ELEMENTS OF AGRICULTURE

FREAM'S
ELEMENTS OF
AGRICULTURE

A TEXTBOOK PREPARED UNDER THE AUTHORITY
OF THE ROYAL AGRICULTURAL SOCIETY OF ENGLAND.
ORIGINALLY EDITED BY W. FREAM, LL.D.

FIFTEENTH EDITION

Edited by
D. H. ROBINSON, Ph.D.

Revised and Metricated by
NEIL F. McCANN, B.Sc., N.D.A.

LONDON
JOHN MURRAY, ALBEMARLE STREET

First Edition 1892
Thirteenth Edition 1949
Revised and Reprinted 1951
Revised and Enlarged 1955
Reprinted 1956
Reprinted 1959
Fourteenth Edition 1962
Fifteenth Edition 1972
Reprinted 1973
Reprinted 1975
Revised and Metricated 1977

Printed in Great Britain by
Butler & Tanner Ltd, Frome and London

0 7195 2579 9

PREFACE TO THE
REVISED AND METRIC IMPRESSION

DURING the eighty-five years that have elapsed since Fream's *Elements of Agriculture* was first published in 1892, no fewer than fifteen editions have been printed. The last of these was published in 1972 and a revised reprint issued in 1973. The rapid changes that are taking place in British farming and, particularly, the introduction of Système International d'Unités (SI) into all aspects of life have induced the Society to make yet a further revision. Apart from metrication, the eclipse of the horse as a working animal on the farm and the almost total specialisation of poultry keeping have resulted in great changes in the sections devoted to livestock.

The form of this revision follows closely the lines laid down by Dr D. H. Robinson, who was editor from 1949 to 1974. Supervision of the 1977 text has been undertaken by Neil F. McCann with the help of the original contributors and a few newcomers, all of whom are named in the list of acknowledgements.

The Society hopes that, with its new text and the addition of a list of " Recommended Reading ", this latest revision of Fream will prove a valuable source of basic information and of references to those interested in British agriculture and, particularly, to students.

JOHN GREEN
Chairman of Education Committee

ROYAL AGRICULTURAL SOCIETY OF ENGLAND
35 BELGRAVE SQUARE
LONDON SW1
1977

v

CONTENTS

PLATES

ix

ACKNOWLEDGEMENTS

Text: The Royal Agricultural Society of England is most grateful to the various teachers and research workers who have brought the text up to date and, particularly, have dealt with the complicated and time-consuming task of converting Imperial measures to those of the International System (SI) to which farming, like most other industries, is now committed. They are : N. Barron, B.Sc., Ph.D., M.R.C.V.S. ; G. D. H. Bell, C.B.E., Ph.D., F.R.S. ; W. J. Bevan, M.Sc. ; B. M. Boyns, B.Sc., Ph.D. ; C. L. Coles ; A. J. Costley, B.Sc. ; V. Cory, B.Sc. ; C. Culpin, O.B.E., M.A., Dip. Agr., F.I.Agr. E. ; S. Culpin, M.Sc. ; D. Bryan Davies, B.Sc., Ph.D. ; R. M. Deakins, B.Sc. ; F. K. Deeble, B.Sc., M.S., N.D.A. ; M. Eddowes, M.Sc., Ph.D. ; A. Eden, M.A., Ph.D., F.R.I.C. ; R. Gair, B.Sc. ; J. MacLeod, B.Sc. ; Miss Ann Martin ; C. G. Pointer, B.Sc. ; J. Rudman, B.Sc. ; J. Tabram, B.Sc. ; R. E. Taylor, B.Sc., Ph.D. ; B. D. Trafford, C.Eng., M.I.C.E. ; D. B. Wallace, M.A. ; J. Young, B.Sc., A.R.I.C.S.

Plates : The Society wishes to thank the following firms and individuals who have kindly permitted the reproduction of copyright illustrations :
   (*a*) Livestock : *Farmer and Stockbreeder* ; *Farmers' Weekly* ; J. Halliwell ; C. Hosegood ; G. S. McCann ; National Pig Breeders' Association ; *Pig Farming* ; Pig Improvement Ltd. ; *Sport and General* ; Wigfield & Pluck Ltd.
   (*b*) Machinery : Alfa-Laval Co. Ltd. ; Allaeys Ltd. ; Alpha-Accord Ltd. ; Armer Agricultural Machinery (ISC Ltd.) ; Bamords Ltd. ; Bomford & Evershed Ltd. ; British Leyland UK Ltd. ; David Brown Tractors (Sales) Ltd. ; I. W. Chafer Ltd. ; Colchester Tillage Ltd. ; John Deere Ltd. ; Farrow Irrigation Ltd. ; I. Gibbs Ltd. ; Howard Rotavator Ltd. ; Imperial Chemical Industries Ltd. ; International Harvester GB Ltd. ; Kongskilde UK Ltd. ; F. W. McConnel Ltd. ; Mann's of Saxham Ltd. ; Marshall-Fowler Ltd. ; Massey-Ferguson (UK) Ltd. ; Richard Pearson Ltd. ; Ransomes Sims & Jefferies Ltd. ; Sperry New Holland Ltd. ; Stanhay (Ashford) Ltd. ; Vicon Ltd. ; John Wilder (Eng.) Ltd.
   (*c*) Scientific : Long Ashton Research Station ; H. W. Miles ; H. C. F. Newton ; P. T. Thomas.

Introduction

THE MAJOR CHANGES
IN POST-WAR BRITISH AGRICULTURE

BRITISH agriculture since 1945 has moved from a labour-intensive to a highly capitalised industry. It showed great strength and resilience in absorbing new techniques, blending them with traditional skills and applying them unhesitatingly to farm production.

With the strengths, weaknesses have also arisen but nevertheless the industry's progress has been impressive. Its labour productivity has shown a consistent rate of increase unapproached by that of any other industry.

WARTIME EFFORTS AND THE 1945 SCENE

Agriculture in Britain in 1945 presented a very different picture from that of 1939. At the start of the second world war a run-down industry which had received only limited attention from governments was called upon to fulfil the role of the nation's chief supplier of basic foods.

The dereliction of much of our pre-war farmland is now hard to visualise. After a brief encouragement provided by the first world war, farming became a victim of the world-wide slump. Industrial production declined and with it demand for both manufacturers and for primary foods. Farmers tightened their belts. They spent less on themselves and on their farms. The result was land depleted of essential plant foods and stock cheaply fed but with low outputs of milk and meat. Worst of all was the depressed outlook of the farmers themselves who felt they had been abandoned by the country at large. When the call for increased food output came the cynicism of a whole generation had to be overcome. That this was achieved reflected credit on British agriculture as well as upon clear governmental directives fairly and firmly administered.

The overall achievements of the war period are difficult to measure but it appears that the country's dependence on imported temperate food was reduced in total volume from around two-thirds to under a half. The index of Gross Agricultural Output of the United Kingdom at current prices stood at 215.9 in 1945-6 against 100 as the pre-war average. This was achieved to a considerable extent by applying proven methods of husbandry and by making available, on a quota basis, feeds, fertilisers and lime. To these inputs labour and machinery were also added. The entire effort was controlled and serviced by the Ministry of Agriculture and Fisheries with the Ministry of Food purchasing virtually the whole of the country's farm produce.

Home food supplies during the war were supplemented by such imports as could be obtained and carried and an important part was supplied in the form of canned foods, by the United States on " Lend-Lease " terms. These American arrangements were linked to the military effort, and when the war in Europe terminated in May 1945 all aid by " Lend-Lease " was abruptly concluded—much to the consternation of the British public.

THE FIRST POST-WAR PHASE 1945-52

Thus the first phase of peacetime agriculture opened with an intensification of the " siege " conditions experienced for the past five and a half years. Added to the difficult food supply at home were the gloomy predictions of Boyd-Orr and others as to the long-term situation. It was confidently predicted at this time that the total world demand for food would be beyond the capacity of production. With a shattered and starving European continent at hand the British people were under no delusions as to the importance of home agriculture. This was accentuated by bread rationing introduced for the first time in 1946–7. It was clear that farming was being provided with a very great challenge.

1945 Efficiency. How efficient and productive was British agriculture in 1945 ? As far as the soil was concerned, though there had been some over-cropping of lighter types, the fertility of most of the land was reasonably high. Soil nutrient reserves, notably of phosphates, had been judiciously improved and much lime had been applied to acid areas. In addition the war had seen a wide application of grassland management principles which had been established at Aberystwyth, Cockle Park and elsewhere during the period before 1939. Much arterial drainage had been improved during the war though much more work was needed. It was also necessary to restore to farm use areas which had been given over to the fighting services. Such military zones were very extensive and exceeded in area all the derelict land which had been reclaimed for farming.

Livestock numbers were severely depleted during the war. Pigs and poultry, being heavily dependent on cereals (vital for human food), became early casualties in 1939. Though cows were reduced in numbers rationing of feeding stuffs sustained the milk production which the nutritionists so greatly valued. Artificial insemination had been established during the war, and after 1945 this was to prove of great benefit in the expansion of the dairy enterprise.

By contrast farm labour in 1945 was a hodge-podge and could by no measure be regarded as satisfactory. Key farm workers were reserved from national service, but supplementary labour, largely unskilled, was employed on a very large scale. The labour force included the Women's Land Army (who were trained), conscientious objectors, Italian and German prisoners of war and various groups of displaced persons from Eastern Europe. Thus in 1946 there were 976,000 workers employed on the land in the United Kingdom. Clearly farming was a labour intensive industry. With the dispersal of the supplementary work force mechanisation moved forward to replace hand labour in every sector. The continuation of the replacement of manpower by machine represents the most consistent single trend in the development of post-war British farming.

Mechanisation. At the heart of farm mechanisation is the tractor—a general-purpose mobile power-unit adaptable to many field and stationary purposes. It has provided a complete replacement both for the horse and for much hand labour.

Manufacturing capacity after the war was at first limited and use was made of the relatively inefficient machines available. Government War Agricultural Committees retained fleets of tractors for contract work, and new machines were allocated on the basis of need. However, notable technical advances in tractors became available in the immediate post-war period. The most important of these was the incorporation of the hydraulic principle for close-coupled implements following its invention by Ferguson in the 1930's. It was progressively employed both for tractors and other implements from that time.

A second significant change in the farm tractor took place in the early post-war period—that of the substitution by the compression-ignition or diesel engine of the existing petrol or petrol-kerosene units. The diesel engine, though heavier, was a more efficient power-source, consuming less fuel and requiring less maintenance than the spark-ignition engines.

The complicated, if picturesque, cornfield scene began to change with the advent of the combine harvester. This machine, though known in Britain since 1927, was rarely seen outside the chalklands of Cambridgeshire, Berkshire and Hampshire before 1939. The development of the self-propelled combine represented a major step forward, and by the mid-1950's this machine, together with accompanying drying and storing equipment had completely ousted the self-binder and stationary thresher. The stooks, ricks and stackyards had disappeared for ever!

Considerable advances were made in the 1950's with a number of new farm machines. These included the forage harvester, the powered rotary cultivator, the one-man pick-up baler, the hydraulic excavator, front- and rear-end loaders, improved FYM distributors, precision drills, auger elevators, sugar-beet harvesters and many more.

Perhaps of equal importance to the development of the machines was the change in attitude of farmer and worker to mechanisation. During the immediate post-war years a lack of basic comprehension of machines retarded their efficient use. Thanks to the efforts of education, advisory and, later, Training Board officers, supported by the manufacturers, a much greater understanding developed.

Fixed Equipment. Fixed equipment on the farm changed more slowly than the machines, and in the late 1940's and early 1950's a somewhat piecemeal picture emerged. While centuries-old structures continued to be in wide use, prefabricated buildings designed for general purposes were pioneered. Specialists and architects strove to develop designs to fit the new shape of agriculture which was emerging. Teams visited the USA and Canada in the search for fresh ideas. Eventually the modern layouts appeared with controlled environment systems for intensive stock and specialised crops and a strong emphasis upon adaptability in buildings for the general farm.

Legislation—the Agriculture Acts. The demand for more food met a good response from farmers but they made it clear that government support would be needed to underwrite their investments in land and equipment. The insistent demand for security of tenancy for rented land made itself felt. On the other hand the government had to take into account a consuming public faced with price inflation and a general shortage of most everyday essentials. The result of wide consultation and careful preparation were the Agricultural Act of 1947 and the Agricultural Holdings Act of 1948—measures agreed by both government and opposition.

The Act, " promoting and maintaining, by the provision of guaranteed prices and assured markets . . . a stable and efficient agricultural industry capable of producing such part of the nation's food . . . as in the national interest it is desirable to produce in the United Kingdom . . . ", also stressed proper remuneration and conditions for workers. It also inaugurated an annual review of conditions in the industry. This review was an occasion when the Ministry of Agriculture met farmers' representatives with the object of reaching agreement on prices and other measures.

This legislation, which lasted until EEC membership, provided for deficiency payments to farmers to meet the differences between the generally lower world prices for commodities and those obtainable in British markets. Thus national food prices were kept to a low level. Accompanying these provisions there were clauses to ensure efficient production, and " supervision " powers were available for use against defaulting tenants and landlords. However, these shades of wartime enforcement proved unpopular and difficult to operate and they were soon dropped.

National Agricultural Advisory Service (NAAS). A significant introduction at this time of rapid growth in farm output was the National Agricultural Advisory Service. Separating advice from teaching the government launched this almost comprehensive service in 1946 with the express purpose of " giving free of charge, technical advice and instruction whether practical or scientific, on agricultural matters ". Its wide coverage replaced the older County Council system which in many cases lacked uniformity in reaching all levels of farmers. The NAAS provided the link between research and practical farming conditions. In the words of its first Director-General, Sir James Scott Watson, " its prime task was to sift the growing harvest of new knowledge and new resources and to commend to the individual farmer such new practices as he might, with advantage, adopt ".

The Service's Experimental Husbandry Farms proved to be most effective test-beds for new technology. Later the NAAS successfully developed applications of the new discipline of Farm Management bringing a practical approach which was acceptable to the farming community.

There is no doubt that the NAAS contributed substantially to the technical progress of agriculture in England and Wales.

In 1971 it was superseded by the Agricultural Development and Advisory Service.

Technical Innovations. In this first phase of post-war farming a flood of technical improvements was released by research institutions and the laboratories and workshops of the manufacturers.

Leading the field were new cereal varieties with outstanding qualities of stiff straw and high grain yields.

Outputs moved steadily upwards :—

UNITED KINGDOM CEREAL YIELDS
(tonnes/ha)

Crop	Pre-1939	1950	1960	1970
Wheat .	2·22	2·82	3·63	4·04
Barley .	2·04	2·59	3·30	3·59
Oats . .	2·02	2·31	2·71	3·43

Source : Agricultural Statistics HMSO (converted)

In the first wave of new wheats Bersée, Nord Desprez and the outstanding Cappelle Desprez were French ; Hybrid 46 was British ; Atle, a fine spring wheat, Scandinavian.

The barley variety Proctor, produced at Cambridge, ousted other types for many years. The Welsh Plant Breeding Station produced several outstanding oat varieties—though this crop commenced a decline in favour of the two more profitable cereals. Plant breeders developed new techniques which were to produce a rapid succession of crop varieties in the succeeding years.

Along with better cereal varieties the chemists made revolutionary progress in two almost entirely new fields.

The New Insecticides The first was in organic insecticides. Leading the new materials was DDT produced in Switzerland in the 1930's.

DDT was very effective against flies, killing by contact and by absorption. It was economical in use since it was long lasting. The material was brought into quantity production after 1945 and was widely used on farms, on market gardens and on fruit plantations. BHC, invented and first used in Britain late in the war, proved to be very successful against the wireworm which had devastated many cereal crops.

These two materials were the first of an extensive family of organo-chlorine compounds—later to be condemned for their persistence. Many other chemicals—notably organo-phosphorus compounds such as Schradan and Parathion—followed and presented the farmer and gardener with an extensive, if complicated, armoury against most insect pests.

The concept of systemic insecticides dates from this period. It was discovered that certain materials could be absorbed by a plant and translocated to all parts without damage.

When the treated plant was attacked by a biting and sucking

insect the toxic plant juice killed the pest. This was to prove a particularly valuable aid against aphids.

Selective Herbicides. The second advance in the chemical
field was equally dramatic. Weeds had been a farmer's adversary
from the earliest times. Most cultivations were based on weed control
and very few aids other than mechanical were available before 1945.
Such materials as sulphuric acid and ammonium sulphate were in use
but, possessing difficult handling properties, they had only a restricted
application.
     During the war two similar chemicals were developed, one, 2,4-D
in the USA and the other, MCPA in England. These materials when
applied to a wide range of broad-leaved weeds produce an exaggerated,
twisted growth pattern followed by death. The chemicals, which
became known as Selective Herbicides, were harmless to cereals.
     Their acceptance by the farming community was immediate and
the effects striking. Weeds such as Charlock and Poppy began to
disappear from the countryside. Chemical weed control became a
routine operation, and a wholly new science complete with its manufacturing industry appeared almost overnight. Many new groups of
chemicals were developed, and by the early 1950's changes in traditional farm practices were everywhere apparent.
     Cereals became a " cleaning " crop and costly hand weeding
became a thing of the past—or almost so—because one or two highly
persistent weeds, notably wild oats, remained to be dealt with.
     Herbicides played a significant part in releasing manpower from
the laborious sugar-beet and potato crops. This was achieved by
both pre- and post-emergent sprays of chemicals which could cope
with the long period when the ground was uncovered in growing
these root crops.

     Fertilisers. The war had seen the development of " Triple-
Superphosphate ", and the first " compounds " incorporating nitrogen, phosphate and potash appeared in the 1930's. In the post-war
period manufacturers did much to develop compounds and to improve their handling qualities. The granulation and pelleting of
fertilisers became common practice and, especially in relation to
nitrogenous manures, greatly facilitated wider usage.

     Grassland. The most typical feature of most British farms—
grass—was traditionally the crop least well managed. Helped by
wartime experience, farmers, after 1945, began to adopt improved
methods. Conservation, however, presented many difficulties. Grass
silage-making, in particular, was slow to be adopted. Only when
improved mechanical equipment, heralded by the Paterson Buckrake,
became available did this type of fodder take its place on the more
intensive dairy farm. The economic need to derive the greater part
of the cow's food intake from grass progressively led the way to better
grazing and conservation methods. Improved herbage varieties
originated with the Welsh Plant Breeding Station but these were
joined by overseas strains from the 1950's onwards. Controlled grazing whether by strip-grazing or by paddocks or related techniques

became the rule—with the considerable help provided by the electric fence. Haymaking, the traditional method of conserving grass, received much attention by agriculturalists and engineers from the 1950's onwards. Wholly new machines for conserving the leaf and reducing the water content appeared on the scene. At the same time methods of drying partly cured hay in the barn were developed to a successful stage. The complete grass-drying techniques developed before 1939 remained however in the hands of large-scale specialists who often used lucerne as their main crop.

The intensification of stocking also owed much to the systematic use of much larger quantities of nitrogenous fertilisers. By the time that Britain entered the Common Market it is probable that its grass-land was better managed than that of most European countries.

THE SECOND POST-WAR PHASE 1952-70

The End of Food Rationing. While productivity on the farm was advancing the pressure of critical shortages of food was beginning to ease.

Food rationing both for the human population and for the animals was brought to an end in 1953. The vast organisation of the Ministry of Food, which had held in its control almost every item of stable foods, was gradually wound up.

With an agricultural economy that was becoming freer the government had already, in 1952, announced an expansion programme. The target for this was " 60 per cent plus pre-war by 1956 ".

This called for increases in grazing stock as well as better yields of grass and arable crops. With the new technical aids at its disposal and with the backing of the Agricultural Acts, British agriculture easily met this modest objective.

The deficiency payment system embodied in the Act was applied to beef cattle and marketing took on a new and more realistic role.

A new marketing organisation was launched by the National Farmers' Union to accompany this developing industry—the Fatstock Marketing Corporation. The FMC provided a dead-weight outlet for beef and lamb which could be negotiated on the farm, so providing an alternative to the traditional live-weight auction system.

New developments continued to flow and farming productivity showed little sign of faltering in the long period up to the change-over in support policy in the early 1970's.

Ironically the period started with a major, though localised, natural disaster at the end of January 1953.

This was the East Coast sea-flood which inundated considerable areas of land and destroyed property and stock in counties lying between Lincolnshire and Kent. That the salt-affected land was restored to cultivation within three or four years is a tribute to scientists and farmers alike.

Mechanisation and Manpower. In this second post-war period the manpower changes which had commenced before 1952 rapidly accelerated.

The farm labour force showed a continuous fall throughout the country over a period of very many years. The gap was filled by mechanisation and it is probable that in many cases machines were obtained in advance of expected labour shortages.

MANPOWER IN UNITED KINGDOM

Year.	Regular Full-Time Labour.	Regular Part-Time Labour.	Total.
1950 . .	718,000	200,000	918,000
1960 . .	505,000	188,000	693,000
1970 . .	274,000	156,000	430,000
1974 . .	233,000	158,000	391,000

Source : HMSO Cmnd 4321 and 4623

Amongst the extended use of machines already known many entirely new concepts appeared after 1952. These included on-floor grain-drying equipment, sealed tower silos, mechanised livestock-feeding layouts and complete potato harvesters.

The following table gives a selection of four machine types which were in considerable demand during the second post-war phase. It will be noted that FYM spreaders and pick-up balers appear to have reached saturation point. Statistics on farm machines over a period of years can be misleading owing to the many innovations which quickly replace older types.

MACHINES REPLACING LABOUR
(England and Wales)

Year.	FYM Spreaders, all types.	Forage Harvesters.	Pick-up Balers.	Complete Potato Harvesters.
1950 .	12,179	1,136	5,023	477
1962 .	94,160	15,260	76,940	3,310 (1964)
1965 .	105,350	19,680	87,410	6,090
1968 .	103,240	21,950	84,070	9,380 (1969)

Source : HMSO Agric. Statistics

It is of interest to note that the proportion of national farm costs accounted for by labour fell significantly between 1950 and 1970 while machinery costs remained steady during the period in contrast to other costs :—

CERTAIN COSTS BY % OF TOTAL COSTS (UK)

Year.	Labour.	Machinery.	Rent/Int.	Feedstuffs.
1950–1 . . .	32·8	15·6	7·95	19·0
1960–1 . . .	24·6	16·9	8·2	27·0
1960–70 . .	18·8	16·0	11·2	29·5

Source : HMSO Cmnd 1658, 4321 ; Review of 1951

Dairying 1952–70. Milk production accounts for about one-quarter of the total UK farm output and therefore ranks of great importance in the industry. In the second post-war period developments in dairying have been associated with concentration in production, yield improvements, breed standardisation and more efficient management of the enterprise. Following the establishment of artificial insemination centres providing a complete national coverage the benefits of improved breeding slowly became apparent. The use of proven bulls by the Milk Marketing Board, along with performance testing and the making of contemporary comparisons, has had the effect of substantially improving milk yields.

AVERAGE ANNUAL MILK YIELD PER COW
(England and Wales)

	Litres.
1938–9 . .	2,546
1944–5 . .	2,355
1949–50 . .	2,832
1954–5 . .	3,068
1959/60 . .	3,341
1964–5 . .	3,545
1969/70 . .	3,750
1973–4 . .	4,000

Source: Dairy Facts and Figures, UKMMB (converted)

Breed changes have been substantial during the period. The Ayrshire breed remains dominant in Scotland.

DAIRY HERD BREED DISTRIBUTION BY %
(Cows and heifers in milk and cows in calf)

Breed	England and Wales				Scotland		
	1955	1960	1965	1970	1960	1965	1970
Ayrshire . .	18·3	19·2	15·7	9·7	79·5	77·8	68·6
Friesian . . .	40·6	51·1	64·2	76·3	14·5	17·8	18·2
D. Shorth. . .	25·3	14·5	6·3	2·5	NA	0·3	NA
Guernsey . .	5·3	6·2	5·7	5·2	0·1	0·1	1·0
Jersey . . .	2·6	3·3	4·3	3·8	1·0	1·0	1·0
Others . . .	7·9	5·7	3·8	2·5	4·9	3·0	2·2
	100·0	100·0	100·0	100·0	100·0	100·0	100·0

Source : " Dairy Facts and Figures ", UKMMB (adapted)

The use of good-type bulls both of beef and dairy breeds and whether by A.I. or natural service is always an important consideration in British agriculture. In the 1960's the introduction of a foreign breed for crossing purposes—the Charolais—took place. Performances were carefully observed before bulls were made available at artificial insemination centres.

The period 1952–70 was characterised by relatively stable total cow numbers and steadily declining numbers of milk producers. The effect of this was seen in the increase in average herd size :

AVERAGE HERD SIZE OF DAIRY COWS (UK)

1960	1963	1965	1968	1974
18	20	23	28	39

Source : HMSO Cmnd 5977 and " Changing Structure of Agriculture "

A radical increase in scale of operation has obviously been taking place. This has been accompanied by the use of many mechanical techniques along with better methods of handling both cows and equipment. The milk parlour began to replace the cowshed fairly steadily in the south though progress has remained slower in the north. The parlour itself underwent change—styles have moved from " tandem " and " abreast " to " herringbone " and " rotary " designs. By 1973 36 per cent of milking systems in England and Wales were of the parlour type and of these a third were " herringbones ".

With parlour systems the cows are normally kept in loose housing conditions, i.e. not tied by the neck Accompanying loose housing was the adoption of " self-fed silage " techniques. Chopped silage in clamps with adequate feeding faces is made available to the cows day and night demanding a minimum of labour. At first the animals were bedded with straw in open yards but with increasing numbers litter costs became excessive. At this point the cow cubicle was introduced by a West Midlands farmer. Cubicles consist of framework standings (without tying arrangements) made of wooden battens or steel tubes with a floor of insulated material. The dimensions are such that the cow may occupy the bed without fouling it. Straw is not required—merely a dusting of sawdust or chalk. However, the price of this improvement has been the burden of disposing of the manure which, without straw, becomes mainly fluid in consistency. In most cases of large herds full liquid handling of this effluent has had to be undertaken. This has entailed underfloor collection, storage tanks or " lagoons " and the means of distributing the slurry on to the fields. By the mid-1970's herds of 120 cows with " herringbone " or " rotary " parlours, refrigerated milk storage vats for bulk collection and full effluent-handling equipment were becoming standard units. Herds of 300 cows became not uncommon.

Thus the labour-intensive dairying farm was passing to capital intensity during this second post-war phase. The technical requirements of the dairyman greatly increased until this highly trained operator often became the farm's highest paid employee. The certain marketing arrangements and standardised prices conferred on milk production a stability which encouraged the heavy capitalisation of this enterprise.

Total production of milk in the UK as a whole has shown a fairly consistent annual increase since 1945. There have been occasional checks, often of a seasonal nature, but the trend has been towards

over-production for the most part. In 1974 however, greatly increased feeding stuff and fertiliser costs have threatened to change the
familiar pattern. Milk for manufacture has also been consistent—
viewed over this period. Some intermittent difficulties in supply have
occurred to interfere with cheese production but these have been of a
temporary nature.

TOTAL SALES OF MILK OFF FARMS (UK)
(Million Litres)

Year.	Total.	Of which, manufactured.
1938/9 . . .	5,878	1,918
1944–5 . . .	6,287	786
1949/50. . .	8,155	1,168
1954–5 . . .	8,905	2,023
1959–60. . .	9,676	2,641
1964–5 . . .	10,706	3,250
1969/70. . .	11,860	4,384
1973–4 . . .	13,312	5,783

Source : " Dairy Facts and Figures ", UKMMB (converted)

Beef Production. A large proportion of the beef produced in
Britain traditionally stemmed from dual-purpose cattle. In recent
years there has been some segregation of our breeds into dairy and
beef types, but even so more than half of the home-produced beef
originates with the dairy herd.

The conversion of food, whether of forage or of concentrates, into
beef is a relatively inefficient process. Consequently progress in this
field has somewhat lagged by comparison with other livestock. Efforts
have been made to improve feed conversion and also to shorten the
rearing and fattening period. Research work since 1950 has included
a study of feeding planes related to growth stages, improved fertiliser
usage on grass, the use of growth-promoting substances and the
utilisation of non-protein nitrogen in diets.

Perhaps the most spectacular system to be introduced in the period
was " Barley Beef ". This technique, developed at the Rowett Research Institute, is based on rolled moist barley with a protein supplement fed to Friesian calves as an exclusive diet to produce finished beef
in about 12 months. The system found a place at a time of barley
surpluses in the mid-1960's.

Not without its problems, e.g. bloat and liver failure in the animals
and calf supply difficulties, the system gave way to " 18-month beef ".
In this technique use is made of well-managed grassland to reduce feed
costs. Efforts have also been made to improve the efficiency of use
of conserved forage for beef production.

Breeding improvements have centred on production assessments of
various crosses—work with which the Meat and Livestock Commission
has been identified. Newly introduced breeds have been the object
of special interest.

A very late development in British beef production is that of
fattening bulls. This practice, making use of the more efficient food

conversion by the entire animal, was impossible under the Sire Licensing Regulations until they were amended in 1970.

Pigs. Important changes occurred in systems of pig husbandry in the second post-war period. Already by 1952 pig-keeping had recovered from the severe setback which the industry had received during the war. At about this time there was a most significant importation of Swedish Land Race stock which provided a stimulus for breeding developments. The crossing of this breed with the old-established Large White produced a stable cross of much improved conformation as well as good feed conversion characteristics.

From now on the "Pig on every farm" policy rapidly gave way to ever-increasing specialisation by a smaller number of highly quali-fied pig farmers. Housing underwent much development, designed both to conserve heat and to reduce disease risks.

The control of disease became a major preoccupation and systems were built around this concept. The revolutionary "High Humidity" technique claimed to ensure a control of respiratory diseases, for in-stance. Early weaning of piglets became an accepted practice and this was followed by many developments in nutrition. These in-cluded the use of antibiotic and metallic copper additives and, later, systems of pipeline feeding.

The traditional pork and bacon animals were joined by the heavy hog outlet which provided for the processors a more versatile product. The Pig Industry Development Association which later became part of the Meat and Livestock Commission gave much guidance to breeders in the fifties and sixties. This organisation pioneered Progeny Testing and also artificial insemination for pigs.

Poultry. The poultry industry made a slow recovery from the heavy blow caused by the war. Not until the mid-1950's did it show a significant recovery. After that date, however, development took place very rapidly. Egg production in 1946–7 had amounted to just over 50 per cent of the country's total requirement. By 1969 99 per cent of the need was supplied by British farmers. The progress in poultry meat production was even more spectacular. From 8 per cent of the total home meat supplies in 1946–7, by 1969–70 poultry constituted 22 per cent of the total. In the case of both eggs and meat much progress was made in breeding techniques making use of the short life-cycle of the hen to select for economic traits. Considerable advances also took place in nutrition and in the environment of poultry houses. A new word joined the farming vocabulary with the start of the "broiler" industry. Modelled on American practice, units were started from the late 1950's which soon rivalled the originals in size and efficiency. Traditional table poultry production had been for larger birds, but the change was accepted readily by the public. They now had a cheap alternative to "red" meat, and the con-sistency and volume of the supply was especially attractive to catering establishments.

Sheep. British farming has long boasted a numerically strong sheep enterprise. However, the contribution of sheep to the whole farm output in the UK has not advanced above 5 per cent since 1945.

Home production in 1973–4 was providing only 54 per cent of the declining total supplies of mutton and lamb. Advances in sheep husbandry have been relatively minor for many years. They have included some improvements in grazing techniques, greater control of some parasitic conditions and a better recognition of some long-standing diseases. Progress in breeding has been quite limited despite some attempts to incorporate factors for greater fecundity. The traditional stratification pattern, geared to mainly extensive production, has remained unchanged.

Home-Produced Meat. The contribution made by various livestock enterprises to the home meat market is as follows :—

HOME PRODUCED MEAT ('000 tonnes)
(United Kingdom)

	1946–7		1953–4		1962–3		1965–6		1969–70	
	Amount	%	Amount	%	Amount	%	Amount	%	Amount	%
Beef/Veal	558·8	64·3	655·3	45·4	910·3	40·2	815·8	34·7	929·6	35·5
Mutton/ Lamb .	137·2	15·7	174·7	12·1	251·9	11·2	253·0	10·8	224·5	8·6
Pigmeat .	103·6	11·9	511·0	35·4	752·8	33·3	870·7	37·0	882·9	33·8
Poultry .	71·1	8·1	102·6	7·1	345·4	15·3	412·5	17·5	579·1	22·1
TOTAL .	870·7	100	1,443·6	100	2,260·4	100	2,352·0	100	2,616·1	100

Source : HMSO Cmnd 4321 and 4623 (adapted)

The declining relative importance of beef and veal and mutton and lamb has been a consistent feature. This was in spite of a continued government exhortation and support. Measures of assistance included Hill Farm Subsidies for cattle and sheep and a Beef Cow Scheme as well as the deficiency payment system.

Taking the index for 1946–7 as 100, beef/veal and mutton/lamb reached an output of 166 and 163 respectively for 1969–70.

The corresponding indices for pigmeat and poultry were respectively 852 and 814.

Marketing Problems. The late fifties and sixties showed the vulnerability of an increasingly efficient industry to the inelasticity of food demand.

Beef production experienced very low prices in the 1960–3 period resulting in unusually large deficiency payments by the government. Barley production increased 366 per cent between 1953–4 and 1967–8, though the government for the most part regarded high grain production as a useful contribution to import saving.

In other sectors of farming the continued expansion brought about changes in the supply chain. This was particularly true of the poultry enterprise. Fierce competition developed within the industry, and with a continuous rise in the prices of concentrate feeds fewer but much larger units emerged. This concentration was especially felt in the hatchery section where both layers and table poultry producers were catered for.

Egg supplies reached high production levels during the period after 1962 and received decreasing government subsidies with almost complete self-sufficiency.

The tendency for supply to exceed demand was also a feature of many other products during this second period.

Potatoes are typical of a finely balanced commodity sensitive to over-production. The Potato Marketing Board controlled the ware crop acreage by an acreage levy and, backed by government support, was responsible for the disposal of surpluses. The position was somewhat eased during the 1960's by the increasing interest in processed potatoes. These outlets catered for a growing proportion of the ware crop and included both the well-established crisping technique as well as dehydrated and refrigerated products.

In vegetables and fruit no price support was provided by the Agriculture Acts, reliance being placed on import restrictions which could be applied when necessary. Greater emphasis than ever was given to quality standards, packaging and marketing of these perishable products, and again increasing specialisation and concentration became apparent.

The government made strenuous efforts to foster co-operation in production and marketing and in 1965 set up the Central Council for Co-operation in Agriculture and Horticulture. Generous grants were made available for the establishment and running of such groups—though the response was somewhat halting. Farmers and growers had shown more interest in forming buying groups, finding it possible to obtain useful discounts for bulk purchases. To concentrate the purchasing of supplies at discount rates the NFU set up " Agricultural Central Trading " for the benefit of its members.

The Pig Industry Development Board was created in 1956 with the object of improving breeding techniques and to foster performance testing, using boars from strategically placed centres. The Board developed artificial insemination for pigs for practical application and sponsored local study groups to improve production methods. However, PIDA did not become involved in marketing aspects. The Meat and Livestock Marketing Commission was set up in 1965 and absorbed PIDA into its work. Although the government was deeply concerned with some of the unsatisfactory features of meat marketing the new commission was precluded from any powers of control over fatstock sales! Instead, it concentrated on production efficiency, carcass classification and certification, research and upon situation and outlook information.

During this period cereal growing and marketing were critically reviewed and a Home Grown Cereals Authority was created. It was charged with the encouragement of forward contracting by making bonus payments from a levy collected from cereal growers. Unlike the Meat Commission the HGCA was given trading powers to intervene in a depressed market. It has provided market intelligence and has an interest in research and development. With the advent of EEC membership the HGCA undertook further commitments.

However, milk remains the only commodity in Britain having a Board that controls all aspects of a single enterprise—with powers of exclusive purchase and disposal of the farm product.

Horticulture. Parallel with the progress in agriculture, horticulture made considerable advances in techniques, new varieties, disease and pest control. Also notable was the better presentation and marketing in the period 1952–60.

Horticulture in Britain, traditionally known for its diversity, has followed the trend towards greater concentration.

The specialisation which this implies has entailed the adoption of new techniques. One crop which saw great changes in the 1950's was green peas. Eastern England traditionally produced both dried and green peas, but with the increased demand for convenience foods the canned and quick-frozen peas brought about a radical change in the production pattern. This entailed new varieties grown on contract to a tightly controlled system, harvested with a completely new range of machinery and delivered to the factory within a specified time. This type of pea trebled in total output between 1952 and 1967.

Better pest control enabled the tonnage of carrots to be increased by 120 per cent in this period. Improved field techniques also permitted increases in the outputs of beans, cauliflower, lettuce and dry bulb onions.

Glasshouse design changed radically between 1950 and 1965 with the rather make-shift Dutch-Light structures giving way to steel-framed houses equipped with automatic ventilation, watering and liquid feeding. Also during this time the heating pipes were changed from a large diameter to those of the " small bore " system incorporating circulating pumps. In the late sixties the use of extra carbon dioxide in glasshouses to improve photosynthesis became common practice.

The mushroom crop provided another example of specialisation. Using standardised composts and advanced control of the environment in the sheds, the growing closely resembled factory production. With the crop concentrated amongst a very limited number of growers, the tonnage for sale increased five times over in 24 years.

Fruit. Fruit-growing has always been a strong feature of British horticulture and it shared in the general flood of new technology. To meet a greatly increased demand the production of dessert apples increased, although from a smaller total area than formerly. This was accomplished by much more effective pest and disease control as well as close attention to the major and minor plant food requirements. Quality control and effective long-term storage techniques completed the cycle.

Of the small fruit, strawberries and blackcurrants deserve mention. Though remaining a labour-intensive crop at harvest strawberry-growing has greatly benefited from the new herbicides and fungicides. Plant breeders have also played a part by producing a better spread, of prolific but high-quality varieties. Blackcurrants have undergone drastic husbandry changes. Here mechanisation has advanced to the stage of mechanical harvesting enabling much larger blocks to be grown. Attention to fertilisers as well as to pest and disease control have evened out current yields—though avoiding frost damage remained a stubborn problem.

Land Pests. Of the larger-sized pests which afflicted the British farmer pride of place went traditionally to the rabbit.

Rabbits were probably introduced to England at the time of the Norman Conquest, and especially on the lighter lands of England they consumed large quantities of grass and growing corn.

It had been calculated that 7 rabbits ate as much green material as one sheep! Then, in 1953, there occurred a totally unexpected event which was to bring a quite fortuitous benefit to British farming. This was myxomatosis. This disease of rabbits was of South American origin. It had been introduced into Australia to control the huge rabbit population and it appears to have been brought from there to France. The disease appeared in the early summer of 1953 in southeastern England and immediately devastated the local rabbit population. It became an offence in law to spread the disease artificially but despite this measure the epidemic developed and gradually affected rabbit colonies throughout Great Britain. The mortality rate was around 99 per cent and though this fell in time the disease recurred throughout the country in localised areas.

The benefits to crop yields were soon apparent, and in order to deliver the *coup de grâce* to this ancient pest, rabbit clearance societies comprising farmers and landowners were officially sponsored.

Rodents and Pigeons. A new weapon in the continuous war against rats and mice became available in the late 1940's. This was the compound " Warfarin " and it acted as an anticoagulant in the blood of rodents which had taken the bait. Harmless to other animals this material became a standard control, ousting older and more dangerous chemicals.

The wood pigeon, capable of causing much damage to green crops and peas has always proved a difficult pest to control.

A close study of the habits of the bird by research workers proved helpful in directing control by nest destruction, while later narcotic baits were employed with some effect.

THE THIRD POST-WAR PHASE—1970 ONWARDS

As with the whole of the UK economy the move of agriculture towards a European base represented a change of fundamental and historical importance. No sector remained unaffected and the period was dominated by rethinking and replanning of many basic details. Britain had been interested in a common agricultural market since the Congress of Europe in 1948. After the failure of the proposed Free Trade Area in 1958, internal discussion commenced with the application of the Macmillan government in 1962 to join the EEC. Soundings in the farming industry were taken and the government kept the leaders informed of progress in the negotiations. Nevertheless no step was taken at that time to restructure agriculture policy so that it might fit into the Common Agricultural Policy (CAP).

This application was vetoed by France but was renewed by the

Wilson administration in 1967. Again no basic changes were made in the form of government policy towards agriculture. The emphasis of this policy was upon " selective expansion " as a means to reduce food imports. This had been a recurring theme in post-war agricultural policy commencing with the " 60 per cent Plus " campaign of 1952–6. The National Development Plan of 1965 featured expansion with improving productivity, and the import-saving merits of agriculture were stressed in successive Price Reviews. As the EEC entry loomed larger on the horizon so discussion increased as to the agricultural implications. Though the second application to Brussels was not accepted it nevertheless remained " on the table "—a decision which met the wishes of both political parties. When the Heath government took office in 1970 it immediately developed its manifesto to adapt British farm policy to the CAP. This accompanied an energetic and finally successful campaign to gain for Britain full EEC membership. It may be asked, " How did the agricultural sector obtain such a prominent position in EEC affairs ? "

When the Treaty of Rome was signed in 1957 the original six countries, France, West Germany, Italy, Belgium, Netherlands and Luxemburg, made agriculture the keystone of Community economic policy. The formation of a common market was the means to an eventual European union, and agriculture was the largest industry and the biggest employer—covering 25 per cent of the labour force. Furthermore, the Six were almost completely self-supporting in temperate food supplies. Another major difference between farming systems of continental Europe and those of Britain was in the structure of the holdings. The Six were characterised by a farm pattern which was the result of farms becoming divided amongst members of the family upon the death of the occupier. Small, fragmented units are frequent and run counter to efficient operation under modern conditions. Nevertheless the Rome Treaty undertook to encourage productivity and better use of resources in agriculture as well as to stabilise markets and to improve the living standards of farmers. Though not without some economic conflict, reasonable prices were to be kept for consumers, mainly by relying upon competition. The objectives, if not the means of their attainment, resemble those of the Agriculture Act in Britain—a fact which probably helped the UK farmer to understand and broadly accept the EEC principles. The table on page xxxii summarises differing features of agriculture in the Nine.

The Six were concerned to reduce trade barriers in agriculture between their countries and had, by the late 1960's, produced much harmonisation through Community procedures. These mechanisms varied slightly between commodities but the case of cereals will provide an example. World grain prices are usually lower than European prices. Therefore a " threshold " price, i.e. one which will allow domestic producers a price advantage, is arrived at. The difference between a lower world price and the Threshold price is made up by Variable Import Levies. A price below the Threshold price known as the Intervention price is established as the level at which the EEC would buy in grain through its agencies. The basic Intervention price for soft wheat (and formerly, barley) is established for the area of greatest deficit, deemed for this purpose, to be Duisburg, in the Ruhr. The surrounding surplus regions have progressively lower Intervention prices based on the cost of transport to Duisburg. The storage of products bought in at Intervention prices has sometimes been a source of embarrassment. Arrangements exist to ease exports of surpluses by means of Restitution Payments where markets can be found.

By 1973 cereals, pigmeat, beef and veal, milk products, eggs, poultry, vegetables, fruit, fats and oils were covered by the EEC regulations. The source of finance for intervention, export restitutions and certain schemes to help member states with structural and technical problems is known as FEOGA (*Fonds Européen d'Orientation et de Garantie Agricole*). In turn FEOGA since 1971 has been one part of the Community Budget. Its finance is to be derived from customs

FARM WORKFORCE AND
CONTRIBUTION TO NATIONAL OUTPUT
1953/5–70

	% of Workforce in Agriculture.		Agric. Output as % of GNP.	
	1953–5	1970	1953–5	1970
United Kingdom . .	4·8	2·7	2·7	2·7
Belgium	9·3	4·1	8·1	4·2
Netherlands	13·7	5·8	12·0	6·1
West Germany . . .	18·9	5·6	8·5	3·3
Denmark	25·4	9·0	19·2	6·4
France	25·9	12·7	12·3	6·6
Italy	39·5	13·1	21·6	9·8
Ireland	38·8	25·7	29·6	16·9

Source: Adapted from Open University, "EEC Economics and Agriculture", P933, 5–6

duties, agricultural levies and 1 per cent of Value Added Tax. This would be implemented by progressively increasing fractions of the total sums, and additional needs would be met by direct contributions of the member states. This transitional stage was in operation at British entry and in practice the special arrangements determined a maximum UK contribution.

How then was the well-established British farm support system adapted to the EEC requirements? The system of deficiency payments, designed for an era of food shortages had already required alterations to meet periods of surpluses. These were becoming pronounced at the end of the 1950's. A depression in the prices of fat cattle, pigs and sheep in 1961–2 resulted in a near-fourfold increase in exchequer cost for these commodities compared with the previous year. By 1964 the deficiency payments on cereals had been moving upwards to a degree which caused the Minister of Agriculture to apply a Minimum Price for imports. At the same time he fixed "Standard Quantities" for the home production of wheat and barley. In 1970 the government extended the Import Levy systems to beef and veal, mutton and lamb and to milk products. Minimum price systems favour overseas suppliers at the cost of high domestic prices. Variable Levy systems ensure that downward movements of prices on world markets are reflected in higher exchequer receipts.

Negotiations between Britain and the Six commenced in 1970. They were conducted by Mr Geoffrey Rippon and involved tough bargaining over a number of issues. Chief amongst these were the transition arrangements. It was obvious that although the ground had been prepared for a radical change in support policy, neither the farmers nor the consuming public could make an overnight transformation. Along with other contentious points final agreement was reached with the old member states on 13 May 1971.

The CAP was to be applied to the new member states but at lower price levels. These prices were to be progressively increased to the full price over a transitional period of 5 years, expiring on 31 December 1977. The difference in prices were to be met by Accession Compensatory Amounts. These were additional to Monetary Compensatory Amounts which had existed since 1971 to cover currency movements. The UK would be permitted to retain its production grants for farmers until the end of 1977 but should endeavour to abolish them as soon as possible. Also to be dropped with maximum speed would be the Deficiency Payments System.

There were some " escape " clauses in the EEC legislation. One of these provided that a state incurring difficulties before December 1977 could apply to the Commission for sanction to take protective measures to " rectify the situation and to adjust the sector concerned to the economy of the Common Market ". There was also an anti-dumping clause referring again to the transitional period. Sugar, for which the UK had made long-term arrangements with the Commonwealth, was the subject of a special dispensation to expire on 28 February 1975. As for New Zealand butter this commodity also received special attention. Imports were to be allowed up to the end of the transition period with some possible arrangements even beyond that time.

During the long negotiation period a number of issues dear to the heart of the farming community were apparently lost to view. Nevertheless these became clearer as time elapsed following the agreement. One of these was the fate of the Annual Price Review. This was a valued activity born of the 1947 Agriculture Act. Where would it fit in future ? It was clear that the main price decisions would henceforward be taken in Brussels where COPA (*Comité des Organisations Professionelles Agricoles*) represented the interests of European farmers, now including the British Farmers' Unions, *vis-à-vis* the Commission. However, the British review might still continue even though it might form no more than a round-table discussion. The position of the Milk Marketing Board was obscure in the face of the EEC's known antipathy towards monopolies. Nevertheless it soon became obvious that this efficient body was highly regarded and that its future was not in question.

Hill farming constituted another source of concern in Britain. The many problems of farming in the uplands had long been sensitive topics since they included social considerations along with those purely agricultural. The continuation, if need be indefinitely, of official aid was demanded by all concerned. Despite the absence of a clear-cut decision on the matter in 1971 it became apparent that many member states shared the problem. Later special provisions were to be made from FEOGA sources of aid for " Less-Favoured Areas ". In the meantime UK hill support schemes continued.

The concept of " Intervention " was novel in the UK. The change-over to employing, with renewed vigour, skills in the market place and at the same time putting pressure on the political front for a high Intervention price were things the farmer must develop. To operate Intervention policy it was first necessary to set up a statutory board. This Intervention Board would enforce in British law the EEC regulations. Within this new organisation cereal intervention was thereupon delegated to the Home Grown Cereals Authority which already had purchasing powers. This body could also supervise in times of surplus the " denaturing " of grain intended for human consumption. By this operation an owner could be paid a premium for treating the grain so as to render it suitable only for animal feed. To deal with meat the Meat and Livestock Commission was given powers (which it had been denied upon its inception !) to buy in surpluses and to store carcass meats etc. However in the UK the decision to invoke Intervention procedures was not automatic and the government did not apply them for the 1974 beef surplus. In Northern Ireland these functions were passed to the Ministry of Agriculture. Milk and milk products were to be handled by the Agricultural Departments. Import levies and similar administration was handed to HM Customs and Excise, while a Seeds Executive was set up to deal with farm seed crops.

In 1972 an unrelated event took place in America which was to have far-reaching effects. The Soviet government made massive and quite unexpected purchases of grain in the USA and Canada. Cereal prices reacted quickly to reach levels considerably higher than the usually high European standards. Prices for cattle and pigs also rose in 1972–3. Thus the impact of the EEC on prices was surprisingly little felt by British producers when, on 1 January 1973 the UK with Ireland and Denmark joined the EEC and on 1 February adopted the Common Agricultural Policy.

The biggest problem facing the operation of the CAP in the mid-1970's was one of interstate monetary relationships. In Britain the " Green Pound " based on the sterling values of early 1973, expressed EEC prices in Units of Account. To balance the difference between the " Green Pound " value and the real value of sterling after serious inflation, Monetary Compensatory Amounts are payable. A reluctance to devalue the " Green Pound " according to the decline in sterling rendered UK domestic food prices lower than they would otherwise have been.

ADDITIONAL FEATURES OF BRITISH AGRICULTURE SINCE 1945

Structure of British Agriculture. By comparison with many European countries and in particular those subject to the Napoleonic

Code, Britain has benefited from a farm structure based on a typical " ring fence " holding. Such holdings, usually remaining intact between generations have been capable of intensification and expansion as occasion has demanded. Thus the great changes of the post-1945 period have been contained within traditional patterns but on

HOLDING NUMBERS BY AREA SIZE GROUPS
(England and Wales)

Size Group Hectares.	1955		1960		1965		1970	
	No.	%	No.	%	No.	%	No.	%
120 plus . .	13064	3·5	13800	4·2	15300	5·1	21174	9·1
40–120 . . .	64029	17·3	62600	19·0	59300	19·6	57313	24·8
20–40 . . .	59556	16·1	57600	17·5	53100	17·6	42413	18·6
8–20 . . .	66222	18·0	61300	18·7	54200	18·0	41964	18·0
under 8 . .	166694	45·1	133400	40·6	119800	39·7	66770	29·5
TOTAL . . .	369565	100	328800	100	301800	100	233408	100

Source: HMSO Agric. Statistics converted and rounded

larger scales and with much concentration of enterprises. Over the period holdings fell in total numbers but size changes increased the proportion of larger holdings.

In terms of area, holdings of over 120 ha in England and Wales

DISTRIBUTION OF HOLDINGS BY SIZE OF BUSINESS
(in SMDs)
UNITED KINGDOM

	SMD.	1969	1974
Number of Holdings ('000) with .	Under 275	142·9	115·4
	275–599	83·2	66·5
	600–1,199	54·6	55·0
	1,200 plus	34·2	41·1
	TOTAL	317·0	278·0
Holdings 275 SMDs and over . .	Average Size—SMD	953	1,114
	Average Area—(ha)	89·5	96·6
	Contribution to output %	92·2	94·3
Estimated No. of full-time farms ('000)	Under 600 SMD	102·0	84·5
	600 SMD and over	88·1	91·6
	TOTAL	190·1	176·1

Source : HMSO Cmnd 5977 (adapted)

rose progressively from 28·2 per cent in 1960 to 45 per cent in 1970. Similar trends occurred in Scotland and Northern Ireland so that farms in this size range accounted for more than half of the major products of British agriculture. In 1974 the simple average of size of all holdings was 44 ha.

In order to express size of business the Standard Man Day is used. Individual SMD values are applied to enterprises on the farm and are totalled for each holding. Holdings with 275 SMDs and above are taken as full-time businesses.

Concentration of Enterprises. Developments involving concentration of dairy cows, pigs and poultry have already been mentioned. In many arable crops the number of growers has been falling for some years. This has been marked in potatoes where growers of less than 4 ha have virtually disappeared. In 1970 52·7 per cent of the crop was grown by 30 per cent of the producers in the size-of-holding group 60–280 ha. The average annual increase in the size of the potato enterprise between 1960 and 1968 was: 3·1 per cent between 1960–3, 8·0 per cent between 1965–8 and 6·4 per cent for the whole period. Similar rates apply to barley and wheat. Thus the tendency towards greater specialisation is very apparent, implying a more efficient use of such increasingly scarce resources as labour and capital.

Full-time and Part-time Farms. It was estimated that in 1968 there were as many as 140,000 part-time holdings in the UK, equivalent to 40 per cent of the total. These holdings accounted for only 7 per cent of total SMDs.

Land Tenure.

TENURE OF AGRICULTURAL HOLDINGS
(England and Wales)

	Numbers ('000)					
	1950		1960–1		1970	
	No.	%	No.	%	No.	%
Owned . . .	138·7	36	157·7	47	107·8	46
Rented . . .	185·0	49	123·7	37	71·6	31
Part Owned . .	56·3	15	51·8	16	53·9	23
Part Rent . .						
TOTAL . . .	380·0	100	333·2	100	233·4	100

Source: HMSO Agric. Statistics

In England and Wales the traditional preponderance of rented over owner-occupied land changed somewhat in the 1950's and 1960's. By 1960 47 per cent of the farms were owned and occupied, 37 per cent rented and 16 per cent of mixed occupation. In 1970 the rented proportion had fallen to 31 per cent—though these farms were, on average, larger than the owner-occupied. The situation in Scotland was much more stable with 61 per cent of farms rented in 1961 and

59 per cent in 1969. Owner-occupied farms tended to be larger than rented holdings. In Northern Ireland the traditional situation obtained whereby almost 100 per cent of the land has remained in owner-occupation—together with " conacre " or annually let land, as a flexible extra.

Contract Farming and Vertical Integration. With the trends moving towards increased specialisation and standardised production techniques there has been a tendency for some enterprises to become identified with processing industries. Thus vined peas have been grown from seed supplied by the freezing or canning company and cultured and harvested to the instructions of the fieldman employed by that firm. Chicken may be supplied to the producer by the company which will process the poultry, payment being made on the headage basis. Though fears were expressed that farmers might lose their freedom in this way and become mere employees of factory interests the trend has not developed to any general extent. The independence and resilience of the farmer has remained unimpaired. Perhaps for the same reasons, repeated attempts to foster close cooperation between operators for particular enterprises or for joint management have never made substantial progress.

ENVIRONMENTAL CONSIDERATIONS AND AGRICULTURE

Chemicals. As we have noted, the progress made by farming after 1945 owed much to new chemical materials. Compounds were produced to control pests, diseases and weeds. In addition, improved types of fertiliser of greater facility in application and composed of more concentrated nutrients (e.g. urea) became available. Among the insecticides was a large group of materials developed from organic acids with chlorine ions. They were notable for being non-toxic to mammals and for possessing great persistence. These were desirable assets as far as the purely farming objectives were concerned—a single application being capable of remaining effective over a long period. For instance the chemical Dieldrin kept a sheep free from the dreaded maggot fly for an entire season with one dipping. DDT was used to control the Anopheles mosquito which is the vector for the malarial parasite. However the degree to which mammalian body tissues absorbed such chemicals appears not to have been recognised until long after these chlorinated pesticides were put into general use. With the development of a new laboratory technique—gas chromatography—residues at very low concentrations became detectable. The presence of organo-chlorine materials was established not only in mammalian body-fat but also in certain species of birds including some birds of prey. When animals consume other species residues may accumulate in body organs. Thus in a " food chain " there may be a progressive concentration of persistent chemicals reaching a peak in flesh-eating fauna.

This sequence is thought to have been the cause for the drop in numbers of certain raptores, e.g. the Peregrine (feeding on pigeons—themselves grain-feeders) and the Golden Eagle. The incidence of low egg viability as well as direct toxicity was supported by much circumstantial evidence. Nevertheless a great deal of public interest was attracted to the subject of chemical usage in farming and around the countryside. This was highlighted by the American authoress and biologist, Rachel Carson, whose book *Silent Spring* was very widely read. Though the subject was often attended by much ill-informed comment, farming in Europe and America became the object of critical attention. The British government took action to reduce and phase out most of the offending materials. Public awareness of the environment, however, had now become a factor which farmers and manufacturers alike were obliged to take seriously. Pressures to reduce pollution of air and water grew during the same period, and on the farm it became an offence to allow effluent to enter main water courses.

The Countryside. Along with the public sensitivity over the use of chemicals came an increasing demand for more access to the countryside. As very extensive holders of the land area of Britain farmers have long been faithful custodians for the community at large.

After the second world war National Parks were officially established. The name does not imply, as in other countries, that the parks are devoted to recreation to the exclusion of farming. They were delineated in areas of natural beauty ; development has been controlled and the character of the region preserved by careful and unobtrusive management. In 1968 a Countryside Act was passed with provisions for improving the means by which the population might further enjoy the open country. Local Authorities were empowered to arrange for parks, recreation sites, footpath identification and other amenities. With the increased usage of the countryside by townspeople using their own transport, farming has benefited in some parts of the country by supplying accommodation, camping sites, fishing, pony-trekking and even conducted visits on the farm. It is an ironical fact that high soil fertility and outstanding scenery are at opposite ends of the scale in Britain. Thus farmers in scenic areas are encouraged to develop natural advantages to offset other limitations.

Intensive Livestock and Public Reactions. Reference has been made to intensive methods of livestock production which were developed after 1950. Poultry were traditionally kept on " free range ", but the use of cages proved to be a more efficient method both for egg production and economy in manpower. By 1960 free range became a rare form of poultry husbandry—there was even a development of roadside sales by which the small producer secured premium prices for eggs produced in this way. Intensive methods were also developed for pig production, including the so-called " sweat house " or high-humidity system. In this, both growth rates and feed conversion efficiency benefited and manual labour could be considerably reduced. In the very specialised system of intensive veal production, the young calves are kept under conditions of restricted lighting and with a diet designed to give the desired white meat.

These and similar farm livestock techniques attracted increased public attention and criticism. The term " Factory Farming " came into use, and to a nation predominantly urban but strongly animal-loving, horrifying images were conjured up of suffering livestock. Ruth Harrison led an active campaign against intensive methods. Allegations of cruelty were voiced with such vigour that a government committee was set up. This resulted in the Bramham Report, following which legislation was introduced to lay down standards to be observed by farmers in safeguarding the welfare of animals. Yet again the agriculturalist was surprised to find himself vulnerable to public scrutiny. No longer a folk character, his new role as a businessman-producer was difficult for his town cousin to understand or accept.

The use of antibiotics to improve the performance of certain animals—notably pigs, poultry and young beef stock was also a subject of inquiry. To avoid a carry-over of residues to the consumer, safeguards were introduced which restricted the use of such materials as terramycin and penicillin.

Throughout the three post-war periods there was a continuing search for aids both major and minor to achieve greater technical efficiency. Highly developed chemical and engineering industries both at home and abroad acted quickly upon research findings and placed in the farmers' hands materials and techniques which he was remarkably quick to employ. Not every innovation was sound and often independent testing was not awaited. Sometimes experience was dearly bought but fortunately there were few calamities.

VETERINARY ASPECTS OF THE POST-WAR PERIODS

The fight against animal diseases, particularly those occurring on epizootic scale, has long exercised the farmer and the veterinarian. With the increased level of intensity of livestock production since 1945 the battle to retain control over both old-established and entirely new diseases became very fierce indeed. The advantages were by no means always on the side of the veterinary surgeon and the pressure on diagnosis techniques, epidemiology and immunology as well as extensive control measures severely taxed the resources of both government and private services.

Foot-and-mouth disease, which affects cattle and pigs, is capable both of sudden appearance and an alarming rate of spread. A number of localised outbreaks occurred between 1945 and 1967 and these were isolated and controlled using the well-known " stamping out " procedure. In 1967, however, following an outbreak of the disease in pigs in the West Midlands an unprecedented spread took place, bringing movements of stock to a halt over the entire country. A total of 2,364 cases in a period of 9 months severely tested the established control measures. The slaughter of 211,825 dairy cows had an immediate, though temporary effect on milk supplies. The cost of the replacement of animals slaughtered was high—government compensation payments alone totalled £27 million and many associated losses had to be made good.

Another example of a severely epizootic disease was provided by the poultry enterprise. Newcastle disease, or Fowl Pest, was first identified in 1926. It reappeared in 1947, and with the development of large poultry units it became easily transmissible. At first the disease had been controlled using the stamping out policy of slaughter and compensation, but by 1963, despite every effort, the disease gained ground and a voluntary vaccination system was introduced. A serious resurgence of the disease occurred in 1971 resulting in 7,000 outbreaks encompassing every English and Welsh county. It involved 42 million birds and appears to have resulted from inadequate vaccination procedures on the part of some owners. As a result redoubled efforts to tighten up farm-control measures were instituted.

Bovine tuberculosis was officially eliminated as an enzootic disease in 1960 after a long campaign. Nevertheless continuous testing and strong action for breakdowns have been continuing commitments. In the west of England badgers were shown to be carriers of tuberculosis—necessitating urgent local slaughter.

Swine fever was virtually eliminated in 1966. This was after a slaughter policy which was started in 1963. Attempts to control this long-established disease by the voluntary use of vaccines proved to be unsuccessful. The much-increased size of modern pig units is a direct result of the disappearance of swine fever. However, two other new pig diseases soon arrived on the scene. Transmissible gastroenteritis has provided intensive producers in some areas with a most difficult disease to control. The second appeared in 1972 as an entirely new disease with symptoms resembling those of foot-and-mouth. It was named Swine Vesicular Disease or SVD. By 1975 a large number of outbreaks of SVD had occurred—most being associated with swill feeding. New legislation to ensure proper sterilisation of waste food sources was introduced.

Brucellosis, causing abortion in cattle and Undulant Fever in man, had long been known when, in 1971, a gradual eradication campaign was commenced. The undertaking has been especially difficult to organise since the vaccine S19 had been used on farms for many years and this tended to confuse the interpretation of blood tests.

As with human medicine the introduction of antibiotics after 1945 proved to be of great benefit. However, changes in the antibiotic sensitivity of various bacteria have necessitated the production and use of more potent preparations in the treatment of infections.

Old problems such as mastitis in dairy cows were still causing much economic loss in the mid-1970's. This was despite the existence of antibiotic treatments and wide-scale publicity on control measures. The introduction of BHC-based sheep-dips in 1948 led to a rapid conclusion to the campaign against Sheep Scab, and the last case had occurred in 1952. However, imported sheep from Ireland were thought to be the origin of a series of outbreaks which began in December 1972.

Investigational work in the animal health field immediately benefited from the establishment in 1946 of Ministry of Agriculture Investigation Centres. The staffs, acting as specialist consultants for private veterinary surgeons, also carried out experiments on local problems. Veterinary research work during this period was also greatly intensified at specialist institutions throughout the United

Kingdom. With increasing world travel and shrinking distances animal disease problems become progressively international in scope. Research stations become part of global referencing and testing procedures, thus contributing to large-scale control measures.

Chapter I

THE SOIL

THE scientific study of the soil embraces a great deal more than a series of chemical analyses, and consequently it is important at the outset to appreciate the distinction to be made between soil as a *material*, and soil as a *natural body*. The difference between a handful of dry earth in the laboratory and the undisturbed soil in the field is just as great as between a handful of hay and the living, uncut grass. Fundamental knowledge of the soil must be sought in three main directions : the chemical, the physical and the biological. Questions relating to the nature and proportions of soil material generally require the chemical approach ; considerations such as the movements of water in soil, and the manner in which soil material is built up to form the soil body in the field, are examples of soil physics ; whilst the biological aspects include studies of soil bacteria, fungi and animals, and the relationship of soils to their natural vegetation. These represent the more significant features about soil in relation to the business of farming, and they will now be considered further.

SOIL FORMATION

Many quarry faces show on inspection a gradual transition between the original solid rock and the surface soil, such as depicted in Fig. 1. As the eye travels upwards cracks and fissures can be observed, which grow more definite until separate masses of rock are apparent. These become smaller, and the fissures more numerous and more open, until soil appears between the rock fragments and the general condition is brashy. At a still higher level the predominant constituent of the mass is obviously soil, and finally the top few hundred mm become darker in colour and are found to contain the root systems of the herbage growing on the surface, with other accumulations of organic matter. Throughout the transition the significance of the original rock and its more obvious characteristics can be readily observed, and it will be appreciated that the rock constitutes the parent material from which the soil has developed through the action of various natural agencies such as climate and vegetation. Any source of information concerning the various kinds of parent materials distributed throughout the country is obviously of considerable importance in the study of soil ; it is primarily in this respect that some understanding of geology, the science which is concerned with the origin and nature of the earth's crust, is of value to the agriculturalist.

Geology and Soils. The geological sequence of events in the earth's history was probably as follows. Presuming the earth to have been formerly a molten mass at high temperature, the first formed rocks arose from the cooling down of the surface with the development of a hard outer crust with a still molten interior. These

I

rocks, because of their association with heat and fire, are known as *igneous* rocks. Subsequently further igneous rocks were, and are still being, formed by volcanic eruptions. Typical of igneous rocks are granite and basalt, which are hard and fairly resistant to the weathering effects of climate ; for this reason, and also because of their often volcanic origin, they are generally associated with mountainous areas.

FIG. 1.—DIAGRAMMATIC REPRESENTATION OF SOIL FORMATION FROM UNDERLYING ROCK.

Once a shell of igneous material began to form, local variations of temperature would provide the beginnings of a climate, and the concentration of water in low-lying areas would establish seas and lakes. From time to time, however, gigantic earth movements brought about an entirely new geography and a redistribution of climate, thus instituting new geological eras. During every such era the natural agencies would be at work on exposed rock surfaces, weathering them down to masses of broken material which might be transported and deposited elsewhere to form new rock masses called *sedimentary* rocks.

Sedimentary rocks are almost all composed of small rock particles which have been carried along by rivers and deposited on the floor of seas, lakes and estuaries. These sediments later become compressed by the weight of overburden and cemented together into new rock forms ; included amongst these are sandstones, shales and clays. A less-abundant but none the less important group is of organic origin ; that is to say, its constituents have been accumulated as the result of living organisms. The most prominent members are chalks and limestones, composed of the shells of marine organisms of various sizes, and coal which has resulted from preserved plant remains. It will be noted that clay is in the geological sense classed as a rock when it constitutes a deposit laid down in a well-defined geological era. The earlier formed sedimentary rocks were later themselves partly weathered and transported to form fresh deposits in subsequent eras each of which may represent a span of millions of years ; thus a whole sequence, formed throughout geological time, has been traced by geologists and embodied in a geological classification of the earth's crust.

In geologically more recent times events during various " ice ages " resulted in vast quantities of surface material being pushed across land areas in front of moving glaciers and deposited where the latter finally melted. Thus in certain parts of the country, formerly subjected to the action of glaciers, extensive sheets of surface material may bear no relationship to the underlying sedimentary or igneous rocks ; these deposits are styled as *glacial drift*.

The distribution of igneous rocks, sedimentary rocks, and drift throughout Great Britain, has been studied by geologists and depicted upon geological maps ; published information, in the form of district memoirs, is also available regarding the actual nature of the various deposits concerned.

The Composition of Rocks and Minerals. Igneous rocks, as has already been indicated, appear to have been formed by the cooling of original molten material. During this process the various chemical elements present become formed into various distinctive minerals. A mineral may be defined as a rock component having a definite chemical composition and a definite crystal structure. Thus a study of the mineral composition of igneous rocks, and of the manner in which they are decomposed by weathering, affords some idea of the kinds of materials which ultimately form the basis of soils.

The make-up of igneous rocks, in general, includes such chemical elements as oxygen, silicon, aluminium, iron, calcium, sodium, potassium, magnesium, titanium, carbon, phosphorus and sulphur, in the form of various minerals. The approximate percentage proportions of these elements present in the earth's crust are oxygen 50, silicon 25, aluminium 7, iron 5, calcium 3·5, sodium 2·5, potassium 2, magnesium 2, titanium 0·5, carbon 0·1, phosphorus 0·1 and sulphur 0·06 per cent ; the actual amounts, however, vary for different types of rock, a factor of significance in relation to the derived soils. Those classes of igneous rocks in which the proportion of silicon, which is non-metallic and therefore acidic, is greater than 25 per cent are known as *acidic* rocks ; where the silicon content is less than 25 per cent, together with a consequent increase in the proportions of calcium, sodium, potassium or magnesium, which are metallic and therefore basic or non-acidic, the term *basic* rock is applied. Granite is an example of an acidic rock, basalt of a basic rock.

If a freshly broken piece of granite be examined it is seen to consist of large white or pink crystals together with much smaller dark or black crystals, firmly set in an irregular mass of whitish crystalline material. The large white or pink crystals represent a mineral known as felspar, which is a potassium (or calcium) alumino-silicate ; the small black crystals are the mineral mica, which contains in addition to alumino-silicates appreciable proportions of iron and magnesium. Mica is thus one of the so-called ferro-magnesian minerals. The whitish crystalline mass in which the other crystals are embedded is a mineral known as quartz ; quartz is a crystalline form of silica, or silicon oxide, a compound composed of silicon and oxygen. It is the presence of this high proportion of quartz which makes granite an acidic rock, and this acidic tendency is transmitted to soils derived from granite. Basalt differs chemically from granite principally in the elimination of quartz, its place being largely taken by ferro-magnesian minerals.

Since sedimentary rocks are in the first place derived from igneous rocks it is not surprising that they are often composed of some of the minerals already mentioned ; owing to the processes of weathering and transportation which have occurred, however, the proportions of each present have been altered and the range considerably restricted. Thus most sandstones have a predominant content of quartz and are more accurately known as quartzose sandstones ; if mica or felspars, reduced to the size of sand particles by weathering, are present in appreciable quantities the sandstones are known respectively as micaceous sandstones and felspathic sandstones. The reason for the predominance of quartz in the sedimentary sandstones will be more fully appreciated after a consideration of the mechanism of weathering.

The Processes of Weathering. Weathering is the result partly of the operation of physical forces and partly of chemical reactions on original rocks. Variations in temperature produce expansions and contractions of the rock minerals and, since the different classes of minerals respond to differing extents, stresses and strains are set up in the rock as a whole, causing it to crack and shatter. Water enters these cracks in the rock mass and, if freezing occurs, the expansion of the resultant ice produces an enormous disruptive force. The result is that the rock is more widely split, and it may finally shatter down to a deposit of angular fragments. The natural agencies already mentioned, such as running water or glaciers, may subsequently transport the fragments, grinding them together and against the surfaces over which they move in the process, thus carrying disintegration still further. All these actions have as their primary effect simply the reduction of a single mass of rock to a large number of small rounded particles which may ultimately collect to form a deposit of sand or sandstone. Since there has as yet been no alteration in chemical composition the processes so far outlined constitute what is termed *physical* or *mechanical* weathering.

Concurrently with physical weathering, *chemical* weathering is also taking place. Here the most potent force is the action of water which in the passage of time very slowly dissolves certain of the materials of which the rock particles are composed and sets up chemical reactions on their constituents. The solvent action of water is considerably enhanced by the fact that rain in its passage through the atmosphere acquires a certain amount of carbon dioxide and becomes weakly acidic ; this solvent action is increased by further contact with carbon dioxide as the water percolates through the surface soil. The general trend of the reactions which take place is the splitting up of some of the complex minerals in the rocks, removing certain of their ingredients dissolved in the drainage water and leaving others behind as insoluble residues. These decompositions are very complicated because of the considerable number of varying factors involved, but the final result will be largely determined by the nature of the original rocks and the prevailing climatic conditions.

The issues may be more clearly understood by considering specific cases. When granite weathers chemically it is found that its quartz is highly resistant and is very little affected ; the felspars lose their potassium as soluble potassium carbonate, leaving behind a residue of hydrated alumino-silicate, which is a variety of clay ; the mica similarly loses soluble constituents, and clay-like residues may accumulate, often coloured brown or red by the liberated iron compounds. Thus the weathering of granite results broadly in the formation of stones, sand and clay, which may go to form a soil on the spot, or may be transported away to become the basis of various sedimentary sand and clay deposits, which may in their turn be subjected to further weathering in a subsequent period.

It will be appreciated that these weathering processes still continue after a soil has been considerably developed from its parent material. They are, in fact, the means by which mineralised chemical substances are continuously being liberated as plant food. This ultimate weathering will be referred to later in the chapter, in relation to soil classification.

Biological Factors in Soil Formation. While the processes of weathering are producing the mass of mineral particles of which most soils are mainly composed, various forms of life have developed on and in the soil, and they in turn have exerted important effects. The most obvious of these living organisms are plants whose stems and leaves are supported in the air by means of an underground root system. Plants normally develop from seeds buried in the soil which, given suitable surroundings of oxygen, moisture and warmth, germinate and produce both a root and a shoot. If the supply of food materials in the seed suffices to enable the shoot to reach the surface by penetrating the upper layer of the soil, the shoot produces small green leaves which assimilate carbon dioxide from the air. At the same time the rootlet absorbs moisture containing small quantities of nutrient material derived from the mineral matter of the soil. The substances thus absorbed—the carbon dioxide, the water and the mineral substances—are, with the aid of energy derived from sunlight, elaborated by the plant into complex organic compounds such as starch, cellulose and lignin with which the plant builds up its structure and makes growth. When the plant dies it decays, and from its remains a decomposition product known as *humus* may be formed in the soil.

In addition to the plants visible upon its surface, the soil teems with living organisms of many different kinds, from earthworms and insects down to microscopic forms such as fungi and bacteria, all of which play their part in the development of the soil complex. In the past half-century micro-organisms have been the subject of much research work so that today soil microbiology represents a very important branch of soil studies. It is now known that many micro-organisms are dependent for their food upon the organic matter accumulated by plants : this food they obtain by attacking and decomposing plant material. Humus is a by-product of processes of this type. Other micro-organisms have the ability to assimilate atmospheric nitrogen, which they accumulate in their own structures and which may ultimately become available as plant food.

As well as enriching the soil with various forms of organic matter, living organisms undoubtedly further the general processes of weathering soil minerals. The mechanical effect of plant roots in disrupting rocks by penetrating into cracks and fissures is fairly well known ; less obvious are the slow grinding effects which earthworms and kindred creatures exert on soil mineral particles. The excretions of various organisms may also assist in chemical weathering.

It is to be expected that what might be termed the living or active part of the soil would be the surface layer, and examination generally shows this to be the case. The content of organic matter, insect life, bacteria and plant roots fairly rapidly decreases with increasing depth. Consequently, from the point of view of its employment for crop production, the top soil has very different properties and capabilities from the subsoil, a fact of great importance in connection with such questions as the depth of ploughing, the value of subsoiling and the most suitable system of cropping.

THE PHYSICAL PROPERTIES OF SOIL

So far the soil has been considered from the point of view of its development and of the natural phenomena which influence this. It is necessary now to consider the composition and properties of the mineral and organic portions : the internal conditions of moisture, temperature, and aeration, which constitute *the soil climate* : and the nature and functions of the associated living organisms.

Texture and Soil Material. Texture is the property which largely determines the ease or difficulty of cultivations, and walking on the land is one of the most impressive ways of appreciating its significance. During wet weather we have at one extreme, slippery, sticky soils where walking is difficult owing to tenacious accumulations on the feet, and in which moisture is obviously excessive ; at the other extreme are non-plastic soils, which exhibit no adhesive tendencies, and on which walking is comparatively easy and where conditions are dry underfoot. By contrast, in times of drought the first set of conditions is replaced by those of hard clods, or of a rock-like surface traversed by numerous cracks, while the latter give place to a loose, dry, friable condition in which the individual soil particles show little tendency to cohesion. These examples of soil texture are due primarily to the nature and quantity of the various mineral particles present. If a handful of ordinary soil be examined closely by the aid of a pocket lens, it can be observed that the mineral particles are of all shapes and sizes, diminishing from obvious pebbles. If a little of the soil is moistened and rubbed between the fingers the gritty nature of the larger particles is revealed, even when to the eye their proportion is extremely small ; at the same time it may be realised that the greasy feel and stickiness of a soil are attributes of its finer particles. In ordinary parlance the terms " sand " and " clay " imply some of these impressions of grittiness and stickiness, the former being responsible for uncohesive and loose conditions and the latter for tenacious and plastic qualities.

Size of Particles. To give precision to such terms, and to facilitate analytical comparisons between different soils, a convention has been established with respect to particle size. On this basis, particles with a diameter exceeding 2 mm are considered as stones, whereas those particles of less diameter than 0·002 mm are considered to have the properties of clay. In the natural condition, as has been previously mentioned, soil particles may not all be separate from one another but may be aggregated into crumbs and clods of varying sizes. For the purpose of analysis of particle sizes any soil sample must undergo a preliminary treatment to destroy the crumbs ; when this is accomplished the larger particles can be removed by sieves of appropriate dimensions. The smaller particles, however, can pass through the meshes of the finest sieves practicable for this work, so recourse must be made to another method of separation. This is based on the fact that particles of the same material, but of varying sizes, sink through a liquid medium such as water at different rates, dependent upon their diameters. By combining the two processes of sieving and sedimentation any soil sample can be divided into a series of arbitrary fractions whose limits are set out below.

Fraction.	Diameter of Particle.
Coarse sand	2·0–0·6 mm
Medium sand	0·6–0·2 mm
Fine sand	0·2–0·06 mm
Silt	0·06–0·002 mm
Clay	less than 0·002 mm

The process of separating the mineral portion of a soil into these fractions is known as a *mechanical analysis*.

The differing textural properties imparted by stones, coarse sand, fine sand and silt are largely the result of their diminishing particle size and the particles may in fact merely represent successive stages in the degree of sub-division of an original rock mass. Clay, on the other hand, as has already been indicated, is derived from the chemical decomposition of larger mineral particles and therefore it can impart additional textural effects which are partly due to its changed chemical nature as distinct from the further reduction in particle size.

Specific Surface. The effect of " specific surface " of soil particles is, however, of very great importance and merits a more detailed explanation. If a cube with 1-m sides be considered, the total area of its six faces is 6 m². If this cube is now sliced up into smaller cubes each with sides of 0·1 m there will be 1,000 small cubes and the total area of each cube will be 6 × 0·1 × 0·1 = 0·06 m². Since, however, there are 1,000 of them their total area will be 1,000 × 0·06 = 60 m² ; thus by subdividing the material its total surface area has been increased. The more finely the material is subdivided the greater will be the total surface area for a given quantity. It has been computed that the total surface area of the particles of a cubic metre of sand is 6,000 m², whilst that for a cubic metre of clay is 100,000 m² ; it is therefore the clay fraction which is responsible for by far the greatest proportion of the internal surfaces of soils. If these surfaces are visualised as covered by, and retaining, a uniform film of moisture, the connection between the water-holding powers of the various-sized soil fractions begins to reveal itself.

THE PROPERTIES OF CLAY

Apart from the retention of water by the surfaces of all the particles of a wet soil, water can also be held internally by the clay particles. This special capacity for water absorption causes clay particles to expand on wetting and shrink on drying. Thus wet clay soils tend to present a barrier of tightly packed particles impervious to the passage of surface drainage water, whereas a dry clay soil is permeated by wide and deep cracks. Furthermore, since clay particles have the special property of cohesiveness, a mass of drying clay will tend to form clods between the fissures. This can be contrasted with coarse sand which shows practically no cohesion and water absorption, and allows free drainage ; and with fine sand which does show some cohesion on drying, but is easily crumbled and permits free movement of water. Silt more closely resembles clay in its physical properties.

As has already been indicated, the great reactivity of clay is not due entirely to its extensive internal surfaces. The clay fraction includes hydrated alumino-silicate compounds that have been formed from the decomposition of minerals present originally in unweathered rocks. Certain of these compounds have themselves characteristic crystalline structures and are known as " clay minerals " ; amongst these can be mentioned kaolinite, beidellite and montmorillonite. In addition to recognisable minerals, such as those mentioned, there are also present in the clay fraction substances that are chemically reactive but of indefinite composition, together with hydrated alumina and free hydrated iron oxides which confer yellowish or reddish soil colours.

Colloidal Behaviour. Many of the properties exhibited by clay are characteristic of a class of substances known as *colloids*, which are substances of extremely small ultimate particle size often showing considerable affinity for water. An important feature of colloids is the ability of their particles to aggregate into larger composites (flocculation) or to be dispersed into separate ultimate particles (deflocculation) according to the surrounding conditions. In the presence of a sufficient concentration of calcium a clay will tend to become flocculated, thus increasing its particle size and consequently decreasing its surface area and surface retention of water ; if calcium on the clay is partially replaced by sodium on the other hand a clay will tend to deflocculate, thus decreasing its particle size and increasing its imperviousness. This has important practical implications in that drainage, aeration and ease of cultivation in a clay soil can be appreciably increased by adequate liming whereas, in the absence of lime, acidity induced by carbonic acid will tend to degrade the physical conditions. It is in this respect that clay shows important differences from silt which, although exhibiting the properties of plasticity and cohesion, is not sufficiently colloidal to be flocculated into an agriculturally more desirable condition by liming.

Cation Exchange. Clay is potentially of an acid character and the clay complex behaves in many respects as a weak insoluble acid ; thus it can combine with various basic (non-acidic) substances to form neutral compounds. The principal soil base in the British Isles is lime (calcium oxide). When a clay has combined to its fullest extent with bases it is said to be " base-saturated ". If insufficient bases are present the clay will make up the deficit by combining with hydrogen derived from soil acids and will reassert its acidic tendencies. Normally a neutral clay is a predominantly calcium clay, with smaller amounts of magnesium, potassium and sodium ; an acidic clay is largely a hydrogen clay. These bases and hydrogen are exchangeable, one for another. Thus if an acidic clay is limed, calcium is taken up by the clay and hydrogen is displaced into the soil solution. Similarly, if sulphate of ammonia is added to a calcium clay, ammonium (a base) is taken up by the clay and calcium is displaced to form soluble calcium sulphate, which is subsequently lost in the drainage water. This phenomenon, which is characteristic of many colloidal substances, is called *cation exchange* ; the ability of a given quantity of clay, or soil, to retain exchangeable bases is called its " cation exchange capacity ". It will therefore be appreciated that two main forms of calcium can exist in a soil, namely calcium directly combined with the clay (exchangeable calcium) and calcium in the form of calcium carbonate, which is not at the time chemically combined with the clay but which may form a potential reserve, later to be taken up as exchangeable base material.

Thus in addition to conferring upon a soil its water-holding capacity, clay is also to a considerable extent responsible for the retention of various essential plant nutrients and soil improvers such as calcium, potassium and magnesium.

THE ORGANIC MATTER OF THE SOIL

Particular emphasis must be placed upon soil organic matter as having an important effect upon the physical properties of soils. Soil organic matter can be considered as (*a*) undecomposed organic matter, and (*b*) decomposed organic matter or humus. In the first category are included recognisable plant remains such as fragments of roots, stems and leaves ; in the second category come dark-coloured, amorphous materials, not readily distinguishable from the main soil mass, which are collectively known as *humus*. It is a common error to regard soil organic matter and humus as being identical, but such is far from the case. It is more accurate to recognise two separate classes, firstly " undecomposed organic matter ", and secondly " humus ", which is derived from the former through decomposition by soil micro-organisms.

Humus. The physical properties of the two classes are notably different. For instance, when straw, old turf, freshly made strawy farmyard manure, etc., are ploughed under they tend to prevent the soil particles from combining with one another ; in other words, undecomposed organic matter keeps the soil " open " ; this is desirable in heavy land but rarely so in light land. Humus, on the other hand, because of its colloidal nature, has those properties of water absorption and base exchange which have already been indicated in relation to clays. Humus can thus impart these desirable qualities to sandy soils, in which they would otherwise be deficient. Furthermore, humus has the property of facilitating the formation of soil crumbs, a function which it can best fulfil when endowed with an appreciable content of exchangeable calcium. The precise mechanism whereby humus enables crumb development is not clearly understood, but it can be considered as forming envelopes of colloidal material which enclose a number of individual soil particles and retain them distinctively from other similar units. Humus, by giving light soils enhanced water-holding and base capacities, and by giving heavy soils a crumb structure which increases porosity and aeration, is consequently an indispensable factor in soil fertility.

Humification. The chemical functions of humus are obscure, as a brief consideration of its mode of formation will indicate. Soil organic matter may decompose either by oxidation or by humification. Oxidation can be regarded simply as akin to burning, in which the organic matter is dissipated as carbon dioxide and water, leaving a small mineral residue ; it is caused by certain micro-organisms and favoured by moisture, warmth, absence of undue acidity and, above all, by a good supply of air. In these circumstances, which are to be found in light, well-drained soils, and under arable conditions, organic matter can disappear very quickly.

Whilst oxidation is destructive, humifying processes are by contrast conservative in their effects.

The commonest kind of humification is an anærobic humification which takes place in the absence of, or with a restricted supply of air. Such conditions are outstanding in water-logged areas, but even in well-drained soils there are times when there is a sufficiently high moisture content and enough compaction for anærobic conditions to be brought about. Humus is derived largely from the fibrous component of plant tissues, called lignin, combined, through the activities of specific micro-organisms, with nitrogenous material obtained partly from nitrogen compounds already existing in the soil : consequently, when fresh organic matter is ploughed-in the level of readily available soil nitrogen may be reduced whilst humification is proceeding.

Another important factor in humification is that of base status ; in the presence of lime the humus formed will be a calcium humus with properties similar to calcium clay ; it is called " mull humus ". But, in the absence of lime, such as occurs on dry heath, pine forest, upland moor, etc., a different product known as " mor humus " results. Furthermore, the mineral content of soil organic matter may vary considerably according to the class of plant material from which the organic matter has been derived. If the original plant material was a product of fertile land its considerable content of calcium, potassium, phosphorus, etc., will be ultimately conveyed to the soil, which becomes enriched in both humus and mineral matter. If, however, the original plant material was itself of low mineral status, then it can do little in this respect ultimately to improve the soil, and its only contribution can be to provide organic matter for possible humification ; its main effect on the soil will probably be the long-term one of physical betterment. It should thus be clearly understood that organic matter is not necessarily the same thing as humus, and that neither is necessarily synonymous with fertility. For example, if an old phosphate-deficient pasture is ploughed-in, succeeding crops will be unsatisfactory in the absence of phosphate manuring in spite of the abundance of organic matter, a state of affairs which has sometimes caused crop failure following reclamation of deficient land.

SOIL STRUCTURE

Everyone accustomed to handling or cultivating soil will have noticed that the smallest particles tend to stick together in lumps which may show great resistance to further breakdown. A soil in which these aggregates have been fully broken down is said to be " dispersed " ; energetic chemical and physical treatment is normally necessary to achieve this. The study of the size grades of dispersed soil particles gives information about soil *texture* and by contrast the study of the aggregates themselves, their nature, stability and size grades gives information about soil *structure.*

The study of soil structure is of importance in practical cultivation and the more intensively the land is cropped the greater will be the importance of maintaining a good structural condition. Tillage operations in general, the trampling of stock, the passage of heavy implements and the impact of rain on bare ground all tend to harm the soil so far as its structure is concerned. A soil of good structure will be well drained and aerated, will provide a good medium for vigorous root growth, and will be resistant to harm done by necessary tillage operations and will tend to restore itself quickly to a satisfactory condition after it has been disturbed.

The unit of study in soil structural observations is the *ped* ; this is the natural aggregate formed in the soil and is contrasted with terms such as " clod " or " fragment " which refer to aggregates imposed by tillage implements and other external forces. The factors which should be noted are the extent to which the soil is aggregated into peds and the stability of the peds in the presence of conditions which tend to break them down. Other factors are the size and shape of the peds and the nature of their surfaces. Peds which are stable and persistent show faces that are well formed, and which are usually smoother than the interior when the ped is broken across ; they may also be of a different colour.

The shape of peds is important and a useful classification can be made as follows. Platy—with greater development of horizontal faces; prismatic—with vertical faces developed to a greater extent than the horizontal ; blocky—with roughly equal development of horizontal and vertical faces. Of these, blocky structures are the most wide-spread ; prismatic structures are most frequently found in clay sub-soils ; platy structure is somewhat less often encountered. The term blocky structure needs further subdivision. This can be " angular blocky " when the faces tend to be casts of the opposing faces, and " spheroidal " when the surfaces are curved and rounded showing little or no accommodation to the adjoining faces. When " angular blocky " structure is well developed the peds have sharp interfaces and vertices ; when some of these are rounded, the term " sub-angular blocky " is used. Spheroidal structure may be " granular " if the peds are relatively non-porous or " crumb " if the peds are porous.

Crumb structure is by no means a frequently occurring condition, but it is found in the topmost layers of old permanent grass where the influence of numerous rootlets and the presence of insects and fungi has produced a layer containing rounded, spongy aggregates of this type. While it persists it forms an excellent medium for plant growth.

Structural defects in the soil are particularly evident on newly restored land after industrial working has taken place. On this type of land proper growth of crops cannot be obtained by use of manures and fertilisers alone however generously these are applied. The land tends to be waterlogged in winter, and cracked and parched in summer owing to poor movement of water through the soil. Aeration is poor and erosion can often be serious on sloping ground.

For scientific work calling for an exact recording of changes in structure taking place under particular systems of husbandry, attempts have been made to devise laboratory methods of measuring soil proper-ties associated with soil structure. This is not an easy task owing to the complex nature of the phenomena, and no fully satisfactory methods have yet been generally accepted. Three factors are involved in these measurements. (1) The strength of the peds and their resistance to dispersive treatment. (2) The particle size grades of the peds as they are found in the soil. (3) The extent and nature of pore space, that is of the space both between one ped and the next and within the ped itself.

SOIL MOISTURE, AIR AND TEMPERATURE

Reference has already been made to various grades of soil moisture, including the moisture held by surface tension round the surfaces of soil particles, and the moisture absorbed internally by the clay-humus complex. Since, however, water supply is probably the most important factor in deciding the cropping abilities of a soil, the subject merits further consideration.

Moisture. The two categories of soil moisture referred to above are sometimes known as *capillary* water and *imbibitional* water : to these may be added a third, water actively draining through the soil and known as *gravitational* water. Given the same distribution of rainfall the proportions of these in different soils will depend upon the factors already mentioned, i.e. size of particle, amount of pore space and quantity of absorptive colloidal material present. Assuming the soil to be porous and the rainfall sufficient, the soil will ultimately become saturated with imbibitional and capillary water— the amount of which is termed the " field capacity " of the soil—and the surplus or gravitational water will drain downwards. Ultimately it will come to rest in a saturated underground zone known as the regional water table, the general level of which is indicated by the level of valley streams, lakes and wells.

It was formerly considered that capillary water could rise up to the surface from the water table, following a period of surface evaporation and serve as a source of moisture for crops, providing that the soil particles were pressed close together to form narrow conducting tubes ; and also that this capillary rise could then be checked and its loss by evaporation prevented by maintaining a loose open surface. On these assumptions the water-conserving effects of rolling and mulching were explained. Modern soil investigators now consider that capillary water is unlikely to rise above the water table for more than 1·5 or 2 m at the most, and they consider that the cultivations mentioned are effective mainly because they suppress weeds. It seems therefore that in many soils crops must rely, during periods of drought, mainly upon the water held at and in the soil particles themselves, since the soil surface may be many metres above the influence of the water table. Under these conditions the imbibitional and capillary water of the subsoil assumes considerable importance since it is more protected from evaporation. This emphasises the outstanding significance of certain factors of soil management. A satisfactory humus content should be developed in the surface soil so as to maintain the field capacity for moisture at the best possible level. Good drainage conditions should be ensured, including those induced by humus and lime, and those brought about by surface and under draining when necessary, so that plants may be encouraged to root as deeply as possible even at times when the soil tends to be water saturated, for a common cause of drought failure is the shallow rooting enforced upon crops by seasonal soil waterlogging. Finally, whenever possible, root penetration to the subsoil should be aided by deep cultivations, sub-soiling, etc.

Air. The amount of air present in the soil will depend upon both the pore space, including the spaces between crumbs and particles and the spaces inside crumbs, and the moisture content. In an ordinary dry soil the pore space may amount to as much as 50 per cent of the total volume, but will be considerably less under wet puddled conditions. In composition the atmosphere within the soil differs from the external atmosphere in two important respects; first, it is usually saturated with water vapour, and second it contains a much higher proportion of carbon dioxide. Whereas normal air has a carbon dioxide content of about 0·03 per cent, that of the soil is often over 1 per cent. This increase originates from the respiration of plant roots and from the activities of soil micro-organisms in decomposing organic matter ; it consequently varies from soil to soil and from season to season. The carbon dioxide thus formed dissolves in the soil moisture to form carbonic acid which, as indicated previously, reinforces the dissolving action of the water on mineral matter and thus plays an important part in rendering available plant nutrients.

Temperature. The outstanding consideration concerning soil temperature is not so much the daily and seasonal variations, which in a general way are related to atmospheric temperatures, but rather the rapidity with which a soil assumes higher temperatures in the spring consequent upon the removal of surplus winter rainfall by drainage and evaporation. It takes ten times as much heat to raise a given quantity of water through one degree of temperature as it does the same weight of dry soil, and it takes over 500 times as much heat finally to evaporate that water as water vapour. It follows that in a soil of high moisture content a very considerable proportion of the heat absorbed from the sun and from the atmosphere is used up in evaporating water when it might otherwise, under drier conditions, have been used in warming the soil. The removal of surplus water by adequate drainage is therefore a very important feature in inducing a more rapid rise in soil temperatures in the spring. Another important temperature factor is the aspect of the soil ; a soil facing to the south will receive a greater intensity of the sun's rays and will therefore warm more rapidly than a similar soil with a northern aspect. The colour of the soil is also not without significance, since dark surfaces are more effective in absorbing heat from the sun than are other shades. On the whole " early " soils are usually light textured and of southern aspect, whereas conspicuously late soils under the same range of climatic conditions are the northward-facing and badly drained heavy loams and clays.

THE CLASSIFICATION OF SOILS

The development of any branch of knowledge based upon scientific research can generally be traced in the increasing precision of its terms of reference, for it is a fundamental requirement of any pursuit claiming to be scientific that all its exponents shall mean precisely the same things when they use the same terms. Soil classification in its essentials simply means giving various distinctive classes of soils suitable names and descriptions, so that when these particular terms are used they will convey a very definite meaning or impression, and so serve to identify a particular soil or class of soil.

TEXTURAL CLASSIFICATION

The earliest attempts to classify soils were essentially elaborations of the farmer's conception of light, medium and heavy land ; or sands, loams and clays. These were subsequently given greater precision by being linked with mechanical analyses, in which the actual amounts of sand, silt and clay present in a sample were determined in the laboratory. Thereafter soils were classified into light loams, medium loams, heavy loams, clays, etc., upon the basis of their ascertained content of sand, silt and clay. The amounts of the various fractions appropriate to the different textural classes cannot, however, be stated absolutely because of the modifying influence of organic matter, shape of particle and type of clay mineral etc. The basis of the present classification is the loam which contains very roughly 40 per cent sand, 35 per cent silt and 25 per cent clay. Soils with less clay and silt than the loam but more sand are called

TABLE I

TEXTURAL CLASSIFICATION OF SOILS

Class.	Properties of Moist Soil.	% Composition of Typical Examples.		
		Sand	Silt	Clay
Sands . . .	Gritty, cannot be moulded	94	3	3
Loamy sands .	Gritty, little cohesion	87	5	8
Sandy loams .	Gritty, moderate cohesion	70	15	15
Loams . . .	Readily moulded, moderate cohesion	40	40	20
Silt loams . .	Readily moulded, smooth and soapy	20	65	15
Clay loams . .	Highly cohesive, polish when smeared	35	30	35
Clays . . .	Very strongly cohesive, high polish	20	20	60
Organic mineral soils . . .	Dark and friable	(8–25% organic matter)		
Peaty soils . .	Black and very friable	(more than 25% organic matter)		

loamy sands and sandy loams, soils with more clay and less sand are clay loams and clays, and those with more silt and less sand are called silt loams. Table 1 gives representative particle size distributions for the USDA textural system and brief comment on the hand textural properties.

The predominance of coarse sand, fine sand or silt within the textural categories is indicated adjectivally as " fine sandy " where a large proportion of the sand fraction is in the form of fine sand or " silty " where the percentage of silt is particularly high, e.g. silty clay loam. Clay loams with significant proportions of sand are termed sandy clay loams. Apart from the difficulties of securing exact standards, this type of classification suffers from certain grave objections when it is used to indicate the distribution of soils by the aid of soil maps. Texture is only one of the many properties of a soil, and as such can have only a limited relevance to the broadest conceptions of soil classification, which should embrace all the important soil properties. If, for instance, the soils of England were grouped into their various textural classes, and these portrayed on maps, it would by no means follow that the soils in any one group would resemble each other in properties apart from texture, such as subsoil and drainage conditions, depth, aspect and elevation.

GEOLOGICAL CLASSIFICATION

Those familiar with the English countryside cannot fail to be impressed by the great influence of geology on soils and on land forms. Some formations such as the Chalk, Old Red Sandstone and Oolitic Limestones are readily recognised by their soils ; others are not so easy to distinguish but, because of the sometimes close connection between soil and geological formation, a geological basis was adopted in many of the earlier soil surveys. For example, Hall and Russell, in their survey of the agriculture of Kent, Surrey and Sussex, distinguished Chalk soils, Wealden soils and Gault soils, amongst others, and within the area studied the method was fairly satisfactory in so far as it enabled the distribution of various classes of soils to be followed by means of a geological map. The system, however, has serious limitations.

A geological map, although often wrongly regarded as such, is not a " parent material " map. The geologist classifies and groups various strata on the basis of their relative ages and not upon the actual nature of the materials ; thus the actual composition of any one formation is by no means constant. For instance, the Forest Marble formation of Dorset, in so far as it is a soil-forming material, is mainly a heavy calcareous clay, whilst the Forest Marble of Gloucestershire is extensively a brashy limestone. Thus it is misleading to speak in the widest sense of Forest Marble soil, however useful such a conception may be in dealing with a small area. Even in the case of " Drifts " the geologist is primarily concerned with their age and origin and only to a lesser degree with the actual nature of the material in question. Equally misleading can be the geological separation of similar parent materials. It is probable, for instance, that little or no difference from a soil-forming aspect exists between the various oolitic limestones—the Cornbrash, the Great Oolite and Inferior Oolite, etc.—or between the Tertiary and Glacial gravels and sands. The use of a geological interpretation of soils, however, tends to foster the reverse impression.

PEDOLOGICAL CLASSIFICATION

Since the beginning of the present century, and mainly through the influence of American and Russian soil workers, efforts have been increasingly directed towards the treatment of the study of soils as a definite branch of natural science rather than as a part of agriculture. For this the name *Pedology* has been adopted.

The Soil Profile. This has led to an important advance in the classification and mapping of soils by the recognition of the *soil profile* as the unit of study. The soil profile is represented by all the distinctive *horizons* or soil strata from the surface down to the parent material, and gives expression to all the processes of soil formation, including physical and chemical weathering, and the vertical movement of soil constituents up or down the profile brought about by soil water. The nature of these processes, and hence the character of the soil profile, is generally governed by climate which also controls the type of natural vegetation, which in its turn influences profile development. The character of the parent material also plays an important part and may under certain conditions be the dominant factor.

Soil Series. Soils with similar profiles, derived from similar material under similar conditions of development, are conveniently grouped together as a *series*. The soil series in Britain is the basic unit of classification and may be subdivided on texture, slope, depth, and other profile features into lower classification categories.

Soil Groups. Avery describes ten major Soil Groups and thirty-nine Sub Groups as occurring in Britain. Since it is outside the scope of this introduction to describe all the divisions and sub-divisions or this system, five of the most important major soil groups, with minor modification, have been selected from Avery's classification. Only the main morphological, chemical and mineralogical features of these soil groups will be described. The groups are :

1. Calcareous Soils
2. Brown Earths
3. Podzolised Soils
4. Gley Soils
5. Peat Soils.

1. **Calcareous Soils.** The occurrence of these soils is closely though not exclusively related to the geological boundaries which demarcate limestone, chalk and other calcareous rocks. Two main sub-groups are recognised : (a) *Rendzinas*. These soils are shallow and of neutral to alkaline reaction throughout. A dark grey brown or black mull humus layer is often the only soil horizon above the parent material, and this contains variable amounts of organic matter and free calcium carbonate. (b) *Brown Calcareous Soils*. Though slightly acid in the surface horizons the soils of this sub-group have calcareous lower horizons. They develop on limestones and highly calcareous sandstones : profiles in Britain are generally brown, freely drained, and of medium to heavy texture.

2. **Brown Earths.** These soils are well, or moderately well, drained. Profiles under natural conditions are leached of soluble salts and carbonates and the tendency is maintained also under cultivation. In the surface horizon, organic and inorganic mineral materials are intimately mixed to form mull humus, and the soils vary from slightly to strongly acid. In some profiles there is evidence of clay movement from upper to lower horizons though, apart from structure differences, there is no marked horizon differentiation. Soils in this group occur in Britain on parent materials from sands and sandstones to clays and shales. Under natural conditions the sandy profiles are usually more acid, and under some conditions podzolise. However, much of

the agricultural land of Britain consists of brown earths, and ploughing and cultivation have tended to stabilise profile development.

3. Podzolised Soils. Podzolised soils develop in Britain in regions where acid raw humus accumulates under heath or woodland. They are more characteristically found where drainage is free, on light-textured parent materials such as non-calcareous sandstones, sands and gravels. Profiles have a strongly acid layer of raw humus below which is a bleached, mineral zone. Organic acid solutions have removed iron and aluminium oxide from this bleached horizon, and through precipitation a bright brown horizon enriched in these components is formed below. Some soils display a zone of humus accumulation, which overlies the sesquioxide-rich layer. These horizons are formed under conditions of acid leaching and all are strongly acid. In Britain podzolised soils occur widely, though generally they are found in areas of low to moderate rainfall. In the lowlands extensive cultivation by man has masked, and possibly restricted, the extent of podzolised soils.

4. Gley Soils. Where profile characteristics are the result of water-logging, soils are termed gley soils. They are generally medium to heavy-textured, but sandy soils which occur in basin sites, or above an impermeable sub-soil layer may be termed gley soils. The morphological features of gleys largely result from the effects of waterlogging on iron-compounds in the soil. Waterlogging leads to anaerobic conditions. When this happens the ferric compounds, which impart brown and red colours to the soil, are changed by reduction to the grey or blue-grey ferrous compounds.

Where alternating oxidising and reducing conditions operate due to a fluctuating water table, grey, and rusty mottlings and blotchings are found together in soil horizons. Gley soils are essentially grey in colour, or at least possess a horizon where grey colours predominate. They are often sub-divided into surface-water gleys and ground-water gleys, though this implied relationship to water tables is not fully understood. Surface-water gley soils display a strongly gleyed layer near the surface, resting upon heavier-textured, less-permeable sub-soil horizons which apart from structure faces do not show features of gleying; in contrast, ground-water gley soils are typified by grey and blue-grey colours, and other features indicative of waterlogging and gleying more or less throughout the profile.

5. Peat Soils. Peat soils develop in areas of excessive wetness and in these conditions plant residues accumulate to give essentially organic soils. Regions of high or moderately high rainfall and low temperature favour peat development as do basin sites where there is a marked excess of precipitation and seepage over evaporation.

Fen peat is formed in basin sites where seepage water is base rich. This neutral to moderately acid peat consists of the humified remains of plants which have inhabited shallow water, intermixed with variable amounts of washed-in mineral matter. With the continued accumulation of humified material, the basin is gradually filled and the free water surface disappears. Plants which demand small amounts of nutrients from the peat below invade the site, and their decaying residues, acid and poor in nutrients, build upwards until a final stage known as *Raised Moss* is reached. Under conditions of high rainfall and low temperatures, acid peats form from a variety of plant associations. Because these peats blanket the uplands, where climatic conditions are favourable, they have been called *Blanket Bog*. They are found principally in western Britain where altitude and rainfall favour this development.

SOIL SURVEY

Despite the fact that the pedological conception of soils has been developed from a purely scientific standpoint largely divorced from agricultural considerations, it quite obviously has important practical applications. This is hardly surprising when one considers the importance to farming of those factors which have been found to determine so largely the evolution of soils, namely climate, parent material, situation and moisture conditions.

When the distribution of the various classes of soil in an area is recorded on a soil map, the convention is adopted of colouring the Podzols various shades of red, the Brown Earths brown, the Gleys blue, Peats purple and Rendzinas yellow. Thus at a glance the map would convey the general locations of acid soils and non-acid soils, wet soils and dry soils, light textured soils and heavy textured soils. A more detailed study, supplemented by reading an explanatory soil memoir of the area, would provide information about less obvious contrasts—those, for instance, which differentiate the various series within a sub-Group, such as elevation, depth and stoniness, the whole being a useful summary of agriculturally significant data.

A large amount of survey work has been completed in this country in recent years, including areas in Lancashire, Yorkshire, the West Midlands, the East Midlands, East Anglia, the Home Counties, southern England, Devon, Cornwall and central southern and northern Wales. A considerable amount of survey work, largely directed from Rothamsted Experimental Station, is in progress at the present time.

The example on page 20 will indicate a typical recording of soil profile observations. It is slightly abbreviated from a description published by E. Crompton.

A sample of each horizon from a number of profiles of each series is taken for laboratory analysis; this enables further data to be obtained as additional criteria for defining the soils. In the laboratory examination attention is directed primarily to those more stable soil characteristics which as a rule can be altered only to an inappreciable extent by the activities of man. These include the percentages of sand, silt and clay, the exchangeable base capacity, and the proportion of silica to iron and aluminium oxides (the silica–sesquioxide ratio) in the clay fraction.

Reference has already been made to man's influence upon soil characteristics, the significance of which is not always appreciated. The mature soils we see are the products of evolutions from earlier forms, due to the varying interactions of parent material, climate, site, etc.; these soils thus each represent a state of equilibrium between the soil body and its environment. Any act of man, such as liming, drainage or manuring, constitutes a change of environment and hence a shift in equilibrium, which may be sufficiently great to alter the soil from one genetic class to another. The conversion of a Podzol to a Brown Earth by cultivation, liming and manuring is

CHARNOCK SERIES

Drainage of profile.	.	Imperfect.
Parent material	.	Till of Carboniferous material.
Vegetation	.	*Agrostis*, ryegrass and white clover.
Topography	.	Altitude 230 m, 5° slope to West.

Profile :—

0–80 mm	.	Very dark brownish-grey, tight root mat, organic silt loam, well developed fine crumb to fine granular, structure, moist. Clearly defined boundary.
80–300 mm	.	Greyish-brown loam, very fine blocky to sub-angular blocky structure; numerous roots and root tracks penetrating down from above. Merging boundary,
300–550 mm	.	Yellowish-brown, with some very pale grey, rather diffused and poorly defined mottling, fine sandy clay loam ; structure is weakly developed but breaks easily to fine crumb ; porous and fissured with some darker old worm tracks penetrating from the horizon above. Merging boundary.
550 mm +	.	Dull brown, pale grey and ochreous ; old root channels with some darker patches (probably manganiferous), fine sandy clay loam with some fine sandstone fragments tending to be laminated, probably reflecting the bedding of the parent material.

a case in point which, in the agricultural sense, constitutes " soil improvement ". The new point of equilibrium is the result, however, of the artificial introduction of an additional environmental factor, and can only be maintained for so long as the new factor persists. Thus the alteration and improvement of land by good farming can rarely be fully permanent and will last for only so long as the higher standard is maintained. If this is at any time discontinued, the soil will tend to revert to the original state.

THE SOIL AND PLANT NUTRITION

Attention may now be paid to the growing plant to see what its requirements are and to what extent they can be supplied by the soil. Both the atmosphere and the soil play a part in the life processes of plants. To begin with, respiration is continually taking place just as in the case of animals, though in a much less obvious fashion. Some of the tissue of the plant is oxidised with the production of the necessary energy for life and growth. This is accompanied by the liberation of carbon dioxide and water, waste products which are got rid of by the leaves.

Photosynthesis. In the daytime, however, and especially in bright sunlight, another vital phenomenon is apparent. This is the assimilation of carbon dioxide from the atmosphere, where it is always present in small amounts. On the under side of a leaf are found numerous pores or stomata, leading into small internal cavities lined by the walls of active living cells. The air containing carbon dioxide finds its way into these spaces and dissolves in the moisture in the cell walls. The green colouring matter in the cells, called chlorophyll, is able to utilise the energy of the sun's rays which fall upon it, to split up the compound carbon dioxide and build up the carbon so obtained into new substances of the nature of simple sugars. At the same time the oxygen which has been split off is set free and finds its way, via the stomata, into the air once more.

It will be seen then that in the daytime a plant utilises carbon dioxide from the atmosphere to build up its tissues more quickly than it produces that compound by respiration. The sugars produced are then distributed through the growing parts of the plant, and from them are elaborated many other compounds, mostly more complex, which play their part in the building up of its structure. Some of these compounds contain elements other than carbon, hydrogen and oxygen, the constituents of the sugars, and it is these which are obtained from the soil by means of the root system.

Absorption by Roots. The very fine root hairs which the growing plant is continually putting out to explore the soil for nutrient material, push their way amongst the soil particles and lie closely up against their wet colloidal surfaces. The soil solution or moisture in the soil diffuses through the membranous walls of the root hairs and along with it go small amounts of the various chemical substances present in it. The manner in which this happens is not fully understood, but there is no doubt about many of the facts connected with it, such as the enormous amount of water which a plant takes up, passes through its structure and respires from its leaves. There is no doubt, also, that the root system can absorb nutrients from the soil selectively, up to a point.

The substances which are taken in from the soil in this way include nitrogen, phosphorus, potash, silica, iron, magnesium, calcium and others. They pass up into the structure of the plant and, together with the products of carbon assimilation from the leaves, form the materials from which all the various organic compounds comprising the tissues of the plant—proteins, carbohydrates, fats and oils—are produced. So for its growth a plant requires, in the main, oxygen and carbon dioxide from the air, and water, nitrogen, phosphorus, potash and lime from the soil.

Some idea of the quantities in which these substances are required is provided by the two following examples, one afforded by a cereal and the other a root crop. The former, if it yields 5·5 tonnes of grain and 3·8 tonnes/ha of straw, removes from the soil 130 kg of nitrogen, 40 kg of phosphate, 60 kg of potash and 50 kg of lime besides other constituents and during its growth will transpire from its leaves 550 tonnes of water. A 60 tonnes per hectare crop of mangels removes 95 kg of nitrogen, 56 kg of phosphate, 280 kg of potash and 28 kg of lime and will transpire 900 tonnes of water during its period of growth.

Water Supply. These figures indicate that the biggest demand on the part of a growing crop is for an adequate supply of water, without which no additions of other plant nutrients can exert anything like their full effect. It is an interesting arithmetical study to see what this means when put against the existing rainfall. Twenty-five mm of rain on 1 ha gives 250 tonnes of water, so that for a successful barley crop, the water equivalent to 140 mm of rain must be actually supplied by the soil and transpired through the plant, which occupies the ground from March to August. In the eastern counties the average rainfall is about 640 mm, of which 200 mm will usually fall within the period mentioned. So that it will be readily seen that the physical properties of the soil and the success of the cultivations designed to conserve the moisture in the soil will have as much effect on the barley crop as the various chemical compounds concerned as nutrients, especially when it is remembered that transpiration from a crop is taking place continually, while rainfall is spasmodic.

PLANT NUTRIENTS

It will be noticed from the above examples that nitrogen, phosphate, potash and lime are the mineral substances needed in greatest amount by the growing plant. Other substances are needed also in much smaller quantities and it is usual to classify the essential plant nutrient elements as follows :—(a) *Major elements*, or those needed in relatively large amounts : nitrogen, phosphorus, potassium, calcium, magnesium and sulphur. (b) *Trace* or *Minor elements*, or those needed in relatively small amounts : iron, manganese, boron, copper, zinc and molybdenum.

A few other elements are taken up by plants with beneficial effects, but do not seem to be essential for growth. Among these are sodium, chlorine, silicon and aluminium.

Mineral Deficiencies. Fortunately most fertile soils in this country are able to supply adequate amounts of all these elements except nitrogen, phosphorus, potassium and calcium, and it is usually necessary to supply these substances only as manures or fertilisers.

It is sometimes necessary to apply other elements as well and those most often required are given below, more or less in order of importance.

Magnesium. Deficiency of this element in crops occurs most frequently on light soils that have been under arable cultivation for some time and have become low in organic matter and have had generous dressings of phosphate and potash. Some soils contain ample natural reserves of magnesium. Symptoms of the deficiency include an intervenal chlorosis but these can be brought on by excessive potash dressings. Root crops tend to show the symptoms on the older leaves as the crop matures ; often there is an unaffected green rim to the leaf. Cereals show the symptoms as a transitory mottling of the leaf when the crop is young ; usually this disappears later and reduction in yield is uncommon. Blackcurrants and other fruit crops are also susceptible to this deficiency. The remedy is to apply dolomitic limestone when the land comes for liming (see p. 34). If the soil does not need liming more costly forms of magnesium may be considered such as kieserite or Epsom salts (magnesium sulphate). Magnesium, deficiency also occurs in stock (hypomagnesaemia). This is a complex problem and not necessarily associated with magnesium deficiency in the soil, but if hypomagnesaemia is feared a useful precaution is to ensure that the soil is not deficient in this element (see p. 678).

Sulphur. Although this element is required by plants in comparatively large amounts, deficiency conditions have hitherto not been observed due to the use of sulphate-rich fertilisers such as ammonium sulphate and superphosphate, and because of the large-scale emission of sulphur dioxide in urban areas which dissolves in atmospheric moisture and falls on the land in rain.

Iron. Cases of iron deficiency in fruit crops arise on calcareous soils or those heavily limed, but are rare among ordinary farm or vegetable crops. The soil invariably contains large amounts of iron, but the strongly alkaline conditions render the element unavailable. The deficiency shows itself as lack of chlorophyll in the growing shoot, and similar symptoms can arise from toxicity caused by contamination of the soil by copper, zinc and other metals.

Manganese. As with iron, all soils contain reserves of this element that are ample for the needs of the crop. If soils containing high amounts of organic matter are heavily limed the manganese becomes unavailable and the crop in severe cases may fail. The deficiency is easily corrected by spraying the crop with a 4 per cent solution of manganous sulphate at 220 litres/ha which is effective if applied early enough. Cereals are especially subject to this deficiency.

Boron. Boron deficiency occurs naturally in a few soils in Britain, but in the main is induced by alkalinity resulting from supplies of free lime. It is particularly noticeable in dry, light-textured soils. While a very wide range of crops may show symptoms of the trouble, more commonly sugar beet, mangels, swedes and turnips are the most susceptible. In mangels and sugar beet the characteristic symptoms are a rotting of the crown of the root (Heart Rot), while in swedes and turnips the interior tissues are affected and turn brown (Raan or Brown Rot). Treating the soil with a small dressing of borax, about 20 kg/ha, will prevent this condition, but compound fertilisers containing boron in the required amounts are available.

Copper. Copper deficiency is uncommon but can result in complete failure of cereal crops where it occurs. It has been recorded on peats, heathland soils and chalk rendzinas. The application of a spray of copper oxychloride (3 kg/ha) to the crop provides a cure. Copper deficiency is also found in pear trees, and the effects are equally dramatic. Only very small amounts of copper are required by the plant and excessive amounts quickly cause toxicity symptoms to appear. Copper toxicity is common in industrial areas and once land has been contaminated in this way it is slow to recover.

Zinc. Although soluble salts of zinc are toxic to plants, sometimes sandy soils have been found where the small amount of zinc

c

required by plants is absent or unavailable. Zinc deficiency has been often reported in America, particularly with orchard trees and maize. Also zinc toxicity, due to zinc minerals in the soil or to zinc-smelting works in the neighbourhood, is of frequent occurrence, and cases are often recorded where excessive amounts of this element have reached the land from applications of refuse materials containing it, such as sewage sludge and the liquid sewage effluent from industrial sources.

Molybdenum. When it is necessary to apply this element it is usual to use sodium molybdate either in solution or at the rate of 100–200 g/ha. Cases of severe deficiency are very rare in the United Kingdom though in other countries they appear to be widespread in some restricted areas. The " whiptail " condition in cauliflowers has been proved to be associated with molybdenum deficiency, and molybdenum is necessary in the nodulation of leguminous crops and hence facilitates the fixation of nitrogen. Responses of legumes to molybdate applications are reported abroad, though if the land is well limed this is less likely to occur. It is interesting to note, however, that excess of this element has caused considerable trouble in a few areas known as " teart " pastures, where the herbage has caused excessive scouring in cattle, the grasses apparently containing more of this element than is normal. Soils on the Lower Lias formation in Somerset and elsewhere have caused this trouble, which has been cured by feeding traces of copper compounds either direct to the stock or through the pastures.

AVAILABILITY OF NUTRIENTS

The utility of fertilisers is, nowadays, demonstrated beyond doubt by the huge dimensions of the fertiliser trade, and most people are personally acquainted with instances of marked responses by crops to their application. Yet when the increases in crop produced, the amount of added fertiliser, the amount of the particular constituent removed by the crop, and the amount of it present in the soil are all considered together, no very obvious connection is apparent.

The average arable soil contains about 0·2 per cent of nitrogen, which, in a hectare of soil 200 mm deep, represents 5,000 kg, an amount enough to supply this constituent for 350 tonnes of grain. But the roots of wheat, for instance, in their search for plant food, go much deeper, so that they come into contact with an even greater amount of nitrogen, probably not exaggerated if put at 10,000 kg/ha. In spite of this, it is found that the average yield of wheat on the classic unmanured plot at Rothamsted, containing an amount of nitrogen of this same order, is steady at about 1 tonne/ha. On a neighbouring plot, receiving annually 40 kg of nitrogen in the form of concentrated fertiliser, an equally steady average yield of 2 tonnes/ha is maintained. This would suggest that the total amount of nitrogen has little direct bearing on the size of crop grown, but that there is probably a factor of great importance in the form in which the plant can assimilate it.

If a similar rough calculation is made of the amount of nitrogen present in the soil in forms similar to ammonium nitrate and sulphate, the figures obtained take on a more ordered aspect. There are found in the soil at different periods of the season varying amounts of two comparatively simple compounds of nitrogen, namely nitrates and ammonia, of the order of 0·001 per cent of nitrogen as nitrates and 0·0001 per cent of nitrogen as ammonia. These represent, per hectare, a total of 28 kg at any moment. This figure is substantially of the same order as the amount of nitrogen in the fertiliser dressings referred to above, so that there is strong support for the view that the amount of nitrogen present in the soil in an assimilable form, e.g. as nitrate, has a direct bearing on the crop yield. It has already been indicated that the bulk of the nitrogen present in the soil is in the form of complex protein-like compounds in the humus or organic matter, and that these are constantly being broken down into simpler compounds by the action of specific bacteria, until they result in the production of ammonia which, in its turn, is converted into nitrate. The organic nitrogenous compounds are mostly insoluble and certainly not able to be absorbed into the root system, while ammonia and nitrate are comparatively simple in composition, readily soluble in water, and easily assimilated by root hairs.

The Nitrogen Cycle. This sequence of changes forms part of what is known as " the nitrogen cycle ", or the way in which the element nitrogen is continually in circulation in nature. In its simplest form it can be set out as follows to illustrate the continuous growth and decay of plants.

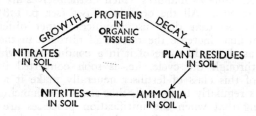

But the activities of bacteria and the processes of decay are not merely the mechanism of this simple cycle ; they have trends of their own leading in other directions, as indicated in the following version :—

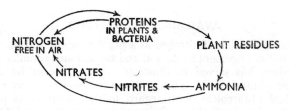

Some of the nitrogen of the proteins is set free and added to the store already existing in the atmosphere, while some of that in the nitrates produced in the soil suffers a similar fate. On the other hand, several different bacteria can feed on the free nitrogen of the air and elaborate it into proteins in their own bodies.

Sufficient has been said, perhaps, to indicate the respective influences of proteins and nitrates, to mention the two most important stages of the cycle, upon the fertility of the soil and the growth of plants. The proteins constitute the reserve in the soil, the material from which a constant supply of easily assimilable nitrates is produced. The rate of this production depends on the temperature, moisture and aeration of the soil. It is at a low level during the winter months, but as soon as the soil warms up in spring it rises rapidly, with the result that the amount of nitrate found in fallow land constantly increases until autumn, when production falls off and drainage commences once more, gradually reducing the amount to its winter minimum. In the case of land under a crop, the nitrate is continuously absorbed, with the result that the amounts found are negligible under growing crops, they then accumulate after crops are removed in late summer and later fall off during winter because of leaching in drainage water.

It will be understood from this brief outline that easily soluble nitrogenous compounds used as top dressings in the early spring, before nitrification is really active, are in a position to exert a very potent influence on the growth of a crop in its early stages. It also follows that the effect of such fertilisers is exerted within a short time of their application and that they are unlikely to have a permanent effect in raising the fertility of soil. But there are many materials of undoubted value as manures which in themselves are more of the nature of proteins. All the organic refuses (see p. 178) come into this category, so that their use produces rather different effects. They do, in fact, increase the reserve of nitrogen in the soil at the same time as they add some of it in a condition in which it easily progresses through the cycle, i.e. is soon converted into nitrates. The users of this class of fertiliser generally make it a practice to add supplies regularly to the soil, thus steadily enhancing its fertility and ensuring that when the nitrification processes are active, the amounts of nitrate produced may be correspondingly increased all through the season. Evidence of advantage from this kind of treatment is not conclusive, but it is probably better for some horticultural crops than a liberal supply of quick-acting nitrogen applied early in the season.

AVAILABLE PLANT FOOD

The consideration of nitrogen as a plant nutrient and the relations between the store present in the soil and the amount actually absorbed by the plant has served to illustrate what is meant by the term " available plant food ". This is an expression frequently employed in technical phraseology with regard to the nutrition of plants, to call attention to the fact that it is not so much the total amount of any one constituent which matters as the extent to which that constituent is in a form assimilable by the plant roots, or, in other words, how much of it is available. In the case of nitrogen there are a number of established facts, as indicated above, which enable a fairly clear idea of the position in the soil to be obtained.

In the case of those other important nutrients, the mineral constituents, especially phosphates, potash and lime, our knowledge is not quite so satisfactory. Looking at them on the same simple quantitative basis, it is found that a common figure for the amount of phosphate in a soil is about 0·1 per cent, which, by calculations similar to those employed in the case of nitrogen, represents over 5,000 kg/ha within root range, and yet by the addition of 250 kg/ha of superphosphate, containing about 40 kg P_2O_5, remarkable effects on the crop are obtained on deficient land.

In the case of potash, a much greater range of total content in the soil is experienced, from 0·1 per cent or less in light, sandy soils to over 1 per cent in heavy clays, representing from 5,000 to 50,000 kg/ha within root range ; but again, when the soil responds to applications of this constituent, it does so to dressings containing amounts of the order of 40 kg/ha.

The phosphoric acid in the soil is present partly as mineral matter, such as calcium phosphate, and partly as a constituent of the organic matter. The potash is there as a constituent of the original minerals or of their partly weathered residues. In the soils of the south-eastern half of England it occurs chiefly in the smaller particles, particularly in the clay fraction, so that the total amount increases with the heaviness of the soil. That all these compounds, whether phosphate or potash, are practically insoluble in water is evidenced by the very small amounts which are found in drainage water or in the soil solution.

METHODS OF ASSESSING AVAILABILITY

At the same time there is evidence that in some soils the phosphate is more " available " to the plants than in others. It is possible to find instances of soils containing the higher amounts of total phosphate which respond to added phosphatic fertilisers, while others with a lower total content show no response. Phenomena of this kind led to the view that there are differences in the state or constitution of the phosphates in soils, making them more or less easily assimilable by plant roots. It was also suggested that the root systems excreted acid juices which could dissolve the mineral nutrients, phosphate and potash, and so render them useful to the plant.

Dyer's Citric Acid Method. After a period of experiment Dyer came to the conclusion that this action of the roots could be very closely imitated in a laboratory by shaking up the soil with a 1 per cent solution of citric acid under certain standard conditions. By means of this test he was able to differentiate between soils with respect to the amounts of " available phosphate " and " available potash " which they contained. His results agreed fairly well with the actual response of the soils in the field to added fertilisers, and to this day similar methods are widely employed by advisory chemists to assess the need of soils for these particular constituents.

In Britain Olsen's method for assessing the availability of phosphate in soils is the most widely used. Phosphate is extracted from soils with a solution of sodium bicarbonate of pH 8·5 at 20 °C. Phosphate concentration in the extract is determined by measuring the intensity of the phosphomolybdate blue colour produced on reduction. In arable soils amounts of Olsen's extractable phosphorus range from 1 to 70 parts per million (ppm) and in glasshouse soils levels up to 300 ppm are encountered. Below levels of 9 ppm there is a risk of failure unless phosphate fertilisers are applied. Available potassium is most commonly assessed by an ammonium acetate/acetic acid extraction and reduction in yields are likely to arise when soils contain less than 50 ppm.

It is obvious that such standards cannot be regarded as hard and fast, but due consideration being given to other known facts, such as history of the field and soil texture, this form of analysis is a valuable aid to correct fertiliser application.

A useful modification of the method consists in studying the ratio of the " available " to the " total " amount of either constituent instead of the absolute percentage. The following examples are from different fields on one farm :—

Per cent.	1	2	3	4	5	6
Total phosphoric acid . . .	0·123	0·092	0·083	0·119	0·116	0·243
Available phosphoric acid . .	0·013	0·011	0·025	0·033	0·031	0·016
Available as % of total . . .	11	13	30	28	27	7
Total potash	0·248	0·332	0·147	0·115	0·164	0·492
Available potash	0·016	0·012	0·023	0·007	0·005	0·017
Available as % of total . . .	7	4	15	6	3	3

Numbers 1, 2 and 6, by this criterion, seemed to be short of available phosphate ; in practice they were the fields where phosphatic fertilisers produced the bigger response. There is not such a marked contrast in the case of the potash. On the ratio of available to total potash, they all look poor except number 3, while in practice numbers 4 and 5 gave the best results for the use of potash.

It is 200 years since that pioneer Jethro Tull expounded his views on availability and they bear close resemblance to some of those expressed by eminent soil scientists of the present day. He realised the tremendous importance of tilth in making a soil productive, and the connection between it and the internal surface area of the soil. He regarded this as the " pasture " area of the soil so far as plant roots were concerned, and devoted all his energies to the improvement of tillage implements in order to keep this area at its maximum. In his day little was known of the chemistry of the soil or of plant nutrition. He was convinced that an important part, if not all, of the necessary nutrients resided on the surface of the soil particles, and so argued that the more of this surface there was made accessible to the searching roots the more nutriment the plant would obtain. Enough has already been said in the present account concerning the soil solution and the action of plant roots on the colloidal surface of the particles to show that his line of reasoning had very much the same trend as that of subsequent workers.

On many arable farms and some grassland farms regular use of fertilisers over several decades has built up substantial reserves of phosphate and potash, so that, except for very responsive crops such as swedes and potatoes, crops grown in the presence of these reserves do not respond to current fertiliser applications. This situation gives farmers much more latitude in their fertiliser planning and the emphasis can be changed from response of the next crop to one of maintaining satisfactory nutrient reserves during the rotation. For example, where reserves are high the total phosphate and potash offtake by a sequence of five cereal crops can be applied in the autumn before the first crop and soil analyses used to monitor nutrient levels before the next application. Levels of potash in sandy soils cannot be economically maintained at high levels and frequent applications are needed to avoid crop losses. Nitrogen reserves cannot be built up in soils by application of nitrogen fertilisers and so this nutrient must be applied for each crop. Responses to nitrogen of most non-leguminous crops are very large and efforts to achieve optimum levels of application are well worth while. Magnitude of crop response to fertiliser nitrogen is inversely proportional to the level of nitrogen residues left by preceding crops.

Other Extractants. Dyer's method is quoted as an example of an available nutrient determination, but many other similar methods have been tried and as these are more convenient in a laboratory where many routine soil analyses are carried out Dyer's method is not now widely used. Extractants can be vigorous or gentle according to whether they extract a high or low percentage of the total plant nutrient, and as the various crops differ in their ability to absorb the nutrient in question, no one extractant is likely to have universal application. Many workers have observed that when a soil sample has reached equilibrium with an extractant there still remains a further smaller amount of nutrient in the soil that could be extracted by means of a second extraction. Thus the use of a large amount of solvent for a given amount of soil extracts a proportionately larger fraction of available nutrient. The following solutions are among those used in available nutrient determinations :

N/2 acetic acid (solvent : soil = 20 : 1)
N/2 acetic acid + 10% sodium acetate : Morgan's solution (solvent : soil = 5 : 1)
N/5 sodium bicarbonate : Olsen's solution (Solvent : soil = 20 : 1)
3N sodium nitrate (solvent : soil = 2 : 1)

Other methods of assessing the availability of the mineral plant foods in a given soil have been tried out by investigators in recent years, especially on the Continent, by a more direct appeal to the plant, instead of by soil analysis. On the grounds that a single nutrient factor, such as phosphoric acid, exerts a specific effect on the crop, as evidenced by the yield, attempts have been made by studying the yields obtained by varying additions of a single manurial constituent, to calculate the amount of this originally present in the soil. This method has been severely criticised because of the difficulty of isolating the effect of a single factor where so many are obviously concerned. This drawback, coupled with the slow and laborious nature of the investigation, precludes its general adoption.

Neubauer's Method. There are, however, investigational methods which are more rapid and more easily carried out which have attained a certain measure of success. That of Neubauer is an example. This worker relies on the ability of seedlings, given favourable circumstances, to extract all the available plant food in a given sample of a soil within a comparatively short time. His procedure is to mix a small quantity of the soil sample with about three times its weight of pure sand, set it up in saucers, plant in each 100 seeds of rye or barley, and keep the whole at carefully regulated conditions of temperature and water supply for seventeen days. At the end of this period the seedlings are lifted, carefully washed free of soil, dried, and analysed for their content of phosphate, potash, etc. The figures obtained are then used to determine the sufficiency or lack of any individual constituent in the particular soil under experiment. This line of attack has been much used, since its discovery, by the fertiliser interests in Germany, in connection with field investigations and for routine advisory purposes.

Larsen's Method. Recently the advent of radioactive tracers has given another type of method to the soil chemist. By adding a known amount of radioactive phosphorus (P_{32}) to the soil and measuring the activity of the plant grown on it and that of the soil a direct measure of the ability of the soil to provide phosphate can be obtained. Fried and Dean have developed a similar method in America. It is harder to apply the method to potassium and other elements because of the absence of isotopes with half lives of convenient duration.

SOIL REACTION AND LIME STATUS

There still remains to be considered one constituent of the soil which ranks as a very important food nutrient but which is the more indispensable because of its effect on the condition and fertility of the soil. This is lime, or the calcium base, which is absorbed by plants along with their other food materials, but which also has a dual effect on the soil in improving its structure and in keeping its reaction suitable as an environment for most of our crops and for the beneficial micro-organisms which so greatly influence fertility.

Soil Reaction. All materials in the presence of water can be either neutral, alkaline or acid, and the degree of acidity or alkalinity is expressed in terms of a scale, known as the pH scale, on which the neutral point is numbered 7. Numbers above 7 indicate alkalinity, and below 7 acidity, the alkalinity and acidity being greater the further the number deviates from 7. Alkaline conditions are due to the presence of bases such as lime, soda or potash, while acid conditions are due to acids, an example of which is citric acid, present in fruits such as lemons. When the soil is well supplied with bases, particularly lime, the soil will have a neutral or even slightly alkaline reaction, with pH values from about 7·0 up to about 8·0. When such bases are lacking, acid conditions set in, and the soil will have pH values less than 7, which in extreme cases may fall as low as pH 3·0. As plants appear to be fairly responsive to the reaction of the medium in which they grow, being able to tolerate only slightly alkaline or slightly acid conditions, the farmer must maintain the soil within the optimum pH range.

Many soils in this country have a reaction on the acid side, i.e. they have pH values less than 7·0, but exceptions may be found in soils on the chalk and limestone formations or in soils which have been limed recently. Plants can be divided roughly into groups according to their natural ability to grow under (1) very acid, (2) moderately acid, (3) slightly acid or (4) alkaline conditions. In the first group are found wild plants such as heather, bracken, bilberry, rhododendron, spurrey (mountain flax, sandweed, dither, yarr, etc.), sheep's sorrel or sour dock, and such agricultural crops as potatoes, rye and oats. In the second group we find other wild plants and agricultural crops such as wheat, swedes, turnips, kale, ryegrass and wild white clover. In the third group are found mangels, sugar beet, barley, carrots, red clover, beans and peas, while the lime-loving plants in the fourth group are the wild plants usually found on chalk lands, such as scabious and burnet and the agricultural plants lucerne and sainfoin.

The above division is not exact, since most plants tolerate a range of reaction sufficiently wide to bring them into more than one group, but the differences are sufficiently marked to enable many of these plants to be used as indicators of the general soil reaction. Thus, on arable land, the presence of spurrey or sour dock may indicate a very acid condition if these plants grow in large numbers to the exclusion of the more sensitive agricultural plants such as sugar beet, barley or red clover. The failure of the sugar beet crop can be instanced as a good example of what happens under too acid a soil condition. The plant grows normally until the hoeing stage, after which growth practically ceases and the plant assumes a stunted appearance, becoming pale in the leaf, while root development is checked and the plant may die. With very acid conditions the plant may die immediately after germination. The failure of the crop allows weeds to grow without competition, and spurrey and sour dock tend to smother any weakly plants which have survived. A very usual symptom of soil acidity is patchiness of the crop, rather than complete failure, since the soil is never uniformly acid all over the area. With barley, excessive acidity results in a rapid diminution of the crop and the consequent patchiness is always very evident. Root growth is stopped and a stunted, dark-coloured root system results. Red clover is the most sensitive of all the common agricultural crops to acidity and usually fails to reach the seedling stage. Wild white clover and alsike clover can withstand much greater acidity than the other clovers.

Crop Tolerances. Different crops require different pH value for optimum growth and can tolerate different degrees of acidity before failure of the crop is likely. Soils differ in their nature in this respect : peat soils can support growth of crop plants at a lower pH than mineral soils and on the latter ample treatment with farmyard manure can go a long way to mitigate the effects of acidity. Highly organic soils, however, are more likely to give rise to trouble at high pH (neutral or alkaline). The common crops can be divided into groups as shown below. The pH figures would be average for a mineral soil of average humus content. They are intended only as a rough guide.

The Nature and Effect of Soil Reaction. It has previously been indicated that the reaction (pH) of a soil is largely determined by the condition of its clay and humus. These may be, in extreme cases, hydrogen saturated and behave as insoluble acidic compounds ; or calcium saturated and behave as insoluble alkaline compounds. Weak organic acids, carbonic and sulphurous acids are constantly being formed in the soil water due to biological activity, and brought into contact with the colloids ; if the latter are themselves acidic then the soil solution will maintain a similar reaction ; but if the clay and humus are calcium saturated some of this calcium will be liberated to neutralise the dissolved acids, and so the soil solution

Group.	Crop.	Optimum pH.	Dangerously low pH.
I	Lucerne Sainfoin	7·0	6·0
II	Sugar Beet Mangels Barley Red Clover	6·5	5·8
III	Wheat Brassicas Wild White Clover	6·5	5·5
IV	Oats Potatoes Rye Ryegrass	6·0	5·0

will become non-acidic. The calcium lost from the colloids is replaced either by the hydrogen from the original mineral acids, in which case the colloids become progressively acidic, or by more calcium if the soil has reserves of calcium carbonate, or is limed.

Although there is often a close connection between crop failure and low pH, as shown in the previous table, the assumption cannot be made that it is the acids themselves, or the exchangeable hydrogen, which inhibit plant growth. It has long been known that plants grown experimentally in sand or water cultures can succeed at pH ranges well below the failure levels of field conditions, suggesting that other factors are involved in addition to that of direct acidity. Some of these factors are considered elsewhere and need only be indicated at present. They include the indirect effects of acidity on the physical condition of the soil, on the activity of soil micro-organisms, on the prevalence of plant diseases and on the competitive powers of weeds.

Two other indirect effects of soil reaction merit further consideration—the effects upon the availability of essential plant nutrients, and upon the solubility of toxic substances. Essential elements such as iron, manganese, copper, zinc and boron appear to become less soluble when the pH rises above the neutral point (pH 7). This lowering in availability sometimes explains the unfavourable results occasionally reported as a result of heavy liming. At the other extreme, increasing acidity enhances the solubility and hence the availability of certain elements, notably manganese and aluminium, to the extent that they may become harmful to plant growth. Investigations have shown that for certain crops the characteristic "acidity" symptoms are probably due to manganese toxicity under conditions of low calcium intake. In other words, a given soil, in relation to certain crops, might be manganese "deficient" at high pH, manganese toxic at low pH, and of a satisfactory soluble manganese status in the intermediate range of perhaps pH 5·5 to pH 7·0.

Causes of Soil Acidity. We must now mention the reasons why soils become increasingly acid year by year unless adequate dressings of lime are applied when they are needed. The chief cause of soil acidity is the fact that the easily exchangeable bases held by the soil colloids (clay and humus) are removed in a soluble form by rainwater as it percolates through the soil. The rainwater is slightly acid due to carbon dioxide gas in solution and various "humic" acids formed by the decay of humus, and these weak acids remove the absorbed or exchangeable bases, particularly lime, so that they are washed out or leached from the soil and lost in the drainage water. Light, sandy soils have the lowest capacity for holding exchangeable calcium and being freely drained lose this lime rapidly by leaching. It is therefore on such soils that soil acidity most commonly occurs, but from their nature, they are the most easily corrected, only small amounts of lime being needed to restore fertility. On the other hand, heavy clay soils, or soils rich in humus, hold much larger quantities of exchangeable calcium so that soil acidity develops much more slowly. But even these soils may in time reach such a state of acidity that crops begin to fail or grass 'and begins to deteriorate and become matted ; in such conditions small dressings of lime are of little use. Large or repeated applications of lime are often needed, and on grass land it may be necessary to plough out and re-seed so that the lime may be mixed effectively with the soil. The rate at which lime is removed from the soil in the drainage water varies, of course, from soil to soil and according to rainfall, but it may be taken to be from 100 to 500 kg/ha (as CaO) per annum.

Besides these losses from the soil by natural drainage, the crops themselves extract small amounts of lime each year, since calcium is an essential plant constituent. Heavy application of nitrogenous fertiliser causes substantial acidification of soil as a result of microbiological transformations of the ammonium ion. To counteract this effect, lime is added during the manufacturing process of nitrogen fertilisers but a more recent trend towards fertiliser with higher analyses has intensified lime losses.

Chemical Tests for Soil Acidity. The presence of appreciable amounts of *free* lime in the soil may be shown by adding a little dilute hydrochloric acid to some of the soil, when a fizzing will result, due to the action of the acid on the calcium carbonate.

Soil Indicator. But soils may be fertile and not in need of lime, even if they are devoid of free lime ; numerous field trials have shown that slightly acid soils (pH about 6·0) are still able to grow good crops. Hence a sensitive test has been devised to measure the pH value of the soil in the field by using the so-called " soil indicator ", which is a solution of various synthetic dyes, the colours of which change according to the degree of acidity of the soil when mixed with the indicator. The value obtained is only an approximate one, but it has proved useful in field work to discriminate between soils which are very acid and those which do not need immediate liming. Accurate pH determination is carried out in the laboratory by electrometrical methods.

Lime Requirement. If a soil has been shown by the above tests to be too acid to grow crops satisfactorily, it is then necessary to ascertain the amount of lime which should be applied. This is done by determining the " lime requirement " of the soil by another chemical method, which again needs the apparatus of a laboratory. The lime requirement figure is stated in terms of the amount of lime (expressed as carbonate of lime, $CaCO_3$) needed in tonnes/ha. It is usually only necessary to determine the lime requirement of those soils which have a pH value less than about 6·5 as it has been found in practice that it is only such soils which need liming. The application of lime to soils having pH values greater than 6·5 would be wasteful, and on some soils might be definitely harmful as it might induce manganese or boron deficiency in the succeeding crops. In any case it must be remembered that some experience is needed in interpreting the results of chemical analyses, because such factors as soil texture and the nature of the crop to be grown must be considered before deciding upon the precise amount of lime which should be applied.

THE LIMING OF LAND

Details of the various forms of lime which can be used to correct acidity will be found in Chapter VII. The time and method of application of these products to the soil will now be considered.

On arable land the lime should be applied after ploughing, as a more even distribution can be made and the material can be well worked into the soil without burying it too deeply. Both the soil and the lime should be as dry as possible. The lime should be spread on the furrow and well harrowed or cultivated into the soil but, for small dressings of up to 2 tonnes/ha, it is best applied through the drill as finely ground material. If the soil has been found to be badly in need of lime, the first opportunity should be taken to apply the lime, whatever the next crop ; but if the dressing is intended only to maintain a reserve of lime and to keep the land in condition, the most suitable time in the rotation is during the first preparation of the land for a root crop other than potatoes. The potato crop is unsuitable because the tubers are likely to be affected with Common Scab after recent liming.

On grass land the lime is best applied in autumn or winter, particularly if burnt lime is used, as large dressings of this form of lime are liable to check growth if added in the spring. The herbage should be dry when the dressing is put on and the turf should then be heavily worked with pitch-pole or other harrows to ensure that the lime is well mixed into the surface soil. Other forms of lime, such as ground limestone or chalk, may be used much more freely on grass land, and at any time of year, since their action is less severe. Grass land which has been allowed to become very acid gradually deteriorates, and the dead material decays so slowly that it tends to accumulate and form a fibrous " mat " on the surface. In these circumstances it is best to plough out the old turf and apply the lime to the exposed soil. The lime is worked thoroughly into the soil, after which the land may be re-sown at once (with application of appropriate fertilisers) or cropped for a few years with arable crops before sowing down to grass again.

The length of time during which the effect of an application of lime will last depends on several factors, such as soil type, rainfall, type of crops and manurial treatment, all of which affect the rate of removal of lime from the soil. As a rough guide under average conditions it is usually assumed that a dressing of 2 tonnes/ha of carbonate of lime (or 1 tonne/ha of burnt lime) will last from four to six years. A condition which needs to be watched with great care occurs in some soils overlying limestone or chalk. Leaching of the lime from the surface soil may have taken place to such an extent that the residual soil is acid, although free lime is present beneath the subsoil ; immediate liming of the land may be necessary.

Other Beneficial Effects of Liming. Besides correcting soil acidity, liming confers other beneficial effects on the soil. Thus heavy soils containing a high proportion of clay are improved in structure by liming. Under the influence of lime the fine clay particles tend to clump together to form larger aggregates, thus imparting easier working properties to the soil. The whole of the bacterial activities taking place in the soil, such as the production of nitrates from nitrogenous organic matter and the fixation of atmospheric nitrogen by free-living organisms such as Rhizobia, Azotobacter and Clostridium, occur more rapidly in soils supplied with lime. The accumulation of undecomposed mat on grass land, as mentioned above, is a direct result of the inability of the micro-organisms to function correctly under acid conditions. An improvement in herbage quality as well as in quantity is obtained by liming grass, more lime being taken up by the herbage and built up into organic calcium compounds. This improves the feeding value of both grass and hay. Finally, the relation of liming to plant diseases should be mentioned. Lime has long been used to combat attacks by Club Root (Finger-and-Toe) on cruciferous plants. The appearance of this disease has usually been taken to indicate soil acidity, but this is not always the case, as the symptoms have been found on neutral soils. There is evidence, however, that dressings of lime assist in preventing the disease, an increase in soil pH to about 7·0 having a beneficial effect (see p. 362).

Chapter II

FIELD DRAINAGE

THE practice of field drainage is almost as old as that of agriculture itself, a recognition of the fact that while a proper supply of moisture is essential to the successful growth of crops, it not infrequently happens that in the soil there is present an excess of water with decidedly harmful effects. The farmer's problem is to preserve the best possible balance between the two, to ensure an equable and constant supply of water and to avoid the presence of an excess. Left to herself, nature produces a balance between soil conditions and plant growth which is a reasonable optimum for the seasonal cycle of the weather. In considering the soils of this country attention was drawn to the great influence of the drainage factor on their general characteristics. Much of our land in its natural condition was ill-drained, and while nature clothed it with a vegetation suited to the circumstances, its conversion to farming was not successful until drainage measures were taken to remove the excess water and prevent the recurrence of water-logging.

The importance of field drainage in this country is evident from the amount of attention which it has received during and since the war of 1939–45 and, indeed, at all times other than those of agricultural depression. Apart from major arterial drainage works by public authorities, farmers have spent almost £100,000,000 gross on field drainage works in the 30 years 1939–69. This includes ditching and the under-draining of 827,000 ha. During the prosperous times of agriculture between the years 1846 and 1872, intensive tile drainage was carried out on at least 1,200,000 ha, with a total of probably 4,850,000 ha being drained in the whole of the 19th century. Indeed, at some time or other, all the farm land of the country which for reasons of soil or site could benefit from artificial drainage has been so improved at least once in its history.

Disadvantages of Wet Land. Wet land is late land and one of the chief objects of field drainage is to remove this disability. Fields which habitually lie wet for a substantial part of winter are slow to dry and warm up in the spring, implements cannot be got on to them soon enough, cultivations are late and their results are less satisfactory. The deterrent effect of high ground water level on root development and the harm which can follow a rise during the growing season are well known. In the development of crops the early stages of growth are of great, even vital, importance and on this growth soil temperature has a potent influence. Its response to increasing sunshine in the spring is greatly influenced by drainage conditions. The higher temperatures attributable to drainage were realised by the famous drainage engineers of a hundred years ago. Josiah Parkes recorded a gain in temperature of 7 °C at a depth of 180 mm, in the middle of June, in drained bog land. Bailey Denton found a benefit of 1 °C from March to May, at a depth of 460 mm in drained clay land. Such records have been confirmed and supplemented in more recent years by American observers.

Soils which are perennially water-logged are structureless, their subsoils in particular are tight-packed, and the water in them is stagnant. Soils which are seasonally water-logged, such as clay land and heavy land in general, have the same disabilities except that the summer, especially in droughty years, causes substantial drying-out from the surface downwards. The effect of this becomes evident in the shrinkage, cracking and structuring of the soil mass, the aggregation of the soil particles into larger units, and the appearance in the soil of a system of larger interspaces than were there before. This is the essence of permeability. Clay land drains better after a dry year than after a wet one. Careful study has led to the opinion that a good drain line leads to a steady improvement in the physical condition of the soil about it. In drainage one thing leads to another. The removal of stagnant surplus water allows root systems to develop in greater depth. The simultaneous improvement in soil structure and permeability also helps, the pore space is increased and aeration benefits accordingly. At the same time, in spite of the removal of much water, the amount of moisture available to the growing plant is substantially increased because the roots can derive their supply from a much greater depth of soil, in much more healthy conditions.

The Movement of Water in Soil. In general, the amount of moisture in soil depends on how much it receives as rain and how much it loses by evaporation and transpiration. The soil acts continuously as an absorbent and storage reservoir, so that losses due to the latter factors are sooner or later made good by the former. In this country the rainfall is more or less evenly spread over the year but the loss of moisture from the soil is distinctly seasonal. At the spring equinox the increasing power of the sun and the lengthening days combine to cause evaporation at a rate which can exceed that at which the rain is received. Based on monthly averages, this is particularly true of the predominantly arable south-eastern half of the country, and is generally applicable to all but the mountainous regions. In the spring and early summer, accordingly, we find that drain flows slacken and in heavy land cease altogether before midsummer. Evaporation and transpiration together deplete the soil of its reserve of moisture, working from the surface downwards.

The measure of this reserve is impressive: 1·5 m of clay soil, for instance, holds moisture equivalent to 760–890 mm of rain, even when drained, though some of this is too tightly held by the soil to be available to the plant roots. An average crop uses 150–200 mm of rain during its growing period. As a result, the normal autumn condition is one of a deficiency of moisture in the soil, varying in amount according to the summer season. Normally, it will be of the order of 75–100 mm, but after a droughty summer may be as much as 200 mm. Such rains as come in summer, of course, interrupt this trend, but as a rule do not obliterate it entirely. At the autumn equinox, when rainfall once more begins to exceed evaporation, there is still as a rule storage room for much of the autumn and early winter rains before an excess appears in the soil and percolation and drainage ensue. From time to time we have a run of years with abnormally low rainfall—sequences of droughty summers with dry winters between them ; the soil may even fail to become remoistened to capacity and so the drains do not run and the springs fail to rise.

If more rain falls than can be retained by the soil the surplus moves downwards under the influence of gravity, provided that there is no obstacle to its percolation. The characteristic of the soil which determines the extent to which this can take place is called " permeability " and it depends on the texture and structure of the soil mass and of the underlying geological material. If this is coarse and open, as in the case of soils lying on deposits of gravel or sand, the water can move downwards through the comparatively large interspaces with fair ease. If, on the other hand, the soil mass consists chiefly of the finest mineral particles, as in the case of clay, water cannot percolate through it easily, if at all. Such material can only attain a measure of permeability as a result of the channels made by earthworms or by decaying roots, or of the development in it of what is known as " structure ", which arises out of the process of weathering and its effect on the clay particles. These particles are of a colloidal nature and display a marked degree of shrinkage on drying and of swelling on re-moistening. The broad effects in field conditions are the well-known fissuring of the ground in summer, and the splitting of the clay soil into structural units with a network of interspace amongst them which persists somewhat even after remoistening and swelling. But this only affects the top 300 mm or so of the soil and its benefit is temporary. Such land is essentially impermeable and, in the absence of a drainage system, wet periods will be accompanied by surface water-logging.

But even in the case of soils of the most open nature we find instances of bad and persistent water-logging, which can arise from two causes. If an otherwise permeable soil lies on an impermeable geological formation, it will be liable to water-logging if the depth to the obstructing material is only a metre or two, because in wet times the surplus water percolating downwards piles up on the impermeable floor as a rising ground water level, which, if it comes within about 900 mm of the surface, has important effects on the soil. The other cause of trouble of this kind is that while percolation is mainly a vertical downward movement, the resulting body of ground water moves in a different fashion, influenced by the shape of the impermeable floor on which it rests and the differences in hydrostatic pressure which arise. So we have to deal with the possibility of trouble due to foreign water, i.e. water which fell as rain or otherwise entered the soil at a distance and then moved underground to come to the surface where it is ultimately found, the result of certain combinations of topographical and geological features. It must be remembered that it is differences in permeability that matter. A layer of moderate permeability underlying one of a very open nature will impede percolation in times of heavy rain. This impedance, acting in conjunction with sloping ground, causes lateral movement of water and accounts for many localised wet spots. The vital importance of the nature and properties of soil in the field is often lost sight of in this connection. We discuss and think of soil as a material, something which is manipulated in the business of farming, and unconsciously tend to regard it as being homogeneous. This is not so. All soil characteristics are liable to vary with depth, and natural variations of this sort, especially permeability, are accentuated by tillages. Thus it comes about that soils which in nature drain well enough are found under farming conditions to acquire a propensity to seasonal drainage troubles. Plough pan is not an uncommon cause of defective soil drainage, and in conjunction with sloping ground aggravates it locally.

Heavy soils and peaty soils have a much greater absorptive capacity for moisture than the light and sandy soils, and so it is common experience that, if markedly different soils are found within the same farm, the light land drains become active earlier than those of the heavy land. Further, the fashion of events is different in the two cases. The drains of open soils start gently, and flow continuously, with gradually increasing rates as the winter advances. The heavy land drains, by contrast, suddenly gush forth after a particular shower, flow at a great rate for a short time and then fall off to a trickle until the next rain.

FIELD DRAINAGE PROBLEMS

The farming community has been wrestling with the problem of surplus soil water from its earliest days, so what more useful background could one have to the study of present-day difficulties than the history of drainage efforts in this country? About the large-scale drainage system of the country no more need be said than that it has consisted in the main of the protection of certain areas from the influx of foreign water, the improvement of the natural river systems and their supplementation with artificial channels to get rid of the surplus water originating in those areas.

On the field scale, one of the oldest practices, dating from the times when the land was farmed in small parcels and the drainage efforts of one individual were liable to be nullified by the lack of co-operation or even the active opposition of his neighbours, was to work the land on the ridge and furrow system, which did at least ensure a modicum of land moderately well-drained, even in the worst seasons. This method lent itself particularly well to the clay-land areas, where it exploited the natural tendency of the impermeable clay to get rid of its surplus water by surface run-off.

The establishment of underground drainage channels, of course, was an obvious early development, but the materials used were those most easily obtained, and they varied as a result. To make such channels in the pre-pipe era required large stones, or boulders, or hewn stone, which were used where they occurred naturally. Where they did not, bricks were the obvious solution of the problem. Where stones were abundant in the soil and in arable farming, on the surface too, it was not surprising to find that stone-filled drains were used. In clay and silt areas, where neither solid rock might be accessible nor surface stones abundant, all sorts of adventitious fillings were employed in conjunction with a method of constructing the drains which exploited to the full the natural plasticity and cohesion of the subsoil material. Thus, in the clay lands of the eastern counties there were in widespread use two main methods, the hand-dug channel, narrow at the base, with a filling of straw rope, thorns, bushes or similar material ; and the mechanically made channel known as the mole drain.

Early Drainage Systems. As the enclosure of land proceeded, a network of ditches developed to the increasing benefit of field drainage. Many of them were natural watercourses, but from them were dug side ditches at convenient intervals to divide the land into suitable plots for working and at the same time to take the drainage water away from them. Many were sited along lines vital from the point of view both of soil and drainage conditions. It frequently happens that such ditches coincide with the boundary between the exposures of two adjacent geological formations, as a consequence of the farmers' realisation that the soil and drainage conditions were different on the two sides of the line. Ditches, whether natural or artificial, are the most important feature of our field drainage system.

Joseph Elkington. During the latter half of the eighteenth century a Warwickshire farmer, Joseph Elkington, became widely known for his success in dealing with wet springy areas lying across slopes. He had studied the problem very closely on his own farm at Princethorp and appreciated the significance of the geological structure of the ground in this connection. It is not uncommon, along the sides of valleys, to find that there are alternating exposures of sand and clay under the surface soil. Where this happens, spring lines are almost certain to be found, brought about through the percolating water being forced out on the surface along those lines to the detriment of a belt of land lying below them. Elkington's method of dealing with them was to site a deep drainage channel across the slope where the water first issued from the ground, i.e. towards the top side of the wet area. Sometimes it took the form of a deep ditch, but often it was a deep, covered channel constructed in stone. It was always deep, however, to prevent the underground water from rising to the surface to be a nuisance. It was not an extensive formal system to remove the water from all parts of the wet area, but a single channel to intercept the water as it moved along the floor which was forcing it to the surface. This method was widely and successfully employed, on Elkington's advice, in many parts of the country.

Smith of Deanston. The method advocated by Elkington was not of universal applicability. James Smith of Deanston applied himself to the problem of the extensive areas of heavy land which because of its poor permeability was unable to get rid of the excess of water falling on it as rain, land which was prone to lie wet all over, in greater or less degree, during the winter and spring months. His solution was to install an intensive system of underground channels in a formal pattern, covering the whole field, to collect and lead the surplus water to a limited number of outfalls in the ditches. The channels were hand-dug and filled with small or broken stones to within 500 mm of the surface. He himself was not satisfied even with this, and proceeded on his own farm to make it work more effectively by carrying out subsoiling operations behind his ploughs. The production of drain tiles cheaply and on an extensive scale in England from the early years of the nineteenth century, enabled his method to be applied in principle all over the country.

Josiah Parkes. Meanwhile, in those regions which were low-lying and occupied by less heavy soils, Josiah Parkes had become convinced of the need for deeper systems of drains, laid to a formal and uniform pattern to prevent the accumulation of ground water and its rise from below.

Thus, a number of different methods of field drainage came into being and were applied, in the main correctly, in those parts of the country to which they were best suited, and by their means practically the whole of the agricultural land of this country was drained, where necessary, before the 1880's.

Diagnosing Drainage Troubles. It is important to remember this in approaching field drainage problems today. Many of them arise out of the neglect to maintain the existing systems, but some are due to faulty planning or faulty work when the job was first tackled. If not regularly attended to, ditches soon begin to be obstructed by weed growth, collapse of the sides, fallen boughs, broken culverts under gateways and so on. They begin to silt up, the outfalls are submerged and silted over, the pipe-lines behind them cease to function, bursts occur and water breaks out over the surface. In some circumstances trouble occurs which cannot be attributed to the state of the ditches. The passage of time or natural factors may put underdrains out of action. Such are silting attributable to the nature of the soil ; the deposition by chemical action of lime or iron compounds from certain soil types ; the settling down of the back-filling over the pipes in clay land ; or the entry of roots into them. Few drainage systems can be said to be permanent in the sense of complete efficiency.

Before carrying out any infield drainage work it is worth while to study carefully the problem of the wetness to avoid unnecessary expense and to obtain the maximum benefit to the land in question. A knowledge of the geological build of the country and the situation of the affected area relative to its surroundings should give a fair idea of the fundamental difficulties ; whether they belong, for instance, to the nature of soil and subsoil or to the site. This done, the condition of past works should come under review, beginning with the ditches. All of them are important. On sloping land the function of the top ditch is to protect the field from the drainage of the land lying above it, that on the sides to carry it safely away, and that along the bottom generally to receive and evacuate the drainage of the field itself. The existence of obstacles, the extent of silting and the location and examination of outfalls is important, so that a reasonable assessment of the efficiency of existing installations can be made as well as of the chances of their functioning effectively once more. Next the internal condition of the pipe-lines should be ascertained, particularly in respect of silting and the possibility of clearing them. The drainage characteristics of the soil, the condition of existing drains and the cause of local wet areas, other than obvious bursts and overflows, can best be ascertained by digging holes at appropriate points.

Proceeding on these lines a reasonable diagnosis can be made and the most effective remedial measures can be planned accordingly. One's natural impulse is to concentrate attention on the wet area and devise means to take the water away. There is no particular difficulty about that, but the real problem is to prevent the wetness from recurring, i.e. to stop the water from getting there. To do that involves a knowledge of whence and how the water comes, which calls for a study of the wet area in its context of soil and site in the field, and of the field relative to its neighbours.

DITCHES

It is not often that new ditches are dug in this country; more frequently, indeed, their elimination is contemplated because the mechanisation of tillages and the general increase in the use of large machines calls for bigger fields and there is often a desire to get rid of these obstacles to cultivational operations. It is not an act of husbandry to be encouraged. It is true that ditches can be success-fully piped, but the job needs to be done carefully and well. The various functions of ditches, such as those of interception, draining and dealing with run-off, have already been described. The first two call for depth rather than width and this can be attained with no loss of ground by means of a piped channel of adequate size. The third function, above all, demands a big carrying capacity, i.e. both depth and width. Accordingly, in making, restoring, maintaining or piping ditches, it is important to appreciate to the full the purpose which any individual ditch serves. In the past it was not uncommon in the systematic draining of large agricultural estates to carry the efflux underground for quite considerable dis-tances to the outfalls by means of commodious brick or stonework sewers in place of open ditches. Many of them are still sound and functioning. But this is a different matter from putting a line of pipes in the bottom of a ditch and filling-in over them. The feasi-bility of piping any ditch should be considered in the light of its function and the amount of water which it may be called upon to carry.

An existing ditch even when " bottomed out " does not make an ideal bed for a line of pipes. These ought never to be less than 150 mm diameter, and the actual size should be sufficient to deal with the rate of flow to be expected from the area of land whose run-off the ditch carries. All existing drains emptying into the ditch will need to be connected up to the new pipe-line, and if they are of major importance they should be led into it at a suitable and properly constructed sump. The inlet end of a piped ditch should be fitted with a grid to prevent the entry of floating debris. One danger associated with piped ditches is that those carrying water the whole of the year round may have a fatal attraction for the roots of plants growing above them and it is not unknown for such drains to become choked by the roots of agricultural crops.

Those ditches which run continuously are for the most part natural streams or their canalised successors, and as a result are of some size, are winding rather than straight and have, as a rule, trees growing alongside. They are spring fed. Others are entirely artificial and straight ; these run along the contours or directly down the slopes, to function as convenient field boundaries, as drains in themselves and as recipients of outfalls from the fields which they skirt. The dimensions of all should be adequate for their duties. Fortunately, other considerations also operate to attain this end.

The lack of stability of an exposed soil face under the influence of weather, treading and erosion, is countered by giving ditch-sides a slope or batter. The amount varies in practice, a common one being to cut the top edge back 500 mm for every 1,000 mm. The bottom of a ditch needs to be wide enough to accommodate the feet of the men working in it and a minimum of 400 mm is desirable. Where possible, the inverts of outfalls should be higher than the floor of the ditch, so that in few circumstances is there justification for depths of less than 900 mm. A good working rule is to make the top width equal to the depth plus the bottom width, the depth being decided by that of the outfalls or the draining effect to be achieved by the ditch itself, and the width by the amount of water to be carried.

Avoidance of irregularities in the conformation of a ditch is obviously desirable. There should be no constrictions, no sudden turns, no marked or rapid changes in grade, because all these lead to localised erosion and silting. The stability of ditch-sides is a matter of importance. Weathering and crumbling are natural occurrences and the most effective preventive is a cover of the indigenous herbage, hedgerow shrubs, grass and weeds. In reconditioning ditches it is as well to bear this in mind and to avoid unnecessarily flaying off such sod as exists on the ditch-sides. This is a point of importance in ditch maintenance, when silt is being removed from the ditch floor and the annual growth of weeds is being " brushed out ". When the job amounts to a thorough bottoming-out, however, the amount of material to be removed in order to uncover the original ditch bottom and the outfalls of old drain systems entails drastic treatment of at least one side. This type of work can now be done quite efficiently in ordinary field ditches by means of one of the varied range of ditch-clearing machines which have been developed in recent years.

In all ditching operations, care should be taken to discover, liberate, and render effective any outfalls which may exist. The disposal of spoil from the ditches presents a problem which, if neglected, may cause an amount of trouble and nuisance which seriously offsets the advantage accruing to the drainage. Its amount, nature and condition may all present problems, but it should be deposited in such a way that it does not fall back, nor get trodden or washed back into the ditch, that it does not interfere with surface drainage into the ditch, that its weight and wetness do not cause the berm to collapse, and finally so that it does not dry and set in irregular heaps and banks which are difficult to spread, which harbour weeds, and which are an obstacle to cultivations and cropping.

Culverts are an important item of equipment in ditches, as is testified by the original and common practice of building them substantially in bricks or masonry. The casual and makeshift measures so common in the immediate pre-war times caused much more trouble and depreciation in the drainage condition of our agricultural land than is generally realised. Culverts should be commodious and well built, with the floor on a level with the original bottom of the ditch, wide enough to carry all the water brought to them, without causing substantial backing-up of water on the high side, and strong enough and sufficiently well finished not to come to grief under the traffic which crosses them.

TILE DRAINING

In considering the practice of tile draining it is essential to have in mind the salient points of soil moisture relationships, and the way in which water moves in the soil, as outlined on pages 13 and 38. The conditions in which harmful water-logging occurs and with which farmers' drainage problems are commonly associated may be grouped into five classes :

1. "Bottom water", or a rising water table due to an impermeable stratum below a soil of open texture or structure.
2. "Top water", or surface water-logging due to the impermeability of the immediate subsoil, e.g. clay land.
3. A combination of (1) and (2), as found in heavy silts and heavy subsoils other than clay.
4. Springs or spring lines at the division between permeable and impermeable geological strata on slopes or hillsides.
5. Ground water rising from below in valleys in permeable geological formations as a result of hydrostatic pressure operating from a distance.

Each of these various problems calls for its appropriate solution. The first is best dealt with by a system of parallel pipe-lines running with the fall if this is slight, or from ditch to ditch if the land is flat. A "gridiron" is a fitting description, whether each line of pipes has its own outfall or whether a number of them are linked by a main along the bottom of the field leading to a common outfall.

The second problem calls for similar treatment in principle, though the drain lines may be either pipes or mole channels. When, however, the fall is pronounced, the pipes should be laid across the fall, thus intercepting the downhill drift of water which is natural in such circumstances. As will be seen later in discussing mole draining, this is a point of some importance.

The third type of problem will be found to be amenable to the principles applied for (1) and (2) with appropriate intensification in the manner so "thoroughly" executed and expounded by James Smith.

The fourth problem constitutes the subject of Joseph Elkington's many triumphs. His method was to elucidate the lie and exact position of the troublesome clay stratum involved and to site a deep ditch or a deep commodious underdrain just on the high side of the critical spring line, reaching down to the impermeable floor and leading to a convenient side ditch, or by means of a main, to the ditch at the bottom of the field. Such deep drains operate by intercepting the water moving along the impermeable floor before it comes out at the surface.

The fifth type of problem is probably the most difficult to deal with, since the water is rising from below rather than moving in laterally, and interception in the simple sense of the term is not easily accomplished. Areas thus circumstanced are fortunately restricted and their subjugation is not often attempted. To do so effectively involves difficult and expensive operations in the way of laying deep parallel drains, which must first start to function as ditches, later to be piped and filled in. Here again an effective outfall is essential to achieve the requisite depth for the pipe lines, no easy matter in the general topographical conditions in which this particular problem is encountered.

Depth and Distance Apart of Drains. A hundred years ago shallow draining meant channels at a depth of about 750 mm and an interval of 5 m; it was practised for the most part on heavy land. In deep open soils, by contrast, depths of about 1·2 m and intervals of about 20 m were frequently employed. Later there was a regrettable tendency in some districts to lay pipes in clay at very shallow depths, 500 mm or even 450 mm. This is excessively shallow work and soils which are so tough and impermeable as to warrant it would be much better dealt with by means of a mole plough, which in good practice is operated at 500–600 mm depth and about 2·75 m intervals.

Depth and distance in draining are decided by what is feasible in the circumstances and what is desirable for particular farming purposes. To be early, land must be drained very well indeed; and when one finds it achieved artificially, it is generally by the deep draining of open soils. As a case in point there may be quoted some light silt soils of the Fens, with drains from ditch to ditch across the field, laid 1·8 m deep and 40 m apart. Such are the circumstances in which permeability can be exploited to the full. But the lower the permeability of the soil mass, the closer and shallower it becomes necessary to lay the pipe-lines, and the less feasible it becomes to compensate for wide intervals by greater depth. A different factor becomes important; the surplus water is more and more a feature of the top soil, in which it moves until it comes to the more open material over a pipe-line, to which it thus obtains easier access. It is in the heavier soils other than true clays that depths of 600–750 mm and intervals of 5, 10 or 15 m are found justifiable.

Drain Pipes. Up to the present time land drain pipes have been made mostly either of clay or concrete—in agricultural field work clay pipes have been almost universally employed. The deciding factor is cost, so long as the pipes are reasonably sound, uniform and durable. Some 150 years of experience and experiment have eliminated a multitude of specialised shapes and forms in favour of the cylindrical clay pipe, 300 mm long, easily handled and comparatively cheap to produce. There is a limit to the diameter and length of drain pipes made with clay, but concrete is successfully used for making larger pipes and is the best material for pipes over 200 mm in diameter and 600 mm in length.

The use of pipes of small diameter should be avoided, because of the ease with which they get displaced in the ground with the result that the continuity of the channel is broken : 65–75 mm is a desirable minimum internal diameter. Otherwise, carrying capacity and the area of land served should decide the size of pipes used. In this country a 75 mm pipe should deal adequately with the run-off from 1·2 ha of light land, or 0·8 ha of heavy land ; a 100 mm pipe should serve 2·8 ha of light or 2 ha of heavy land ; a 150 mm pipe should serve 7·3–8 ha of light land.

Laying the Pipes. At the outbreak of war in 1939 there were not many skilled hand drainers in existence, but the amount of drainage work then required produced a new generation of drainers and also led to the appearance of a number of machines for excavating drain trenches, laying pipes and back-filling over them. Both hand work and machine work require skill to produce the best results, and in the execution of work a number of points should be kept in mind. The system should be marked out on the ground, the grades established, and the pipes laid out in such a way that they will not interfere with the digging and pipe-laying. The grades should be set up above ground by means of three or more horizontal bars for each length of drain line so that the level of the trench floor can be checked against them.

Work should begin at the outfall and proceed in such a fashion that the developing system can function if need be in case of heavy rain or tapping pent-up water. Excavated material should be placed tidily and systematically with the view of its easy replacement. The digging should be the minimum possible for the aim in view and the trench bottom should be completed by the scoop, with the floor undisturbed, and in such a way as to provide a bed, semicircular in cross-section, and of the right size to fit the pipes snugly. The pipes should be laid in succession from the outfall, properly butted-up end to end, and given a first cover of material calculated to remain open enough to allow percolation but close enough to prevent silting. The top soil may be the best material for this purpose and it is certainly the most easily obtained. The operation of blinding, or giving the pipes their first few inches of cover, is particularly important before back-filling by machine, and in trench bottoms which do not fit close to the pipes, to avoid displacing them. Junctions of minors with mains, and of submains with mains must be made with care to avoid silting, erosion, collapse or bursting. According to their importance and to the loads anticipated they may range from purpose-made clay ware junction pipes, second quality sanitary ware, or brick chambers with a covering slab for more important junctions.

The chief difficulty with outfalls is to render them stable and secure. The best cheap method is to use a scrap-metal pipe, offcut pitch fibre or plastic pressure pipe, placing it so that it achieves stability by the combination of a smooth tight bed and a well-tamped back-fill over it. Alternatively, the pipe-line can be made to terminate in a socketed glazed pipe laid in concrete, with a surrounding brick or concrete headwall built to the batter of the ditch-side and supplemented by an apron of similar material under it and also on the opposite side of the ditch to prevent erosion.

It has been pointed out that most fields have been drained before and that frequently the drains can be brought into action again. Where, nevertheless, it is decided to redrain and it is found that old pipe-lines are being broken into by the new, the old line should be appropriately joined up with the new so that if it still functions it will have an adequate outlet.

MOLE DRAINING

Clay land farmers, struggling against the seasonal wetness which is a natural characteristic of their soils, have for centuries exploited the plasticity and tenacity of clay in their field drainage operations, producing by hand or by machine underdrainage channels which would last and function for a number of years without the aid of pipes or other rigid lining materials. An account of one of these, known as plug draining, was given by Stephen Switzer in *The Practical Fruit Gardener*, published in 1724. Machines called mole ploughs were not uncommon in Essex at the beginning of the nineteenth century, and much that is of importance in the practice of mole draining is on record in Arthur Young's *Agriculture of Essex*, 1807.

The Mole Plough. The mole plough consists essentially of a chisel-pointed, horizontal, cylindrical mole affixed to the foot of a stout knife-edged vertical coulter, which in its turn is mounted in a long heavy beam so designed that it can be drawn along the ground by suitable power, with the mole moving through the subsoil at a set depth. At the rear of the mole is attached an expander, of greater diameter than the mole by 13–20 mm (Plate IIIc). The beam is usually elaborated and slung from a frame mounted on wheels, so that it can be lifted out of action or lowered into position by a power unit on the tractor or a self-lift mechanism on the plough wheels. Mole ploughs are built to different designs and dimensions, but common agricultural use aims at a 75–100 mm channel 500–600 mm below the ground surface.

A mole plough moving over an uneven ground surface tends to produce an uneven channel unless there is some steadying influence or mechanism for smoothing out the irregularities. This is the function of the beam whose length and weight, in conjunction with a flexible attachment to the power unit and with the chisel-shaped leading end of the mole, combine to cause the mole to move horizontally through the subsoil and produce a smooth even channel. The action of a mole plough, which should travel at the rate of

3–4 km/h, is to prize up the ground in its track in a systematic fashion and to leave behind it a smooth channel, circular or ovoid in vertical cross-section. The soil mass above the channel is left with frequent vertical fissures branching at an acute angle from the slit made by the coulter and joining it in the direction from which the plough has come. The ground is thus moved and opened on both sides of the slit to a total width of 1·2–1·5 m in the case of heavy mole ploughs. The slit and accompanying fissures let the water down to the mole channels which in turn lead it away.

It is important to ensure that the seating of the coulter in the frame and the set of the mole relative to the beam are maintained in such adjustment that the mole proceeds through the subsoil with its axis parallel to the ground surface, otherwise the action will be a tearing one and the channel will be distorted vertically, with rough broken walls and the roof much too close to the ground surface : in such conditions rapid and early collapse of the system will ensue.

Land Suitable for Mole Draining. Mole draining is eminently successful in clay land, and can be used effectively in most heavy land with stiff subsoils if carried out in the right way. The major clay formations in Great Britain are well defined and are mapped in detail by the Geological Survey. The land most suitable for mole draining is found on the exposures of the Gault, Ampthill, Oxford, Lias, Kimmeridge and London clays and on the Boulder Clay of the south-eastern half of England. There are also isolated areas of Boulder Clay and other clay formations suitable for mole draining in other parts of the country.

Land which is suitable for mole draining is characterised by even, unbroken surfaces and gentle slopes. It is hard and cloddy when dry, with marked fissuring in times of drought, sticky and tenacious when wet, and very prone to surface water-logging. The subsoils, on examination to 600–750 mm in depth, will be found to be dull in colour, generally buff, grey or blue-grey, with faint mottling of brown or buff. They contain few or no stones, and not much sand. In short, they show the properties of shrinkage, tenacity, plasticity, and impermeability to a marked degree. Sandy clays and heavy silts are less suitable subjects for mole draining. mainly because of their lack of stability which renders the mole channels prone to silting and to premature collapse.

Mole drains should be 75–100 mm in diameter, circular or nearly so in cross-section, 500–600 mm deep (to the floor of the channel), 2·5–3·5 m apart, and not much more than 200 m in length. They should run with the fall as a rule, except where this is excessive. They should be drawn over a properly covered main or system of sub-mains, and should cover the whole of the field or area concerned in a uniform manner. In this way the subsoil will be rendered open and permeable (some $\frac{4}{9}$ of the area is actually moved or fissured) and surface water-logging is thus prevented, all excess of rain being rapidly carried away by the mole channels below. Mole systems in action work swiftly : the run is in the form of a flush and the outfall rates are very high—some 6 litres/ha/sec have been recorded. Hence, to get rid of all this and to safeguard the mole system, it is advisable to use

100 mm pipes for the mains and to provide at least one outfall per 2 ha of land.

Obviously, the layout of mole drain systems depends on the size. dimensions and topography of the fields concerned. The simplest are those of roughly rectangular shape with a receiving ditch along the bottom side. A piped main is laid parallel to the ditch, about 14 m away from it, with leads to the ditch every 80–100 m run. The moles are drawn over this main, parallel to the sides of the field, and any length up to 200 m. For lengths exceeding this it is advisable to divide the run into two, by means of a second main across the field, so sited that it leads to a convenient side ditch, or to a carrier main to the ditch at the bottom of the field. The second main may have only one outfall and therefore may not be able to serve as much as half of the field. The aim in planning should be so to site all mains that it is possible for the mole plough to traverse the field, to and fro, without interruption, as simply as the ordinary plough. In land with poor falls, this may be attained by siting the pipe-lines in such " lows " as do occur, so long as they run more or less in the same direction. A field with a pronounced depression through it may call for moling either parallel to the trough and across the slopes, or at right angles to the trough, in which are laid two parallel mains some yards apart.

It is possible to mole drain at any time of the year, but circumstances render some months much more favourable than others. The surface should be dry enough to take the tractor without its slipping or digging in, and the subsoil should be moist enough to take the mole without excessive " bursting " effects. In this way the right degree of fissuring of the ground will accompany the production of a smooth-walled mole channel, likely to last its full time. There should not be water standing about when moling is in progress, nor should the land even be wet, if it can be avoided. On the other hand, a drought-hardened subsoil, though maybe desirable for drastic subsoiling or bursting operations, is not in the best condition for successful moling. Corn crops and leys suffer little harm from moling carried out in favourable soil conditions, so long as they are not in full growth. The best months for moling will be found to be March to June, with July to October also good, except in droughty summers. In the winter months the soil conditions are frequently unsuitable.

Mole Drains and Main Drains. The provision of a proper main is essential to success in mole draining. It is possible to draw moles direct from a ditch and have some of them function, but they soon collapse and go out of use. The ends can be piped to keep them open, but this practice is wasteful and inefficient and does not overcome other weak points such as the damage done to the ditch by the mole plough, the loss of time, and the cost. With a moderate expenditure per hectare on a really good piped main, a durable foundation can be obtained which will serve for an indefinite number of repetitions of the moling at intervals of a few years at its own characteristic low cost. The main should be laid according to sound tile draining practice at such a depth that the mole plough has easy clearance of the pipe. The excavated trench, being in clay, should be as narrow as it can be made, both to provide a tight firm bed for the pipes, and to economise in the covering material which must be put over the pipes up to a level well above that of the roof of the mole channels. The function of this cover is to let the water from the moles down to the pipe-line. It generally takes the form of the most economic material of an open nature which is to hand, e.g. coarse clinker, coarse gravel, broken tiles, rubble, shells, bushes, hedge cuttings. Bushes and clippings should be carefully trodden in ; the other materials should not contain any quantity of very fine particles, to avoid risk of silting.

When using gravel or similar material on the mains the moling may take place as soon as the surplus trench spoil has been levelled, assuming that the sub-soil moisture conditions are suitable. With fills such as bushes it is sound practice to lay the mains a year before moling so that the trench-filling becomes consolidated before moling takes place.

Occasionally circumstances arise in which moled mains can be used. This is not very good practice and is only justified on a short term view. Points to bear in mind are (1) to draw three or four channels for the main, at least 3 m apart and at sufficient depth to give them 25–50 mm clearance of the floor of the minors ; (2) to " eye " all of them into tiled leads on the same level, one lead per 1·2 ha ; (3) to draw the mains before the minors.

The use of Plastics and other Recent Developments. Plastics first began to be used experimentally in field drainage during the 1950's. Not only are the strength characteristics very different from those of clay tiles, but for economic and other reasons plastics pipe is manufactured in smaller diameters than those traditionally used for clay ; 50 mm is a common size in plastics. During the late 1950's and early 1960's the Ministry of Agriculture ran a series of field trials in which " side by side " comparisons were made between clay and plastics. These trials established a basis for the strength required in a plastics pipe so that it can withstand handling stress and also support the long-term loads for the soil. They also showed that the drainage effect was the same provided that the pipe diameter was adequate for the situation. In general, pipe diameter is not an important factor in either water entry to the pipe or drain spacing, but it is important in transporting away the water. Hence when plastics pipes became acceptable for grant aid in March 1964 the only major limitation on their use was that closer attention had to be given to scheme design to avoid the pipes being overloaded. Tables have been prepared which allow the maximum safe length of pipe to be found for any given circumstances. Despite the fact that a 50 mm pipe only carries some 33 per cent of the water that would be carried by a 75 mm pipe, various surveys have shown that it is only in a few large, flat fields of eastern England that drainage design must be materially altered to allow the use of plastics.

The first plastics available were extruded in semi-rigid pipes some 4·5–6 m long. These fit together by a simple spigot and socket type joint and are equally suitable for either hand laying in a previously prepared trench or for passing through the laying chute of a specialised mechanical trencher. From the beginning it was realised that one of the most important aspects of plastics was the degree to which they would allow further mechanisation. Early experiments concentrated on the use of a machine developed from the mole plough to allow the continuous laying of a drain lined by a plastic strip. In one development this strip was zipped together at the edges to form a tube. As the whole process was completed underground, out of sight, the process was not readily acceptable ; however, these experiments paved the way for the development of the technique of " trenchless " drain laying which is now widely used in conjunction with flexible, coilable, plastic pipe.

Currently (1976) the major part of the plastic pipe used is of the corrugated coilable type although about 20 per cent is smooth rigid tube. Both may be in either PE (Polyethylene) or PVC (Polyvinyl chloride). At present the percentage of the total drainage in this country done in plastics is well under 20 per cent. This is largely due to economic factors in that the price differential between work done in plastics and clay has been small. Where this is not the case, as for example in Northern Ireland, the percentage is much higher (currently about 60 per cent). It is quite possible that future price movements and the development of trenchless working will greatly increase the percentage of plastics used in this country.

Another recent development with a wide future potential is the production of light-weight aggregates for trench backfill on tile-mole work and schemes of similar nature. Currently no plastics based material is economically viable, but various synthetics such as expanded clay, pulverised fuel ash, shale and similar industrial waste materials are being increasingly used.

SURFACE DRAINING

The old and widespread practice of setting up the land in ridge and furrow or in high-backed lands was essentially surface drainage. It survives still in much of our permanent grassland, but is rarely used on new drainage work. However, surface methods are employed on moorland grazing and in forestry work. The usual practice is to cut surface drains by means of a specially developed plough. The layout is usually based on a herringbone layout leading to a natural burn or beck. By these means substantial improvements in the grazing are possible at a cost of £5–£10 per ha.

Chapter III

BUILDINGS AND OTHER FIXED EQUIPMENT

MANY of the farmsteads and their buildings in existence today were first erected at some date between 1800 and 1875. The enclosure of the open fields, which had been going on gradually for several centuries, became much more rapid in the last quarter of the eighteenth century until well on in the nineteenth century and resulted in the creation of large numbers of new farms which had to be equipped with buildings. At the same time farming methods made great advances, and the ramshackle, draughty buildings of the eighteenth century, constructed of local materials such as timber, stone and thatch, no longer met requirements, particularly for livestock which were being housed through the winter.

Development in the design of farm buildings went rapidly ahead, encouraged by such writers as J. C. Loudon with his *Encyclopaedia of Cottage, Farm and Villa Architecture*, published in 1836 and J. Bailey Denton's famous *Farm Homesteads of England* which appeared in 1863. New and more durable materials were used, since with improved communications building materials could be transported further afield. Bricks became more widely used, iron roof trusses were developed and roofs were covered in tiles or slates brought from Wales or the north. The buildings were substantially constructed and for this reason so many have survived until the present day. By modern standards, however, the livestock buildings appear cramped and stuffy, wagon sheds do not meet the needs of modern machinery; the large enclosed barns are awkward for storing hay and straw and too little account was taken of the ways in which labour might be saved.

When the so-called Golden Age of Farming had come to an end about 1875 rents slumped and had fallen by about a third at the end of the century. Landlords were no longer willing to invest capital in farm buildings and progress stopped. With farming in a depressed state rents remained low for the next forty to fifty years. Few new buildings were erected and they were mainly such inexpensive structures as dutch barns and Scandinavian piggeries which first appeared about this time. There was a marked swing to dairying, bringing a large increase in the number of tie-up cowhouses, but these were provided mostly by adapting existing buildings as cheaply as possible. Two relatively inexpensive materials also first started to be used in the 1920's, concrete for floors and walls, and asbestos-cement sheeting for roofs. With expenditure on fixed equipment being so limited there was little opportunity for the development of new ideas in design and in materials for farm buildings.

By 1939 farm buildings as a whole were out of date and were often in poor condition as well. The wartime National Farm Survey of England and Wales of 1941–3 revealed that 61 per cent of farmsteads had buildings which were technically unsatisfactory. Although in the 1939–45 war there was rapid progress in new farming methods little

could be done to improve farm buildings.   Since the war the revolution in farming methods has continued and has had a counterpart in
the design and construction of farm buildings.

The development of new designs and materials for farm buildings,
testing them out under farm conditions and then getting the new
developments widely adopted, is inevitably a slow process.   It is a
far slower process and involves much higher expenditure than the
bringing into use of most new techniques in farming.   Because of this
the progress with fixed equipment of farms, though very considerable,
has been inclined to lag behind farming progress generally.   Metrication will greatly affect the fixed equipment of farms and Chapter XXV,
" Metrication ", should be studied.

To encourage the injection of capital into the improvement of
fixed equipment various forms of Government assistance have been
available since the war, including a scheme which in 1946 provided
grants of half the cost of reconditioning hill farms, later extended to
all upland farms depending on livestock rearing.   The Farm Improvement Scheme of 1957 granted one third of the cost of new and improved buildings and other fixed equipment, except housing on all
types of farm.   Changes in 1967 and 1970 made grants available for
a wider range of improvements and also altered the rate.

Planning and Building Controls.   Farm buildings, but not
farm houses and cottages, may be put up or altered without the need
to obtain local planning authority permission provided that, in the
case of a new building, it does not exceed 465 m^2 in ground area
either by itself or added to other buildings erected within 90 m during
the preceding two years, or is not more than 12 m high.   There are
other exceptions and special arrangements apply in areas of high
scenic value.   Planning law changes from time to time and in case of
doubt the local planning authority should be consulted.

Farm buildings also enjoy some exemption from the Building
Regulations Act, 1965, but plans with details of new building work
or reconstruction need to be deposited with the local authority before
starting.   There are also restrictions on the sinking of wells and boreholes and the storage of petrol.

There are particular requirements affecting farm buildings.   Regulations made under the Health and Safety at Work Act, 1974, prescribe the structural precautions needed to safeguard apertures, stair-
ways, pits and stationary machinery in farm buildings.   There are
also statutory Codes of Recommendations for the welfare of livestock
which can affect layout and design.   The divisional offices of the
Ministry of Agriculture can give advice on these.

MODERN DESIGN AND BUILDING MATERIALS

With building costs rising steadily until they are now many times
what they were in 1939, the aim in design has been to have durable
buildings suitable for their purpose, but at the lowest possible capital
cost and constructed of materials requiring little or no maintenance.
As a result, traditional buildings of brick and tile, or stone and slate,
constructed to fit the site are being superseded by framed structures

bought prefabricated from the manufacturer. This type of building comprises basically a framework spaced at 3–4·8 m intervals, supporting a roof which is covered in light sheeting. Walls, where required, are in the form of infilling panels, carrying no load from the roof. The framework can be either stanchions or pillars to which built-up roof trusses are fixed, or a " portal " or arch frame (see Fig. 16). The portal frame has the advantage of requiring no tie beams or other obstructions. Wide spans are possible with framed buildings, without the need for intervening stanchions to obstruct the floor space. Upper floors are more costly to provide than ground floors, and two-storey buildings are not often erected, except for such purposes as granaries where gravity can simplify the movement of foodstuffs and other materials. Roof lights are preferred to windows in walls because overhead lighting is usually more effective and less likely to be damaged. Flat roofs are also being used instead of lean-to or pitched roofs for small buildings such as dairies. With a simple structure of timber, covered with asphalt or felt laid by a specialist firm, this type of roof need not be expensive.

Roofing. Corrugated sheeting of various materials is used for roof coverings and walls ; it has the advantage of being light, weatherproof and reasonably durable. Asbestos-cement requires no maintenance but is brittle, and should not be used in places where it may be knocked, such as the lower parts of walls, or where the building itself may move. Galvanised steel costs about the same and is better in such places though the galvanising wears off in time, particularly if used in livestock buildings where condensation occurs, and the sheets then have to be painted periodically. Galvanised sheets may now be obtained pre-painted and for a more durable coloured finish plastic-coated galvanised sheets are marketed. Transparent plastic and resin bonded glass fibre sheeting, in corrugations to match other types of sheet, provide weatherproof lighting.

Concrete. For livestock buildings, reinforced concrete is probably the most suitable material for the structure, taking into account cost and durability without maintenance. It is available in the form of structural frames, roof trusses, window frames, beams and joists for upper floors, water troughs, rain-water gutters, drain pipes and chambers. Concrete components tend to be clumsy and heavy, but pre-stressed concrete is an improvement. It is reinforced concrete in which high tensile wire takes the place of mild steel rods, with a considerable saving in weight.

Steel. Steelwork is lighter and smaller in section than concrete and is easier to transport and handle on the farm. It needs painting, except in smaller components such as window frames which should be bought galvanised.

Timber. Timber is expensive and should be used economically. With modern techniques using graded timber and metal connectors and plates to simplify and strengthen joints, lighter and shorter lengths of timber can be used for structural purposes. The Timber Development Association has developed a number of designs of simple timber frame which can be constructed on the farm (see Fig. 2).

D

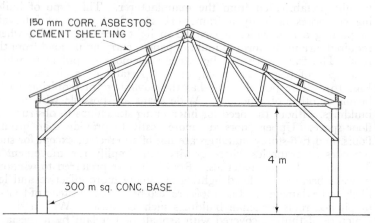

150 mm CORR. ASBESTOS
CEMENT SHEETING

4 m

300 m sq. CONC. BASE

Fig. 2.—Timber Implement Shed designed by the Timber Development Association.

Protection. Metal must be protected from corrosion, and timber from the effects of damp, fungi and insects. A film of paint will prevent damage by damp though repainting is essential before the film perishes. Oil paints are expensive for farm buildings: tar, and bitumen paints are much cheaper for timber and metal but equally satisfactory though they are dark in colour. The best forms of these paints are bitumen solution paint and black paint (tar base). Timber can also be treated with preservatives, and whilst these will not prevent it becoming damp and swelling they stop any attack by fungi, such as dry rot, and insects. Preservatives are cheaper than paints and should be used for timber farm buildings and all timber in contact with the ground, such as fencing posts. Creosote is the cheapest form of preservative but it has a smell causing taints and it is unpleasant to handle. There are inexpensive water-soluble preservatives which can be used instead ; they are usually sold under proprietary names. It is often possible, and an advantage, to buy timber which has already been treated with preservative.

Walls. For walling there is little to choose between concrete blocks and brickwork ; blocks are probably easier for unskilled labour to use but to avoid trouble with cracks there should be very little cement in the mortar so as to allow for shrinkage movement. To be weatherproof, walls should be 225 mm thick and should have a damp-proof course just above the ground to prevent rising damp. Rainwater falling on roofs should be kept away from walls by using gutters and downpipes. Asbestos-cement is often used as it is inexpensive and requires no maintenance.

Floors. Concrete is the most suitable material for floors in most farm buildings, though in livestock buildings insulation is needed, as will be explained later. All concrete floors are improved by being made damp-proof by the laying of a polythene membrane under the concrete.

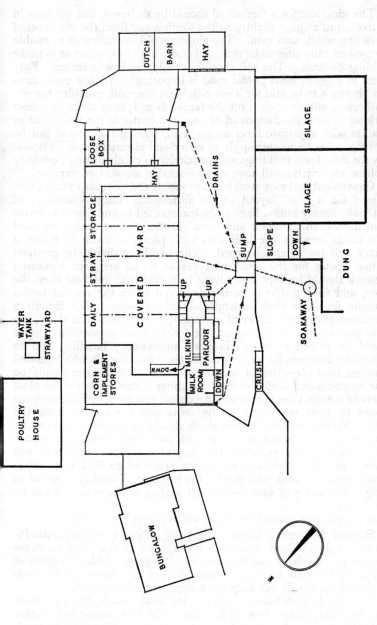

Fig. 3.—Dairy Farmstead for 28 ha Holding designed by the *Farmers' Weekly* at Bulborne, Hertfordshire.

57

THE SITING AND LAYOUT OF THE FARMSTEAD

The ideal site for a farmstead should be sheltered, but not shut in by trees, and sloping slightly to the south away from the direction of prevailing winds and rain. The slope should be sufficient to enable sewage and rainwater to be removed easily but not so great as to make building difficult. The soil should, ideally, be free draining. Easy access to a good, hard public road is important. With a small farm there is much to be said for a site adjoining the road, provided the site is otherwise suitable : but on the larger farm it may make for easier working to have the farmstead as near the centre of the farm land as possible with a private farm access road, though this should not be too long. An adequate supply of water and electricity will be necessary for most farm buildings and proximity to existing lines, especially 3 phase electricity, will save a considerable amount of expense.

Convenient and easy working is probably the most important point making for a good layout of the farmstead. Large quantities of materials, mostly bulky, have to be transported to and from the farmstead from both the fields and from outside. There should be good wide roads between the buildings with plenty of room in which a tractor and trailer can be used. The buildings should be grouped so that routes for the different types of traffic are kept separate ; livestock should be kept away from transport with produce from the fields, and field traffic in turn should not block the way of lorries bringing fertilisers and foodstuffs or taking away milk. Buildings concerned with the delivery of materials, whether the farm dairy, the grain storage and drying plant, the fodder store or the slurry store should be placed so as to eliminate unnecessary handling and to reduce the distances for moving materials. Buildings in which related processes take place should be close together. Food stores should be near or combined with livestock buildings ; dairies should be close to where the cows are milked and accessible to a road; milking parlours should be near the yards, to the cows and the way in from the fields. Tractor and implement sheds should be together, and can be away from other buildings. Livestock buildings should face south : tall buildings, such as dutch barns, placed on the north side will provide shelter. Buildings should be arranged so that mechanical handling can be used, whether for the bringing of food or removal of dung. The siting of silos for the self feeding of silage should not be overlooked.

A layout for a small farm is shown in Fig. 3.

Saving Labour. Continually rising labour costs and a diminishing supply of farm workers are emphasising the need to make efficient use of labour and to eliminate drudgery. Although study of labour use has long been a common practice in industry the methods employed have only recently been applied to farm work.

Work Study, also known as *Time and Motion Study*, has two main branches likely to be of use in farming. The first, method or motion study, involves analysing and recording in a standard form methods and equipment used or to be used in the performance of work, followed by a critical examination in a specified form aimed at showing where improvements can be made. These improvements are generally found to be self-evident, though it is only after such a dispassionate study that they become so obvious. The alterations required in buildings and in equipment are frequently small though the resulting increase in efficiency may be large. For the best results, training and experience in work study are required ; but common sense and an elementary knowledge of the basic theory can often suggest valuable improvements in work routines.

While method study is probably most useful in showing improved ways of working in existing buildings, some general principles for saving labour have emerged which are of value when designing new buildings. The reduction of unnecessary movements, particularly walking about, is probably the most important. This can be achieved in such ways as :

(a) handling such materials as foodstuffs and manure in the largest possible quantities (e.g. barrows and trolleys instead of buckets) and having buildings suited for bulk handling with wide doors and level passages, and containing bulk tanks as for milk, grain, fertilisers and feeding-stuffs ;

(b) having close together those buildings and work areas where related operations are carried out (e.g. food stores close to the animals to be fed) ;

(c) arranging buildings for circular travel and avoiding " back tracking " ;

(d) keeping tools, machinery and services, such as water, close to where they are used.

The use of gravity and simple machinery can do a lot to reduce work and drudgery.

Work measurement, or time study, is another branch of work study of use in agriculture. It is concerned with determining the time required under specified conditions for carrying out different operations. The preparation of standard times is primarily a matter for experts but these times when available—and they are very few as yet—provide a valuable basis of comparison between ways of carrying out the same task using different specified equipment and buildings, e.g. milking in different types of milking parlours. Standard times can also show how efficiently a task is carried out on a particular farm judged by the time taken compared with the standard time.

GENERAL PURPOSE AND TEMPORARY BUILDINGS

Most farm buildings are constructed of materials which will last very many years—often longer than the need for which the buildings were first erected. There is, therefore, much to be said for having buildings which can be altered without excessive expense to meet changing requirements and changing methods, or to have buildings suitable for a number of purposes without the need for any structural alterations. From the point of view of adaptability there are many advantages in the wide span building which has few intervening stanchions to support the roof, and partition walls independent of the main structure which divide the building up as required. A building which provides merely a roof and walls, leaving the floor free for any use, is a variation. The covered stockyard, which can be used for cattle, pigs and poultry as required, is one example. Another is the dutch barn which, particularly if sheeted on one or more sides, is very adaptable and can be easily and cheaply extended by the addition of lean-tos. It can be used for storing hay and straw, covering silage pits or housing implements and livestock. Loose boxes, particularly on a small farm, will serve many purposes.

The building suitable for more than one purpose may suffer from the disadvantage of not being ideal for any one purpose, and in many circumstances the special purpose building is to be preferred since, being specially designed, it will provide better conditions for livestock or other uses and will be more labour saving. Against this must be put the initial cost, which is usually higher, and the fact that the capital invested in it may have to be written off over a much shorter period if it cannot easily be adapted to other uses.

With the high price of timber and many other materials, and costly labour, light temporary buildings are almost as expensive to erect as more permanent ones, particularly those of the prefabricated type. It is possible, however, by using rough timber and straw to make some substantial economies with home-made buildings such as open or partly-covered cattle yards, lambing pens and shelters for breeding and store pigs. There are no fixed designs, but a little ingenuity and a frame of coppice poles or woodland thinnings, with straw in bales or sandwiched in netting, will make useful buildings. Fire risks are high and vermin can be a nuisance ; such buildings should be taken down before becoming derelict and an eye-sore. More ambitious buildings of the *pole-barn* type are also possible. A building tried out in this country comprises a framework of telegraph or similar poles and railway sleepers, supporting a roof of wire netting and roofing felt with walls of straw bales. Units of 12 m × 12 m have been found suitable for a wide range of livestock, including fattening and breeding pigs, and poultry.

VENTILATION AND INSULATION

Any enclosed building, particularly if totally enclosed, where livestock are housed for long periods, should have the air changed from time to time if the conditions in the building are to be satisfactory. Animals emit large quantities of carbon dioxide and moisture into the air and moisture may also be taken up by the air from floors, wet from washing down, from animal excreta, from drinking water and food. As a result the air in a livestock building will become humid and foul unless it is changed. Farm animals will not thrive in such an atmosphere, and the structure of the building may be seriously affected by the condensation which occurs when air, saturated with moisture, comes in contact with colder surfaces in the building.

Temperature. In addition to the affect of temperature on condensation it is important that the air in a building should be warm enough but not too warm for the animals. For cows, a temperature of 10–13 °C is considered to be the best though their milk yield is little affected at much lower temperatures provided there are no sudden fluctuations. Pigs are less adaptable and need a more closely regulated temperature if they are to thrive. A temperature of 13–19 °C is usually recommended for fattening pigs, and 19–24 °C for sows and young pigs. Farm buildings are not often heated artificially and any heat is supplied by the animals themselves. The amount can be considerable, and a full-grown dairy cow will give off the equivalent of 1 kilowatt-hour of electricity. In winter the heat can, however, disappear rapidly from the building through excessive ventilation or by too rapid conduction through walls and roofs.

The cubic air space per animal should also not be too large since otherwise the animal will give out too little heat to warm the air in cold weather : 14–17 m³ per cow and 1·5–2·25 m³ per fattening pig are common standards.

Ventilation. Natural ventilation is all that is usually provided in farm buildings to change the air : it depends partly on wind which blows air in on one side of a building and sucks it out of the other, and partly on the fact that the air on being warmed expands, becomes lighter and therefore tends to rise. In a building it will rise to the highest point, normally the ridge, and if there is an outlet there it will pass out of the building and fresh, colder air will be drawn in from outside to replace it. The fresh air should be admitted through inlets in a position where it can quickly mix with the warm air without causing draughts. The quantity of air passing through the building should be sufficient to remove enough moisture to keep the relative humidity within the limits mentioned. If too large a quantity of air passes through during cold weather the temperature in the building will fall because the animals will be unable to heat the air fast enough.

Cowhouses. Experience has shown that to preserve satisfactory conditions in a cowhouse, each cow needs 65 m³ of air per hour under normal conditions, increasing to about 170 m³ in mild weather. This can generally be achieved with ventilation *outlets* at the ridge of 0·5–0·8 m² per cow. The type of outlet will vary with the roof construction : for asbestos-cement sheets there is special ventilating ridge capping, or a 100 mm gap can be left between the sheets at the ridge, though this may cause down draughts in windy weather ; with slates or tiles, alternate ridge tiles can be raised. Patent extractor ventilators are also available. *Inlets* for fresh air need not be more than 0·1 m² per cow at the most since a great deal of air will enter the building through windows and the opening of doors, which are also often badly fitting. Inlets should have a baffle to prevent wind blowing in and are best set at least half-way up an outer wall, spaced one to every two cows. Windows with hoppers which can be closed are an advantage, particularly in a gable end, as they provide a means of regulating the quantity of air entering the building. The quantity required will, of course, vary with the relative humidity and the temperature of the external and internal air, but with cattle, which can tolerate a fairly wide range of conditions, a close control of the ventilation is probably unnecessary.

Piggeries. As pigs require higher and more constant temperatures, piggeries should have a smaller cubic air space and the ventilation should be more controlled. It is found from experience that one of the most satisfactory designs of piggery has a flat, false ceiling over the pigs at eaves level with ventilation by means of a duct or flue, with an outlet which can be regulated in size, in each compartment of the piggery. A suitable design of flue is given in Fig. 14A. The quantity of fresh air recommended for a fattening pig of 90 kg is 20–23 m^3 per hour in a totally enclosed piggery, or 8·5 m^3 in a house with outside runs for dunging. This can be provided by outlets or a flue of about 0·1 m^2 per 15 pigs and inlets about a half less. In partly enclosed buildings, such as covered yards or piggeries with outside runs, no provision for inlet ventilation is required but outlets are needed at the ridge. Natural ventilation requires suitable climatic conditions to be effective and electric fans are now sometimes used as a supplement when conditions are unfavourable. Fans are also used to provide all the ventilation in highly specialised piggeries.

Insulation. All building materials allow heat to pass through them to some extent. If a livestock building is to have a constant temperature in winter and summer and the ventilation system is to be satisfactory, with little condensation, either the materials of which the building is constructed must be resistant to the rapid transfer of heat or materials with good thermal insulation properties must be added to the structure as a lining. The materials in general use for roofs—asbestos-cement and corrugated iron sheeting—have a very poor insulation value and often need a special lining.

Air is excellent for insulation, and most materials used for insulation contain a large number of small air cells. The most suitable type of material will depend on the position in which it is to be used.

For roofs, the insulation can be in the form of boards or slabs of materials such as expanded or extruded polystyrene or polyurethane, fixed as an underlining just under the external covering of the roof or, as in piggeries, as a ceiling at eaves level. An alternative is to use loose materials, some available in a quilt, as asbestos fibre, glass and wood wool, or tightly packed straw supported on building boards or wire netting. To prevent the insulating materials being damaged by moisture passing through them and condensing on the cold surface of the outer roof, it is essential to seal the underside against moisture penetration either by painting it with bitumen or aluminium paint, or by lining it with polythene or waterproof building paper : any joints need to be watertight.

For walls a single leaf of 225 mm thick, hollow blocks, rendered externally if porous, or two leaves of 112 mm thick brickwork with a 50 mm air cavity between provide a wall with adequate insulation properties for a livestock building. With thinner walls the insulation can be improved by adding an inner leaf of concrete blocks made with a lightweight aggregate.

Insulation is essential in concrete floors on which animals lie if they are to have a warm bed. The insulation is best provided by having a layer of material containing air spaces close to the surface of the concrete. Suitable materials are a 75 mm layer of concrete made with lightweight aggregate or with no fine aggregate as sand, known

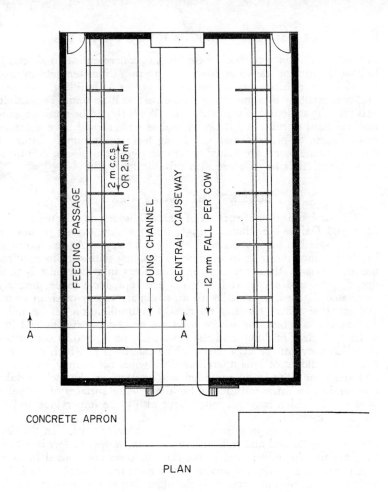

PLAN

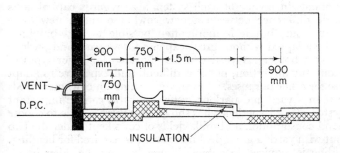

SECTION A—A

Fig. 4.—Double Range Cowhouse for 24 Cows

as " no fines " concrete (both these types of concrete contain air cells), hollow flooring or partition blocks, or 75 mm diameter field drains.

Moisture reduces the insulating properties of building materials and consequently it is important to see that as little damp as possible gets into the structure of the building. Walls should be made reasonably resistant to rain, and floors as well as walls should have a damp-proof course to prevent moisture rising from the ground. Bitumen and polythene sheets are valuable for damp-proofing floors, and bituminous materials in rolls are also suitable for walls.

HOUSING THE DAIRY HERD

The design and construction of dairy buildings is affected by the Milk and Dairies Regulations, 1959, under which anyone producing milk for sale has to be registered by the Ministry of Agriculture and has to produce the milk in premises complying with certain requirements. Broadly these are that any buildings in which milk is produced, processed and stored, or cows are kept, have to be properly ventilated, lighted, provided with an adequate supply of clean water and situated where there is no risk of contamination of the milk. The access to a building where cows are housed but not milked has to be kept clear of dung, and the cows must be housed in conditions in which they can be kept clean. The buildings in which cows are milked should have an impervious floor wherever it may be soiled, and the floor sloped and provided with gutters or channels to take any liquid to a drain outside the building. The surfaces of the parts of the walls which may get soiled have also to be impervious and to be easily cleaned.

There are two main systems of keeping a dairy herd. In the first the cows are housed and milked in a *cowhouse, shippon* or *byre* in which they are tied in stalls : and in the second, cows are housed in loose yards or cubicle sheds, and are milked in a separate building, a *milking parlour*. The cowhouse system is the older and so is more common in the traditional dairying areas of the country, but the loose housing and parlour system is almost universal in new buildings. There are many arguments in favour of each system : as regards capital costs there is little difference between a new cowhouse and a new loose house and parlour, but it is more often possible to find buildings which can be adapted without great expense into a yard and parlour than into a cowhouse. A yard can also be used for other types of stock without any alteration, but it is difficult and expensive to adapt a cowhouse for other purposes. On the other hand, a cowhouse needs less space than a yard and parlour system and this can be important on a congested site. Experience has shown that there is little to choose between food consumption and milk yields of cows kept under the two systems, but with yards a great deal more straw is required for bedding. Little bedding is required in cubicle houses but, on the other hand, individual attention and rationing of food is more difficult. Considerably less labour in milking, feeding and cleaning out is used in a yard and parlour than in a cowhouse ; the difference, however, depends a great deal on the efficiency of the layout of the building

A

B

PLATE I
A. Helicopter spraying potatoes against blight
B. An automatic auger-feeder distributing maize silage to fattening cattle

PLATE II

A. Abreast two-level parlour with one stall, one unit, direct-to-churn milking plant
B. Tandem milking parlour showing operator's pit

and in the use of labour, but savings of 25–33 per cent are often quoted.

The Cowhouse. A cowhouse should be sited to be convenient for the dairy and food store and to provide a short and dry approach for the cows coming from the fields. There should be adequate space round the building so that the cows can get in and out without trouble and manure can be easily removed.

Cowhouses can be *single range* or *double range*; in the latter, the cows stand back to back, separated by a central passage used when milking and for the removal of manure. Sometimes the cows face inwards on to a central feeding passage. For herds over 15 a double range cowhouse is to be preferred as it requires less labour and is cheaper to build, and for herds over 30 feeding passages are a worth while addition.

The following are the usual internal dimensions and details of construction of a cowhouse: they are also illustrated in Fig. 4.

Standings. These are the raised part of the concrete floor on which the cows are kept. The concrete should be insulated for comfort and slightly roughened to prevent slipping. The standings are usually 1·5 m long and arranged in pairs, 2·15 m wide for each pair. These dimensions can be varied slightly according to the size of the breed of cow.

Stall Divisions. The divisions separating each pair of cows can be of galvanised tubular steel or solid, to prevent draughts, in which case precast concrete, concrete blocks or cement rendered brickwork are the usual materials.

Mangers. There are wide variations in what is provided, from no mangers at all but a tiled area of floor in front of each cow to individual glazed troughs. It is essential that a manger should be capacious enough for the feeding of bulky fodders. The usual manger is of precast concrete or of concrete made *in situ* at least 750 mm wide and lined with a section of glazed stoneware pipe for durability and easy cleaning. Permanent or movable divisions in the mangers between each cow are desirable to prevent poaching of food. Automatic drinking bowls should be provided over the mangers.

Tyings. There are two types of tying, the central and the side tie. The central tie can either be a yoke or a double chain, which is usually preferred as it gives the cow more freedom. Side ties, fixed to stall divisions, are either single or double chains which can be adjusted to fit the cow. All ties should have a quick release arrangement.

Feeding Passages. Where provided passages should be 1·1–1·2 m wide to allow the use of a food trolley. There should also be no piers projecting from the outside walls into the passage and, to save walking, passages should go round the cowhouse without blind alleys at the end.

Dung Channels and Passages. The dung channel behind the standing should be not less than 750 mm wide and 900 mm is better. It should also be 150 mm below the standing. The passage adjoining it should be as wide as possible and in a double range cowhouse at least 1·5 m, so that a tractor and trailer or a tractor with a scoop can be used for cleaning out the building. In a single range cowhouse the passage should be 1·2 m wide. Passages can be continuous with the dung channel but it is better for them to be 50 mm higher. For the drainage to work satisfactorily, the whole floor should be carefully laid to proper falls. The standings should have a fall of 25–37 mm towards the channel and the passages a fall of 25 mm. The whole floor should be sloped lengthwise towards an outside drain at the rate of 12 mm per cow. The channels should have no fixed covering and where they cross a passageway they should be covered with removable iron gratings or concrete slabs. The channels should run to a trapped gulley just outside the building, and it is an advantage if the gullies are fitted with strainers to intercept straw, etc. In mucking out by hand, labour can be saved if there is a ramp outside the cowhouse up which the manure can be taken in a barrow and tipped into a trailer or manure spreader. If a pit is used for the manure it should be at least 18 m from the cowhouse and it should have a low retaining wall and a concrete floor. In some cowhouses, to save labour in mucking out, a scraper driven by a small motor is fitted in the dung channel.

Walls. The walls can be of brick, stone, concrete blocks or panels and there should be a damp-proof course. Internally the walls are rendered up to 1·4 m in cement to provide an impervious, easily cleaned surface. Walls should not be too high ; 2·25 m to the eaves is ample.

Roofs and Lighting. Roofs can be of slate, tile or asbestos-cement but should not be of corrugated iron. Ventilation should be provided. Natural lighting should be at the rate of 0·4 m² of wall window for each cow or 0·3 m² of roof lighting. Electricity should if possible be used for artificial lighting. If filament lamps are used, 80–100 watts per cow is the usual allowance, or one-third of this rate for fluorescent lighting. The lighting is most effective if provided at eaves level behind the cow.

Doors. For single range cowhouses the doors used by the cows should be 1·2 m wide and 1·8 m high, in the clear. For double range cowhouses the doors should be at least 2·4 m wide in the clear ; they can be double-hung or it is an advantage to have sliding doors although they are inclined to be draughty. Where possible, sliding doors should be hung internally to protect the sliding gear, but if hung externally the sliding gear should be protected by a hood.

It will be seen from the dimensions given above that the internal width of a single range cowhouse without a feeding passage is 4·2–4·5 m and a double range cowhouse 7·9–8·2 m or 9·7–10·6 m if there are feeding passages.

The Dairy. Milking in a cowhouse is normally by either a bucket or pipe-line milking plant. The milk has to be taken by hand or on a trolley to the dairy for cooling, though in large cowhouses overhead runways are sometimes used. To avoid unnecessary walking the dairy should be placed as close as possible to the cowhouse. With small cowhouses, for up to say 16 cows, the best position for the dairy is at one end : but for larger cowhouses it is better to have the dairy at the centre of one side. Other considerations affecting the siting of the dairy in relation to the cowhouse are that it should be in the coolest possible position, that is, on the north side. The Regulations also prohibit direct access from the cowhouse to the dairy ; there has to be either a small ventilated lobby between them or the dairy has to be approached from outside the cowhouse, often under a short covered way. For easy working the food store should be placed in much the same position as the dairy, that is at one end in the smaller cowshed and either centrally or at one end in large sheds.

The Yard and Parlour System. There are several variations in the yard and parlour or loose housing system. In its simplest form the cows graze out of doors all the year round and are milked in a movable shed or bail, taken from field to field with them. Sometimes the bail is kept for the winter in a sheltered spot on concrete paving to provide easier conditions for milking. The cows may also be housed for the winter in temporary or permanent yards and milked in a bail on an adjacent site. Permanent accommodation is, however, provided most frequently and comprises essentially housing for the cows, either in kennels or cubicle house or covered yards or a tie-up cowhouse unsuitable for milking, an assembly area where the cows are kept while waiting to be milked, and the milking parlour itself.

The Milking Parlour. A milking parlour is a highly specialised building used only for milking, a skilled operation which takes up about 40 per cent of the time of looking after a dairy herd. It is essential, therefore, for the building to be as labour saving as possible ; it should also be easy to keep clean, arranged so that the cows can be got in and out quickly and the cowman can do his work efficiently and with the least amount of effort. There are many designs of parlour and the most suitable for a particular farm will depend on such factors as the number of cows, their average yield, the milking routine to be used, the labour and time available for milking, whether concentrates are to be fed in the parlour, the amount which can be spent on the parlour and its equipment and the size and shape of the site or building available for the parlour. The milking equipment is also important. New developments in machines and in the technique of milking are continually occurring and this affects the design of parlours. There are three main types of milking machine. In a bucket plant the milk flows to a bucket which has then to be carried to a cooler. This is probably the simplest and cheapest form of machine. With an in-churn plant the milk goes direct into a churn in which it is also cooled and dispatched from the farm : the churns are taken from the parlour on a trolley or overhead runway. The in-churn plant saves labour by handling the milk in bulk and in the same vessel. The third is the releaser type of plant in which the milk from the milking unit is carried by a common pipe to the dairy where it passes under vacuum over an enclosed cooler into churns or into a bulk tank. This is the most expensive plant but the most labour saving and now that bulk milk collection is widespread and will shortly become universal the pipe-line system is the only one to consider in new parlour installations.

Choice of Parlour. There are basically two types of milking parlour, those with two milking stalls to each milking unit and those with one stall to each unit. With two stalls per unit a throughput of 10 cows per unit per hour is usual but with one stall per unit 8 cows per unit can be milked in an hour. The greater throughput of the two stall to one unit type is because one cow is being fed and prepared for milking while the other cow of the pair is being milked. As a result the unit is more fully used and the cow has longer to feed in the parlour. The advantage of feeding concentrates in the parlour is that it avoids the need to have expensive yokes and mangers for individual feeding in a yard. A cow will only eat about 0·35 kg of food per minute in a parlour and it will require some minutes longer than it requires for milking to get through a reasonable quantity of concentrates.

With one stall to one unit a more regular routine is possible with each cow and this encourages a quicker release of milk : for herds with high yielding or slow milking cows one unit to each stall is to be preferred if the concentrates can be fed elsewhere.

The layout of the stalls in a parlour can be abreast where the cows stand side by side, tandem where the cows stand head to tail, or herringbone where they stand in echelon. These formations can be used either in static parlours or in rotary parlours, and in all cases the cows are at a higher level than the milker, although with an abreast parlour the cow cannot be raised more than about 450 mm.

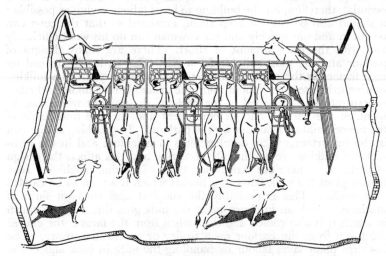

FIG. 5.—ABREAST-TYPE MILKING PARLOUR.

The Abreast Parlour. The abreast parlour can merely be some stalls separated from the rest of a cowhouse, to which the cows come only to be milked, entering and leaving by the same door.

This is not the most labour saving arrangement and it is usual to have a special building with 2–12 tubular steel stalls arranged in pairs. The layout of a typical abreast parlour is shown in Fig. 5 and Plate IIA. It will be seen that a length of 5–5·35 m is required, and a width of 4·8 m for four stalls. To reduce the amount of stooping, stalls are sometimes raised 350–400 mm above the level of the rest of the floor. Cows can step up this height but for anything higher an intervening step is required, which can be an obstruction.

The Tandem Parlour. In this type of parlour the cowman works in a pit about 1 m below the stalls. The pit can either be sunk below ground level or the stalls can be built up. A tandem parlour can be arranged with the stalls in the form of an L or U, or in one row on one side of the pit, or two rows on either side of the pit. The cows enter separately from a common passageway 900 mm wide. Details are shown in Fig. 6. There are two simplified forms of tandem

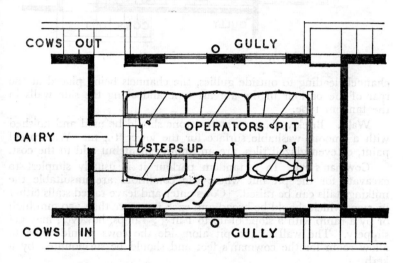

FIG. 6.—TANDEM-TYPE MILKING PARLOUR.

parlour without access passages, in which the cows are brought in and out in batches and there is one milking unit to each pair of cows on either side of the cowman's pit (Plate IIB). In the *chute* parlour the cows enter in pairs up steps or a ramp either side of a narrow pit, and the stalls are placed one in front of the other without separate passages so that both the cows on one side have to enter and leave together. By omitting passages a narrow building 3·1 m wide and 4·8–6 m long is large enough for four stalls, and this is probably the cheapest and simplest form of tandem parlour for a herd of medium size. In the *herringbone* parlour the layout is the same but there are no separate stalls : two tubular rails on either side of the pit holds batches of 6–10 cows in echelon (Fig. 7). A very high rate of operation is possible but little individual attention to the cows is possible. A recent development is the *rotary* parlour with the stalls on an elevated moving platform with the cowman in the centre.

The principal points in the construction of a parlour are as follows :

Floors. Concrete is the only suitable material. For drainage, the floors should have a fall of 1 in 40 towards shallow, open, dished

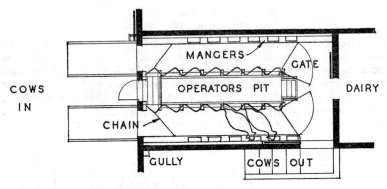

FIG. 7.—HERRINGBONE MILKING PARLOUR.

channels leading to outside gullies, the channels being placed at the rear of the parlour in the abreast type and along the side walls in the tandem type.

Walls. Brickwork, concrete or stone should be used and finished with a smooth washable surface for the lower 1·4 m. Waterproof paint, or even glazed tiles, are an improvement but add to the cost.

Cowman's Pit. In a tandem parlour it is usually simplest to excavate for the pit but where site conditions are unsuitable the milking stalls can be raised. Cows enter and leave raised stalls either by steps which should be broad and each not more than 250 mm high or by a ramp with a slope of up to 1 in 3 ; ramps, however, may get slippery. The wall of the pit alongside the cows should slope to allow room for the cowman's feet and should be surmounted by a kerb.

Stalls. Tubular steel is used. In an abreast parlour stalls are 1·7 m long and each pair is 2·1 m or 2·55 m wide depending on whether bucket or in-churn plant is used. Tandem stalls are usually 2·25 m or 2·4 m long and 750 mm at the widest point with gates at each end.

Lighting and Ventilation. Natural and artificial lighting are most effective if provided overhead. Ventilation should also be arranged through the roof.

Doors. A width of 1–1·1 m in the clear is usual. To save walking it should be possible to open and close entry and exit doors to the parlour as well as doors to the stalls by lever or cable from the washing area or cowman's pit.

Services. Water, hot and cold, should be laid on ; in a tandem parlour it is worth while laying on water to each stall.

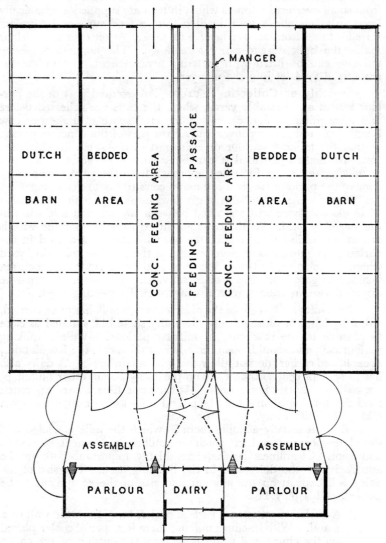

FIG. 8.—YARDS AND MILKING PARLOURS (C. L. A. McTurk Design No. 1).

Feeding. Feeding in the parlour can be highly mechanised. Mangers can be fitted in all types of parlour either to the stalls or to the parlour itself, depending on the type. Mangers may be filled from small overhead hoppers which in turn are supplied mechanically from an auger above them leading from a food store or by gravity through tubes from an overhead bulk tank. A loft over the parlour makes the bulk storage easier to arrange. The hoppers supplying mangers can be fitted with metering arrangements to regulate the amount of food let into the manger.

Assembly or Collecting Yards. An essential part of the parlour system are assembly yards, where the cows are collected before and after milking (Fig. 8). An uncovered space of 2 m² per cow is sufficient and it can conveniently form part of the concrete paving around the milking parlour or even part of the yards in which the cows are housed provided it can be separated by gates or a chain at milking times. A funnelled approach from the yard to the entry door of the parlour prevents too many cows trying to get in together. An " electric dog " also helps to get cows into the parlour : it consists of an electric fence wire stretched between two parallel side wires in the assembly yard. As the cows go through the parlour the wire is drawn up behind those left in the yard by means of a cord in the parlour. A more recent development is the circular collecting yard where a backing gate brings the cows into the parlour and the area behind the gate is used as a dispersal yard. This saves the separate dispersal yard necessary with a conventional collecting yard.

The Dairy. To reduce the distance which milk has to be carried, either by hand or mechanically, the dairy should be situated as close as possible to the cowhouse or milking parlour. While a milking parlour can, and should, open into the dairy, the Regulations, as explained earlier, do not allow this with a cowhouse. A dairy also has to be kept cool so that a site on the shady side of other buildings, normally the north, is required. At the same time a hard approach road and turning room should be provided for lorry or other transport taking the milk away.

A dairy is usually a single room in which the milk is cooled and stored temporarily and in which the milking equipment is washed and kept. Cleanliness is, therefore, highly important both in the approaches to the dairy and in the building itself. The size of the dairy will depend on the amount and size of the equipment to be housed. This may be :

(a) A cooler, most often a surface cooler, hung from the wall or a stand. With in-churn milking there is a special cooler placed on the churn and water flows over the outside of the churn into a tray. Where there is insufficient water (3 to 4 litres per litre of milk to be cooled) or it is too warm in the summer, some form of refrigeration may be necessary involving additional plant and possibly a chilled water storage tank or a tank in which churns may be immersed.

(b) An electric water heater or a boiler to provide hot water for washing.

(c) A steam raiser and sterilising chest where steam sterilisation is practised.

(d) Hot and cold water wash trough.

(e) A hand wash basin, if not provided elsewhere.

(f) Racks for clean equipment.

(g) The milking machine power unit, if small and electrically driven.

Where electricity is not available, water will have to be heated by oil : a separate boiler room and fuel store are then necessary and space outside the dairy will also have to be provided for the milking machine power unit. A dairy should not be unnecessarily large, and where milk is not retailed a room of 10 m² is sufficient for a small herd, 15 m² for herds of about 30 cows and 20 m² for larger herds. Fig. 9 illustrates an all-electric dairy for 30 cows.

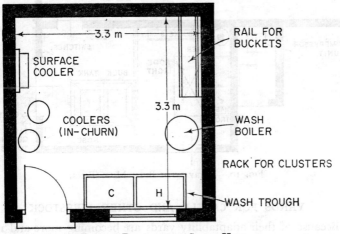

FIG. 9.—DAIRY FOR A SMALL HERD.

To provide clean conditions in the dairy the walls should be cement rendered and if possible finished with a waterproof paint ; white glazed tiling is better still, though expensive. A flat ceiling of asbestos-cement sheets is desirable as it is easily washable, but ventilation outlets through it will be needed. A well-made concrete floor is best, reinforced with steel chequer plating where there is most wear from churns, although blue bricks well pointed in cement mortar can be satisfactory. The floor should have a slope for drainage to open channels leading through a hole in the wall to an outside gulley. Good natural lighting by windows in the wall and overhead artificial lighting are also important. The labour of handling churns can be much reduced by using overhead hoists and rollers inside the dairy and outside by a platform for loading on to lorries about 1 m high, approached by steps or a ramp.

Bulk Collection. A recent development affecting dairy design is the introduction of bulk collection of milk in a tanker lorry calling once a day or every two days. The milk is stored and cooled in a stainless steel tank suspended in a water bath and surrounded by an insulating jacket with an outer case of stainless steel or resin bonded fibre glass. The tanks are large—the smallest size now in common use is 560 litres requiring about 3 m² of floor space with an extra 600 mm around it for cleaning—so that there should be adequate room in the dairy and an opening wide enough to get the tank installed. The tanks are also heavy so that the floor should be well constructed. Tanks are being developed in which the refrigerating unit and water bath are separate from the milk vessel so that they can be more easily fitted into existing buildings. A good access road right up to the dairy is required for the heavy tanker collecting the milk (Fig. 10).

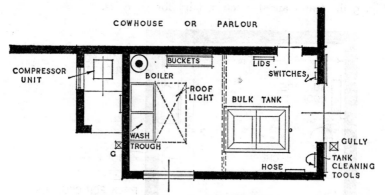

FIG. 10.—DAIRY WITH BULK MILK TANK.

YARDS FOR CATTLE AND OTHER LIVESTOCK

Because of their adaptability yards are becoming more and more popular for housing all forms of livestock but a good deal of care needs to be taken over their design and layout if some of the advantages are not to be lost by poor design, especially when improving or adapting existing buildings. The saving of labour and the comfort of the stock are the most important points to be considered. So far as labour is concerned the main problem is one of movement, partly of large quantities of bulky materials such as hay, roots, silage and other fodder being taken in and manure coming out, and partly of animals, particularly dairy cows, going in and out. It will obviously help if the distance over which the materials have to be carried is kept as short as possible by having storage for the fodder and the yards close together or even incorporated in the same building. Some of the carrying can even be eliminated if self-feeding of the silage can be arranged. Mechanical appliances will reduce the labour required —a tractor and trailer or a trolley for filling mangers and loaders, bulldozers and other implements for removing the dung. Buildings need to be suitable for the use of machinery. Mangers should be along the outside of the yard so that they can be filled without entering the yard, and wide spacing of gates and stanchions will enable tractors and machinery to get in for cleaning. For the comfort of the animals, generally cattle, the important points are freedom from draughts, shelter from bad weather, dry lying space and an adequate length of manger.

Yards can be covered or partly covered. With partly covered yards orientation is important and the open side should face south-east and away from the rainy quarter ; more shelter is given if other buildings, particularly a tall building like a dutch barn, are on the north side. To prevent draughts, at least two adjacent sides, and not opposite sides, of a yard should be enclosed with a wall, not necessarily to the eaves, or with other buildings ; and the eaves should be kept as low as possible compatible with getting tractors into the yard when full of manure.

Size and Shape. The size and shape of a yard are largely dictated by the length of manger and floor space per animal to be provided. The minimum length of manger required is 600 mm for adult de-horned stock and slightly less for fattening animals. Full-grown animals require approximately 4 m^2 of lying space and 2 m^2 of concrete feeding space. With partly covered yards the lying space is generally covered and the feeding area left open. To prevent driving rain penetrating too much of the covered area it should be at least 6 m from back to front and not more than 3 m high at the eaves.

Mangers should be situated where there is short and easy access without sharp corners from fodder stores and along one or two sides of a yard so that they can be filled from outside. Where there are several yards a central feeding passage with mangers and yards on either side is a convenient arrangement. The passage should be wide enough for a tractor and trailer, that is about 2·7 m and in some designs it is raised with ramps at either end as this makes it easier to fill the mangers. Where the passage is between two covered yards the convenience of the shelter often justifies the extra cost of having a cover over the passage, but with partly covered yards it is usually more difficult and costly to cover the passage. The mangers will be supplied from a dutch barn which if possible should be alongside the yard and can even form one wall. Bedding will also be supplied from the barn. A food room for concentrates and roots should be near by.

Silage. If silage is fed the silo should be close to the yards or alternatively the cattle can be allowed access to the silage to feed themselves as this saves the labour of handling bulky material. For self-feeding the silo may either be in the yard or between two yards or in an adjacent dutch barn but if the layout of existing buildings does not allow this kind of arrangement or if partly covered yards are used the silo can be a short distance away, sometimes along one side of an open yard. There should be a hard approach to the silo and the area around it should be concreted. Where cattle have access to the silage all the time it is sufficient to allow 150–200 mm of the face of the silage for each adult beast either at the end or along one side of the silo. The silage should not be more than 1·8 m high and special feeding barriers or an electrified wire along the face of the silage will prevent the animals wasting it by pulling out more than they need.   Where *ad lib* self-feeding is not required, labour in feeding can be saved by having a movable feeding barrier or fence adjoining the silo in front of which loose silage, hay or roots can be put for the cattle to feed themselves.

Litter.   To keep the cattle reasonably clean it is important to have sufficient straw or other litter available for bedding.   In a fully

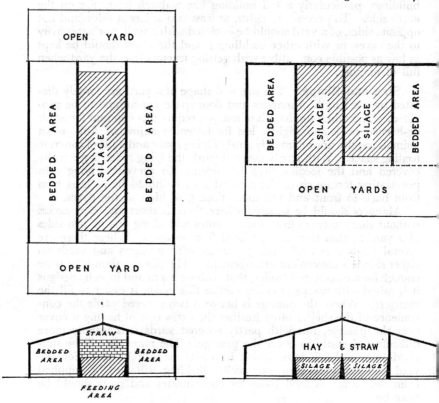

FIG. 11.—PARTLY COVERED YARDS WITH FODDER STORAGE.

covered yard 6–8 kg of straw per day should be provided for each fully grown beast, equivalent to 1·5–2 tonnes for the winter.   Where straw is in short supply a much smaller quantity is needed if the yard is divided into a bedded area for resting and a concreted feeding area.   The bedded area should have an area of 4 m² per head for adult stock and should be well covered with straw as soon as the stock are put in.   The manure pack which results soon heats up and keeps the animals warm.   To prevent the waste of straw the bedded area should be kept free of movement, mangers and water troughs should be kept outside and access to a milking parlour should not be from the bedded area.   The concrete feeding area should be at least 3 m wide and should be kept clear of dung and straw by the daily use of a scraper. The feeding area can either be covered or open but, if it is open, it is usually larger and includes a loafing or exercise area. The design of this and other types of yard are illustrated in Fig. 11.

Cow Cubicles and Cow Kennels. An alternative way of saving straw and at the same time solving the problem of dealing with strawy dung, when it does not suit the farming system, is to divide a covered yard into cubicles arranged in rows like stalls in a cowhouse. The cows are not, however, tied but are able either to lie in the cubicles or to move about the yard. The beds in the cubicles are separated from the passage behind by a raised kerb and are covered with bedding which can consist of sawdust, shavings or chopped straw ; only very small quantities of bedding material are required.

To keep the cows clean the cubicles should be only just large enough for the cow to lie down in comfort : for a Friesian cow a total length of 2·1 m and a width between divisions of 1·2 m is sufficient. Dunging passages behind the cubicles should be wide enough to allow two cows to pass comfortably, that is, 2–2·4 m, but if these passages are too wide cows may lie down in them. The passages can be of solid concrete, which can be scraped clean with a tractor, or can have a slatted floor, normally with concrete slats 75–150 mm wide, with a tank underneath into which the dung is trodden. The gap between the slats should be about 37 mm. Slatted floors increase the capital cost considerably and provision must be made for the storage of large quantities of slurry until it can be removed.

To reduce the cost of constructing new covered yards and installing the necessary cubicles, lighter forms of structure have been developed known as " cow kennels ". In these structures the stall divisions form part of the main structure carrying the roof and walls; the passages are also often left unroofed as a further measure of economy.

Materials. A wide range of buildings is suitable for yards but a framed structure with the stanchions spaced as widely apart as possible is the most economical and adaptable. Materials which are not affected by corrosion from moisture should be used and reinforced concrete with asbestos-cement roof covering is very suitable. The building should not have too wide a span as otherwise the height at the ridge becomes excessive, and the building may be cold and draughty. The height at the eaves should be about 3 m or 3·6 m if there are tie beams as this allows sufficient, but not too much head-room when the manure has risen in the yard. Walling can be of brickwork, concrete blocks or rough timber, but to reduce cost the upper part of the walls can be sheeted in asbestos-cement or corrugated iron. With open fronted yards the walls should be carried up to the eaves. Divisions between yards should be removable for cleaning out and high enough when the yard is full of manure. Timber or tubular steel rails are usual and there should be wide gates which can be raised. Around the yards the same type of rails may be used instead of walls, but solid walls may be necessary to provide shelter even in an open yard. Floors of covered yards can be of earth, stone or hardcore but for open yards concrete is very desirable to keep the yards dry, though efficient drains to dispose of the rain-water are essential.

Mangers.   In the same way that removable rails round at least part of a yard make for greater flexibility of use, so do portable or removable mangers make a less rigid design and save the expense of fixed mangers.   There are various types of movable and adjustable mangers in timber and in steel : they can be free standing, suspended on chains or on hooks on a rail or bolted to parts of the main structure.   Where there is a fixed feeding passage part of the passage can be used as the manger on which fodder is placed, the cattle being kept from the passage by a feeding barrier.   Where fixed mangers are provided they should be 750–900 mm wide from back to front and deep enough to provide plenty of room for bulky food.   The back of the manger adjoining the feeding passage should not be more than 900 mm high as otherwise it is more difficult to put the food in.   The front adjoining the yard should be 1·1 m above floor level to allow for the build-up of manure but 375 mm is sufficient if there is a clean concrete area in front of the mangers.   Some form of barrier through which the cattle feed is required to keep them out of the mangers and to prevent bullying.   Where concentrates are fed in the mangers closable yokes are usually installed so that the feeding can be controlled.

Water troughs should also be provided and fixed at the same height as the mangers ; a 150 mm length of trough for each adult beast is sufficient.

LOOSE BOXES

Loose boxes are some of the most useful general purpose buildings on a farm and there should be sufficient of them, conveniently placed. A box about 3·6 m × 4·2 m to give a floor area of 14–16 m² is suitable for general purposes.   Boxes are best grouped together in ranges, facing as near south as possible : division walls between boxes need only be up to eaves level.   A covered feeding passage is an advantage, though it adds to the cost.   Each box should have a door leading on to an outside concrete causeway.   The doors should open in two halves or alternatively a sheeted gate will allow for mechanical mucking out.   The internal fittings should include rings for tying stock, and a manger and water bowl or standpipe, adjustable in height if different types of stock will use the box.   The floor should be of concrete, insulated and sloped to an outside drain.   Roof lighting is preferable, though hopper type windows set high up can be used. Ventilation should be provided at the ridge.   One or more boxes in the farmstead will be required as isolation boxes and should be sited as far away as possible from other buildings.

BULL PENS

Bulls are too often housed in unsatisfactory conditions, such as in a stall in a cowhouse or in a small dark loose box, and this aggravates their natural bad temper and makes their handling unnecessarily dangerous.   A bull should have a properly constructed box with either a yard or a small, well-fenced paddock. The box should be in a sheltered spot, facing south, and where the bull can see what is going on ; it should also be close to where the herd is kept. It should be soundly constructed with a floor area of about 16 m² and a length of not less than 3 m. There should be a manger and water bowl on one wall, arranged so that the manger can be filled from outside, either through a hatch or from a feeding passage : it is

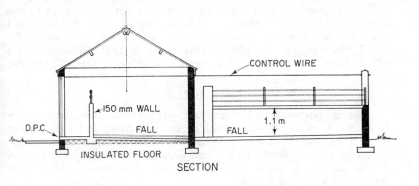

SECTION

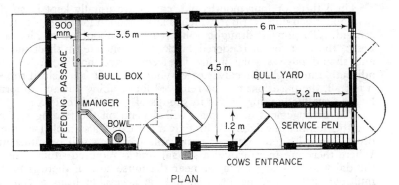

PLAN

Fig. 12.—Bull Box with Open Yard.

In this plan provision has been made for two escape gaps between the bull box and the wall of the yard

an advantage to have a catching yoke or ring fixed over the manger. The manger and water bowl should be built up from the ground in solid brickwork or concrete to prevent the bull damaging them by getting his horns underneath. Doors should also be stout and with a smooth surface, lined with sheet metal, as this makes them more difficult to damage. The doors and manger should be placed so that the bull has room to lie down out of the way of draughts, and the door between the box and the yard should be hung so that it can be operated from outside when it is required to keep the bull in the box or in the yard for cleaning out (Fig. 12).

The yard needs to be at least 6 m long to provide sufficient exercise, and 9 m is better. The floor should be of concrete roughened with a brush to prevent it becoming slippery. A bull is more contented if he can look out from the yard, but if solid walls are not provided it is important to have the gaps between the fencing rails or tubes so small that a bull cannot get his legs or head trapped. The best arrangement is to have a solid wall not less than 225 mm thick and 1·1 m high. Above that, tubular rails are sufficient and enable the bull to see out. It is an advantage to have a service pen built in one corner of the yard. There should be a stout gate from the yard and some escape arrangement for use in emergency, either in the form of a gap in the wall 300 mm wide or two upright steel stanchions across a corner forming a refuge until a man can climb over the wall behind.

CALF HOUSING

Calves, particularly for the first few months, need warm, dry housing. Concrete floors should be well insulated and have plenty of bedding ; alternatively, a false floor of timber slats or expanded metal on a timber framework is sometimes used. Doors and windows should be well fitting to prevent draughts, and the roof should be fairly low and well insulated.

Until three or four months old, calves are usually kept in single pens about 900 mm × 1·8 m which is large enough for the calf to turn around. To prevent draughts solid partitions about 1·2 m high are best : they can be of rendered brickwork, concrete or sheet metal, and should have a yoke fitting for feeding and holding a bucket, adjustable in height for calves of varying ages. It is a help if some, if not all, the partitions can be removed for cleaning out and to form larger pens for housing calves in groups of 3–6 when older.

The arrangement of the pens will depend on the feeding system and the number of calves. With only a few calves a suitable loose box divided up with temporary or removable partitions will be sufficient. Where there are more calves a special unit is required situated near the dairy and mixing room, or near the nurse cows if there is to be multiple suckling. The pens should be accessible from a central passage containing shallow channels into which the pens drain : the channels discharge into gullies outside the building. Adequate lighting should be provided either by roof lights or hopper windows high up the walls and there should be outlet ventilation in the ridge. As calves sometimes die from lead poisoning from sucking or chewing paintwork, paints not containing lead or other preservatives should be used on timber or metal work in buildings where there are calves.

HOUSING FOR PIGS

Unlike other farm animals the domestic pig has almost no coat to protect it against extremes of temperature—in this country cold being the most important. The exercise which pigs get when kept out of doors is sufficient to keep them warm in cold weather but this uses up much of the energy obtained from their food. While this does not matter with breeding stock it is important on grounds of cost that fattening pigs put on weight quickly and with the minimum consumption of food. Warm indoor housing has therefore to be provided for them ; young piglets have no layer of body fat and so are very susceptible to the cold, and warm housing is even more important. Experience has shown that provided the pigs are kept warm high humidity in the air does not affect their health. High humidity causes condensation in the piggery which in time will damage the structure of the building, particularly the insulation. High humidity should therefore be avoided by good ventilation and the rapid drainage of liquids from the interior of the building. Outside runs in which the pigs dung provide a simple and effective way of reducing humidity and drainage problems.

Outdoor Systems. Where there is sufficient land available and it is dry enough to use all the year round there is a great deal to be said for running breeding pigs out of doors. They are healthier and housing problems are simpler. Gilts or in-pig sows can be kept together in temporary or permanent enclosures : they are best in groups of 15 or 18 to the hectare. Water should be provided and a shelter where the pigs can go in bad weather. A temporary building with a low roof and made of straw bales, or a timber hut, is sufficient if erected in a dry, sheltered spot. Near farrowing time some breeders consider it desirable to provide individual feeding arrangements for the sows so that each can get a proper share of food without bullying.

For sows with litters, portable arks on skids, which are moved frequently, are the most suitable. They can be put in a small fold surrounded by hurdles, or the sows may be tethered on a long chain. The ark should have a floor area of 2·4 m × 1·8 m or be rather larger if there is a creep for the young pigs. There are various designs of arks, but generally the construction is a double-skin wall coming to an apex with insulating material between. The walls can be of timber tongued and grooved boarding but, to reduce cost, sheet metal or plywood is sometimes used for one thickness of the wall. Other essentials are a stout timber floor, a draught-proof attendant's door and an entrance door placed off centre or with a baffle wall inside. It is also an advantage to provide a creep feeder for the piglets in one corner.

The main disadvantage of the outdoor system is the work involved in looking after a number of pigs in scattered units, often in bad weather during winter. To some extent this can be overcome by arranging the runs so that the housing or arks are close together, alongside a hard road, with the runs radiating out from them. A separate food hut filled once a week or so from a central store and kept near the arks will reduce the carrying of food. In winter the arks can be brought on to concrete paving, if possible where water and electricity for heating units in the arks are available.

Store and fattening pigs are sometimes kept in covered or partly covered yards in lots of about 15-30, with or without cattle. They should have sufficient straw and a separate feeding place in one corner of the yard where they can be fed apart from the cattle. If there is any risk of the pigs not being warm enough they should also have a warm sleeping place with a low roof. With more specialist enterprises, however, it is generally found that pigs fatten more quickly when kept in smaller lots and with less space to move about.

Indoor Housing. Although full scientific information about the requirements of pigs when housed indoors is lacking, experience has shown that if the pigs are to thrive the housing has to meet certain requirements. These are :

1. Roof, walls and floor properly constructed and well insulated to prevent excessive heat loss. The cubic capacity of the building should not be too large (1·5–2 m³ per pig) so that the pigs themselves, as the only source of heat, can keep it at a reasonable temperature. A headroom of 1·8–2·1 m is ample, and where there is a lofty roof the air space can be reduced by a ceiling at eaves level.

2. A warm dry sleeping space away from draughts. In some designs of piggery the roof over the sleeping quarters is kept as low as 1–1·2 m to form a warm " kennel " for the pigs ; an alternative is to have a low false ceiling over the sleeping place. Pigs are naturally clean animals and to encourage them to dung elsewhere the area of the sleeping place should be kept small, 0·5 m² per pig being sufficient.

3. A separate place for dunging which may either be an outside run or a passage placed next to the outside wall. The passage should be 1·2 m wide so that there is enough room for the pigs, but 0·3 m² per pig is sufficient floor space.

4. Sufficient trough space for each pig to feed without overcrowding and bullying. This is one of the most important factors in the design of pens as it controls their size and shape. A length of 300 mm of trough for each fattening pig and 450 mm for sows and boars is usual ; the trough should be lined with 300 mm wide, half-round, salt-glazed channels.

5. Draught-proof partitions between pens, at least 1·1 m high and sometimes carried up to the roof. Partitions can be of 112 mm brickwork or concrete blocks. The pen front adjoining the feeding passage can either be solid (fixed or swinging) or open, consisting of rails of galvanised steel tubing. The bottom tube is often a water pipe, either with holes at intervals through which the water flows into the troughs below, or supplying automatic water bowls.

6. Impervious but insulated floors sloped so that drainage runs quickly from the pigs to the outside to prevent excessive humidity. A fall of 1 in 40 is usual. To save labour, dung can be removed from the enclosed type of piggery as a sludge either by the use of automatic scrapers or by having a slatted floor in the dunging passage leading to a storage tank outside the building or to a dung cart.

7. Adequate but not excessive inlet and outlet ventilation as described in the earlier section on ventilation. Forced ventilation by fans is used in some piggeries. In one design fans draw the air at each end into a central overhead duct from which it passes over the pens before reaching the outlet. In others extractor fans are placed in the outlets.

Several designs of piggery meet these requirements with varying efficiency. They can broadly be divided into two types, first, those with outside runs and second, totally enclosed houses.

Piggeries with Outside Runs. There are two designs of piggery with outside runs for dunging. The first is the *McGuckian*, in which the pens are in pairs with feeding troughs arranged along narrow side feeding passages leading off a main passage from which they are separated by a draught-proof door. The pens are long (4·5–6 m) and narrow with all walls carried up to the roof. There is a small doorway with a baffle leading to the outside run and the roof of the pen is carried down over part of the run to form a verandah.

The *Harper Adams* house (Fig. 13) is a single range building sometimes constructed as a lean-to off an existing wall, with a single feeding passage on to which the pens and feeding troughs face. The pens are normally 3·3 m long and 2·4 m deep. The roof is low, being 1 m at the eaves and 1·8–2·1 m over the feeding passage. The outside run is approached by a small doorway with an outside baffle wall which has a roof over. The pen walls are carried up to the ceiling and the feeding passage acts as a horizontal ventilating shaft, regulated by the entrance doors at each end which open in two halves. Six to ten pens

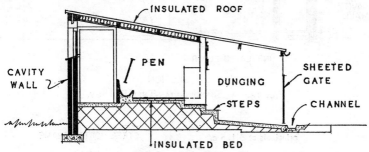

Fig. 13.—Harper Adams-Type Piggery.

form a suitable length of building, otherwise the range becomes too long.

With both these types of house it is important that the runs should face south and have wide gates so that they can be cleaned out by a tractor with a fore-end loader. If plenty of straw is used the yards need only be cleaned out at long intervals. The food room can be placed at one end or in the centre.

Totally Enclosed Piggeries. The best known is the *Scandinavian* or *Danish* type (Fig. 14A), which at times has had a poor reputation because it has been incorrectly copied with inadequate insulation. It comprises two ranges of pens along either side of a central feeding passage, and dunging passages along each outside wall. The pens are 3–3·6 m in length along the feeding passage and 2·4–3 m wide, separated by a partition from the dunging passage which is about 1·2 m wide. To conserve heat a flat insulated ceiling over the whole building at eaves level is an improvement; a cheaper alternative is a false ceiling only over part of the pens. There are several variations of this type of house. In one, the divisions separating the dunging passages from the pens are carried up to the roof, and the passages have separate and higher roofs so that a tractor with a scraper blade or loader can be used to clean them out. In another, the pens are placed against the outside walls where the headroom is kept low and the dunging area is next to the feeding troughs and central passage.

For farms with plenty of straw there are designs of piggery with the pens covered in deep straw in which the pigs keep themselves warm

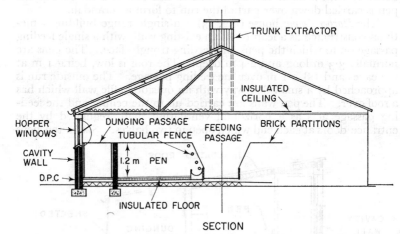

FIG. 14.—A. SCANDINAVIAN-TYPE PIGGERY.

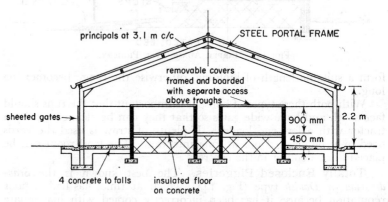

FIG. 14.—B. SOLARI-TYPE PIGGERY.

without the need for so much insulation and ventilation as in other types of house. The pens are arranged on either side of a central passage and are separated by partitions which can be removed for cleaning out periodically. There is no dunging passage but the floors of the pens are at a lower level towards the outer wall to contain the straw.

The *Solari* piggery (Fig. 14B) is another design adaptable to other uses. It comprises a steel dutch barn with double pens arranged on either side of a central feeding passage. The pens are divided by 100 mm block walls 1–1·2 m high : the inner sleeping pens have hinged timber boarded covers which can be raised or lowered according to the temperature in the pens. The outer dunging pens are 300–450 mm lower and are deeply strawed. They can be cleaned by a front-end loader through large doors.

Farrowing Pens. Accommodation for indoor farrowing can be arranged in loose boxes, in small pens provided in a fattening house or for larger units there may be a special farrowing house. Requirements are the same as for other houses except that warmth is even more important since young pigs require a higher temperature and a sow and litter gives off less heat. Other points are a floor area of 6–9 m² with separate dunging place or small outside yard ; farrowing rails round the pen to prevent the sow crushing the piglets and a creep to which the piglets but not the sow can go for feeding. Artificial heating reduces losses of young pigs. It can be either an infrared radiant heater hung over the creep or electric heating wires incorporated in part of the floor.

Farrowing crates are sometimes used in the pens for the first few weeks as they make the sow easier to handle and protect the piglets. The crates are either in the form of a long narrow box with slatted sides to contain the sow, or they form a circular (Ruakara) structure with a central safety zone for the piglets.

ACCOMMODATION FOR SHEEP

In the past sheep have had little in the way of fixed equipment, but to make flock management easier and less laborious for the shepherd the recent trend has been to provide more.

The accommodation for lambing depends a great deal on local circumstances. The essentials are a shelter from the worst of the weather and a place where the ewes can be gathered reasonably close together to make supervision easier. A temporary shelter yard of hurdles covered with straw, or of straw bales, with a pen or two for lambing, may be sufficient ; when lambing has finished the yard can be destroyed, so simplifying the control of disease. More permanent accommodation on hill farms can be in the form of small walled enclosures or on lowland farms, in lambing sheds and yards at the farmstead. When permanent accommodation is used it needs to be well disinfected after lambing.

The dipping of sheep is an important item in good management, and in many counties it is still compulsory. Coupled with the dip should be drafting pens and a footbath. A well-designed layout can make the handling of a flock more economical of labour and more efficient. Convenient siting is important, namely a central position with good access. An adequate supply of water should be near by and somewhere to dispose of liquid from the dip without causing pollution. A common arrangement is to have a collecting pen where the sheep are gathered. From this they pass through a sorting race either to a catching pen leading to a dipping bath or to drafting pens.

A short swim bath is the most popular ; it should have a capacity of 2 litres of dip for each sheep to be dipped. Baths can be bought ready made of galvanised steel from 630 litres capacity upwards or they can be constructed on site from concrete or brickwork. Sheep are lowered into the bath tail first, usually on their backs. After passing through the bath they climb up an exit ramp into draining pens. For large flocks a long swim bath or a circular ring bath may be preferred. Post and rail fencing for the pens is satisfactory with the rails of hardwood spaced not more than 150–175 mm apart to a

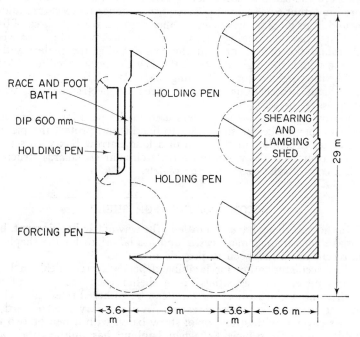

RACE AND FOOT BATH

DIP 600 mm

HOLDING PEN

FORCING PEN

HOLDING PEN

HOLDING PEN

SHEARING AND LAMBING SHED

29 m

3.6 m   9 m   3.6 m   6.6 m

Fig. 15.—Sheep Dip and Pens.

height of 1·1 m. Pen floors should be of concrete. A convenient installation is illustrated in Fig. 15.

A shearing shed may also be provided adjoining the pens. This can merely be a roofed area to shelter the workers, or it can be an enclosed building in which the wool is stored as well. An electricity supply makes the use of electric shearing machines possible.

The wintering of hill lambs in buildings (as an alternative to sending them to lowland farms) is an old-established practice which has recently been extended to lambs and ewes on lowland farms. On hill farms there may be a special wintering shed, for which an existing building may be adapted, with a slatted floor ; 0·5–0·75 m² of floor area per head is sufficient, depending on the breed. A shallow pit is needed under the slats which should be of hardwood 37 mm square with tapered sides about 15 mm apart and supported on timber joists. The building should be well ventilated and not totally enclosed so that the sheep do not become too pampered as this can lead to management troubles in spring when they are let out. On lowland farms more improvised arrangements can be made with existing buildings such as dutch barns and cattle yards into which slatted floors are temporarily fixed, though straw can be used as an alternative.

ACCOMMODATION FOR POULTRY

Conditions in the poultry house vary from the strictly controlled environment for broiler production to the simple pole shed for growing turkeys. In an era of intensive methods of poultry husbandry, with stock housed at great density and growth rate accelerated by improved breeding methods and nutrition, our basic knowledge of the optimum environment for the different classes of poultry is rather deficient. Modern poultry houses are in consequence designed very much on a trial and error basis. If more were known of the physiological processes involved, then the needs of the chicken or laying hen for warmth, ventilation and light could be better safeguarded in the design of modern poultry houses.

Poultry are adaptable creatures and can thrive under a wide range of conditions but with less latitude in the chick stage than for the well feathered growing or laying stock. Critical temperatures, below which food is diverted from production to maintain animal heat, are especially important because costly foods are used by poultry. In modern housing the temperature and the length of day is carefully controlled at all times and variation in temperature and length of day will govern the laying patterns of the birds. Similarly for broilers the temperature is carefully controlled from birth to slaughter.

Ventilation rates based on the replacement of 0·25–0·5 m³ of air per hour per kg body weight appear optimal. The technique of controlled lighting for growing and laying stock, together with more exacting regulation of heat and ventilation, has led to the production of windowless or " controlled environment " houses, incorporating automatic heating and ventilation systems.

ACCOMMODATION FOR IMPLEMENTS

Most modern implements are expensive and when not in use need protection from damage and deterioration caused by the weather. Near-by workshop accommodation where they can be easily adjusted and maintained is very desirable.

Implement accommodation need not be close to other buildings though it is usually more convenient to have it in the farmstead. The main requirement is good road access and a site where the buildings can face north, away from driving rain and the sun. It is difficult to prescribe the amount of storage space which should be provided on a farm as the number and size of implements vary widely from farm to farm : the following are typical figures of accommodation

E

requirements for farms over 40 ha obtained from a survey by Cambridge University :

Type of Farm.	Area per 40 ha.	
	m² of Accom.	No. of Bays of 30 m².
Rearing, grass, sheep and large arable farms . .	45–90	2
Dairying, mixed and heavy arable	65–160	4
Intensive arable and market gardens	90–250	6

For a small farm with a limited number of implements a shed with an open front is probably the most satisfactory. One or two bays can be enclosed with doors to house tractors and a workshop. The shed should be at least 6 m deep—6·5–8·5 m is better—to give full protection from the weather, and in bays 4·8 m wide. A height of 2·4 m to the eaves is enough for most implements. For small implements such as ploughs, harrows and cultivators, a simple lean-to building 3–3·6 m deep and with 1·5–1·8 m of headroom is adequate.

For larger farms with more implements and a number of them with engines and many moving parts, a more expensive totally enclosed shed is probably justified. A useful building is one with bays 10 m deep and 4·8 m wide ; the walls can be in concrete blocks or bricks for the first 1·2 m with sheeting above. One bay with a headroom of 3·6 m at the eaves makes the housing of combines and loaders easier. Doors should be 3–3·6 m wide ; sliding doors, though more expensive, are less likely to get damaged. Good overhead lighting is important, and concrete floors make servicing easier.

Workshop. The workshop can be sited either in part of the implement shed or in a separate building. It should be fitted with well-lighted work benches, racks for spare parts and tools, and an inspection pit ; water and electric power should also be provided. Overhead lifting tackle is required in a large workshop. The size of the workshop and the amount of equipment will depend on the type and number of jobs to be undertaken. It is very important that the workshop should be warm so that the operators can work in comfort.

Fuel Storage. There should be fuel storage near the tractor sheds with good road access for the tractors and tankers, but the storage should be away from wells which may be contaminated by the seepage of fuel. For diesel fuel and vaporising oil overhead tanks of 1,200 litres capacity for one tractor and 3,000 litres for two to three tractors are usual sizes. The tanks should not be galvanised internally as the zinc may be dissolved by the fuel. Tanks are generally rectangular and if they are mounted 1·2–1·8 m high tractors can be filled by gravity. The supports can be of brick, concrete or angle iron but must be substantial since the weight of a 3,000 litre tank with fuel is 2·5 tonnes. A timber or bituminous fillet placed between the bottom of the tank and the supporting pier will prevent rusting. Tanks should be tilted backwards at least 75 mm from the filling taps to form a sludge trap with a drain cock for emptying. A licence from the local authority is needed if more than 19 litres of petrol are to be stored, and for larger quantities underground storage is a usual requirement. This is expensive and probably not justified except on a large farm.

STORAGE BUILDINGS

Dutch Barns. A dutch barn is one of the most useful storage buildings on a farm. Although usually provided to store hay and straw it can be turned to many other uses when necessary, such as a cover for grain silos and silage pits, the storage of large implements like combine harvesters, and temporary yards for livestock. An all-steel barn with a curved roof is the most popular, but reinforced concrete barns with asbestos-cement roofs are often found. There are also designs of barns in timber covered in corrugated sheeting.

Barns are usually manufactured in standard bays of 4·5 m or 4·8 m with widths in 1·2 m modules and a height to the eaves of 4·8–6 m. One bay 7·2 m wide and 4·8 m high would hold about 20 tonnes of loose hay or 8–10 tonnes of loose wheat straw.

To preserve the feet of the stanchions the concrete bases in which they are set should be brought well up above ground level and sloped away from the stanchions. One or more sides of a barn can be covered in sheets to give more protection from the weather. Lean-to extensions can be added, separated if required from the main barn by concrete or brick walls. Upper floors are sometimes inserted, but expert advice should be obtained before structural additions are made which put extra load on the stanchions.

A recent development is the use of one or two bays of a barn for hay drying by blowing warm air through loose or baled hay to provide high quality hay in small quantities (see p. 345). The bay has to be enclosed with airtight walls up to a height of 4·2 m and air is blown by a fan through a false floor of wire mesh. It is sometimes possible to use the fan and heating system of a grain drying installation for supplying the air for hay drying as well as for ventilating potato stores. The possibility of having a combined installation should always be considered when designing a layout for one form of crop drying.

Silos for Silage. Another way of conserving green crops is by ensiling them. While much silage is stored out in fields less wastage occurs if the silage is made in a pit or trench (which can be above or below ground level) with permanent sides, a roof and drainage. A minimum width of 6 m is required for consolidation by a tractor but a width of between 9 m and 16 m is most satisfactory for the use of modern machinery. Grass silage occupies about 1·5 m³ per tonne.

Walls can be of precast concrete staves or plywood panels available from manufacturers, or of timber planks with a supporting framework, of concrete blocks, of reinforced concrete or of bricks made up on the site. On free-draining soils it is an advantage if the pit or trench is excavated below ground level so that the soil and excavated material will give support to the walls. Otherwise care should be taken to have the walls strong enough to resist the pressure on them when the silage is being consolidated. There should be a row of land drainage tiles along either side of the floor leading to a suitable drainage outlet. The best form of roof is a dutch barn, but if something cheaper is required a low roof just above the silage and removable in sections can be used. Corrugated steel sheeting fixed to the silo walls or to a timber framework is satisfactory (Fig. 16).

The use of tower silos for silage, a common practice in the USA, is now being re-introduced into this country. Chopped silage is blown into the top of the tower. It is removed for feeding by a mechanical cutter in the top of the tower, down a conveyor to the ground where it is passed by means of an auger to cattle in yards. Concentrates can be added to the silage as it passes through the auger. Large numbers of cattle can be fed mechanically by an installation of this type.

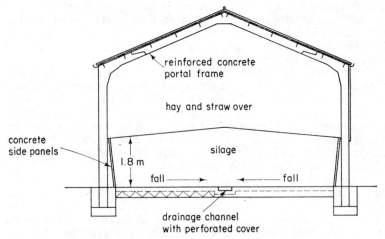

FIG. 16.—COVERED SILO WITH PORTAL FRAME ROOF.

Potato Stores. The indoor storage of potatoes has many advantages over a clamp in the fields. Existing buildings can generally be made suitable for shallow storage up to 2·4 m high at no great expense ; for greater depths forced ventilation may be required and loading may be more difficult.

The main points about a building for potato storage are that it should be big enough for handling the potatoes—a width of about 6 m is usually considered necessary with 900 mm of headroom when full —and it should have wide lofty doors, preferably hinged to avoid draughts. There should also be a space of about 4·5 m in front of the potato stack when full for sorting and loading the potatoes. Walls should be strong enough to resist the pressure of the potatoes, which is about 5 kg/m^2 per foot of depth, or should be strengthened, and be thick enough to keep out frost. Where there is any doubt about the insulating value of the walls they should be lined with a single layer of straw bales stacked closely together, or with loose straw tightly packed behind wire netting, or some commercial insulating materials should be used. If the storage is to be in a building with open sides such as a dutch barn, walls of at least two layers of straw bales suitably supported should be provided. Potatoes in a store should be well covered in straw, particularly if there is no roof insulation.

A tonne of potatoes will occupy about 1·5 m³, so that with shallow storage just under a square metre of floor space is required for each tonne of potatoes.

Granaries. The granary is the principal building on the farm used for storing and processing food for livestock ; on larger farms the labour of constant journeys to the granary can be saved by having subsidiary food stores attached to the larger livestock buildings such as the cowhouse, piggery and poultry buildings, filled up periodically from the central granary. As large quantities of heavy and bulky foodstuffs are brought to the granary, stored and processed and taken away, a convenient central site and machinery for handling the food will reduce the work. The site should have a good hard approach for lorries and farm transport delivering and taking the feeding-stuffs away. The loading of sacks can be made easier if part of the granary floor is at lorry level, about 1·1 m, approached by a ramp so that a sack barrow can be used for moving the sacks without the need for lifting them. Where home-grown grain harvested by combine is used, it is convenient to have the granary close to the grain store or even amalgamated with it.

A weatherproof building is essential for a granary and the ground floor particularly should be dry. A concrete floor with a damp-proof membrane is best. Upper floors should be designed to take the necessary weights and can be of timber or pre-cast concrete units. It is convenient sometimes to have root storage attached to the granary and the root cutter and cleaner can be driven from the same shafting. A building with wide entrance doors and stout walls is required, and it is usual to allow 2 m³ per tonne of roots to be stored.

Grain and meal are usually stored in bins which can have sides of concrete blocks, brickwork, timber boards, plywood, galvanised metal sheets or asbestos-cement. The front of the bins should not be more than 900 mm high so that sacks can be tipped without much effort. If the front is made of removable boards sliding in slots the bins are easier to clean out. The number and size of the bins will depend on the requirements of the individual farm, and the following figures show the space required per tonne for different grain :

STORAGE SPACE FOR GRAIN AND MEAL

	Whole grain.	Meal.
Wheat . . .	1·4 m³ per tonne	2·2 m³ per tonne
Barley . . .	1·5 ,, ,,	2·0 ,, ,,
Oats. . . .	2–2·3 ,, ,,	4·0 ,, ,,

The bins can be on a ground floor, but handling is sometimes easier to arrange if they are on an upper floor above the processing machinery into which the grain or meal can fall by gravity. The ground floor is then free for the machinery and the storage of cake, etc., in sacks.

EQUIPMENT FOR GRAIN DRYING AND STORAGE

The general use of combine harvesters in Britain has resulted in a high proportion of the wheat and barley being threshed during August and September, and has now made it essential for most farmers to be able to store a proportion of their combined grain. In the past a seasonal rise in the average price of wheat helped to repay the cost of drying and storing this crop. Moreover, modern marketing conditions make it advantageous for farmers to be able to hold other crops if necessary. Fortunately, research and development work have provided many satisfactory solutions of the farm grain storage problem, though some of them are high in capital cost, and some unattractive to tenant farmers. In the following pages the main systems of farm drying and storage are briefly discussed and compared.

Variations in Need for Drying. The proportion of grain that needs drying, and the amount of moisture that needs to be removed to ensure safe storage, varies considerably from farm to farm and from season to season. There are some farmers in favoured parts of the country who have never had a drier and who are quite sure that the only insurance they need is adequate combine harvester strength in relation to the amount of corn to be handled. For most farmers, however, having a spare combine harvester is neither the cheapest nor the most reliable insurance against unfavourable harvest weather, and it is necessary to have available, either on the farm or elsewhere, some means of getting at least a proportion of the grain dried. Grain harvested at 20 per cent moisture content or above will readily heat, and no practicable amount of turning can be relied on to keep it good if it is put into a bin. If a good drying wind blows through it while it is well exposed in hessian bags, it will rapidly lose some of the moisture ; but this is a form of drying that is too unreliable to be advocated. Grain containing 18–20 per cent moisture that is clean and has been harvested without excessive damage can be kept for short periods in sacks which are spread out in a well-ventilated building, but it can rapidly give trouble in warm, damp weather. At 16 per cent moisture content sound grain can be stored fairly safely in sacks, and may be put into unventilated bins if it is watched and turned occasionally. At 15 per cent moisture content, sound grain will keep safely in bulk, and needs only an occasional check to ensure that there are no troubles arising from damper patches or from insect infestation. For unventilated long term storage, drying of both wheat and barley should be to 14 per cent m.c., and occasional turning is recommended. Most progressive farmers who can afford to do so, while hoping for and making the most of a season that enables them to harvest most of the grain at a moisture content suitable for storage, will plan for an average or bad season that makes conditioning of some of the grain essential.

The amount of drying needed, while depending to a great extent on the season, can also be considerably influenced by management of the combine harvester. On a summer day in good weather, grain that is fully ripe and has not recently been wetted by rain may have a moisture content of 19 per cent in the early morning. Moisture will then evaporate fairly rapidly from a standing crop, and by late afternoon the grain may contain under 16 per cent moisture. This emphasises the need for adequate combine harvester strength, so that the work can usually be carried out in favourable conditions. Another important factor in management is choice of early-maturing varieties of cereals, which ripen when the days are long during the early part of the season.

In any mass of grain as it comes from the field, the volume of air between the individual grains in a bag or in a bin is almost as great as the total volume of the individual grains. This air, if stagnant, very rapidly takes on a degree of dampness corresponding to that of the grain itself—a characteristic made use of in one form of grain moisture meter, which measures the relative humidity of the inter-granular air. All farm grain driers operate on the principle of blowing away this moisture-laden air and replacing it by relatively dry air which can take up more moisture from the grain before it is in turn expelled. The vast differences between grain driers arise from the considerable variations in the thickness of the grain layer, the dryness of the air used, and the speed at which the air is blown through. Choice of drying system depends on a large number of factors some of which are mentioned below in describing the characteristics of common types of plant.

Continuous-flow Driers. A characteristic feature of these machines is a grain layer only 75–150 mm thick, with arrangements for moving it continuously through the drier. The most common type is in the form of a short tower down which the grain flows by gravity ; but horizontal conveyors and inclines are also employed. Some horizontal conveyor machines are suitable for either grass or grain drying. Drying is rapid, the grain usually staying in the machine for less than an hour, during which time 5–6 per cent moisture might be removed, at a throughput ranging from 0·5 to 20 tonnes per hour. In order to achieve maximum throughput, continuous driers usually employ the highest temperatures that can be maintained without damaging the grain. These maximum safe temperatures are as follows:

	Max. Temp. °C.
Corn for stock feed	104
Milling wheat	65
Malting barley or seed corn of under 24 per cent m.c.	49
Malting barley or seed corn of above 24 per cent m.c.	43

Continuous driers are usually fairly expensive in capital cost. Running cost depends partly on such factors as the fuel used, oil-fired heaters being most popular ; but depreciation and interest charges inevitably outweigh both fuel and labour costs if an expensive installation is little used.

With a continuous drier it is usually advisable to arrange not only for pre-cleaning of the grain but also for some pre-drying storage of grain, so that the machine can operate with little labour for long hours if necessary. A drier with a throughput of 5 tonnes per hour when removing 6 per cent moisture at 65 °C will be capable of dealing with the produce from two large combine harvesters in most areas and seasons.

With all " high-temperature " driers, including some tray machines, it is essential to cool the grain to within 5 °C of atmospheric temperature before storing it in bulk.

Batch Driers. Horizontal tray driers with trays having perforated floors are suitable for drying grain in layers 300–600 mm deep. Dual-purpose (grass and grain) driers employ temperatures similar to those used in continuous driers, but owing to the fact that the grain is not moved during drying it is advisable to keep maximum temperatures 5 °C lower than those given for continuous-flow machines above. Drying at the higher temperatures tends to be uneven, but subsequent mixing of the grain during emptying minimises the disadvantage.

Tray driers employing temperatures not more than 14–17 °C above atmospheric temperature dry the grain slowly and safely. Handling the grain into and from the tray may be facilitated by the use of a tipping mechanism or by having the tray permanently inclined at an angle of 15–20 degrees. Throughputs of portable tipping tray low-temperature driers are from 1½–2 per cent moisture extraction per hour from 1·25 to 3 tonnes of grain, while a typical output of a high-temperature dual-purpose drier is 5 per cent removal per hour from 1·5 to 3 tonnes.

More complex batch driers may be fitted with automatic controls to regulate filling, drying and emptying. Length of the drying period may be controlled either by a time clock or by the relative temperatures of the drying air at inlet and exhaust.

Platform driers for bagged grain consist essentially of a fan and a platform with air ducts beneath and openings over which single bags of grain or seed are laid. Adoption of bulk handling methods rapidly made such driers obsolete for grain, but they are useful for drying small quantities of seeds of types which are difficult to handle in most driers.

Vertical-flow Ventilated Bin Plants. In this type of installation the grain is dried in the storage bins, and the plant serves the dual purpose of providing for both conditioning and storage. For this reason it is more expensive to instal than a simple drier. The drying bins are provided with perforated or porous floors, and slightly warmed air is blown through the grain from the bottom to the top of the bin. Grain depth during drying normally ranges from 1·5 m to 4·5 m and an air temperature rise of 1–5 °C above atmospheric is generally employed. The aim is to keep the relative humidity of the drying air at 55–65 per cent and this results in the grain having a moisture content of about 14 per cent. There is no need to keep the relative humidity absolutely constant, since slight over-drying by air of low relative humidity during the day is balanced by the higher relative humidity at night. Occasional tests of the moisture content of the grain near the bottom of the bin suffice to indicate the need for any adjustment of the temperature rise, and such adjustments can easily be made in the last few days of drying.

A properly designed ventilated bin plant is a sound method of providing grain storage in any part of the country where the bulk of the grain can normally be harvested at a moisture content of 20 per cent or below. Ventilated bins can deal effectively with very damp grain, but only in relatively small quantities which are treated in shallow layers. It is advisable to provide all bins with ventilated floors and to be able to provide an air flow of 25–34 m³ per hour simultaneously through at least half of the bins when they are filled with grain to a 3 m depth. In wet regions, provision of 34 m³ per hour through two-thirds of the bins simultaneously may be advisable.

Typical performance in a well-designed plant ranges from removal of 5 per cent moisture in 10 days from a single 20-tonne bin filled 3 m deep, up to removal of 5 per cent in 5 days from half a dozen or more 20 tonne bins where the grain depth during drying is limited to 1·5 m. The first requires a small fan which can also do the pneumatic conveying for a 40–60 tonne plant, while in the latter case a more powerful fan is required and conveying must be done by other means.

Drying in a vertical-flow ventilated bin is effected from the floor upwards. When a fairly damp batch needing 5 per cent moisture content removal is being dried, the situation after 2–3 days is that the bottom 600 mm is perfectly dry, drying is proceeding in the layer 600–900 mm from the bottom, and the upper part is substantially unchanged. Ventilation should not be interrupted until the grain at the top is dry, and it is then advisable to turn the grain once in order to secure thorough mixing. This turning is unnecessary for grain put in the bin fairly dry, but desirable if any grain was put in very wet.

There is no limit to the size of a ventilated bin plant, but it becomes desirable to consider the alternative of a continuous drier or of using a floor drying system at storage capacity above the region of 400–500 tonnes. The system can be adapted in the smaller sizes to suit a wide range of existing buildings, but it is often preferable to put the plant on a clear site, owing to the extent of the strengthening and alterations needed.

Any farmer who considers installing a plant is well advised to seek up-to-date information on such matters as general layout, bin and floor construction, methods of warming and controlling the air and efficient operation.

Other Ventilated Grain Storage Systems. In the radial-flow ventilated bin system the drying air is introduced into the centre of the cylindrical silos through a drying cylinder, with perforated walls. The air then flows outwards through the grain, and leaves the silo through perforations in the walls, which may be made of expanded metal, hessian supported by a suitable steel mesh, or louvres of various kinds. A typical drying silo is 2–3 m in diameter and has a central cylinder about 450–600 mm in diameter. The relatively thin walls of grain through which the air has to pass can result in high efficiency and a drying rate appreciably greater than that usual in vertical-flow ventilated silos. For example, the removal of an average of 2 per cent moisture per day from a 2·2 m diameter silo is fairly easily achieved.

In one type of ventilated bin plant the warm air is introduced and the damp air exhausted at various levels through the silo. The thickness of the grain layer between inlet and exhaust ducts is thereby limited to about 1·5 m, and a fairly high temperature rise (e.g. up to 14° C) or chemically dried air may be effectively employed. This makes it possible to remove from 2 to 5 per cent moisture content per day from a single 20 tonne bin, and with such an arrangement it is advantageous to have a few self-emptying drying bins, together with plain bins for subsequent storage. Uneven drying may be partially avoided in self-emptying bins by arranging a limited amount of grain circulation.

On-floor Driers consist of a general-purpose building with ventilating ducts laid over the floor in such a way as to provide the equivalent of a large, vertical-flow ventilated bin. It offers a low-cost system in which the major investment is in a building which can, if necessary, be converted to other uses. In the most common type the drying equipment includes a fan, a main duct running along the centre of the building or along one side according to the width, and portable laterals at right angles to the main duct and spaced at 1–1·2 m centres. Air supply to each lateral is positively controlled by shutters which are usually fixed from inside the main duct. The laterals are laid down as filling of the store proceeds, and grain is piled over them to a height of about 2·4 m. In a building 18 m wide it is convenient to employ a high centre duct to divide the store longitudinally. A few subdivisions at right angles to this can be provided by means of free-standing wall sections of suitable design.

Though lateral ducts are usually portable and above-ground, some farmers are sufficiently keen on easy handling to distribute the air by building a series of shallow trenches in the concrete floor, and covering the tops with strong flat prefabricated metal grids. The covers are rebated accurately into the tops of the ducts so as to form a level floor which will support the weight of tractors, trailers and trucks.

Another method of avoiding the use of portable laterals is by means of a " centre-duct " drying system in which the grain is piled high above the single large central ventilating duct, and is left at a natural angle of repose. The drying air flows from the duct in a more-or-less radial pattern. Though this sytem has the merits of simplicity and low cost, it is less positive than a typical flat-top floor drying system.

Grain Aeration. There are many reasons for cooling grain to not more than 15–18 °C as soon as possible after storage. For this purpose small, easily moved ventilating units providing about 10 m^3 per hour per tonne of grain can be effectively used.

Moist Grain Storage. Grain intended for stock feed can be safely stored at moisture contents considerably higher than those recommended above, either by means of stores which can be effectively sealed against entry of air, or by the careful use of preservative chemicals such as propionic acid.

Whatever method of moist storage is chosen it is advisable to aim at only a moderate moisture content, preferably of about 18–20 per cent m.c., and not over 24 per cent. At higher moisture contents the grain is more difficult to handle, while the consequences of a break-down of the preservation system are likely to be more serious.

Grain Cleaning. In weedy crops, or where a short-strawed barley is undersown with grasses and clovers, a considerable amount of greenstuff may be taken into the combine harvester with the straw, and broken pieces of leaves, thistle heads and many insects are often present in the grain sample. The greenstuff, along with dust and broken pieces of straw, tends to collect in patches that will quickly heat if the grain is stored in bins. Pieces of straw will also cause trouble in some types of drier by preventing an even flow of grain through the machine. It is therefore essential with tower driers and also with ventilated bins to remove most of the rubbish by passing the grain through a " pre-cleaner ".

The term " pre-cleaner " is applied to various types of machines designed to remove the impurities referred to above, and not necessarily expected to make a properly graded commercial or seed sample. Nevertheless, there is no fundamental difference between some types of pre-cleaners and a complete corn dresser. A pre-cleaner usually employs one or more of the following devices :

(*a*) An inclined reciprocating screen with large holes to remove pieces of straw, thistle heads, string and other roughage.

(*b*) An air blast to remove dust, chaff, and light or broken grains. This may be provided by a simple winnowing device or by a controlled aspiration in what is termed an aspirating leg.

(*c*) A screen with small holes to separate small and broken grains and weed seeds.

A combined cleaning and grading machine may employ all the above devices and it may also incorporate certain other mechanisms such as :

(*d*) A rotary screen to separate grain according to its width.

(*e*) Discs or indented cylinders with pockets to separate grain according to its length.

A good pre-cleaner should be capable of operating for long periods without blockage of any screens that are employed ; points to look for are ample capacity and a fairly steep angle of slope on the screens, together with suitable arrangements for disposing of the large amount of dust that may be removed in the cleaning operation. The latter requirement is often best met by machines which employ one or more aspirations and deliver the light liftings through a duct to the outside of the building.

On many farms it is advisable to select a machine that is capable not only of carrying out any necessary pre-cleaning, but also of preparing a commercial sample after drying or storage.

Grain Conveying. It is possible on a small scale to treat grain in batches without the use of any form of conveyor, but where the amount to be handled is considerable it is usually essential to employ power operated conveying equipment to move the grain from a receiving hopper to and from cleaning or drying equipment, and into or out of bulk storage.

Bucket Elevators are simple in design, easily repaired and adjusted ; they are widely used for vertical lifts for feeding cleaners and driers, for filling silos near the elevator, and to link upper and lower horizontal conveyors used for delivering to distant storage silos. They need a separate pit when used to empty underground receiving hoppers. The foot is not self-cleaning.

High-speed Auger—some designs are suitable for vertical or horizontal use, but generally augers are employed either fixed or portable, for conveying in inclined positions. They occupy little space, and need little or no extra height at head or foot. With some models the foot can be dropped directly into the mass of grain.

Chain and Flight, and Belt Conveyors are used for horizontal runs. The chain and flight type is cheaper, and satisfactory for most farm stores, where the amount of annual use is small. Short belt conveyors can be movable where necessary. Some chain and flight types and belt and flight conveyors are suitable for use in an inclined position.

Pneumatic Conveyors can provide a complete conveying system from a single fan, and are often suitable where electrical power is not available, or where long and complicated horizontal runs are involved. Power requirement is high, the conveyor is noisy and tends to create a dusty atmosphere. The drive unit can be made easily portable.

In choosing a conveyor for a small or medium-sized installation, the advantages of a completely automatic press-button arrangement must be considered in relation to the cost. The annual usage is always likely to be small, and a considerable saving in capital cost can sometimes be made by accepting the need to spend a little time in moving a part of the equipment.

EQUIPMENT FOR PREPARING RATIONS

The introduction of new grain-harvesting methods, the extension of rural electrification and the development of new types of equipment for corn grinding and meal mixing, have combined to make many old granaries unsuitable or at least inconvenient for modern needs. Where the basis of the stock rations is to be home-grown corn harvested by combine, it is very desirable to site the food-preparing plant adjacent to the grain store, so that a short extension of the conveying system provides for easy transfer of the grain for grinding. It is often possible to house a new bulk grain store in an existing barn adjacent to the granary, but where the grain store has to be built on a new site, the provision of grinding facilities there should be considered.

An existing two-storey building can be put to good use for food preparation, and valuable covered space at the end of a high barn used for grain storage can also often be most efficiently utilised by putting in an extra floor ; but with modern equipment sited adjacent to a bulk grain store there is no other important advantage in a two-storey arrangement, and a single storey layout is likely to be preferable. In either case it is usually advisable to adjust floor and road levels so that there is one loading and unloading point which has the granary floor at the level of the platform of a lorry standing out-side the building. This facilitates loading and unloading with a sack barrow.

Where mains electricity is available it is always better to use it for all farm machinery, equipping each machine with its own motor, and making use of the freedom that this bestows, to locate each machine in the best possible place. In order to keep electrical loadings and capital costs within reasonable limits it is advisable to choose machines of small capacity and to arrange them for unattended operation wherever this is practicable.

Grinding and Crushing Mills. The main types of mills now used on farms are :

(a) Plate mills in which the corn is ground between the surfaces of two corrugated plates made of chilled cast iron.

(b) Crushing mills in which the grain is rolled between the surfaces of two iron cylinders that rotate close together.

(c) Hammer mills in which the corn is disintegrated by steel hammers which rotate at high speed inside a casing.

A popular machine in the past has been one incorporating both chilled-iron grinding plates and crushing rollers, but the use of small automatic hammer mills has extended rapidly with the provision of electricity supply. These machines have a low power requirement, can easily be made to operate unattended for long periods, and are low in capital and maintenance cost. The usual method of control is to put the required quantity of grain into the hopper and set the mill running. The mill switches itself off when the grain hopper is

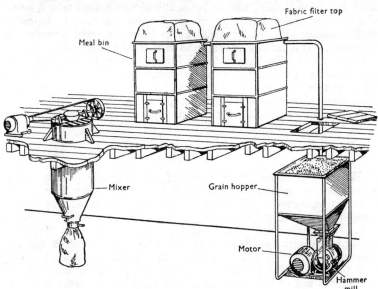

FIG. 17.—TWO-STOREY ARRANGEMENT FOR GRINDING AND MIXING RATIONS.

[*Christy and Norris*]

empty. By the use of interchangeable screens the hammer mill can do either coarse or fine grinding. Where barley or oats are to be fed to cattle then rolling is preferred because it cuts down on dust and is more palatable to the cattle.

The mill needs a grain hopper of at least 0·5 tonne capacity, and this may be bought from the mill manufacturer or made up on the farm. Where it is necessary to feed grain into the hopper from sacks, the top of the hopper can be about 450 mm above floor level and the mill itself sunk into a shallow pit. Mills of up to 5 h.p. are normally arranged to blow the grain into one or more meal bins equipped with a filter-cloth top. When larger mills are used it is almost essential to have a cyclone, but this can be arranged to deliver into bins if required. It is difficult to make meal flow freely out of storage bins, and the most satisfactory form of construction is to have parallel sides and a gently sloping bottom which leads to a large vertical bagging-off slide. The slide is protected from the weight of the meal by an inclined board fixed above it, and a short rake can be used to bring the meal forward to the bags if necessary. Fig. 17 shows a typical installation which is capable of an output of up to about 5–6 tonnes of meal per week.

Food Mixing. A mechanical food-mixing machine is rapidly becoming a regular item of equipment, and the type now generally preferred is a bottom-feed machine which employs a high-speed auger to take in the ingredients and throw the mixture over the top of the mass inside a conical casing. Where the drive is by electricity the motor may be mounted at the top of the machine itself, and this helps by leaving the floor space clear. The mixer is started up when the first sack of ingredients is tipped in, and by the time the whole of a batch has been loaded, mixing is usually almost complete. A mixer of 0·5 tonne capacity is large enough for many farms. If desired, a small hammer mill can be arranged to blow the meal directly into the mixer. It is usually advisable to provide the latter with an outlet at the top which leads to a fabric filter of adequate size, but some manufacturers provide for complete re-circulation of air back to the mill. In a well-arranged food-preparing plant, ample floor space is provided for storing bought-in feeding-stuffs that are to be mixed with the home-grown meal, and also for the temporary storage of mixed rations.

SERVICES

Water. To make the fullest use of a farm an adequate supply of pure water to buildings and fields is essential. Apart from public water mains, sources of a farm supply are deep and shallow wells, bore holes, springs, streams, ponds, ditches and rainwater. Wells and bore holes require pumping plant and some form of storage. Where a supply by gravity from springs and streams is impossible a *ram* and sometimes storage will be necessary, though in the fields properly constructed watering places in a stream can be satisfactory. Ponds and ditches are unreliable sources of water and are liable to spread disease through pollution. Rainwater rarely provides an adequate supply and is more useful as a supplement to other sources.

Pumps. Before deciding to use any source the adequacy and purity of the water should be fully tested, and with shallow sources steps must be taken to prevent pollution by surface water. For farm supplies a reciprocating pump (suction, lift or force) is generally the most suitable as it usually requires no priming and will work with a high suction lift. For bore holes, a submersible form of turbine pump, working in stages, is required. Pumps can be driven by electric or internal combustion engines or a windmill. Where a supply of electricity is available an electric motor has many advantages on grounds of cost and ease of operation as it can be controlled automatically by switches in the storage tank, and only a small pump house is required to house them.

Pumping can be either to a storage tank or reservoir situated at the highest point in the system, with distribution from it by gravity or into a small pressure tank close to the pump. Elevated storage tanks are often costly and a reservoir at or below ground level is to be preferred ; they should be capable of holding two or three days' requirements and to make them larger is expensive and unnecessary with an adequate supply. Requirements can be calculated from the following average figures of consumption :

WATER REQUIREMENTS PER HEAD OF LIVESTOCK

Cows in milk	70	litres per day
Cooling and cleaning utensils, etc.. . .	70	,, ,,
Other cattle and horses at work . . .	45	,, ,,
Pigs	14	,, ,,
Sheep	7	,, ,,
Poultry in batteries	0·5	,, ,,
Domestic houses sewered	90–135	,, ,,
Domestic houses unsewered	70	,, ,,

Pipes and Troughs. The pipes of the distribution system should be laid 750 mm deep to avoid frost damage and there should be sluice valves to all main pipes. The materials most used for pipes are plastic (polythene), galvanised steel, copper and asbestos-cement. Plastic and copper pipes are available in long coils and can be most easily laid by a mole plough which avoids the expense of excavation work. Galvanised steel can also be laid with a mole but it is sometimes damaged ; it is cheap, but on the other hand it corrodes rapidly in some soils and with some waters. Copper corrodes much less readily but is more expensive. Asbestos-cement pipes may also be affected by corrosion at the metal joints unless they are protected, but otherwise they are cheap though only available in 37 mm sizes upwards. The expense of a distribution system to each field can sometimes be saved if only a central main is laid : it is provided with points to which troughs, placed in each field being grazed during the season, can be connected by a length of plastic tubing laid along the ground.

Pre-cast concrete troughs for livestock are the most durable. To prevent damage the ball valve should be in a separate locked compartment at the end or centre of the tank with the supply pipe entering from the bottom. The pipe should be fitted with a stop cock so that the water can be turned off when necessary. A surround of hardcore or clinker prevents poaching of the soil by cattle. The number of troughs can be reduced by placing them in the dividing fence to serve two fields.

WASTE DISPOSAL

Changes in legislation to prevent the pollution of rivers and the greater concentrations of livestock on many farms have in recent years required a new approach to the disposal of farm wastes. The Rivers (Prevention of Pollution) Acts, 1951 and 1961, made the consent of a River Authority necessary before any discharge into a stream is made or continued and this consent is not given if the waste does not reach a satisfactory standard in terms of parts per million of B.O.D. (Biochemical Oxygen Demand) and of suspended solids. On some farms it may be possible to discharge wastes into a local authority's sewers, but as farm wastes are classified as " trade effluent ", the local authority is entitled to make charges for receiving and treating them. Many farm wastes are expensive to treat and the local authority's charges may make it more economical to dispose of the wastes on the farm itself rather than into a public sewer.

Surface Water. The disposal of wastes from the farmstead is often made unnecessarily difficult because large quantities of relatively clean water are allowed to mix with them. All buildings should have eaves gutters and down-pipes leading by underground drains or open channels to soakaways or to a nearby ditch. Paved areas around buildings should also have proper drains which can be connected to the same system if they are kept clean and free from livestock droppings, and when higher ground slopes towards the farmstead there should be open channels or drains to divert the storm water.

Disposal Methods. Possible methods of waste disposal for a particular farm will depend on such factors as the type and number of livestock being kept, the kind of soil and the cropping programme. The simplest system is to handle the manure as a solid by providing adequate bedding, normally straw, to absorb as much liquid as possible : any excess liquid should be drained by appropriate falls in the floor to channels and pipes leading to a tank which is pumped out periodically over the land. The tank should be large enough to contain at least one month's flow as more frequent emptying can be inconvenient. It is usual to reckon the daily quantity of urine at 14 litres for each cow, 7 litres for other cattle and 2 litres for pigs. The solids can be stored in a compound or dungstead, the size depending on the frequency with which they can be spread on the land.

Where little or no bedding is used the waste will be in the form of a slurry with a consistency varying with the livestock and the amount of bedding. The slurry is stored in a below-ground tank constructed of reinforced concrete or brickwork or, in suitable conditions, of a simple earth retaining wall : pumping the slurry to a prefabricated above-ground storage tank is an alternative. The daily quantities of slurry to be stored are about 40 litres for cows ; 5 litres for pigs of 70 kg body weight ; and 900 litres per week for 1,000 hens. These figures should be increased to allow for washing and waste water : in the case of dairy cows a total of 70–90 litres a day is usual. The number of days storage for which a tank should be provided will depend on the frequency of emptying possible under the most difficult conditions likely to be encountered. The slurry can be spread on the land either by means of a mobile tanker, in which case water will need to be added in the ratio of 1 : 1 to get it to flow, or through an organic irrigation system (see p. 174) for which additional water in the ratio of 3 : 1 is needed.

Purification of farm waste to a condition in which the effluent may be discharged into a stream is possible but often difficult owing to the high B.O.D. and solids content. Basically the systems used rely on oxidation of the slurry by some form of agitation giving greater contact between the air and the waste and so encouraging bacterial action.

Where there is insufficient suitable land available for the distribution of wastes, either as a solid or a slurry, as with an intensive pig or poultry unit, removal elsewhere by a contract tanker or installation of a special drier may be necessary.

Silage Effluent. The effluent from silos can be particularly difficult to deal with, largely owing to its high B.O.D. content. The amount of effluent coming from the silage can be minimised by protecting the silage from rain and by ensiling only wilted crops of high dry matter content. In this way the effluent can be reduced from as much as 350 litres per tonne of silage to as little as 5 or 10 litres. In all cases the effluent should be collected in a sealed tank from which it is pumped after dilution and spread over the land. It is advisable to allow a storage capacity of 20 litres per tonne of silage stored.

ELECTRICITY

Electricity from the public supplies is normally at 240 volts or with a current alternating at 50 cycles a second. The supply can be single phase or three phase ; the single-phase system is the cheaper to extend to a farm as it requires fewer wires at the transformer and throughout and so it is what electricity boards usually provide. A single-phase supply is adequate for most farm purposes such as lighting, heating and driving small motors. Three-phase motors are cheaper than single phase owing to the simpler starting but the difference only becomes appreciable in sizes over 10 h.p. Where only a single-phase supply is available it is cheaper to use a number of small automatically controlled motors rather than to do the job more quickly with a large motor. For example, a 3 h.p. small hammer mill running for five hours to grind a ton of meal will cost no more to run than a large 20 h.p. mill grinding the same quantity in 1½ hours. For crop drying where heavy loads are involved, or where large motors are essential, a three-phase system would be required.

Where a mains supply is not available, engine-driven generating sets can be used. A plant generating alternating current at 240 volts is the most suitable for farm use and has the advantage that no changes in wiring and appliances are involved when a mains supply becomes available. It is, however, impossible to store alternating current so that the plant has to run all the time that electricity is required. In a direct current system storage batteries, recharged by the generator, can be used ; batteries are, however, expensive particularly for a high voltage. A 3–5 K.V.A. diesel plant will supply lighting and drive most small motors on a farm, especially if arrangements can be made not to use the larger motors at the same time. Starting can be automatic whenever the current is switched on, remote control or manual. Water and wind power can also be used to drive small generators.

Wiring in farm buildings is mainly done with plastic conduit, plastic cable or with self-contained tough-rubber-sheathed (T.R.S.) cable of which there is a special form with an impregnated outer braiding known as farm wiring cable. It is fixed by special wiring clips and cleats and being flexible there is no difficulty with irregular surfaces. The wiring is less likely to be damaged if it is fixed as high up as possible and out of reach of livestock and other sources of damage. In positions where abrasion may occur the cable can be enclosed in conduit, preferably the plastic type. Electrical wiring, is, however, a matter for the expert and advice should be obtained on the most suitable form of installation. All farm wiring installations should be carried out to the standards set out in the current regulations for the wiring of farm buildings, published by the Institution of Electrical Engineers.

ROADS

Traffic on internal farm roads is relatively light and infrequent so fairly simple methods of construction are all that is required. It is important to appreciate that it is the soil which supports the weight of the traffic, and everything possible should be done to improve the load-bearing strength of the soil. The soil, however, needs to be kept dry if it is to bear the weight of traffic without sinking. Rain falling on the road should be removed quickly by making the surface as impervious as possible, and by sloping the road so that the water will run off to the sides, preferably into grips or channels which will carry it well clear of the road and its base. Shallow side ditches close to the road are ineffective and may do more harm than good. In places where the water table is high, deep side ditches may also be necessary to keep the ground water away from the road. Except for clay, the soil should be well rolled or compacted to increase its bearing capacity by removing air voids, a heavy roller (not less than $2\frac{1}{2}$ tonnes) being used after vegetation or turf have been removed ; it is unnecessary to remove any soil. Road work should be carried out when the soil is as dry as possible, that is, during the summer and autumn.

Some soils, mainly gravels and sands with well-graded particles, will, with shaping and consolidation, form a satisfactory farm track provided it is reshaped periodically when it gets worn. Roads of this type will give better service if the surface is waterproofed by dressing with bitumen, blinded with 6 mm stone chippings. Normally the bearing capacity and wearing qualities of the soil have to be improved by the addition of several layers of stone or similar material to form a *flexible pavement*, or by putting down a slab of concrete.

In a *flexible road* the choice of materials for the various layers will make a lot of difference to the cost. Hard, crushed stone or macadam make the best road, but it is usually the most expensive and local materials should be used wherever possible to save cost. The first layer lying on the soil is the *sub-base*, about 50 mm thick, which levels out the surface of the soil and acts as a cushion between the soil and the main layers of the road. It can be of clinkers, gravel or cinders. The next layer is the *base*, of any thickness from 75 mm upwards, though anything over 100 mm is unnecessary on most farm roads, and costly. It is usual to specify that the thickness of the base should not be less than twice the size of the largest stone in it. Crushed stone may be used or such local materials as flints, chalk, builders' hardcore, cinders, clinkers or slag. The top layer is the *surfacing* or *wearing course* which provides a smooth surface for traffic and protects the layers below from wear and the effects of the weather. The simplest way of surfacing is to " blind ", or scatter and roll over the base, a layer of fine material such as stone dust, gravel or ash. An improvement is to waterproof the surface by binding it with a water slurry of stone dust or of sand and clay. An alternative is to surface the road with fine material of not more than 37 mm size, grouted with tar, bitumen or bitumen emulsion, followed by a blinding of fine grit. The tar and bitumen have to be applied hot and need special equipment. Bitumen emulsion, although more expensive, can be applied cold without special equipment. Stone can also be bought already coated with tar or bitumen. Although the various layers have been described separately, they are often combined and laid in one or two layers.

The flexible road needs maintenance if it is to last. Mud and animal droppings must be removed and pot holes and depressions filled and grouted with bitumen. Complete resurfacing will be required eventually dependent upon the condition of the road. At the worst, the base may have to be renewed in places and a complete new surfacing applied. In other cases only the surfacing will need filling up and shaping, though if a heavy roller is available the surfacing should first be scarified or broken up.

Concrete is generally used for entrance roadways and internal roads where large numbers of cattle are moved regularly. For internal roads 100 mm of concrete is normally sufficient but entrance roads carry heavy traffic and (20 tonne lorries are now common on farms) usually require 150 mm of concrete. A concrete mix of 1 part of cement to 2 parts of sand and 4 parts of stone is required, laid on a suitable base. A polythene sheet laid immediately under the concrete will allow the road to expand and contract without cracking and will prevent too rapid drying.

The strength of the concrete road depends more upon the soundness of the base than on the thickness of the concrete. Where concrete is laid on an existing hard road only minor levelling will be required. In most cases it will be necessary to excavate the top soil and any soft places and fill with well consolidated hardcore. The hardcore should be compacted in separate layers of a maximum 100 mm depth and the surface should be blinded before the polythene membrane is laid. Where the base is suspect reinforcement of the concrete is advisable and is a better guarantee of strength than increasing the thickness of the concrete.

The choice of road will depend largely on cost, and the cheaper the road the more maintenance it will need. The approach road to a a farm should be as smooth as possible, and if traffic can be restricted mainly to vehicles with rubber tyres a flexible road with a bituminous finish or the more expensive concrete road is often the most suitable. Around the buildings concrete is required as it provides the only surface which can be kept clean without damage. Tractors and implements not on rubber tyres, cattle and sheep will damage roads with bituminous surfacing so that for roads leading from the farmstead to the fields a simple form of flexible pavement is required, or concrete, if the extra cost can be justified.

Farm roads are generally 3–3·6 m wide, but this is not strictly necessary and a road 2·7 m wide gives a track wide enough for most traffic and saves cost ; an occasional passing place is desirable and fencing should be set well back from the road.

FENCES

There are four main types of farm fence—post and wire, post and rail, hedges and walls ; the last two are valuable also for the shelter they give, particularly in exposed districts.

Post and wire fences are the cheapest and can be either of single wires or of woven wire. To be satisfactory the wire fence must be strained very taut and for this there should be stout, well-strutted straining posts at intervals of not more than 135 m and at all changes of direction. The posts should be fitted with eye bolts or ratchet strainers so that the wires can be tightened when necessary. Straining posts should be 2·7–2·8 m long and 125 mm × 125 mm in section, with one-half to one-third embedded in the ground. Intermediate posts or stakes are spaced at 3–3·6 m intervals, though this can be increased to 9 m if droppers (spacing bars of steel or timber 37 mm square to which the wires are nailed at 1·8 m intervals) are used. A pointed stake of 75 mm diameter and 1·6 m in length driven 450 mm into the ground is usual.

Timber posts if not of oak, larch, chestnut or cedar should be treated with creosote either under pressure or by the hot-and-cold open tank method. With the open tank method the timber is placed in a tank of cold creosote which is heated to 80–90 °C. After keeping this temperature for 1 to 2 hours the creosote is allowed to cool and the timber is then removed. One hundred and fifty litres to each cubic metre is required but very absorbent timber will take up to 450 litres. This is unnecessary and adds to the cost, but some of the creosote can be recovered if the tank is again heated and the timber is removed before the creosote cools. Any watertight tank is suitable, and for fencing posts an oil drum large enough to take the butts up to ground level will suffice.

Angle iron and concrete posts are satisfactory but concrete should not be put in places where it may be knocked by implements. A tractor driven post-hole borer reduces the labour of putting in posts and there are also devices for making it easier to drive them in.

Single wire fences can be of three to seven wires, of which the top, bottom and not more than one intermediate wire need be barbed. Other wires should be plain : 3·25–4 mm is sufficient thickness except in areas where corrosion of steel occurs. The top wire should normally be 1·15 m above the ground. Woven wire fencing is available in various widths and spacings of the wires for pigs, sheep and cattle ; it is more stock proof if a strand of barbed wire is fixed at the bottom and top. Woven wire fencing is more expensive than plain wire but it is easier to erect and lighter posts are sufficient.

An inexpensive form of dropper fence is the *lightning* or *spring* fence, common in Australia. Strainers are spaced 200 m apart and the intermediate posts are 30–50 m apart. Light gauge high tensile wires are used, strained very tight and kept apart by cleft chestnut palings every 2 m or 3 m.

Post and rail fences last longer than wire fences but are two or three times as costly. The most durable type consists of three or four rails morticed into posts spaced at 2·7 m centres with an intermediate or prick post in between. The timber should be well treated with preservative. Cleft rails nailed to split round poles provide a cheaper fence, and cleft timber lasts longer than sawn.

Hedges are three or four times as expensive as wire fencing but have a much longer life. Many varieties of plants are suitable but hawthorn (quickthorn or *Crataegus oxyacantha*) is the most used. The hedge should be planted on the flat with the plants 225 mm apart in a double row and a good ditch provided where necessary. Wire fences to protect the young hedge against livestock are often required. The cost of planting can be reduced by using farm implements instead of hand digging to prepare the bed. Tractor-operated hedge trimmers make subsequent trimming cheaper ; they are no substitute for the laying by hand every ten to fifteen years required for a good stock-proof hedge previously laid.

GATES

The traditional timber five-bar gate 3 m wide and 1·2 m high is still very popular on the ground of cost and durability. Many kinds of timber can be used but they should all be treated with preservative, even oak and larch. The important parts of a gate are the top rail or *bar*, the *heel* and the *head*, which should be of oak or larch well morticed together. Only the top, bottom and one other bar should be morticed through the heel and the head to avoid weakening them, and the gate should be well braced with diagonal struts and braces bolted rather than nailed together ; a diamond pattern for the bracing is often preferred.

Hanging posts of timber or concrete should be 200 mm square in section and 2·4 m long, set at least 1 m in the ground. A gate will sag in time and it helps to have adjustable hinges. Gates should be hung to be self-closing, and it is an advantage to have a latch which will support them when closed.

Wider openings of 3·6 m upwards are required for modern implements but timber gates of over 3 m are too heavy. The solution is to have two gates up to 3 m wide which either meet at the head or else overlap and can be locked together by a swinging bar. An alternative is a wide tubular steel gate which is lighter than a timber gate. A steel gate should be well galvanised after manufacture and painted with bitumen if it starts to rust. Angle iron gates are also satisfactory.

GRIDS

A cattle grid is a convenient way of providing a barrier to livestock in a road where a gate is liable to be left open or is a nuisance. It consists of a number of bars fixed across the direction of the road over a shallow pit about 300 mm deep. The bars are spaced 150 mm apart, which is sufficient to deter animals from crossing but does not hinder

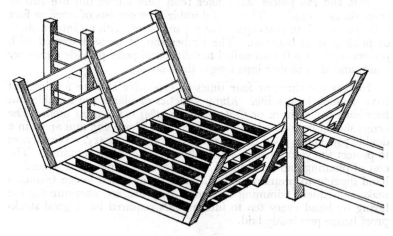

FIG. 18.—CATTLE GRID.

wheeled traffic. A grid should be the full width of the road and for farmstock a length of 2–2·4 m is sufficient. There should be fences along either side of the grid to prevent animals getting round (Fig. 18). A number of inexpensive proprietary grids in concrete or steel are available. Grids can be simply constructed of 50 mm water tubing or 75 mm × 75 mm rolled steel joists fixed to concrete or timber bearers.

Chapter IV

THE USE OF MECHANICAL POWER IN FARMING

THE farmer uses mechanical power and machinery either to increase food production or to maintain a given volume of production at a smaller cost. The problem is essentially the economic one of how to organise use of his labour and equipment in the most profitable way. Power and equipment costs today represent an important and increasing proportion of total farm costs, and vary widely from farm to farm. It is clearly important that farmers should have such a knowledge of power and machinery as will enable them (1) to choose and buy equipment well suited to the needs of the farm ; (2) to understand the working principles of the machines so that they may be kept in good running order ; and (3) to understand the applications of the equipment in order that it may be operated efficiently

The power of animals has been used in farming from the earliest historical times. In Britain, use of animal power is now rare and on most farms the working horse has been replaced by mechanical power. As mechanical ingenuity developed, and Britain became industrialised, increasing use was made of water and wind power on farms, especially for the driving of grinding mills. Water power is still used where circumstances are especially favourable, and there are also many wind-powered pumping plants in constant use. When the steam engine was developed it was, naturally, harnessed to agricultural jobs, and was successfully used for heavy cultivations and threshing ; but steam power on the farm has had its day, and is now generally superseded by the much handier power of the internal combustion engine and the electric motor.

Internal Combustion Engines. Internal combustion engines are today of prime importance in British agriculture. Three main types of engine are used on farms, viz. petrol, paraffin and Diesel engines. The smallest sizes, giving $\frac{3}{4}$–3 h.p. (0·6–2·3 kW), are generally run only on petrol. For larger sizes, paraffin (or V.O.—tractor vaporising oil) may be used, owing to the greater importance of fuel cost. Petrol is in many ways a more satisfactory fuel than vaporising oil, and but for the artificial difference in cost caused by the heavy excise duty on petrol, there is no doubt that petrol would be the standard fuel for all small- and medinm-sized farm engines. An increasing number of new engines of 5 h.p. (3·8 kW) or more are of the compression-ignition or Diesel type. The Diesel engine has advantages in fuel economy, and is thoroughly reliable ; but it costs more than the petrol or paraffin type. The farmer should understand the inter-relation of initial cost, depreciation and fuel economy for the various types and sizes of engine, in order that the most suitable type for a given application may be chosen ; but neither this aspect nor the working principles of the engines can be further discussed here.

Stationary internal combustion engines are used for driving barn machinery, water pumps and milking machines. Small portable engines are used for driving machines such as elevators and sugar beet cleaner-loaders, while trailed machines may occasionally be powered by a small engine mounted on the machine. Many large self-propelled harvesting machines are driven by multi-cylinder engines incorporated in them. The greatest and most important use of internal combustion engines on farms is, however, as the power unit of the farm tractor. It seems unlikely that there will now be any great increase in the use of fixed engines on farms, owing to the increasing use of electric motors for all kinds of stationary work —especially the lighter jobs—and of tractors for all field work and such heavy stationary work as driving hammer mills or large fans for crop drying.

Use of Tractors in British Farming. The modern farm tractor is a very versatile power unit. Equipped with pneumatic tyres, it can in most conditions haul loads along the highways or over the fields ; and in addition to providing power at the drawbar it can supply it at the belt pulley for such work as driving a saw bench, or at the power take-off in order to transmit steady power to such trailed machines as forage harvesters. Moreover, many farm machines have been re-designed to operate with tractors on the "unit" principle : i.e. with the implement mounted on the tractor in such a way that the tractor is an essential part of the equipment, and provides power through a hydraulic lift for raising and lowering the implement. It is clear that developments in this direction are still only in their infancy.

As the number of tractors employed on British farms has increased, so the range of work that they can profitably undertake has been continuously extended. Small farms with low tractor use requirements tend to have more tractor power than larger farms in relation to the amount of work to be done. The number of tractors required depends not only on the total annual need but also on the peak seasonal demand. Average annual use is 700–800 hours annually, but on light-land farms some tractors may do up to about 2,000 hours annually. Specialised tractors such as large 4-wheel-drive tractors and tracklayers are usually heavily used for short periods in autumn and spring, but relatively little used on an annual basis. Modern tractor equipment, such as the hydraulic farmyard manure loader, makes it possible to eliminate manual work from an ever-increasing range of farm tasks.

THE PRINCIPAL TYPES OF TRACTORS

Tractors range in size from minute single-wheeled hoeing machines of about 1–2 h.p. (0·7–1·5 kW) up to giants of 200 h.p. (150 kW) or more, while the majority of those used in Britain are medium-powered four-wheeled general-purpose machines which are able to deliver about 40–60 h.p. (30–45 kW) at the drawbar in good working conditions. There is no simple system of classification which is entirely satisfactory, but a general picture of the range of types and sizes is given by a grouping based mainly on the power of the engine. Farmers are more concerned with drawbar power than with engine power as such, but comparable figures for drawbar power are difficult to obtain in practice, owing to wide variations in operating conditions. Drawbar power is readily calculated if the speed and drawbar pull (measured by means of a dynamometer) can be determined. It is given by the formula

$$\text{Drawbar h.p. (Imperial)} = \frac{\text{Speed (m.p.h.)} \times \text{drawbar pull (lb.)}}{375}.$$

$$\text{Drawbar power (SI Units)} = \text{Speed (m/sec)} \times \text{pull (kilonewtons)}$$
$$= \text{kilowatts (kW)}.$$

The power exerted at the drawbar is always appreciably less than the net engine power. It may be only a little less on a test track, but can be very much less on loose, deeply cultivated land.

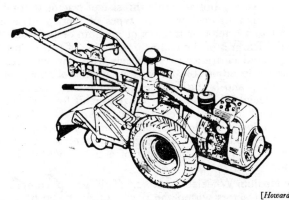

[Howard]

FIG. 19.—A TWO-WHEELED ROTARY CULTIVATOR.

Market Garden Tractors. These are generally two-wheeled machines of up to about 10 h.p. (7·6 kW), but single-wheeled machines, single-track machines, very small tracklayers and very small three-wheeled and four-wheeled tractors may be included. These tractors are used mainly for hoeing and light cultivation, though some machines can deal successfully with a single-furrow plough. The two-wheeled machines have a 1- or 2-cylinder engine ranging from a very small 2-stroke up to a robust 4-stroke giving 8–10 h.p. There is usually some adjustment of wheel track width, and sufficient ground clearance to permit straddling crops of moderate height. Methods of steering two-wheeled machines, type of gear-box, and ease of handling generally, vary between different makes. The usefulness of two-wheeled tractors depends largely on (a) mechanical reliability, (b) ease of handling and (c) the range and quality of the equipment available for use with the tractor, and the ease with which various units can be interchanged.

Equipment includes ploughs of various types, including " one-way "; general and row-crop cultivators; drills; hoes; harrows; rolls; mowers; spraying machines; belt pulley; and a bogie which converts the tractor into an articulated four-wheeled outfit suitable for light transport work with a two-wheeled trailer.

Pedestrian-controlled tractors are commonly used on small family farms in some parts of Europe and the Far East; but use on commercial holdings in Britain tends to be limited to situations such as work in glasshouses where space restrictions prohibit use of larger tractors. For out-door work cropping is usually adapted to suit use of the smaller "conventional" four-wheeled tractors. Special-purpose rear-engined tool carriers are sometimes used for accurate drilling and hoeing.

Small General-purpose Row-crop Tractors. These are generally four-wheeled and occasionally three-wheeled machines, specially adapted for work in growing crops, and with engines of about 12–45 h.p. No hard line can be drawn between the groups. On some small farms a machine of this size constitutes the main power unit for field work; but as a rule this size of tractor is not considered powerful enough to deal with all types of work and is used as a supplement to a tractor or tractors of medium or large size (Fig. 20).

Essential features of the conventional type of row-crop tractor are (a) high ground clearance; (b) wheels with narrow tyres or rims; (c) wheels easily adjustable for various widths of row; (d) a small turning circle; and (e) fittings for the attachment and easy operation of various special row-crop tools. A power lift for the mounted tools and rear-wheel steering brakes are fast becoming standard fittings on row-crop tractors. While most of the small row-crop tractors used in Britain have four wheels, all adjustable, others achieve a very small turning circle and easier mounting of a front tool-bar by having a single front wheel.

Small–Medium Wheeled Tractors. This group comprises mainly four-wheeled tractors within the power range 45–60 h.p., which are capable of doing most kinds of farm work and may be described as "all-purpose". Tractors of this group possess the desirable "row-crop" features already mentioned, and they are particularly well suited to lighter operations such as distributing fertiliser and light transport work. They are, nevertheless, capable of doing the heavy work that needs to be done on smaller farms, such as deep ploughing, heavy cultivating and forage harvesting, though at a work rate appreciably lower than that of large–medium or high-powered tractors. Though normally rear-wheel-driven, a minority incorporate 4-wheel-drive, usually with only medium-sized front wheels. The main features involved in choice depend on the particular needs of individual farms. On some, ease of adjusting wheel track width and ease of fitting specialised (e.g. forward-mounted) row-crop equipment may be important, while on other farms such features may be irrelevant.

Large–Medium Wheeled Tractors. Tractors in this group have general characteristics similar to those of the small–medium size, but with 60–75 h.p. available such tractors are capable of appreciably higher work rates on heavy tillage, forage harvesting, etc. For such work it is particularly advantageous to have a wide range of gears, preferably with a change-speed mechanism of the semi-automatic

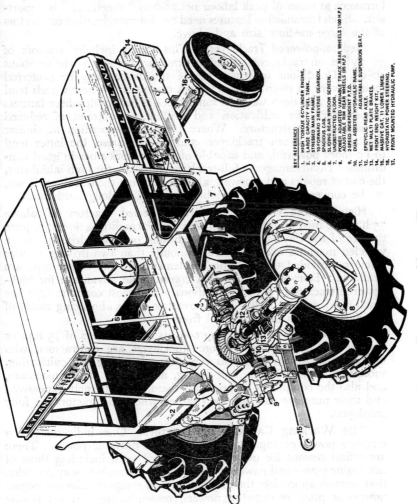

KEY REFERENCE:

1. HIGH TORQUE 6-CYLINDER ENGINE.
2. LARGE CAPACITY FUEL TANK.
3. STRESSED MAIN FRAME.
4. 10 FORWARD 2 REVERSE GEARBOX.
5. SPACIOUS CAB.
6. SLIDING REAR WINDOW SCREEN.
7. UNOBSTRUCTED FLOOR.
8. POWER ADJUSTED 'VARIABLE TRACK REAR WHEELS (100 H.P.)
AND/OR FIXED REAR WHEELS (85 H.P.)
9. 2 SPEED INDEPENDENT P.T.O.
10. DUAL ASSISTER HYDRAULIC RAMS.
11. ADJUSTABLE SUSPENSION SEAT.
12. EPICYCLIC REAR AXLE.
13. WET MULTI PLATE BRAKES.
14. FRONT END WEIGHT KIT.
15. MASSIVE CAT. 11 LOWER LINKS.
16. HYDROSTATIC POWER STEERING.
17. FRONT MOUNTED HYDRAULIC PUMP.

FIG. 20.—HIGH-POWERED WHEELED TRACTOR.

type, that permits gear-changing " on the move " and under load. This enables a conscientious driver to keep the tractor working reasonably near to its potential work rate, and can result in improved performance at times of peak labour need when " timeliness " is important. Such transmission features need careful consideration on tractors of the large–medium size and above.

Medium-powered Tracklayers. This group includes tractors of 50–75 h.p. on tracks, corresponding drawbar horse-powers being about 40–65. Conditions in which use of tracklayers may be preferred include work on steep hillsides ; work on soft land where wheels tend to dig in ; and deep work on some types of heavy silt where farmers choose tracks for cultivation and drilling, on account of reduced damage to soil structure. Where there is competition at similar capital cost between tracklayers and 4-wheel-drive, the latter tend to be more powerful, and capable of higher work rates in good conditions. Disadvantages of full-track machines are the high initial cost, the cost of repairs and renewals, and the fact that the tracklayer is not, in the conventional form, suitable for running along hard roads.

High-powered Wheeled Tractors. Wheeled tractors of above 75 h.p. are used on many large farms where conditions do not require the use of tracklayers. In such conditions, one man driving a large wheeled machine may be more economic than two driving smaller tractors, provided that suitable implements for the bigger type are available. Four-wheel drive is a considerable advantage for high-powered tractors in British conditions. The most common type has equal-sized drive wheels ; but increasing use is also being made of auxiliary drive to medium-sized front wheels.

Large Tracklayers. This group consists of tracklayers of 75 h.p. or more. The chief users of large tracklayers are contractors, who undertake such heavy work as very deep ploughing and cultivating, subsoiling, mole-draining, bull-dozing, and so on. These machines, and also the equipment for use with them, are extremely expensive ; and their purchase can only be justified where the tractors are fully employed.

The Working Capacity of Tractors. One indication of a tractor's power or capacity for work is its drawbar power. There are sound reasons for using other standard tests, including those of net engine power and power at the p.t.o. This is satisfactory provided that farmers appreciate the differences, and compare *either* net engine powers *or* p.t.o. *or* drawbar powers, remembering that net engine power does not include any transmission losses, and that these will always be appreciable, even without the use of power-consuming devices such as hydraulically assisted gear-change. In choosing a tractor the farmer should have a clear picture of what various machines will do on his own land and with the equipment that will be available. Few things are more annoying to a farmer than to send a tractor and implements to do a job and find that power unit and implement are not suited to one another or to the work in hand. While small tractors can often tackle heavy work if equipped with tools that they can manage, they may fail completely if the tools are too wide. On the other hand, a big surplus of power is wasteful owing to the higher running costs.

If it is known how many plough furrows a tractor will normally pull on a certain type of land, a fair estimate may be made of the width of other standard implements that it should handle. Thus, a tractor capable of pulling a given width of plough at normal depth can generally handle three times the width of cultivator or disc harrow, six times the width of heavy harrow and five times the width of drill. Within limits, power needed is proportional to forward speed.

THE ECONOMICS OF CHOOSING AND USING TRACTORS

Cost of Operation. The cost of operating a tractor, like many other costs connected with farming, varies considerably from farm to farm. Important factors influencing cost, in addition to the obvious ones relating to size and type of machine, are the number of days worked annually, the types of work performed, and the care the tractor receives in maintenance and operation. Annual use of tractors varies from a few hundred hours on small farms to 2,000 hours or more on large light-land farms where field work can be carried on throughout the year. Naturally, a high level of annual use generally results in a low hourly cost, owing to the fact that some of the overhead charges are little affected by the amount of use. While due regard may be paid to a low hourly cost, neither this nor a low total annual cost necessarily gives a measure of efficiency, since true efficiency can be arrived at only if the results of the work done are taken into account. The average cost of operating a medium-powered diesel tractor on a farm where 1,000 hours of work per annum are done may be of the order of £1·60 per hour, of which the labour cost amounts to £0·80 while fuel and lubricants account for £0·20 and depreciation and repairs for £0·50 per hour. The balance is made up of licence, insurance and sundries.

Choice of a Tractor. The choice of a tractor requires careful study. To be over-powered is wasteful, while to be under-powered may be disastrous. The most important work for the average British tractor is still ploughing and heavy cultivations, and where only one tractor is used it must be able to do this work in the limited time available. Where several tractors are needed it is usually best to have a range of sizes. The cheapest power is often provided by medium or large–medium tractors of the conventional 4-wheel type with rear-wheel-drive only ; but on many farms 4-wheel drive can be justified for the higher-powered tractors.

When a British farmer buys a new tractor there are some parts of the specification where little choice can be exercised, since statutory regulations apply. Such regulations control details of safety cabs, and of the maximum noise level perceived by the tractor driver. Another feature concerning the driver's welfare is a seat that helps to reduce harmful vibration. Control systems are important because they can strongly influence the driver's ability to make effective use of the tractor's power. The simplest hydraulic implement control systems can be ineffective with long implements.

An important need is a transmission that provides easy matching of speed of work to the power available. " High-Low " or more complex semi-automatic change-speed systems naturally cost more than a simple gear-box ; so the benefits have to be compared with the extra costs. In making comparisons, the effects on yields of getting work completed in good time must be estimated. Most tractor manufacturers aim to use many of the same basic components and also to follow similar layouts for control systems for the whole range of models. There are generally advantages of easier operator training and interchangeable spares in " standardising " on one make where practicable.

Economic Loading. One of the most important factors concerned with economic operation of a tractor is the choice of implements suited to the tractor's capacity. A tractor works most economically when provided with a load round about 75 per cent of its maximum power capacity. Light loads often lead to inefficient use of the driver's time, and always show a poor specific fuel economy. Where practicable, a low engine speed and high gear should be chosen. In some instances it is possible to achieve both speed and economy of operation by hitching together two or more implements in tandem. For example, where a tractor is capable of pulling a drill and a set of harrows together, it is worth while to devise a hitch so that both operations may be done in one journey across the field.

Working Speeds. The most economic operating speed depends partly on the job and partly on the wheel or track equipment. Tracks and wheels with spade lugs are not designed for fast work, and the maximum drawbar power is developed at speeds in the region of 4·8 km/h. Pneumatic-tyred tractors, which cannot exert such high pulls, need to go faster in order to develop their maximum drawbar power. Some rubber-tyred tractors develop their maximum drawbar power at speeds of about 9·6 km/h. Most modern tractors are equipped with one very high gear giving a road speed of about 32 km/h ; but this cannot often be used for field work.

Optimum speed is, of course, much influenced by the implement. With ploughs, where draught increases with speed, there is a practical limit to speed at well below 10 m.p.h. (16 km/h) ; and above about half this speed plough bodies specially designed for high-speed work are required. Most cultivation implements, drills and manure distributors work best at fairly low speeds, while hoeing cannot be efficiently done at above about 4·8 km/h. A spring-tine cultivator works better at high speed, and this may be achieved economically by running in high gear with the engine throttled down. Modern mowers need a speed of 9·6 km/h while combine harvesters need a range about 1·6–6·4 km/h to suit different crop conditions. Hay windrowing is one of the few field operations calling for a fairly high speed. At the other end of the scale, such a machine as a transplanter requires speeds of as little as 0·4 km/h. Thus, an all-purpose tractor requires a wide range of gears, with several fairly close together at 3·2–9·6 km/h, and a governor that is effective over a wide range of engine speeds.

Tractor Maintenance. Tractors and their equipment are now of such prime importance on British farms that systematic attention to care and maintenance is essential. All tractor manufacturers provide a handbook which gives detailed instructions on such matters as adjustments, lubrication and minor running repairs. These books are of interest to intelligent drivers, and full use should be made of the information contained in them.

Adequate maintenance of tractors and implements is facilitated if a fuel and spares trailer is provided, so that everything required for everyday use is ready to hand, even when the tractor is working a long way from home.

At home there should be a tractor shed, and a workshop where the tools necessary for running a farm can always be found. Manufacturers now tend to make machines so that they can be repaired by replacement of worn or broken parts; so effective maintenance requires care rather than great skill. Simple workshop equipment makes the tasks easier.

ELECTRICITY ON THE FARM

The use of electricity on farms in Britain has steadily increased in recent years, and with much attention having been devoted to electrification of rural areas it may be expected that progress will be much more rapid in the future. In addition to farms provided with a public electricity service, there are a few which obtain electricity for lighting by means of small generators driven by an internal combustion engine. The modern Diesel-electric generator is an efficient unit, and is worth consideration on isolated farms where there is no possibility of a public service.

The supply of electricity to farmyard, farmhouse and cottage is of great importance to the well-being of country people generally, supply to the houses [being as beneficial as that to the farm buildings. Much daily drudgery is removed when lighting, heating, water-pumping, washing and ironing can be done by electricity. In short, electricity is probably more important than anything else in raising the standard of living of country people, and so encouraging men and women to live and work on the land.

The main applications of electricity on farms are concerned with quite orthodox lighting, heating and power devices. The advantages of electric lighting in farm buildings, especially where dairying is carried on, are immense. Heating applications include sterilisers for dairy utensils, where the cleanliness and saving of labour often more than offset any extra cost; brooder heaters, where ease of control and safety are important factors; grain drying, where ease of control and efficiency of utilisation of the heat to some extent offset extra fuel costs; and incubators, where thermostatic control leads to fuel economy, as well as ease of regulation. The special advantages of electrical methods often outweigh considerations based on the relative cost of heat from electrical and alternative sources, and it is likely that heating applications of electricity on farms will extend considerably.

The electric motor has many advantages over other stationary power units for small or moderately small power requirements, and its use on farms is steadily increasing. The standard industrial "protected" types are quite suitable for most farm purposes. The larger motors require special devices to reduce the starting current, and occasionally the more expensive slip-ring type has to be used for large power applications. Typical tasks, with the size of motor used, are as follows : milking (1–5 h.p.) ; cooling ($\frac{3}{4}$–5 h.p.) ; grinding (3–50 h.p.) ; crop drying (3–50 h.p.) ; and water pumping (1–3 h.p.). For large power applications such as drying and driving a large hammer mill the farm tractor may be just the right size, and where demands for such work do not interfere unduly with essential field work it may be better to use the tractor than to install a large electric motor. A characteristic of electric power of which good use can be made is the ease with which automatic control can be arranged. For example, when a pressure tank is used in a water supply, pressure control may be employed to switch the motor on and off; and the motor used in a farm grinding installation may be switched off after a certain time, or when the flow of grain to the mill ceases. Other advantages are the ease of starting, reliability, and very small amount of time required for maintenance.

While the major uses of electricity on farms are concerned with straightforward heating, lighting or power applications, there are a few special uses where characteristics peculiar to an electrical device are employed. Examples of such uses are the electric fence, by means of which a sharp but harmless electric shock is used for fencing against livestock ; the electric insect trap, which destroys flies alighting on a wire mesh window screen by electrocuting them ; and the treatment of livestock by ultra-violet and infra-red radiation. Such special uses of electricity will be multiplied now that electric power is widely used ; and use of automatic controls is already commonplace.

Electric Fences. Most British electric fence units are battery-operated and are powered by a 6-volt accumulator of the type used in cars. A battery of 20 amp.-hours capacity will electrify up to 24 km of fence and will last for about 20 days of continuous use. Mains-operated fences are widely used in some countries, e.g. Denmark, and are becoming popular here ; they are safe and efficient if properly designed and tested control units are employed.

Chief points to observe in operating electric fences are :

One wire at 600–750 mm is satisfactory for cattle and horses ; sheep and pigs need two wires at 300–450 mm. Proper insulators must be used, and the fence wire must be kept clear of the crop and of trees, etc. The unit must have an effective "earth" connection. The stock must be trained to respect the fence, since it is fear rather than the wire which keeps them back. When the fence is first erected the animals must be allowed to approach it quietly. A good way of tempting them to try its effect is to place some food just beyond it.

PLATE III

A. Medium-powered tractor and 4-furrow plough fitted with discs, skimmers, and overload-release general-purpose bodies

B. 4-wheel-drive tractor with articulated chassis operating 4-furrow reversible deep digging plough fitted with "Diamond" bodies and large skimmers

C. Tracklayer and 4-furrow semi-mounted reversible plough fitted with semi-digger bodies

PLATE IV

A. General-purpose plough bodies with discs and skimmers set for work on grassland

B. 6-furrow semi-mounted plough with automatic depth regulation at both front and rear. (Load monitor system.) Work of the semi-digger bodies is assisted by skimmers

C. "Chisel plough" fitted with rigid tines designed for deep work in hard unbroken soils

A

B

C

PLATE V

A. "Chisel plough" fitted with C-shaped spring-steel tines

B. Spring-tined cultivator-harrow with depth wheels and spring-tine harrow

C. Trailed tandem disc harrow with depth/transport wheels and scalloped discs

PLATE VI

A. Rotary cultivator fitted with spike rotor and crumbler roller, making a coarse seed-bed

B. Reciprocating harrow used with " giraffe-neck " coupling and drill. Note cage-wheel extensions

C. Rotary cultivator combined with seed drill and crumbler roller for " once-over " reduced cultivation system

PLATE VII

A. Combined seed and fertiliser drill with large hoppers and transport
    wheels, and fitted with disc coulters
B. 6-row single-seed unit drill with belt feed
C. 2-row automatic high-speed potato planter with fertiliser and in-
    secticide placement attachments, and mouldboard-type covering
    bodies

PLATE VIII

A. 4-row transplanter planting cauliflower

B. "Triple-disc" drill being used for direct drilling of cereal crop into burnt stubble

C. Direct drill for row crops, especially kale and maize

PLATE IX

A. Tractor-driven centrifugal pump, above-ground slurry store and large-capacity slurry distribution tanker

B. Hydraulic power loader with double-acting auxiliary ram to roll back and tip the manure fork. Trailer type spreader is also shown

C. Side-delivery flail type manure spreader

PLATE X

A. Spinning-disc type mounted fertiliser distributor
B. Trailed fertiliser distributor with oscillating-spout distribution mechanism
C. Liquid fertiliser distribution by 2·5 tonne trailed sprayer

Chapter V

TILLAGE AND TILLAGE IMPLEMENTS

THE OBJECTS AND PRINCIPLES OF TILLAGE

TILLAGE is the practice of working the soil with implements in order to provide conditions favourable to the growth of crops. It is based partly on a knowledge of soil science, but possibly to a larger extent on an acquired skill which comes by practice and experience. It has been truly said that the cultivation of the soil remains today more an art than a science.

Not only do different soils vary greatly in their reactions to implements ; but an individual soil varies from day to day according to such factors as its moisture content (which is easily recorded) and the exact sequence of moisture changes in individual clods (which is almost impossible to record accurately). While it is not possible in the present state of knowledge to lay down any hard-and-fast rules about tillage, there are certain fundamental principles which every student of farming must understand. In general, " light " or sandy soils are easy to work, and can be tilled at almost any time of the year, while " heavy " or clay soils are more difficult to manipulate and can be satisfactorily tilled only at certain seasons, when the moisture content and weather are suitable. The effects of natural agencies on soils is a matter of great importance to the farmer, since it is only by working with these natural agencies that the most economic methods of working can be achieved. For example, frost is an agency of vast consequence in the farming of heavy land. When such land is ploughed early in winter it often has a texture similar to that of plasticine. When it is well frozen, however, the water in it expands and splits the soil into planes of partition which later assist in the formation of soil " crumbs " or tilth. As the soil dries out in spring the clods break down, and if worked by implements at the right time a very fine tilth can be secured on land which a few months earlier seemed to defy any attempt to secure a seed-bed. If, however, the farmer is in too great a hurry and a soil which has been frozen and is lying well is cultivated too soon, i.e. when still too wet, all the advantage gained by weathering can rapidly be lost. Even very heavy rain coming after frost may sometimes so " puddle " the land as to destroy the crumb structure and leave in its place a nasty sticky clay which is very difficult to handle. There are, of course, all gradations of soils between the heaviest clays and extremely light soils ; and some which behave as light soils for most of the year have just sufficient fine particles to render their cultivation a tricky job.

The other great ameliorative weathering agency—even more important than frost—is alternate wetting and drying of the land. Tillage can make use of this effect in many ways. A furrow or clod of clay soil, when thoroughly dried, contracts considerably and becomes hard, almost like concrete. When re-wetting takes place expansion occurs, and the outer layers of soil become split off into small particles. The process is repeated when re-drying and re-wetting occur, so that in time a tilth, or layer of fine friable soil, is formed.

These two examples should make it clear that the farmer, especially he who decides to farm on heavy land, cannot possibly use his tillage implements to best advantage without a knowledge of what is likely to happen to the soil in changing weather conditions.

The main objects of tillage may be conveniently discussed under the following headings :

1. The production of a suitable " tilth " or soil structure ;
2. The control of soil moisture, aeration and temperature ;
3. The destruction of weeds ;
4. The destruction or control of soil pests ;
5. Burying or clearing rubbish, and incorporation of manures in the soil.

1. The Production of Tilth. When soils are left uncultivated for long periods, the soil particles become stuck together, the air spaces become small, and the friable structure of the soil is lost. In such conditions it is impossible to sow seeds satisfactorily, and tillage implements must first be used to restore a " tilth " favourable to the crop. The tilth required and the implements used vary according to the type of crop, some crops needing a fine, firm seed-bed, while others do best where there are clods, and only sufficient fine mould to ensure covering the seeds.

In the preparation of seed-beds, ploughing is usually, but not always, the first tillage operation. The plough cuts the soil into rectangular slices or furrows and turns them over so that the rubbish is buried and the soil surface is exposed in a rough condition to the weather. There may then be a considerable interval before the next operation, but later on the land is generally cultivated by one or more of the tined implements, by disc harrows, by rolls, or other tackle described later. The farmer must choose those implements which will give him the desired tilth with a minimum of effort. Very often it will be necessary to try out various implements in the field before a satisfactory effect is produced, for it is not always possible to forecast in advance which of the cultivation implements will work best.

2. The Control of Soil Moisture, Aeration and Temperature. Tillage is concerned in many respects with adjustment of soil moisture content to the needs of the crop. For example, the land may be deeply cultivated and left rough at the surface in autumn, so that winter rains may easily penetrate, and the surplus pass down to the drains. On some soils, if the surface is left too fine and smooth in autumn, the winter rains will cause the soil to " run together " on the surface and prevent the percolation of water. When it is necessary to produce an early seed-bed on a soil that lies wet in spring, a rough surface must first be produced, generally by the plough, in order to induce a partial drying.

On the other hand, at seeding time in spring it is often necessary to produce at the surface a finely divided crumb structure which will act like a sponge and keep any rain that falls in the upper layers, so that the seed will germinate. Consolidation of the land sometimes has important effects on movement of water through the capillary spaces in the soil. For example, if the soil is very moist below and has a dry " fluffy " layer on top, rolling will cause moisture from below to pass into the surface layer.

Aeration of the soil, like many other effects of tillage, is still not perfectly understood. It is well known that oxygen is required for the germination of seeds, the respiration of plant root systems and the activity of beneficial aerobic bacteria in the soil. The few scientific experiments recorded, however, suggest that it is only rarely, as when the soil is waterlogged, that lack of aeration as such becomes a limiting factor in plant growth in normal farming conditions.

Soil temperature is closely bound up with soil moisture. Soils which are wet in spring are cold because water absorbs much heat in warming up, while large amounts of heat are absorbed in the evaporation of moisture from the soil.

3. The Destruction of Weeds. Weed destruction is one of the most important of all the objects of tillage, and the methods employed vary widely according to the type and condition of both soil and weeds. Annual weeds can be killed by completely burying them, or by dragging them out and leaving the roots exposed in dry weather. Most weeds are quite easily destroyed in the seedling stage, and one further object of tillage is to create soil conditions which will make weed seeds germinate, so that the seedlings may be destroyed by subsequent cultivations before the crop is sown. On some light soils where annual weeds are troublesome, farmers make a regular habit of killing successive crops of weed seedlings before sowing their roots in spring. One of the objects of autumn cleaning, referred to again later, is to germinate the seeds of such weeds as speedwells, chickweed, wild oats, black-grass, poppy, etc. There are, unfortunately, a few weed seeds, such as those of corn buttercup, fat hen and knot-grass, which do not readily germinate till late in the year.

Some perennial weeds, such as couch grass, occasionally need to be combated by special methods, as described in the section dealing with combined tillage operations for cleaning (see p. 137).

Many experiments on the inter-cultivation of root crops show that the benefits of tillage between the rows are almost entirely due to weed control. This has been proved by tests in which weed control has been achieved by cutting at ground level, without stirring the soil in any way. Other experiments have shown that additional hoeings beyond the amount required to control weeds may result in lower crop yields, because of damage to roots of the crop close to the surface.

Other aspects of weed control are dealt with in Chapter VIII.

4. The Control of Soil Pests. While tillages sometimes react directly on soil pests by exposing them to the attacks of birds, by drying them out, or by crushing them, they are more generally employed to resist pest attacks indirectly, by improving conditions for plant growth. For example, when wheat growing on a " fluffy " soil is attacked by wireworms or by the larvæ of the wheat bulb fly, firming the soil by the use of a roll may often be effective in minimising the damage. The improvement produced in such circumstances is due more to the better conditions provided for the plant roots than to any direct effect on the insects themselves, though thorough consolidation may to some extent restrict the insects' activity.

5. **Burying Rubbish and Incorporating Manures in the Soil.** Crop remains and farmyard manure must be buried in order that the organic matter may be incorporated in the soil, and that the surface of the land may be cleared of rubbish which would interfere with the action of drills and hoes. The plough is normally used for burying rubbish, and one of the chief requirements of good ploughing for British conditions is that all vegetation should be effectively covered by soil. It may be observed, however, that there are important " dry-farming " areas in other parts of the world, notably in North America, where a primary aim in cultivation is to prevent wind erosion by leaving as much rubbish as possible anchored in the soil but sticking out from the surface. In the " dry-farming " areas of the United States and Canada the plough which buries rubbish completely has been found worse than useless, and disc cultivators, sub-surface cultivators, and other special implements are now used in those areas. It has been argued from this, though without much justification, that ploughs which bury vegetation completely are unnecessary. Well-planned use of chemical herbicides has, nevertheless, made it practicable on many British farms to substitute tined implements for ploughs, or even to do without any cultivation. (See " Direct Drilling ".)

Cultivators and harrows are often used for mixing fertilisers with the soil before crops are sown. In other instances drills and planting machines are so designed as to distribute the fertiliser in bands alongside the rows of the crop. Fertiliser placement is further discussed in connection with manure distributors and combine drills.

PLOUGHS AND PLOUGHING

The plough separates a layer of soil from the underlying subsoil and inverts it, so that any vegetation or manure present on the surface is buried and soil from below is exposed to the action of weathering agents and to other implements. The ploughed land is laid up in parallel furrow slices, the type of slice depending on the shape of the plough body and on the nature of the soil. Modern ploughs are the result of a process of gradual development extending over thousands of years.

The Plough. Ploughs may be broadly divided into two main types, viz. the lea plough (Fig. 21) and the digger plough (Fig. 23). The chief difference between the two types is that the lea plough turns long, unbroken furrow slices while the digger has a tendency to pulverise the soil.

The *coulter*, which makes the vertical cut, is fastened to the plough beam. The other soil-working parts, viz. the share, which makes the horizontal cut; the *mould-board*, which turns the furrow; the landside, which slides against the furrow wall; and the slade or sole, which slides along the furrow bottom, are all attached to the plough frame which is a casting or pressing of complicated shape fastened beneath the beam.

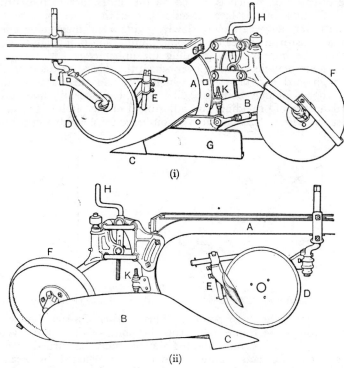

FIG. 21.—THE PRINCIPAL PARTS OF A TRACTOR PLOUGH BODY (LEA TYPE). (i) LAND SIDE. (ii) FURROW SIDE.

A, beam. B, mould-board. C, share. D, disc coulter. E, skim coulter. F, rear wheel G, landside. H, rear wheel adjustment. K, pitch adjustment. L, coulter tilting adjustment.

Coulters are of three main types, viz. knife, disc and skim. The knife coulter was the normal type on horse ploughs, but on tractor ploughs the disc type, which can work in difficult conditions without blocking, is used on all except a few special implements designed for very deep work. The skim coulter is used to assist in the complete burying of manure, rubbish or crop remains. It acts as a miniature digger plough, and is fastened to the plough beam just behind the disc coulter or just in front of the knife coulter. On rare occasions disc, skimmer and knife coulters may be used simultaneously. The combined disc and skimmer has become a very popular fitting on tractor ploughs.

The *share* (Fig. 22) fits over the front or " nose " of the frame.

On the lea plough it has a definite " neck " which has the effect of placing the cutting edge well ahead of the mould-board. The " lead " thus obtained assists in keeping the furrow slice unbroken, enabling it to pass gently from share to mould-board. The digger plough, on the other hand, has the cutting edge of the share close to the mould-board and often almost at right angles to the line of draught. Most shares are of steel, but many used on smaller tractor ploughs are made of cast iron which is chilled on the under face of the cutting edge. Chilling of this face makes it harder than the top ; and in use the lower side wears more slowly, giving a self-sharpening effect. If chilled cast-iron shares do become blunt when only slightly worn they may be improved by grinding the upper edge.

On some ploughs the nose over which the share-box fits is carried at the forward end of a lever which pivots in the frame casting. This type of plough body is called the " lever-neck ", and it has the advantage of providing easy adjustment of the pitch of the share. In districts where rocks are often encountered in ploughing, a popular

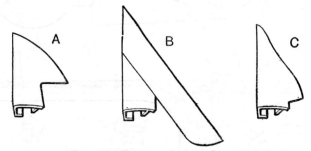

FIG. 22.—FORMS OF PLOUGH SHARES.
A, normal type. B, for skimming stubbles. C, pointed share for stony land.

type of share is the " bar-point ". This consists of a steel bar which passes through the frame and forms the point of the share. When the bar-point has become blunted the bar may be turned over or reversed end for end. Later, it may be sharpened by heating, hammering out to a chisel point, and hardening.

The mould-board is attached to the frame by bolts with counter-sunk heads. The lea type, which may be up to about 1·1 m long, is convex from top to bottom throughout its length with a gradual twist which turns the furrow slice very slowly and finally presses it against the previous one. The digger type of mould-board is much shorter and has an abrupt concave curvature which bends the furrow slice sharply on itself and throws it aside. There are many intermediate general-purpose bodies between the extreme lea and digger types. Digger and semi-digger bodies are becoming increasingly popular for general work. They have the advantage that deep work can easily be done when required, whereas with lea ploughs depth of work is very limited. Even where land is being set up for the winter it is generally found that the rough clods left by a digger body weather as easily as the unbroken furrow slices left by a lea type.

The *landside* takes the side thrust caused by the pressure of the furrow slice on the mould-board. It is often in one piece with the slade, which supports the weight of the plough. On some types of wheeled tractor ploughs, where these thrusts can be taken on suitably disposed wheels or the hydraulic lift linkage, the landside and slade can be almost or entirely dispensed with.

Plough Types. Early tractor ploughs were almost invariably trailed implements with three wheels and a number of parallel beams, each carrying a body of the general type already described. The two large wheels, i.e. the furrow wheel and the land wheel, were carried on cranked axles ; and the depth of work was regulated by screws or levers which controlled the setting of the cranks. The plough was lifted out of work by a mechanical lift, usually operated from the land wheel. Modern ploughs are almost invariably attached to the tractor's 3-point linkage, and both working depth and lifting depend on the tractor's hydraulic implement control system. The plough frame now often consists of a single diagonal beam to which all bodies are attached. The number of bodies can often be easily varied from 2 to 6.

Where practicable, ploughs are fully mounted, as this facilitates short turns. Long ploughs that are fully mounted, especially heavy reversible types, require very high lift forces and can create difficulty with tractor balance and steering. There is therefore a tendency for very long ploughs to be semi-mounted. With long semi-mounted ploughs, however, depth regulation by means of simple top-link draught control systems can be unsatisfactory. One solution is to provide at the rear of the plough a depth-regulating hydraulic cylinder that works in unison with draught control through the 3-point linkage. Sensing for load control with high-powered tractors is often more effective with systems working through the lower links, or by one that senses torque in the tractor transmission, e.g. " Load Monitor " sensing. One reason for wide spacing between adjacent plough bodies is to avoid the blockage caused by surface trash if parts of the furrow being turned come in contact with the body ahead. An advantage of the " Diamond " plough is the absence of disc coulters, a design that enables large skimmers to put surface rubbish easily into the bottom of the preceding furrow, and a furrow shape suited to turning in a very short distance. This facilitates design of a plough that is short in relation to its working width, and makes it practicable to apply the fully mounted principle to large, deep-working reversible ploughs.

Reversible ploughs are used in Britain mainly for deep work in preparation for root and vegetable crops, where their ability to leave a level top leads to easy seed-bed preparation. Modern implements are usually of a single-beam type, the bodies being mounted on opposite sides of a beam that is rotated through 180°.

Some deep and highly productive soils such as fen silts are occasionally ploughed about 400 mm deep and subsoiled a further 150 mm. Experiments have shown, however, that such deep ploughing cannot be justified as a regular practice. About 300 mm can be considered deep enough for normal deep-cultivation purposes. Very deep ploughing requires use of exceptionally powerful tractors and expensive equipment.

Interest in " reduced cultivation " systems has led to the development of mould-board ploughs specially designed for shallow work at high forward speed, and in particular for the first operation on cereal stubbles.   Such ploughs, being limited to work at only about 100 mm deep, can be lightly built ; so it is practicable to have fully mounted implements with 8–10 furrows on high-powered tractors.   In suitable conditions such ploughs work very fast and can bury rubbish effectively.

Subsoil Ploughs.   On certain types of land it is desirable occasionally to stir the soil very deeply—below normal ploughing depth—in preparing for root crops.   It is particularly necessary to do this where repeated ploughing at the same depth has caused the formation of a " plough pan ", or where precipitation of iron salts in the soil has formed a rock-like " iron pan ".   In these circumstances there is little doubt that subsoiling once every few years leads to better crops and is well worth the extra cost.

The methods used for breaking up the subsoil vary, but the general aim is to stir the lower layers without bringing unweathered subsoil

[*Ransomes*]
FIG. 23.—DEEP-DIGGING PLOUGH WITH SUBSOILING TINES.

to the surface.   Subsoiling, even with a single tine, requires a high drawbar pull if carried out in the hard, dry soil conditions in which it is usually most effective.   Subsoiling is sometimes carried out by tines fitted behind the bodies of digger ploughs in such a way that they burst up the bottom of each furrow to a depth of about 150 mm below ploughing depth (Fig. 23).   This is an economic way of subsoiling, but the higher pull needed slows down the rate of ploughing.   A more usual method is to carry out subsoiling or " bursting " as a separate operation, usually on stubble, when both surface and subsoil conditions tend to be suitable.   The operation is well suited to contract work by a powerful 4-wheel-drive or tracklaying outfit, fitted with a toolbar carrying two or more deep-working tines.

Subsoilers for mounting directly on the 3-link hitch of medium-powered tractors are also available, and are capable of working effectively down to about 380 mm deep.

Ridging Ploughs. Ridging ploughs have a narrow share and double mould-boards, and are used to form ridges in preparing the land for planting potatoes and, in some districts, roots. The ridger is used after the land has been worked to a deep tilth. The width between ridges is generally 600–900 mm, but is varied to suit the soil and the crop. Ridgers are much less used than formerly because potatoes are now almost invariably planted by a machine which works on level land and incorporates the necessary bodies for both opening the furrows and raising a ridge to cover. Ridgers are still required for earthing up ; and the main trend is towards short bodies that throw the soil lightly over the ridges and avoid pressing, which may result in clod formation in damp soil conditions.

THE SETTING AND OPERATION OF PLOUGHS

All types of plough are similar in essential components, and all must be correctly set for the performance of a satisfactory standard of work. Bad setting can cause an unnecessarily high drawbar pull and a low rate of work. It also inevitably results in uneven work, which makes subsequent cultivations more difficult, and often adversely affects crop yields.

Adjusting the Share. The plough body should run level and smoothly, and correct adjustment of the share plays a large part in securing this result. The point of the share is given, in manufacture, a slight pitch in two directions—downwards and towards the landside. Faults in the fitting or setting of the share can sometimes be masked a little, but can never be completely remedied by adjustments elsewhere. On many modern ploughs the pitch of the whole plough body in relation to the beam can be adjusted. There is usually no special provision for adjusting the pitch of the share, except in the leverneck type of body ; but care must be taken to avoid upsetting the correct pitch when, for example, steel shares are sharpened.

Adjusting the Coulters. The knife coulter is set with its point sloping slightly forward, the point itself being a little above the point of the share, and about 13 mm towards the unploughed land. As with all cutting tools, there must be a clearance behind the cutting edge of the coulter so that the side which is towards the unploughed land does not rub hard against the furrow wall. When the disc coulter is used to make the vertical cut, as on most tractor ploughs, it should be set for average conditions about 13 mm to the landside of the share point, and with its centre just above the share point when viewed from the side. Bad coulter setting is one of the commonest faults of tractor ploughing. Owing to the method of adjustment by a cranked stalk, discs always tend to run too narrow, and settings should be checked occasionally to see that they have not slipped. Too narrow setting of discs leads to increased draught and wear and produces work of bad quality.

Modern disc coulters for tractor ploughs have a fitting which allows them to be tilted. A good effect may often be produced in rough conditions, where a skimmer will not work, by setting the disc over two to three notches so that the lower edge undercuts the unploughed land, thereby producing a crested furrow.

On some types of wet, heavy land, where any form of coulter chokes, "wing" shares which have a vertical fin attached to the share sometimes work well.

Skim coulters are of two main types, viz. that with a straight shank which is attached directly to the plough beam, and that which is attached to the arm of the disc coulter and swings with the disc. In either case the skimmer must be set close to the disc to avoid blocking. In bad conditions the independent skimmer sometimes works best. The skimmer is generally set to cut a small "neck" of soil from the top left corner of the furrow slice, so that when the furrow is turned, weeds or turf are prevented from growing up through the cracks between adjacent furrows. When ploughing turf skimmers should generally be set well forward, so that the cut is made before the top of the furrow is disturbed by the share and mould-board. With digger ploughs it is often best to use a large skimmer that pares off a high proportion of the top of the furrow slice. Such equipment is suitable for burying arable crop remains or a dressing of farmyard manure.

Adjusting the Mould-board. It is not often necessary to interfere with the setting of the mould-board. Where the furrow slices will not turn over properly there may be a temptation to try to complete the turning by setting out the tail of the mould-board ; but this generally makes matters worse and causes increased wear and draught. The most common cause of furrows failing to turn is too narrow a furrow width in relation to the depth. As a rule, a furrow slice which is not pulverised in ploughing must have the width about one-and-a-half times the depth if it is necessary, as it generally is, to turn the furrow well over. The width of furrows on most tractor ploughs is adjustable within small limits, though the adjustment is sometimes not an easy one.

Adjusting the Hitch and Wheels (Trailed Ploughs). With plough body and coulters correctly fitted and set there remains the adjustment of the hitch and wheels. It should be emphasised that though the depth levers or screws provide for final setting, the required depth and width of work are achieved primarily by adjustments of the hitch. The depth of work is increased by raising the hitch on the front of the plough, and the width of cut is increased by moving the hitch point away from the unploughed land. The wheels are used ta steady the plough in the required position.

With ploughs that are attached to the tractor's 3-point-linkage there are only limited possibilities of moving the implement sideways in relation to the tractor centre-line ; and because much ploughing is done with one pair of tractor wheels running in the furrow, a vital adjustment that needs to be about right before getting down to plough-setting details is the tractor wheel track width. This must be adjusted if necessary to ensure that when the plough runs straight the front furrow width is approximately correct.

One of the advantages that 3-point-linkage systems provide is a means of securing rapid penetration of implements to full working depth. This is achieved by designing the linkage geometry so that the pitch angle of the implement is steep as it enters the soil and is reduced as working depth is reached. When it has reached working depth a plough should run level. If it runs nose down and leaves a gouged furrow bottom, drawbar pull is excessive. Some tractors provide a choice of systems of implement control. For most small ploughs straightforward draft control is satisfactory. The soil forces acting through the linkage provide a " weight transfer " effect that helps to ensure good wheel grip.

Setting of Mounted Ploughs. The principles of setting tractor-mounted ploughs without depth-regulating wheels are easily understood. Pitch and sideways movement of the plough are controlled entirely through the linkage which connects the plough to the tractor. The pitch of a mounted plough is readily varied by altering the length of the top link. Sideways movement of the mounted plough is a little more difficult to understand. The usual method on ploughs connected to the tractor by a 3-point linkage is to employ a mechanism which in effect makes one of the lower connecting links longer than the other. This has a tendency to make the plough run at an angle to the direction of travel ; but the soil forces immediately tend to make the plough run straight ahead again, and the net result is that as the outfit moves forward after the adjustment has been made, the implement swings to one side and then runs more or less straight ahead in its new position. A common method of achieving this result is to mount the plough's hitch pins on the ends of a cranked cross-shaft which can be rotated by means of a screw mechanism.

Mounted ploughs without a depth regulating wheel depend on the tractor's automatic draught control mechanism to ensure that the plough runs at the chosen depth. In normal soil conditions in Britain such mechanisms can give entirely satisfactory results, and this method has the merits of lightness and simplicity of the implements, coupled with very easy alteration of working depth through the medium of the hydraulic control lever.

Some plough-setting faults caused by incorrect hitching, and the remedies, may be summarised as follows :

Fault.	Remedy.	
	Trailed ploughs.	*Mounted ploughs.*
Plough will not penetrate.	Raise hitch at hake.	Shorten top link. Set top link and draught links in lower position on tractor.
Plough will not run shallow.	Lower hitch at hake. Fit extension rim to plough if necessary.	Lengthen top link. Set top link and draught links in upper position on tractor.
Uneven depth of ploughing.	Re-set hitch at hake.	Re-set top link.
Front furrow too wide or too narrow.	Re-set hitch on tractor.	Check setting of tractor wheels Adjust cross shaft on plough.

METHODS OF PLOUGHING

Most ploughs in use today turn the furrow slices to the right, and with such ploughs it is impossible to plough a field by starting along one side and working to and fro until the other side is reached. The method commonly adopted for ploughing in "lands" is briefly as follows : The headland is first marked off by turning a shallow furrow about 7·3 m from the boundary of the field. Ridges are then drawn parallel to one another across the field, the distance apart being normally about 20 m but being varied according to local requirements. The distance between the headland furrow and the first ridge is made one-quarter of the normal distance beween ridges.

There are various methods of setting up a ridge, the quickest being to turn a shallow furrow and then, turning right at the end, to turn another furrow towards it so that the two just meet, leaving an unploughed strip beneath. If this job is really well done and the land is clean this method is satisfactory for many crops, but a more thorough method is to split the ridges before setting up the " top ". To do this, a left turn is made after drawing the first shallow furrow, and the second furrow is thrown away from the first. On the third trip across the field the second furrow and the undisturbed soil beneath it are turned together towards the first furrow ; and on the fourth trip the ridge is completed by turning the first furrow and the soil beneath it so that it just touches the other pair of furrows.

Ploughing then proceeds by working clockwise round and round the first ridge until a strip 10 m wide has been ploughed. The same process (called " gathering ") is repeated round the second ridge, and then the 10 m stretch left between the strips is worked out by turning anti-clockwise, this process being termed " splitting " or " casting ".

When the last furrow between the lands is finished, a wide open furrow is left. It is usual to leave the field thus, but sometimes a deeper drainage or " mould " furrow is drawn in the bottom, while in other conditions the land may be levelled a little by ploughing back the last two furrow slices. When all the lands are finished the headlands are finally ploughed by working round and round the field.

With high-powered tractors similar methods may be followed, but headlands are generally made wider. The ridges are set farther apart except where drainage considerations rule otherwise, and short turns are avoided by working two ridges at once.

Ploughing round and round the field is quite common in some areas, especially where there are large fields, as on the Downs. There are various methods of dealing with the turns, the commonest being to have the corners rounded and to keep the plough at work all the way. A more thorough method is to mark out headlands along the diagonals, lift the plough at the turns, and plough these internal headlands last. Where a field has some gradual turns and some sharp corners it may be advisable to make a headland and lift the plough at the sharp corner, but to plough round the gentle curves.

A popular method for deep work is called " square ploughing ". The field must be fairly regular in shape, and a land is ploughed out in the middle of such a shape that the boundaries of the field are equidistant from the adjacent sides of the ploughed part. The field is then ploughed by working clockwise round this central part, the plough being raised at corners and an anti-clockwise loop made to bring the outfit round ready for dropping the plough into work along the next side. With careful attention to raising and lowering the plough, extremely good work can be done, all the land being ploughed and troublesome open furrows eliminated.

Essentials of Good Ploughing. It is now generally agreed that some of the fine points of style that used to decide ploughing matches are of little practical significance. The more important factors today may be listed as follows :—

1. All the land should be ploughed, including that under the " top " or ridge, so that subsequent cultivations may be carried out without trouble.
2. All rubbish and manure must be completely buried.
3. The plough furrows should be left in the condition desired. This may be rough, for easy weathering ; or well pulverised for rapid preparation of a seed or plant bed.
4. The depth of work should be uniform.
5. Unless drainage conditions require something different, the land should be left as level as possible, the ridges not being too high and the open furrows not too deep. Attention to this point greatly facilitates seed-bed preparation.

CULTIVATORS

Cultivators are implements which are commonly used for breaking clods and working soil to a tilth in the production of seed-beds. Other uses include the killing of weeds and mixing of fertilisers with the soil. Cultivators and harrows are tined implements, cultivators usually having wheels and being heavier than harrows, and used for deeper work. The terms cultivators and harrows are, however, loosely used, and some implements which are called harrows are equal in effect to cultivators.

Various types of tine may be fitted, according to the kind of work to be done. Rigid tines, spring tines and spring-mounted curved tines of sickle shape are commonly used, the tines being mounted in two or more rows with successive rows " staggered " to allow clods to pass through easily. Springs allow the tines to yield a little when they strike large clods or very tough land, and it is claimed that the vibrating motion assists in pulverising the soil and reducing the draught. This effect is most marked at high forward speeds, and modern implements must be strong enough to work at 8–10 km/h in normal conditions. Penetration is achieved mainly by virtue of the pitch of the shares, rather than by the weight of the implement. Various types of share can be fitted to the tines, some being narrow and designed for deep working, while others are broad and designed for such jobs as cutting all weeds just below ground level.

With any type of wheeled tractor there is usually every advantage in using the mounted type, rather than a trailed implement. The mounted implement is simple, cheap and sturdy, and it is only where heavy tracklayers are used that trailed implements are sometimes preferred.

Many types of cultivator/harrows designed for work at high forward speed employ a large number of spring tines.

The heaviest cultivators are often called " chisel ploughs ", and are suitable for bursting up unploughed land. On many farms mould-board ploughs are normally no longer used, their place being taken by tined implements, assisted by the action of herbicides as necessary. Such equipment is available in sizes designed to utilise all the power available in high-powered tracklayers or 4-wheel-drive tractors. Rigid-tined implements may be used with a few tines at low speed for deep bursting of hard land : this is beneficial if done in suit-ably dry conditions. A more common aim is to produce a tilth by work at high forward speed with closely spaced tines at shallow depth: for this purpose various types of strong spring or spring-mounted tines are suitable. The inclined points of tined implements often have fluctuating effect on penetration, and for this reason it is advisable to use depth wheels to assist draft control.

HARROWS

Harrows are generally used for shallow cultivations, in operations such as the preparation of seed-beds, covering seeds, destroying seedling weeds, and so on. The commonest form of harrow is called the *Zig-Zag* harrow. It has the tines staggered so that the rear ones do not follow the same tracks as the front. The lightest type has short, straight tines : several sections totalling 4·5–6 m wide are often attached to a tractor-mounted frame. It is called a *Seeds* harrow, because it is used for the final touches in preparation of seed-beds and for covering seeds after the drill, as well as for very light weed-control operations. When used for covering seeds, the harrows are now often hitched directly behind the drill.

The heaviest kind of zig-zag harrows, called the *Duck-foot* or *Drag* harrow, has curved tines and often wheels, and is capable of quite deep work. The action of the drag harrow is, in fact, very similar to that of the cultivator, the main difference being that it has more tines and does not, as a rule, work quite so deep. On difficult land it is often found that the drag harrow produces a better effect than the cultivator in the preparation of spring seed-beds, as it does not bring so many unweathered clods to the surface.

The *Spring-tined* harrow consists of a number of sickle-shaped spring tines mounted on frames 0·9–1·2 m wide. It can be adjusted to produce very variable results, the tines being set at various angles and depths by means of a lever. Its great advantage is that it can be set to do all kinds of work from that of a very light harrow, when the points of the tines are vertical and only scratching the surface, to that of a light cultivator, when work up to 150 mm deep can be done. As with most other types of harrow, the easiest method of use is to have several sections attached to a hydraulically lifted harrow frame.

Chain harrows have no rigid frame, being built up of a series of flexible links. As the name suggests, the original chain harrows consisted simply of a number of chain links joined together to form a rectangle. The main use of this type of harrow was rolling up weeds which had been pulled free from the soil by other implements. Spiked link harrows were soon found more generally useful, and most of the flexible harrows in use today have teeth of one sort or another. One common type has the teeth formed by an extension of the link. Another type, primarily for pasture work, has knives carried in cast-iron sockets which are flexibly linked together. Some spiked flexible harrows have on one side long tines, suitable for use on arable land, and shorter ones, suitable for use on grassland, on the other.

On uneven land flexible harrows follow the surface irregularities better than rigid types. On such work as harrowing pastures, semi-rigid harrows, such as the zig-zag, jump about too much to be of any use, while spiked flexible harrows work steadily if the surface is not too hard.

With the lighter harrows a full load for the tractor often necessitates an implement 6–9 m or more wide for the lightest types. It is often not economic or convenient, however, to buy such a wide harrow ; and it is frequently found better to do two operations, e.g. harrowing and rolling, with the implements arranged in tandem. The development of effective herbicides for couch and other grassy weeds of arable land has greatly reduced the need for harrows to be able to deposit rubbish that has been dragged out of the soil. An increasing need is for equipment that is easily transported from field to field, and can produce a tilth rapidly, without the need to run over the land several times. In this connection, ground-driven rotary harrows and power driven rotary and/or reciprocating harrows need to be considered.

An effective type of harrow for many purposes is a *rotary* type with blades on spools driven by contact with the soil, and arranged in gangs at an angle to the direction of forward travel, as in disc harrows. Such implements are self-cleaning and work well at high forward speed.

An even more effective way of ensuring that tractor power is effectively applied to cultivation is by use of power-driven *rotary reciprocating harrows*. By choice of a suitable tractor gear it is possible to achieve in one journey across the field a better tilth than would be possible by repeated use of a ground-driven implement.

Disc Harrows have a series of saucer-shaped discs mounted on two or more axles, which may be set at an angle to the line of draught. The discs are generally 450–600 mm in diameter, the larger sizes being capable of deeper work. The action of the discs depends largely on the angle at which they are set. If the gangs are set perpendicular to the line of draught, penetration is at a minimum. Penetration increases with the amount the discs are " angled ", until the position is reached at which the forward edges of the discs lie tangential to the direction of travel.

Disc harrows are particularly valuable for the preparation of seed-beds after old turf or leys, where the use of tined implements would bring turf to the surface. They are also often selected for preparation of spring seed-beds on all kinds of arable land—especially where it is undesirable to cultivate very deeply. An advantage of disc harrows is that when the soil is in suitable condition to be reduced to a tilth, the discs will do the work more rapidly than the common tined implements ; the fewer operations necessary result in a smaller loss of moisture—an important consideration in preparing spring seed-beds.

The draught of disc harrows is heavy, but a medium-sized general-purpose wheeled tractor can generally pull a medium-weight tandem disc harrow 2·4 m wide. The tandem disc harrow is, in fact, one of the few types of harrow that will provide such a tractor with an adequate load. Light tractor mounted discs are now widely used.

Like all tillage implements, disc harrows have some limitations and disadvantages, chief among these being the rather high first cost and a greater rate of depreciation than comparable tined implements. Road transport difficulties have been resolved by direct mounting of lighter implements and provision of hydraulically controlled combined depth-regulating/road transport wheels for heavier implements. Wheeled discs can make effective use of the weight-transfer advantages available with " pressure-control " hydraulic systems. They are also well suited to depth regulation by means of transmission torque-sensing (" Load Monitor ") hydraulic control systems.

ROLLS

Rolls are used to consolidate the soil, to crush clods, to smooth the surface of the soil, to prepare a surface for the drilling or broadcasting of seeds, and occasionally to break a surface crust. Consolidation is necessary in some soil conditions to give plants a firm roothold and to ensure continuity between subsoil and top-soil. This is particularly necessary in spring, after winter frosts have raised the soil surface. The crushing effects of rolls need no explanation, apart from emphasising the importance of choosing the right time to do the work. The effects of rolls, with regard to both consolidation and clod crushing, are largely bound up with the soil moisture content. Rolls cannot be used when the soil is wet ; and when it is dry the effects are often negligible. There is, however, an intermediate stage, when the clods have been wet and have started to dry, at which the optimum results from rolling can be achieved.

Rolls must, of course, be used with caution on clay soils. It is seldom advisable to roll such soils in autumn owing to the danger of getting too fine a tilth, so that the surface runs together in winter. In spring, rolling must be delayed until the soil is reasonably dry underneath as well as on top, especially in conditions where the soil is short of humus.

Spring rolling of autumn sown corn is a common practice, and experiments show that the small expense is well justified, quite apart from the important matter of levelling the ground for the combine at harvest. Such rolling sometimes produces markedly good effects in conditions where the soil is loose and wireworms or other soil pests are active.

Many types and sizes of rolls are used on farms, but two types predominate, viz. the *Flat* roll, which has a smooth surface, and the *Cambridge* or *Ring* roll which has a ribbed surface. The flat roll usually has two or three sections mounted loose on an axle in an implement about 2·5 m wide. Division of the cylinder into sections aims to avoid excessive scrubbing of the soil surface when the roll is turned. Diameters cover a wide range, with a majority about 450–750 mm ; average weight is 500–750 kg, but rolls weighing 3–4 tonnes are used in spring for pressing stones into grassland on which forage harvesters will be used.

The Cambridge roll consists of a number of rings, generally 75 mm wide, and having a rib about 37 mm high. The rings are mounted loose on an axle, so that turning is easy, and intermotion of the rings keeps the surface clean. This roll is generally heavier than the flat type, and is more effective as a clod crusher. It leaves a characteristic soil pattern in which narrow, well-compacted grooves about 37 mm deep and 12 mm wide are separated by small, rounded ridges which are only lightly compressed. This surface is a very suitable one for many needs. It can be left after drilling without fear that it will run together as badly as a flat-rolled surface often does ; and it is often convenient to broadcast small seeds, such as grasses and clovers, on the corrugated surface, covering them by a subsequent harrowing or flat-rolling.

While single swing-type implements are often used, the width of work is limited, and triple rolls are commonly used where a wide strip must be covered. Three rolls, each about 2·5 m wide, are provided with a hitch which permits them to be run in tandem for transport, and so that they just overlap when at work.

The Furrow Press is a very heavy type of roll with wedge-shaped press wheels which are designed to break down the lower parts of the furrow slices after ploughing. The implement is generally of a multi-ring type and is trailed behind the tractor plough. It is sometimes used on light land when ploughing out leys in preparation for wheat.

The Planker or Scrubber is a simple implement which is sometimes used for breaking down clods on land where it is impossible to use a roll. It is constructed of planks approximately 2 m × 250 mm, fixed together so that they overlap about 50–100 mm × 50 mm, the corner thus formed being shod by an iron bar. When the implement is pulled over the ground it scrapes the surface and rolls the clods over, creating a tilth and smoothing the top. It can occasionally be used when the soil is still wet below and use of a roll would be inadvisable.

The Float is an implement which is used in the preparation of very smooth seed-beds for root crops. It consists essentially of a framework carrying two or three boards or metal plates, which can be set at a slight angle so that they work on the soil rather like a plane. The float picks up soil from the high places and drops it in the hollows.

Some farmers drag a single plank just in front of the drill coulters, in order to fill up the ruts made by the tractor wheels.

ROTARY CULTIVATORS

Rotary cultivators of various types were invented and used at an early stage in the application of steam power to agriculture. None of the early steam-powered rotary cultivators, however, achieved any substantial success. The internal combustion engine is a more satisfactory power unit than steam for all types of field work, and in recent years several satisfactory types of rotary cultivator have been produced.

Machines now available include very small two-wheeled hoeing units with a cut of about 250 mm, sturdy two-wheeled units with engines of 5–8 h.p. designed to do all the basic cultivations in small gardens, and wide tractor-mounted or trailed outfits designed for operation from the power take-off of ordinary farm tractors. The engine drives a horizontal rotor situated at the rear of the machine, and the outfit is equipped either with " spike-tooth " or spring-steel chisel-shaped tines, or more usually with sturdy hoes which successively enter the ground as the machine travels forwards. The rotor is covered by a hood, and working depth ranges from 100 mm with the smallest machines to 250 mm with larger models.

The rotary hoe has been found effective for the preparation of seedbeds and for such work as stubble cultivation on many farms ; and now that units which can readily be fitted to any size of tractor have been developed, a steady increase in the use of the power-driven cultivating tools on farms as well as on market gardens, may be expected.

INTER-ROW HOES

When root crops became extensively grown during the nineteenth century horse hoes were widely used to assist in keeping the land clean. The primary object of hoeing is to kill weeds, and experiments show that hoeings additional to the amount required for weed control are seldom justified. Competition from weeds checks the growth of the crop right from the first week or so after germination, and it is necessary to start hoeing root crops at a very early stage. The surface mulch created by hoeing appears to be of some value in improving soil moisture conditions (by providing an insulating layer between the moist soil below and the atmosphere) ; but it has been shown that the effects on moisture conservation are not, as a rule, so great as used to be supposed.

It is important to be able, at an early stage in the growth of root crops, to hoe close to the crop plants without smothering them. The best type of hoe blade for such work has a vertical knife which is continuous with the L-blade. In some conditions, plain or concave discs running close to the plants opposite the L-blades permit very close work ; but the extra cost and complication of such equipment is a disadvantage.

TRACTOR TOOL BARS FOR INTER-CULTIVATIONS

The main features of row-crop tractors have already been described. The cultivating tools may be attached at the rear, between the wheels, or in front of the all-purpose tractor, and they are generally mounted on tool-bars in such a way that they can be raised or lowered by means of the hydraulic lift. For inter-cultivation of potatoes after the crop has been planted, rear-attached tools are generally most satisfactory. With root crops, on the other hand, when it is necessary to hoe close to the rows, the best position of the hoes for a one-man outfit is on a tool-bar lying ahead of the rear wheels. Individual hoes or groups of hoes working each row should float independently in order to work efficiently on uneven ground. Good work can be done by rear-attached hoes which are independently steered ; but this method is wasteful of manpower. Some light tractors which are primarily designed for hoeing have the power unit and transmission mounted over the rear axle, and have the hoes mounted on a frame in front of the driver in such a way that they can be seen clearly and adjusted with ease.

A disadvantage of most hoes is that blockages easily occur in moist soil, and that weeds easily grow again if rain falls soon afterwards. An effective inter-row hoe for such conditions, or for use where weed growth is heavy, is the inter-row rotary hoe. Hoeing problems can be reduced by the skilled use of herbicides in crops such as sugar beet and potatoes. There is advantage in using a wide hoe. For example, it is quite practicable on flat fields to hoe 12 rows of sugar beet at a time, where the same tool-bar has been used for drilling ; and the use of 4-row potato planters, scufflers and ridgers in place of the customary 3-row has advantages in addition to the extra speed of work. With 4-row equipment, potatoes do not have to be planted in tractor wheelings ; and later on, none of the rows has tractor wheelings on both sides.

COMBINED TILLAGE OPERATIONS FOR CLEANING LAND

The Bare Fallow. Bare fallow is the term given to land which is left uncropped throughout a whole year, during which time it is cultivated with the object of killing weeds. The system was formerly more generally adopted in England than it is at present and it formed an essential part of the three-field rotation under the Manorial system. It is also part of the normal system in some dry-farming areas of the world, notably in some parts of North America and Australia, where the collection and storage of rainfall in the soil during the fallow year is one of the main functions. In Britain today the bare fallow has been largely discarded in favour of shorter spells of cleaning. Improvements in tillage implements and increased use of tractor power have made it possible to keep the land clean without resort to an expensive and entirely unproductive year. There are, however, still a few farmers on heavy land who employ the system, and a brief outline of the methods employed in fallowing such land is given below.

The operation normally follows a corn crop, and the first point to note is that the land must be left undisturbed all through the winter and early spring. The object is to kill the weeds (e.g. couch) " in the clod " by desiccating them, and the first operation is to plough after all risk of severe frost has passed. The first ploughing should be done to full depth, using a lea type plough, preferably when the soil is wet. When dry weather follows, the furrows dry out hard and do not crumble. When the buried vegetation has been well smothered and the furrows are fairly dry they are ploughed back again in order to separate them completely from the subsoil and to dry out the other side. The object at this stage is to keep the soil in large clods and avoid the formation of tilth.

The next operation depends on the equipment available. The field may either be cross ploughed, or cross-cultivated with an implement having very few tines, the object being to break the furrows into very large clods. After this the clods should be stirred and turned over when the weather is hot and dry. If the field is kept on the move during a dry spell of 10 days or so the fallow will have been " made " and the rest of the job is easy. A deep bursting of the subsoil, carried out in dry conditions, will often have long-lasting beneficial effects on drainage.

Towards harvest time or earlier the fallows begin to crumble to a tilth, and then the weed seeds begin to germinate. These may be killed by a final ploughing, and this also serves to cut thistles, which are sometimes still quite strong after a fallow.

The efficient farmer today generally plans his cropping and arranges his tillages so as to avoid any necessity for a bare fallow. If the land becomes very foul it means that either the farming system is at fault or that the tillage operations are not fully effective. If, however, the land should by some mischance become so foul as to need a thorough cleaning it becomes advisable to make a bare fallow to remedy the situation. It should be mentioned, in passing, that a bare fallow has good manurial effects on the soil, as explained on page 168.

The Bastard Fallow. The bastard or half-fallow is a cleaning operation which is carried out after a crop has been removed from the land in early summer. For example, it may be carried out after mowing the first cut on a dirty piece of " seeds ", or after an arable silage crop has been removed in June or early July. In general, this is a cleaning method which can be recommended, especially for heavy land. The shorter period of the fallow naturally provides fewer opportunities of prolonged spells of dry weather ; but if advantage is taken of good weather that does occur it is seldom impossible to achieve satisfactory results. The method is to plough and kill the weeds in the clod, as in a bare fallow.

The main difficulty about a bastard fallow is that the land may be too hard to plough properly after the crop is removed. There are occasions when only a heavy tractor and a very heavy cultivator will make much impression on the hard-baked land. There are also years when July and August are persistently wet, and in this event the bastard fallow is not likely to succeed. There is also little opportunity for germinating and killing annual weed seeds.

The advantages of a bastard fallow compared with a bare fallow are that a crop is obtained, and that if this crop is a satisfactory one the weeds are well drawn up and smothered before cultivations begin, while the total cost of cultivations is less.

Autumn Cleaning. With both bare and bastard fallows cleaning the land depends mainly on drying out the perennial weeds " in the clod ", and no weeds are pulled out and burnt. In autumn or stubble-cleaning, however, though killing in the clod may be employed in especially favourable dry spells, the normal practice is to employ quite different tactics. The weather is often unsettled after harvest, and the object in autumn cleaning is generally to break the soil down into a fine tilth so that the weeds may be shaken free from soil, pulled out, and burnt. In addition, an effort is made to make weed seeds germinate, so that they may later be killed.

The first operation after harvest has been cleared—or even between the rows of bales—is to separate from the soil below a layer of soil 80–100 mm deep, in which most of the weeds are growing. This may be done by ploughing, or by use of heavy cultivators, rotary cultivators or sometimes disc harrows. Speed of working is essential, as the time available before the land becomes too wet is restricted. The land is then harrowed or worked with any convenient tined implement until a tilth is obtained and the weeds come free from the soil. They may then be collected by chain harrows, gathered into heaps and burnt. In carrying out these operations a fine tilth will have been created, and after a week or two a crop of weed seedlings may appear. Further cultivations will destroy these seedlings and help to germinate a second crop.

The success of stubble-cleaning depends mainly on starting early while the land is still dry, and before the weeds have time to recover from being drawn up and shaded by the crop. The method is particularly useful when early-sown roots or spring corn are to follow, for in such cases there is seldom time to do any effective cleaning in spring without unduly delaying the sowing date. If full advantage is taken of opportunities of autumn cleaning, the standard of cleanliness of the farm is improved and better crops are secured.

Spring Cleaning. Spring cleaning is a technique which is normally employed just before sowing late-planted roots. It commences with a winter ploughing which is deep enough to be below the couch or similar rubbish, and the land is then left until it begins to dry out in spring, and the rubbish begins to grow through. At this stage the field is ploughed again, in order to dry the land and bring the weeds back to the surface. Cultivators and harrows are then used, especially in dry weather ; and if the couch is dense it may be pulled out, collected and burnt as in autumn cleaning. Often, however, if the land is not excessively foul and a dry spell of weather ensues, the weeds may be so thoroughly dried out that it is safe to plough them under or, if there are not enough to cause hoes, etc., to block, they may be left on the surface. Spring cleaning may sometimes be quite successful if a prolonged dry spell is followed by nice rains when the crop is seeded ; but in some seasons the rain comes in the period when cleaning is being attempted, or the land remains dry so long after sowing that germination and subsequent development of the crop are impaired. For these reasons spring cleaning is not a practice that should be regularly followed. Where autumn cleaning is possible it is generally to be preferred.

The Influence of Mechanisation on Tillage Practices. In general, the mechanisation of tillage operations has taken place without many fundamental departures from traditional methods. The only outstanding exception to this generalisation is the rotary cultivator, which has become well established. Power harrows show great promise, but most tractors pull ploughs, cultivators, harrows, etc., which are fundamentally little different from horse implements.

Mechanisation makes it possible to telescope tillage operations, as by ploughing, pressing the furrows, drilling and harrowing in one trip across the field. It is, however, only in rare conditions that such complex operations can be advocated, though more limited combined operations such as harrowing and rolling together may often be undertaken with advantage.

Adequate tractor power and suitable implements make it easy to undertake operations which would be difficult to accomplish without them. New developments in agro-chemicals make effective weed control by use of herbicides possible, and these can quickly lead to new techniques such as " direct drilling " on suitable soils for crops that do not require deep cultivation.

Suitable implements and tractor power have made it possible to cultivate the land deeper than hitherto, and in many instances the occasional deep ploughing and subsoiling have unquestionably resulted in improved crop yields.

Chapter VI

FARM MACHINERY

BROADCASTING MACHINES AND SEED DRILLS

SEED is now most commonly sown by means of a drill, which places the seeds in the soil in rows and covers them. Broadcasting by hand is, however, still practised for certain crops and field conditions, while broadcasting machines are used in preference to drills for the sowing of particular crops. The advantage of broadcasting for such a crop as a grass and clover ley is that a more complete covering of the ground can be achieved than is possible with a drill. The disadvantage is that the depth of covering is irregular and that much of the seed may be left on or very near the surface, even after harrowing to cover it. In conditions of soil and climate where the upper layer of the soil dries out badly it is essential to place small seeds well into the soil, and this can only be economically and efficiently done by use of a drill.

Broadcasting Machines normally consist essentially of a long hopper, with rotating feed units placed at intervals along it. The simplest type is the seed barrow, in which the seed box is mounted transversely on a light barrow frame and the feed mechanism consists of a revolving brush which is driven from the wheel, and which sweeps the seed through adjustable holes in the rear of the hopper. Similar machines, some with fluted roller feed, are mounted on two wheels and fitted for operation by tractor. While broadcasting machines are used on bare ground some form of marker or system of sighting rods is required to avoid gaps or overlapping.

The *Fiddle* is a small broadcasting device which is popular in some districts, especially for sowing grass and clover seeds on hilly land. It consists of a hopper and a light mechanical distributor, which are strapped to a man's shoulders. As the operator walks along, the seed pours in a thin stream on to a horizontal ribbed disc, which is rotated by a rack and pinion mechanism, the rack (or bow) being pushed to and fro as the operator walks. The seed is thrown out a considerable distance by the rotating disc. The chief disadvantages of the fiddle are that some experience is required to secure an even cover, especially if the weather is not perfectly calm ; and the heavier seeds tend to be concentrated at the edges of the bout.

Seed Drills. Most of the drills used in Britain are of a general-purpose type, being capable of sowing almost any kind of seed. There are, in addition, various special types of drill for such jobs as sowing roots on the ridge.

The general-purpose drill consists of a seed box, carried on a pair of land wheels, and having a feed mechanism driven by gearing from the wheels. The seed passes down seed tubes to the coulters, which cut grooves in the soil. The seed rate is adjusted by variations of the feed mechanism and the gearing. The depth of sowing and the width between rows are controlled by the setting of the coulters. A simple device is provided for raising the coulters clear of the ground and putting the feed mechanism out of gear. Most modern drills are basically of semi-mounted type, with remote hydraulic cylinders to raise and lower the coulters and to put the drill in and out of gear.

Provision of a large-capacity hopper facilitates use of effective bulk handling methods for seed. Comparisons show that the time taken to refill small hoppers can seriously restrict overall work rates. Where large hoppers are used big wheels are needed to carry the load, and to avoid excessive soil compression in wet conditions. Satisfactory crop establishment requires action of the seed drill to be thoroughly reliable : finding undrilled strips when the crop emerges is completely unacceptable. For this reason it was a long time after reasonably

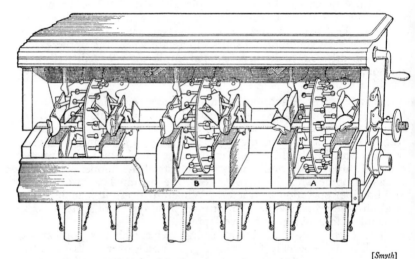

[*Smyth*]

FIG. 24.—MECHANISM OF A CUP-FEED DRILL.

At A the hopper chutes are shown turned back ready for the cup barrel to be removed. At B the hoppers are shown in the drilling position.

reliable tractor-operated drills were introduced before it became common practice for drilling to be done without an extra man riding on the drill. Some farmers still consider this a worth-while insurance. The improvements in reliability have been brought about by providing feed mechanisms that can hardly fail to work properly ; by development of coulters that in normal conditions are free from blockages and by the use of seed that contains no impurities.

Modern general-purpose drills need to be able to work well in fairly rough seed-beds at a forward speed in the region of 8 km/h. Depth of drilling must be reasonably uniform and easily adjustable. Seed rates need to be easily adjusted and to correspond well with the manufacturer's setting instructions. Good drilling of cereals is usually the main need since " single-seed " drills are now generally chosen for row crops. Feed mechanisms include :

The *external force feed* consists of a number of fluted rollers rotating in the bottom of a V-shaped hopper. Regulation of seed rate is partly controlled by varying the length of fluted roller in contact with the grain. The best drills of this type will sow a wide range of seeds satisfactorily. The force-feed mechanism is steady over clods and sows uniformly on hilly land.

The *internal force feed*, which is commonly used on combined grain and fertiliser drills of American design, has the corrugations on the insides of flanges which form the rim of small disc wheels (Fig. 25). This type is very efficient for sowing cereals but is not adaptable to a wide range of crops. The *studded roller feed*, which is somewhat similar in action to a simple force feed, but less positive, is widely used on drills of Continental origin. It is adaptable to almost the whole range of farm seeds, and is normally used in conjunction with a fairly complex gear-box giving a very large number of sowing rates.

A drill of Scandinavian origin with a *centrifugal feed* unit employs a single high-speed rotary conical feed which delivers the seed by centrifugal force down the required number of coulters, which have the tops of their seed tubes arranged in a ring round the feed mechanism. The practical sowing width of a single unit is limited to about 2·7 m, but two units can be linked side by side to form a wide drill.

Most drills with pneumatic delivery employ a large centrally mounted bulk hopper, and a row of mechanical metering mechanisms, one for each coulter. An air stream provided by a fan delivers the seed along smooth tubes to the coulters. Some pneumatic drills have light shoe-type coulters that are best suited to use on fine seed-beds.

The main types of coulters used on general-purpose drills include Suffolk (shoe) single-disc and hoe or tine types. The *Suffolk* coulter has a V-shaped shoe which cuts a groove in the soil, and the seed is delivered into the groove just before the soil falls back. This coulter works satisfactorily in most conditions, but is apt to block in rubbish and on sticky land, and fails to penetrate in hard land. The *single-disc* coulter consists of a saucer-shaped hardened steel disc which cuts a groove in the soil, and has a seed boot attached to the convex side of the disc, just below and behind the centre. This coulter works well in all kinds of unfavourable conditions, but it does not do quite such a good job as the Suffolk in good conditions, since it does not place all the seeds at the same level.

Tine or hoe coulters use narrow reversible double-ended shares, generally mounted on C-shaped spring tines. These are often a strong type capable of direct drilling in reasonably soft soil conditions. Tine coulters are usually arranged in three or more rows to avoid blockages. They are well suited to work in stony soils or for drilling ploughed heavy land without other cultivation.

The Operation of Drills. Whether the coulters are power-lifted, operated by a self-lift or hand-operated, the action of lowering them into the soil automatically puts the drill in gear. Control of seed rate is achieved by adjusting the feed mechanism and also the drill gearing. While the chart supplied with the drill provides a good guide, it is necessary to check the quantity being sown at an early stage of the work owing to variations in seed samples. When the land is at all sticky it is advisable to drill the headlands first. There is always a lag between the operation of the feed mechanism and arrival of the seed at the coulters, and this must be allowed for whenever work is re-started after a halt. The coulters must always be raised if the drill is backed.

Care is necessary to ensure that adjacent drill bouts are parallel, and that the distance between bouts is equal to the distance between the drill rows. Most modern drills are designed for operation by one man, and have a row of spring tines fitted at the rear for covering the seed. If a foot-board with a harrow hitch is fitted, the hitch should be designed to offset the harrows, so that the outside wheel-mark is not covered, and to allow the harrows to swing over to the other side as the drill turns. With wide drills, fitting of a marker is necessary to secure accurate work. A large hopper which minimises stops for filling, and large-diameter pneumatic-tyred wheels, are valuable features in a corn or " combine " drill.

Spacing Drills. Though many root and vegetable crops are still sown by means of versatile multi-purpose drills of the type mainly used for corn, these have been quickly superseded by special-purpose drills on many farms where sugar beet or vegetable crops are grown on a large scale. Essential requirements of such special-purpose drills are that they should sow at a very precise depth, and should deposit the seed, either singly or in small clusters, at regular intervals. Seed rate must be accurately adjustable, and the drill must be capable of use in conjunction with equipment for application of a pre-emergence herbicide.

The drills for such needs are almost invariably of " unit " type, with a positive metering mechanism which may consist of a vertical cell wheel with indentations in the periphery, a horizontal plate with notches at the edge, or an endless belt with holes in it. Seed picked up by these metering mechanisms from the hopper is positively ejected into the soil. Accurate grading of the seed and choice of the right size of aperture to suit the grade are necessary, and pelleting assists handling of the more difficult types of seeds such as lettuce.

The objective is usually to drill at a spacing which requires subsequent thinning of the crop, either by hand or by machine ; but as techniques of cultivation and sowing are improved and labour available becomes more scarce and expensive, the trend is towards " drilling to a stand ", for crops such as sugar beet sown in good conditions of soil and climate.

PLANTING AND TRANSPLANTING MACHINES

Potato planting is now almost invariably carried out with the help of a planting machine, which is normally tractor mounted and may also apply fertiliser, either with or, preferably, slightly away from, the seed.

During the development of modern planters many simple machines were devised to help seated operators to place seed potatoes directly into the soil. They had only the virtues of cheapness and simplicity and soon became obsolete, being superseded by machines that need less labour and provide a more positive control of plant spacing.

Semi-automatic planters require one operator for each row to place one tuber in each of a succession of compartments in a mechanism that is basically a rotating wheel. Some types can handle chitted seed without excessive damage to sprouts. Achievement of a good work rate in potato planting may be restricted by one or both of two basic handling factors : (a) the rate at which workers can each ensure that the receptacles in semi-automatic machines can be filled, leading to a low forward speed ; and (b) the time taken to handle the seed in its movement from store to headland and thence to the planter. When new potato planters are designed, one of the important aims is, inevitably, removal of the forward speed restriction associated with even the best semi-automatic machines. Another objective is to control movement of potatoes within the machine to minimise damage to sprouts. A third aim is achievement of the desired average spacing, with a reasonable limit to the variability of the spacing.

In positive-feed automatic planters individual potatoes are picked out of the hopper by an endless conveyor ; and an auxiliary feed mechanism is provided to ensure that " misses " in the primary feed are detected and remedied. The operator's main task is to keep the auxiliary feeds operative.

The second type uses vibration, reciprocating mechanisms and/or rollers to provide an endless flow of tubers, final delivery being by endless belt.

The best method of fertiliser distribution for potatoes is still the subject of experimental work. It depends on a variety of soil and climatic factors, as well as on the types of equipment used. In general, however, distributing the dressing before seed-bed preparation and " side-band " placement are most likely to give good results.

Transplanting machines include a simple type with which the operator sits very near the ground and places the plants directly into a furrow opened by the planter. A pair of inclined press wheels follows and firms the soil around the plant roots. An intermediate type has a pair of driven sorbo rubber discs to grip the plants and convey them into the furrow. In another type the operators place the plants in fingers on an endless conveyor belt. The belt takes the plants down to a furrow which is opened by the machine, and releases them as the press wheels firm the soil alongside. All types require nimble-fingered operators for successful working. The work done with planting machines which have been correctly adjusted is often of better quality than that done by hand.

FERTILISER DISTRIBUTORS

Fertilisers are, as a rule, applied to the soil surface by means of distributors which are designed to operate as broadcasting machines. Uniformity of distribution is important where very small amounts of concentrated fertiliser are to be applied, and distributors must be capable of sowing evenly various sorts of fertiliser at rates varying from 125 to 2,500 kg/ha. In some circumstances, as when fertiliser is being sown along the ridges for potatoes, it may be advantageous to concentrate the manure in bands, so that none is applied to the tops of the ridges. Special devices to facilitate such a distribution may be obtained for fitting to some distributors.

A common type of distributor employs a long low hopper carried on two wheels, with a feed mechanism, which is driven by the wheels, operating in the bottom of the hopper. There are many different types of feed mechanism, the most noteworthy being the following :

Roller feed. A delivery roller rotating in the bottom of the hopper carries the manure through an adjustable opening. The roller may be plain or corrugated, and may be driven at variable speeds by means of the gearing. An agitator is generally required in the hopper to prevent the manure " bridging " over the roller, and a cleaning brush may be used to sweep off the manure when it has been carried outside the hopper. In one type, soft-coated contra-rotating rollers carry the fertiliser out between the two rollers.

Agitator feed. The roller may be replaced by an agitator which pushes the manure to-and-fro over a number of adjustable holes in the bottom of the hopper. This is a simple and efficient type for sowing heavy fertilisers which are in good condition.

Reciprocating Plate feed. The bottom of the hopper consists of two slotted plates with the holes not opposite one another. Sowing rate is controlled by the speed of reciprocation of the moving plate and by the distance it travels.

Star Wheel feed. A number of star or finger wheels rotate in the bottom of the hopper, and push the manure through slots of variable size. This type of feed is commonly used in conjunction with an internal force-feed grain mechanism in combined grain and fertiliser drills.

Revolving Plate feed. Saucer-shaped discs are mounted along the bottom of the hopper and driven by bevel gearing. As they rotate, a layer of manure is carried to the outside of the hopper, where it is broadcast by fingers on a high-speed shaft. Advantages of this type are that it can handle almost any type or condition of manure. Many simpler types work faster and distribute as evenly.

Conveyor-and-brush feed. The bottom of the hopper is formed by an endless canvas belt which moves slowly to the rear and takes a layer of manure to the outside, where it is swept off by a high-speed cylindrical brush. This type is made of non-corroding materials.

Spinning Disc delivery. A popular distributor has a hopper of V- or truncated cone shape, from which the manure is fed on to a rotating disc. The disc throws the fertiliser over a fairly wide strip. When correctly adjusted good machines of this type work well in calm weather, but distribution is somewhat irregular in windy weather. The hopper is frequently mounted on the tractor's three-point linkage and driven from the tractor p.t.o. Such machines have a high working rate and are particularly useful for applying light top dressings to corn and grassland. A pair of contra-rotating discs may be used.

Oscillating Spout delivery. The fertiliser passes into a tapered pipe which swings rapidly from side to side. Distribution pattern over the limited swath width tends to approximate in character to that achieved by distributors with " full-width " hoppers.

Pneumatic. Pneumatic-delivery distributors have a large hopper suited to bulk fertiliser handling ; a row of metering mechanisms, often of fluted roller or similar type ; and a wide boom with evenly spaced curved plates for spreading. The fertiliser is delivered on to the plates by individual plastic pipes.

Bulk Handling of Fertiliser. Work studies show that filling the distributor hopper often takes more time than spreading. Moreover, heavy lifting may be involved. These factors lead to development of mechanised handling methods. If the distributor has a large hopper (1–2 tonnes), bulk handling direct into the machine at the store is often practicable. If the hopper is only of small or medium size, pallet handling from store to field or use of a " high-tip " trailer may save time and effort.

Combined Grain and Fertiliser Drills. Combined grain and fertiliser drills are now commonly used for corn crops which require

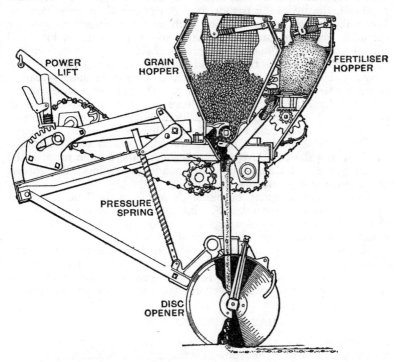

FIG. 25.—MECHANISM OF COMBINED CORN AND FERTILISER DRILL (M.M.).
Grain and fertiliser are in separate hoppers but are delivered down the same spout.

a dressing of phosphatic or compound fertiliser. Most combined drills sow the corn and fertiliser down the same seed-tube, and are fitted with single disc coulters (Fig. 25). Excellent results have been obtained throughout Britain from phosphatic manures applied with the seed in such combined drills. Some injury to germination may occur if any considerable quantity of fertiliser is placed in close contact with spring-sown crops, especially if a period of dry weather follows. Some crops, such as peas, cannot tolerate fertilisers close to the seed, and should be drilled either through ordinary drills, or, better, by means of fertiliser placement attachments which deliver the fertiliser to one side of the row and at about the level of the seed.

One of the chief faults of the combined drill is the difficulty of cleaning it out effectively, and the harm that results when it is not thoroughly cleaned. Use of rubber or plastic materials can minimise corrosion, but cleaning with water before storage is advisable.

Farmyard Manure Distributors. The need to be able to handle distribution of farmyard manure effectively and expeditiously with a limited labour force has led many British farmers to equip themselves with mechanical loaders and spreaders. The hydraulically operated tractor loader is a sturdy and fairly inexpensive device which has many other uses, and though several types of loader are available, the hydraulic tractor loader is preferred by most farmers. When this is used in conjunction with an easily-hitched trailer-type spreader, one man can, in reasonable working conditions, load and spread just over 3 tonnes per hour in a field 550 m away from the yard. Alternatively, a tractor loader that does nothing else can keep busy a pair of spreaders working fairly near to it, or more spreaders or trailers working over a long distance. The more expensive tractor-mounted hydraulic slew loaders have considerable advantages in situations where bad ground conditions hinder tractor movement. These, like front or rear loaders, have other uses, such as ditching.

One type of farmyard manure spreader consists essentially of a trailer with a slatted conveyor running along the bottom towards the rear, together with a shredding and spreading mechanism that throws the manure out at the back.

The mechanisms may be either driven from the land wheels or by the tractor's power take-off. P.t.o.-driven machines are preferable where very heavy dressings need to be applied, and where the going is difficult, as on hillsides or slippery land.

A simple flail type manure spreader delivers the manure to one side by means of sturdy chains which gradually unwind on the p.t.o.-driven central shaft as the cylindrical spreader body is emptied. Absence of exposed chain conveyors avoids maintenance difficulties.

HAY-MAKING MACHINERY

Finger-bar Mowers. Most mowers used for cutting grass crops have a cutting mechanism consisting of a finger-bar in which a steel knife, made up of triangular sections, reciprocates at high speed. The cutter bar is flexibly mounted, so that it can lie close to the ground and conform to the irregularities of the surface. The knife is usually driven though a crankshaft and connecting rod by the tractor p.t.o. V-belt transmission between the p.t.o. and the crankshaft provides both a speed increase and a safety device. Crank throw is often 76 mm but may be more. The fingers on the cutter bar are set 76 mm apart, and knife sections are the same width at the base. The cutting action is like that of a pair of scissors, the reciprocating knife representing one blade and the stationary ledger plates on the fingers the other. For clean cutting the following conditions are necessary : (*a*) each knife section must move at least from the centre of one finger to the centre of the next ; (*b*) the knife must have a slight lead at the outer end, so that it runs square to the work when actually cutting a heavy crop ; (*c*) the cutter bar must be straight and the fingers in line, so that the knife slides evenly over the ledger-plates ; (*d*) the knife sections must be sharp ; (*e*) the edges of the ledger plates on which the knives slide

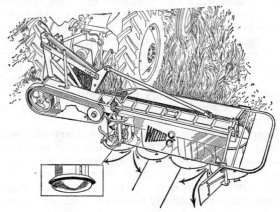

FIG. 26.—VERTICAL SPINDLE 4-DRUM MOWER (Bamfords Ltd.)

must not be rounded ; (*f*) the malleable clips which press the knife sections on to the ledger plates should be adjusted to ensure that the knives press very gently and evenly on the ledger plates along the whole length of the bar.

The cutter bar is carried on two shoes, the outer shoe carrying a divider and swath-board, and sometimes a small wheel for easy adjustment of the height of cut, and the inner shoe a bearing on which the knife-head slides. The inner shoe also carries the drag link and lifting mechanism. The lifting mechanism may generally be operated by either hand or hydraulics ; and a further lever controls the " pitch " of the cutter bar, for close cutting when the ground is smooth, or elevation when this is undesirable.

Rotary Mowers. Though finger-bar mowers are widely used and have the merits of a low power requirement and low capital cost they are far from ideal as regards maximum practicable forward speed and ability to keep going when cutting very dense laid crops. Flail forage harvesters and flail mowers, which are considered later, can cut laid crops effectively but rate of work is often restricted by the high power requirement and care must be taken to ensure that the crop is not cut into short lengths which are lost in subsequent swath-treatment and harvesting operations. Mowers of the *vertical spindle rotary type* can operate at high forward speeds, need not chop the crop excessively, and have only a moderate power need in relation to rate of work.   It seems advisable to seek simplicity of design and soundness of construction.   In this respect, 2-drum types may have advantages over 4-drum types.   Similarly, drive to the tops of drums allows more latitude in design than is possible with the disc type, where the drive has necessarily to be slim in order to run beneath the lower edges of the discs.   Disc types are capable of cutting a reasonably short stubble by tilting the discs down a little at the front.

Tedders.   When a heavy crop of hay is cut by finger-bar mower, the swath lies dense and compact, and it is advisable to lighten it up so that sun and wind can penetrate it.   This operation should be carried out very soon after mowing, so that drying of the whole swath proceeds uniformly.   Tedders are rotary forking machines which are designed primarily for this first operation.   Most machines are able also to turn and windrow.   Some are of the forward-action type in which the tines rotate in the opposite direction to the wheels and carry the hay over the cylinder beneath a curved hood.   Many designs have one or more rotors with a substantially vertical axis, which can easily be set, in conjunction with adjustable guide cages, for swath turning or collecting.

For a long time the development of tedders was neglected by manufacturers because farmers thought that such machines stripped off too many leaves.   This can, indeed, happen if tedders are used too late in the haymaking operation, after the leaves have become brittle.   If, however, treatment is begun early, the whole crop dries more uniformly, and can be effectively teased out without losing its leaves.

Swath-Turners and Side-Delivery Rakes. Swath-turners are designed to turn swaths or windrows gently, and to move them sideways slightly during the operation.   This often has the effect of turning the hay on to dry ground, in addition to exposing the lower part of the swath to the action of sun and wind.

There are many multi-purpose haymaking machines that can be used for swath turning, windrowing or scattering.   The rotating-head type has two or three sets of curved spring-steel tines which rotate and gently roll the swath over to one side as the machine advances.   Rapid change of purpose is a major aim of design.   This type, which has a gentle action, can be used as a collector by making adjacent heads rotate in opposite directions.

The inclined rake-bar machine usually has four horizontal bars attached to a pair of discs by a linkage which keeps them horizontal. The machine acts as a swath turner if the centre section of each rake-bar is removed.

The finger-wheel type has a number of ground-driven wheels mounted obliquely on a light frame so that they can be set so as just to touch the ground.   They are rotated by their contact with the ground and crop.   Disadvantages are a tendency to roll damp swaths tightly and to pick up loose stones.   Modern multi-purpose hay-making machines are normally mounted on 3-point linkage and are designed for use at high forward speed.   The one machine can effectively carry out all essential operations after cutting ;   but drying rate

B

A

C

PLATE XI

A. 4-disc rotary mower cutting a heavy crop
B. Wide flail mower with hydraulic controls
C. Small rotating-head type haymaking machine arranged for collec-
    ting

PLATE XII

A. Mounted 4-wheel finger-wheel multi-purpose machine shown swath-turning

B. Pick-up baler and 8-bale accumulator shown setting down group of bales

C. Handling flat-8 group of bales with fork-type loader

PLATE XIII

A. Forming 20-bale group with accumulator towed behind the baler.
    Bands of twine hold the group together after setting down and
    subsequent handling

B. Big bale of rectangular press type being set down

C. Big-roll baler at work in straw

PLATE XIV

A. Double-chop forage harvester
B. Metered-chop forage harvester picking up wilted swath
C. Self-propelled forage harvester fitted with reciprocating cutter bar
    collecting grass for drying

PLATE XV

A. Large-capacity self-propelled combine harvester working in a heavy partially laid cereal crop
B. Small self-propelled combine harvester working on hill land
C. Single-row potato harvester side-loading

PLATE XVI

A. " Bunker " type single-row potato harvester
B. 2-row trailed tanker sugar beet harvester tipping to field heap
C. 2-row self-propelled tanker sugar beet harvester

C

A

B

PLATE XVII

A. Mobile irrigation sprinkler operating on grassland
B. General-purpose mounted crop sprayer
C. Contractor's self-propelled field crop sprayer

PLATE XVIII

A. Cutting overgrown hedge with high-power heavy-duty flail type hedge cutter
B. Hydraulic digger-loader with ditching bucket working across a narrow ditch
C. Self-propelled trencher/pipe layer laying clay tiles

can be improved if a fairly severe swath treatment is carried out during cutting.

Roller Crushers and Crimpers. If the hay swath is effectively bruised in a roller crusher or crimper during or soon after cutting, the drying rate in good weather conditions can be considerably faster than that of swaths that are only tedded. As a general rule, one drastic bruising treatment is carried out by a combined machine, and thereafter tedding or swath-turning is done according to the condition of the crop.

Another method of applying drastic swath treatment is to use a flail mower for cutting the crop. The machine must be equipped with flails to suit the crop and the purpose. For hay, it is essential to avoid chaffing, otherwise much of the crop may be lost during subsequent operations. This requires a low rotor speed, high forward speed, and a wide clearance between the flails and the shear-bar of the machine.

Pick-up Balers are used for harvesting both hay and straw. The common type is a p.t.o.-driven trailed machine with a retracting-tine pick-up, and a ram, reciprocating in a rectangular chamber, which forms bales having two bands of twine running the long way of the bale. A common bale section is 450 mm × 350 mm, but larger sections are available in some high-capacity balers which are intended mainly for use on dry straw. Bale length is variable : for most bale handling methods a length of about 0·9 m is used, but for special purposes length may be shortened to about 0·6 m or extended to about 1·1 m. Baler capacity depends on the machine's ability to keep going without blockages or other faults. It is dependent on detailed design of the combination of mechanisms for handling the crop from the pick-up into the bale chamber, and also on uniformity of the windrows.

Twine-knotting mechanisms are surprisingly similar to those originally developed for grain binders, though the mechanisms and twine are heavier. Typical consumption of sisal twine of average thickness is usually about 1·5 kg per tonne of hay or 2 kg per tonne of straw. Polypropylene twine has some advantages over sisal, notably resistance to rotting if the bales are stored in damp conditions. Medium-sized balers should have a potential output of about 6 full-length bales per minute, equivalent to 6–10 tonnes per hour in hay or 3–5 tonnes per hour in straw. Overall working rates are usually considerably lower.

Curved bales are a fairly common fault. The cause may be simply bad adjustment of the chute or deflector at the rear of the bale chamber, but is more likely to require re-adjustment of the feed into the bale chamber. If too much crop is fed to one edge of the bale, no other adjustment will remedy the fault.

When hay is being baled a fairly accurate control of bale density is necessary, especially if appreciable further drying of the crop is necessary. In the usual type of bale chamber the crop is compressed until the force exerted on the partly-formed bale is sufficient to move it along the bale chamber. Resistance to movement is adjusted by means of tension controls which resist the tendency of the bale chamber sides to move apart. Unfortunately, the dampest hay always tends to be baled tightest at a given setting. Moreover the increase in bale density which may take place if no adjustment is made when dew falls can be considerably more than that due solely to the small change in moisture content.

Knotter faults can waste time, especially if over-winter maintenance has been neglected, so that parts such as the knotter-bills are covered with rust.   When overhauling, it is advisable to check that the needles place the twine accurately in the retainer, and that the tensions required to pull the twine from the ball and from the knotter-bills accord with figures specified by the manufacturer.

Bale Handling Equipment.   Bales are far from ideal packages from a mechanical handling viewpoint. The shape and size were originally chosen to provide a package which one man could lift and carry ; but as mechanical handling methods replace human muscles there is interest in systems more suited to the use of tractor power and electric motors.   In spite of its drawbacks the bale retains its position on many farms because the buildings were designed for hand feeding, and new buildings suited to power-handling systems can be costly. In the field, hand lifting of bales on to trailers is becoming less common, though some farmers prefer specially designed long and narrow low-loading trailers to the many tractor loaders and endless chain elevators that are available.

When hydraulic loaders are to be used, the first requirement is to get individual bales assembled into rectangular blocks. One type of wheeled automatic collector uses a " flat-8 " group, loaded by means of curved forks which penetrate the bales from the top.   With this type it is possible to make a good bonded load if the front ladder of the trailer is inclined forward at about 45 degrees, and the successive groups of bales are pushed up to the front. A further advantage of the " flat-8 " fork-type loader is that it can also unload effectively, and can stack directly into a well-located barn.   Another type of automatic collector makes rectangular blocks of 16 or 20, which are held together by two bands and are well suited to handling by hy-draulic loader or fork-lift.   Automatic bale pick-up trailers can form loads that are either designed to deliver single bales for stacking, or to be set in position in rectangular blocks where needed for storage.   On some farms where reduction of time on field work is a major con-sideration, machines capable of forming rectangular or round bales weighing 0·5 tonne or more may be chosen.   Such big bales are well suited to farms where straw for bedding needs to be moved into well planned yards.   Use of big bales for hay in British conditions generally requires either a suitable barn drying system or careful use of a chemical additive to assist preservation.

Even when bale handling from field to barn has been effectively mechanised, mechanisation of subsequent handling often leaves much to be desired, compared with the flow-handling methods which can be used with short-chopped silage.

SILAGE EQUIPMENT

The development of improved equipment for silage has made it possible to eliminate practically all hand work from both making and feeding. Ability to do the ensiling speedily, when the crop is at the right stage of growth, helps to ensure a reasonably consistent product, capable of providing the basis of a high-production ration for both beef and dairy cattle. Aims vary considerably according to the facilities available and the system of feeding. A common objective is to produce a medium-quality silage for self-feeding. This can be done with a minimum expenditure on equipment, and suits many farms where shortage of capital rules out higher-cost systems. Essentials for success with a simple direct-cut system include harvesting the crop at a dry matter content of not less than about 20 per cent. Wilted silage made in horizontal silos at a medium dry matter content of about 25–30 per cent usually requires more equipment, especially if it is to be fed by forage box. The making of high dry matter silage of 30 per cent D.M. or over is normally practicable only with tower silos or in large, quickly-filled bunker silos where effective sealing against entry of air is ensured. Use of towers inevitably involves high capital cost but has the potential advantage of a press-button feeding system capable of putting a measured amount of high-quality feed into the mangers.

Any silage system which aims to provide mechanical feeding works better if the crop is short-chopped. With tower silos and feeding by fixed equipment, failure to achieve a short chop can cause much unnecessary trouble with some types of unloaders and feeders. With high dry matter silage a short chop is essential for effective packing and exclusion of air. Even with self-feed silage a reasonably short chop helps in securing a good fermentation ; and cattle are able more easily to achieve a high intake of silage where this is desired. The necessity to exclude air from all types of silage should be emphasised. The higher the dry matter content, the more difficult this becomes ; but plastic sheeting can help to improve sealing.

Flail Forage Harvesters. A simple flail forage harvester consists essentially of a high-speed horizontal rotor with a large number of knives mounted on swinging arms. A hood partially covers the flails. Direction of rotation of the flails is opposite to that of the land wheels. The hood is extended at the top into a chute. In the simplest (" in-line ") type the outlet from the hood is a simple one to the rear, and the trailer has to be hitched directly behind the harvester. This simple arrangement works well in practice. Only one man is needed, and the in-line arrangement makes it easy to fill trailers effectively— an important consideration in getting a high output from the harvester. Where transport distance is short the tractor, harvester and trailer may run to the silo without disconnecting the hitch. For high-speed work two harvesters may cart to the same silo. For longer hauls along public roads it may be better to use a hydraulic pick-up hitch on the harvester, and to use the second man, tractor and trailer for transport. The main disadvantage of the " in-line " system is the fact that a tractor wheel has to pass over the crop before it is cut. This is not a serious disadvantage in most conditions. By working to and fro the harvester is able to cut and pick up the pressed-down crop. In adverse field conditions, however, soil from the tractor wheels may contaminate the crop ; and even small amounts can spoil silage quality.

There are slightly more complex versions of the flail harvester to suit other working systems. Offset harvesters of either the trailed or semi-mounted type avoid a tractor wheel travelling through the crop before cutting ; and when fitted with a swivel delivery head they may be used either with the trailer hitched behind the harvester or towed alongside it by a separate tractor. The semi-mounted type has a considerable advantage for use on hillsides because it allows the trailer to be hitched directly to the tractor. When the crop has to be transported along narrow roads, the harvester is easily disconnected and left in the field.

The principles of setting and operation of flail forage harvesters are simple. The primary adjustments are height of cut and the setting of the clearance between the flail tips and a shear-bar at the front of the machine ; but performance is also greatly influenced by the tractor used to drive the harvester. Both rotor speed, determined by the tractor p.t.o. speed, and forward speed have marked effects. The higher the rotor speed, the shorter the chop length and, within limits, the greater the power requirement. The higher the forward speed, the greater the chop length and the more ragged the stubble. Though flails depend for their cutting effect more on peripheral speed than on sharpness, it is wasteful of power to use blunt flails. When flails are replaced, attention must be paid to maintaining the balance of the rotor, since excessive and dangerous vibration may be set up if new flails are fitted to one side only.

A good flail harvester is capable of picking up a wilted swath left by a mower. If difficulty arises, fitting of an extension plate at the front of the hood on some machines helps to concentrate the air and lift the swath.

"Double-chop" Harvesters. These are trailed or semi-mounted offset machines which use a flail rotor for the primary cutting or pick-up operation : an auger running in a trough behind the rotor then takes the crop sideways and delivers it to a flywheel type chopper mechanism. Paddles attached to the flywheel deliver the crop up a curved chute suitable for loading trailers. Length of chop is influenced not only by flail speed in relation to forward speed, but also by the number of knives used on the secondary chopper. Though a double-chop harvester can normally produce a considerably shorter chop than a simple flail if both machines are operated in a normal manner, chop length is not uniform, partly because the crop is fed to the secondary chopper in a somewhat tangled form. Double-chop harvesters cost more to buy and to maintain than simple flails, and they require a greater power input per ton of product where a shorter chop is achieved. Nevertheless, many farmers consider that the improved results easily justify the higher costs.

Metered-chop Harvesters. Considering the world as a whole, " metered-chop " harvesters are the most widely used type. They are expensive in both capital and running costs, but experience proves that the uniformly short chop which they can achieve is a considerable advantage in many conditions, and that cheaper and simpler harvesters are incapable of giving the desired level of performance. The heart of a metered-chop harvester is the short-chopping mechanism and a feeder. In addition, the harvester for British conditions usually needs a tined pick-up. Alternative collecting equipment is a cutter-bar or a maize head for handling one or two rows of maize.

Two main types of chopper mechanism are used, viz. " flywheel " and " cylinder " types. Choice of harvester is often decided by such factors as ease of ensuring that the chop does not deteriorate due to a steadily increasing gap between knives and shear-plate. Sharpening and adjustment, using a semi-automatic device to touch up the blades as they are slowly turned under power, is quick and easy on some cylinder choppers. Most flywheel choppers require considerably more skill and take more time. Some cylinder choppers can work at a nominal chop length of 5 mm. The shorter the chop, the greater the power requirement. P.t.o. powers of well over 50 h.p. can be effectively utilised. For very high outputs over a long season, such as is needed for supplying a large grass drier, self-propelled harvesters are often preferred. The main technical advantage, apart from sheer engine power and chopper size, is ease of manoeuvring and trailer hitching.

With any kind of metered-chop harvester, large stones or pieces of metal which get into the chopping mechanism can cause serious damage. It is therefore worth while to adopt practices likely to avoid such accidents. For example, a very heavy roller should be used in spring when the soil is still moist ; and any swath treatment machines used should be of such type and condition that steel tines are unlikely to break and find their way into the swath. Because chopped-up metal can have unfortunate effects on livestock some manufacturers offer powerful electro-magnets for fitting to harvesters and forage blowers at a point where the iron may be collected.

Equipment for Silo Filling. The common method of filling a horizontal silo is to tip the load of silage on to a concrete apron and transfer it to the silo by means of a rear-mounted *buckrake*. It is easier to get a good load on the buckrake and to place it accurately if the device is designed for this specific purpose, with slim, spring-steel tines. Where a large amount of silage has to be made quickly, the extra cost of a hydraulic push-off type is easily justified. Rear mounting of the buckrake improves traction ; but light-industrial type front loaders used on some farms have obvious advantages.

Blowers can be effectively used for silo filling provided that the crop is chopped reasonably short and feed to the blower is effectively regulated by means of a forage box or dump box. Blowers are the normal equipment for filling towers, and can also be used with advantage where high-dry-matter silage is to be made in very deep horizontal silos. A combination of a semi-mounted forage harvester, a forage box and a blower provides a practicable fully mechanised alternative to buckrake filling on small farms. Some hand manipulation of the blower spout is required when filling horizontal silos. For filling towers, automatic spreaders can greatly reduce or eliminate hand work. A common type of spreader is electric motor-driven, and located high up in the centre of the tower. For good results the crop should be delivered vertically on to the spreader, which then spins it off equally in all directions. Another type of spreader used for towers has a swan neck which oscillates and at the same time moves slowly up and down ; these two actions are adjusted so that the crop is delivered successively all round the circumference of the silo. Careful attention to even filling results in better silage and a much easier job for the unloading equipment.

Forage Boxes and Dump Boxes. The essential feature of a forage box or dump box is ability to receive a load of forage and then to unload it in a uniform stream, either directly into a manger or into some other forage handling equipment. A common method of delivering the load to the unloading mechanism is by an endless chain and slat conveyor running along the floor of the box. An endless apron of rubber may also be used. Two or three spinners mounted above one another throw back surplus forage. The layer delivered between conveyor and spinners from one end of the box is concentrated into a narrow band and delivered positively to one side by means of a cross conveyor.

Dump boxes are designed primarily for standing beside the silo and receiving loads from tipping trailers. Much of the advantage is lost if the trailer load cannot be tipped quickly and completely. It is therefore wise to choose a box of ample width, and if necessary to build a good ramp to assist load transfer.

Forage boxes usually have to deliver forage from the field to the silo filler, so their basic needs are adequate size and no time lost on turn-round. For their other main duty—feeding into stock mangers—good manœuvrability is needed. Forage box feeding is advantageous where short-chopped silage has to be fed to a number of different herds in scattered premises. Where this feeding method is chosen it is often worth while to provide a labour-saving method of delivering components such as rolled barley and minerals on to the forage. The supplements are then automatically mixed in during unloading.

Silo Unloaders. Electrically operated tower silo unloaders make it practicable to feed forage crops to cattle by press-button systems. In ideal conditions a single " start " button sets in motion the whole operation of feeding a pre-determined ration to a yard of cattle, and when the job is completed all of the equipment involved is automatically switched off. Most farms employ systems making less use of automatic controls, and needing some supervision during operation, but all of this can usually be done from a single point and with only minor adjustment of the unloader.

Top unloaders scratch or cut the silage from the surface by means of rotating augers or endless chains. A common method of transfer of forage from tower to feeder is to use a centrifugal impeller to throw it into the chute which surrounds the line of doors, but some unloaders drop it down a central opening, formed by a cylinder which is pulled up through the silage as the silo is filled, while others drop it down a special chute just inside the doors. All top unloaders necessitate a climb up the tower for adjustment and maintenance.

Bottom unloaders undertake the more difficult task of cutting successive layers of forage from the lowest layers in the tower ; but they have the advantage that the unloader works in a fixed position at ground level. One type of bottom unloader has an endless chain which pivots about the centre. A second chain running in a trench at a lower level picks up the silage and carries it outside the silo. Another endless-chain bottom unloader pivots at a point outside the silo, and reciprocates to and fro across the base, running into specially formed lobes at the extremities of its sideways travel.

COMBINE HARVESTERS

The combine harvester cuts or picks up a mown crop and threshes in one operation. Combines are most used for direct-cut cereals, but most can be fitted and adjusted to handle a wide range of crops. The main types are (1) small machines driven from the tractor power take-off ; (2) machines which are pulled by a tractor but have the threshing mechanism driven by an independent engine mounted on the machine ; and, (3) self-propelled machines. The self-propelled type, which has the cutter bar at the front, has many advantages over other types, and is popular on all types and sizes of farms. The main disadvantage of this type is the high cost. Engine-driven trailed machines have

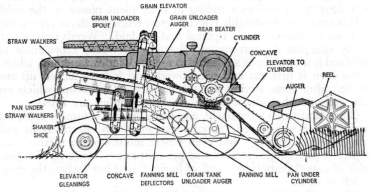

FIG. 27.—SECTION OF A COMBINE HARVESTER SHOWING FLOW OF GRAIN AND STRAW.

marked advantages over machines driven from the power take-off, especially on undulating land ; but the latter have the merit of relative cheapness.

Combine harvesters can usually leave a stubble from about 75 mm to 600 mm long in a standing crop ; and with crops which are not laid it is usually advisable to cut a fairly long stubble, so as to ease the work of the threshing mechanism and to keep above the worst of the weeds. Where it is necessary to cut very low the speed of working is reduced, since threshing capacity is the limiting factor.

The threshing drum in combines used in Britain is usually of the beater type. Both drum speed and concave clearance on modern machines can be easily adjusted. Recommended setting procedure is to start with both set according to the manufacturer's instructions, and if the combine does not thresh cleanly the first adjustment should be to increase drum speed.   If this is not effective, concave clearance may then be reduced.   Similarly, if threshing is clean but grain is being cracked, the first step should be to widen the concave clearance. Drum speed may then be reduced if damage to the grain continues.

The sieves are often of a louvred type.   The fans need frequent adjustment to ensure that the grain is cleaned but that none is blown away with the chaff.   There is a device for returning to the threshing drum any pieces of ear which have not been effectively threshed. Some combines separate and collect the weed seeds, but others either scatter the very small seeds or deliver them mixed with the grain.

Combines are normally equipped with a grain tank to facilitate bulk grain handling and operation by one man.   Bagging types are sometimes used where adequate facilities for receiving and handling grain in bulk are not readily available, but where an efficient grain conditioning and storage plant is installed the bulk-handling system is always preferable.   It requires the provision of suitable grain-tight trailers that can be easily emptied into a receiving hopper.   The usual method of transport is now to employ large hydraulic tipping trailers, which are also suitable for many other kinds of farm transport work.

Where a bagger type combine is used it pays to arrange to drop the bags in windrows, so that they are easily picked up, or to transfer them directly to trailers standing at convenient places in the field.

It is worth a considerable effort to grow crops which are reasonably free from weeds, uniformly ripened and not laid.   Weeds reduce output by clogging the screens and shakers.   If weedy crops must be combined it is sometimes advisable to use the windrow method, whereby the corn is first cut and windrowed, and is later picked up and threshed after some drying has taken place.   Laid crops which are clean can often be satisfactorily dealt with if lifters are fitted.

Crops that ripen irregularly, e.g. peas, linseed, mustard, rape, are usually better windrowed.   Combines can thresh effectively at well over 30 per cent m.c. ; but even where moist storage methods are adopted it pays to aim at about 20 per cent m.c.   There is, of course, a risk of oats " shattering " and of barley " necking " when the crops are left to ripen.

The chief problems connected with use of the combine harvester in Britain are drying of the grain where necessary, and handling of the straw after the machine.   It is almost essential in most parts of Britain to have available drying facilities for dealing with grain that is too damp for safe storage, and the main farm drying and storage systems are briefly discussed on pages 92–98.   Grain cannot be safely stored in unventilated bins if moisture exceeds about 14 per cent and while it is often possible to combine corn as dry as this, many samples may contain 20 per cent or more of moisture.   It is therefore necessary in Britain to have drying facilities available if the combine is to be used to best advantage.   Farm grain driers vary from simple, home-made machines to elaborate and expensive ones.

Where the straw has to be collected after combining, pick-up balers are almost invariably used.   Where the straw contains much greenstuff it is usually advantageous to leave the bales out for a day or two in fine weather, but otherwise they may be carted and stacked immediately.

The main alternatives to baling are burning, or the more laborious chopping of the straw and working it into the soil. Chopping unfortunately needs a high-powered tractor at a time when it cannot easily be spared, and experimental work has so far shown little advantage from returning the straw to the soil. Burning, however, provides ideal conditions for starting stubble cleaning and destroys some noxious weed seeds.

ROOT HARVESTING MACHINERY

Potato Diggers. The extent of mechanisation of potato harvesting depends on availability of casual labour for hand-picking as well as on suitability of the soil type to use of mechanical harvesters. The crop continues to be grown on some soils where lifting conditions may be extremely adverse and the only effective equipment capable of uninterrupted use is a simple potato raising plough or a spinner.

Elevator Diggers are preferred wherever they will work effectively, since they expose more of the crop and leave it easier to pick. The main principles of operation are to lift all of the potatoes but as little soil as possible, and to get the crop freed from the soil with the least possible shaking in order to do this. Two-row diggers have considerable advantages where they can be used. Digging can proceed ahead of the picking gang, and putting two rows into one reduces the amount of work to be done. Effective systems based on handling the crop in pallet boxes have been developed, but on most farms baskets are tipped directly into trailers.

Potato Harvesters. The simplest harvesters are essentially elevator diggers with a platform to carry the pickers, a sorting table to make their job easier, and conveying mechanisms to deliver the crop for transport. Such harvesters have a low rate of work if all potatoes have to be handled. The most that can normally be expected of a total gang of 5–6 is about 0·4 ha per day. On more complex harvesters stone and clod separators take over most of the necessary sorting and the main function of the operators is to correct mistakes. A typical work rate with a gang of 6–8 is 0·1 ha per hour.

Some types of 2-row harvesters do not attempt final sorting. The roughly-sorted crop is loaded into large-capacity self-unloading trailers and the crop is passed through a stationary pre-cleaner before being stored or transferred to the final grading and packing machines. The most complex type of harvester employs automatic electronic sorting devices which can distinguish between potatoes and contaminants such as stones and clods. The sorting mechanisms are made self-levelling, the whole harvester is designed for one-man operation, and the machine may be self-propelled, with a storage tank to facilitate a quick turn-round. Use of such equipment will inevitably supersede hand picking methods.

Sugar Beet Harvesters. There are several ways of carrying out the mechanised harvesting which has completely superseded hand harvesting. One of the cheapest harvesters is a single-row side-loading type which tops, lifts and cleans the beet and delivers it into a trailer drawn alongside. This is not a good method because it needs 2 men and 2 tractors to work at no more than 0·1 ha per hour. The 1-row tanker harvester overcomes this disability. One man can work alone and deliver the beet to a headland heap. Alternatively he can transfer loads to a trailer for longer transport.

Multi-row harvesters may split the topping, lifting and loading into two or more jobs. With 3-row machines side-loading is fast enough to be practicable, though a very large transport fleet may be needed for continuous working.

It is necessary in planning operations to consider the factory's loading requirements and also to take care in adjustment to avoid unnecessary top and dirt tares.

FIELD CROP SPRAYERS

In recent years, spraying for weed control has become a common-place operation on almost all farms where corn crops are grown (see p. 215). The widespread adoption of this practice has followed the development of new chemical sprays which can be effectively applied as a fine mist at low rates, e.g. down to 112 litres/ha, and has been aided by the production of cheap and simple machines designed to apply the solutions. Spraying machines for field crops may be broadly classified into the following groups :

1. Low Volume (112–448 litres/ha). Such sprayers are designed primarily for applying true solutions such as MCPA and 2,4-D. They are almost invariably tractor-mounted, and have a small tank and a simple p.t.o.-driven pump, usually of the gear or vane type. The machine works at low pressures (normally 2·7–2·8 bar) and employs simple fan jets.

2. Low/Medium Volume (112–670 litres/ha). The tank usually has a capacity of 225–450 litres, and the larger capacity pump may sometimes be of a type suitable for spraying suspensions. Usually, however, these machines, while capable of handling materials like proprietary potato fungicides for short periods, are unsuitable for spraying more difficult types of suspensions. As in low-volume machines, agitation of the fluid in the tank is usually effected by arranging for surplus liquid from the pump to be directed through jets back into the tank.

3. High-Low Volume (112–1120 litres/ha). These machines usually have a tank capacity 360–1140 litres, and in the larger sizes are trailed machines. The pump is a type capable of handling any sort of spray material at pressures ranging up to 6·9–13·8 bar. A mechanical agitator is used to stir the tank. Low-volume applications are handled by using fan jets, while swirl nozzles are usually fitted for high-volume application.

The simplest and cheapest sprayers cannot be expected to handle effectively a full range of field-crop spray materials, but may be quite suitable for farmers whose chief or only need is to spray corn crops. At the other extreme, a contractor or a farmer who regularly has to spray potatoes or to use difficult suspensions is well advised to buy one of the more expensive machines that are capable of effective high-volume work. Machines of the intermediate (low/medium volume) group can be used to a limited extent for applying suspensions, but there may be difficulties due to insufficient agitation of the liquid, or to the rapid wear of an unsuitable type of pump.

For fertiliser suspensions use may be made of widely spaced " flood jets " which employ a curved metal surface to break up a solid jet. Wide booms can help to achieve high work rates and to minimise damage caused by tractor wheelings. Whipping and yawing of wide booms can lead to uneven distribution : such faults may be minimised by use of stiff booms with damped spring mountings.

Pneumatic sprayers do not employ hydraulic pumps but distribute the spray by using an air compressor to exert a pressure on the liquid contained in a strong airtight tank. This method is particularly suitable for spraying but use is mainly confined to small machines, generally of knapsack type.

Some of the chief operating points to observe in using any sprayer are:

1. Correct dosage. With modern p.t.o.-driven machines this is simply arranged by choosing a set of nozzles of the correct throughput, seeing that the operating pressure is set to that recommended by the manufacturer, and taking care to drive the tractor at the correct speed—usually 6·4 km/h.

2. Avoidance of spray drift. The finely divided spray produced by low-volume nozzles can easily drift on to neighbouring crops. This may be a serious matter if weed killers are being used. Drift dangers can be minimised by choosing suitable weather and by using more dilute material at a higher application rate.

3. Cleanliness of machine and of water supply. All filters must be used and cleaned regularly, and all parts of the machine should be washed clean immediately after use. When changing over from weed spraying to the spraying of susceptible crops, use of a detergent such as washing soda is advisable.

4. Danger to operators. Some spray materials are very poisonous, and where these materials are used, certain regulations designed to safeguard health and life must be complied with. Fortunately, the " growth-regulating " sprays such as MCPA and 2,4-D are relatively safe to use, and the special safety regulations do not apply to the handling of these less-dangerous materials.

EQUIPMENT FOR IRRIGATION

The starting-point for any scheme of irrigation must, of course, be the availability of a suitable water supply, and the quantity of water needed to do a worthwhile job should not be under-estimated.

It requires 250 cubic metres ($\frac{1}{4}$ million litres) of water to apply 25 mm to one hectare and during a severe drought it may be necessary to apply 25 mm every 10–14 days in order to secure best results. Nevertheless, on average, only 50–150 mm per year are needed for most farm crops, though much more may be applied with advantage in exceptionally dry years.

When a stream is to be used as the source of supply, its flow, or the amount that may be extracted, must be reckoned in dry, summer conditions. Other sources of supply include boreholes, wells, ponds, and artificially constructed reservoirs. In all cases it is advisable to check the legal position concerning water extraction rights, as well as to ensure that the supply is technically suited to the need.

Having cleared the situation concerning water supply, it is advisable next to consider details of the crops to be watered. It is necessary to assess the total area to be covered in any one season, and also the extent to which the needs of the crops overlap, so that an estimate can be made of the total area to be covered during 10–14 days in a dry spell. When this is known, the area that must be covered in a day is easily reckoned, and knowing the time taken to put on say 25 mm, the area that must be watered at any one time can be decided. For example, a typical farm plant may cover 0·6 ha at a given moment, and may be capable of 3 moves daily. At this rate, 20 ha are covered in about 11 days.

Types of Distribution Equipment. The main types include :

1. Large-diameter rainers,
2. Small to medium diameter rotary sprinklers,
3. Oscillating spray-lines.
4. Mobile rainers and/or sprinklers.

Large rainers employ a large nozzle and high pressure to throw part of the water a long distance. This leads to the production of large drops which, applied at a high rate, can do considerable damage to the structure of unprotected soil. Rainers are best suited to crops that provide a full cover, e.g. established grassland and top fruit.

Small-diameter sprinklers are versatile equipment which can be used for a wide range of crops, including vegetables. They operate at a low or moderate pressure. Application rates can be varied by choice of nozzles to suit the soil and crop conditions. Quick couplings and aluminium sprinkler lines can make moving quick and easy.

Spray-lines, which water rectangular strips, are particularly suited to vegetable cropping. They cost more and take more time to move than rotary sprinklers, but can provide small drops and a uniform distribution suited to the watering of delicate crops such as lettuce.

Mobile equipment includes water-operated vehicles carrying either boom and sprinklers or an oscillating rainer ; and transportable (e.g. tractor-mounted) rotating booms with sprinklers. Mobile equipment can greatly reduce labour for moving.

Pumps. The type of pump and distribution mains employed depend on the source of water and the layout of the fields. Where the water is at a fixed point and the fields are so arranged that the first part of the main is always in more or less the same position, there is much to be said for a permanent pump-house with an electrically driven pump and a buried permanent main. The pump used for distribution is usually a single-stage centrifugal type, cheap to buy and largely self-regulating. A diesel engine may be used where electricity is not easily available. Where there are several scattered sources of water, it may be preferable to have a mobile engine-driven or tractor pump and a temporary main made of aluminium alloy.

It is necessary to make a careful calculation of the pump capacity needed, and account must be taken of such factors as the difference in levels between the supply and the highest field, and the friction losses in pipes, couplings, etc. It is essential to plan the whole scheme not forgetting such accessories as a good strainer, the necessary gate valve and pressure gauge near the pump, take-off hydrants, elbows, end stops, and so on. A plan should be made of the best method of working, with a view to economising in capital cost and time taken to move equipment. Time taken for moving varies considerably from farm to farm and from crop to crop. Whereas one man can move 0·4 ha of sprinkler equipment in about half an hour on short, level grassland, a typical time for moving the same equipment in a well-grown potato crop is about 2 man hours.

The total cost of irrigation varies considerably, according to the capital cost, which may range from under £75 to over £250 per hectare. When account is taken of depreciation and maintenance cost as well as the running cost of all equipment, and allowance is made for labour involved, a typical cost of applying 25 mm/ha of water by sprinkler equipment is in the region of £20 but may range from about £10 to well over £25, according to the type of plant and the amount of water applied. Thus, the cost of applying 75–100 mm to a crop of potatoes or sugar beet may be in the region of £50–75 per ha, excluding the cost of the water.

HEDGE CUTTERS

Tractor hedge cutters fall into three classes, viz. : those with a circular high-speed saw, those with a cutter bar of the reciprocating sickle type, and the flail type with a high-speed cylinder-cutter.

Machines of the circular saw type usually employ a heavy blade similar to that of a typical wood saw. The saw is usually on an articulated arm and is manœuvred by a number of hydraulic rams. The drive may be from the p.t.o. through a number of universal joints. Some cutters of this general type are driven by a hydraulic motor, and this effectively gets over the difficulties in the transmission of power to the cutter head. This type of machine is capable of dealing with overgrown hedges, and can cut branches several centimetres thick.

Reciprocating cutter-bar machines are usually p.t.o.-operated, and the best of them can tackle growths over 25 mm thick. The cutter bar is a heavy type with blunt-nosed fingers. There are variations between different makes in such factors as the reach of the machine, ease of operation, and the time taken to put it on the tractor.

The flail type is usually hydraulically driven, a p.t.o.-driven power pack being incorporated where necessary ; but cheaper, mechanically driven machines are also available. Flail hedge cutters have the advantage that young growth is well chopped up ; so the work of clearing up trimmings is minimised. For this reason flail machines are tending to replace other types for regular trimming.

For work on hedges that are very difficult of access, there are hand-held machines. In one type the light cutter-bar is operated by compressed air, and in another by electricity. Hand-held machines are for regular trimming of light growths, and have a considerably lower output than tractor-mounted machines.

DITCHING MACHINES

A wide range of ditching machines has been developed generally for ditch maintenance, rather than for ditch construction work that contactors' excavators can undertake. There is, however, no sharp distinction between farmers' and contractors' machines, for developments in the use of the farm tractor's standard hydraulic system have resulted in relatively inexpensive ditchers which can do very effective work in field ditches of the normal size.

Farm-scale ditchers are almost all of the same general type, but there are differences in cost, construction and performance. Some of the less expensive machines are designed for easy attachment to and removal from the tractor; but the connection must be rigid: hydraulically controlled weight-carrying pads to provide stability are desirable. Work rate of these machines is often limited by the power available in the tractor hydraulic system. A slewing boom enables the tractor to remain stationary during the digging, lifting, and emptying operations. The tractor runs parallel to the ditch and moves forward a short distance at a time. Types of bucket must be chosen to suit the work. Long, half-round types are suitable for working at high speed across the bottom of the ditch. The digging, lifting and slewing are controlled by independent rams. With some machines the basic elements may be employed in a wide variety of machines for digging, lifting, loading and hedge cutting. Such multi-purpose machines are normally operated by a control unit situated inside the tractor safety cab and used with the seat reversed. A one-tractor unit consisting of tractor, loader and trailer can be used for loading without unhitching.

More expensive ditchers are operated by a more powerful external hydraulic system, with an independent oil supply. The machine may employ 6–7 hydraulic rams, and is usually of the " back-acter " type. In such machines the boom pivot may be offset so that the bucket is drawn along the line of the ditch. In this type, a tapered bucket which automatically batters the ditch bottom to a suitable shape may be used.

TRENCHING AND PIPE-LAYING MACHINES

Comprehensive under-drainage schemes can only be carried out at a reasonable cost if all unnecessary hand work is eliminated. Digging of trenches for laying either tile drains or PVC drainage pipes is easily mechanised, and an increasing amount of this work is done as a combined trenching and pipe-laying operation. Two main types of trenching mechanism are used, viz. a large digging wheel and an endless bucket mechanism on a light boom. Both types are used complete with attachments for laying either ordinary tiles or coils of PVC drainage pipe. A porous back-fill may be delivered on top of the pipe before back-filling, if required.

The main difficulty with any pipe-laying machine is connection with existing drains. This applies even more to " trenchless " pipe-layers than to those which first excavate a trench and lay the pipe, and subsequently re-fill it. Trenchless pipe-layers use a very powerful tractor or winch to pull a broad blade through the soil, and the pipe is introduced behind the blade before the slot closes again.

With all pipe-layers an accurate grade is essential. This is usually achieved by manual/hydraulic control after sighting rods have been set up along the track of the drain. There are, however, radio-control methods by which the depth of work of the implement is adjusted by a distant radio operator, who takes sights on to the implement and adjusts its working depth via a radio signal to a solenoid valve in the implement's hydraulic system. Fully automatic depth control is achieved by use of a " laser-plane " system. A portable controller establishes a plane at the slope required ; and a unit on the pipe-layer regulates depth according to this plane.

Chapter VII

MANURES AND MANURING

In the section dealing with soils the meaning of the term fertility, as well as the various properties of natural soils and the constituents which contribute to it, were considered. It is the purpose of this section to consider what effect the growing of crops and husbandry generally can have upon fertility, what practices may lower it, and particularly by what agencies it can be maintained or augmented.

The Fertility Balance. Fertility of the soil has been shown to depend partly on the physical conditions or tilth, partly on moisture conditions, partly on bacterial activity, and partly on chemical composition. Man attempts to influence these various factors by suitable tillage operations, by rotations of crops, and by the addition of a number of manurial substances which will be considered in detail. Some of these are added for direct nutritional reasons while others fulfil a dual purpose of feeding the plant and improving or maintaining physical condition, thus increasing moisture retention and making the soil a more suitable medium for crop growth.

Leaving aside for the moment the question of tilth, it is a useful starting-point to consider manuring from the standpoint of plant nutrients and to view the soil of the farm as a unit containing a limited amount of organic matter, nitrogen, phosphoric acid, potash, lime, magnesium and some other substances required in greater or lesser quantities—the trace elements. Of the trace elements the soil usually contains all that successive crops require except under certain exceptional circumstances already referred to in a previous chapter (see p. 22). Before the farm is established, vegetation of sorts occupies the land. This vegetation is supplied with nutrients from the soil and from the atmosphere, and when it dies upon the ground it adds to the stock of humus and nitrogen, and returns the mineral matter which has been drawn from the soil. By this process the soil has gradually arrived at a certain level of fertility. If now man takes possession and grows crops to sell away, he at once disturbs the balance of this system, and very rapidly lowers the fertility level of the soil by removing soil constituents in the crops and also by causing other constituents to be lost in drainage. This loss may not be appreciable when considered in relation to the *total* reserves present in any soil, but it must be remembered that the loss falls primarily on the *available* portion of the total reserves, and so the soil becomes unable to provide anything like the amounts of some of the individual constituents necessary to keep up a high level of production. If the crops are used solely for feeding the farm stock or for the production of animal products sold off, the inroads on fertility are less severe than when the crops themselves leave the farm since some part of the nutrient constituents are retained and returned eventually to the soil in the animal's excreta. Nevertheless, even under the

166

best-managed conditions of mixed farming, some losses of soil constituents are bound to occur. A compensating effect is brought about when food, especially of the more concentrated kind, is bought in for the stock, because a good portion of its nitrogen, phosphates and potash remains on the farm, thus augmenting the natural reserves. Finally, the balance of fertility is greatly affected by the deliberate importation of fertilising substances from extraneous sources.

The main lines of movement of nutrient materials are shown in the following diagram :

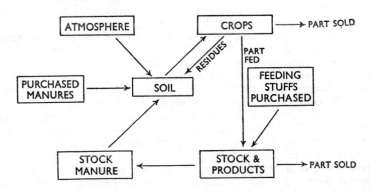

From this it will be seen that a certain amount of nutrient material is constantly being removed from the farm in the form of crops or animal produce ; a certain amount is circulating—from the soil, through the crops, stock and farmyard manure, back to the soil ; and a certain amount is added from the atmosphere (i.e. nitrogen and carbon), and from purchased feeding-stuffs and manures. Thus it will be realised that it is possible to balance gains and losses by adjusting the farming system to the inherent fertility of the soil, or to raise the level of fertility by additions of fertilising materials and so increase its production or turnover. It is equally possible, also, by specialising in one direction, to impoverish the soil in one or two of the main constituents unless measures are taken to replenish them from outside sources.

Many different systems of farming are practised or can be designed to make the most of the fertility of the soil ; but the decisive factors in adopting any one system are mainly economic, so that it is more useful to consider various prominent farming operations and the fertilisers procurable than to discuss the subject from the point of view of any particular system of farming. Any farming project must have a plan behind it. The maintenance of fertility will need to be thought out in relation to all parts of that plan. In this way it will be possible gradually to increase the soil's store of nutrients or, in certain special circumstances, to depress it. Nowadays there is a tendency towards supplying only those amounts of mineral fertilisers thought to be necessary for the particular crop to be grown. This opinion is based partly on experience of the use of nitrogenous fertilisers, where any excess beyond the present crops' needs may not be available to the succeeding crop, and partly on the somewhat spectacular results obtained by the placement with a combine drill of very small quantities of fertiliser close to the seed.

Fallowing.   The benefits of fallowing as a means of restoring the fertility of the soil were forced on the attention of the farming community at an early date.   When agriculture consisted mostly of corn growing and sparse grazing, man soon found that he inevitably exhausted the soil to a marked extent : he was forced to adopt a system whereby the land was left uncropped for a period of months or even years.   He next discovered that it was helpful, not simply to leave the soil to its own efforts at recuperation, but to assist the process by tillage during the resting period.   Thus the systematic fallowing of land became an established practice and in some systems of farming, particularly on clay soils, remains so to this day.   The explanation of the benefits produced is not even now fully understood, but it is known to a certain extent.   Some of the more obvious beneficial effects of the fallow have already been referred to in the chapter on tillage, such as the killing of all kinds of weeds by the actual drying out of the soil, the improvement of soil structure by alternate drying and wetting, the effects due to temperature changes, and the accumulation of nitrates through the work of nitrifying bacteria in the absence of a growing crop.   There are, however, other benefits beyond these.   It may be that the drying or heating effects on the colloidal coverings of the soil particles are such that, on re-wetting, a larger amount of the phosphate and potash in the soil comes into solution, or at all events is rendered more suitable for absorption by the plant roots.

Green Manuring.   Occasions arise in farming practice when it is preferable, instead of leaving ground unoccupied for an interval, to grow a crop with the express purpose of ploughing it in later. This process, known as green manuring, is a means of adding humus to a soil and has undoubted advantages in some circumstances.   If it is desired to follow a main crop with a green manure crop in the same year, it is essential to get the latter in quickly before the bare compacted surface left by its predecessor becomes totally dried out, a condition which follows rapidly, in dry weather, on the removal of the covering crop.   Green manuring is also useful as a means of conserving the nitrates accumulated in the soil during a fallow period, if a rotation crop is not planted before winter.   The winter rains would wash out the nitrates from an uncropped soil, but a green manure crop planted after a fallow absorbs the nitrates and holds them in its own tissues until it is ploughed under.   Various crops are employed, such as mustard on chalk soils, lupins on light sandy soils, tares on heavy soils and Italian ryegrass on a variety of soils.   The process is not always successful for, unlike an application of farmyard manure, it does not seem able simultaneously to raise the organic matter content and at the same time to increase the supply of available nitrogen.   If the green material is old and fibrous it increases the soil's store of humus but adds little available nitrogen.   On the other hand if the green crop is young and succulent it may increase the amount of available nitrogen but add little to the soil's store of organic matter. Apparently this latter effect is due to enhanced activity of the soil organisms, causing them not only to attack the easily decomposable succulent green material but also the stores of more stable organic matter. The factors which make green manuring successful are much the same as those which operate in ley farming or alternate husbandry, and no doubt are partly physical, partly biological as well as partly chemical. The land, when under a three or four years' ley, is considerably enriched in humus and nitrogen, whilst at the same time producing both pasturage and crops of hay. When the ley is broken up, and ploughed under, beneficial results become apparent both in the ease with which a tilth is established and in the enhanced yields of the arable crops which follow. Increased results can often be obtained by judicious treatment of the green crop with fertilisers, which not only increase the bulk of the organic material available for ploughing in but also build up the elements, nitrogen, phosphate and potash into more complex compounds later to be used as slow acting reserves.

FARMYARD MANURE

In spite of the growth of the fertiliser industry, farmyard manure is still one of the mainstays of British crop production and is worth all the care and attention which can be bestowed upon it. It is the product of the intermingling of the fæces and urine of various farm animals with straw or some other form of litter. There is obviously great variation in its composition, but in spite of this the properties of the final product are strikingly constant. A consideration of some of the factors involved in the making of dung is of use in explaining the points of practical management of this commodity.

The chief fertilising value of the material fed to animals obviously lies in its content of nitrogen, phosphoric acid and potash. Table 2 shows the approximate percentage of these elements in different fodders :

TABLE 2

FERTILISING ELEMENTS IN FODDERS (PER CENT.)

Fodder.	Nitrogen (N).	Phosphoric Acid (P_2O_5).	Potash (K_2O).
Decorticated cotton cake . . .	6·7	2·8	1·8
Groundnut cake	7·5	1·3	1·5
Beans	4·0	1·2	1·3
Oats	1·6	1·3	0·7
Meadow hay	1·5	0·6	2·0
Oat straw	0·5	0·3	1·5
Swedes	0·2	0·09	0·3

Working animals, fattening adult stock and store animals retain little of the constituents of their food : grazing stock, pregnant animals and animals in milk retain a good deal, especially the nitrogen and phosphoric acid. It should be observed that while most of the potash excreted occurs in solution in the urine, much of the phosphoric acid appears undigested in the faeces.

Different foodstuffs are of varying digestibility ; some of their constituents pass through the animal unchanged, and it seems reasonable to assume that what proves indigestible to the animal is not likely to be available to the plant until it has been broken down by some suitable agency. Of the portion of the food digested, some is used to provide energy for the animal and is in effect burnt up, some is stored as fat, and some is used to replace body tissues. The undigested portions are voided as fæces. The proteins are the nitrogen-bearing constituents, and much of the proteins may be used in producing growth (flesh), milk or young, thus ultimately leaving the farm. But a certain amount of the protein is used to replace similar compounds consumed by wear and tear in the animal's body, and this is the portion whose nitrogen appears in the urine as comparatively simple compounds, such as urea, which have been collected in a soluble form by the kidneys and thence passed out dissolved in water. Of the ash constituents, a little of the phosphoric acid and most of the potash also appear in the urine. The main trends of these various processes are set out graphically below.

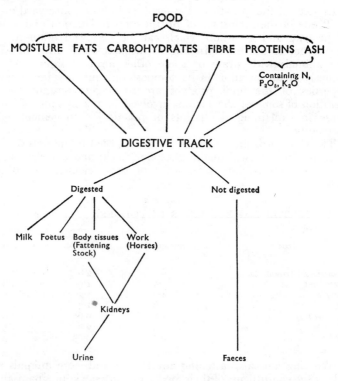

The litter also has a definite content of manurial constituents, as is shown by the analysis of oat straw in the table on page 169. The first function of litter is to absorb the urine, although conditions in which it accomplishes this to any considerable extent are rare. Straw will absorb two or three times its own weight of water, while peat moss can assimilate as much as ten times its own weight of water and in addition can absorb considerable amounts of ammonia gas, a compound which is formed in considerable quantities during the decomposition of urine. Its advantage over straw in absorptive powers, however, is offset by its stability and its resistance to decay, so that it does not rot down in the same desirable manner as farmyard manure made from straw litter.

Chemical Changes occurring in Dung. At the time of production, farmyard manure consists of a crude mixture of straw, fæces and urine, commonly termed " long dung " ; but this at once begins to undergo various changes which result in it ultimately producing a very uniform material in which many of the original differences of composition, due to the type of animal, richness of food, amount and type of litter, have been considerably mitigated if not obliterated. To begin with, the bulk of the soluble manurial constituents are in the urine, so that to the extent that this is allowed to drain away, they become a total loss. Liquid manure tanks and suitable drainage systems avoid this loss but, with the increased use of water for washing purposes in cowsheds, the difficulty of using them to the best advantage has been aggravated. The soluble compound of nitrogen, urea, is very quickly converted by bacteria into ammonia. Its presence, particularly in stables, is usually obvious. Ammonia appears as a gas whenever the litter or the floor dries out, but so long as conditions are moist it remains in solution. This change of urea into ammonium compounds is only one of many which are brought about by bacteria in the manure heap. Each change comes into prominence as the conditions of moisture and aeration in the heap become suitable.

The most important chemical changes which take place in the manure heap are as follows :

1. The conversion of urea into ammonium compounds.
2. The fermentation of the carbohydrates of the litter and fæces with the production of heat, various gases (such as carbon dioxide, methane and hydrogen) and a decayed mass of organic matter richer in nitrogen and darker in colour than the original straw.
3. The breaking down of the proteins of the litter and fæces into simpler compounds of nitrogen such as ammonia.
4. The assimilation and fixing of nitrogen as protein in the bacteria.

These changes become manifest in the gradual disappearance of any recognisable structure, the whole heap tending to become uniform in texture and colour. The raw, soluble compounds of nitrogen gradually disappear, and drainage from the heap takes on a dark brown or black colour. This is " dung liquor ", and its appearance is due to the presence of soluble compounds of ammonia and organic matter. When all these changes are well advanced, the heap is in the condition known as " short dung ". There is inveitably a great amount of wasting in the heap, both as regards total weight and amount of fertilising constituents. Observations indicate that some 15 per cent of the nitrogen is lost in the first few days, and that this loss steadily increases to as much as 40 or 50 per cent as storage continues. Yet, in spite of this, the final product as a rule is richer in nitrogen than the original components owing to the comparatively greater loss which falls on the non-nitrogenous constituents. In the diagram below the main changes which take place in a manure heap are set out :

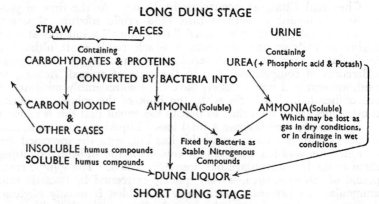

LONG DUNG STAGE

STRAW   FAECES   URINE

Containing   Containing
CARBOHYDRATES & PROTEINS   UREA(+ Phosphoric acid & Potash)

CONVERTED BY BACTERIA INTO

CARBON DIOXIDE   AMMONIA (Soluble)   AMMONIA (Soluble)
&   Which may be lost as
OTHER GASES   gas in dry conditions,
                                                 or in drainage in wet
INSOLUBLE humus compounds   Fixed by Bacteria as   conditions
SOLUBLE humus compounds   Stable Nitrogenous
                             Compounds
                     DUNG LIQUOR

SHORT DUNG STAGE

The initial losses of nitrogen may be minimised by allowing manure to accumulate beneath the animals, a practice carried out with success in the case of fattening animals in suitable covered yards. The daily removal of manure leads to the heaviest initial losses, especially if it is thrown out into a yard in casual fashion. Losses in yards and heaps may be cut down by collecting the manure in a well-compacted heap on a concave surface kept free from surface and roof drainage. The utmost saving in constituent nutrients is obtained by immediate removal to the field and ploughing into the land in fresh, long condition.

The Composition and Use of Dung. Enough has been said to describe the nature of the fertilising constituents in farmyard manure, both in the "long" and in the "short" dung stage. An average sample contains about 0·5 per cent nitrogen, 0·1 per cent phosphoric acid and 0·5 per cent potash, so that the amount of phosphoric acid is low in comparison with the other constituents. If these figures are converted into their equivalents of common nutrients, 25 tonnes of farmyard manure will contain 125 kg nitrogen, 25 kg phosphoric acid and 125 kg potash. The presence of these nutrients does not mean, however, that they are in an easily available condition. An average figure for their availability has been deduced from an examination of the results of many experiments in which the effect of dressings on yield were being examined. Most of the nitrogen in dung is contained in organic forms and some only becomes readily available for crops. Likewise, the phosphates and potash are more slowly available than those in fertilisers.

By comparing field experiments conducted both with and without farmyard manure, it was found that on dunged land fertiliser dressings can be reduced by 38 kg N (equivalent to 125 kg ammonium nitrate), by 25 kg P_2O_5 (equivalent to 63 kg triple-superphosphate), and by 75 kg K_2O (equivalent to 125 kg muriate of potash) per hectare to allow for the amounts of these nutrients in 25 tonnes of farmyard manure.

Farmyard manure, however, does not owe its value solely to the nitrogen, phosphoric acid and potash it contains : although in the case of dressings normally applied, experimental work has invariably shown that any increased yields of farm crops can be accounted for solely by the amounts of these nutrients it contains. There are five points with regard to the humified organic matter in it which must be borne in mind. Firstly, this constituent exerts a profound influence on the tilth of the soil, whether it be light or heavy. It opens up heavy soils and gives body to light soils. It opens up heavy soils by a mechanical loosening effect, it increases pore space and stabilises the soil crumbs, improves drainage and ventilation thus favouring the activity of micro organisms. Light soils cohere through the presence of colloidal humus matter, the pore space and permeability are reduced and water-retaining power increased. The increase in water retention is illustrated by the data coming from Rothamsted experiments. A plot treated every year with farmyard manure showed 3–4 per cent higher water content during the experimental period than the neighbouring plot to which no farmyard manure had been applied. From the same station comes further information that the water-stable crumbs in the soil of more than 0·5 mm. diameter were increased from 28 to 50 per cent following an annual application of farmyard manure over many years. Secondly, its content of nutrient compounds makes an almost permanent addition to the fertility of the soil which, by its continuous use, is considerably increased. Thirdly, it provides a source of energy for the beneficial micro-organisms in the vicinity of the plant roots. Fourthly, either by reason of its own composition or by some action on the soil as yet not understood, it makes available major nutrient elements like magnesium as well as trace elements to overcome conditions of deficiency affecting the growing crop. Finally, considered as a whole, the use of farmyard manure is probably the best means of maintaining fertility, for it mitigates to a great extent the effects of seasonal variations in weather, and in any one year it gives the most equable effect.

In the making of farmyard manure the great importance of the urine lies in the fact that its soluble nitrogen compounds enable the straw-rotting organisms to do their work, a point which has a bearing on the quantities in which straw should be supplied as litter. To prescribe a definite figure, such as 1 tonne of straw to every 50 kg of digestible protein in the food supplied, may be a counsel of perfection, but it is based on facts which are well worth attention in the making and management of farmyard manure.

Liquid Manure. Reference has been made to the urine of farm animals, to its composition, and to the loss involved in the absence of any liquid manure tank system. In the last century, when most livestock farms had storage tanks for collection of urine and drainage from the buildings, the liquid was either distributed on the fields or pumped over the solid dung heaps. The increased use in the present century of open yards and improved hygiene in cowsheds, with the consequent dilution of the drainage by both rain and washing water, has made the profitable utilisation of what, in many cases, has become little more than polluted water, a very difficult proposition. Any analysis of such an effluent, so varying in dilution, would not be meaningful, but if it be assumed that on most mixed farms the main contribution will be from cows' urine which has an approximate composition of 0·8 per cent nitrogen, 0·01 per cent phosphoric acid and 1·8 per cent potash, and that each cow would contribute some 13–14 litres (say 14 kg) of such urine per day, it is possible to get an idea of the total amounts of plant nutrients produced. Furthermore, it will be seen that such an effluent is in effect a nitrogen–potash stimulant very suitable for grassland so long as it is not so concentrated as to scorch the vegetation, although to make the best use of its high potash content it would be better balanced with extra nitrogenous fertiliser. At present prices the manurial value of this effluent is such that it is no longer cheaper for farms to allow it to run to waste and replace the plant nutrients by bought fertilisers.

Slurry. This form of liquid manure is a worrying responsibility to the farmer firstly as a result of The Rivers (Prevention of Pollution) Act of 1961, which has made it more difficult to allow effluent to run to waste ; and secondly because of the difficulties of dealing with farmyard manure in modern methods of intensive management of dairy cattle, pigs and poultry, where the costs of obtaining bedding and of handling manure in the solid state have resulted in the conversion of faeces and urine, sometimes with a minimum of litter material, into a slurry.

Such a slurry diluted with sufficient water to give a dry matter content of from 10 to 12 per cent, known as " Gülle " on the Continent, can be handled by pumping and spread on the land as a liquid. The addition of the faeces to the urine produces a better balanced manure and a typical winter slurry from dairy cows might contain 0·23 per cent nitrogen, 0·10 per cent phosphoric acid and 0·24 per cent potash. The potash content is still rather high for use on grassland and where this extra potash might interfere with magnesium uptake, balancing with extra nitrogen fertiliser might be advisable. The handling of slurry is, however, not without its problems. On many intensive livestock farms sufficient land is not always available to take the large volumes of slurry (45 litres per cow per day) produced by large numbers of animals housed together. In wet weather there is the difficulty of getting on to heavy land to dispose of slurry daily. Storage facilities are, therefore, desirable and the cost of producing these and of the handling equipment, whether handling be by tanker or by organic irrigation, can be appreciable. On the other hand a dairy cow produces some £22 worth, and a fattening pig some £2½ worth, of available nitrogen, phosphate and potash per annum at present fertiliser prices.

Poultry Manure. It is estimated that 1,000 light-to-medium type hybrid layers will produce about 1 tonne of fresh droppings, containing about 75 per cent moisture, per week. A similar number of broilers would produce some 400–450 kg of fresh droppings per week of their 10–13 weeks life. Modern methods of poultry management demand high egg production from the layer and high growth rate from the table bird, to meet both of which requires a high input of protein-rich food. Some 40 to 50 per cent of this food protein is excreted, giving droppings rich in both nitrogen and phosphoric acid. An average figure for the analysis of such fresh droppings would be 1·5 per cent nitrogen, 1·2 per cent phosphoric acid and 0·7 per cent potash. Whilst age and type of stock, kind of food and efficiency of utilisation will affect the composition of the manure, a far greater influence will be exercised by the method of conservation of the droppings and the kind and amount of litter mixed with them. Thus the litter from one crop of broilers on wood shavings could be three times as rich in nitrogen, phosphate and potash as the fresh droppings, due to the reduction in the moisture content. The addition of dry earth or absorbent material improves the physical condition and lessens the loss of nitrogen as ammonia from the stored material. Almost all of the fertiliser ingredients will be conserved if the poultry are kept on deep litter, but will be diluted by the absorbent material. Compared with farmyard manure, fresh poultry droppings contain about three times the quantity of nitrogen and phosphoric acid : a dressing of 7½–9 tonnes/ha will be the equivalent in these nutrients of 25 tonnes of farmyard manure. Experiments have shown that the nitrogen is from 50 per cent upwards as available as the nitrogen in inorganic fertilisers and the phosphate, being largely insoluble, acts slowly. It probably should not be the sole source of phosphate for arable crops (especially where there is a soil phosphate deficiency). The potash will nearly always require supplementing for root crops like potatoes and sugar beet. Being essentially nitrogen-rich there is always a danger of scorch damage to crops if the dressings are too heavy, this being especially so where very high nitrogen content broiler house manure (containing 4–5 per cent nitrogen) is applied at normal farmyard manure rates instead of at the more realistic 2½–4 tonnes/ha rate. The physical condition of the fresh droppings makes it difficult to handle and store ; conversion to slurry is practised and has the same problems as slurry produced from other farm stock, with the extra one that specialist poultry farms rarely have sufficient land to handle the slurry themselves. Processes have been elaborated for drying and grinding poultry manure, and the product appears to have an outlet as a fertiliser in horticulture as well as a nitrogen feed supplement to ruminant livestock.

Compost. Some years ago work at Rothamsted resulted in the elaboration of a process for the production of compost from straw and other vegetable refuse. This followed on discoveries made on the nature of the changes proceeding in the manure heap. Briefly, it was found that two sets of organisms, working independently, brought about two main results, the one the rotting of the straw and the other the fixation of nitrogen. It was also found that the former needed a supply of readily available nitrogen compounds to enable it to go on. By building up a heap of straw, layer by layer, each of which was well watered and given a sprinkling of chalk, and by washing in some easily soluble nitrogenous fertiliser such as sulphate of ammonia, a complete rotting down of the heap to a product which was very like short dung, and which gave similar results in the field, was effected. This material has been variously called artificial farmyard manure or compost.

On farms where there are few or no cattle the replacement of dung by other forms of humus is an important problem. The use of artificial farmyard manure or straw compost is virtually undeveloped. Green manuring, or ploughing-in of straw, or a combination of both practices may be a solution.

Sewage Sludge. Large quantities of sewage sludge are produced in all the large towns. Urban sewage consists of about 1 part of solid matter in 2,500 parts of liquid, and the bulk of the solid is separated by sedimentation and partially dried in lagoons freely exposed to air. Practically all the water soluble salts in the original sewage pass into the effluent, and much nitrogen, phosphate and potash are thus lost. Most of the sewage sludges undergo aeration during their production and in some cases are later allowed to heat in heaps. At a moisture content of less than 55 per cent a sludge will crumble and can be easily handled. Sewage sludge does not have the same beneficial effect on soil texture as farmyard manure.

The percentage composition of raw sludge, digested sludge and farmyard manure is set out in Table 3 :

TABLE 3

COMPARATIVE PERCENTAGE COMPOSITION OF SEWAGE SLUDGES
AND FARMYARD MANURE

	Dry Matter.	Organic Matter.	Ash.	Total Nitrogen.	Inorganic Nitrogen.	P_2O_5.
Raw sludge	40	20	20	0·9	0·05	0·5
Digested sludge . . .	52	23	29	1·4	0·06	1·1
Farmyard manure . .	20	16	10	0·6	0·06	0·4

Contrary to older opinion, digested sludges (those in which an aerobic fermentation has taken place) are more active than raw ones. The process of digestion improves the physical condition and also increases the proportion of available nitrogen.

Heavy dressings of sewage sludge furnish large amounts of phosphoric acid, half of which may be regarded as being as effective as the phosphoric acid in superphosphate. By comparison with farmyard manure, sewage sludges are poor in potash, a fact which is borne out by field experiments. Materials present in certain industrial sludges (zinc, etc.) may be directly injurious to crops.

Some municipal authorities compost sewage sludge with other

forms of town refuse such as dust-bin siftings. The resultant material is of the nature of compost but contains a high ash content. From the point of view of the authorities the production of such material partly solves the difficulty of disposal of voluminous and often unwanted materials by providing a substance which can readily be incorporated with the soil for crop and vegetable production.

Seaweed. Seaweed has long been used in coastal districts for bulk application to the soil. It contains about 80 per cent moisture, has more potash than farmyard manure and has the following approximate percentage nutrient content in the fresh condition: 0·6 nitrogen (N), 0·2 phosphoric acid (P_2O_5), 2·0 potash (K_2O).

FERTILISERS

In ordinary farming the purchase of fertilisers consists of buying commodities containing one or more of the principal plant nutrients, namely, nitrogen and phosphoric acid and potash. Each of these substances has its particular effects on plant growth, and as the form in which it is offered to the plant may also have specific effects, it will be profitable to consider individually the materials which are obtainable. Of these there is a wide choice available, and for each one there can be claimed special advantages in certain circumstances, so that a knowledge of the properties of the various kinds may enable the farmer to come to a decision as to which is most likely to serve him best for any particular purpose. It is useful, also, to bear in mind that some kinds are available only in comparatively limited quantities, a fact, among others, which has its effect on the price at which they can be bought. In the majority of cases a grower will use fertilisers to obtain a reasonably rapid effect, but in some cases he may be content to get a slower effect while gradually building up the fertility of his soil. In the former case he is best served by a material which is in a form immediately assimilable by the plant, or readily and quickly converted by bacterial or other action to an available form. Generally speaking, the more soluble it is in water the quicker it will act. In the latter case, substances of a more durable nature are desirable, such as will release the desired constituent slowly but evenly over a long period, so that regular additions continually increase the reserve in the soil, and hence the supply at any particular moment. Other points which are of practical importance are the keeping qualities of the manure, its concentration, the ease and comfort with which it can be handled, and the evenness with which it can be spread

NITROGENOUS FERTILISERS

The effects of nitrogen-containing manures upon plant growth are amongst the most spectacular which can be achieved by manuring, for nitrogen is the food material which is especially responsible for growth, particularly of foliage and stems. Rapid growth of these organs is to be desired on occasion, as in kale, but it may become a handicap if, for instance, the production of grain is the main object of the crop. The excessive use of nitrogenous fertilisers encourages the growth of the normally less valuable parts of some crops and may influence the proportion of straw in a grain crop, or of leafage in a root crop.   Late season applications (mid May–mid June) of nitrogen fertilisers to cereals tend to increase the yield of corn rather than straw.   The use of nitrogen has a marked effect on the appearance of herbage, giving to it a luscious green appearance ; indeed, the lack of nitrogen is often shown by a stunted growth, and also by pale or yellowish-green leaves.   There are two things to beware of in the use of a quick-acting nitrogenous manure—its tendency to retard the ripening processes and its tendency to produce succulent and flabby growth.   These effects become apparent more quickly when the soil is deficient in phosphates and potash.   Flabby growth due to this cause has two drawbacks ; it may be unable to support the weight of its own structure and so become laid, or it may more readily fall victim to the attacks of fungoid diseases.

Organic Refuse Materials.   The   important   organic   refuse manures and the percentage nitrogen they contain are as follows :

	Percentage Nitrogen.		Percentage Nitrogen.
Animal Residues—		*Animal Residues (continued)—*	
Shoddy	3–15	Fish manure	6·8
Hoof and horn . . .	13–14		
Dried blood	13	*Vegetable Residues—*	
Leather refuse . . .	12	Castor seed meal ⎫	
Fur, hair, skins, feathers .	8–12	Rape ,, ,, ⎬ . .	5–8
Meat meal	10	Malt culms or dust . .	3–4

Of those mentioned in the list above certain fertilisers, viz. meat meal, fish manure, oil seed refuse, etc., contain phosphoric acid and potash in addition to the nitrogen.

The Animal Residues originally contain large quantities of protein, and unless they become contaminated with dirt or are mixed with other substances during the manufacturing processes, they may contain up to 15 per cent nitrogen.   Their utility and value depend partly on their nitrogen content but more on their physical state. For instance, the barbs of a feather or finely ground fragments of hoof and horn have an obvious advantage over pieces of shoe leather when considered as fertilisers.   All these materials may be regarded as comparatively slow-acting manures, useful for raising the fertility of the soil, i.e. they have residual values.   They are therefore in demand by intensive growers for use in market gardening, hop and fruit growing.   This, coupled with the rather limited supply, probably explains the fact that they are usually the most highly-priced forms of nitrogen.

Shoddy is a by-product of the woollen industry.   Pure wool is wholly protein and contains 15 per cent nitrogen, so that the fragments which are discarded in woollen manufacturing as shoddy will approach that figure.   But in some processes the fabric manufactured contains an admixture of cotton, a substance which contains no nitrogen, so that it is possible to have almost any percentage of nitrogen in shoddy according to the nature of the material being made up.   For this reason shoddy receives special mention in the

Fertilisers and Feeding Stuffs Act, which exempts it from the statutory declaration of its nitrogen content. Shoddy is, on the whole, a finely divided material, but it is of such a nature that it is bulky and not easy to spread in small quantities. It is highly esteemed by intensive growers, who often apply it at the rate of 2·5 tonnes/ha.

Hoof and Horn also contains a high proportion of protein and consequently of nitrogen. It is dried and ground so that an effective spread can be obtained with quite low dressings per hectare.

Dried Blood also is marketed in a very good physical condition. Its content of nitrogen varies as sometimes the drying process is helped by the addition of lime which, of course, has its own value but must inevitably reduce the percentage of nitrogen in the final product. This material is classed as a moderately quick-acting fertiliser.

Leather refuse is occasionally available as a fertiliser, as are also fur, hair, skin, feathers and meat meal, etc. All these commodities vary so much from one consignment to another that it is impossible to describe them precisely. They all have potential fertilising value, but their value can only be computed with difficulty. Sometimes the material is of such a tough nature that it can be recognised in the soil for years ; in this case it is obviously of poor value as a manure. It is necessary that it should be finely divided or that it should be chemically treated in order that it may easily undergo the changes by means of which its nitrogen content becomes available for plants.

The Vegetable Residues form a group by themselves. They are usually the result of processes which have extracted certain constituents of seeds for technical purposes, leaving a residue which for some reason or other is unsuitable for use as a foodstuff. It will thus be realised that these residues cannot as a rule be very rich in nitrogen. The average seed contains carbohydrates, fibre, ash, moisture, oil and protein. In the case of rape and castor seed the last two are the most abundant constituents. The oil is extracted for various purposes, leaving a residue in which the protein provides nitrogen to the extent of about 6 per cent of the bulk. Small amounts of phosphoric acid and potash are also present.

All these organic manures owe their chief value to their nitrogen content, this often being of a form which becomes available gradually. In some instances the conversion is so rapid that an immediate response can be seen in the crop, and the manure is practically used up in one season : this happens with dried blood and seed refuses. With others, such as shoddy, hoof and horn, fur, leather, etc., there are greater or less residual effects, so that their use results in a semi-permanent improvement in the fertility of the soil. This property is reflected in the practice of awarding compensation for their use on the termination of a tenancy.

Inorganic Nitrogen Fertilisers. The merit of these fertilisers lies in the fact that they provide nitrogen in a concentrated form, convenient to handle and distribute, and readily assimilable by plant roots. Nitrates are the compounds which in the soil form most of the plant's immediate source of nitrogen, and ammonium compounds are only one stage removed in this respect, being easily converted to nitrates by the action of certain bacteria in the soil. Each kind has its merits and demerits, and to make the best choice it is necessary to consider these in connection with the soil and crop on which it is intended to use them. For instance, ammonium compounds are better suited to well-limed soils, while fertilisers containing a lime base may have an advantage for general purposes on soils which are, or tend to be, acid. Ammonium compounds are generally regarded as more suitable for potatoes, while nitrates have an advantage with such crops as wheat or mangels. It is well to remember, however, that these differences are more relative than absolute, as ammonium compounds soon undergo nitrification when conditions are favourable and are converted into nitrates. In fact, it may well be that the ammonium ion is itself absorbed by some crops and utilised therefore as a source of nitrogen. The main characteristics of our chief nitrogenous fertilisers will now be described.

Sulphate of Ammonia. At first produced as a by-product of the manufacture of coal gas, sulphate of ammonia has been used as a fertiliser in this country for over a hundred years. The conversion of the ammonia gas liquor into sulphate of ammonia has long ceased to be economic and practically all the fertiliser produced now in Britain is of synthetic origin. However produced it has the same properties and is composed of small needle-like crystals which are perfectly dry when fresh, are easily distributed, and can be kept for a reasonable time without losing condition. Like all ammonium salts which do not contain a metallic base, its conversion in the soil into nitrate entails the loss of a certain amount of the reserve lime which combines with the sulphate radical and is washed out of the soil. The effect of this action soon becomes apparent, especially on light soils poor in lime, when continued use of this fertiliser will bring about a decrease in soil pH. Sulphate of ammonia has a nitrogen content of 20·8 per cent, and this relatively low nitrogen content precludes its use in the manufacture of the present-day high-analysis compound fertilisers ; furthermore the sulphate radical is costly to produce and thus sulphate of ammonia can no longer compete economically with ammonium nitrate and the other more concentrated nitrogen fertilisers. Whilst at the present time it only constitutes about 2 per cent of the straight nitrogen market, sulphate of ammonia will continue to be available as a by-product of the synthetic fibres industry and to be compounded in the lower analysis compounds used, for example, in horticulture.

Ammonium Nitrate. Ammonium nitrate is probably the most extensively used nitrogen fertiliser in Britain at the present time, both as a source of straight nitrogen and as a component of mixed or compound fertilisers. Containing 35 per cent nitrogen, it is only exceeded in concentration by urea and anhydrous ammonia, with the advantage that half the nitrogen is in the rapidly available nitrate form and half in the slightly more slowly available ammonium form. Ammonium nitrate has for long been used in the manufacture of explosives and the widespread use of the undiluted material as a fertiliser is a comparatively recent development due partly to the unstable properties

of the material formerly produced and partly to its extremely hygro-
scopic nature and tendency to cake.   Since 1965 a safe, commercially,
pure, salt has been produced in the form of prills (cooled and hardened
droplets) packed in waterproof polythene bags and marketed in this
country under the brand names of " Nitram " (34·5 per cent N),
" Nitra-Shell 34 per cent " and " Nitro-Top " (33·5 per cent N).
Provided the bags are kept intact and the contents dry until immediately
before use, these materials are in excellent physical condition and easy
to distribute.

Ammonium Nitrate/Chalk Blends.   Prior to 1965, the only
forms of ammonium nitrate on the British fertiliser market were blends
of the synthetic salt with either chalk or ground limestone.   Such
material is still produced and, being in the form of even granules, is
in excellent physical condition, safe to handle and supplying both
nitrate and ammoniacal nitrogen.   Furthermore, the chalk or limestone incorporated in the fertiliser neutralises the acidifying nature
associated with all ammonium salts and make such fertilisers suitable
for all soils.   " Nitro-Chalk " was one of the earliest of these materials :
formerly marketed containing 15·5 per cent nitrogen it is now made
with 25 per cent nitrogen.

Nitrate of Soda, extracted from the vast deposits of " Caliche "
which occur in Chile, has been imported into Britain since the beginning
of the nineteenth century.   Despite improvements in its physical condition and its rapid action in the soil, the relatively low nitrogen
content coupled with freight costs no longer enable it to compete with
home-produced synthetic ammonium nitrate and its limited use is
largely confined to horticulture.   The sodium radical has the value
that it can replace potassium as a plant nutrient to a limited extent,
but it also has the less desirable effect, when used intensively on heavy
soils, of producing a sodium clay, whose presence becomes evident
in poaching, deflocculation, stickiness and loss of tilth.   Nitrate of
soda contains 16 per cent of nitrogen.

Urea.   The world use of urea as a fertiliser has shown a rapid
increase in the past ten years.   Its attraction lies in its ease and cheapness of manufacture and its high nitrogen content ; it contains 46 per
cent of nitrogen.   Whilst, like ammonium nitrate, it does tend to
absorb moisture and become difficult to handle, prilling the commercial product has improved it in this respect.   Due to the presence
in soil of an enzyme called urease, urea is rapidly broken down to
liberate ammonia gas.   If this change occurs on or near the surface,
ammonia may be lost to the air, which not only results in waste of
applied nitrogen, but can also lead to severe ammonia scorch damage
to any crop that might be growing.   This problem can be overcome
by working the urea into the soil where, on heavy soils at least, the
ammonia liberated is adsorbed by the soil colloids and becomes available for plant use.   As it is often impracticable to work fertiliser into
the soil, until research has found methods of preventing this loss there
will always be the risk of loss of efficiency when urea is used.   Urea
itself is much less likely to scorch foliage than either ammonium or
nitrate fertilisers and this together with its high nitrogen content,

makes it a very useful substance for foliar sprays. The high mutual solubility of urea and ammonium nitrate enables a mixture of the two in water to be made with a higher nitrogen content than either separately would give when dissolved. Such mixtures are commonly used in the preparation of liquid nitrogen fertilisers. Urea is also used in stock feeding as a source of non-protein nitrogen.

Ammonium Phosphate. Synthetically produced ammonium phosphate is a very important source of both nitrogen and water-soluble phosphate and will be described under the heading Phosphates.

Anhydrous Ammonia. This is the most concentrated of all the nitrogenous fertilisers : it contains 82 per cent nitrogen and consists of pure, water-free ammonia. Although a gas at normal temperatures, for purposes of transport, storage and application as a fertiliser it is maintained as a liquid under pressure. Its commercial use as a fertiliser was first developed in the U.S.A. after the second world war and at present totals 50 per cent of all the straight nitrogen used in that country. Commercial application of anhydrous ammonia in Britain commenced in 1965 and some 38,000 tonnes are now applied annually.

Because it reverts to a gas as soon as it is released into the atmosphere, anhydrous ammonia must be injected into the soil at least 150 mm below the surface, where the ammonia is held by the soil colloids (see p. 9) within a radius of 100 mm of the point of release. Ideally, ammonia should be injected when soil moisture and tilth permit rapid and complete closure of the injection channel, otherwise both loss of nitrogen and damage by scorch to any crop present could occur. Heavy, excessively wet or dry cloddy, coarse textured or stony soils can present problems of this nature. The specialised equipment necessary for both injection and transport of anhydrous ammonia is costly : this confines its use to very large farms, syndicates or contractors. Despite the fact that the nitrogen in anhydrous ammonia can be produced at about one-third of the cost of that in ammonium nitrate, the higher cost of application and the fact that the soils and crops in this country are not so suited to its use make it doubtful if it will ever reach the same proportions of the straight nitrogen market as it has done in the States.

Aqueous Ammonia. Ammonia in aqueous solution has been used for many years in the form of crude gas liquor. When it no longer remained profitable to convert this by-product from gas works and coke ovens to sulphate of ammonia, it became common practice to transport the relatively dilute liquor in bulk and apply directly to the surface of the ground. Such liquor containing about 1·7 per cent nitrogen, provided it is free from such toxic substances as tar, phenols and thiocyanates (all of which are produced when coal is gasified), can be a useful fertiliser for farms within a close radius of the source of production. The liquor tends to scorch green crops such as grass and, to reduce such damage to a minimum, it should be dribbled rather than sprayed on the grass as soon as possible after the crop has been cut or grazed. The centralisation of coal gas production has been responsible for a decline in the use of this liquor.

More concentrated solutions of ammonia are used in the U.S.A. and are being tried in this country. Whilst solutions containing up to 28 per cent nitrogen can be made at atmospheric pressures, to obtain higher concentrations the ammonia must be kept in solution under pressure. With both types of solution loss of ammonia gas by volatilisation means that the solutions are unpleasant to handle and must be injected into the soil to prevent scorch damage and loss of nitrogen.

Non-pressure Nitrogen Solutions. The high mutual solubility of ammonium nitrate and urea enables relatively high analysis fertiliser nitrogen solutions to be made, higher in nitrogen than solutions of either one of the component salts singly. Such solutions containing from 28–32 per cent nitrogen compare very favourably in analysis with solid nitrogen fertilisers with the advantage of ease of handling and distribution by spraying machine.

Soot. Soot is widely used, but not in very considerable quantities, as a nitrogenous fertiliser. It contains its nutrient material in the form of sulphate of ammonia. The best material comes from domestic chimneys ; boiler soot is of limited value. Its continued use in old gardens probably accounts, in part, for the dark colour of the soil. It is said to have a beneficial effect on the structure of clays.

PHOSPHATIC FERTILISERS

Equal in importance among the plant nutrients commonly added to the soil by means of fertilisers comes phosphoric acid. This has its own peculiar effects on the growing plant. It fosters the development of the seedling and enables it to produce a more vigorous, fibrous root system which, in its turn, leads to a healthy growth above ground. It counteracts the weakening effects of excessive supplies of nitrogen, and helps to produce herbage of a much more nutritious kind. Finally, it quickens up the ripening processes in the plant, a useful and very real help with cereals in some circumstances.

All natural forms of phosphate are insoluble in water and therefore slow in action. There are two ways in which this drawback can be overcome in fertiliser practice. The raw material may be treated chemically in order to render it water soluble ; or it may be ground very finely by suitable mills in order to give it a greatly increased surface area on which the dissolving or assimilating influences in the soil and roots of plants can operate.

For the water-insoluble forms of phosphate, solubility in a 2 per cent solution of citric acid has proved to be a very good indication of their expected rate of action.

Mineral Rock Phosphates. Phosphorus occurs naturally in many parts of the world in forms of the mineral apatite. These deposits vary considerably not only in phosphate content but also in the hardness of the rock and hence the availability of the phosphorus. The majority of these rock phosphates are not suitable for direct application as fertilisers, even when finely ground : they have to undergo acid treatment to be converted into one or other of the water-soluble phosphates.

H

The mineral rock phosphate obtained from the Gafsa district of Tunisia in North Africa is, however, much softer than other naturally occurring forms. When this material is finely ground so that not less than 80 per cent passes the 100 mesh B.S. sieve, it is a useful slow-acting source of phosphate for grassland and swedes and turnips on the more acid soils of the higher rainfall areas of the west. Whereas many of the harder rock phosphates, unsuitable for direct application, contain more than 30 per cent P_2O_5, the Gafsa rock is more commonly found containing 27–29 per cent P_2O_5.

In addition to the normally ground Gafsa phosphate a proportion of this material is ground to a superfine condition (not less than 80 per cent through the 0·053 mm mesh sieve) and marketed in the United Kingdom. Although it would be expected that such fine grinding would improve the availability, results of experimental work in the field comparing this material with normally ground Gafsa phosphate would not warrant the extra cost of fine grinding.

Superphosphate. For many years this was the most widely used phosphate fertiliser, chiefly on account of its high solubility in water and its rapidity of action. It has been largely superseded by triple superphosphate and ammonium phosphate, both of which are more concentrated and therefore more suitable for use in the modern high analysis compounds. Superphosphate is made by treating ground rock phosphate with the calculated necessary amount of sulphuric acid to convert the insoluble tricalcium phosphate (in the ground rock) into water-soluble monocalcium phosphate. The resulting superphosphate is a bulky, greyish, powdery or granular material consisting mainly of this monocalcium phosphate and calcium sulphate in approximately equal proportions, and containing from 18 to 21 per cent water-soluble phosphoric acid. In the soil it probably passes rapidly to the diacalcium salt which, although no longer so water-soluble, is in a sufficiently fine state of division to be easily available to the plant. The calcium sulphate could be a useful source of sulphur in soils where the latter is deficient.

Triple Superphosphate. This fertiliser is made by treating ground rock phosphate with orthophosphoric acid instead of sulphuric acid. It is composed mainly of monocalcium phosphate with no calcium sulphate and contains 46–48 per cent water-soluble phosphoric acid. In its behaviour it resembles superphosphate, of which it may be considered to be a more concentrated form.

Ammonium Phosphates supply the water-soluble phosphoric acid as well as some of the nitrogen in many of the high analysis compound fertilisers on the British market. They are normally made by passing ammonia into orthophosphoric acid which itself is produced from ground rock phosphate. Whilst two ammonium phosphates, mono-ammonium phosphate containing about 12 per cent nitrogen and 60 per cent water-soluble phosphoric acid, and di-ammonium phosphate (21 per cent nitrogen and 53 per cent water-soluble P_2O_5) can be prepared, at the present time little ammonium phosphate is sold as a straight fertiliser ; the bulk is used in the manufacture of high analysis compound fertilisers which are obtained by adding the necessary amounts of nitrogen (either as ammonium nitrate or ammonium sulphate) and potash (as muriate of potash) to an ammonium phosphate slurry.

Superphosphoric Acid and Ammonium Polyphosphates. The use of more concentrated phosphoric acids than orthophosphoric acid, e.g. metaphosphoric acid, pyrophosphoric acid and the polyphosphoric acids, enables the production of more concentrated fertiliser salts. Both calcium metaphosphate (64 per cent P_2O_5 in the commercial fertiliser) and potassium metaphosphate (57 per cent P_2O_5 and 37 per cent K_2O) have been produced and have been shown experimentally to be valuable sources of phosphorus. However, neither is commercially available in Britain at the present time.

A mixture of the above three acids, along with some unchanged orthophosphoric acid, is produced when the latter is dehydrated, and such a mixture is known commercially as superphosphoric acid : it contains about 70 per cent P_2O_5 (of which some 45 per cent is as orthophosphoric acid). Use of this superphosphoric acid in the formulation of both liquid and granular fertilisers permits the production of higher analysis compounds than can be produced using conventional orthophosphates, with the extra advantage that superphosphoric acid is a sequestering or chelating agent allowing the incorporation of heavy trace element metals in solution fertilisers, something which is not possible when normal orthophosphoric acid is used.

Ammonium polyphosphate, containing from 56–61 per cent phosphoric acid and 10–16 per cent nitrogen, is a stable, non-hygroscopic material which could find a place in the fertiliser industry in Britain where higher analysis compounds are required. As yet, however, little is known as to whether fertilisers based on ammonium polyphosphates are as effective as those based on ammonium phosphates (ammonium orthophosphates).

Basic Slag. This material accounts for approximately one-fifth of all the phosphorus fertiliser used in Great Britain and is the most widely used straight phosphate. For long associated with the encouragement of wild white clover, it still maintains its popularity as an improver of grassland, especially on the heavier soils and in the wetter regions of the country.

Basic slag is a by-product of the manufacture of steel from pig-iron. Many, but not all, iron ores contain phosphorus, and when such iron ores are smelted in the blast furnace this phosphorus is left in the pig iron. In the conversion of pig iron to steel the phosphorus is oxidised and extracted with lime or limestone in the form of a molten basic slag which, being lighter than steel, can be floated off. Such slag on cooling is ground to a fine powder and constitutes a very useful source of fertiliser phosphorus. The phosphorus in basic slag is in the form of calcium silico-phosphate which is not water-soluble ; it is, however, soluble to some degree in a 2 per cent solution of the weak acid citric acid and the extent of its solubility in such a solution has long been recognised as a useful indication of the availability of the phosphate to the plant. In assessing the value of a basic slag not only is the total phosphate content (referred to as the Grade of the slag) important, but also the amount of that phosphate which is soluble in 2 per cent citric acid (the citric solubility) and the proportion of the material which will pass through a 0·5 mm sieve.

The amount and availability of the phosphorus in slag vary with the phosphate content of the original iron ore and the actual steel-making process. The bulk of the home-produced basic slag is from British ores rich in phosphorus which have been converted to steel by either one of the basic oxygen " Bessemer " type processes or in modified " open hearth " plants. Such high-grade slags contain from 10–16 per cent total phosphoric acid, are of high citric solubility (80–95 per cent of the phosphoric acid being soluble in citric acid) and are ground so that at least 80 per cent will pass the prescribed test sieve.

The presence of free calcium oxide and calcium silicate confers a neutralising value on slag, making it almost equal, weight for weight, with ground limestone in its effect on soil acidity.

Basic slag contains appreciable quantities of magnesium, manganese and iron : there is evidence that a proportion at least of this content is available to plants, as would also seem to be the case with such elements as copper and cobalt present in trace amounts.

One disadvantage of the finely powdered basic slag is its dirty, dusty, nature which makes handling the material unpopular on the farm : the spreading of slag is therefore commonly a contractor's job. Furthermore, there is the drift hazard to nearby dwellings, or to adjacent land being grazed or carrying vegetable crops ready for harvesting. The introduction of a " mini-granular " basic slag, of particle size range 0·15–0·60 mm., which eliminates all the fine dust, has provided the farmer with a free-flowing material which is clean to handle and virtually free from drift, which advantages, it is claimed, outweigh the inherently slightly lower initial availability of the phosphate.

Bone Products. Raw bones contain fat, gelatine, calcium phosphate and calcium carbonate. The fragments are extracted by various means, such as steam or solvents, to obtain the valuable fats and gelatine, and *Bone meal* is the residue after the extraction of the fat. It contains 21 per cent phosphoric acid and 4 per cent nitrogen, the gelatine not having been removed. The bone structure is still recognisable in the fragments. *Steamed bone flour* is the result of more drastic treatment ; the gelatine being largely removed, the amount of nitrogen falls to 1 per cent, while the phosphoric acid rises to 28 per cent. At the same time the product is finer and more dusty than bone meal. The presence of the nitrogen and the difference in composition between the two forms are reflected in their prices. Much of the steamed bone flour produced nowadays is used by the manufacturers of animal feeding-stuffs. The costs of bone products relative to the nitrogen and phosphoric acid they contain are high. The phosphoric acid of bone fertilisers is only slowly available to plants.

Fish Guanos. The so-called fish guanos (or fish meals) are manufactured from unmarketable fish and their offals. Their composition is variable, with 7–8 per cent nitrogen and 4–8 per cent phosphoric acid. They are too expensive for general farm use, and are employed on horticultural holdings in the same way as meat and bone meals.

POTASH FERTILISERS

Potash as a plant nutrient nowadays receives much more attention that it used to do. In the plant it is found in stems and leaves more than in the grain, and its function is to help to establish and maintain the plant in a healthy growing condition. It is intimately connected with the assimilation process in the leaves and in the production of starch and sugars. This helps to explain its beneficial effect on quality of fruit, grain and tubers. Plants suffering from lack of potash show a dull blue colour in the leaves and often a characteristic marginal scorching. In cereals it often results in softness of straw and lodging. The range of potash fertilisers is small and the points of difference between the various kinds are not numerous. They are all easily soluble in water, but in the soil they are fixed by chemical reaction with the mineral complex and so are not liable to be washed out by rain, except probably in the lightest of sandy soils. The chief features of commonly used potash fertilisers are described later.

Chlorides, usually called muriates, constitute nowadays the chief bulk of the potash fertilisers. In addition there is still some sulphate of potash and small amounts of kainit containing sodium or magnesium or both. Since for several high grade crops, such as tomatoes, the sulphate radical seems to produce consistently higher quality than the muriate, sulphate of potash is the form preferred in horticulture. Potatoes show the same preference, but cost and supply must be taken into account ; there is no doubt that custom and use have veered almost completely in the direction of the muriate.

In the past all the potash fertiliser used in this country was imported, mainly from East and West Germany, France and Israel. However, a deposit which exists some 1,300 m below ground-level in north-east Yorkshire is now being worked, and this will not only supply this country's requirements but also a sizeable surplus for export.

Muriate of Potash. This material contains a minimum of 60 per cent K_2O and is now granulated when sold for use as a straight potash fertiliser. It has a common salt content of up to 3 per cent.

Sulphate of Potash. This material, which reaches the market in the form of a white crystalline powder, contains 50 per cent of K_2O and is now used almost exclusively for fruit, market gardens and for crops under glass which are susceptible to chloride damage when muriate of potash is used. Sulphate of potash produces potatoes with a higher dry-matter content than is the case when muriate of potash is used, and this might justify its use in compound fertilisers for potatoes grown for the processing trade.

Kainit. A small amount of kainit, containing from 15–20 per cent K_2O and about half its weight of common salt, is on the market and is used mainly in East Anglia on sugar beet. Kainit may also contain up to 10 per cent of magnesium and is thus a useful source of this element.

Sulphate of Potash—Magnesia. An undoubted benefit which arose from the use of the Stassfurt kainit was due to its magnesium content. For this reason the re-introduction of potash fertilisers containing magnesium could be of considerable benefit. Sulphate of Potash—Magnesia (28 per cent K_2O and 5–6 per cent Mg) is obtainable on our market.

Nitrate of Potash. This fertiliser has the advantage of supplying both potassium (44 per cent K_2O) and nitrate-nitrogen (13 per cent N) in the one compound. It has long been recognised as an excellent fertiliser, although hitherto its use in agriculture has been restricted due to high costs of production. With costs of production coming more into line with those of the other potash fertilisers its fertiliser advantages which include absence of the harmful chloride ion, presence of nitrate nitrogen, and higher solubility than sulphate of potash, could well warrant its use in both solid and solution compounds, especially where the chloride ion in muriate of potash is a disadvantage.

Flue Dust from industrial sources (5–15 per cent K_2O), **Kelp** the ash of seaweed (12–16 per cent K_2O) and **Wood Ash** (up to 5 per cent K_2O) are all sources of potash occasionally available. Bracken is notoriously high in potash (40 per cent K_2O in the ash) and its value as compost should compensate for the trouble taken to cut and prepare it.

MAGNESIUM FERTILISERS

Magnesium, like nitrogen, phosphorus and potassium, is an essential major nutrient removed in crops from the soil. Whilst most farmers know how important it is to replace nitrogen, phosphate and potash in the soil if maximum production is to be achieved, it is only in recent years that the need for magnesium fertilisers has been recognised.

Magnesium deficiency is associated by most farmers in this country with the disorder *Hypomagnesaemia*, or grass staggers, in livestock, but the direct effect of magnesium deficiency in cultivated crops is rapidly becoming of equal importance.

In the plant, magnesium is an essential constituent of the chlorophyll responsible for the green colour in leaves and necessary for the production of sugar and starch. Thus magnesium deficiency shows itself by non-green patches (yellow, brown or sometimes red and purple) on the leaves. Magnesium is necessary for the movement of phosphorus in the plant and for the development of seed, consequently a deficiency can have a severe effect on crop yield. As the plant is capable of transferring magnesium from the older to the younger leaves, deficiency symptoms generally occur in the older leaves first. Whilst crops like barley, oats, sugar beet and potatoes exhibit the above visual symptoms when deficient in magnesium, grasses commonly do not and thus the grazier is deprived of any helpful pre-warning of the possibility of hypomagnesaemia.

In the past the small but steady supply of available magnesium resulting from the weathering of the soil minerals was augmented by the addition of farmyard manure, a fairly rich source of magnesium (25 tonnes would supply 75 kg magnesium element) or by such magnesium-rich compounds as kainit and low-grade potash salts. The use of these sources of magnesium has declined ; at the same time, not only have increased crop yields, consequent on higher usage of nitrogen phosphate and potash, meant greater requirements for magnesium from the soil but the more intensive use of lime- and ammonia-based fertilisers has resulted in greater loss of magnesium in the drainage water. Furthermore, high rates of nitrogen, potash and sometimes lime, can have an antagonistic effect on what magnesium is available and so induce magnesium shortage in the crop. Such induced magnesium shortage is recognised as being associated with luxury uptake of potash in young spring grass.

Fertiliser sources of magnesium, in addition to those mentioned above, include basic slag, ground magnesian limestone and magnesite (54–55 per cent magnesium). There may, however, be cases where basic slag is not the most suitable form of phosphate or where the soil pH is already high and the use of magnesian limestone or magnesite could lead to over-liming troubles.

Kieserite is a product of the refining process in the production of muriate of potash ; it is a form of magnesium sulphate and contains 16–17 per cent magnesium. It is in the form of a greyish-white crystalline powder.

Epsom Salts are also a form of magnesium sulphate, containing 9–10 per cent magnesium. Easily soluble it may be used as a quick-acting spray ; it is, however, relatively more expensive than keiserite and being purgative may have an adverse effect on stock if sprayed in quantity on herbage being grazed.

COMPOUND FERTILISERS AND SPECIAL MIXTURES

Compound or mixed fertilisers are fertilisers containing two or more of the essential plant nutrients. Whilst at one time these were made by simply mixing compatible straight fertilisers in the proportions required for any particular purpose, there is also on the market a range of compound fertilisers produced by complex chemical synthesis from such raw materials as rock phosphate, ammonia, nitric acid and muriate of potash, resulting in final products of much greater concentration than could be attained in the old mixtures of straights.

Most fertilisers, straights and compounds, are now available in a granular or prill form and have many advantages over powders. They are much more convenient to handle, and they flow more evenly through the drill. They do not absorb moisture so rapidly, and in consequence they do not cake or form hard lumps so readily during storage. When used as a top dressing they tend to bounce off plant foliage rather than adhere to it. In the case of water-soluble ingredients these dissolve more slowly in the soil giving a steady flow of nutrients over a longer period. For insoluble phosphates such as ground rock phosphate and basic slag, granulation, because it reduces the surface area on which the soil solution can act, is not desirable. Two commonly used examples of mixed fertilisers marketed in the powdered form are Potassic Basic Slag ("K-slag") and Potassic Mineral Phosphate (P.M.P.), both of which contain their phosphate in non-water-soluble form.

The value of a compound or mixed fertiliser obviously depends on its content of valuable plant nutrients. The purchaser of a fertiliser is fairly adequately safeguarded in this respect by law. The Fertilisers and Feeding Stuffs Regulations ensures his being supplied with an adequate description of the nature and analysis of most fertilising materials, and provides him with the means of obtaining satisfaction in cases of dispute. Thus in the case of a compound fertiliser the vendor must state the amounts, if any, of nitrogen, phosphoric acid soluble in water, phosphoric acid insoluble in water and potash contained in the material, and it is usual to express the constituents in this order.

Plant Food Ratios. From this analysis the ratio of the three plant nutrients can easily be worked out by taking the percentage of the nutrient present in least amount as unity. Thus a fertiliser containing 22 per cent N, 11 per cent P_2O_5 and 11 per cent K_2O would have a plant nutrient ratio of 2 : 1 : 1, and one containing 13 per cent N, 12 per cent P_2O_5 and 20 per cent K_2O would have a ratio of approximately 1 : 1 : 1$\frac{1}{2}$.

Designation of Compound Fertilisers. A large number of compound fertilisers of varying analysis is offered to farmers and growers. Although different soils, crops and climatic conditions influence plant nutrient requirements, it has been found that a simplified list of plant food ratios can meet the majority of crop–soil–climate combinations that arise in this country. Such a list is given in Table 4, together with some of the needs they meet and examples of formulations at present commercially available.

Fertiliser Recommendations. In discussing the manuring of crops in Chapters IX and X recommendations have all been expressed in terms of kilograms of plant food per hectare. This replaces the use of the unit (one per cent of a cwt. or 1·12 lb) which was the customary method of recommending both compound and straight fertilisers before metrication was adopted. The number of kilograms of a particular nutrient in 100 kg of fertiliser will be the same as the percentage of nutrient in the fertiliser. The standard 50 kg bag of fertiliser will therefore contain $\frac{\text{percentage nutrient}}{2}$ kg of that particular nutrient. Thus a 50 kg bag of ammonium nitrate (34 per cent N) contains 17 kg N ; a 50 kg bag of a 12 : 12 : 18 compound contains 6 kg N, 6 kg P_2O_5 and 9 kg K_2O.

The following example shows how the system can be used in choosing a suitable compound fertiliser to replace a home mix. Suppose a farmer has been in the habit of using 625 kg sulphate of ammonia (21 per cent N), 687$\frac{1}{2}$ kg superphosphate (18 per cent P_2O_5) and 312·8 kg muriate of potash (60 per cent K_2O) all per hectare on his potato crop.

TABLE 4

COMPOUND FERTILISERS

The principal classes of plant nutrient ratios in compound fertilisers, the main needs they meet and examples of commercial sources.

Class	Plant Food Ratio	Commercial Examples	General Use
High nitrogen with phosphate and potash	$2\frac{1}{2} : 1 : 1$	25 : 10 : 10, 25 : 9 : 9, 23 : 10 : 11	Spring-sown cereals, kale and others
	2 : 1 : 1	22 : 11 : 11, 20 : 10 : 10*†, 16 : 8 : 8†	Forage crops, intensive grassland
	$1\frac{1}{2} : 1 : 1$	21 : 14 : 14, 20 : 14 : 14, 15 : 10 : 10	Root crops where soil P_2O_5 and K_2O high
High phosphate with nitrogen and potash	$1 : 1\frac{1}{2} : 1$	⎰16 : 24 : 16, 13 : 20 : 13 ⎱12 : 18 : 12, 10 : 15 : 10	On soils very deficient in phosphate
High potash with nitrogen and phosphate	1 : 1 : 2	12 : 12 : 30, 10 : 10 : 15*	⎰Potatoes, sugar beet, horticultural crops.
	$1 : 1 : 1\frac{1}{2}$	15 : 15 : 23, 15 : 15 : 21, 13 : 13 : 20*, 12 : 12 : 18	Most row crops on potash-deficient soils
High phosphate and high potash with nitrogen	$1 : 2\frac{1}{2} : 2\frac{1}{2}$	9 : 25 : 25, 10 : 25 : 25, 10 : 24 : 24	⎧Crops grown on land well supplied with nitrogen, e.g. following good leys, on peats. Autumn-sown corn, and all crops where land deficient in both phosphate and potash
High nitrogen and high potash with phosphate	2 : 1 : 3 $1\frac{1}{2} : 1 : 2$	17 : 8 : 24, 17 : 11 : 22 18 : 12 : 18	⎰On soils well supplied with phosphate
Equal nitrogen phosphate and potash	1 : 1 : 1	17 : 17 : 17, 15 : 15 : 15, 10 : 10 : 10	⎰Grassland and fodder crops on soils of average fertility
Nitrogen and phosphate only	2 : 1 : 0 1 : 2 : 0	20 : 10 : 0, 26 : 14 : 0 12 : 24 : 0	⎰On soils of very high potash
Nitrogen and potash only	$1\frac{1}{2} : 0 : 1$	25 : 0 : 16	⎰On soils of high phosphate status, in conjunction with slag or ground rock phosphate
Phosphate and potash only	0 : 1 : 1	0 : 20 : 20	⎧In seed bed for autumn-sown cereals, peas, beans, lucerne. As basal dressing for grassland (e.g. K-slag, potassic mineral phosphate)

* Also with added Boron.
† Also with added Magnesium.

He wishes to save the trouble of mixing these straights by using a compound fertiliser. How does he choose a compound which will supply the same total amount of nutrients in the same proportions? The quantities of the respective nutrients expressed in kg would be:

$$N\frac{625 \times 21}{100} = 131 ; \quad P_2O_5\frac{687 \cdot 5 \times 18}{100} = 124 ;$$

$$K_2O\frac{312 \cdot 5 \times 60}{100} = 187 \cdot 5.$$

He then examines his fertiliser price lists and chooses a compound having the three nutrients roughly present in the same proportions (or ratio) viz : $1 : 1 : 1\frac{1}{2}$. He may find several quoted in this ratio, viz : (a) $15 : 15 : 23$; (b) $13 : 13 : 20$; or (c) $12 : 12 : 18$.

It only remains to calculate the particular one he chooses, in this case (c), in kg/ha ; thus the amount of compound containing 131 kg of nitrogen will be :

$$\frac{131 \times 100}{12} = 1100 \text{ kg or } 22 \times 50 \text{ kg bags.}$$

This quantity will at the same time contain approximately the amounts of P_2O_5 and K_2O he requires.

The system is flexible and the requirements of crops can be stated directly in terms of kg/ha required. It can also be utilised for stating the amounts of ingredients in farmyard manure and other similar products : for example 25 tonnes of average F.Y.M. can be said to contain 125 kg N, 25 kg P_2O_5 and 125 kg K_2O.

Comparative Values. This system provides a useful criterion of values for comparing different fertilisers. In the case of fertilisers supplying only one plant nutrient, if the price in £ per tonne is divided by the number of kilogrammes of plant nutrient per tonne (or the percentage plant nutrient multiplied by 10) this gives the price of one kilogramme of plant nutrient in that fertiliser. Thus :

Ammonium nitrate (34 per cent N) is quoted at say £58·00 per tonne, therefore price per kg $N = \dfrac{5800}{34 \times 10} = 17p$. Nitro-chalk (25 per cent N) is quoted at say £44·00 per tonne therefore price per kg $N = \dfrac{4400}{25 \times 10} = 17 \cdot 6p$.

However, it must be realised that cost per kg may not be the only criterion of value and, in this case, it could be that the higher cost per kg of Nitro-chalk is offset by its greater suitability for the conditions prevailing.

When assessing the value of compound fertilisers containing several nutrients it is necessary to adopt kg values of the constituent nutrients by reference to well-known straight fertilisers.

Mixing Fertilisers on the Farm. In the comparison of proprietary compounds versus home-mixed fertilisers, the farmer must assess whether, after allowing for cost of bagging and transport, the difference in the value as calculated on price per kg of constituents is enough to outweigh the merits of the thorough mixing, granular condition and convenience for use (including higher analysis of concentrated proprietary compounds than can be achieved using straights at the farmer's disposal) of the proprietary compound.

Solution Fertilisers. Liquid fertiliser solutions are essentially compounds made up of fertiliser chemicals all of which are soluble in water. Thus they contain nitrogen, phosphorus and potassium in the same water-soluble forms as are found in most solids. Their main advantage is ease of handling, for all transfer can be done by pump. Because they are in solution, both placement and broadcast coverage are uniform. They have the disadvantage of limited concentration but with the use of the more concentrated superphosphoric acid and its compounds this limitation may be overcome. Furthermore, concentration in granular fertilisers has of necessity tended to restrict range or flexibility in plant nutient ratios ; this flexibility is available with solutions and individual batches can be made up more cheaply, to which batches minor or trace elements can be added where required. Many farmers are already applying compatible herbicides and solution fertilisers in one operation.

Unexhausted Manurial Values. Provision is made in the Agriculture (Calculation of Value for Compensation) Regulations 1975 and their amendments for the recompense of an outgoing tenant for fertilisers (other than inorganic and dried blood nitrogen) and liming materials used on the farm, in the previous 1, 2 or 3 growing seasons in the case of the former and up to 7 growing seasons in the case of the latter. In the same way the manurial value of purchased feeding stuffs is catered for.

MATERIALS FOR LIMING THE SOIL

In the section dealing with soils the subjects of soil acidity and calcium deficiency were dealt with at some length. There are several commodities available for correcting these deficiencies and their value is now generally expressed in terms of their neutralising value (N.V.). In agriculture the word " lime " is used to include all those materials which will have the same effect as " burnt " or " quick " lime in neutralising the acidity in the soil.

Some 7–8 million tonnes of lime in one or other of its forms are applied annually in Britain, and it is estimated that some $3\frac{1}{2}$–4 million tonnes are removed from the soil annually, either used in neutralising soil acidity, lost in the drainage water or removed in crops.

The agricultural value of a liming material depends upon its N.V. and its availability or solubility in the soil, a property which depends upon its softness and fineness of grinding. The harder the material, the more important is this fineness of grinding.

The bulk of the liming material used in British agriculture is obtained from natural rock deposits of limestone and chalk which are widely, although rather unevenly, distributed over the country. The different forms vary in purity and hardness and each presents its own particular problems in the production of a material suitable for agricultural application.

Ground Limestone. The harder limestones have to be quarried and passed through crushers to break them down to a size suitable for grinding in hammer mills where they are ground to a fine powder. A product which will entirely pass through a 4·75 mm sieve, with 95 per cent passing through a 3·35 mm sieve and at least 40 per cent passing through the B.S. 100 mesh sieve, is considered satisfactory for agricultural use. Such material is officially designated *ground limestone* and normally has a neutralising value between 50 and 56 depending on the source and purity of the rock. In general the harder, drier and purer quality limestones are easier to grind to the required standard than the softer, wetter, and impure limestones. Limestones which do not come up to the specifications given above are often designated *agricultural limestone* or *limestone dust* : their usefulness depends on the fineness of the material.

Ground Chalk. Chalks are quarried in the south and east of England. Chalk consists of almost pure calcium carbonate but many of the natural deposits contain bands or pockets of flints, which flints should be removed before grinding. Chalk is much softer than limestone and quarrying can often be carried out by digging with an excavator ; but the ease of quarrying is more than counteracted by the costs of removing flints and reducing the high moisture content, which latter often necessitates drying in rotary driers or kilns or the addition of burnt lime, before the material can be ground to pass a 6·7 mm screen and be designated *ground chalk*. Most ground chalks have a N.V. of at least 50 and many are ground so that 100 per cent will pass the 3·35 mm sieve.

Screened Chalk. With some of the softer chalks it is possible to produce a somewhat coarser material without quarrying or grinding. On suitable level sites the soft chalk can be loosened and broken up with disc harrows, scraped up and passed through a screen. A 1 in. screen is often used, giving a material varying in particle size from 1 in. downwards. Such material is relatively cheap and somewhat heavier dressings are normally used than with the finer ground material.

Lump or Dug Chalk. Some raw chalk is still dug and spread on the land in dressings of about 50 tonnes/ha. Normally this is applied in the autumn, making use of the winter frosts and spring cultivations to break the material down to a fine enough state.

Burnt Lime is still produced extensively and represents the form with the highest neutralising value. It is composed of calcium oxide containing residues which were present as impurities in the limestone from which it was derived ; these do not amount to more than a few per cent in the purest samples. Little is now used on the land in the coarse lump form. Instead much of it is kibbled, the product having a maximum diameter of about 25 mm ; in this form it can be spread mechanically. Ground burnt lime in bags is still produced but is more expensive than the kibbled form. It must not be stored long since it rapidly absorbs moisture and bursts the bags. In most cases a ground carbonate of lime is cheaper than, and equal to, burnt lime, unit-for-unit of neutralising value.

Hydrated Lime. High-grade slaked lime in a finely divided condition is usually available to, and popular with, horticulturists. It is rather expensive compared with other forms of lime, but can be intimately mixed with dry soils with consequent rapid action.

By-product and Waste Limes are usually in the form of calcium carbonate in a very fine state of division, being products of various industrial processes. They are not scheduled under the Fertilisers and Feeding Stuffs Act. In many of them the moisture content is both high and variable making the value difficult to assess and making them sticky to handle and expensive to transport. Some, like the dumps of greyish-green product of the old Chance process, and the waste lime from tanneries and tar distilleries, may contain unweathered materials and care should be taken to avoid such samples. Some, like the sulphate of ammonia waste lime in the north-east of England, contain about 1 per cent of nitrogen. A large quantity of waste carbonate produced at sugar beet factories finds its way back to the land. Most of these waste limes are cheap but all should be bought and applied on their Neutralising Value, which varies considerably according to moisture content.

Ground Magnesian Limestone and Lime. There is a growing production both of ground magnesian limestone and its burnt equivalent. Its increased demand probably arises from at least two causes. Firstly, considerable areas of soils now exist where lime is not only needed to neutralise acidity but also where added magnesium is required as a nutrient for the crops ; secondly, for application to grassland where it is thought that its presence will help to overcome hypomagnesaemia in animals. When purchasing ground magnesian limestone as a source of magnesium it should be remembered that any ground limestone containing 3 per cent or more of magnesium, 40 per cent of which is fine enough to pass the B.S. 100 mesh sieve, can be sold as ground magnesian limestone : a magnesium content statement should therefore be requested.

MISCELLANEOUS FERTILISERS

In addition to the materials already described, some mention must be made of those supplying trace elements, and of salt.

Manganese Sulphate. This is usually applied in a calcined form to supply manganese on areas where manganese deficiency is established. It may be applied in the ordinary way at the rate of 60–125 kg/ha, or as a solution in the form of a spray ($11\frac{1}{2}$ kg/ha dissolved in 450 litres of water).

Borax. This material can be applied in the form of a solid dressing to the soil before seeding (6·3–9 kg/ha), or in the form of a spray. In either case its application is to overcome boron deficiency in soils leading to such diseases as " Heartrot " in sugar beet and " Raan " in swedes. Boronated fertilisers can now be obtained to be applied at a rate designated to supply about $22\frac{1}{2}$ kg borax per ha.

Copper Sulphate and Copper Oxychloride. Copper in the form of copper sulphate or copper oxychloride can be used as a spray or as a solid application. Care must be taken when using copper sulphate solution as this is phytotoxic.

Salt. Agricultural salt has long been known to produce yield responses on mangels, sugar beet and carrots. It increases the yield of sugar beet even when potash is applied, and field experiments in England have shown that an extra 630 kg sugar per ha may be expected from the application of 630 kg of salt. Salt should always be applied some 2–3 weeks before sowing the crop. Some crops show a strong dislike to applications of salt, probably due to the deleterious effect of the chloride radical rather than to the sodium present, and indiscriminate application of salt should be avoided.

FUTURE TRENDS IN FERTILISER PRODUCTION

The last thirty years have seen a large increase in the consumption of fertilisers in British agriculture accompanied by marked changes in form and concentration. Granulation has become almost universal, and improvements in the conditioning of both straight and compound fertilisers have removed many of the problems of storage without caking, and distribution through drills without clogging and corrosion. The same period has seen the replacement of sulphate of ammonia by ammonium nitrate, and superphosphate by triple supers and the ammonium phosphates, as the main sources of nitrogen and water-soluble phosphate respectively; whilst improvements in methods of manufacture have increased the concentration of K_2O in muriate of potash from 50 to 60 per cent. Thus the demand for more plant nutrients to produce ever-increasing crop yields has been met without recourse to higher rates of fertiliser application.

This need for higher yields, with its demand for more plant nutrients, will continue; and the increasing cost of labour in production, handling and application of fertilisers, together with the need to reduce the amount of compaction of the land by the passage of heavy implements, will call for yet increased concentration of fertilisers. At the same time, however, the recent phenomenal rises in the price of energy feed-stocks and rock phosphate will certainly call for the more careful use of fertilisers, especially phosphates, and the value of soil analysis in this connection coupled with a better appreciation of the fertiliser value of the organic by-products of the farm is to be expected.

Of all the plant nutrients nitrogen will continue to give the best returns more especially on grassland. The relatively lower cost of production and higher nitrogen content of gaseous anhydrous ammonia and solid urea make them both attractive forms of nitrogen, and will, no doubt, encourage further research into methods of handling the former and overcoming too rapid breakdown and consequent loss of ammonia from the latter. Of the phosphate fertilisers, the replacement of phosphate-rich iron ores by cheaper imported phosphateless ores has resulted in a decline in the production of high grade, high citric soluble basic slag and a substitute for this popular grassland fertiliser will have to be forthcoming. Possibilities would appear to be the fortification of such phosphorus-free ores with rock phosphate or the heat treatment of aluminous phosphate rock—like that from Senegal—in order to increase the availability of the P_2O_5.

The increased use of polyphosphoric acid and the polyphosphates, especially ammonium polyphosphate, in compounds is to be expected, not only to increase the concentration of P_2O_5, but as carriers of " available " trace elements, the need for which must increase as our main fertiliser ingredients become purer, and larger yields take more trace elements out of the soil.

Chapter VIII

CROPPING, IRRIGATION AND WEED CONTROL

PRIMITIVE tillage farming in all parts of the world is, and in the past has always been, associated with a simple sequence of cropping. In the typical case a section of grass or light scrub-covered land is cleared and cropped with the same or similar crops until it ceases to yield profitable returns, either because of the exhaustion of fertility or because of the accumulation of weeds. In the former case the cultivators move on and break up another virgin area ; in the latter the same practice may be adopted or a bare fallow may be introduced to kill the weeds, after which the land is cropped as before. The former practice was probably adopted in primitive times in Britain. The latter practice was typical of the farming in California when wheat growing was first developed in that country during the latter part of last century. It is equally typical of certain types of farming in the Prairie provinces of Canada where continuous cereal crops are interrupted by bare fallows. As the needs of a community increase and farming becomes more intensive some definite sequence is developed ; thus in the Manorial period of this country, " when each man had his rood of land ", the three-field system was developed. In this, autumn corn was followed by spring corn, followed in the third year by a bare fallow after the stock had grazed the stubble of the previous crop through the winter and spring periods.

At this period of British farming, when the fields were divided into a series of narrow strips, each occupied by a different cultivator, the need for a rigid yet simple rotation was paramount, since every field would have become a patchwork of crops if each cultivator had been allowed to crop his " lands " as he pleased. When Enclosure followed, the need for a rigid system of cropping for purposes of maintaining fertility still persisted, though to a lesser degree, and with the introduction of clover by Sir Richard Weston, turnips by Lord Townshend and the drill husbandry by Jethro Tull, the four-course system was developed and thereafter formed the basis of English farming until the end of the nineteenth century. During the latter part of this period the introduction of many types of artificial manure, and the knowledge of how to use them, rendered the rigid four-course rotation no longer necessary for the purpose of maintaining fertility, and the Agricultural Holdings Acts of 1908, giving tenant farmers freedom of cropping, finally overthrew the idea that a rigid rotation was essential to arable farming. The introduction of the farm tractor, giving much greater speed of manipulation of tillage operations, further reduced this necessity.

At the present time most arable farmers follow no fixed rotation. Tremendous developments have taken place in British agriculture in the last two decades following the stimulus of the second world war. Better varieties of crops and their improved protection by seed dressings, herbicides, fungicides and insecticides ; widespread use and better understanding of chemical fertilisers ; the decline in manpower ; and above all the vast impact of mechanisation, have rendered a fixed sequence of cropping less necessary, and often less desirable. Farming has become more specialised, and the term " sequence of cropping " implies a greater flexibility to meet changing conditions in competitive farming, than does rotation of crops. Nevertheless, the main objects of rotation still apply.

OBJECTS OF ROTATIONS

The objects to be attained by a definite sequence of cropping may be considered under the following headings :

(1) The farming system in relation to land, labour and capital.
(2) The increase or maintenance of fertility.
(3) The economical distribution of labour through the year.
(4) The suitability of crops to the farming system.
(5) The degree of specialisation.
(6) The control of weeds.
(7) The control of plant diseases and pests.

1. **The Farming System in Relation to Land, Labour and Capital.** Farming conditions vary widely and the sequence of cropping must be varied accordingly. Climate is an important factor to be considered in deciding what crops to grow. The dry climate of the eastern counties is favourable to tillage farming and especially to the cereals, wheat and barley. The moister and cooler districts of the west and north are more suitable to the growth of temporary and permanent pasture and, if cereals are grown, to oats rather than wheat or barley. Smaller differences in local climate, as for example those associated with altitude, aspect, liability to frost, etc., may favour the growth of one crop and make it advisable to vary the sequence.

The character of the soil is largely responsible for the success or failure of different crops. On light soils with easy drainage the land warms up quickly in spring, growth is rapid and harvest is early. Tillage and intercultivation are comparatively easy and root growth is facilitated. On the other hand, fertility is not generally very high, and in periods of dry weather crops may suffer from lack of water. These conditions are specially favourable for the growth off malting barley, provided the land is not too dry. They are not so favourable for wheat and oats, but rye grows to advantage on the lightest and driest of soils. Root crops of all descriptions can be easily cultivated and easily harvested or folded if desired. Such soils are specially favourable for sugar beet, turnips, and to early potatoes. They are not so suitable for main crop potatoes because lack of fertility and of moisture frequently result in small crops. Light soils, especially in moist districts, are favourable for catch cropping because seed-beds can be quickly obtained, and growth is rapid. Chalky soils are frequently light and easy to cultivate, in which case they favour the growth of barley and roots and in addition such leguminous crops as peas, sainfoin and many clovers.

If the soils are shallow they are not very suitable to wheat, beans or potatoes.

Heavy soils are difficult and costly to cultivate. Their texture does not facilitate the production of good seed-beds nor the easy growth of plant roots. They are not favourable for the harvesting of root crops in bad weather. As a result of these conditions relatively few crops can be successfully grown upon them, and the crops which can be successfully grown are generally possessed of a vigorous root system. Wheat and beans both have strongly growing roots and these crops grow to advantage on clay soils. Mangels have deeply penetrating roots and grow well on clays, but potatoes cannot be grown successfully because the soil is unsuitable for growth and because it is impracticable to harvest them in wet seasons. Red clover and white clover, together with certain grasses used in temporary pasture such as rye grass and cocksfoot, grow freely and abundantly provided the land is not allowed to become waterlogged ; since such pastures entail little annual expenditure for labour and management they can be grown economically upon heavy land. Loamy soils are suitable for the growth of nearly all crops because both fertility and water supply are generally good and the crops produced are generally large. Both barley and wheat grow well, as do beans, and large crops of good quality potatoes can generally be produced. Very large crops of turnips and mangels can be grown, though the folding of the former in wet weather sometimes damages the texture of the land. Red clover and other leguminous crops grow freely and many vegetable crops can be incorporated in the sequence of cropping. For these reasons it is necessary in planning a sequence of cropping to give careful consideration to the suitability of the crops to the soils on the farm.

The stock policy adopted on the farm will make varying demands for food and bedding, especially for the winter months, and the cropping sequence will need to be framed to provide for these. If sheep predominate, a sequence of folding crops, roots, kale or temporary leys must be provided. If dairy cows, a winter supply of hay, mangels, silage and such winter-green crops as marrowstem kale may be required. Provision must be made to guard against periods of shortage of grass during summer droughts by the growth of lucerne, green maize, cabbage, kale, rape or a reserve of silage. Conversely, the amount of stock kept and the amount of farm-yard manure produced may influence the sequence of cropping by rendering it advisable to introduce a larger area of potatoes, green vegetables, etc., to make more profitable use of this manure.

When the farm is considered as a unit the amount of land, labour and capital available will determine the farming system. If land is the limiting factor, intensive cropping is likely to be practised. Where labour is scarce, grass and cereal crops will be preferred to the more labour demanding root crops.

Local markets and local demands for special produce may make it desirable to vary the sequence of cropping. Proximity to a large provincial town or to a seaside resort will create a good demand for vegetables of all sorts which can be cheaply delivered if the distance is short. In such cases these crops should find a place in the sequence. Proximity to racing stables will create a good demand for hay and straw. Proximity to a sugar factory reduces the cost of transport on sugar beet, especially if the fields on the farm are close to a main road. In such cases the sequence of cropping should be planned accordingly. In other cases, where farms and fields are inaccessible for one reason or another, provision should be made in the sequence of cropping for a greater proportion of the produce to be fed to stock, or for the growth of crops which are not bulky in character so that transport is facilitated. These and many other considerations peculiar to the circumstances must be taken into account when planning the sequence of cropping on any farm.

2. The Increase or Maintenance of Fertility in the Soil. Whether it be considered from the point of view of the nation, the owner, the occupier or the farm worker, there is no condition (except that the farming system is attended with profit) of greater importance than that the fertility of the land be maintained or increased, because upon this mostly depends the power of the soil to produce future crops.

Fertility of the soil, in the sense in which the word is here used, is governed firstly by the supply of humus and the nitrogen which it contains, and secondly by the supply of certain available minerals, especially lime, phosphorus and potash. All soils, whether they be cultivated or uncultivated, are continuously undergoing change brought about by chemical, physical and biological actions. Some of these changes lead to an increase of fertility, others to a decrease. The sum of the changes in ground covered with vegetation and not cultivated generally leads to an increase of the humus, and therefore to increased fertility. Under arable conditions the sum of the changes generally results in loss of humus and therefore in reduced fertility.

Most tillage operations tend to increase the aeration or ventilation of the soil so that oxidation is encouraged. By this the humus is decomposed and the nitrogen which it contains converted from an insoluble to a soluble form. In this condition it is available for absorption by plant roots. At the same time, tillage and the consequent aeration of the soil tends to encourage the solution and availability of the mineral plant food in the soil. Plant foods, when in this soluble condition, are liable to loss in two ways ; either they may be absorbed by plant roots to form part of the structure of roots, stems, leaves or fruits of the plants growing in the soil, or failing this, they may be carried away in solution by the excess rain-water draining through the land. In the former case such parts of the plant as are removed from the field at harvest result in loss to the field of the plant-food elements which they contain, whilst other parts, such as roots, stubble, broken leaves, etc., are left on the field to be reincorporated in the soil, together with the plant food they have taken up from it.

When crops are consumed by stock on the field a comparatively small portion of the elements of plant food is retained in the structure of the animals, and the greater part is returned in the faeces or in the urine to the land. In the same way when harvested crops are consumed in the yards by stock bedded upon straw, the resulting farm-yard manure contains a large portion of the plant food removed in the crops consumed and used as bedding, and if the manure is carted back to the land it returns to it the plant food which it contains. In the making and storage of farm-yard manure, however, it must be remembered that considerable losses of nitrogen and smaller losses of minerals are liable to occur, so that the efficiency of the return of plant food is not so great by this method as when the crop is consumed on the land where it is grown. It will thus be seen that loss of fertility is greatest in the case of those crops such as cereals or potatoes which may be sold away from the farm, intermediate in the case of crops which are consumed in the yards and made into dung (provided the dung is subsequently returned to the land), less in the case of crops consumed on the land, and least of all in green crops which are ploughed back in to the land where they are grown.

Some crops, either because of the greater penetrative power of their roots or because of the more efficient tillage which they receive, extract from the soil larger amounts of plant food than others. A comparison in this respect between barley and mangels shows that the latter extracts from the soil very much larger amounts of plant food than the former and is to this extent more exhausting. Root crops generally require, and do in fact extract from the soil, considerably larger quantities of plant food than cereals. Root crops, however, are generally heavily manured and frequently consumed by sheep on the fold, so that the fertility of the land is increased during the root break. For this reason root crops are generally regarded as renovating crops ; whilst cereals are exhausting crops because they are not heavily manured and because the whole crop is removed from the field. Leguminous crops in like manner extract from the soil more mineral plant food than the cereals and in this respect are more exhausting crops. They also obtain very much larger quantities of nitrogen than the cereals—a 5 tonne crop of clover hay will contain approximately four times the amount of nitrogen contained in a wheat crop consisting of 3·5 tonnes of corn and 2·5 tonnes of straw—but the nitrogen of the leguminous crop is mainly obtained by the nodules on its roots, not from the soil itself, but from the air in the soil. This greater accumulation of nitrogen, therefore, by leguminous plants does not result in loss of fertility, but on the contrary is a most potent means of accumulating fertility. When the roots and stubble of such crops are ploughed back into the soil, and still more when the stems and leaves are consumed on the land, the nitrogen content of the soil is increased. The growth of peas and beans under arable conditions, and of white clover under pasture conditions, thus enhances the level of nitrogen in the soil.

3. The Economical Distribution of Labour through the Year. In framing a sequence of cropping the economic use of labour requires to be considered not only from the point of view of the distribution of labour through the year, but also with regard to the elimination of unnecessary or unproductive work.

In farming practice it is generally necessary to employ the greater part of the labour continuously throughout the year. It is not so feasible in farming as in other businesses to discharge men when a busy season of work is finished and expect to find them available for work when the busy season comes on again. If such methods are attempted it will be found that casual labour is either very inefficient or very costly. Continuous employment must, therefore, be found for the greater part of the labour employed. Now, no single crop requires continuous labour throughout the year, but each crop requires labour, and sometimes a great amount of labour, at special seasons. Thus the wheat crop in this country is planted in autumn and harvested in late summer and therefore demands labour mainly during October and November for planting and in August and September for harvesting. Very little labour is expended on the crop at other seasons. The potato crop has requirements for labour at other seasons ; the land may be ploughed in winter, planted in spring, cultivated and sprayed in early summer, harvested in autumn and marketed in winter. It will be seen that these two crops require labour at dissimilar periods and for this reason are suitable crops to combine in a sequence of cropping. Other crops have their own seasonable requirements of labour. In framing a sequence of cropping, therefore, great care must be given to see that too great a congestion of work does not arise at any one period of the year, for nothing prejudices successful farm management so badly as being unable to do work at the right time. Conversely, things must be so arranged that the periods when little or no labour is required are not too protracted. A certain amount of labour can generally be economically utilised upon matters of upkeep such as drainage, fences, roadways, etc., which do not require attention at very precise times, but these are limited in quantity, and any excess of labour beyond these requirements is unproductive.

Another point in the economical utilisation of labour is illustrated in the cartage and use of farm-yard manure. At the present time the cost of labour involved in the cartage and spreading of this product is very high. If the dung is utilised for the production of crops which are saleable at good prices all is well, but if it is utilised in the maintenance of fertility for the production of ordinary farm crops such as wheat and beans the cost may bear no relation to the benefits received, especially when the field lies a long distance from the site where the dung is made. A sequence of cropping which involves the carting of farm-yard manure for ordinary farm crops must so far as possible be avoided, and in its place the fertility of outlying fields should be maintained either by the use of green manuring or by the use of temporary pastures which are grazed during part of the year.

4. The Suitability of the Crops to the Farming System. Technical and scientific developments have led to a much greater freedom in cropping. Inherently infertile light land may be improved by the use of lucerne and cocksfoot leys, irrigation and subsequent heavier stocking. Heavy land may yield to powerful mechanisation and produce crops previously considered unsuitable. None the less there remain important differences in the response of crops to environmental conditions of soil and climate.

Each successive crop should allow sufficient time to carry out the necessary tillage operations in order to prepare an adequate seed-bed. The interval should not be too long nor too short. If too long, not only is time wasted but soluble plant food may be drained away whilst the land is bare. If too short, there may not be sufficient time for tillage and weathering to enable a good seed-bed to be prepared, and the following crop has to be planted either in an unfavourable seed-bed or too late. Thus wheat planted in November can be conveniently taken after potatoes harvested in September or October, but it cannot be advantageously taken after roots folded in November because December is not a favourable month for planting wheat. Similarly it would be unwise to grow forage crops planted in May or June after potatoes harvested in October because of the waste of time and of plant food during the interval.

The growth of, and the tillage for, some crops leave the land in a suitable condition for the growth of other crops. The deep cultivation generally given to all root crops leaves the subsoil mellow yet firm for the subsequent growth of corn crops. Clover and other leguminous crops enrich the soil with nitrogen which provides plant food for subsequent crops, thus wheat can be advantageously grown after clover or beans. The fibrous roots left in the land after the growth of a ley consisting of mixed grasses and clover provide, when ploughed up, an open texture which is very favourable for the growth of potato roots, which do not have great penetrative power.

5. The Degree of Specialisation. It may be reasoned that a simple system enables a farm to intensify production from two or three enterprises and therefore to reduce unit cost of production more easily than a traditional mixed farm with half-a-dozen enterprises. However, prices may fluctuate considerably and seasonal effects on crops can be critical. The present trend is towards simplicity.

6. The Control of Weeds. Different crops have different habits and are planted and managed in different ways. Some crops such as wheat are planted in autumn, others such as barley are more frequently planted in early spring, whereas the root crops are generally planted in late spring or early summer. Each of these seasons of planting is favourable to the growth of some weeds and prejudicial to the growth of others.

Some crops, such as the cereals, are generally planted in narrow rows so that their foliage may cover the surface quickly and thus tend to crowd out annual weeds. Other crops, such as beans or turnips, are planted in wide rows so that intercultivation of the crop for the purpose of killing weeds may be carried out whilst the crop is young. These different conditions favour or prejudice different weeds. Some crops have a dense foliage which tends to shade the ground and others have a much more open foliage amongst which weeds grow freely. Barley is illustrative of such open foliage and is liable to contain many annual weeds. Oats make a much denser foliage and tend to smother weeds. A crop of marrow stem kale with its broad horizontal leaves forms a perfect canopy of foliage when well grown, and completely smothers small weeds. A crop of beans, potatoes or kale may form a dense smothering canopy.

Some crops occupy fields for a longer or shorter time than others. These different conditions also favour some weeds and prejudice others. A wheat crop, growing from October to August, favours the growth of perennial weeds such as couch and thistles much more than a root crop planted in April or May and harvested in October. A long ley, during which the land is uncultivated for three or four years, is very disadvantageous to the growth of annual weeds but may lead to the spread of perennial weeds. On the other hand, production of two crops during the year with the additional cultivations which are entailed may, if the cultivations are well executed, help to eradicate both annual and perennial weeds.

In addition to these general considerations certain crops are introduced into the sequence of cropping with the special purpose of enabling weeds to be killed, or a bare or bastard fallow may be taken. The crops specially suited to the killing of weeds are the root crops, including turnips, swedes and other brassica crops, as well as mangels and potatoes. Such crops facilitate the killing of weeds in three ways : they are generally planted late in spring or early in summer, so that there is a long interval of time after the previous corn crop, which gives time for cleaning operations either after harvest or in the spring before planting ; they are planted in wide rows so that both tractor and hand hoeing for the destruction of weeds can be easily carried out ; they produce horizontal foliage, which, when the plants are established, shades the ground and checks the growth of weeds. Methods of controlling weeds by the use of chemical substances are described later in this chapter.

7. The Control of Plant Diseases and Pests. Diseases and pests are likely to accumulate in land continuously cropped with the same crop. Thus land is said to become " clover-sick," owing to the accumulation in the soil of one or more of several disease-producing organisms, when cropped too frequently with red clover. Turnips and other brassica crops in the same way are likely to fail with club root, and potatoes to suffer from wart disease and eelworm if grown too frequently. Some crops, for example, barley and mangels, can be grown more frequently and yet continue to yield satisfactory crops. But with wheat, continuous cropping may lead to the accumulation in the ground of take-all, eyespot and other soil-borne diseases, and the practice is, therefore, undesirable.

If an interval of time is allowed to elapse between the taking of two similar crops, the diseases and pests find no host plant and die out or migrate. The interval of time necessary to complete the process must, of course, depend upon the life cycle of the disease or pest, and this and other points are discussed in Chapters XIII and XIV.

THE NORFOLK FOUR-COURSE SYSTEM

From the middle of the 18th century until the beginning of the 20th, this system formed the basis of a very large portion of English farming, and for the conditions prevailing during the larger part of this period was a wellnigh perfect system, especially in the drier districts.

The rotation in its original form was :

Roots—Barley—Seeds—Wheat.

The roots consisted first of all of turnips and swedes, but later on mangels, and later still kale, came to occupy part of the root break.   The seeds were principally red clover, with or without rye grass, but trefoil, sainfoin, peas or beans could be substituted to widen the interval between successive crops of red clover.   Where necessary, oats displaced some of the barley.   The folding of sheep on arable crops and the winter feeding of bullocks in yards were essential parts of the system of farming.

At a time when artificial manures were non-existent or but little understood the system provided for the maintenance of fertility at a reasonable standard, so that the owner was insured against depreciation and the occupier was assured of good crops in the future. It provided continuous employment of labour throughout the year : the autumn was employed in harvesting some of the roots, in preparing and planting wheat, in ditching and hedging.   The winter was employed in threshing corn, in feeding stock in the yards and folding sheep, in ploughing for roots and for barley.   The spring was employed in completing the ploughing, in drilling spring corn and clover seeds, in cleaning the root land, and in spring-tillage of corn crops.   The summer was occupied in drilling and hoeing root crops, in the hay harvest and the corn harvest.   With one-fourth of the land devoted to roots, this system required a relatively large staff of labour, but during this period farm wages were very low so that this drawback was not very serious.   The system provided two cash crops, namely, the wheat and the barley, and at the same time an abundance of food for livestock, namely, the roots and the clover hay as well as the cereal straw for litter, so that ample farm-yard manure could be produced to maintain fertility.

The system was specially suitable to light land farming in dry districts where the trampling of the sheep on the fold caused no detriment to the soil texture and where the sheep manure, provided the folding was properly controlled, improved not only the manurial but also the textural condition of the soil.

Where the system was applied to heavy land, sheep-folding in winter and early spring was liable to injure the texture and so prejudice the growth of the barley.   In this case the roots needed to be folded early or carted from the land for consumption in the yards.   On the heaviest types of clay land the root crop was eliminated and its place taken by a bare fallow.

In wet districts of the north and west, which are less favourable to barley, an oat crop was taken in its place.   If potatoes were to be grown they could be planted in place of some of the roots.

MODIFICATIONS OF THE FOUR-COURSE SYSTEM

The four-course rotation broke down on economic grounds.   A world depression, coupled with imported grain from the American prairies, and frozen and chilled meat from abroad, undermined the home market. The period of prosperous " high farming " ended abruptly and much land in the arable areas " tumbled down to grass ". The system was modified to meet the changing needs of agriculture and emphasis shifted to the more perishable commodities, milk, eggs, fresh meat, potatoes and vegetables. By the end of the 1914–18 War the Norfolk four-course rotation, in its original state, had ceased to exist. Most modern cropping sequences are, however, seen to be modifications of the old Norfolk system.

The Repeal of the Corn Production Act in 1921 resulted in a severe agricultural depression. The introduction of the sugar beet crop, and the sugar beet subsidy first granted in 1923, were undoubtedly a vital stabilising factor which buffered, especially, the East Anglian farmer against the worst effects of the depression. Some farmers changed their farming system radically, but others incorporated this profitable crop into the structure of the Norfolk system.

Four-Course Rotations. The simplest modification to meet modern economic conditions is by replacing all or part of the fodder roots by direct, high value, cash crops—sugar beet and potatoes :

Sugar Beet—Barley—One Year Ley—Wheat.

Five-Course Arable Rotations. The proportion of cash crops is raised still more by including both sugar beet and potatoes to give the following widely practised sequence :

Sugar Beet—Barley—One Year Ley—Potatoes—Wheat.

Another method of increasing the saleable crops is to extend the cropping sequence to include three cereal crops, thus:

Roots—Barley—One Year Ley—Wheat—Barley (or Oats).

This five-course system increases the acreage of cereals from one, half to three-fifths of the cultivated area and consequently increases the amount of saleable corn. At the same time the area of roots is reduced from one-fourth to one-fifth, so that a saving of labour can be effected on this crop. This means that a smaller proportion of the land is cleared each year and weeds may accumulate. This difficulty may be overcome by stubble cleaning and by the use of chemical weed-killers.

The recent introduction of efficient forage harvesting and handling machinery is likely to result in the more widespread use of high-yielding forage crops. Maize and other cereal crops grown for silage and kale may replace fodder roots and one-year leys in intensive arable cropping sequences, and on heavy land upon which it is difficult to fold stock, and where traditionally most of the roots are carted.

Winter-planted crops such as wheat and winter oats can be grown to better advantage than spring crops on heavy land, and roots are less suitable.

Six-Course Rotations. In districts and conditions peculiarly favourable to the growth of cereals, the Norfolk system may be extended to a six-course system by the introduction of corn crops in both places in the sequence as previously discussed :

Roots or Fallow—Barley—Barley—Seeds or Beans—Wheat—Oats.

In this case roots are reduced to one-sixth of the area so that the labour requirements on these is much reduced, and the cereal area increased to two-thirds. The liability to weed accumulation is further increased and must be dealt with by efficient stubble cleaning after harvest and by chemical weed-killers. In times when the only source of cultivating power was that provided by horse labour this concentration of work might have been difficult of accomplishment, but, with tractor-power available, cultivations can be so rapidly executed that this difficulty need not now arise. In cases where this rotation is applied to heavy land the roots may be in part or in whole substituted for by a bare or bastard fallow, the latter following either grass or some other crop cut for hay or silage, or folded before late summer.

When cereal prices are high this system provides good returns under suitable conditions, but is relatively unprofitable when cereal prices are low.

A six-course system adopted for mixed cropping in some districts runs as follows :

Roots—Barley—Seeds—Potatoes—Wheat—Oats.

In this case potatoes are taken after clover, which is a very favourable arrangement, partly because of the accumulation of nitrogen by the clover, but especially because the fibrous roots and stubble of the clover produce humus and a good soil texture in which the potatoes revel. It also provides a longer period of growth for the clover, and mixed grasses if any, in autumn than in the case where the ley must be ploughed up for wheat sowing in autumn.

The rotation provides for two cleaning crops, roots and potatoes, in six years so that the land can be easily kept clean, and the potatoes can be cashed to advantage, as well as the three cereal crops.

Another six-course system suitable for mixed cropping is that called the East Lothian system, which runs as follows :

Roots—Barley—Seeds—Oats—Potatoes—Wheat.

In the district of East Lothian and other parts of Scotland to which it is applied, conditions are often more favourable to the growth of oats than of other cereals, and the crop can commonly be marketed to advantage for oatmeal. This crop is therefore taken after the seeds, where it benefits from the accumulated fertility of the clover roots. As the oats are sown in the spring a longer grazing period is available on the clover stubble. In other respects its advantages are similar to the previously described six-course system.

The cropping sequences discussed above are typical of most arable farming districts in Britain, but some notable exceptions must be mentioned. The relative simplicity of growing and harvesting cereals, and their relative profitability, has resulted in repeated and continuous cereal cropping in some districts. In high-fertility areas successive wheat crops, and in lower fertility chalk land areas barley crops, are often successfully grown. With wheat such practice involves added risk of the build-up of certain diseases such as take-all and eyespot, although the use of resistant varieties of winter wheat, and the extended use of the less susceptible, improved spring wheats counters eyespot to some extent. Barley is less susceptible to these risks. Fertility is maintained by the liberal use of chemical fertilisers and by ploughing-in the larger crop residues obtained. Meanwhile interest in oats has declined due to lower average yields and added harvesting problems compared with wheat and barley.

Silt and Fenland Rotations. These soils are deep and rich and the land is very valuable. Consequently high value crops predominate in the rotation which takes the form :

Potatoes—Sugar Beet—Wheat

or

Sugar Beet—Potatoes—Wheat.

These crops are frequently accompanied by market gardening crops such as bulbs, celery, peas, brassicas and root crops for seed.

EXTENDING THE SEEDS LEY

In the Norfolk and similar systems the clover and other " seeds " crops have been described as one-year leys, growing for one year only. These systems have proved specially suitable to the drier districts of eastern and southern England in times when cereal prices were remunerative, because in these districts the second growth of red clover in the first year can often be profitably harvested and cashed as seed, and because weeds and wireworms tend to accumulate in second and third year seeds to the serious disadvantage of subsequent crops.

In moister districts the leys are commonly kept down for two, three or more years. The first extension gives a six-course arable-with-grass rotation :

Potatoes—Wheat—Sugar Beet—Cereal—Two Year Ley.

The proportion of cash crops remains high and provision is made for livestock to consume the grass and arable by-products. Expressed in this way it represents a further extension to the Norfolk system.

ALTERNATE HUSBANDRY

A further extension of the " seed " course to three or more years, or the inclusion of a long ley, leads to a practice more graphically described by the term " alternate husbandry ".

In alternate husbandry the land, having been cropped for some years, is laid down to ley or temporary pasture for a number of years, after which it is again broken up and cropped. Whilst the land is under ley many advantages are obtained. The land becomes filled with fibrous roots of grasses, clovers and other plants which, together with the surface vegetation, greatly help the texture of land when next the ley is broken up. The fertility of land, measured in terms of humus and nitrogen, can be rapidly increased in this way, and in minerals also if lime, phosphates and potash are added by way of top-dressings.

Under proper management the increase of fertility is largely associated with the presence of white clover. The annual cost of management is greatly reduced after the initial seeding, whether in the preceding corn crop or by direct seeding on bare ground, since little further cultivation is required. The cost of harvesting is relatively small in the case of silage and hay-making, and much less when the ley is grazed.

During the period of temporary ley, production of herbage under good management is generally very great, and often on inferior land much greater than that of permanent grass grown upon similar land. This is due mainly to the establishment of more prolific species of plants under more favourable environmental conditions than those occurring in the old pasture. Coincident with this is the fact that stock thrive better upon new leys than upon old pasture, not only because of the more abundant herbage but also because the new temporary ley is relatively free from pests and diseases. A further advantage is that the system permits better weed control since the same weeds are unlikely to thrive both on grassland and arable land.

When the ley is broken up and recropped the fertility stored up by the nodules of the leguminous plants, by the organic residues of the grass and clover roots and stems, by the top-dressings applied to the pasture and by the droppings of the animals grazed on the pastures, is utilised for the production of crops. This accumulated fertility is mainly in an organic condition which, under arable conditions, becomes progressively available for plant growth. In this way the fertility stored during the ley period can be utilised in the production of cash crops, and when this fertility is partially exhausted the land may be again laid down to ley and the process repeated. It will be noticed that under this system fertility is maintained without the use of farmyard manure, the carting of which, especially to outlying fields, involves the use of much labour and power. Any farmyard manure produced on the farm can, if desired, be used upon fields near the buildings or devoted to special cash-producing crops, such as fruit and vegetables.

There are, however, some drawbacks to alternate husbandry. The capital required for fencing and water supply may be high, although at present grants are available towards such costs. Good seeds mixtures are expensive, while the purchase of stock to graze the leys demands still more outlay. New leys suffer much more from " poaching " by livestock in wet seasons and in winter than permanent grass, a disadvantage on some stock farms.

Alternate husbandry is basically a higher cost system, but it has the great advantage of flexibility to meet changing conditions in competitive farming. When arable farming is relatively profitable leys can be ploughed out earlier ; where livestock farming is relatively more profitable leys can be left down longer. With the introduction of the ploughing grant in 1939, the State encouraged the system and grants of several pounds per acre were available until 1970 to encourage maintenance of the tillage area and regular ploughing-up of leys.

On heavy land, and land unsuited to root crops, cereal crops may alternate with grassland as follows :

Wheat—Wheat—Barley—Long Ley.

It may be convenient on a mixed dairy and arable farm to practise two sequences of cropping ; an inner sequence near the buildings consisting of fodder crops such as Italian ryegrass, cereals, maize and kale ; and an outer sequence made up of cash crops and long leys.

This system of ley farming or alternate husbandry has been practised in the moister districts of England and especially in Scotland for many years, and has received special attention since the discovery of the outstanding value of wild white clover as a nitrogen-collecting plant, and the use of basic slag and close-grazing in stimulating its growth. It has now spread to most parts of the country, due partly to the stimulus of war-time farming, and partly to the development of improved techniques for establishing and managing leys. These techniques are discussed in Chapter XI.

CATCH CROPPING

Catch cropping is not a system of farming in itself, but consists in the snatching of an extra crop between two of the main crops of a rotation. An example of a catch crop is the broadcasting of Italian ryegrass or rape on a scarified corn stubble in autumn to provide keep during early spring, after which the land is prepared for a late sown crop. Similarly, winter rye or a mixture of winter rye and Italian ryegrass sown in early autumn are frequently used to supply " early bite " on dairy farms. It is a common practice in some districts where early potatoes are grown to plant a catch crop of Italian rye grass immediately the crop is lifted : the forage can be grazed off or ploughed in as green manure. Mustard is frequently planted as a catch crop after a bare fallow, to be ploughed in during the preparation of the ground for autumn wheat. The mixing of small quantities of rye-grass and trefoil with spring sown oats or barley is also catch cropping, for after the removal of the cereal the forage plants can either be grazed or allowed to grow unchecked before being turned under during cultivations prior to the planting of the next major crop in the rotation.

Opportunities for catch cropping on heavy land are limited, and it is essentially a light land technique for intensifying land use and raising the gross output. The chief advantages are : provision of stock feed at difficult times of the year ; replacement of main crops that have failed ; conservation of nutrients, especially nitrates ; checking of weeds ; and additions of crop residues when ploughed in. The main disadvantages of catch cropping are possible detrimental effects on the yield of the following main crop by the removal of nutrients and water from the soil, and the temptation to leave a successful crop down too long, with subsequent delay in seed-bed preparation.

IRRIGATION

Irrigation is the artificial application of water to the soil to ensure an adequate supply of moisture to meet crop needs. Under natural conditions a small part of the crop's requirements is available from the moisture reserves stored in the soil from rain falling on the land before the crop is sown ; the remainder, and by far the greater portion, is supplied by rain falling directly on the soil during the growing period.   Thus the amount of rainfall and its distribution during the growing season largely controls the availability of soil moisture and considerably influences crop yields.   Over large areas of this country the amount of rainfall during the growing period is generally inadequate, because its distribution is poor in relation to crop needs: lack of soil moisture frequently reduces crop yields.   In these conditions irrigation is used to supplement natural rain as and when necessary to provide adequate moisture for the crop.   Used correctly, irrigation will pay handsomely over a wide range of conditions.

The amount of water that is available to the plant from the reserves of moisture in the soil depends on two things : the amount of water that is available per unit depth of soil, and the depth of soil from which the particular plant can extract moisture.   It is usual to refer to the amount of water available per unit depth in terms of *mm of water per mm of soil*.   Similarly, the amounts of irrigation water applied to, and the quantities of moisture used from, a soil are also referred to in terms of *mm depth*, 25 mm of water being 25 mm depth over the prescribed area.   Generally, light textured soils have smaller moisture reserves than heavy soils : a very approximate guide to the water reserves of various soils is given below.

APPROXIMATE MOISTURE RESERVES OF SOILS

Soil Texture	mm per 300 mm of soil
Coarse Sand	20–25
Sandy Loam	25–35
Medium Loam	35–45
Clay Loam	45–55

The depth of soil from which a plant can draw water is largely dependent on the type of crop, provided normal root development is not restricted by soil or other conditions.   For irrigated farm crops the rooting depths are given in the next table.   The depths given are not the maximum rooting depths of the crops, nor the extreme depths from which moisture extraction occurs, but they indicate the depth of soil to be considered for irrigation purposes.

"ROOTING DEPTHS" OF CROPS UNDER IRRIGATION

200–375 mm	375–500 mm	500–600 mm
Early potatoes	Peas	Sugar Beet
Short grass	Beans	Long Grass
	Mangels	Lucerne
	Brassicas	

From the information given in the above tables it is possible to estimate the total soil moisture reserves available to the crop growing on a particular soil.   Thus, for sugar beet growing on a sandy loam the total soil moisture reserves will be 50 mm ; the reserves are 25 mm per 300 mm of soil, the depth of rooting being 618 mm.

The amount of water and the rate at which it is used by a crop are determined by the weather conditions so long as there is a supply of available moisture in the soil, and are independent of the crop itself providing the crop gives reasonable ground cover. On average, in this country water will be used at the rate of 25 mm every 14 days in April and September, 25 mm every 10 days in May and August and 25 mm every 8 days in June and July. These figures vary little year by year except during periods when there is an unusual excess or lack of sunshine. If rainfall is insufficient to provide water in the above quantities, the moisture reserves of a cropped soil will be depleted and exhausted after several days of dry weather. Irrigation should be applied before the soil reserves have been depleted to the point where plant growth is checked.

The amount of available water per unit depth is determined by measuring the volume of water held by the soil at two reference points, *field capacity* and the *permanent wilting percentage.*

Field Capacity. On many well-drained soils the moisture content of the soil falls to a fairly definite value after rain or irrigation has ceased, provided evaporation is prevented. In this condition the soil is holding the maximum quantity of water against free drainage, and the soil moisture reserves are at the maximum. In this state the soil is referred to as being at field capacity. Any additional water applied at this stage will not be retained by the soil and will be lost through drainage.

Readily Available Moisture. A plant growing in the soil removes water, until the water remaining in the soil is held so firmly that the plant is unable to obtain sufficient water to meet its needs and the plant wilts. A soil in this condition is at the permanent wilting percentage, and the moisture reserves of the soil available for plant growth are exhausted. The volume of water in the soil held between these two reference points is known as the *readily available moisture.*

It is now generally accepted that as a soil dries from field capacity towards the permanent wilting percentage there is a falling off of the growth rate and a reduction in marketable yield of a crop. The amount of available water that can be used from the soil without seriously affecting the marketable yield depends on a large number of factors and it is not possible to use the same standards for the irrigation of all crops. Recommendations based on existing experimental data and practical experience are given later.

Soil Moisture Deficit. As the timing of the irrigation and the amounts of irrigation water to apply must be related to the amount of water used from the soil, it is necessary to estimate this. In practice a reasonable estimate can be obtained by comparing the quantities of rainfall or irrigation for the period in question with the water-use figures for the same period. When the water-use exceeds the amounts of rainfall or irrigation, moisture will have been used from the soil and a soil moisture deficit will exist. The *soil moisture deficit* is the amount of water required to restore the soil to field capacity. The size of this deficit and the crop conditions will determine whether or not irrigation is necessary. More details and the figures of water-use for specific areas are obtainable from the Ministry of Agriculture's Technical Bulletin No. 138 " Irrigation ".

Irrigation Practice. For most farm crops sufficient water should be applied at each irrigation to restore the major portion of the root-zone to field capacity—in other words to satisfy the soil moisture deficit completely. On heavy soils it may be advisable to leave 13 mm deficit to accommodate any immediate rainfall. More water than the quantity required to restore the major portion of the root-zone to field capacity should never be applied.

It is usually unsound to irrigate for the purpose of hastening germination of seeds. If the seed-bed is likely to be dry the soil should be irrigated during seed-bed preparations : irrigation should not be carried out immediately after sowing, nor in the young seedling stage of the crop. During dry weather, late sown crops of kale, peas, etc., often benefit from water before sowing, and an application of 40–75 mm of water should be applied before the land is ploughed or before final seed-bed preparations are made. The smaller application is usually sufficient for light soils, but 75 mm may be required on heavier soils after a long dry period.

Early Potatoes. Irrigation should be applied when 50 per cent of the total soil moisture reserves have been used, the water used from the soil being calculated from the time when 80 per cent of the plants have emerged and have produced small bushy plants 100–125 mm high, each having 3 or 4 stems. Irrigation should stop some 5 days before the crop is to be lifted to allow the soil to dry out and the crop to benefit from the water applied. On light soils it may be more economic to withhold irrigation until 25 mm is required than to apply smaller quantities more frequently. Where water or equipment is restricted, a single application of 25 mm (if needed) some 14 days before lifting will be beneficial.

Maincrop Potatoes. Irrigation should not be considered until the tubers have formed and are about the size of small marbles. Water should then be applied when 50 per cent of the soil moisture reserves have been used, and continued until leaf drop due to blight or maturity prevents any further yield increase. Again, it may be more economic to delay irrigation until 25 mm is required than to give more frequent light applications. On farms where there are limited irrigation facilities, the most critical period for water shortage is early July. Infrequent and irregular irrigations may reduce the quality of the crop because they may induce secondary growth.

Sugar Beet. The use of irrigation should be considered from shortly after singling, when the plants are just meeting in the row, until about the middle to the end of August. Irrigation is required when 50 per cent of the soil moisture reserves have been used up during this period. July and early August appear to be the critical periods, if irrigation facilities do not permit continuous irrigation. Irrigation after the end of August may reduce the sugar content of the crop. Under dry conditions increased root weights should compensate for this sugar reduction.

Cereals. On light soils and in a dry spring, a single application of 25–40 mm is frequently well worth while. Cereals are most responsive to irrigation when they are starting to run to ear.

Peas. Normally, irrigation is not necessary before flowering except in very dry summers, and water should only be applied from flowering onwards when 60 per cent of the soil moisture has been used. At the first irrigation the aim should be to restore to field capacity or give 50 mm of water, whichever is the less ; but smaller quantities are beneficial. Late-sown crops of peas are likely to respond to seed-bed watering.

Grassland. Maximum production is obtained if the soil is maintained as close to field capacity as possible. In practice it is most convenient on grazing fields to restore the soil to field capacity when 25–40 mm have been used, and when 40–50 mm have been used from hay and silage swards, regardless of soil type. A typical schedule for a grazing field would be to graze, top dress with fertiliser and irrigate, to give a second irrigation in 7 to 10 days if necessary, and graze again in a further 8 to 10 days.

Kale and Mangels. No irrigation is necessary before there is reasonable ground cover, and then irrigation should start when 60–70 per cent of the soil water reserves have been used.

Haphazard watering may give rise to some benefits but is usually wasteful of water and equipment, the best results being obtained from controlled irrigation. Benefits other than yield increase are frequently obtained from controlled irrigation, because the quality of the produce is often improved and the time the crop reaches the marketable stage can sometimes be influenced favourably.

WEED CONTROL BY CHEMICAL METHODS

The primary purpose of weed control is to prevent weeds competing with a crop for nutrients, light and water and so reducing the yield. Weed control often benefits the farmer in other ways, for it brings about easier singling in root crops, easier harvesting of cereal crops and improved crop quality.

For very many years weeds have been kept under control by good rotations, ploughing, cultivations, hoeing and hand pulling and these methods are still of value today. The control of weeds by chemicals is of recent origin. At the beginning of this century a limited use was made of calcium cyanamide and finely powdered kainit, and also 3–4 per cent solutions of copper sulphate in water, to destroy charlock and wild radish growing in cereal crops. The broad, hairy, more or less horizontally held leaves of the weed retained the dusts or liquids, which were held to a much less degree by the narrow, vertical leaves of the cereal. In the 1920's, sulphuric acid at concentrations of between 7 and 10 per cent solution of B.O.V. (brown oil of vitriol) was used. Later it was discovered that certain types of mineral oil, such as the vaporising oil used as tractor fuel, have little effect upon the growth of umbelliferous plants, including carrots,

I

parsnips and parsley, but are toxic to many non-umbelliferous annual weeds. In 1939 compounds of dinitro-ortho-cresol (DNOC) began to be widely used. Since 1942, when the synthetic growth regulating substances were discovered, selective weed control has developed at a tremendous pace.

Selective Weed Control. A selective weed killer, or selective herbicide, is a substance which checks or destroys weeds without seriously harming the crop in which the weeds are growing. The word herbicide, meaning plant killer, is used partly to avoid the suggestion that these substances will kill only weeds : crop plants may be killed just as readily if the chemical is used on a crop or in a way for which it is not recommended. To control weeds effectively it is not always necessary to kill them ; sometimes sprayed weeds do not die but remain stunted and fail to set seed, which may be quite as beneficial to the farmer.

The selectivity of a herbicide is only a matter of degree : very few chemicals have absolutely no effect on the crop and yet will kill weeds. Most herbicides check the crop very slightly, and some weeds, even of those species regarded as susceptible, recover and manage to survive. Each herbicide has a range of weeds which have varying degrees of resistance and some which are quite unaffected by it and are termed "resistant ".

The selectivity of any herbicide depends on the amount applied : a high rate will often kill both crop and weeds. Selectivity is also very dependent on the stage of growth both of crop and of weeds. For each herbicide and crop the stage of growth at which it is safe to spray must be carefully defined : at other stages there may be crop damage. As weeds become older they become more resistant to herbicides so that satisfactory control is often dependent on spraying as soon as the crop reaches a safe stage.

Selectivity may also be affected by other factors, such as the volume of spray applied per acre, air temperature at time of spraying, and cultivations before or after spraying.

Herbicides can be classified according to their mode of action, as shown below :

Contact Herbicides.	Translocated Herbicides.	Residual Herbicides.
Dinoseb	CMPP	Ametryne
Diquat	MCPA	Chlorpropham
Ioxynil	2,4-D	Lenacil
Paraquat	MCPB	Linuron
Sulphuric acid	Amitrole	Monuron
	Dalapon	Pyrazone
	Dicamba	Simazine

Contact herbicides have a rapid scorching action on those plant tissues with which they are brought into contact. They are not moved about inside the plant to affect other parts untouched by the spray. If the scorching is sufficiently extensive the plant will be very severely checked or it may die, but it will recover from a slight scorching. Therefore good coverage of the plant by the spray is very important with contact herbicides and thus a volume of spray of 225–450 litres per hectare is generally applied. Contact herbicides will often kill the shoot systems of perennial plants but regrowth takes place from the unaffected root system.

Translocated herbicides have normally no scorching (or contact) effect but are absorbed by the leaves of the plant and moved about within it, causing a slow but profound disturbance of the normal processes of growth and eventual death of susceptible plants. In their effect on growth many of these translocated herbicides resemble auxins, the natural growth regulating substances present in all plants, and they are sometimes described as " growth regulators ". In the group shown above CMPP and MCPA produce characteristic twisting of the stems and leaf stalks and general severe distortion of the normal growth habit. Because of their movement within the plant these herbicides can be carried down to the root system of perennial plants and can effectively control many, though not all, perennial species. In contrast to the contact herbicides, complete coverage of the plant by the spray is not necessary and low volume sprays are generally used.

The most familiar type of treatment is to dilute the chemical with water and spray when the seedlings of both crop and weeds have emerged from the soil. This is known as " post-emergence " (i.e. after the emergence of the crop) and is the most common way of using contact and translocated herbicides.

Residual herbicides are applied to the soil and are taken up by plant roots. They persist in the soil for some weeks and are intended to kill weeds either when the seeds are germinating or in the very young seedling stage. They are often used for horticultural and root crops with the pre-emergence technique described below.

Residual herbicides are usually applied soon after the seed of the crop has been sown but before the crop seedlings have appeared above the ground, i.e. the " pre-emergence " technique, before the emergence of the crop. Young weed seedlings and germinating weeds are killed and the soil kept clear for some weeks by the residual activity of the herbicide. In some uses the selectivity depends on the inherent tolerance of the crop to the herbicide ; in others the crop may be liable to damage by the chemical but is protected by being sown deep enough to escape the full effect. The results with pre-emergence treatment are very much influenced by soil type, rainfall after application and temperature ; the method calls for precision and skill and is now widely used. Contact and residual herbicides may also be used pre-emergence, e.g. linuron in the carrot crop.

Non-selective or Total Weed Control. It may be necessary to kill weeds on land which is not cropped, such as paths and around yards and buildings or on waste land which has become overgrown. Sometimes weeds can be killed in arable land in between crops, such as couch or twitch or bindweed in cereal stubbles. Here the herbicide need not be selective as between crop and weed, and a chemical can be used such as Dalapon for couch or twitch, which would be severely damaging to the corn.

Often, in addition to destroying existing vegetation, it is desired to keep the land free from weeds for as long as possible (soil sterilisation). Here a very persistent residual herbicide is needed—a suitable application of monuron will keep land free from all vegetation for a year or more.

The chemicals used for these purposes may be the same chemicals which can be used selectively on crops, simazine for example, but they would be applied at considerably higher rates. Or they may be chemicals which have no selective use—dalapon, sodium chlorate or borax compounds.

Herbicides are sometimes used where no weed problem arises. The burning-off of potato haulm is one example, using sulphuric acid or one of the newer chemicals such as diquat. Clover seed crops carrying a lot of leaf can usually only be direct combined by first burning off the leaf with similar chemicals, a technique known as pre-harvest desiccation. Recently a method has been devised of re-seeding grassland without ploughing by first killing the existing grass sward with a herbicide such as paraquat and then sowing the seed after light surface cultivations or rotavating.

The nature of herbicides. The main groups of the large and expanding list of available herbicides are classified in the Ministry of Agriculture, Fisheries and Food booklet, *Approved Products for Farmers and Growers*, which is published annually.

Commercial products may be supplied either as aqueous concentrates or as wettable powders. If the basic chemical is an acid, as in MCPA, the commercial product may be supplied as a " salt " (usually sodium, potassium, ammonium or amine) which is water-soluble ; or as an ester, a compound dissolved in oil with an emulsifier so that it forms an emulsion when diluted with water.

Rates of application are quoted in terms of active ingredient (a.i.) : for example, 0·3 kg a.i. per litre for a liquid, or 50 per cent wettable powder for a solid.

The way in which the active ingredient is made up commercially is known as the *formulation*, and the type of formulation may modify the activity of a compound.

Detailed information on herbicides, their uses and application rates, is given in the official booklet mentioned earlier.

Chapter IX

THE CULTIVATION OF CEREAL AND PULSE CROPS

THE CEREALS

THE cereals wheat, barley, oats and rye are species which produce enormously enlarged grains used extensively for human consumption and for feeding to farm livestock. Although they have been cultivated for a very long time and are familiar to most people, yet occasions do arise when it is not easy to distinguish between the cereals, especially if the ear is not visible.

The Recognition of Cereals. By paying special attention to certain details of structure it is possible to identify the cereals with certainty at any stage of growth.

I. Seed. As in all the Gramineæ the grain of the cereals is a one-seeded fruit, or caryopsis, in which the bulk of the grain is occupied by starchy food reserves (endosperm) ; at one end of the grain is the embryo, or germ. In wheat and rye the grain is naked, but in barley and oats the grain is wrapped round by two chaffy leaves called the paleæ, or pales. The appearance of the four grains is shown in Fig. 28.

2. Seedling. When germination takes place in wheat and rye both the plumule (or shoot) and the young radicles burst forth from the embryo end of the grain as shown in Fig. 28A. But in barley and oats the plumule is at first hidden by the pales, and it is unable to burst its way out until it has grown at least half as long as the grain. Consequently, it appears that the rootlets and plumule of barley and oats arise at opposite ends of the seed, though this is not really the case.

In the very young seedling stages it is possible to identify cereals by digging up the plants and examining the sprouted grains.

3. Established Plant. As the plants get older, however, the grain rots away and it is no longer possible to use its characteristics as a means of recognition. Attention must then be devoted to the appearance of the leaf. As in all grasses, the cereal leaf is in two parts. There is a portion called the sheath which clasps the stem between one node of the stem and the next, and there is the blade itself. Where sheath meets leaf there are, in wheat, barley and rye, extensions of the sides of the blade called auricles, or claws. In wheat the claws are of moderate size, and carry a few long, scattered hairs. In barley the claws are relatively very much longer and overlap considerably ; they are bare, or quite without hairs. In rye the claws are very tiny, but the sheath is purplish. In oats there are no claws whatsoever (Fig. 29).

4. Ear. The recognition of oats when in ear presents no difficulty, because the ear is so obviously branched, each branch or

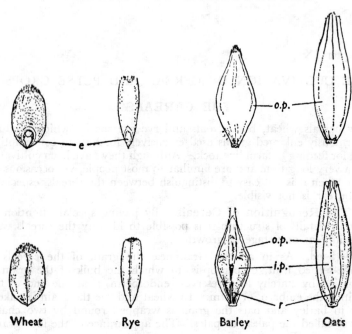

Wheat   **Rye**   **Barley**   **Oats**

FIG. 28.—GRAINS OF THE CEREALS.

e, position of embryo. *o.p.*, outer pale. *i.p.*, inner pale. *r*, rachilla.

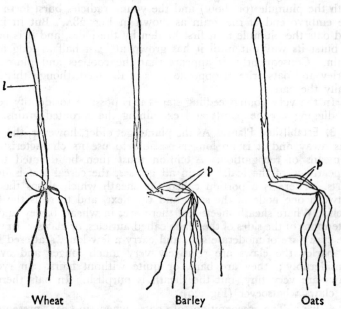

Wheat   **Barley**   **Oats**

FIG. 28A.—SEEDLINGS OF THE CEREALS.

c, coleoptile or sheath. *l*, first leaf. *p*, pales

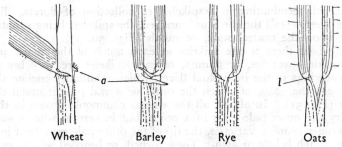

Wheat   Barley   Rye   Oats

FIG. 29.—AURICLES AND LIGULES OF THE CEREALS.
a, auricle. *l*, ligule.

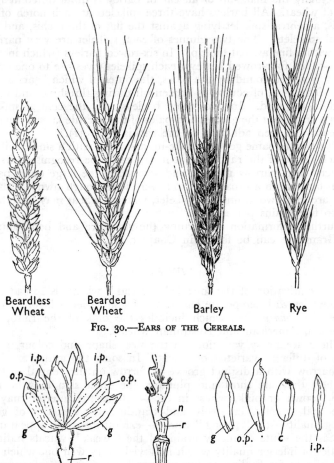

Beardless   Bearded
Wheat   Wheat   Barley   Rye

FIG. 30.—EARS OF THE CEREALS.

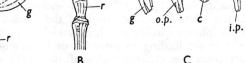

FIG. 31.—SPIKELET OF WHEAT.
A, single spikelet. B, rachis (much enlarged). C, detail of floret and glume.
r, rachis. *g*, glume. *o.p.*, outer pale. *i.p.*, inner pale. *c*, caryopsis or grain. *n*, notch of rachis.

sub-branch terminating in a spikelet, or collection of florets. But in the other cereals the ear is unbranched, the spikelets being situated direct upon the central stalk, or rachis (Figs. 30, 33).

In wheat there is one spikelet at each notch of the rachis and the outside coverings, or glumes, of each spikelet are very broad. The coverings to the individual floret, which eventually enclose the grain, are the pales, of which the outer or dorsal pale is much the larger (Fig. 31). In almost all the wheats commonly grown in this country the outer pale ends in a point, but in certain wheats such as the older bearded varieties, the tip of the outer pale grows out into a long, rough bristle or awn. These awned, or bearded wheats may be mistaken for barley on casual inspection (Fig. 32).

Actually the make-up of an ear of barley is quite different from that of wheat. All barleys have three spikelets at each notch of the rachis, a central spikelet lying against the flat of the rachis, and two lateral spikelets. The two glumes of each spikelet are very narrow, each with a fine awn at the tip. In six-rowed barleys (which include the so-called four-rowed barleys) each spikelet gives rise to one grain, tightly wrapped round by the pales, the outer of which bears a long awn. The awn, of course, is broken off during threshing and is not found in the seed. In two-rowed barleys only the central spikelet forms a grain, but the glumes and pales of the infertile lateral spikelets can easily be seen adhering to the rachis (Fig. 32).

Rye has the same general make-up as wheat, with a single spikelet to each notch of the rachis. But it can easily be recognised, first by the very tiny, narrow glumes, and second by the large outer pales, which taper from a broad base to a very long, rough awn. Usually there are only two grains per spikelet, whereas wheat may have from one to five grains per spikelet.

Further information concerning the structure and behaviour of the Gramineæ can be found in Chapter XI.

WHEAT

The cultivation of the so-called " bread " wheats is spread over all countries with temperate climates. These varieties belong to the species *Triticum vulgare*, which includes the bulk of the European, Canadian, American and Australian wheats.

There are many variations in the size, shape and colour of the grain of different varieties of wheat. In some the grains are long and narrow with a distinct groove or furrow on the ventral surface, in others they are short and plump ; the outer skin may be red, white, brown or pink, whilst in cross-sections the interior may be white and starchy, or steely and translucent. Varieties of good baking quality—the " strong " wheats—exhibit the steely appearance, whereas the white or mealy grains of the " weak " wheats indicate varieties of inferior quality which may yield very well but which are suitable only for feeding to livestock or for biscuit manufacture.

Apart from the characteristics of the grain already described, such factors in wheat as earliness and lateness in ripening, resistance or non-resistance to disease, adaptability to conditions of soil type and degrees of soil fertility, and the capability to withstand adverse winter weather, are of great economic importance. Yield of grain is the result of a large number of contributory causes, and it is not until they are all operative in the same direction that maximum results can be obtained. For instance, a crop of great promise may be so badly attacked by rust or mildew that the resulting grain is only half its normal weight and size, leading to a corresponding reduction in the final weight of crop. It is not uncommon to experience serious loss in yield as a result of the crop becoming lodged due to the inability of the straw to support the heavy ears until harvest time. Varieties also vary from season to season so that conclusions as to the value of any particular variety should not be reached until experience over at least three seasons has been gained

Quality in Wheat. The large-scale buyer of wheat takes into consideration three main items before making a purchase of a particular lot of grain. These are its condition, its milling characteristics and the quality of its flour. The condition of the grain has little to do with variety, though some varieties are naturally of better appearance than others and white wheats are usually preferred to red wheats. Grain in good condition is undamaged, free from screenings, clean and dry enough to store well. For normal storage in bulk, the moisture content should not exceed 14 per cent. The farmer can influence, and is responsible for, the condition of the grain he offers for sale.

The miller naturally desires a wheat of good milling quality. A " free-milling " wheat is one in which the flour is easily and completely separated from the bran, while the flour itself is in a granular or gritty condition, in which state it can easily be screened or sieved. Milling quality is quite distinct from baking or bread-making quality. For example, the variety Bouquet has good milling characteristics but its flour is of poor bread-making quality. The variety Maris Huntsman mills poorly, but has fairly good flour quality. The spring variety Sappo has both good milling and good baking qualities.

Quality in flour is usually judged from the bread-making angle. A good bread-making wheat yields a flour whose dough is elastic and which bakes into large loaves of even texture and a good shape. Such a wheat is often said to be a " strong " wheat in contrast to a " weak " wheat whose flour bakes into a smaller, denser, less-even-textured loaf. Wheats imported from Canada and USA are " strong " wheats whose capacity for making large, well-shaped loaves of good texture make them popular with bakers. In contrast, most of the wheat varieties grown in this country yield a " weak " flour, but Maris Widgeon, Bouquet, Flinor and Maris Freeman, amongst the winter varieties, and the spring varieties Maris Dove, Sirius and Sappo, are all good bread-making wheats which also mill easily.

For biscuit-making, however, wheat flour should yield a dough which is not so much elastic as viscous. It so happens that most of the accepted biscuit-making wheats are of only moderate milling quality : they include Maris Nimrod, Atou and Champlein.

Quality for both bread-making and biscuit-making depends very much upon the nature and amount of the gluten in the grain. Gluten is a protein which can be increased to some degree by the amount of nitrogen given to the plant as a fertiliser, though nitrogen cannot alter the nature of the gluten and change a poor-quality wheat into a good-quality wheat. Nitrogen can, of course, very markedly increase the yield of grain of both autumn-sown and spring-sown wheat crops. Certain varieties, such as Maris Huntsman, Atou and Champlein, have been specially bred with very short, strong straw, capable of supporting the ears even when heavily manured with nitrogen. Yields of well over 5 tonnes/ha of grain are not uncommon with these varieties; unfortunately, the flour quality of these very productive varieties is not high. Much effort is being made at the present time by plant breeders to combine the high yielding characteristics of these varieties with the good milling and baking characters of varieties like Holdfast and Yeoman, which are now obsolete.

Varieties. Only two species of wheat have been grown commercially in this country in recent times. These are *Triticum turgidum* in which the normal or $2n$ number of chromosomes in the cell nucleus is 28, and *T. vulgare* in which $2n = 42$ (see page 467). The *T. turgidum* types grown were known as Rivet or Cone Wheats. An improved higher yielding variety known as Rampton Rivet was produced some years ago and was moderately popular for a time. All these *T. turgidum* wheats were late maturing and quite unsuited to the modern techniques of growing and handling the crop. They are now obsolete.

The species *T. vulgare* embraces varieties which are sown in autumn and varieties which are planted in spring. If true autumn varieties are planted after about the middle of February they yield very much less than when drilled at the proper time. If planted after the end of March they may even form no ears at all. Generally speaking, winter wheats yield more grain than spring wheat, though spring varieties like Kolibri and Troll can yield up to 5 tonnes/ha in fertile conditions. The number of varieties offered for sale to farmers is very large, and new varieties are continually being introduced. There is no point in adopting a new variety of wheat unless it has some advantages not possessed by an existing type, and the best general guide to varieties is the list of recommended and provisionally recommended varieties published annually by the National Institute of Agricultural Botany, whose current list of recommended varieties should be consulted. It should be realised that the list is drawn up as a result of a comprehensive programme of variety trials carried out at a number of centres throughout the country and for a minimum of three years at each centre.

Early successes with wheat resulted from the selection of individual plants of superior character from the mixed population of types which constituted the old 'land race' varieties. Later work involved the hybridisation of two varieties and the selection of desirable types within the progeny of such crosses (see pp. 477–80). A land-mark in the hybridisation technique of breeding was the production of the two varieties, Yeoman and Little Joss, by Sir Roland Biffen at Cambridge. Yeoman was the result of a cross between the, by former standards, high yielding variety Browick and the high bread-making-quality Canadian wheat, Red Fife. It combined high yield and good baking quality in the one variety. It is of historic importance also in that it showed that baking quality, although influenced by environment, was inherited on Mendelian lines. Little Joss was the result of a cross between Squareheads Master and Ghurka and it combined in the one variety good yield with resistance to Yellow Rust (*Puccinia striiformis*) (see p. 384). It also is of importance historically in showing that disease resistance could be inherited on Mendelian lines. Although both these varieties are obsolete they have been used as parental types in many subsequent crosses.

During the last twenty-five years three main developments have taken place. The first is the great success achieved in France, Britain and Belgium in breeding for straw strength. As a result of reducing the danger of lodging, the varieties produced have permitted higher levels of fertiliser to be used—which in turn has made possible a general increase in wheat yields. Varieties like Atou, Maris Fundin and Val are tributes to their success.

The second development has been the realisation that Yellow Rust exists in the form of several biological races. Varieties showing resistance to one race are not necessarily resistant to other races. It is, therefore, necessary to breed for resistance to all known races of the rust.

The third major development has been in relation to spring wheat varieties. The last twenty-five years have seen a spectacular improvement in yielding ability and straw-strength, although in neither attribute are they yet the equal of the better winter wheat varieties. Varieties like Sirius, Sappo and, to a less extent, Maris Dove, in addition show high milling and baking quality.

Along with the developments mentioned attention has been given to resistance to other diseases, such as Loose Smut (*Ustilago tritici*) and Mildew (*Erysiphe graminis*), to resistance to shattering of the ear at harvest, to resistance to sprouting in the ear, and to date of maturity.

Position in Rotation. Wheat is frequently grown after a one-year seeds or longer ley on medium to heavy loam soil; on the really heavy clay land, it may be taken after a fallow, bastard fallow or after peas or beans. Because the straw is relatively strong and resistant to lodging, wheat on many farms is taken after potatoes or after roots when the level of fertility is often too high for the other cereals. Moreover, under rich conditions the quality of the grain does not suffer as in the case of barley, nor does the yield suffer at the expense of straw production as often happens in the case of oats. It is not customary to take winter wheat as a second straw crop, although in recent years two, or even three, winter wheats have been taken after ploughed-out grass and leys. Spring wheat quite frequently is taken as a second straw crop. There is, of course, the example of continuous cropping with winter wheat over the last 122 years on the famous Broadbalk field at Rothamsted. A succession of wheat crops, or indeed too many wheat and barley crops, in a rotation can give rise to disease troubles. The chief of these are eyespot and take-all (see pp. 393 and 377). Where such trouble has been encountered a two- or three-year rest from wheat and barley is necessary in the case of eyespot, but a one-year gap is all that is needed in the case of take-all, provided the land is free from Couch grass and Yorkshire Fog, which can also carry the disease. Three-year breaks in the rotation may be made either by introducing medium-duration leys or by re-arranging cash crop sequences to give, for example, oats—seeds—potatoes, or other three-year non-wheat-non-barley sequences (see p. 208).

Manuring. As with other crops, the yield of wheat depends to a very large extent upon the availability of adequate quantities of the three main plant nutrients, nitrogen, phosphate and potash. In former times these were applied mainly in the form of farmyard manure. In rotations where wheat followed seeds the F.Y.M. was usually applied at rates from 15 to 25 tonnes/ha on the seeds before ploughing. This practice is still followed at the present time in some areas. In others, the F.Y.M. available tends to be reserved for crops such as potatoes. After heavily manured and fertilised crops, residues are available for the wheat crop, and this reduces the quantity of other fertiliser required.

According to the Crowther and Yates system of calculation of average optimal fertiliser rates, those for the wheat crop, under the price structure of the last few years, and taking account of subsidies, have been as follows :

Nitrogen.	Phosphoric acid (P_2O_5).		Potash (K_2O).
90 kg/ha	S.E. England N + W ,, Wales and Scotland	18 kg/ha 36 ,, ,, 42 ,, ,,	20–30 kg/ha

The average optimal for phosphoric acid, unlike that for nitrogen and potash, varies according to climate regions and is greater in the wetter and lower in the drier areas. These figures represent a basis from which reasonable fertiliser dressings for the crop can be worked out.

The 90 kg of nitrogen represent the average optimal rate, provided the crop will stand such a dressing without lodging. Crowther and Yates originally suggested that an arbitrary upper limit of 50 kg/ha only should be used because the varieties in use 20 years ago were insufficiently strong in the straw to stand up to the full dressing. The position has changed with the availability of stronger strawed varieties. Extensive experimental work in recent years has shown that for varieties like Cappelle Desprez and Hybrid 46, 90 kg/ha is optimal, but for other varieties 60–80 kg/ha are more appropriate for average conditions. Three factors other than varietal strength of straw affect the standing ability of a crop and therefore its ability to make use of high levels of nitrogenous fertilisers. One is seed rate, a low seed rate giving greater straw strength ; a second is in infection with the fungus causing eyespot which reduces straw strength ; a third is the moisture factor as influenced by climate and soil, the moister conditions giving rise to a relatively longer straw.

Residues of plant nutrients applied to previous crops, and still available to the wheat crop, influence the actual rates of fertiliser required. Where wheat is taken after a well-manured potato crop even Cappelle Desprez only requires 50 kg of nitrogen and certainly no phosphorus or potassium. Farmyard manure applied when ploughing out one-year clover seeds or a sequence in which the crop follows a good ley, also reduces the fertiliser requirement for wheat. The type of soil modifies the fertiliser requirements. For instance, wheat grown on true fen peat soil will require less nitrogen than the average optimal owing to the nitrogen content of these peats. On soils deficient in phosphorus or potassium supplementary dressings of the appropriate fertiliser are required over and above the average optimal.

Methods of application of the fertiliser to the crop require special mention. In most cases little or no nitrogenous fertiliser is required in the autumn with winter wheat. The optimal rates refer to that applied as a top dressing in the spring in the form of ammonium nitrate or other 'straight' nitrogen fertiliser. These top dressings may be applied at any time between March and mid-May according to weather and soil conditions. In the dry eastern counties there is evidence that a split dressing, half early and half late, can be an advantage. If all the top dressing is given in early spring excessive growth of straw is liable to occur. In the case of the spring crop the nitrogenous fertiliser may be incorporated in the seed-bed or part placed in the seed-bed and the rest given as a top dressing.

Phosphatic and potassic fertilisers, usually in the form of compounds, are applied to the seed-bed prior to sowing, both in the case of the autumn- and spring-sown crop. With the phosphate, and to a less extent the potash, component of the fertiliser, the practice of combine drilling has been shown to give considerable advantages over broadcast applications. A large series of experiments carried out in many English counties, showed that 35 kg of phosphoric acid placed in contact with the seed gave equal crop responses to 70 kg/ha broadcast. With wheat, as with other cereals, the type of combine drill where seed and fertiliser go down the same spout is satisfactory provided ordinary rates of fertiliser are used.

The yield of grain in wheat depends to a great extent on the number of tillers which produce ears at harvest. Blind tillers, i.e. those which fail to produce ears, are useless, and those which develop late in the season usually yield badly or not at all. Production of the maximum number of ear-bearing tillers is not only dependent upon early sowing but also upon the supply of plant food : for this reason provision of an adequate supply of nitrogen, phosphorus and potassium in the seedbed is essential. A further application of 75 kg/ha of a nitrogenous fertiliser in late spring (i.e. the end of April or the beginning of May) will help to increase the size of the ear and hence the yield of grain, without increasing the risk of lodging. Phosphates, in contrast to nitrogen, help to encourage root development.

Cultivations. The type of seed-bed considered desirable for cereals differs considerably depending upon whether the cereals are spring or autumn sown. With a spring-sown cereal the standard aimed at is one where the soil is reduced to a fine crumby tilth, free from clods and firm enough to prevent the heels of one's boots sinking in when walking over it. For the sake of economy and also to minimise the loss of moisture from the surface layers, the specified seed-bed should be achieved with the minimum cultivations possible. To do this it is necessary on all but very sandy soils to plough the land sufficiently early to obtain the mellowing effects of alternate freezing and thawing and alternate wetting and drying. " Forcing " a tilth refers to the excessive cultivations required to produce a fine tilth on the heavier soils ploughed too late to receive the benefit of weathering. Where the necessary standard of fineness and firmness is achieved with minimum loss of moisture, the resulting germination of the seed and early growth of the crop are uniform and as rapid as the prevailing temperature permits. Where the tilth is coarse, where it lacks firmness or where there is undue loss of moisture due to forcing a tilth, uneven germination results and there is usually a reduced plant population established from the seed sown.

In the seed-bed for an autumn-sown cereal firmness is also a desirable characteristic. This kind of seed-bed requires a certain amount of fine crumby material in it, but in addition it is considered that an appreciable quantity of fist-sized clods are required. The latter have the effect of preventing the " panning " or " capping " which occurs when a uniformly fine seed-bed is exposed to the winter rains. Where capping occurs, free gaseous interchange between the soil air and atmosphere is checked, often to the detriment of plant growth. In addition it is often considered that the clods give a measure of protection to the young plants from cold dry frosty winds of the winter.

The spring cereal type of seed-bed can usually be prepared easily enough by harrowing, or a combination of rolling and harrowing, provided the plough furrow has been sufficiently mellowed. In the special circumstances where a spring sown crop follows ploughed-out grass or ley, it is advisable to use disc harrows for the first cultivations following ploughing.

For autumn seed-bed preparation fewer cultivation operations are required. After ploughing, drag harrows only are used in most cases, but discing is also common. Rolling is not necessary and in fact is likely to crush too many of the clods. On soils liable to " puff out " over the winter, as for instance the chalk soils, it was formerly customary to use a press drill or a furrow press. These are still used in some districts. Disc harrows are used when taking wheat after ploughed-out grass or ley. Where the crop follows clover seeds, an aftermath growth is often ploughed in and this needs to be done with care so that the green material is completely buried. Skim coulters and drag chains may help to bring this about. After peas, beans or another cereal crop, stubble cleaning may be necessary prior to ploughing. After peas, clover seeds, or a cereal-legume silage mixture, a bastard fallow may be possible. After potatoes the land is usually clean and the soil fine and loose in texture. In many cases no further preparation is required beyond a turn of the cultivator before drilling. Frequently this results in a seed-bed which is too fine. For this reason the land is often ploughed and re-worked down ; an old practice was to broadcast the seed on the surface and plough it in with a very shallow furrow—a practice known as " shelling in ".

Sowing. The sowing is usually carried out with a drill having coulters spaced 150–200 mm apart : the seed is buried to a depth of about 35 mm. Sowing too deep leads to delayed emergence and, sometimes, to unsatisfactory establishment. Seeds placed at considerable depths (say 100–120 mm) may completely fail to emerge. Seeds sown too shallow may remain in a relatively dry layer of soil and show irregular germination. Sowing by broadcasting on an incompletely prepared seed-bed, such as may be necessary under adverse circumstances in the autumn, leads to some seeds being too deep and some too shallow. As a precautionary measure against seed-borne diseases the seed should be dressed with one of the many organo-mercurial seed powders (see p. 387). Where there is reason to suspect a high wireworm population, such as when wheat is taken after ploughed-out grassland, it is advisable to use one of the combined insecticidal-fungicidal seed dressings. The latter form of seed dressing can also give a measure of control against wheat bulb fly (see p. 436).

In practice the quantity of seed sown depends upon a number of factors such as time of the year, the quality of the tilth, the district and the germination percentage of the seed. There is, however, a definite relation between seeding rate and yield. Up to a point, increases in seed rate increases the grain yield ; beyond this point further increases in the seed rate lead to decreases in grain yield, although straw yields may continue to increase. This point where the change in response takes place is called the physical optimum seed rate. Experimental work in recent years has shown that the average physical optimum seed rate for grain yield of winter wheat is 190 kg/ha and 250 kg/ha for spring wheat. The economic optimum seed rate is less than the physical optimum and will vary according to the relative value of seed and crop, weight for weight. Where the farmer uses his own seed and therefore the seed and the crop are both of the same value, the economic optimum seed rate is 156 and 190 kg/ha for winter and spring wheat respectively ; where the seed value is twice that of the crop the optimum will be 140 and 168 kg/ha, and where the seed value is three times the crop value the optimum will be decreased to 125 kg/ha for both wheats. These average optima can be taken as working guides. With abnormally late planting, with seed-beds of unsatisfactory standards, where the germination percentage is low or where broadcasting is used in place of drilling, the actual seed rate needs to be greater than the average optima.

Date of sowing is an important matter with the wheat crop. It is not advisable to sow autumn wheat too early. The middle of October is the time to commence sowing in many areas, and it may be continued up to mid-November. Sowing earlier than mid-October may, where followed by a mild winter, lead to excessive vegetative growth or " winter proudness ". Such excessive growth is liable to cause subsequent lodging of the crop, either directly or indirectly, by facilitating the spread of eyespot disease (see p. 393). A true winter variety should not be sown in the spring although the earlier maturing types may be sown up to mid-February. Sowing later than mid-February may mean a complete failure of the crop to produce ears.

With spring wheat, drilling as early as possible is the rule. This usually means as soon as the land is dry enough to work down to a tilth during the period January to March. If weather delays sowing the crop until April, one of the early ripening varieties, such as Sirius, should be used.

Post Drilling Cultivations. When wheat is examined in the early spring, it may be discovered that frost has lifted the surface soil on some fields and that the plant is loose at the root. On other fields the soil may be panned down and set into a hard, compact crust. Frequently seedlings of annual weeds will be growing. To rectify these conditions either rolling or harrowing or both will be necessary : but these operations must not be carried out until the soil is dry enough to carry the implements. If the weed infestation is likely to affect the corn crop a suitable weed killer should be used. The choice of weed killer should be determined according to the dominant weed species to be controlled. Information on this matter is given in the *Weed Control Handbook* and in Ministry of Agriculture short-term leaflets.

If " winter proud " growth has occurred the excessive leafage may be grazed with sheep or cattle. Grazing of excessive growth should not be done later than the end of March or severe yield depressions are likely to occur. The feeding value of the leaf is by analysis similar to that of young grass, but deteriorates both in feeding value and palatability during April. Winter proudness is best regarded as a condition to avoid by attention to sowing date, and spring grazing should only be considered as a means of alleviating the worst effects of an undesirable condition.

Harvesting. Although almost all cereal crops in this country are now harvested by combine the binder can occasionally be found working in some areas and for special purposes : for this reason the technique is still described here.

The Binder. With the binder the crop is cut before it reaches the dead ripe stage whilst the grain is still in the " cheesey " condition. Immediately after cutting, wheat should be stooked 8–12 sheaves to a stook. The sheaves should be set with the butt ends firmly on the ground, the ears forming an acute angle so that the rain is quickly shed. It is customary to run the stooks North–South for the sake of even drying. The stooks need to remain in the field from 7 to 14 days, during which time grain maturation is completed. The time in stook depends upon the condition of the crop when cut, whether there is much green stuff in the sheaves and also upon prevailing weather conditions. Ideally it is carted when perfectly dry, although once it is properly mature, carting may proceed even if some surface damp from dew is present. The stacks should be constructed on a dry base and preferably one which allows air to circulate freely below the rick. In building the stack it is important to keep the centre higher than the walls, with all the sheaves sloping downwards and outwards ; in this condition the rain, if it beats against the side of the stack, does not run in and cause damage. The roof should be particularly well built, keeping the centre well hearted and placing the outside sheaves so that the butts overlap like slates on a roof. If the stacks are made in the open they should be thatched soon after completion.

In favourable seasons, more especially in the south of England, threshing may be done directly from the field with a considerable saving in man-hours on the harvesting operations. In the stack the grain may sweat slightly and, should it have been carted in bad order, it may be necessary to leave it to dry out in the stack until the spring before threshing. The few threshing machines still used in this country are usually of the beater-bar type which gives an efficient separation of grain from straw, chaff and weed seeds. The straw is not damaged appreciably and can be used for thatching. The peg-drum American and Canadian types of machines also give an efficient separation of grain, and there is, in addition, a higher output with less labour than from the English machines. The straw, however, is badly shattered and useless for thatching.

The Combine. The binder, commonly used only 20 years ago, is obsolete, and grain cereal crops are now almost all harvested by combine, with emphasis on speed and efficiency. The main advantage of the system is that the whole of the harvest operations are completed quickly with a considerable saving in terms of man-hours compared with the binder-harvest system. As the crop must be dead ripe before combining it is desirable to use a variety which is not prone to shattering. Although the crop has to stand uncut for a longer period than when the binder is used, it is not on that account any more vulnerable to bad harvest weather. Field experience, in fact, has shown quite the opposite. In a wet harvest a standing crop will dry out rapidly in the short drying spells whilst a stooked crop will remain wet and unfit for carting. Lodged crops, although always a nuisance, can be harvested more efficiently with the combine, particularly the self-propelled type, than by the binder. As the speed of the combine is determined to a large extent by the volume of straw passing through it, it is usual to leave longer stubbles than with the binder. This factor also re-emphasises the necessity for cereal varieties with a high ratio of grain to straw to obtain the best field efficiency.

The facilities available for handling and drying, or for direct sale of the wet corn from the farm, dictate the moisture content at which the crop is combined in the field. Data for safe moisture contents for storage as well as other field aspects of the technique of using combine-harvesters are given on page 92. The combine harvester, together with the grain drying and storage equipment, does of course involve considerably greater capital cost than binder equipment. On the other hand the saving in man-hours is approximately 50 per cent. (Figures quoted from economic surveys in the Ministry of Agriculture publication entitled " The Farm as a Business " are shown on page 232.)

MAN HOURS PER HECTARE FOR HARVESTING WHEAT

Combine Harvesting (includes handling of straw).		Binder Harvesting (includes Threshing).	
Self-propelled. 4 m cut.	Towed. 2 m cut.	Stacked and Threshed.	Threshed in Field.
25	30	60	36

Straw collection after the combine harvester is usually done from the windrow by pick-up balers. Any green material in the windrow should be allowed to dry before the pick-up baling is done. Some combine harvesters are fitted with straw trusses which eliminates the need for a separate pick-up baler. Trusses of this type are not so suitable in a crop which contains green material in the straw. It is convenient to tow a bale sledge behind the pick-up baler so that the bales can be collected into 6- or 8-bale stacks. These small stacks render the bales less vulnerable to wet weather, they allow wet bales to dry out, and they facilitate carting.

The combine harvester system has introduced rather a different attitude to straw in the economy of the farm. Under the binder systems, except where threshing was done in the field, the straw was transported to the buildings along with the grain. The straw so collected had to be disposed of in some way, either to stock or for sale from the farm. With the combine harvester technique the straw need only be collected if it is required either for stock or for sale. If it is not required it may either be burnt *in situ* or ploughed in according to the beliefs held as to its value for soil fertility. If it is to be ploughed in, a spreader fitted to the combine harvester and " wavy " disc coulters fitted to the plough facilitate the operation.

Yield. The average yield of wheat in this country has been steadily increasing due to the introduction of stronger strawed varieties with a higher yield potential, and to higher rates of nitrogenous fertilisers applied to the crop. The average yield in the 1930-40 period was 2·25 tonnes/ha : the average yield in the five years 1970-4 has been about 4·3 tonnes/ha, an increase of about 90 per cent. It is by no means exceptional to have grain yields of 5 tonnes/ha and more. Wheat growers in the good cereal-growing soils in the east of the country expect to have yield averages of about 5 tonnes/ha according to season. In field practice and in some experiments, yields of over 8 tonnes/ha have been recorded.

The straw yields accompanying these grain yields will vary between 2·5 and 4 tonnes/ha.

BARLEY

In barley the glumes are relatively small and insignificant and the spikelet contains one floret only : the paleæ, which form the " skin " of the grain, are fused with the caryopsis, or true kernel in most of the varieties grown in this country. There are, however, species of barley with naked grains. As with the other cereals there is a large number of species and varieties but, from the British farmer's standpoint, two main species only are of importance, Six-rowed barley (*Hordeum polystichum*) and Two-rowed barley (*Hordeum distichum*). The area devoted to the six-rowed barleys, however, is very small compared with that sown with the spring varieties of two-rowed barley.

In the six-rowed barleys three fully fertile spikelets arise from each node of the rachis; the same arrangement is found in the two-rowed barleys, but here only the middle or median row of the three spikelets is fertile and forms a grain. The median rows of spikelets in both species have their ventral surfaces, i.e. the surface with the furrow, opposed to the broad side of the rachis. The other spikelets are attached to the node nearer the edge of the rachis, and in six-rowed forms the grains are consequently not absolutely symmetrical in longitudinal section, but slightly twisted (Fig. 32). Since to each set of three spikelets there is only one median grain, six-rowed varieties have one row of median and two rows of lateral spikelets on each side of the rachis, the complete ear having two rows of median and four rows of lateral spikelets. In two-rowed varieties there is one row of fertile and two rows of infertile spikelets on each side of the ear, the complete ear having two rows of fertile and four rows of infertile spikelets.

Varieties. Six-rowed barleys are not much grown in these islands, although in Scotland both winter-sown and spring-sown types are still grown locally under the names of " Bere " or " Big ". Astrix is a new, very early ripening 6-row winter barley for feeding; it is resistant to Rhynchosporium. Another true winter barley is Maris Otter, which has good malting quality combined with early maturity.

Malting quality and general agronomic characteristics have both had their influence upon the barley breeding programmes. As in the case of wheat, the first success of the plant breeder was in pure line selections from the old land race varieties. In this way the varieties Archer, Chevallier, Plumage and Spratt arose and dominated the scene for many years. The next phase of development was hybridisation between these types in order to combine desirable characteristics. This led to the two famous varieties—Spratt-Archer, produced by Dr. Hunter in Ireland, and Plumage-Archer, produced by Dr. Bevin at Warminster. These varieties were both brought out in the early 1920's, they both had good malting qualities and, by the former standards, were high yielders. These two varieties dominated the acreage grown through the late 1920's and the whole of the 1930's. They remained upon the recommended list of the N.I.A.B. until 1960. Few varieties of any cereal have enjoyed such a long period of success as these (see also Chapter XV).

Early in the 1930's the Danish variety, Kenia, was introduced and this was followed by Maja, Freya, Herta, Rika and Carlsberg. All these varieties had shorter and stiffer straw than Plumage-Archer and Spratt-Archer. Their yield at low levels of fertiliser use was not much superior to the English varieties, but by virtue of their stiffer

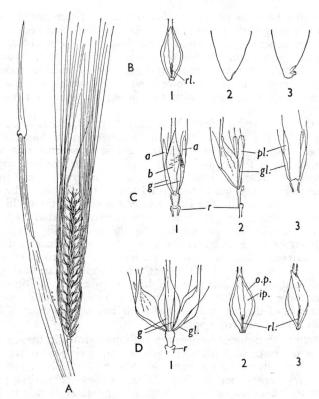

FIG. 32.—EAR OF BARLEY.

A. Typical ear of two-rowed barley.
B. Grains of two-rowed barley, ventral and side views.
　　B1, ventral view of grain.　B2, chisel-edge base to grain of drooping-eared type.　B3, "lumpy"
　　base of erect-eared type.
C. Structures at rachis notch of two-rowed barley.
　　C1, C2, central (fertile) and lateral (sterile) spikelets.　C3, glumes and pales of lateral spikelets
　　(central spikelet removed).
D. Structures at rachis notch of six-rowed barley.
　　D1, triplet of fertile spikelets.　D2, symmetrical grain of central spikelet.　D3, twisted grain
　　of lateral spikelet.　r, rachis.　rl., rachilla.　a, sterile lateral spikelet.　b, grain fertile of spikelet,
　　g, glumes of central spikelet.　gl., glumes.　pl., pales of lateral spikelet.　o.p., outer pale,
　　i.p., inner pale.

straw more fertiliser could be used and, as a result, their yield potential
was much greater. Their malting quality, however, was considered
to be inferior to the English types. Because of the superior agronomic
merits of these Scandinavian varieties, crosses were made between
them and the English varieties in an attempt to combine the virtues
of each into new varieties. This programme led to the production
of Maythorpe, Proctor and Provost, all products of the Cambridge
Plant Breeding Institute. Of the three, Proctor became the most
successful and it once dominated the market. Another feature of the
Scandinavian varieties was their earliness of ripening compared with
Plumage and Spratt-Archer. A further development due to hybridisation has been the combination of high malting quality with winter
hardiness in Pioneer and Maris Otter.

A present trend in barley breeding is directed to producing a malting variety with the agronomic characteristics of Proctor but with earlier ripening. A further trend is to produce varieties resistant to cereal mildew (*Erysiphe graminis*). It has been shown that this disease can give greater yield depressions than was formerly supposed, and thus breeding for resistance assumes greater importance. In the future, no doubt, work will be directed to breeding for the physiological components of yield.

For the description of the recommended varieties the leaflets produced by the National Institute of Agricultural Botany at Cambridge should be consulted. The recommended varieties change from time to time as new varieties are produced and some now popular may be obsolete in a few years.

Malting Quality. Less than one-fifth of the barley crop produced in this country is used for malting and the remainder goes for stock feeding. Since there is usually an appreciable price differential for malting barley the characteristics of a good malting sample need to be outlined. Unlike wheat, quality barley is to some extent incompatible with high yield. The new Cambridge hybrids have reduced this incompatibility to some extent but have not eliminated it, as less fertiliser nitrogen must be used to obtain the highest malting quality. Malting barley must contain a high proportion of starch and a small proportion of protein because it is from the starch that the brewer ultimately obtains maltose, and it is this material that later, when in solution and acted on by yeast, produces a fermented liquor. A cross-section of a good malting sample shows a white " mealy " interior, whereas one with a high nitrogen content will have a translucent or " steely " interior.

The physical features of the grain are usually the guide as to whether a sample of barley is fit for malting or not. In the first place the grain must have been well harvested, be free from excessive moisture and have a sweet smell with no hint of mustiness. Great care is also necessary in threshing, for any skinning or fracture of the grain will ruin its germinating capacity and hence its value for the maltster. It should be even in size, with an evenly coloured and finely wrinkled skin indicating good harvest weather and the right stage of maturity. The sample should also be clean and free from weed seeds or any trace of fungus disease.

To attain this degree of perfection necessitates a good climate, avoiding such extremes as very dry, sunny weather, and wet, sunless weather. Excessive drought after flowering leads to premature death of the plant and small, shrunken and often steely grains, resulting from incomplete translocation of the starch from the straw. Cool wet conditions delay the grain maturation process and lead to discoloration of its surface. Before the grain is fit for malting it must undergo a complete drying-out process (i.e. down to about 14 per cent moisture), otherwise it is likely to exhibit dormancy and give delayed germination on the malting floor.

Soil, too, is a factor of great importance in obtaining malting quality and here, irrespective of rainfall, it is found that heavy soils do not yield the best malting samples. The ideal is a fairly rich medium loam in a district of moderate rainfall. Very light soils produce excellent barleys when the precipitation is adequate, but usually fail in dry years. Soils of reasonable depth derived from chalk, or other calcareous rocks, have the highest reputation as good malting barley soils.

Position in Rotation. Barley was usually taken at two points in the traditional rotations : as the spring cereal crop following a folded root crop ; and as a second straw crop, usually after wheat. In many parts of the country barley is now taken after a sugar beet crop. It was formerly inadvisable to take barley after potatoes, but with the stronger strawed varieties this is now a much safer procedure though it is a sequence only suitable for feeding-quality barleys. Barley is a crop not able to stand up well to weed competition and this was one of the reasons for taking it after turnips or swedes. With the modern range of chemical weed killers there is now less need for this sequence, and the practice of taking barley as a second, or even third, straw crop has increased considerably. There are examples of successful continuous cropping with barley to be seen on Hoos field at Rothamsted, and formerly on the chalk land farms belonging to the late Mr. Chamberlain of Oxfordshire. Barley is probably the safest of the cereals to be cropped at a high frequency in a rotation because of its relative freedom from serious rotation diseases. Eyespot, although able to attack the crop, rarely becomes a serious trouble except with winter barley. Barley is also susceptible to Take-all, but a one-year rest from wheat or barley is sufficient to control the disease provided there is no Couch grass, Yorkshire Fog or volunteer wheat or barley plants. In recent years cereal root eelworm has attacked barley, and this pest must be regarded as a potential threat to intensive barley cropping.

Seed-bed Preparation. The seed-beds prepared for the autumn sowing of barley are similar to those for wheat, but for spring sowings much greater care is necessary in their preparation—partly because dry weather conditions may follow the sowing, during which a badly prepared seed-bed is liable to dry out in parts and give rise to an irregular plant. A further factor necessitating additional care in cultivation is the shorter interval of time between sowing and harvesting, which means that everything possible conducive to even germination and rapid development must be done. In the case of autumn sown crops, time and weather help to ameliorate irregularities in the germination of the seed, but in the case of spring-sown crops this latitude is not possible. In the latter case the typical spring cereal type of seed-bed is required with its fineness and firmness achieved with minimum loss of moisture. For this purpose the initial ploughing does not need to be very deep but, on soils other than the lightest, it requires to be done early enough to obtain the mellowing effect of alternate freezing and thawing. Subsequent working down with harrows, or harrows plus Cambridge roller, will then achieve a seed-bed of the desired standard with the minimum of operations.

Great care is necessary when preparing the seed-bed after late folded roots, for at this time of the year dry weather is likely to prevail and there is a danger of the seed-bed drying out completely during its preparation. Time does not permit of two ploughings, whilst one deep ploughing would turn the damaged surface soil down to the subsoil in an unweathered condition. In the circumstances the normal procedure is to plough shallow, and roll and harrow each day's ploughing immediately afterwards so that the furrows are left covered with a little fine soil which prevents them drying out before ploughing is completed. Alternatively, disc harrows may be used on the trampled surface before ploughing in an attempt to ameliorate the poached surface. This operation should be carried out when the soil begins to dry. Then the land is ploughed to a shallow depth so that the friable surface soil falls upon the subsoil as the furrow is turned. The area ploughed each day must be rolled and harrowed whilst still retaining sufficient moisture for the soil to be crumbly. Final touches to the seed-bed can be given with disc harrows and roller, when the result should be a uniform tilth to the full working depth.

When barley is taken as a second, or third, straw crop every opportunity should be taken to destroy perennial weeds by stubble cleaning operations. This may be achieved by broadshare cultivation or shallow ploughing. If done immediately after harvest, in many years a kill of the weed rhizomes may be obtained by desiccation. In a very foul stubble these operations may be followed by the use of the drag and chain harrows to pull the rhizomes to the surface and collect them for burning—in fact the traditional set of " wicking " operations. Where soil conditions are suitable many farmers prefer the technique of deep ploughing as a control of perennial weeds, and in some circumstances rotovating has been found to be a useful stubble cultivation technique. If the problem is an annual weed instead of perennial infestation, the stubble cultivations should be aimed at producing a seed-bed to encourage the buried weed seeds to germinate. These are subsequently killed as seedlings by ploughing. This is a very necessary practice in the case of annual weeds which cannot be controlled by selective weed killers, and it is one of the few techniques that can be used in the case of wild oats.

Manuring. Barley is the most sensitive of the cereal crops to soil acidity. Failure is likely to occur at pH 5·3, and reduced yields and lowered quality may be expected at acidities below pH 6. Thus any suspicion of soil acidity should be rectified by liming before a crop of barley is sown. It is advisable to do this well before the crop is drilled so that the lime may be given sufficient time to act. Undoubtedly the best procedure is to apply the lime to the preceding crop ; failing this, every attempt should be made to apply in the autumn before the barley is sown.

The review of experimental data made by the late Dr. Crowther and Dr. Yates suggested that the theoretical average optimal fertiliser requirements (see p. 226) for the crop was similar to that obtained for the wheat crop, namely 90 kg of nitrogen ; 20, 30 or 40 kg/ha according to district of phosphoric acid ; and 20–30 kg/ha of K_2O. There are, however, two features of importance in which barley differs from the wheat crop. One is that the resistance to lodging due to high levels of fertiliser nitrogen is less in the strongest strawed barleys than in the equivalent wheat varieties.   There are circumstances where the full theoretical optimal of 90 kg/ha of nitrogen may cause lodging, even when lower rates of seeding than the normal are used.   The second is that high soil nitrogen has a retarding influence upon grain maturation and hence upon malting quality.   The customary dressings tend to be 50–150 kg/ha of nitrogen for strong strawed varieties where feeding samples only are required, and 25–50 kg/ha where malting samples are aimed at.   The new hybrid barleys, as well as the Scandinavian varieties, not only have a higher average yield, but also have an intrinsic ability to respond to a greater degree to nitrogenous fertilisers than the older varieties, Plumage-Archer and Spratt-Archer.   To achieve good malting samples from them, however, the grower must be just as cautious with levels of nitrogenous fertilisers as with the latter varieties.   Where barley is taken after a crop of roots folded off, the level of nitrogen applied should be reduced by about 40 kg, and the P_2O_5 and K_2O by 25 and 30 kg respectively.   In many cases this means no P_2O_5 or K_2O need be applied after a folded crop.

Time of application of the nitrogen is also of importance.   With spring barley intended for malting all the nitrogenous fertiliser should be applied before sowing and worked into the seed-bed.   With a feeding barley some farmers put half in the seed-bed and half as a top dressing later in Spring.   Recent experimental work, however, has shown that the best responses are obtained when all the nitrogen is placed in the seed-bed ; it has also shown that it is safe to combine-drill this relatively high rate of nitrogen.   With winter barley intended for malting, the nitrogen, usually not more than 25–30 kg/ha of nitrogen, should be applied as a top dressing early in April.

Where the soil is known to be deficient in phosphate or potash it is necessary to increase the quantities given over and above those specified as the average optimal dressing.   It is usual to give the barley crop greater rates of phosphoric acid than the theoretical optimal unless it follows a folded root crop or sugar beet crop.   This is probably sound practice in the case of malting barley on soils other than the lightest in the low rainfall areas.   It has been shown that phosphate has a very beneficial effect upon grain maturity and hence upon malting quality.

Barley is the most resistant of the cereals to a low potash status of the soil.   Yet in general practice it is the cereal which receives the greatest amount of potash fertiliser.   One reason for this apparent anomaly is that it is the cereal most commonly grown on the chalk and limestone soils, many of which have abnormally low potash status. Care is needed with potash levels for the malting crop ; excessive quantities have an effect upon grain maturation, and hence malting quality, similar to excessive amounts of nitrogenous fertiliser.   Combine drilling of phosphate and potash leads to similar increases of efficiency of the plant nutrients as in the wheat crop (see p. 227).

Organic manures are not normally applied directly to the barley crop.   In many rotations the crop receives the benefits of organic residues from folded root crops and from folded or ploughed-in sugar beet tops.

Sowing. Seed barley should consist of a plump, even grained sample of good germination, and it should be free from impurity. It should be dressed with an organo-mercurial powder as a precautionary measure against Leaf Stripe and Covered Smut diseases (see p. 388). If there is a high wireworm population in the soil a combined organo-mercurial-insecticidal seed dressing should be used.

Drilling should be to a depth of 25–35 mm below the surface, on a firm bottom : it is normally carried out with drills having 175 mm coulter spacings.

Winter varieties, such as Senta, should be drilled in late September or early October ; that is earlier than the winter wheat crop. Spring varieties may be drilled from January to April according to the state of the ground. The crop is often drilled in February in the drier south-eastern counties, but in the north it is rarely possible before March. Drilling as early as is consistent with attainment of suitable seed-bed conditions is the rule. Both yield and malting quality benefit from early sowing. In sheltered areas near the sea, spring varieties are sometimes sown as early as December : there is some risk attached to this practice as these spring varieties are not truly frost resistant.

The usual rate of seeding is 190 kg/ha but, as with wheat, recent trends have been to reduce seed rates. As with wheat, barley shows a point of physical optimum seed rate ; seed rates above and below this optimum give reduced yields. Recent summaries of experimental work have shown that the average physical optimum seed rate occurs at 156 kg/ha. Where the farmer is using his own seed 156 kg of seed is also the economic optimum ; where the cost of the seed is twice the price received for the crop, weight for weight, the economic optimum is 125 kg/ha. Again, these figures should be taken as general guides only. They need to be increased where drilling is late or where the seed-bed is below standard. With malting barley it is probably inadvisable to decrease the seed rate below 156 kg/ha. With feeding barleys the lower seed rates quoted have the advantage of making the crop less liable to lodge with high rates of fertiliser nitrogen.

After drilling, the crop is harrowed in and, on the lighter soils, is usually rolled. If a weed infestation problem arises the crop may be sprayed with one of the selective herbicides which have already been mentioned.

Harvesting and Threshing. The application of full mechanised harvesting has been particularly successful with barley and almost the whole of the crop is dealt with in this way. The use of the combine harvester has greatly speeded up and simplified the harvesting of barley. At first there was considerable prejudice on the part of maltsters and seedsmen against samples of barley harvested by combine but this has now disappeared because of the good quality of barley grain harvested by modern combining techniques.

When combine harvesting, both feeding and malting barley must be left until dead ripe before cutting commences. Where malting barley is harvested by the binder it similarly must be dead ripe before cutting. There is, however, greater latitude concerning the date to cut when a feeding barley is harvested by binder. Barley varieties are not so liable to shed grains by shattering when dead ripe as are wheat varieties. On the other hand severe winds may shear off whole ears at this stage causing severe loss of crop. This is more likely to occur in those varieties having a long " neck "—that is to say, a long length of straw protruding above the uppermost leaf sheath. Crops which, after the dead ripe stage, develop a form of false lodging characterised by a sharp bending over of the straw half-way between ground and ear, are less vulnerable to such damage. Great care is required in the storing and drying of the grain of malting barley after combine harvesting. Further reference should be made to page 92 for details. Many maltsters prefer to buy malting barley direct from the combine and do their own drying and storage.

When harvested by binder the crop is stooked in the field and subsequently stacked. Where a crop of malting barley is cut when dead ripe there is no need for a period of maturation in the stook. The crop can therefore be carted and stacked as soon as dry enough.

In the threshing of the barley crop intended for malting great care is needed in the setting of the threshing machine or combine harvester for the reasons already discussed in detail on page 235. A little of the " beard " left on the grain is not a drawback and is regarded as an indication of careful threshing.

Yield. The extended use of the stronger strawed varieties, together with the high rates of fertiliser their use permits, has resulted in a gradual increase of the average yields of barley in recent years. The mean yield between 1930 and 1940 was 2 tonnes/ha ; the average yield in the five years 1970–4 was about 4 tonnes/ha ; an increase of about 80 per cent. On good soils and in good seasons yields of 4·5 tonnes/ha are common and yields considerably in excess of 5 tonnes/ha have been recorded.

The yield of straw in the barley crop is usually about equal to that of the grain with the older varieties, but it is less with the newer varieties.

OATS

The inflorescence of oats differs markedly from those of wheat, barley and rye, in that the divisions of the rachis are measured in inches instead of fractions of an inch and the spikelets are not borne directly on the rachis but either on branches of varying length arising directly from the nodes of the rachis, or on secondary branches arising from the primary branches. This type of ear is called a spreading panicle (Fig. 33). The manner in which the spikelets are joined to the pedicels, or small branches, is important, and is one of the features used to distinguish different species. Thus, in some species there is a distinct cavity at the base of the grain, the edges of which are well defined, and frequently bear a ring of strongly grown hairs. The extremity of the pedicel is expanded into a knot which fits into this basal cavity, the two thus simulating a ball-and-socket joint. Such an arrangement is found in *Avena fatua*, the wild oat of cultivated land, and in *Avena sterilis*, another wild form, indigenous to the

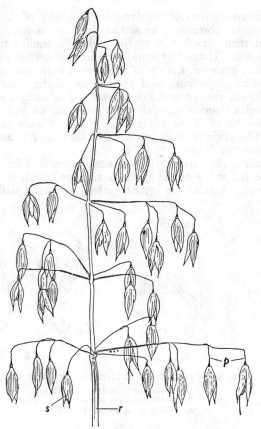

FIG. 33.—SPREADING PANICLE OF OAT.
r, rachis.   p, pedicel.   s, spikelet.

countries bordering the Mediterranean. On ripening, the tissue connecting the pedicel and grain separates naturally, and the grain thus detached from its pedicel is shed on to the soil, thereby assuring continued propagation.

Most of the cultivated varieties found in the British Isles belong to the species *Avena sativa*, in which there is no sharp division of the tissues connecting the spikelet and the pedicel, and the grain when fully ripe separates from the pedicel by a distinct fracture of the tissues uniting the two.

The wild oat can easily be distinguished from cultivated forms by the strongly grown, twisted and kneed awn, which is invariably present on all the grains of the spikelet. An awn is developed in many varieties of cultivated oats, but then only on the lower floret of the spikelet (Fig. 34).

Avena strigosa, another species, must be mentioned here, for some varieties of this group are the oldest forms known in Britain, and are still cultivated in mountainous districts in Wales and in some of the western districts of Scotland. The grains of all the varieties of *Avena strigosa*, commonly known as the " bristle-pointed " oat, are small, and their attachment to the pedicel is similar to that found in *Avena sativa*. Instead of a more or less indeterminate end to the dorsal palea, however, the tip of the husk is extended into two long, fine, awn-like projections, and each grain of the spikelet carries a well-developed, twisted and kneed awn. The various forms of oats of this species flourish in Wales and Scotland on soils where none of the varieties of *Avena sativa* can be grown with any degree of success. The grain is small, but the straw is long, fine and abundant, and makes good fodder.

The bulk of the oats grown in this country is used for stock feeding and hence the value of the grain is adjudged from this angle. In barley the coverings of the grains (paleæ) are fused with the kernel

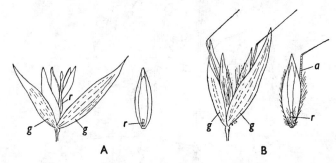

FIG. 34.—SPIKELETS OF CULTIVATED AND WILD OAT.

A. Spikelet and grain of cultivated oat.
B. Spikelet and grain of wild oat.
a, dorsal awn. *g*, glume. *r*, rachilla.

(caryopsis) and it is impossible to detach them from the grain when it is dry : even after soaking in water separation can only be effected with difficulty. In wheat, on the other hand, the kernel is separate from the paleæ and may be removed readily on rubbing the ear through the hand. In oats, although the paleæ envelop and effectively protect the caryopsis against the weather, they are not actually attached to it in any way. There is one species of oat, *Avena nuda*, the so-called naked oat, in which the palea open, as in wheat, leaving the kernel free and exposed at maturity.

If the true kernels of wheat and oats are examined chemically the largest differences in composition are found in the quantity of crude fibre and ash, both of which are higher in wheat ; the fat is invariably higher in oats, and the amounts of crude protein are similar in both cereals. But oats when used for stock food are usually consumed unhusked, and hence the composition of the whole grain is of great importance. Comparing wheat and oats on this basis the differences in composition are considerable, but they are centred mainly around the crude fibre, which frequently amounts to as much as 10 per cent of the weight of the whole oat grain—

the husk forms 20–30 per cent by weight of the grain and of this
about one-third is fibre. After the fibre, the next largest difference
is found in the crude fat, which is naturally much lower in the
whole grain than in the kernel of oats, but is still higher than that
found in wheat. The crude protein also suffers some reduction,
whilst the ash is increased.

Varieties. Cultivated oats may be broadly classified into two
main groups—winter varieties and spring varieties. Within these
groups, however, are further important variations which enable
sub-divisions to be made. The colour of the grain, for instance,
whilst white in the majority of varieties, may be yellowish, black,
brown or grey. Some varieties have short plump grains, others
produce a much longer and thinner type ; some have a spreading
panicle, others a one-sided panicle giving a banner-like appearance,
and a large number have an intermediate type of ear.

It is believed that for several centuries, up to the first half of the
18th century, various forms of naked oat (*Avena nuda*) were extensively
cultivated. They were known as Pilcorn or Piley or the Skinless Oat.
They did well under relatively poor conditions and were often the
last crop grown in systems in which land was cropped for a period
and then allowed to go out of cultivation for a rest or recuperative
period. During the 18th century a variety called Poland was imported
and grown extensively in the south of England : various forms of
Dun or Brown oats, and Black oats in the Midlands, replaced the
naked varieties. Later in the same century Black Tartarian and
White Tartarian oats spread from the east into Europe and subsequently into England. The Black Tartarian became a very popular
variety and replaced the other forms of black oats.

In 1788 the famous variety, Potato, was discovered growing as an
isolated plant in a Cumbrian potato field ; it was probably introduced
from south Europe amongst packing straw. This variety became of
importance in the north, and even today the variety, or one of its
many selections, is occasionally grown in the north of Scotland.
Certain other varieties came into being in a similar way by observant
farmers selecting chance occurring plants or useful looking types
within existing crops, and multiplying them to produce new varieties.
For instance, the variety Sandy originated from an odd plant found
growing on ditch cleanings in Aberdeenshire, whilst the varieties
Hamilton and Hopetown were pure line selections out of the
Potato oats.

Many of the above varieties of oats were grown as much for their
straw as for their grain. This particularly applied to the varieties
Potato and Sandy and their derivatives, which are referred to as *straw
producers*. An outstanding advance in the more predominantly *grain
producers* took place in the last twenty years of the 19th century as a
result of hybridisation work by Dr. John Garton. He introduced a
number of improved varieties, amongst which the most famous was
the variety Abundance, marketed in 1892. These varieties had a
great effect upon oat growing in the country, and Abundance could
still be found in use as late as 1940. Waverley, Yielder, Goldfinder
and Bountiful were other successful types produced. The varieties

Supreme, Onward and Forward have been successful varieties in more recent times.

The Swedish plant breeding station at Svalöf has contributed a number of successful hybrids to the British farmer in the last 40 years. Victory, Star, Eagle, Sun II and Blenda have all been successful varieties. These oats have all been reasonably strong in the straw, good and reliable yielders, with a high kernel-to-husk ratio in the grain and with average date of ripening.

In the last 30 years the Welsh Plant Breeding Station has produced some important varieties. S.84 was a strong strawed spring oat, rather late in maturing and with only an average yield. S.147 and S.172 were two highly successful winter oats; the first had a high yield but rather long straw which made it unsuitable for highly fertile soil or for use with high rates of fertiliser; the second had an extremely short and strong straw but had not the intrinsic yield potential of the former. Powys, a hybrid between S.147 and S.172, to some extent combined the features of the two parents. Its yield potential is actually better than S.147, but the straw strength, whilst good, fails to equal that of its S.172 parent. Each of these winter oats is far superior to the old Grey winter and Black winter varieties they replaced. Another important variety was the spring oat Milford (S.225), a cross between Victory and S.172. It had similar straw characteristics to S.172 and was in a class by itself amongst spring oats for this feature, in the same way as its parent is amongst winter varieties. Again, however, it is not capable of high yields.

An effort has been made at both the Welsh Plant Breeding Station and at the Cambridge Plant Breeding Institute to breed for resistance to Stem Eelworm, not to be confused with Cereal Root Eelworm (see p. 437). The Welsh varieties S.172, Pennant, Milford and Manod, and the Cambridge winter variety, Maris Quest, all show this resistance. Breeding for mildew resistance has also received attention, and the Welsh variety Peniarth is outstanding in this respect.

For hill land areas, earliness of maturity is a desirable characteristic. Winter oats are in general early maturing, but are unsuitable for hill conditions. One of the merits of Supreme, mentioned earlier, was its earliness. Ayr Commando and Ayr Everest produced in this country, Primus II from Svalöf, together with Clinton from USA are all early maturing varieties.

As with wheat and barley, many plant breeders are striving to produce better oat varieties designed to meet modern conditions. An essential feature of this is to combine high yielding potential with extreme straw strength, so that the capacity for yield may be exploited by high fertiliser use without the risk of lodging. The plant breeder has now achieved some degree of success with oats as with wheat and barley. Straw strength is possessed by the varieties Maris Quest and Peniarth plus the capacity for yield. Another feature that requires improvement is loss of grain from the panicle (shattering) when it is left until dead ripe, as is necessary when using the combine harvester.

The current N.I.A.B. list of recommended varieties mentions four winter varieties and five spring varieties.

Position in Rotation. The position of oats in the crop rotation is by no means so fixed as that of wheat or barley, and it varies greatly with the extent to which other corn crops are cultivated on the farm. In Scotland and in many parts of Northern Ireland where neither wheat nor barley is extensively grown, oats follow roots, leys or, if necessary, oats themselves. A high frequency of oats in the rotation can lead to severe trouble with two eelworm pests : Stem Eelworm and Cyst Eelworm. The varieties mentioned above have considerable resistance to the stem eelworm, but all varieties seem to be susceptible to cyst eelworm (see p. 437). In the predominantly cereal-growing districts of the country the sequence of oats undersown with one-year seeds, or a two-year ley, is useful as a disease break in cropping systems containing a high percentage of wheat and barley.

Manuring. Although the oat crop is more tolerant to conditions of soil acidity than wheat and barley, it is likely to give the best yields on soils approaching neutrality. The crop is, however, very sensitive to manganese deficiency, a condition common on soils where a high organic matter content is associated with a high pH. The condition is known as " grey leaf " or " grey speck " because of the characteristic leaf symptoms. On some of the fen peats this deficiency makes it almost impossible to grow oats. If the condition can be recognised early enough, applications of manganese sulphate at rates of 25 kg/ha either as a spray or, diluted with a filler, as a dust can control the condition. By the time the characteristic leaf symptoms appear it is too late to effect a full and complete control.

As regards the major plant nutrients the average theoretic optimal rates are approximately the same as those for wheat (see p. 226). New varieties of oats like Maris Quest and Peniarth will stand more nitrogen than the older weak-strawed varieties. In most cases applications of 100 kg per ha only are used. With phosphate, adjustments for the wetter areas and deficient soils as discussed for wheat (see p. 227) are necessary. With potash the adjustment for the deficient soil only is required. In some rotations oats are taken as the first crop after well-fertilised leys, and in these circumstances the crop is often grown without any fertiliser. The usual economy of phosphate and potash can be effected by combine drilling the fertiliser with the seed. On farms with a high proportion of stock and a low percentage of arable, farmyard manure is often available for the oat crop in which case fertilisers may be omitted. With winter oats the nitrogen is given as a top dressing in the spring as in the case of winter wheat.

Cultivation and Sowing. In the west and the north the crop is mainly spring sown but, in the drier districts of the south and east, there is a greater preference for winter varieties. The winter varieties are less prone to damage from frit fly attacks than the spring varieties (see p. 435). They are also earlier ripening and thus of value in spreading the harvest period. It is important to sow spring oats early in the spring as a protective measure against frit fly attacks ; trouble with this pest may be expected in most areas in crops sown later than the first week in April.

Spring varieties and winter varieties of oats require the typical spring and autumn cereal type of seed-bed respectively (see pp. 228

and 229). Where the crop is taken as the first crop after ley or ploughed-out grass it is advisable to use the disc harrows in the initial stages of preparation of the seed-bed.

The seed sample should be of high germination, plump seeded, free from weed seeds and immature grains. The usual rate of drilling is 125–250 kg/ha. Recent summaries of experimental data place the physical optimum seed rate for oats between 250 and 300 kg/ha, which is appreciably higher than that for winter wheat or spring barley. From the same data the economic optimum seed rate where the farmer sows his own seed is 250 kg/ha ; when the purchased seed is worth twice and three times the same weight of grain from the crop the economic optima are respectively 225 and 150 kg of seed per ha. There are no equivalent data for winter oats.

It is advisable to dress the seed with an organo-mercurial seed dressing to control seed-borne diseases, and this should be replaced by an insecticidal-fungicidal seed dressing where high populations of wireworm are present. After drilling the seed should be harrowed in and, if necessary, the seed-bed rolled with a Cambridge roller. Where injurious populations of annual weeds develop the appropriate type of herbicide may be used. Greater discretion in choice of herbicide needs to be exercised for oats than for wheat or barley ; for instance MCPA should be used in preference to 2,4-D. For further information see Chapter VIII.

Harvesting. The oat crop is cut at an earlier stage of maturity than either barley or wheat, because oats are liable to shed the grain when fully ripe. Moreover, the grain will ripen satisfactorily in the stook and the straw is of much higher feeding value when cut at this early stage. Thus the crop is cut whilst the straw is still slightly green and before the chaff flies.

The crop is left in the field usually for at least a fortnight— tradition decrees that the stook should be out for three Sundays— and to prevent heating in the stack. In districts of higher rainfall it is usual to make relatively small stacks and to arrange for through ventilation of the stack. Faggots laid out in the middle of the stack are helpful.

The liability to shedding when dead ripe makes the oat crop less suitable for combine harvesting than wheat or barley. In addition a dead ripe straw is less palatable and of lower feeding value for stock. Nevertheless, great economy in the man hours needed to harvest the crop can be gained using the combine harvester technique. The standard man hours per ha for producing and binder harvesting the crop are taken as 72 whereas the equivalent figure, using the combine harvester is much less. Consequently most of the crop is now harvested by combine.

Yield. In common with the other cereal crops oats have shown increasing yield trends in recent years. The official average figures show that whereas the pre-war yields averaged 1·9 tonnes/ha, the average yield in the five years 1970–4 has been about 3·5 tonnes/ha, an increase of about 80 per cent. The yield increases have not been as great as those with wheat and barley over the same period. Crops of 4–5 tonnes/ha are not uncommon on good land in good seasons, and yields of over 5 tonnes/ha have been recorded.

Straw yields tend to be about one-third greater than grain yields under most conditions. Grain to straw ratios are greater in dry areas than in wet areas ; they are also influenced by variety and quantity of fertiliser used. High nitrogen levels decrease grain-to-straw ratios.

RYE

Rye as a grain crop is of very minor importance in this country, its cultivation being confined mainly to very poor light land where the other cereals would not prove remunerative. The public taste for " crisp bread " has in recent years tended to extend the cultivation of the crop, and during the food production campaign of 1939–45 the crop received a mild fillip.

The crop bears a strong likeness to the other cereals botanically, but has additional features of agricultural value (Fig. 30). It is, for instance, the hardiest of all the cereals and can over-winter under conditions of low temperature to which even wheat will succumb. It does well on sand and light peat soils under conditions of considerable soil acidity, and its rapid germination and quick establishment make it a useful forage catch crop.

The grain is similar in composition to wheat and may be used for stock feeding under comparable conditions, but the plant is used chiefly as a fodder crop for sheep. Sown in the early autumn it makes abundant growth before the end of the year, and by the succeeding February furnishes a good bite for lambing ewes. This early bite can also be used to good advantage for the dairy herd, for it provides an invaluable stimulus to milk production by affording a change of diet from the monotony of the winter feeding programme. Provided the sheep and cows are not kept on the crop too long, rye re-establishes its growth rapidly and will then produce a fair yield of grain. Frequently, however, it is ploughed in after folding, and the land is left in that condition until it is time to prepare it for roots. On light land this method of dealing with the crop is an excellent preparation for roots.

Varieties. Very little work has been done on the problem of varieties of rye. Rye, unlike the other cereals, is cross-fertile, and is therefore more difficult to maintain in a pure condition. It exists in two more or less distinct forms, winter rye and spring rye. It is inadvisable to sow winter rye too late in spring, for there is then a danger of an almost complete failure through the non-production of stems and ears.

Certain continental varieties of winter rye are available, including Dominant and Otello. These have shorter, denser ears, and shorter straw, than the older types. Stocks of genuine spring rye are not plentiful.

Position in Rotation. Where rye is grown it usually replaces the other cereals in the rotation.

Manuring. Although rye will succeed on poor acid land this is not to say that much better results will not be obtained when the

K

crop is suitably fed. Applications of lime, phosphate and potash give remunerative returns when applied in appropriate quantities. Care must be taken in the application of nitrogenous fertilisers lest excessive growth is induced and lodging follows : top dressings for grain production should be made in late spring.

Cultivation and Sowing. The preparation of the seed-bed follows the lines already indicated for cereal crops.

The usual rate of seeding varies from 125 kg/ha in early autumn to 190 kg/ha for winter and spring sowings. The seed of winter varieties should be drilled in September or early October and of spring varieties in February to early March. The seed is drilled, harrowed in and rolled as with other cereal crops. Little or no subsequent cultivation is given to the crop.

Harvesting. Rye is the first of the cereal crops to ripen and cutting takes place when the straw is dead ripe. Thus little fielding is necessary.

Yield. The national average yield of rye in the five-year period 1950–4 was 2·2 tonnes and during 1955–9, 2·5 tonnes/ha. It is thus a lower yielding cereal crop than the others mentioned above. These yields were obtained from soils which were much poorer than those from which the wheat, barley and oat yields were obtained. Straw yields from such crops will be of the order of 3–5 tonnes/ha.

PULSE CROPS

Certain leguminous plants which are cultivated mainly for their seed are called pulse crops. In this country the only pulse crops are beans and peas ; vetches are not usually included because as often as not they are grown for cutting in the green state and not for the seed. Sometimes these pulse crops are called " black " straw crops, to distinguish them from " white " straw crops, which are the cereals.

Pulse crops owe their importance largely to the nitrogen present in the seed. Beans, for instance, contain double the quantity of protein found in wheat and oats, and almost three times the quantity found in barley. This protein does not result from the extra liberal use of nitrogenous manures, but is due to the ability, shared by all members of the Leguminosæ, of the plant to make use of the free nitrogen of the air through the medium of bacteria which inhabit the nodular swellings on the roots (see p. 403). Not only the seeds but all parts of leguminous plants are relatively rich in nitrogen. When the roots of these plants decay the soil is enriched with nitrogen which becomes available to the next crop. This is the main reason why beans, or a clover ley, are recognised as a good preparation for a cereal like wheat.

BEANS

Beans are distinguished from the clovers, vetches and peas principally by the square, hollow stems which vary in length with the variety. The leaves are pinnate, and usually consist of two, four or six leaflets, but the petiole or stalk of the leaf instead of terminating in a tendril, as in vetches and peas, ends in a short, very much reduced leaflet. The flowers arise as short racemes in the axils of the leaves (Fig. 35).

Up to the time of ripening the pods are soft and green, with

FIG. 35.—FIELD BEAN (*Vicia vulgaris*).
fl., detail of flower. *p*, pods.

a soft, white, woolly lining. On ripening they dry up considerably, become black and much tougher, the soft lining disappears, and if left until completely dry the pods eventually split along both the upper and the lower lines of division, and the seeds are ejected.

The bean is characterised by a strongly developed tap-root which is capable of penetrating to considerable depths and eventually when ploughed in of adding large amounts of organic matter to the soil.

Varieties. Beans are classified into two main groups, winter beans and spring beans. The former are sown in the autumn and are hardy and able to withstand the cold of a moderately severe winter, though they may be killed during spells of exceptionally severe weather, especially if the land is wet from want of drainage. Winter beans develop more secondary stems and usually produce a fuller crop than do spring varieties. They are also less subject to attacks of black aphis which do not, as a rule, appear until June, by which time an early sown crop of winter beans may escape serious injury (see p. 438). They are, on the other hand, more susceptible to the fungus disease known as Chocolate spot which, in some years, causes severe loss of yield.

There are three main forms of spring beans ; *Tick beans* are small, round or flat-oval seeded types with a 1,000 corn weight of 284–568 g ; *Horse beans* are larger seeded, flat-oval in shape, with a 1,000 corn weight of 568–852 g ; the *Large Seeded beans* are flat-oval with a 1,000 corn weight over 852 g. Tick beans may be grown as a cash crop for the pigeon trade ; the others are used for farm stock. Horse beans, and even more so the Large Seeded beans, are earlier and shorter strawed than the Tick beans. Some of the larger seeded types are related to horticultural broad beans.

Beans are a crop that are partly cross-pollinated and, unless a stock is grown in isolation, admixture can occur. Up to recently the variety situation had got into a thoroughly unsatisfactory state. In the last few years the N.I.A.B. have examined some 510 stocks, 110 of them in full field trials. As a result of this work they have listed two varieties of winter beans for general use, and another variety for special use in the west of the country ; and five varieties of spring bean. It is evident that there is room for considerable research by plant breeders on the bean crop.

Position in Rotation. On heavy land beans are sometimes taken as a cleaning crop in the place of roots, and the success of the crop may then be taken as a test of skill in cultivating clay land. In some rotations they replace the clover break, and in those districts where the three-course system including a bare fallow was practised, beans followed wheat after fallow. Often beans may precede wheat.

Manuring. To grow beans well, a soil whether heavy or light must contain a plentiful supply of lime. Being a leguminous plant and thus able to obtain supplies of nitrogen from the air, the crop does not as a rule need nitrogenous fertiliser. Most growers believe in the value of farmyard manure and experimental work confirms that substantial responses are usually obtained to a 25 tonnes/ha dressing.

Use of fertilisers is not based upon such comprehensive experimental information as in the case of the cereal crops. In the past high rates of phosphatic and moderate rates of potash fertilisers have been used ; 100–120 kg of phosphoric acid plus 60–70 kg of potash in addition to F.Y.M. are examples of fertiliser practice formerly recommended. It is known that the crop is exceedingly sensitive to a low soil potash status, the plants becoming stunted, showing a typical black scorch on the margins of the leaves, and giving very low yields. Most of the experimental work has shown greater responses to potassic and less to phosphorus manuring on ordinary soils. The response to potash is considerably reduced where F.Y.M. is used. The bean crop is liable to give unsatisfactory germination if seed and fertiliser are placed in direct contact, as occurs with ordinary combine drilling. On the other hand greater responses to both phosphorus and potash are obtained where the fertiliser is placed in a band 50 mm to the side and 75 mm below the seed than with broadcast application. A dressing of 0 : 50 : 100 or 0 : 50 : 50 kg/ha is probably the most satisfactory fertiliser for the crop, particularly where a placement drill is used.

Cultivations and Sowing. Beans require a deep but not necessarily a very fine seed-bed, and unless the clods are so large that they prevent the seedlings piercing through to the light a rough surface is not harmful ; in fact, with winter beans it is often desirable, since it provides some shelter for the young, tender plants, during the winter.

After wheat on heavy land, or barley on the medium loams, the dung is applied as soon after harvest as possible and ploughed in. After the furrows have been harrowed to a rough seed-bed and the fertilisers applied the beans are drilled and covered by harrowing. A distance of 300 mm or more is allowed between the rows to facilitate steerage hoeing and weeding and sometimes the rows will be spaced as much as 600 mm apart, but with the widespread use of Simazine for weed control in beans, these wide spacings are no longer necessary and row widths of 200–400 mm are commonly used and beans are sown in finer seed beds.

The seed rate varies from 190 to 290 kg/ha : as with the cereal crops there is a definite physical optimum seed rate. The optimum seed rate varies with environmental conditions and there is evidence to suggest that it is at a higher seed rate on the more fertile soils. There is insufficient experimental evidence to draw reliable conclusions concerning the economic optimum seed rate. The sowing of winter beans should be completed in October if possible, spring beans being sown during March.

In the north of England it was formerly a practice to ridge the land in the same way as for turnips, the ridges being 600 mm apart. Farmyard manure and the fertilisers were put in the ridges and the seed was either sown broadcast before, or drilled after, splitting the ridges. Another method of sowing, particularly winter beans, was by fitting a small seed hopper or drill on the plough beam in such a way as to sow down each second or third furrow opening, the seed being covered by the turning furrow. The spring sown crop is usually drilled.

Post-drilling Cultivations. Where beans are sown in rows of appropriate width inter-row cultivations as cleaning operations can be carried out. Recent trends are to use suitable herbicides for the crop. Simazine is commonly used as a selective herbicide for the control of annual weeds, and Dinoseb may also be used.

Harvesting. Beans should be cut when the leaf has fallen, at which time the hilum or scar, or point of attachment of the bean to the pod, will be black. It was usual to cut and tie the crop with a binder. The sheaves were stooked in fours and remained in the field until perfectly dry. A little outside moisture is not detrimental, for the stack is open and the wind can blow through. The stack should be thatched immediately, because beans cannot shed the rain-water as can wheat or rye. It is unusual to thresh beans until they have been in the stack at least four months, and it is better to allow a stack to stand over the summer if possible because new beans are not suitable for feeding.

The bean crop can be harvested direct with a combine harvester provided it is dead ripe. Under these circumstances, however, some shedding of the seed is liable to occur before the crop is cut. There is a considerable need for a variety of beans having pods resistant to shattering when dead ripe so that the labour saving of combine harvesting may be obtained.

The straw has limited feeding value but may be given to stock to pick over. The upper part of the plant is very palatable but the fibrous portions are generally discarded by the stock.

Yield. The average yield of the bean crop over the last seventy years has been about 2 tonnes/ha and during the last ten years there has been a slightly upward trend. Since records have been kept the yields have fluctuated violently from one year to another with often 100 per cent differences between successive seasons. In good seasons yields of over 3·5 tonnes/ha quite frequently occur. Straw yields are 3–3·5 tonnes/ha.

PEAS

The largest acreage of peas is found in the east and south-eastern counties of England, where the relatively dry summer and autumn are favourable for growing and harvesting the crop. The stem of the plant is squarish and generally hollow, as in the bean, but it differs from that plant in its trailing habit. At the base of the leaf are two large, fleshy stipules, in some cases much larger than the leaflets themselves. The leaves are compound with one, two or more pairs of leaflets placed opposite to each other on the leaf-stalk, which is finally extended to form a series of tendrils, these being in reality leaves modified into thread-like structures ; the tendrils are able to attach themselves to any object they may touch, thus acting as supports for the weak succulent stem (Fig. 36). The straw length varies considerably between varieties, from 300 mm to over 1·5 m in length.

The flowers arise on stalks varying in length with the variety ; the stalks bear one, two or more flowers, the number again being a varietal characteristic. The flowers are papilionaceous and the fruits, again two-valved pods, contain a varying number of seeds. As in the case of many other plants of the family, the fruits when fully ripe open by a violent fracture of the two halves of the pod, which twist on themselves and in so doing eject the seeds.

Expansion of the food-processing industries has led to great changes in the cultivation of peas on a farm scale. Most green peas are now grown for canning, quick freezing and dehydration, and there has been a steady decline in the proportion marketed in pods. Dry peas are used for canning and packeting. Similar cultivations are practised for pea production for these purposes, but choice of variety depends upon the market requirement.

Varieties. Varieties of peas may be grouped as follows :

Green peas for canning, quick freezing and dehydration. The main quality requirements are colour and flavour after processing. Varieties producing dark green peas are required for quick freezing and dehydrating. Pale-seeded varieties are usually preferred for canning and colour can be added to the covering liquid. Varieties include : Sprite, Scout, Dark Skinned Perfection and Lincoln.

Dry harvested peas for canning and packeting. Suitable colour is required for the packeted product, but this factor is less important for canning. Uniform texture is very important for canning.

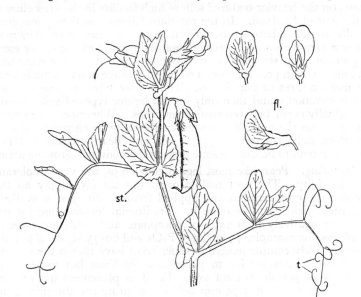

FIG. 36.—FIELD PEA (*Pisum arvense*).
fl., detail of flower. *st.*, stipule. *t*, tendril.

Two types of peas are grown in Britain for harvesting dry : marrow-fats with large dimpled seeds, and blues which have smaller round smooth seeds. Marrowfat varieties include Vedette and Maro. Blue varieties include Lincoln Blue and Dutch Blue.

Green peas for marketing. The main requirement for green peas for market is a range of varieties with attractive-looking pods for size and colour which produce high yields.

The acreage of peas grown for the fresh vegetable market has declined rapidly due to the scarcity of casual labour for picking, competition from processed peas, and consumer preference for " convenience " foods.

Varieties are classified usually into round- and wrinkled-seeded groups and again into early, second-early and main crop varieties. They include : *First early :* Sprite, Sparkle ; *Second early :* Vida ; *Maincrop :* Dark Skinned Perfection, Lincoln.

The choice of peas for picking green is very extensive and the names given are simply those of very popular sorts, all of which are garden peas as distinct from field peas. A good variety for this purpose must not only give a high yield of well-filled pods per acre, but in addition the pods must retain their freshness when bagged for market ; they must be easy to shell and resistant to mildew. Special characteristics are needed for varieties intended for canning or deep freeze. These are usually grown on contract, and the contracting firm supplies the seed of the varieties they require.

Position in Rotation. Threshing peas as a cash crop are usually grown on the heavier textured soils of high fertility in the drier climate of the east of England. In the rotations followed in these areas peas usually are taken between two cereal crops. In some cases they may follow a cereal crop and precede potatoes in place of the clover seeds in popular crop sequence—seeds, potatoes, wheat—of the warp and silt soils. Green peas are grown on a wider range of soils usually near the main centres of population. The fodder types of peas are not grown commonly and then only on the lighter types of soil. Again they usually occur between two cereal crops. Where peas are grown frequently on the same land severe trouble may be encountered with Pea Root Eelworm (see p. 443). Pea and Bean Weevils and Pea Moth also tend to be troublesome pests in the main pea-growing areas.

Manuring. Peas, like most leguminous crops, are very intolerant of soil acidity. The root nodule organisms usually supply all the nitrogen the crop requires and applied fertiliser nitrogen is as likely to depress yield as to increase it. Traditional fertiliser use for the crop was relatively high rates of phosphate and moderate rates of potash, as, for example, 75 kg/ha of P_2O_5 and 60–75 kg/ha of potash. Recently fairly comprehensive fertiliser trials have shown the crop to be relatively unresponsive to phosphatic fertilisers but much more responsive to potash on most soils. Fertiliser placement trials where the fertiliser was drilled 75 mm below and 50 mm to the side of the seed have given much better response than broadcast applications. In common with beans, the germination of the seed is liable to be severely damaged by direct contact placement of seed and fertiliser such as is obtained by ordinary combine drilling. The experimental work suggests that 50–100 kg/ha of potash, applied where possible with a special placement drill, would be ideal for the crop on most soils. Since straight potash fertilisers are not too easy to handle a 0 : 10 : 20 or similar compound is to be preferred. Where the soil is *acutely* deficient in phosphate, up to 74 kg of P_2O_5 may be required. On some soils maximum yields of peas may be obtained without any added fertiliser.

Cultivation and Sowing. The cultural operations are simple and consist of early ploughing of the stubbles in the autumn and a good cultivating as early in the New Year as possible, aiming at keeping the frost mould on the surface. Drilling is carried out from the end of January to early June, dependent upon suitable conditions for seed-bed preparation and the type of peas being grown. Experiments with threshing peas have shown that earlier drilling (in this case early March) gives better crop yields. With the earlier drillings, however, the use of organo-mercurial or, better still, TMTD seed dressings are particularly necessary in order to protect the seed from the fungal rots liable to occur when a cold wet spell follows drilling. With the green pea crop a much wider range of drilling dates is used. To catch the early market, early varieties such as Sprite, are sometimes sown as early as December. Late April and May drillings using later maturing varieties are also practised, as the very late green pea market may be remunerative.

The seed-bed required is similar to that required for spring-sown cereals. With very early drilling it is not advisable to work it down too fine otherwise capping may result. Drilling the seed about 50 mm deep and at a rate of 250 kg/ha is the normal practice. Experimental work has confirmed that 250 kg/ha is both the physical optimum and the economic optimum, except where the seed costs three times that of the same weight of crop. The row spacing may vary from the ordinary 200-mm drill spacing to 400 or 600 mm attained by stopping alternate, or two coulters out of three, respectively. A 400-mm spacing is probably ideal as it gives a well-spaced plant population, and at the same time allows steerage hoeing to be done.

Post-drilling Cultivations. Peas may be harrowed just as the shoot begins to peep above the ground. If used at this time the harrow destroys some small weeds without hurting many of the pea plants. The crop may be harrowed and rolled when about 75 mm high, and inter-row cultivations should be continued until the rows meet and hoes can no longer be worked. The limited time available for this operation makes a clean soil a very necessary condition of pea cultivation, and accounts for the fact that the pea crop is regarded as one of the worst weed-promoting crops we have. In recent years the selective herbicides, Dinoseb, Prometryne and Cyanazine are being used extensively for weed control in the crop.

Harvesting. When the crop is required for grain it should not be allowed to ripen before being cut or many of the pods burst, and the peas fall out during the turnings to which it is necessary to subject them whilst drying in the field. It is sufficient that the haulm should be yellow, and the pods tough and of the same colour. The crop is usually cut with a mowing machine which may be fitted with extensions (" pea-lifters ") to the fingers of the cutter bar. Special pea cutters with better performance than the ordinary mower are also in use in the main pea-growing areas. After cutting the crop is windrowed and turned, either by hand or by side delivery rake type of swath turner until it is completely dry. An alternative procedure is to let it dry partially in the windrow and then complete the process on tripods or tetrapods. In the latter case the crop is protected from the worst effects of wet harvest conditions in which severe losses occur due to shedding and lowered quality due to discoloration.

After drying out thoroughly peas in windrows may be harvested with a combine harvester fitted with pick-up attachments, or they may be carted, stacked and subsequently threshed. Peas on tripods or tetrapods may be swept to a stack by means of buckrakes, combines may be taken from tripod to tripod and the crop forked on to the cutter bar table, or, the more labour-absorbing method of forking from tripod to trailer, stacking and subsequently threshing may be employed.

Finally, peas should be stacked when perfectly fit, and not before, in long, narrow stacks in a position which exposes them to the maximum amount of wind. When carted in doubtful condition it is advisable to stack on wooden frames or tripods, or to adopt other means to permit free circulation of air through the body of the stack. When threshing, it is necessary to set the drum correctly because split peas are useless for seed purposes or for canning. Pea haulm is extremely valuable as winter fodder.

Peas are picked green for human consumption, usually by gang labour on contract at a fixed price per 20 kg bag. Good judgement is necessary to determine the correct stage of the crop when picking should begin. Flat pods give rise to loss of crop whilst over-mature pods are equally unsaleable. The general procedure is for the pickers to pull the halum as they go and strip off all the pods.

The green crop intended for canning or quick freeze is usually cut green and windrowed prior to harvesting and threshing in the field. This method has replaced cutting and transporting to static viners. The whole operation between cutting and processing needs to be carried out in the minimum time if quality is to be retained. Location of the processing factory thus, to a large extent, determines the areas where peas of this type are grown. This type of crop must also be harvested at exactly the right state of maturity, that is to say, when the seeds are a reasonable size but are tender and retain sufficient sugar to impart the sweet taste. The processing factories make graded payments according to quality, which is determined by tenderometer and, sometimes, sugar tests.

Yield. The average yield of grain is 2·5 tonnes/ha with 2·5-3 tonnes/ha of straw. Green peas yield from 4 to 8 tonnes/ha.

Chapter X

THE CULTIVATION OF ROOT, FORAGE AND MISCELLANEOUS CROPS

ROOT CROPS

ALTHOUGH potatoes are often referred to as root crops and have been thus classified in this chapter, they are botanically distinct. The true root crops are those whose principal edible portion is the true root plus a certain amount of what the botanist calls the hypocotyl, which is a transitional structure intermediate between root and shoot. All the true root crops are biennial in habit. As usually grown they spend the first season of growth in developing the root, whose store of food they would normally exhaust the following season in producing stem, flower and seed.

The chief food reserve of the common farm roots is sugar, though the cellulose and other non-woody portions also serve as food materials to ruminant animals. In addition, the "tops" (i.e. the foliage) of the plants is a valuable fodder except in mangels and carrots.

With the exception of the kales and similar crops grown as folded crops, the acreage of fodder root crops has declined steadily in this century. In the 1870's the combined area of turnips, swedes and mangels grown was more than 800,000 ha ; by 1939 this had declined to 240,000 ha and by 1958 to 150,000 ha. By contrast the combined area of kale, cabbage, savoy, kohl rabi and rape was 70,000 ha in the 1870's, 78,000 ha in 1939 and by 1958 had reached 184,000 ha. The reason for these changes is the increased cost of labour during the period which has made the true fodder root crops a relatively expensive source of starch equivalent (i.e. some 50 per cent more per kg of S.E.) than that in the form of hay or silage ; the folded forms of the leafy fodder crops on the other hand can produce starch equivalent at about the same cost as conserved grass products.

Parallel to the decrease in the fodder root crops has been an increase in the cash root crops. Sugar beet was only introduced as a farm crop in the 1920's and the area of potatoes grown has also increased to some extent. The combined area of these crops in the 1870's was 200,000 ha ; by 1939 it had increased to 320,000 ha and by 1958 it had reached 400,000 ha.

The root crops traditionally have been regarded as the principal cleaning crop on the farm. This reputation was based chiefly upon the turnip and swede crop which, being late sown, permitted pre-sowing cleaning operations as well as inter-row cultivations. Where early sown root crops are grown the pre-sowing weed control needs to be done in the autumn. Farm mechanisation makes autumn cleaning a more practical proposition than formerly. The modern trend of thought is that cash root crops are too valuable and too costly to produce for them to be sown on weedy land. This particularly applies to crops like sugar beet and carrots where starting with clean land benefits yield and has a considerable effect upon the costs of production.

THE POTATO

Doubt exists as to the exact date when the potato was introduced into this country, but there is ample evidence to place the period towards the close of the 16th century. It was cultivated in Chile and Peru for hundreds of years before it was grown in Europe and the date of its introduction to the Continent synchronises with the activities of the Spanish discoverers and conquerors of that portion of South America. We are probably indebted to the same agency for the initial distribution of the potato in North America and in Europe. Two 16th-century potatoes have been described, one in *Gerard's Herbal*, 1597, and the other by Clusius (*Rariorum Plantaram Historia*, 1601). Gerard's potato was a white-skinned, round type, while Clusius' was reddish purple and long.

The United Kingdom pre-war area of potatoes averaged about 250,000 ha ; during the war this increased to 580,000 ha, and by 1974 it had decreased once again to about 200,000 ha. With an average yield of approximately 30 tonnes/ha the crop becomes one of the most important cash crops grown in the country.

Varieties. The potato is usually propagated vegetatively by means of a tuber or swollen underground stem, and because of this the character of the original plant may be reproduced, without alteration, for many generations. Sometimes, however, changes capable of transmission do occur in one or more portions of the plant and may give rise to a new variety. This is one of the ways in which new varieties have arisen. Most new varieties, however, are produced from seedlings. Many varieties of potatoes are capable of forming true seed, and the potato flower, although self-fertile, is capable also of being fertilised by pollen of other varieties. In practice it is found that whether self- or cross-fertilisation takes place, the seedlings produced from the large number of seeds of even a single potato fruit or " apple " exhibit an extraordinary diversity.

The whole history of the potato in this and other countries is a long story of struggle with various diseases, with the result that plant breeders have frequently had in the forefront of their work the desirability of producing resistant varieties : yield and earliness of ripening are also factors of prime commercial importance. The series of attacks of potato blight (*Phytophthora infestans*) which swept over the British Isles from 1845 to 1847 acted as a great stimulus to breeding disease-resistant varieties, whilst in more recent times Wart disease (*Synchytrium endobioticum*) has resulted in the introduction of many new varieties capable of resisting the fungus responsible for the disease (see Chapter XIII).

In practice potato varieties are divided into three groups ; (i) First Earlies, (ii) Second Earlies, (iii) Maincrops. The first earlies are harvested as soon as, in the judgment of the grower, they have produced a yield which at the prevailing price will give a satisfactory profit margin. As prices are high in the earliest part of the season it is essential to obtain this moderate yield (say 12 tonnes/ha) as early as possible. Thus a good early variety is one which " bulks up " early : because the crop is lifted before it has completed its growth, yield at maturity is of secondary importance. The early " bulking up " of these varieties is due to their ability to start the formation of tubers under conditions of shorter day length than main crop varieties. As a result of this character some UK early varieties are used as main crops in lower latitude countries. Most early varieties are susceptible to tuber rots during winter and this can cause seed storage difficulties.

The second earlies are heavier croppers than the first earlies and should have better cooking quality. One first early variety, Arran Pilot, has, in addition to early bulking characteristics, a heavy mature yield of reasonable cooking quality ; it is therefore a useful dual purpose variety.

In general, all main crop varieties are characterised by their later maturity, heavier cropping capacity and better cooking and keeping qualities in comparison with varieties in the other two groups. The British market has a preference for the white-fleshed types ; by contrast the American market prefers the yellow-fleshed varieties. The mealy textured potatoes are considered to be the best cooking types ; the " waxy " textured varieties are preferred for chipping because they absorb less of the cooking oils. Blight resistance is a much more important characteristic in the main crop varieties than in the other two groups as a result of the time of the year when this disease occurs.

Briefly the history of potato variety development can be summarised as follows. The severe potato blight attacks of the 1840's resulted in the loss of many varieties. In the re-establishment of the industry four varieties played an important part. These were Paterson's Victoria (produced in 1826), Nicoll's Champion (produced 1865), Clark's Magnum Bonum (produced 1876) and Findlay's Up-to-Date (produced 1886) ; each showed some degree of blight resistance and other desirable agronomic characteristics.

During the latter part of the 19th century and the first two decades of this century many varieties were raised and marketed. In addition many stocks of individual varieties were given different names. The situation became so chaotic that by 1922 out of 1011 named varieties there were only 452 distinct types ; the variety Up-to-Date, for example, appeared under 152 different names. Intensive work on potato synonyms by the Scottish Board of Agriculture and by the N.I.A.B. clarified the situation. This period nevertheless saw some important introductions. The increasing importance of the early potato market led to the development of several first early varieties, the most important of which were Epicure, Sharpe's Express, Eclipse and Duke of York. Some of these are still grown on a limited scale. Of several second early varieties Great Scot was the most important and is still grown to some extent. Amongst the main crop varieties produced during the period mention must be made of Majestic (produced in 1911) and King Edward VII (produced in 1902), which still dominate the market; of Golden Wonder, which though a poor yielder is still regarded as the supreme variety for cooking quality; and Kerrs Pink which has both good yielding and good cooking propensities.

The early part of this century also saw the serious development of potato wart disease soil infestations. It was found that some of the existing varieties showed immunity to the disease. Eventually the disease was scheduled as a notifiable disease by the Ministry of Agriculture. Under the regulations only immune varieties were permitted in the scheduled areas. Since that time it has been necessary to breed for this immunity in the production of new varieties, and this attribute is required before a new variety can appear on the recommended list of the N.I.A.B.

During the late 1920's and early 1930's the Empire potato collection was established at Cambridge. This included a number of potato species other than *Solanum tuberosum*, to which all our existing varieties belong. Some of these species show features such as blight immunity, virus resistance, eelworm resistance, frost resistance, and tuberisation reactions to differing day length, which make them a valuable source of genetical material for breeding programmes. As their chromosome numbers differ from existing varieties, preliminary work on the production of suitable polypoids has to be undertaken before hybridisation can be carried out. Intensive work on these lines is at present in progress at the official plant breeding stations in England, Scotland and Ireland.

For a list of recommended varieties the reader is referred to the appropriate Farmers Leaflet issued yearly by the N.I.A.B.

Position In Rotation. Potatoes are of especial importance in certain counties, as for instance parts of Yorkshire, Lincolnshire, Cheshire, Cornwall, Ayrshire, Norfolk, and parts of Scotland and Ireland. They are grown where soil and climatic conditions are suitable for the production of large yields of high-quality tubers and where the cost of production by virtue of congenial conditions for cultivation and harvesting is not excessively high.

Potatoes flourish best on deep, loose, friable soils, especially when these contain a fair amount of sand and organic matter. On the other hand, poor sandy soils if heavily manured will grow good crops provided the rainfall is adequate; even heavy clays, if suitably treated, may be made to grow potatoes successfully. Some of the choicest potatoes are raised on the Old Red Sandstone, the marls and sands of the Trias and certain limestone soils; those grown on the Greensands, Silts and Warp soils are generally of good quality whilst the rich black fens yield immense crops which suffer somewhat in cooking quality. Big crops are grown on recently broken up grassland or young leys, for the tubers revel in the mass of decaying vegetable matter found under such conditions. Under these circumstances wireworm may ruin the saleability of the crop, and a suitable insecticide incorporated in the fertiliser, or applied in other ways, is a satisfactory control.

If potatoes are grown too frequently in rotations a high population of Potato Cyst Eeelworm (*Heterodera rostochiensis*) may be built up (see p. 444). This will reduce yields to very low levels and a long period (up to 10 years) may be needed before the population is reduced to safe dimensions. As most commercial soils are infected it is necessary to limit the frequency of cropping ; experience on various soils suggests that they should not be grown more frequently than once in four or once in five years. A high eelworm population is such a serious matter that it is advisable to err on the side of safety. Even the frequencies mentioned above are only just safe and a single instance of cross-cropping can lead to trouble. With first early potatoes the crop has completed its growth before most of the eelworms have reached maturity : in consequence the population build up is much slower and does not reach such high levels.

The potato crop occupies part of the root break in many rotations ; in others it is the only root crop grown. The old silt and warp land sequence of seeds—potatoes—wheat has in recent years been incorporated into many rotations in eastern England. This sequence facilitates the autumn application of F.Y.M. to the crop, allows the aftermath growth of clover to be ploughed in and permits early ploughing.

Manuring. In view of its liking for organic matter the crop frequently follows peas, beans, or grassland. Often in potato-growing districts the seeds' aftermath is ploughed in during the autumn in preparation for the following potato crop. Farmyard manure forms the basis of manuring on some soils. The traditional practice was to apply the dung in the row in the spring and plant the tubers on top of it. In recent years, however, this practice has been replaced on a widespread scale in favour of applying the dung in the autumn and ploughing it in. The advantage of this procedure is a considerable saving of labour and an easing of the pressure of spring work. The rate of FYM dressing is usually 25 tonnes/ha or thereabouts.

Whether or not dung is used the potato crop will respond to heavy fertiliser dressings. In 1941 Crowther and Yates summarised all the available fertiliser experiments carried out since 1900. Using prices received for the crop and costs of the fertilisers in recent years, they suggested that the average optimal dressing would be N, 125 kg/ha, P_2O_5, 125 kg/ha, K_2O, 180 kg/ha. On soils where rotations containing a high proportion of cash root crops (i.e. 2 cash root crops in a 5-course or a 7-course rotation) have been the rule, recent experimental work has disclosed a build-up of P_2O_5 to a point where the average optimal dressing is 60 kg instead of 125 kg of this component of the fertilisers. It would be safer to make this reduction only where the results of soil analysis confirm the higher expected average soil status of P_2O_5. The fen peat soils contain more nitrogen than ordinary soils ; as a result the optimal dressing of nitrogen is below that for the average mineral soil ; a reduction from 125 kg to 60 kg is appropriate in these circumstances. This reduction does not necessarily apply to peats which were originally acid bog peats. Where potatoes are grown on chalk or limestone soils it may be necessary to increase the potash dressing to 220–250 kg.

The fertiliser rates given above apply where no FYM is used ; where 30 tonnes/ha of FYM are applied the rates may be reduced by about one-third, although there is some controversy as to whether this should apply to the nitrogen component of the fertilisers. On the soils of very high productive potential farmers often use fertiliser rates greater than those quoted above.

To obtain the greatest efficiency from the fertilisers applied they need to be placed correctly in relation to the seed. Where the land is ridged before planting the fertiliser should be broadcast over the ridge immediately prior to planting. The practice of using deflector plates on the broadcasting drills to concentrate the fertilisers in the bottom of the furrows can check sprouting of the tuber in a dry season. A baulk of timber dragged behind the drill at right angles to the ridges, will pull down sufficient soil to cover the fertiliser and so minimise any damage to sprouting. Where crops are planted by machine without previous ridging, the fertiliser should be placed 50 mm to the side and 25 mm below the seed by means of a fertiliser attachment. Application of fertiliser in solution is beginning to be used in this country ; simple injection equipment using liquid fertiliser facilitates correct placement.

Excessive application of fertilisers can reduce the dry matter percentage in the tuber and so reduce cooking quality. Potash applied in the form of sulphate is rather better than muriate for tuber quality.

Cultivation and Planting. At the present time the potato crop is almost invariably grown on the ridge system. In former times potatoes have been grown on the flat or in beds separated by shallow trenches. The Lazy-bed system was a method of growing potatoes on old grassland, on ley or on reclaimed peatland, formerly used in Ireland and N.W. Scotland. Ridge planting offers three substantial advantages ; it facilitates mechanical weed control, it minimises tuber exposure with the consequential greening and liability to blight infection, and it eases harvesting operations. In the preparation of the seedbed a primary objective is the attainment of a deep loose tilth from which the ridges may be formed. With the increasing popularity of elevator diggers and complete harvesters it is necessary that the tilth should be as clod-free as possible. Creation of new clods by subsequent cultivations should be avoided.

Two old-established techniques of seed-bed preparation exist and are used on different soils. On the light loamy soils ploughing and cross ploughing produces a deep mellowed condition which can be converted into a seed-bed with the minimum use of tined implements. The initial ploughing is done at a moderate depth before Christmas ; the cross ploughing takes place at a shallower depth when conditions are dry enough in January or February. Weathering of the two soil exposures produces the deep, mellowed condition.

The second technique applies to the heavier classes of potato soils. The land is deep ploughed (250–400 mm) as early in autumn as possible using a full digger plough body. If this is done whilst the soil is relatively dry a thorough shattering of the furrow slice is obtained. The shattered furrow allows a deep penetration of the mellowing action of alternate wetting and drying, and freezing and thawing. No further operations are carried out until spring, when tined implements break up the deeply mellowed furrow and produce a tilth. It is advisable to use relatively straight-tined cultivators for this purpose ; curved-tine cultivators bring to the surface unmellowed clods from the deeper layers of the soil.

A third technique of growing popularity is the use in the spring of a rotovator on land deep ploughed in the autumn. On heavy land and in seasons with mild winters, this tool can be of inestimable value in preparing a tilth from the poorly mellowed land.

Each of these techniques leads to tilth production with economy of cultivation in the spring ; economy of cultivations also minimises the loss of moisture from the surface layers of the soil profile.

Once the seed-bed is prepared ridging, fertilising, planting and splitting the ridge should be completed with the minimum of delay. A drying out of the ridge surface before planting is considered to be very detrimental for early growth by most experienced growers. Any perennial weeds in the field should be dealt with in the autumn ; for couch grass deep ploughing itself is an effective weed control operation.

Choice of Seed. The selection of seed is a matter of great importance. Such factors as size of seed, the health and source of the stock and the treatment of the setts prior to planting are of paramount importance. Without good seed it is quite impossible to grow a good crop of potatoes.

For many years now it has been customary for potato growers in the more southern parts of the country to obtain their " seed " from northern districts—and it was assumed the farther north the better. Sometimes after one, more often after two, and almost always after three years' growth in the south the produce of the imported stock exhibits progressive degeneration until finally, in extreme cases, the plants are not worth growing at all. As a result of a great deal of research it has now been established that the cause of the trouble is virus (see pp. 398, 399). Virus diseases occur in a large number of cultivated plants, such as the tomato and raspberry, and their exact nature is still undefined. They are known by their effects rather than by their physical properties. A very important feature, however, is that they can be communicated from plant to plant by insects. In the case of the potato, if the virus passes from the leaf down the stem into the tuber, the plant eventually arising from such a tuber will contain the virus and its tubers in turn will also contain it—most probably in an intensified form. The effect of the virus diseases is a general decrease in the vitality of the plant, exhibited in the form of lessened vegetative development and therefore of tuber production. The degeneration may be extraordinarily rapid, and once started cannot be arrested by improved cultivation or manurial treatment.

Amongst the insects capable of transmitting viruses from plant to plant is the green fly or aphid, which is very common in the more southerly parts of England but less widely distributed in some northern districts. Thus it is that seed from the north is usually healthier and produces better crops, at least until it becomes badly infected with virus. But the mere fact that the seed originates in the north is no guarantee of its freedom from disease, for this depends first on the initial health of the stock, and then on the extent to which it has been subject to infection.   These conditions rather than geographical boundaries demark good seed from bad seed, and provided the initial stock from which a bulk of seed is produced is virus-free, excellent stock can be raised in Northern Ireland and on land lying above the 150 m contour in England and Wales.

Seed Potato Certification.   The health of potato seed is most impor-tant.   Certified seed is of high varietal purity and is substantially free from virus disease.

In some districts new seed should be used each year.   Elsewhere stocks may remain healthy for longer periods, the length of time depend-ing on the frequency and activity of aphid vectors which spread serious virus diseases.

Certificates are issued by the Agricultural Departments in the United Kingdom on the results of inspection of the growing seed crop. The certificates are now standardised and are in two categories each of which is in one or more grades as follows :

Category	Grade
Basic seed intended primarily for further multiplication as seed	Virus-tested stocks propagated from stem cuttings (VTSC)
	Foundation seed (FS)
	Stock seed (SS)
	First-quality commercial (AA)
Certified seed intended for production of ware crops	Healthy commercial seed (CC)

The letters N.I. after the grade letters indicate varieties non-immune to wart disease.

Seed Certificates refer only to the health of the growing seed crops and do not relate to the size grading or condition of the seed tubers derived from such a crop.   These aspects are covered by the Seed Potatoes Regulations, 1965, and by the Sale of Diseased Plants Orders.

In many parts of the country it pays the grower to use fresh seed each year but, where the incidence of virus is not so great, new seed every second year may suffice.   In all other cases it is generally advisable to purchase each year one-third of the seed potato requirements as new certified stock.

Sprouted Seed.   Sprouting or chitting the seed before planting has been shown to have considerable advantages.   With early varieties the growing period may be shortened by as much as two weeks ; with main crops an increased yield of up to two tons per acre may be obtained.   The sprouting is carried out under well-lighted con-ditions in special glasshouses or in buildings fitted with hanging fluorescent lights.   Sufficient light is required to prevent spindly sprouts being formed.   A temperature between 5 and 10 °C is to be aimed at ; below this temperature damage by chilling may occur ; temperatures above 10 °C lead to excessive growth in length of the chits.   Chits of excessive length are liable to be broken off at planting time.   The seed tubers are placed in trays, 12–16 kg per tray, and these are stacked one above the other, the handles providing a spacing arrangement to allow light access. The benefits from chitting are of two forms : diseased tubers can be more readily identified and discarded than when seed is planted directly from the clamp ; but more important is the speedier emergence and early growth of the crop.

Time of Planting. The usual time of planting is from March to the beginning of May, according to the season, a start being made with the early varieties and finishing with main crops. The rule is to plant as early as soil conditions permit and when, according to locality, there is a reasonable chance of avoiding severe spring frosts. Earlies in very sheltered and favourable localities may be planted as soon as January. Whilst the use of chitted seed will to some extent counteract the loss of yield resulting from late planting, the full advantage of chitting will only be obtained when planting is carried out at the normal time.

Rate of Planting. Customary planting rates are from 2 to 3 tonnes/ha. Recent work has shown that although the total yield of tubers increases with rates of planting up to 5 tonnes or more per hectare, the yield of ware has an optimum seed rate well below this figure. The economic optimum seed rate for ware yield is lower still and depends upon the relative value of crop and seed. Where the seed value is 1·5 that of the crop, weight for weight, the average economic seed optimum is at 3 tonnes/ha ; where the seed is twice the value of the crop the economic optimum is 2 tonnes/ha ; where the grower uses his own seed, and thus seed and crop values are the same, the economic optimum is nearer to 3·5 tonnes/ha.

The actual seed rate may be varied either by altering the seed size or by altering the spacing. For yields of ware grades above 45 mm, recent work has shown that, so long as the economic optimum seed rate is used, there is little to choose between planting small seed at close spacing or large seed at wide spacing.

The usual seed size is a 56–84 g tuber obtained by top and bottom-riddle sizes of 45 mm and 50 mm respectively. The specialist seed grower often uses riddle sizes of 58 mm and 32 mm to increase the proportion of seed extracted from his crop. This results in a wider range of seed tuber weights than those mentioned. With the conventional spacing of 350 mm on 750 mm rows an average seed size of 70 g requires 16,000 tubers, or 2·5 tonnes/ha.

In the potato crop the true plant population is the number of sprouts which develop to form independent plants ; it is therefore the product of the number of tubers planted and the average number of shoots developed from each tuber (i.e. the number of hills × the shoots per hill). Some varieties (for example King Edward) produce more sprouts than others (for example Majestic). The effect of this is to increase the number of tubers in the crop, to decrease their average size and therefore to decrease the ware percentage. These high shoot-producing varieties need to be planted at less than the average seed rate to obtain high yields of ware. Very early chitting, starting as soon as the crop is lifted, improves matters. The seed should be placed in a warm chitting house until the sprouting commences and then maintained at a low temperature (i.e. below 6 °C) until planting time. In this way seed with only one to two chits per tuber may be obtained. Single-chitted tubers are also of considerable advantage in producing first early crops.

The seed producer's main objective is a high yield of seed-sized tubers. This can be obtained either by planting normal seed at close spacing or ware-sized seed at normal spacing. As ware-size tubers are the by-product in the seed-producing industry, the second procedure is more appropriate.

The planting can either be done by hand, in which case casual labour is frequently used, or by machine. Under suitable conditions machine planting can be operated with a considerable saving of man hours. Automatic potato planters are now used which speed-up planting considerably and also reduce labour requirements. They require the use of closely graded seed.

Post-planting Cultivations. These are designed to control seedling weeds. About 10 days after planting the ridges are harrowed down, so destroying a crop of seedling weeds. This operation may be followed by a further earthing up and harrowing down at suitable intervals to destroy two further crops of seedlings. After the potatoes have emerged further weed control is done by inter-row cultivations. The final operation is to earth up the crop once again before the tops meet in the row. Contact and residual herbicides are now widely used to destroy annual weeds in potato crops, and they may sometimes replace cultural methods for weed control. Perennial weeds should be dealt with in the autumn.

With seed crops " roguing " is necessary. This involves the digging and removal of any plants affected by virus disease or any plants that are not true to type. This is carried out in May and again before the end of June.

In districts subject to potato blight, spraying is a sound insurance. The first spraying should be done when the haulms are well grown, in dry weather and before the disease appears, for in dealing with blight the measures are preventive and not curative. A second spraying is usually given about one week later but the correct dates for spraying vary from district to district and season for season. Full details of the procedure are given on pages 367–71.

A late attack of blight may call for spraying with 10–12 per cent sulphuric acid to kill the haulms which carry the disease spores. In seasons when the haulms remain green they are troublesome to the lifting implements and it is now becoming quite common to spray such crops with sulphuric acid to kill them off. Tar oil fractions (T.O.F.54), and diquat at 1·12 kg a.i./ha can also be used.

Irrigation. In the dryer parts of the country and particularly on the sandy soils in these areas, irrigation is considered to be a profitable operation provided the water can be applied at a cost of about £8–£10 per ha for 25 mm (i.e. per 250,000 litres/ha). The operation is carried out before the excess of evaporation over precipitation results in a soil-water deficit harmful to the crop. Precision in the operation involves the preparation of water-balance sheets. This may be done by the method described in the Ministry of Agriculture's Technical Bulletin No. 138 " Irrigation ". Certain data for the calculations are given in the Bulletin, but current season's rainfall and sunshine data are required from the nearest meteorological station. Simple instruments which can be used on the farm to give direct readings of irrigation need are being devised and will soon be available commercially (see also p. 211).

Harvesting. Lifting of early varieties commences towards the end of May. Much of this task is hand work and the tubers are packed in bags and despatched to the market immediately. The maincrops, however, are mostly lifted by machinery, which may be of the spinner type, the elevator digger or the complete harvester which lifts the crop and elevates it into carts or bags. Some maincrops are stored in long clamps or pits to be sorted and marketed at favourable opportunities during the winter. The clamps are first given a good covering of straw and finally about 200 mm of soil to render them both frost- and waterproof. Many potatoes are now stored between walls made of baled straw. The practice of storing under cover in sheds or barns is gaining ground since a good deal of labour is thereby saved and the sorting can be carried out under more congenial working conditions. The sorting is into " ware " or sale potatoes ; " seed " which may or may not be used a further time ; and " chats " which are so small as to be fit only for stock food.

After the potatoes have been lifted the ground may be harrowed to expose any tubers still left in the ground, and these can then be collected and stored separately or left exposed to weather and rot.

The production of a potato crop by present techniques involves the use of approximately 350 man hours and 75 tractor hours per ha. Of this 170–200 man hours are required for harvesting and a further 60 for storing. Under suitable conditions fully mechanised lifting can, at a conservative estimate, save 50 man hours per ha with an increase of tractor hours from 32 to 46. Elevator diggers are rather more efficient than spinners ; where two-row models, which deposit two rows into one, are used improved efficiency is obtained ; where two-row lifters are used on the " batch lifting " system instead of the " round and round " system still further improvement occurs. The spinner type of harvester is capable of working under wetter conditions than elevator diggers or complete harvesters. Similarly, a high population of clods or stones will interfere with the work of a spinner to a less extent than the other types. With all harvesting methods care is needed to avoid bruising or cracking the tubers.

Yield. The average yield of ware is 30 tonnes/ha, but on good potato land in favourable years crops of 40–50 tonnes/ha are not uncommon. The average yields of potatoes have increased to some extent in recent years. The increase has been nothing like so great as that with the cereals ; wheat yield for instance has increased by approximately 100 per cent since the period 1934–43 ; whilst potato yields over the same period have only increased by 50 per cent. This smaller increase can probably be attributed to the slower progress in the production of better varieties compared with the cereals.

MANGELS AND FODDER BEET

Mangels belong to the genus *Beta* of the Family Chenopodiaceæ, which also includes sugar beet, fodder beet and red beet. It was some time after turnips became a recognised farm crop in this country that mangels were introduced and their cultivation is more or less restricted to the east and south-east, where the higher general mean temperature is more suited to the successful growth of the crop. In the north of England and Scotland swedes and turnips are much more widely grown than mangels.

Mangels produce a much higher yield of dry matter per hectare than swedes and turnips. The dry matter content of the mangel varies with different classes and is of the order of 7–15 per cent. Whilst this is not appreciably higher than that frequently obtained in swedes the weight of mangels produced per hectare is much greater and, under good soil and climatic conditions and with generous treatment, yields of 125 tonnes/ha are not uncommon. Fodder beet, although giving lower yields of roots than mangels, have a higher percentage of dry matter and will produce as much, or even more, dry matter per hectare than mangels.

Varieties. The number of varieties of mangels in general use is very considerable, and wide variations in cropping capacity, dry matter content, root shape and skin colour occur. The varieties of fodder beet are more restricted and are mainly of continental origin. As these crops are normally cross pollinated they have to be grown in isolation for seed production and, unless strict selection within the stock is maintained, a named variety may change in character over the years. Mangels, fodder beet and sugar beet are best regarded as a continuous series. Mangels have a larger size of individual roots but a lower dry matter percentage than sugar beet, whilst fodder beet is intermediate for both attributes. Similarly, mangels have a smaller proportion of the root in the soil than fodder beet, whilst sugar beet has only a relatively small crown of the root above soil level. This character determines the ease or otherwise of lifting.

Varieties of mangels and fodder beet can be grouped under four main headings :

1. Low dry-matter mangels (D.M. range 9–12 per cent)
2. Medium dry-matter mangels (D.M. range 13–15 per cent)
3. Medium dry-matter fodder beet (D.M. range 14–18 per cent)
4. High dry-matter fodder beet (D.M. range 19–20 per cent)

The low dry-matter mangels are further divided into globe and intermediate types, the latter containing the tankard-shaped varieties. The chief difference between these two groups is that the former have rather more of the root above the ground surface. The skin colour of all groups can be white, yellow, orange or red although the two varieties listed in group 4 are both white skinned. There is little to choose between the dry matter yield per hectare in the two mangel groups ; the two fodder beet groups produce on the average approximately 10 per cent more dry matter per hectare than the mangels. Further information about varieties is given in NIAB leaflets.

The amount of dry matter per hectare and the dry matter percentage are greatly dependent upon conditions of growth. Dry matter percentage is reduced in a wet season although the total dry matter yield per hectare may be increased. Similarly a high plant population will give rise to roots of a relatively high dry-matter content within limits set by the variety. Extensive trials in several countries have shown that the feeding value of these crops can for all practical purposes be estimated from the quantity of dry-matter present.

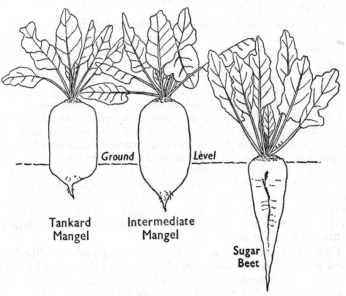

Fig. 37.—Mangels and Sugar Beet.

Position in Rotation. Mangels occupy a part of the root shift of the rotation and thus usually follow, and are succeeded by, a cereal. It is not uncommon, however, on many farms to grow the crop on the same piece of land for a number of years in succession. Usually an area near to the homestead is selected to ease the burden of carting the crop home and the farmyard manure out to the land. Generous manurial treatment year after year builds up the fertility and large yields result ; the repeated cleaning of the land eradicates weeds and enables the cost of hoeing to be reduced year by year. It is because mangels are relatively free from pests and disease that this practice is possible.

Manuring. In common with sugar beet both mangels and fodder beet are very sensitive to soil acidity. Suitable applications of lime should be given to correct the condition where it occurs. This should be done early enough for the lime to act before the crop is sown. In practice it is preferable for it to be applied either to the preceding crop or, at the latest, in the autumn before mangels are sown. The three crops also respond to N.P.K. fertilisers and, in addition, to sodium. The latter is usually given in the form of agricultural salt at the rate of 0·5 tonne/ha. Many fertiliser experiments have been carried out on mangels and sugar beet but a smaller number on fodder beet. Data exist for the economic optima of N, P_2O_5 and K_2O, but a difficulty arises in attaching a value to mangels and fodder beet for the purpose of the calculation. Where an arbitrary value of £4 to £6 per tonne is given to mangels and an equivalent value, in proportion to their dry matter content, to fodder beet, the average economic optima may be taken as being N—125 ; P_2O_5—125 ; K_2O—180 kg/ha.

Where agricultural salt is applied the potash may be reduced to 125 kg ; where FYM is used the optimal may be reduced by 40, 50 and 90 kg of N, P_2O_5 and K_2O respectively per 30-tonne dressing of F Y M. Other adjustments should be made where the soil by analysis shows deficiencies or excesses of individual nutrients. Where the nitrogen is supplied in the form of nitrate of soda (up to 0·5 tonne/ha) there is no need to apply agricultural salt.

Fertilisers for these crops are usually broadcast on the seed-bed before drilling. Agricultural salt needs to be applied about four weeks before drilling or, where this is impracticable, it may be applied the previous autumn. Combine drilling of fertiliser and seed is dangerous ; side band placement is safe and gives a response in the early stages of growth but has no advantage over broadcasting in the final yield. Some farmers top-dress mangels with nitrogenous fertilisers soon after singling. On very light soils in wet seasons this practice probably has a distinct advantage over that of placing all the nitrogen in the seed-bed.

Cultivation and Sowing. A fine, firm and moist seed-bed is required to obtain a satisfactory establishment of the mangel or fodder beet crop. With medium and heavy land this means autumn ploughing so that mellowing of the furrow may be obtained. This in turn leads to an economy of seed-bed cultivations in the spring. It is customary these days to sow the crop on the flat and so a deep tilth such as that described for potatoes is not necessary ; where the crops are still sown on the ridge similar techniques for obtaining the depth of tilth as described for potatoes apply. It is necessary to obtain control of perennial weeds the previous autumn either by stubble cultivations, deep ploughing or by the newer herbicidal methods. In preparing the seed-bed an even and level tilth is required to facilitate precision inter-row cultivations. " Round and round " ploughing or one-way ploughing, helps in the achievement of this objective by the elimination of ridges and furrows. The use of the " scrubber " or " float " in the final stages of seed-bed preparation is also an advantage.

The crop is usually sown at the rate of 10 kg of seed per ha from early April to the end of May, the earlier the better. The depth of sowing is 10–20 mm below the surface. The final plant population is obtained by singling. For high yields of dry matter per ha and for high percentage dry matter a plant population of 60,000–75,000 plants should be the aim. This can be achieved by drilling on a

500-mm row and singling to 240 mm. Where the crop was sown on the ridge row spacings 600–660 mm were common; the low plant population so obtained, whilst giving large individual roots, gave a lower percentage dry matter and lower dry matter yields. Economy in achieving the plant population by mechanisation is described in more detail in the case of the sugar beet crop. Singling should be done at the 4-true leaf stage; in practice it usually extends from the 2-leaf to the 6-leaf stage.

The steerage hoe should be used in the crop as soon as the seedlings are visible and should be continued as necessary until the plants are large enough to smother further seedling weed growth. Residual herbicides like pyrazon and lenacil applied as pre-emergence sprays have been successfully used for the control of annual weeds in mangels.

Harvesting. The mangel crop is harvested some time in October or early November before severe frost sets in. The roots should be injured as little as possible in lifting to prevent bleeding; the tops are either cut or twisted off and the roots are left untrimmed. The practice of clamping mangels with the tops left on has increased in recent years and does not appear to affect adversely the keeping quality of the roots and, being more economical of labour, has much to commend it. After remaining a day or two on the surface to dry, the roots are carted to the pit or clamp, where they are carefully stacked together, and then covered first with a generous layer of straw well laid in position and then with a layer of soil to protect them from frost during the winter. A ventilating shaft, consisting of a bunch of straw or a drain-pipe, should be inserted at a distance of every 6 or 8 m. Mangels are not fed to stock until the New Year to permit full maturation to take place. Failure to observe this is likely to cause severe scouring of the stock. They should not be fed to male sheep since this brings about urinary trouble. Several of the mechanical root harvesters now on the market can be adapted to deal with the mangel and fodder beet crop.

Yield. The average yield of mangels is around 50 tonnes/ha. Under favourable conditions of soil and climate, and especially following generous manurial treatment, yields of 75–125 tonnes/ha can be obtained.

SUGAR BEET

The wild beet is found in its normal habitat along some of the more southerly European coasts. There it develops a fine slender root and may be an annual, biennial or perennial. Contrary to general belief, individual roots of this species may contain more than 14 per cent of sugar.

As an agricultural crop sugar beet has been known on the Continent, particularly in France and Germany, for over a hundred and fifty years, but its cultivation in Britain on an extensive scale did not begin until 1920. In that year some 1,200 ha were under the crop; today the area, about 200,000 ha is an indication of the tremendous rise to popularity of the crop, which has now a secure place in British farming systems.

The cash value of the sugar-beet crop depends very much upon the total quantity of sugar produced per hectare, i.e. weight of crop multiplied by the percentage of sugar in the root. The percentage of sugar in the root is subject to considerable fluctuation, due to season, variations in the character of the soil, to a minor degree to the feeding of the plant, and finally to the variety itself. In general it is reasonably safe to assume that sugar content is largely outside the farmer's control ; the farmer should therefore concentrate on the attainment of a high yield per hectare in order to make the crop remunerative. The grower has, however, valuable by-products in the shape of spent pulp from the factory and the crown and leaves of plants left in the field, both of which are excellent for cattle or sheep and replace roots in the dietary of these animals. Since its introduction many farmers have ceased to grow feeding roots and have replaced them by sugar beet ; this not only provides a direct cash return but also enables about the same head of stock to be maintained, the by-products of the beet replacing swedes and turnips.

The plant is a white-skinned, white-fleshed biennial producing a more or less conical-shaped root weighing on the average about 1 kg and possessing a root capable of penetrating the soil to a depth of 300–400 mm (Fig. 37). The plant may send up its flowering stem in the first year, when it is said to " bolt " : this tendency is hereditary and some strains are more liable to it than others. In all strains, however, too early sowing or a severe check during early growth are conditions which predispose to " bolting ".

Provided there is an adequate supply of lime present and that they are deep and relatively free from stones or a plough pan, there are few soils which will not grow sugar beet. The suitability of any particular soil, however, is determined largely by various economic aspects ; light soils on the whole are to be preferred because the roots are easier to lift and are relatively free from soil when lifted. Under such conditions yield is dependent upon an adequate rainfall. Heavy soils, on the other hand, have the drawback of needing additional cultivation to bring them to a suitable degree of tilth for the crop, and unless early autumn ploughing is carried out and a good frost mould obtained, cloddy seed-beds are likely to result. Without a fine friable tilth it is very difficult to obtain an even plant, the first essential for good yields. On heavy land considerable difficulty is experienced in lifting the roots and carting them off, whilst the amount of soil adhering to the roots, and for which a deduction (dirt tare) is made at the factory, is likely to be unduly high. Fen soils produce good yields of beet but the sugar contents are low : in general, therefore, the ideal soils for growing this crop are the medium loams.

Varieties. The efforts of plant breeders have been directed towards securing the greatest yield of sugar per hectare, and to a very large extent this has been accomplished by gradually raising the percentage of sugar in the roots. It is by no means uncommon for this to register 20 per cent or over, although the average is in the region of 15–16 per cent. Gross yield, too, has been greatly improved, whilst by selection better conformation of the root, which facilitates easier lifting, has been obtained. Branching and fanged roots have largely been eliminated, and in the best varieties the root is now cone-shaped. Further but smaller modifications are the small-leaved tops of varieties produced for growing on rich soils, and non-bolting strains for the earliest sowings.

The higher price paid for beet of high sugar content should be considered when choosing a variety. Moreover, the higher the sugar content of the roots the less the bulk that has to be transported for the same cash value.

The amount of white sugar that can be produced from beet is directly affected by the impurities in the juice. Differences between varieties in impurities present may seem small but have a significant effect on the factory process. Varieties with high sugar content and low juice impurity allow increased sugar extraction and greater processing efficiency.

For some years it has been suspected that breeding programmes based upon intensive selection for the attributes of high sugar percentage, high potential root yield, satisfactory root shape and top characteristics, together with low bolting propensities, have just about reached the limit of their potential. In consequence further substantial improvement must involve a new breeding technique. In recent years the breeding of polyploid varieties has been undertaken. This involves increasing the normal two sets of chromosomes in each cell to four sets or more, a process that can be achieved by treatment of the fertilised ovule with colchicine, or other chemical compounds, at a suitable stage. Polyploid susually show greater vigour and reach a greater size than diploids in most plants. So far two such polyploid varieties have been tested and marketed. One is about average for yield of sugar per hectare and the other is 2–4 per cent above average. It seems likely these modest improvements will be exceeded in the future. Further developments have been directed to the production of mono-germ (i.e. single seeded fruit) varieties. Such types of beet are now grown extensively and produce similar yields to the multi-germ varieties. The development of mono-germ varieties has made an outstanding contribution to the reduction in the spring work in sugar beet production.

For many years the National Institute of Agricultural Botany has tested the behaviour and comparative yield of different strains in accurate field trials situated in the principal beet-growing areas. The current list of recommended varieties, together with a great deal of information about them may be obtained from Farmers' Leaflet No. 5 of the N.I.A.B.

Position in the Rotation. On most farms sugar beet occupies part of the root break, and as already indicated may in some cases replace roots completely. Sugar beet is always grown on contract for the factory which is ultimately to handle it. As the grower is responsible for delivering the roots to the factory, which he does by rail, road, or water, wherever possible fields are selected in close proximity to a hard road. The carting of such a bulky crop long distances over soft tracks in wet weather is very expensive at best, and sometimes impossible. In order to prevent the danger of sugar beet eelworm (which causes the condition known as " beet sickness ", menacing the growth of the crop in this country as has happened on the Continent) the sugar-beet factories do not allow growers to follow beet with another beet crop or a brassica crop at a closer interval than two years (see p. 439).

Manuring. Sugar beet is very sensitive to soil acidity, and where it occurs it should be corrected by lime applications given early enough for it to act before the crop is sown, as described for mangels. Where farmyard manure is available 25–30 tonnes/ha may be ploughed in in the autumn. Spring applications tend to promote fanging of the roots ; many growers apply dung to the previous crop rather than direct for the beet itself. Whether or not manure is given, the crop requires N.P.K. fertilisers and also benefits from an application of sodium. The latter is applied in the form of agricultural salt either one month before the crop is sown or, alternatively, in the autumn before sowing. Although 375 kg are regarded as adequate, experiments have shown that 625 kg of agricultural salt in the absence of potash leads to an average increase of 625 kg/ha of sugar ; in the presence of potash the response is approximately 300 kg sugar.

The experimental data for the calculation of optimal rates of nitrogen, phosphoric acid and potash are more extensive than those for any other British crop. With fertiliser and crop prices of recent years the average optima are : N, 120 kg ; P_2O_5, 60 kg ; K_2O, 180 kg/ha. Where 120 kg/ha of salt are used the potash may be reduced to 125 kg.

These average optima need modification where the soil is below or above average in P_2O_5 and K_2O. On fen peats the nitrogen needs to be reduced by 60 kg, and the P_2O_5 and K_2O increased by 25 kg. Rates of nitrogen application above the optima, whilst not giving an economic increase of root or sugar yield, will lead to an increase in yield of tops. Where 30 tonnes/ha of F Y M are applied, the N, P_2O_5 and K_2O may be reduced by 40, 50 and 90 kg respectively.

Sugar beet on certain soils may suffer from deficiencies of the two micro nutrients boron and manganese. The former causes a condition known as brown-heart and is liable to occur on sandlands in dry years particularly if the land has been over-limed. It may be rectified by careful applications of 25 kg/ha of borax. Manganese deficiency is common on neutral or alkaline soils with a high organic matter content. It often occurs on fen peats. The deficiency causes a condition known as " speckled-yellows " and it is most effectively controlled by spraying 12 kg/ha of manganese sulphate on to the leaves as soon as the deficiency makes its appearance in the crop (see p. 192).

As with mangels, sugar beet will show a response in the early stages of growth to side-band placement of the fertilisers. The final yield is, however, no better than where the fertiliser is broadcast. Direct contact placement of fertiliser and seed can cause severe injury to germination. Sugar beet on very light sands in a wet spring may benefit from a top dressing of nitrate of soda at the rate of 125–250 kg/ha.

Cultivation and Sowing. The seed-bed for sugar beet should be very similar to that described for mangels. The field should have been cleaned in the autumn, the dung applied, and the ground well and deeply ploughed before the severe winter weather sets in. Subsoiling is advantageous, especially on soils which have a hard pan at or about normal ploughing depth. The elimination of ridge and furrow by the use of one-way ploughs is an aid in the production of the level seed-bed required for precision inter-row cultivations. In spring the fine and firm seed-bed needs to be prepared from the frost-mellowed plough furrow with the minimum number of operations, both from the point of view of general economy and in order to conserve moisture. The first operation is designed to disrupt the plough furrow and is usually done by a tined implement working to a depth of 100–120 mm. Subsequent operations are usually carried out by harrows and Cambridge rollers. On the heavier soils a " float " or " scrubber-harrow " is often a considerable help in achieving the desired standard of seed-bed. Badly prepared seed-beds result in poor and irregular seedling establishment which, in turn, hinders singling and inter-row cultivations.

Normally, the sooner sugar beet is drilled after the middle or end of March the better the chances of obtaining high yields and high sugar content. The actual date of sowing is, however, very dependent upon the condition of the soil for seed-bed preparation. It is often a practice to stagger the sowing at two or three dates separated by ten-day intervals, in order to spread singling operations. For the earlier sowings of the crop, low bolting varieties should be chosen.

The natural " seed " is really a cluster of fruits each of which may contain one viable seed. As the true seed is quite small it needs to be sown in a moist tilth not deeper than 20 mm. Where the natural " seed " is used about 16 kg/ha are normally sown. More than one seedling can develop from the fruit cluster and this makes the singling operation relatively slow. Rubbed and graded seed is produced by passing natural " seed " between a rubber pad and a revolving emery wheel and grading the reduced fruit clusters so obtained. This has the effect of increasing the proportion of mono-germ clusters. Pelleted rubbed seed is sown at the rate of 8–16 kg/ha.

The operation of drilling can be carried out in a number of ways using multi-purpose drills, drill units on the tractor tool-bar or precision drill units. The precision drills deliver seed down a very short coulter tube at regular spaced intervals of 70–150 mm. Rubbed and graded seed is essential with this kind of drill and the quantity of seed used per ha is very much less than with ordinary drills.

Covering the seed may be done by harrows, by flat roller wheels behind each seeder unit or by hollow cage-type rollers. The latter are a considerable advantage on soils which have a tendency to form surface " caps " or " crusts " and so impede seedling establishment.

Post-drilling Operations. These take three distinct forms: the production of the required plant population; weed control; and the protection of the crop from disease or pest damage.

The plant population is achieved by the combination of the coulter spacings at drilling and by the singling distance within the row. A large number of experiments in this country has shown that beet yields increase up to and beyond 75,000/ha plants. The yield response to plant population levels off when the latter increases much above 75,000/ha plants and, as the cost of handling the larger number of plants also increases considerably, it is considered that the economic optimum lies somewhere between 60,000 and 75,000 plants per ha. Plant populations of this size may be obtained by row widths of 450 mm with 320 mm singling distance, up to 550 mm with 250 mm between plants. An increase in plant population achieved by making the rows narrower has a greater yield benefit than the same population achieved by closer spacings between plants within the row. Narrower rows increase the total length of row per ha and, as both the labour of singling, seconding and lifting is largely determined by the length of row per ha that has to be dealt with, costs rise accordingly. In practice, row widths are determined by their relation to the width and the positioning of the tractor wheels. Row widths of 450–550 mm are most commonly used and these require singling to 280 mm and 250 mm respectively.

The operation of singling needs careful timing. The ideal time is considered to be at the 4-true-leaf stage. Delay up to the 6-true-leaf stage has little effect upon final yield particularly when rubbed seed is used. Serious loss of yield (i.e. up to 3·5 tonnes/ha) occurs when the operation is delayed, by as little as one week, after the 6-true-leaf stage. The loss of yield is smaller where rubbed seed is used : bunching also reduces the loss.

It takes approximately 360 man hours per ha to produce and harvest a crop of sugar beet ; of this, 100 man hours per ha may be used in singling, seconding and other forms of cleaning the crop. Much attention has been directed to reducing the labour used in these operations. This has been attained by using rubbed seed, by precision drills and occasionally by mechanised thinning. The smaller proportion of " doubles " obtained where rubbed seed is used itself reduces the singling time by approximately 10 per cent. The wider spacing between individual seedlings obtained by using the precision drill still further speeds the operation. On exceptionally level seedbeds and with even brairds the use of the tractor hoe at right angles to the drill row produces a series of evenly spaced bunches which are subsequently singled. This is the process of cross blocking which can, under suitable conditions, lead to appreciable saving in labour. Developments in down-the-row thinners did, however, achieve even greater success in reducing labour without prejudice to the final plant population. The thinner consists of a number of hoe blades attached to a rotating head. As the machine travels along the row the rotating hoe blades strike and leave alternate lengths of row. The number and size of hoe blades used are adjustable. The number and size selected depends upon the beet stand and this is determined by counting, using a special frame, the number of metres in 100 occupied by seedling beets. This is expressed as the percentage of beet-containing metres. After thinning, 10 beet plants per 2·5 m of row will normally be required. Where there are, say, 36 per cent of beet-containing metres, a first thinning using eight blades of a size to cut and leave alternate 45 mm will reduce the number to approximately 18 per cent. A second thinning a few days later choosing double the number of appropriate smaller sized blades will reduce the beet-containing metres to the desired 10 per cent. In each case the size and number of blades are selected by reference to calibration charts using the figure of percentage beet-containing metres obtained by counts, and the final figure required. This system of mechanical thinning is known as the Windsor system. A final trimming of the thinned stand of seedling beets by hand hoes is required. The success of the method is determined by a level seed-bed, an even braird of seedlings and correct adjustment of the thinners. Labour saving of approximately 25 per cent is claimed with twice-over thinning plus hand trimming, compared with hand operations only.

Weed control in sugar beet may be carried out by either herbicides or cultivations or both. Pre-emergence herbicides, e.g. pyrazone or lenacil, should be applied, preferably to a moist soil, within a few days of drilling the sugar beet. A contact herbicide, phenmedipham, may be applied after sugar beet emergence to control small broad-leafed weeds. Side-hoeing may be carried out as soon as the beet seedlings are visible and continued until just before the crop foliage covers most of the ground between the rows.

Amongst many diseases and pests that can attack the sugar beet crop Mangel Fly and Black aphis are two which occur with some frequency according to season. They are described in Chapter XIV.

Irrigation. In some of the drier parts of the country irrigation is now considered to be a profitable operation on the sugar beet crop. The same remarks apply to irrigation on this crop as on the potato crop to which reference should be made (p. 266).

Harvesting. A start with lifting is usually made at the end of September, although maximum sugar content may not be reached until mid-October. Good weather greatly facilitates the work and most growers are willing to sacrifice a little weight per ha or some sugar content in order to get well ahead while conditions are favourable. The rate at which beet harvesting can proceed on individual farms depends very much of the quantity of beet the beet factory will accept for processing.

In recent years great strides have been taken in the mechanical harvesting of sugar beet. In 1946, less than 1 per cent of the area was harvested in this way, but by 1970 it had increased to 99 per cent. There are two systems available, complete harvesters which top, lift, clean and elevate the beet into trailers ; and two-stage harvesters in which one unit tops and a separate unit lifts and cleans the beet. The complete harvesters are suitable either for the growers or for contractors, and on the average they handle 30–40 ha per year although some are capable of doing up to 60 ha in one year. Of the 375 man hours formerly required to produce and harvest a ha of sugar beet approximately 170 man hours were needed for the hand pulling, knocking and topping system of harvesting. Complete harvester systems can reduce this to 65 man hours per ha with an increase in tractor hours of 5 per ha.   The two stage harvesters, where the beet is left in the windrow, can reduce operations to approximately 80 man hours with an increase of about 10 tractor hours per ha.   The efficiency of mechanised lifting varies with the type of machine used, the skill in adjusting it for the particular crop, and the soil conditions. They all work best in crops where an even population of well-developed beets occur.

After the sugar beet is harvested the tops which remain in the field make excellent food for either sheep or cattle : they should be allowed to wilt for 10 days or so.   For sheep the tops are usually folded on the land.   They are generally carted off for cattle and fed on grass or in yards.   When more tops are grown than can be comfortably fed in this way, the surplus, whilst still in sound condition, can be made into silage.   The pit or clamp method is generally the simplest and most satisfactory to use.   When feeding tops to cattle it is always advisable to feed some ground chalk in the ration to counteract any tendency to scour.

Yield.   The crop is purchased by the sugar factory at a fixed price per tonne of washed beet registering 16 per cent sugar.   For every 1 per cent above, or below, the average sugar content a fixed amount is added or deducted as the case may be.   An average yield of washed roots is 32–36 tonnes/ha with a sugar content of slightly less than 16 per cent, but upwards of 50 tonnes/ha may be obtained under ideal conditions.   According to season and other factors the sugar content may be as high as 20 per cent.   There is approximately the same weight of tops per acre as of roots.   These are equal in feeding value to swedes.

TURNIPS AND SWEDES

Turnips were introduced to this country as a farm crop early in the 18th century and swedes towards the end of the same century. The incorporation of roots into the farming system is generally ascribed to the enthusiasm of Lord Townshend, who demonstrated on his estate in Norfolk the value of roots for feeding cattle and the beneficial effect they had on the following cereal crop.   By utilising Tull's turnip drill Townshend was able to grow the crop in rows and to make use of inter-row cultivation in a growing crop as a means of weed control other than by the expensive bare fallow.   Eventually and as a result of his experiments, Townshend formulated a rotation of crops, subsequently known as the Norfolk Four Course, which consisted of turnips, barley, clover and ryegrass, wheat.   This system formed the basis for all crop rotations on light to medium loam soils.

Swedes are distinguished from turnips by the extended stem (generally called a " neck ") from which the leaves arise, and by having smooth, ashy-green leaves in contrast to the hairy, grass-green leaves of turnips which arise direct from the bulb itself (Fig. 38). The most important economic difference between turnips and swedes, however, is the amount of dry matter they contain : in turnips it ranges from $7\frac{1}{2}$ to 10 per cent, and in swedes from 10 to 13 per cent. White turnips contain the smallest quantity of dry matter and green-top swedes the largest. Finally, swedes have a longer period of growth and are much hardier than turnips, and can be stored satisfactorily for feeding during the winter in conjunction with coarse fodders for which they form an excellent supplementary feed.

Both turnips and swedes are more suited to the humid areas of the North and the West compared with the drier regions of the

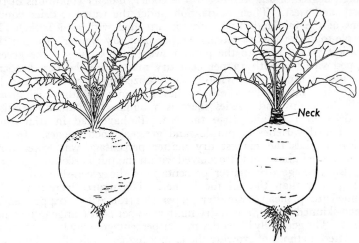

FIG. 38.—TURNIP AND SWEDE.

country. In the latter areas they tend to suffer severely from mildew and late sowing of the crops has long been used as a means of, at least, partial control of the disease.

" Pot " swedes and turnips are produced in some areas. They need to be uniform globe-shaped varieties grown at a relatively high plant population so that the individual roots do not become too large. In addition good " pot " swedes have short necks.

Varieties. Turnips are classified under three main headings white, soft yellow and hardy yellow turnips. Swedes can be grouped into purple, bronze and green-topped varieties.

White turnips are so described because they are white-fleshed, but the outer skin may be white, green, purple or mottled green and purple. These roots grow very rapidly and when sown in late spring are ready for use in early autumn, the leaves and bulb being consumed together. If they happen to be left in the ground and encounter a mild winter they will continue to grow, and a late sown crop may then be consumed in the spring. White turnips can thus be utilised to provide keep over an extended period and are consequently used extensively on sheep farms. They contain 7½–8 per cent dry matter.

The soft yellow turnips and the hardy yellows are yellow-fleshed and contain from 8 to 10 per cent dry matter, the hardy yellows being almost equivalent in feeding value to the purple-topped varieties of swedes. They are somewhat hardier than white turnips and are

L

generally regarded as intermediate in value between soft white turnips and swedes.

Swedes may be white-fleshed or yellow-fleshed, but the latter are the more commonly cultivated. The outer skin of the yellow-fleshed types may be purple, bronze or green. As a class they have a much higher dry-matter content than turnips, a longer growing period and a high degree of adaptability to the cooler, moister conditions of the north and west. Their cultivation under the warmer, drier conditions of the midlands and south is attended with some risk. The purple-topped varieties having 11 per cent dry matter produce the heaviest yields, but are the poorest swedes for keeping ; the green-topped varieties with 13 per cent dry matter are the best keepers, but give a much lower yield ; the bronze-topped varieties can be regarded as intermediate between these extremes. Within each group of both turnips and swedes there is a number of named varieties.

Between 1953 and 1959 the N.I.A.B. have had in trial some 130 named varieties of purple- and green-skinned swedes. In this series of trials the highest dry matter percentage was lower than those quoted above and it occurred within the purple-skinned group ; but the average dry matter percentage of the green-skinned group was greater than that of the former. The figures were : purple-skinned group, mean dry matter 9·3 per cent (range 8·3–10·4 per cent) ; green-skinned group, mean dry matter 9·9 per cent (range 9·7–10 per cent). The generally lower dry matter percentages may be a reflection of altered methods of growing the crop : higher levels of available soil nitrogen, decreased plant populations and later dates of sowing, all tend to decrease the dry matter percentage. The N.I.A.B. data also indicated that each of the green-skinned and two of the purple-skinned varieties listed showed resistance to the club root disease. Also, two of the purple-skinned types showed some resistance to mildew. For further details see Farmers' Leaflet No. 6 of the N.I.A.B.

Position in Rotation. Turnips and swedes are the principal cleaning crop in the rotation in those districts where they are grown extensively ; hence they always follow a cereal crop and in turn are followed in most cases by a cereal crop. Occasionally when the land is very filthy two root crops may be taken in succession in an endeavour to clean the land thoroughly before sowing to corn again.

Manuring. Turnips and swedes show unsatisfactory growth under conditions of soil acidity. Not only is a low pH directly detrimental to the crop, but also such a condition increases the chances of severe attacks of the club root disease. Soil acidity needs to be corrected by suitable dressings of lime applied early enough to act before the crop is sown. Excessive applications of lime on light soils may lead to boron deficiency which produces a condition known as " brown-heart " or " raan ".

Traditionally, turnips and swedes are given large dressings of phosphate but only moderate dressings of nitrogen and potash. As a result of pre-war experiments, 100 kg of N, 180 kg of P_2O_5 and 160 kg of K_2O are calculated to be the average optimal applications assuming present prices and a crop value of £5 per tonne. Revised estimates suggest that 100 kg of N, 100 kg of P_2O_5 and 100 kg of K_2O are more appropriate. Where swedes are to be left in the ground for winter folding 100 kg of N is too high and 50 kg would be safer. High nitrogen reduces the dry matter percentage and decreases the frost hardiness of the crop.

Where 30 tonnes/ha of farmyard manure are applied to the crop, the fertiliser applications may be reduced to approximately half of the revised average optimal. Where deficiencies in soil phosphate or potash are known to exist, supplementary dressings are required.

In areas where turnips and swedes are sown on ridges the fertilisers can be applied on the flat before ridging. The act of ridging then concentrates the fertiliser in the ridge. Where the crops are sown on the flat it has been shown that side-band placement of the fertiliser 50 mm to the side and 25 mm below the seed gives a better yield than broadcast applications. Direct contact placement of fertiliser and seed can cause severe injury to germination and establishment.

Cultivation and Sowing. Turnips and swedes, being small seeded, require a fine and firm seed-bed which needs to be prepared with minimum loss of moisture. In Scotland the crop is often sown on ridges, but in most other areas it is sown on the flat. Where ridge sowing is adopted a deep fine tilth is required to permit the ridges to be formed ; where sown on the flat a shallow tilth only is required. Cultivation operations are arranged to produce either of the two types of tilth mentioned. The deep tilth is produced by early ploughing in the autumn, followed by a cross ploughing later in the year. The second ploughing is done late enough to allow for frost mellowing of the first ploughing and whilst there is still time to obtain similar mellowing of the newly formed furrows. The first dry spell after the beginning of January is the usual time for the operation. The double mellowing then allows the necessary deep tilth to be obtained with the usual tined implements. The fertilisers are applied in the tilth preparation stage. The ridges are then drawn and usually consolidated with ridge rollers, after which drilling of the seed is done. Where sowing is done on the flat only one ploughing is necessary, but this should be early enough to obtain the necessary frost mellowing.

Before the crop is sown the land needs to be cleaned from perennial weeds. This may be done in the autumn by the traditional sets of " wicking " operations, by deep ploughing using skin coulters, or nowadays by the use of either contact or translocatory herbicides. As these crops are usually late sown it has been the custom in some areas to carry out the " wicking " operations in the spring. Whilst the weather in late April and May is often more suitable for these cultivations than that in autumn, the process leads to a drying out of the seed-bed which is to the detriment of germination and establishment.

Farmyard manure, where it is used, is usually ploughed in in the autumn at the present time. Formerly it was often placed in the ridge.

Sowing of swedes begins in the early part of May in the north but may be delayed until the second week in June in the south, where early-sown swedes are specially liable to attacks of mildew if grown on soils which soon become parched in seasons of hot, dry weather.

When the land is ridged it is inconvenient to make the ridges of less width than about 600 mm, but on the flat it is usual to drill roots as near as 450 mm from row to row. Graded seed is precision drilled at 0·5–1 kg/ha and often the crop is not singled, but further cultivation may be needed for weed control.

Turnips are usually sown about a fortnight after the swedes at a similar seed rate, but seeding may continue until the middle or even the end of August.

It is advisable to dress the seed with one of the insecticidal seed-dressings to give the seedlings protection from early attacks of the turnip flea beetle, or " fly " (see p. 439).

Post-drilling Cultivations. Inter-row hoeings should begin as soon as the rows can be seen. When the " fly " is troublesome, dusting or spraying with one of the standard insecticides may be resorted to. Prevailing weather conditions at this time greatly influence the severity of an attack, for with good growing conditions the plants will soon pass the susceptible stage.

Singling of turnips is now seldom practised because of the cost of the operation and because so many crops are precision drilled. It was usually carried out with long-handled hoes. In crops still drilled in the traditional manner the practice has been adopted of using a gapping machine to " bunch " the plants which later can be singled by hand. For success with this method a full plant from one end of the row to the other is essential. Where the crop is to be folded by sheep there is no need to single the bunches as, although the individual roots are relatively small, there is no loss of dry matter yield. Where the crop is to be pulled and carted or chopped for sheep, the larger number of relatively small roots to be handled increases the labour required. Recent developments in the growing of turnips and swedes include the use of precision drills, down-the-row thinners, and graded seed. As a consequence it is now possible to grow good crops either entirely without, or with very reduced, hand singling. This leads to reduced costs and a lessened demand on labour at a critical period.

Harvesting. Swedes are lifted in the autumn for storage in the field where they are grown, or they are carted to the homestead. In favourable districts they are left growing until they can be consumed. When stored, the roots should be topped and tailed and carefully covered with a layer of straw several inches in thickness, and over this a layer of earth at least 150 mm thick should be placed to help in keeping out frost and wet.

Turnips are not usually stored as, being soft, they are easily gnawed by sheep. When required for cattle they are pulled in the same manner as swedes, although they are generally thrown to the stock with the tops left on.

Yield. In good root-growing districts not less than 50 tonnes/ha of turnips would be considered a reasonable yield, although the average for the country as a whole is in the region of 35 tonnes/ha. Swedes crop less heavily than turnips, though the weight of dry matter per ha may be about the same. Even where a 50-tonne crop is obtained the dry matter yield, at 10 per cent dry matter, is only 5 tonnes/ha. As the starch equivalent of the dry matter is approximately 60 per cent the yield of starch equivalent is approximately 3 tonnes/ha. Traditional methods of turnip and swede production involved a heavy labour demand. Developments with precision drills, graded seed, herbicides and mechanical harvesting have greatly reduced labour and power requirements.

CARROTS

Carrot production is now highly specialised and most of the crop is grown to specification under contract.

These roots grow best on light soils free from stones. If grown on heavy land the roots are difficult to lift, and so much soil adheres to them that washing becomes tedious and expensive. Moreover, shapely roots which are easy to market are only obtained on deep, light land. These matters are so important in the case of main crop carrots that the growing of them is confined to the sandlands, the very light silts and the extremely light classes of fen peat. The sandland carrots usually have the superior quality but the yields on the fens are often higher.

As the young plants have very small tops the crop is not suited to weedy land, for in cold seasons the weeds overtake the carrots which are smothered.

Varieties. Varieties of carrot may be divided into three main groups according to shape : Long, Intermediate and Stump-rooted. The long varieties are only suitable for the very deepest soils ; the stump-rooted varieties are favoured for the early market and bunching trade, whilst the intermediates are good general-purpose varieties. Well-known varieties within each group are Long Altrincham, St. Valery and Long Surrey ; James' Intermediate and Autumn King ; and Early Nantes and Early Market stump-rooted varieties. The red-cored Chantenay type of carrot is becoming popular in some districts : in these carrots the root is of an even reddish colour throughout, lacking the contrast between yellow core and red outside so characteristic of most carrots. Cattle-feeding carrots such as White Belgian and " cattle " carrots are not very much grown.

Position in Rotation. It is the main crop carrots which are usually grown as large scale farm crops. The early varieties are more frequently grown on specialist horticultural holdings where cloche culture may be used to obtain accelerated earliness. In the latter case they are often followed by late transplanted crops such as a brassica crop or leeks. In the former case they occupy part of the root break. On the " cash root " producing sandlands they often occupy part of the sugar beet break in a sequence Seeds, Potatoes, Wheat, Sugar Beet plus Carrots, Barley. On the fens they are more commonly grown as a complete break in a four-course sequence of Potatoes, Wheat, Sugar Beet, Carrots. A close sequence of carrots in a cropping system can give rise to trouble with violet root rot and carrot root fly, although control by insecticides of the latter is now possible (see p. 447).

Manuring. Carrots will not grow satisfactorily upon acid soils. Where acidity occurs suitable applications of lime should be given so that the condition is corrected before the crop is sown. It is generally conceded that an application of farmyard manure given direct to the crop lowers the quality of the root. During the 1939–45 war a number of fertiliser experiments were carried out on the crop. On the experience of these experiments fertiliser rates of 55 kg N, 55 kg of P_2O_5 and 100 kg of K_2O per ha are recommended as suitable. In commercial practice rather higher rates than these are usual. The same experiments showed that the crop also responded to applications of 375 kg/ha of agricultural salt and where this is used the potash can be reduced from 100 kg to 55 kg/ha. Side-band placement of fertilisers has been shown to have no advantage in the carrot crop ; direct contact placement is harmful.

Cultivations and Sowing. The seed-bed needs to be very fine, firm and moist. On the light soils upon which the crop is grown little difficulty is experienced in obtaining the required standard of seed-bed. As with sugar beet, levelness of seed-bed is required to facilitate inter-row cultivations. The seed-bed must be free from perennial weeds.

The seed is sown from March to May on the flat in rows 100–450 mm apart. The narrower drilling is more common on the fens ; on other soils the row width is the same as used for the sugar-beet crop. From 1·5 to 4 kg of seed are required, but growers try to sow as little as possible in order to obviate the need for singling.

Before sowing, the seed is sometimes well rubbed and mixed with dry sand or ashes, so that it will not clog in the drill, but will fall freely into the drill-rows. It should be harrowed in with light harrows, and not covered with more than an inch of soil.

The carrot crop is usually grown without singling. The early growth of the seedlings is very slow, and formerly a great deal of hand work had to be used to keep it clean. The use of light mineral oil sprays for weed control has reduced the labour required by at least 40 man hours per hectare. A pre-emergence application of the oil at the rate of 500 litres/ha, followed by a post-emergence spray of from 500 to 1,000 litres/ha at the four-true-leaf stage, is the usual practice. Inter-row cleaning is carried out with the steerage hoe.

The roots were formerly pulled by hand and the tops were twisted off at the same time. It is now more usual to lift the main crop with a potato harvester but the early carrots are still hand lifted. The main crop is usually lifted from October to December or even later. When not sold direct to the consumer carrots are stored in small heaps protected from the weather with straw and soil. For many markets it is necessary to wash the carrots before sale and for this purpose the large growers have special washing plants. The practice of leaving a portion of the crop *in situ* is becoming increasingly popular. The roots keep better in this way and on lifting in the spring usually command a readier sale than those which have been stored in clamps.

The saving in labour is also appreciable. Unfortunately since carrots are susceptible to hard frost the practice involves some risk, and for this reason most growers prefer to secure a portion of the crop under cover.

Yield. The average yield is 25–30 tonnes/ha but under very favourable conditions of soil and season upwards of 50 tonnes/ha may be obtained. The crop is an expensive one to produce in terms of labour, but useful savings have been made through developments in mechanisation and the use of herbicides. A very considerable reduction in labour costs occurs if the crop is sold direct from the field without clamping. Where clamping or pieing is involved a further 125 man hours per ha are required. Mechanical harvesting reduces the harvest labour considerably.

FORAGE CROPS

Included in this section is a variety of plants which are consumed in the green state by farm livestock. Many of these plants belong to the Family Cruciferæ, while others form part of the Family Leguminosæ, and are dealt with here because, unlike the clovers, they are usually cultivated by themselves. Maize, which in this country may not ripen its cobs on a farm scale, is used as a forage and silage crop : it is more closely related to the grasses than to the other forage plants mentioned.

BRASSICA CROPS

Marrowstem Kale, Thousand-headed Kale, Rape Kale, Hungry Gap Kale, Kohl-rabi and Cabbage, although very dissimilar in appearance, are nearly related botanically and are all included in the cabbage group, *Brassica oleracea*. This species also embraces a number of important vegetable crops grown for direct human consumption. Brussels sprouts, cauliflower and broccoli as well as the horticultural varieties of cabbage, savoy, kale and kohl-rabi are included under this heading. These are high value crops which are grown both on a field scale and on more specialised horticultural holdings. There are important economic differences between brassicas grown for forage and as vegetables. In the former case methods of production and utilisation must be such as to provide cheap sources of starch equivalent for stock ; in the latter case the high cash value of the crops allow for techniques of production which are economically not justified for the forage crops. For example, a folded kale crop can be produced and consumed by stock for the expenditure of only about one-third of the man hours needed for the production and harvesting of cabbages and savoys. The acreage of brassica forage crops, including rape and mustard, is approximately three times that of brassicas grown as vegetables for human consumption. Only those brassicas grown as forage crops in this country are considered in the following pages.

Marrowstem Kale is capable of producing very large quantities of greenstuff greatly relished by stock. It is characterised by a thick main stem containing a nutritious pith and bearing large leaves. It thrives on a wide range of conditions but requires generous feeding to produce big yields (Fig. 39).

Thousand-headed Kale is a much-branched plant, growing two, three or more feet in height with numerous plain, uncurled leaves ; tee plant can withstand more severe winter conditions than marrow-stem kale. For this reason it is usually reserved for feeding in the New Year after the marrowstem kale is finished. It is an excellent food for sheep or cattle (Fig. 39).

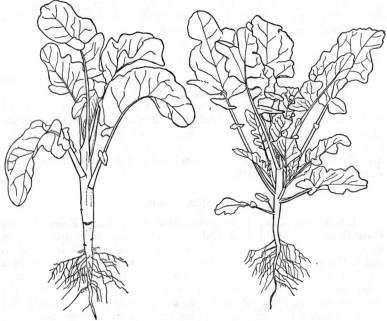

FIG. 39.—MARROWSTEM KALE AND THOUSAND-HEADED KALE.

Rape Kale and Hungry Gap Kale. These are hybrid varieties which find their greatest usefulness in the late spring when the other kales are finished, but the full flush of spring grass is not available for the stock. In April, May, and even as late as June, these varieties send out a profusion of young growth which is extremely valuable for sheep or cattle. Rape kale has a short, thick stem with crinkled leaves. Hungry Gap Kale has a thinner, longer stem with numerous side shoots in the axils of the leaves.

Kohl-rabi is a crop confined mainly to the eastern and south-eastern counties of England. In this plant the stem immediately above the cotyledons or seed leaves becomes much swollen and bears large scars, which are the marks of the junctions of earlier leaves with the stem (Fig. 40). In the comparatively dry eastern counties the cultivation of swedes is particularly difficult owing to their liability to mildew if sown early, and to attacks of turnip fly if drilled later in the season. Moreover, swedes seldom yield well under low rainfall conditions. In these circumstances kohl-rabi provides an excellent substitute, and as it does not suffer severely from drought its cultivation can be extended to lighter soils than those on which it would be possible to grow swedes.

Cabbages are characterised by the shortened main stem and the smooth, glaucous leaves which are closely folded over each other, thus producing a well-compacted, firm head (Fig. 40). In shape, size, colour and hardiness the group furnishes examples of wide differences. Thus, there is the small ox-heart cabbage, the rounder ball-head and the drum-head ; the first, and to a large extent the second, by reason of their more tender and better-flavoured leaves, are used mainly for culinary purposes, whilst the drum-heads provide a valuable green food for stock.

In colour, cabbages vary from various shades of green to red or purple, and in hardiness from forms which can be grown in summer only to those, like the Savoy, which are capable of withstanding severe winter weather.

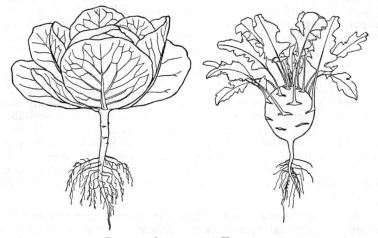

FIG. 40.—CABBAGE AND KOHL-RABI.

Named varieties of each of these brassica fodder crops exist.

Position in Rotation. All the crops in this group occupy part or the whole of the normal root break on the farm. In addition to being grown in rows, thereby permitting cleaning operations to be carried out in the summer they are, with the exception of kohl-rabi and cabbage, foliage crops which effectively smother out the weeds once they meet in the rows.

Manuring. Brassica crops are gross feeders and liberal manuring is essential if maximum yields are to be obtained. They respond to farmyard manure, and when this is available it may be applied in quantities up to 50 tonnes/ha.   Usually the dung will be applied in the autumn and ploughed in.

The evidence that is available suggests that 125 kg of N, 125 kg of P_2O_5 and 125 kg of K_2O per ha are appropriate under most circumstances.   Where 30 tonnes/ha of farmyard manure are applied the P_2O_5 and K_2O may each be reduced to 60 kg/ha.   Where kale is taken as the first crop after good ploughed-out grass or leys, 60 kg of N will be enough but, particularly after ploughed-out permanent grass, more phosphate may be needed.   Other adjustments of P_2O_5 and K_2O may be needed in the light of soil analysis data.   The fertilisers are usually broadcast in the seed-bed ; placement 50 mm to the side and 25 mm below the seed gives better early results than broadcasting, but final yields show little difference between the two methods. Direct contact placement of fertiliser and seed can lead to a severe depression of germination.   In some cases 60 kg of N are top dressed on to the crop before it has met in the row.

Cultivation and Sowing.   These crops require cultivations similar to those given to root crops such as turnips and swedes.   After a fine seed-bed has been prepared, the seed may be drilled and the young plants thinned as desired.   The seed rate is 2–4 kg/ha usually, but when the plants are grown in a nursery bed 1 kg of seed will supply enough plants for 1 ha.   Transplanting is occasionally used for forage cabbages.   It has the advantage of allowing pre-planting cleaning operations to be continued to a later date.   Normally it cannot be justified economically for kale, and is a doubtful proposition for cabbage.   To obtain the highest yields of dry matter per hectare kale needs to be drilled early in April.   When this is done the base of the stem becomes woody in the autumn.   The woody stem bases are unpalatable to stock and have a reduced digestibility. For this reason kale is often sown in May so that reasonable yields are obtained without undue woodiness.   Also with late sowings the crop is not so tall and folding controlled by the electric fence is facilitated.   Kale may also be sown broadcast.   This is sometimes done late in the season after an early harvested crop such as first early potatoes.

Post-drilling Cultivations.   Once the plants are established the subsequent cultural operations are similar to those practised with turnips and swedes—hoeing by tractor and hand hoes being kept going until the crop has arrived at a stage when a continuance involves physical damage to the plants.   For cabbages, singling is necessary to obtain a full development of the head.   With kale, singling, gapping or leaving the drill unthinned are each common practices.   It has been shown that there is no advantage to be gained in terms of dry matter yield by reducing the plant population by either singling or gapping the crop, provided weeds are not a problem. The unthinned crop does, however, have a larger stem : leaf ratio, a lower protein content and a higher fibre content.   On the other hand it is cheaper to produce.

The relatively inexpensive mechanical gapping would appear to be a useful compromise.   This assumes that the crop is either folded or harvested with the forage harvester. Where it is hand cut and carted there is a considerable advantage in having a low population of large plants obtained as a result of singling. In the case of Thousand-headed kale, singling or gapping increases the yield of secondary shoot which is a valuable part of the crop when consumed after the turn of the year.

Harvesting and Yield. Kohl-rabi, which makes excellent feeding for sheep or cattle, may be folded off but it can also be lifted and stored like swedes. It is more resistant to frost than swedes, and if necessary can be left much longer without lifting. Its keeping qualities are remarkably good and the yield varies from 36 to 60 tonnes/ha.

The kales may be folded off by cattle or sheep or the crop may be cut and carted home for feeding fresh to housed cattle. The advent of the flail type of forage harvester has greatly facilitated the cutting and carting of the crop. Whereas kale folded by cattle, using an electric fence, may be produced and used for the expenditure of 100 man hours per ha, at least 250 are needed where hand cutting and carting is the harvest method. Marrowstem kale may be made into silage, which is very useful for feeding in late winter when the crops have lost their freshness and have become fibrous and of lower feeding value. The average yield is 50–70 tonnes/ha. These gross crop yields will give almost 5–7 tonnes of starch equivalent per ha which, if folded, represent one of the cheapest sources of starch equivalent for winter feeding. Marrowstem kale gives the largest yields per ha, followed by Thousand-headed kale, with Rape kale and Hungry Gap kale giving smaller yields.

Cabbages are invariably cut and carted home for feeding to stock In yards or out-wintering on grass. Under good soil and climatic conditions and with liberal manuring, the cabbage crop can yield 100–125 tonnes/ha and is then regarded as one of the most useful crops for feeding to stock.

Rape. This is a valuable forage crop, especially for sheep. Rape is a leafy, succulent plant very similar to a swede except that no bulb is formed. It grows to a height of 600–900 mm and may be cultivated as a main crop or used for catch cropping. Sown in April it is ready for grazing off in July and August, and will yield 30–40 tonnes/ha of greenstuff. The varieties in general use are The Giant and Essex Dwarf; the former is usually regarded as more suitable for poorer and lighter classes of soil and the latter for the richer soils.

Sowing is invariably done on the flat and the seed may be drilled or broadcast. The latter is the usual procedure and 10–12 kg of seed per ha are required; if drilled, 4 kg of seed are ample. Following broadcasting no further cultivations are necessary and, indeed, the manner of seeding makes such impossible. On account of its rapid leaf development rape will smother weeds most effectively. When land is being sown to grass without a cereal cover crop, 2 kg of rape are frequently included in the seeds mixture to act as a nurse crop and provide early grazing for the stock.

As a catch crop following early potatoes or peas, rape is also very

useful. After early potatoes harrowing and rolling are all the operations necessary for the preparation of the seed-bed, but after peas
it may be necessary to plough, or at least disc harrow thoroughly,
and then harrow and roll to obtain the necessary fineness of seed-bed.

MUSTARD

Two species of mustard are cultivated in England, namely White
Mustard (*Brassica alba*) and Black Mustard (*B. juncea*). The tall-
growing species known as Brown or Black Mustard (*B. nigra*) which
was previously grown for condiment has been virtually replaced by
B. juncea and persists now largely as a weed on land on which it was
previously cultivated. All these species are annuals and in many
respects closely resemble the crops dealt with already in this group.

White mustard is largely grown as a catch crop in the same
manner as rape : it differs from the latter in respect to its susceptibility
to frost. A mixture of rape and mustard in equal parts is frequently
grown for sheep feed, whilst in some parts mustard is sown after early
potatoes for the purpose of green manuring. The rate of sowing is
about 20 kg/ha. White mustard is also cultivated for its seed in the
eastern counties ; the seed forms one of the components of mustard
flour for condiment.

The pungency of condiment mustard is imparted by Black Mustard
which contains as much as 22 per cent of volatile oil. *B. juncea* is a
more adaptable plant than was *B. nigra* in that although it will give
a larger yield on more fertile soils it can be grown successfully on
quite light soils. It is also shorter than *B. nigra*—growing up to
950 mm high—and holds its seed better and can therefore be harvested by combine. The crop responds well to liberal dressings of
nitrogen but phosphate and potash are only required where a known
deficiency exists. The crop is frequently grown after a straw crop
and helps to lengthen the rotations. Seedrate is 4–5 kg/ha in drills
180–500 mm apart, sown early in March.

Harvesting takes place in early August. The crop may be cut
with a binder and stacked ; windrowed with a pea-cutter and subsequently picked up by the combine ; or even combined direct. It
should not be cut until the seed is almost ripe and the seeds are a
rich brown colour. Seed yields vary between 1·5 and 2·5 tonnes/ha.

LEGUMINOUS FORAGE CROPS

Vetches or Tares. This crop is used principally to provide
keep for sheep, and green fodder for cattle and horses in spring.
Tares are extremely useful to include in a mixture with oats and peas
or beans for making into hay or silage, whilst in some cases this
mixture may be allowed to ripen to provide grain which can be
ground into a balanced meal for dairy cows. Vetches generally
follow a white straw crop of some kind, and in light land districts
are often grown as a catch crop.

The plant is characterised by slender, square, trailing stems
frequently as much as 2 m in length ; the leaves are pinnate with

numerous obovate leaflets, and the leaf stalk usually terminates in a simple or branched tendril. The flowers, which arise either singly or in pairs on short stalks, or as racemes in the axils of the leaves, may be blue or purple in colour. Because of the weak, trailing stems, tares are more frequently grown in mixture with a cereal which then acts as a support (Fig. 41).

Tares do not exhibit a strong partiality to any particular class of soil ; they may be grown on stiff clays if adequately drained, and equally well on sandy soils, provided the supply of moisture is sufficient and there is no shortage of lime. In their ability to thrive on poor soils they constitute a valuable means of increasing the stock-carrying capacity of this type of land.

Varieties. In the British Isles the most widely grown variety is the Common Vetch (*Vicia sativa*) which is subdivided into spring and winter sorts. These are similar plants botanically, but differ in

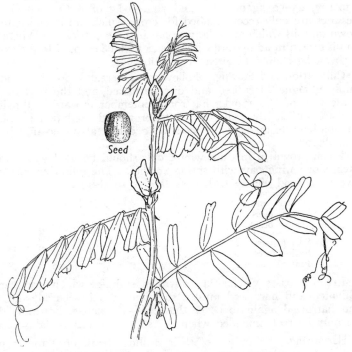

Seed

FIG. 41.—CULTIVATED VETCH (*Vicia sativa*).

their ability to survive the winter. In some districts the Goar or summer vetch is grown extensively and this resembles the common vetch in its flower characteristics, but is a stronger growing plant with larger leaves, stems and pods. The Goar is apparently as hardy as the common vetch, but arrives at full growth somewhat earlier. By sowing portions of the land devoted to vetches with the common variety and also with Goar, the period for cutting the silage crop can be extended without incurring loss in the nutritive value of the silage due to over-ripening and the attendant development of fibre.

Position in Rotation. The crop does not occupy a regular position in the crop rotation, but when grown as a pure crop it is the usual practice to sow after a cereal. When used in mixture with cereals to cut for silage the crop may replace roots completely, in which case as soon as the crop is removed in July the opportunity is taken to bastard fallow in readiness for the next crop.

Manuring. When the crop is grown for seed nitrogenous manures should be used sparingly, if at all, for they over-stimulate the production of haulm. The same objection does not apply in the case of crops intended for soiling or for silage. In the absence of precise experimental data on fertilisers for vetches it can be assumed that the requirements will be similar to those for peas and beans. That is to say, 25–50 kg of P_2O_5 and 50–100 kg of K_2O. The larger dressings are only recommended for crops which are intended to be grown on soils which are relatively low in these nutrients. Where the crop is grown in admixture with oats as a silage crop, N to the extent of 50–75 kg should be used in addition.

Cultivation and Sowing. Following a cereal crop the land is simply ploughed, fertilised and disc harrowed, and the seed drilled at the rate of 150–160 kg/ha from September onwards. Provided they are sufficiently well covered to prevent depredation by birds, it is inadvisable to bury the seed too deeply; about 30 mm is the ideal depth.

When grown in mixture winter oats should be used with winter vetches, or spring oats with spring vetches. The following mixtures have been used widely with considerable success :

(a)	Winter oats	.	.	.	.	.	.	. 160 kg/ha
	Winter tares	.	.	.	.	.	.	64 ,, ,,
	Winter beans	.	.	.	.	.	.	32 ,, ,,
(b)	Spring oats	.	.	.	.	.	.	. 160 ,, ,,
	Spring tares	.	.	.	.	.	.	64 ,, ,,
	Peas	.	.	.	.	.	.	32 ,, ,,

In some cases vetches are sown on stubbles which are merely cultivated and harrowed until a sufficient tilth is secured, and the land harrowed again to cover the seed—a practice, however, that can only be recommended when the stubbles are reasonably clean.

Harvesting. By reason of their trailing stems, once the crop is sown little can be done in the way of subsequent cultivations. In districts possessing a suitable climate vetches are made into hay, but they require a long period of dry weather to cure the succulent stems and may suffer considerable damage if subjected to heavy rain. For hay or silage purposes the crop should be cut when still in flower, for with the development of seed the stem becomes more fibrous and less valuable for feeding. The yield of hay varies from 4 to 8 tonnes/ha.

When grown for seed production the most satisfactory method is to include a small quantity of beans as a support for the vetches. The beans can be readily separated from the vetches by riddling after threshing. The yield of seed varies from 1·5 to 2 tonnes/ha.

For folding with sheep the autumn-sown crops will be ready the following May and the spring-sown later in the summer, according to the date of sowing.

Lucerne. Lucerne is a perennial plant known in some countries as Alfalfa, and botanically as *Medicago sativa* (Fig. 42). It is a very important forage crop used mainly for cutting purposes although with proper treatment it can also be grazed. Its chief characteristic is its deep root system by means of which it has the capacity to remain productive under severe conditions of drought.

In some respects lucerne is an exacting crop, for although it will thrive on a fairly wide range of soils, one essential feature with all of them must be an adequate supply of lime. The land must be well drained and, if not of high natural fertility, then care must be taken to ensure adequate supplies of plant nutrients. Clean ground is also essential, this being important in the early phases of establishment of the lucerne stand. Under reasonable conditions lucerne will give a succession of crops throughout the growing season, resulting in three to four cuts annually and a total dry matter yield of at least 10 tonnes/ha. This production can be sustained for 5 or 6 years or even longer: when finally ploughed up, lucerne is extremely valuable in maintaining arable soils in good physical condition.

The area devoted to lucerne in Britain is comparatively small, some 16,000 ha at the present time, and this is largely restricted to the lower rainfall areas in the eastern and southern counties. Some

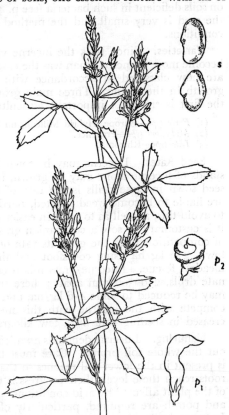

FIG. 42.—LUCERNE (*Medicago sativa*).
s, seed. *p₁*, newly formed pod. *p₂*, mature spirally coiled pod.

effort has been made to extend the area grown to some western and northern counties. Difficulty arises, however, from the fact that the more normal grass and clover seeds mixtures do so well under the higher rainfall conditions in these parts of the country and are less exacting in their management, so that lucerne growing spreads only slowly if at all.

In its early period of establishment from seed, lucerne develops slowly and does not attain its full production capacity for some two or three years after sowing. Conditions of establishment must therefore be of a high order in terms of soil tilth, freedom from weeds and availability of plant nutrients. Given these conditions the young lucerne seedlings, which are extremely vulnerable to competition from other plants, will get away to a quick start and so avoid any initial check to growth. The introduction of a method for inoculating lucerne seed with an appropriate strain of bacteria capable of effecting the development of nitrogen-fixation nodules on the young seedling root system has done much to ensure better establishment of lucerne on soils deficient in such bacteria (see p. 403). The cost of inoculating the seed is very small and the method simple to apply under farm conditions.

Varieties. Since 1945 the lucerne varieties used in Britain have become more numerous than was the case previously. These varieties are now classified in accordance with the time they begin active growth in the spring. Three main groups are recognised, of which the first is the most important agriculturally :

(1) *Early type*—Europe, Hybrid Milfeuil, du Puits, Eynsford and Flamande.
(2) *Midseason type*—Provence.
(3) *Late type*—Rhizoma.

Seed Rates. Lucerne may be sown in a number of ways. Pure stands of lucerne alone may be grown, in which case 14–15 kg/ha of seed sown in close drills is the rate of seeding. These pure stands are liable to become weed infested, particularly on the heavier soils : to avoid this, as well as to assist in easier conservation as hay or silage, it is customary to sow a companion grass. To ensure the minimum of competition from the grass its rate of seeding must be kept fairly low : 2–3 kg/ha using cocksfoot and timothy and 4–5 kg/ha using meadow fescue. Lucerne/grass mixtures may also be sown in alternate drills, 250–300 mm apart ; here the rate of seeding of lucerne may be reduced to some 4–6 kg/ha, and the grasses, which no longer compete with the lucerne with this method of sowing, can be increased in seeding rate by a few kilograms if necessary.

Manuring. Lucerne demands an adequate supply of lime throughout the whole soil profile. Care must be taken to ensure that lime is present in the lower soil reaches so that when the deep penetrating roots reach these levels their progress is not retarded and the growth of the plant affected by acid conditions. Initial supplies of phosphate and potash are required, particularly of potash on the lighter soils. In these latter circumstances supplies of potash are again essential in the later years. Seedling establishment may, in some situations, be improved by the incorporation of a small amount of nitrogen in the seed-bed fertiliser dressing, while backward stands later on may also benefit from a judicious top-dressing of a nitrogenous fertiliser.

Sowing. The time of sowing will depend on the preceding crop and the state of preparation and cleanliness of the land. Perennial weeds, particularly those of a creeping type, should be removed before sowing, and where annual weeds are present it is advisable to reduce their incidence by encouraging them to germinate before the final pre-sowing cultivations are carried out. Many of these weeds in young lucerne stands can now be controlled by the application of selective herbicides based on 2,4-DB. These must be applied after the lucerne has developed its first trifoliate leaf, and preferably between this time and the appearance of the fourth trifoliate leaf.

The seeds may be sown in spring from late March to early May or sowing may be left until late July. Later sowings than this are not possible as winter frosts may severely damage and kill seedlings not properly established and developed. Spring seeding may be under a nurse crop or direct ; in the former method a stiff-strawed early-maturing variety of spring wheat or barley is preferable. July seedings are direct sown.

One particular pest that must be controlled in the seeding stages to prevent failure of the lucerne stand is the Pea and Bean weevil (*Sitona spp*) (see p. 441). This may be done by applying BHC as a dust or spray in the early seedling stages as soon as any leaf damage is observed.

Management. After seeding the management of a lucerne stand should be as lax as possible in terms of mowing and grazing for the first eighteen months or so. This will allow the stand to become firmly established, and once a strong plant with a good root system has developed, the lucerne will give consistent yields over a number of years. Sown under a cover crop in the spring, the lucerne in the stubble should be allowed to develop during the autumn and die back in the winter without being cut or grazed. This is also the treatment for a July sown seeding, whilst a direct sown spring seeding should be cut over once at a height of 150–200 mm during the summer to control weed growth. The stand may then grow on unchecked into the autumn and die back in the winter. Under such treatment lucerne stands will develop strong root systems by the first harvest year, when two cuts may be taken at the early flower bud stage of growth. Subsequent growth in the late summer and autumn of this year should be allowed to grow on until growth ceases, after which it may be grazed. In subsequent harvest years, three to four cuts may be taken, but here again as much autumn rest as possible should be given to maintain a strong plant.

Lucerne may be grazed rather than cut, particularly the growth following the first cut of the season. When grazing, it is essential to use the electric fence to control the period of grazing. This should not be prolonged over too many days to avoid grazing the new growth.

Lucerne hay may be made on tripods. In this way the leaf, which is the more nutritious part of the plant, is prevented from shattering and the hay ensuing will be of high quality. Lucerne may also be conserved as silage, and here a grass companion is useful to allow proper fermentation of the silage and to avoid some degree of un-palatability that may occur with lucerne silage.

In an established stand, heavy harrowing when the lucerne plant is dormant during the winter is of advantage in controlling both broad-leaved weeds and in checking the growth of any companion grass. Harrowing will also open up the stand and result in a more vigorous growth of the individual lucerne plants.

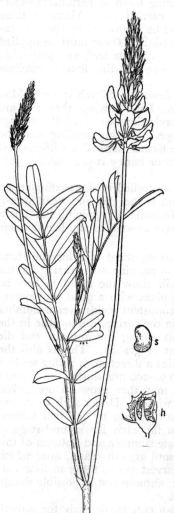

Sainfoin occupies a limited area both as a hay and a pasture crop, more particularly in the eastern and southern counties of England, especially on chalk land. The root is long and strongly developed and capable of penetrating to considerable depths. Sainfoin may be included in " seeds " mixtures but is more commonly sown as a pure crop (Fig. 43).

Varieties. There are two commercial varieties in general use : the Common sainfoin which may be left down for five to seven years or even longer, and the Giant or double-cut sainfoin which usually occupies the ground for one year only. Common sainfoin is better suited to medium soils whilst the giant type does well on light land. There is a number of local strains of common and giant sainfoin which appear to be superior to the ordinary commercial strains.

Manuring and Cultivation. Lime is essential for sainfoin which, like lucerne, needs phosphate and potash in addition. The manurial recommendations for lucerne apply equally to sainfoin. Sainfoin is usually sown with a cereal, which on light land is invariably barley, and on heavy land, wheat. In the former case, and especially in dry districts, it is usual to drill the sainfoin at the same time as the barley, but at right angles to the rows of that crop. As with other

FIG. 43.—SAINFOIN (*Onobrychis sativa*).
s, seed. *h*, husk.

smaller seeded leguminous crops, a firm seed-bed and a fine tilth are essential. The seed rate is 25 kg/ha for milled seed and 40 kg/ha when seed with the husk still on is used. This is sown from February to May, usually in rows about 180 mm apart and not deeper than 25 mm.

From giant sainfoin the average yield of hay is about 4 tonnes/ha, whilst after the second year yields of double that amount may be obtained from common sainfoin.

Trifolium, also called Crimson Clover, is an erect, tall growing, annual clover whose cultivation is confined to the south and south-east of England because of its susceptibility to frost. There are early, late and medium strains of the crimson form, and a late white-flowered variety (Fig. 44).

The crop is chiefly grown as a catch crop, being sown in the autumn on corn stubbles which have been lightly cultivated, usually with disc harrows, and to which a light dressing of phosphatic fertiliser may be applied. The seed rate is 20–24 kg/ha, which is usually simply broadcast and harrowed in.

The crop is fed off by sheep the following April and May in sequence, according to the strain sown, care being taken to consume it before it becomes advanced in growth and woody. At this stage the flowers become hairy and may cause hair-balls to form in the intestines of the stock.

Lupins. The cultivation of lupins is largely restricted to poor light land in the eastern counties. The fact that the crop does well on land deficient in lime and humus is made use of by growing this crop for ploughing in as green manure, for which purpose it is ideally suitable.

Three types of lupin are available, white-, yellow- and blue-flowered, the blue proving the most reliable. In recent years the Sweet Lupin, from which the bitter, toxic principle has been eliminated, has been introduced as a forage crop, but so far it has not been cultivated on a very large scale.

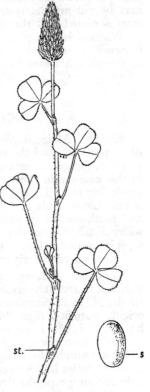

st. ———— s

FIG. 44.—TRIFOLIUM (*Trifolium incarnatum*).
s, seed. *st.*, blunt stipule.

Place in Rotation. On poor light land the crop may constitute part of the rotation, being grown in sequence with rye, potatoes and buckwheat. It may also be grown as a catch crop for green manuring after early potatoes or peas or a bastard fallow. Lupins do best on land that is distinctly acid, and free lime in the soil appears to be a disadvantage.

Cultivation and Sowing. Ploughing followed by harrowing and rolling is all that is likely to be necessary to prepare the seed-bed on typical lupin land. Seeding takes place in April, 60–120 kg of seed, according to variety, drilled in rows 180 mm or 500 mm apart. Broadcasting is quite satisfactory. In most cases nothing in the way

of after-cultivation is likely to be carried out, though the crop may
be horse hoed if necessary.

Harvesting.   For seed purposes the crop must be cut before it is
fully mature or losses due to shedding may be considerable.   Frequently
the crop is cut with a binder, even though the canvases are subject
to considerable wear, for when the crop is cut with a reaper the
strong spines on the ends of seed pods make tying very unpleasant.   It
may be cut and harvested loose.   After fielding until quite fit the
crop is carted and threshed in a comparable manner to beans.

When a big crop of lupins is ploughed in for green manure it
is usually necessary to roll it down first and to attach a drag chain
to the plough beam to ensure complete burial of the plants.

Yield.   The yield of seed is from 1·5 to 2 tonnes/ha.

MISCELLANEOUS CROPS

Maize.   Maize is grown extensively in America and Canada for
silage purposes, and the recent introduction of early-maturing hybrid
varieties, forage harvesters and suitable herbicides has revived interest
in this country in the use of the crop for silage.   It is essentially a
sun-loving crop, flourishing during periods of drought, and hence it
is valuable as a silage, forage or even a grain crop in the dry eastern
and south-eastern counties.   The newer varieties have resulted in a
wider distribution of the crop, particularly in the Midlands, but even
so, it is unsuitable for the north of England.   It is susceptible to frost
and more suited to early soils ; heavy soils are not suitable for it.

Varieties.   A large number of varieties has been tested by the
N.I.A.B. in recent years from European and North American sources.
In the N.I.A.B. recommended list varieties are classified under their
origin, maturity, digestibility and relative yields of dry matter per
ha (see N.I.A.B. Farmers Leaflet No. 7).

Early maturity group varieties include LG 11 and Dekalb 202
which are suitable for forage and grain production.   Medium maturity
group varieties like Kelvedon 33, Austria 290 and Orla 264 are suitable for forage production.

Manuring.   Generous manuring is essential in order to give the
rapidity of growth necessary to ensure a full crop.   A suitable fertiliser
dressing would be about 125 kg of nitrogen, 60 kg of phosphate and
60 kg of potash per ha.   This should be worked into the seed-bed.
The crop responds well to dung application, and when it is applied
the above fertiliser dressing can be reduced accordingly.   Usually all
the nitrogen is applied to the seed-bed before planting, but under some
soil conditions a nitrogen top dressing is beneficial.

Cultivation and Sowing.   Maize requires a clean seed-bed.   Suitable autumn stubble cleaning operations are followed by the application of dung, when available, and the land is then cross ploughed
and left in the furrows over winter to weather : with the aid of harrows
and rollers in the spring the requisite fine tilth is obtained.   During
the cultivation operations the fertilisers can be applied and mixed in

with the surface soil. Sowing takes place when the danger of frost is past, which is about late April in the south and mid-May in the Midlands. The seed, which should be dressed with an appropriate seed dressing, is sown at the rate of 35–40 kg/ha at a depth of 50–80 mm, usually with a corn drill : a precision drill gives a better spaced seed distribution. Row width varies from 500 mm upwards, but a minimum width of 700 mm is necessary when the crop is to be harvested with special maize harvesting machinery. For maximum yields an average plant population of 10/m² should be aimed at. Bird, and particularly rook, damage at the time of germination and emergence is often troublesome and the main cause of crop failures. At sowing time no seed should be left exposed, and whenever possible the crop should be sited near the homestead. In cases of a determined bird attack the crop can be protected by black cotton or string 1 m above the ground and attached to poles set out at 8 m apart. This, however, precludes mechanical methods of weed control. Weeds in maize are now controlled more efficiently by the application of 2 kg (3 kg on medium or heavy soils) of 50 per cent wettable powder of atrazine or simazine in 225–550 litres of water per ha which should be applied during the first 7 days after sowing. Sufficient surface soil moisture is necessary for the latter treatment to be effective : 2,4-D at 0·5 kg acid equivalent in 225 litres/ha of water applied at the four-leaf stage can also give an adequate weed control.

Harvesting. Cutting for forage begins in August, particularly in a dry year when the crop is fed green on bare pastures or in the byre. For ensiling, the crop is cut during late September/early October when the cobs have reached the cheesy stage ; ensiling must be completed before the first severe frost. Harvesting is most easily carried out with a maize harvester, or chopper-type forage harvester fitted with a maize harvesting attachment, blowing into trailers. Care should be taken to ensure that the cobs, which contain a high percentage of the food value of the crop, are not wasted.

Yield. On land in good heart and in a favourable season, 75 tonnes/ha of green crop is not uncommon : 50–60 tonnes/ha may be regarded as a satisfactory yield. It is not difficult to ensile maize when the material is cut with a forage harvester, and the high starch content helps fermentation. In feeding value maize closely resembles kale and cabbage, but on the larger farms where bigger areas are grown it may be cheaper per unit of food. Its high starch content makes it particularly suitable for fattening cattle who can consume efficiently up to about 40 kg of silage per day per adult animal. For milking cows up to 30–40 kg per head per day may be fed, supplemented with a suitably balanced concentrated ration.

Linseed. The linseed plant (*Linum usitatissimum*) may be grown either for the production of its seed, which contains a high proportion of oil, or for flax fibre. In either case the crop succeeds best in districts with a low rainfall and warm climate, and when grown on kindly friable loam soils which are well drained but never likely to dry out. The crop is unable to withstand drought owing to its shallow rooting habit of growth (Fig. 45). Linseed and flax as farm crops in this country have almost entirely disappeared, and the following notes are included only because of the occasional enquiries that are made concerning the culture of these plants.

Varieties. For seed production, Royal and Valuta are suitable, and can be sown early. For late sowings, Dakotor and Redwing should be chosen : the former is considerably more productive, but a day or two later in ripening. Flax is grown only on contract for processing factories and the seed is supplied direct by the factory to the growers. Since about 1950, the cultivation of flax in the British Isles has virtually ceased.

Position in Rotation. Provided the land is clean and extremes of poverty or richness are avoided, linseed may occupy any convenient place in the rotation. As a general rule it will follow a straw crop: it may be used as a nurse crop for seeds, for which purpose it is excellent, since there is little foliage to smother the young grass and clover seedlings and the crop is harvested early in the season.

Manuring. In both cases the crop is generally grown on reserves of plant food in the soil, and little return can be expected from any fertiliser except nitrogen. Not more than 35 kg of nitrogen are needed as a rule.

Cultivation and Sowing. A clean, fine, firm seed-bed is essential and to this end all the cultivation operations should be directed.

Seeding can take place as soon as soil and climatic conditions are favourable in the spring. For seed production the linseed can be sown broadcast at 100 kg/ha or drilled at 40–60 kg, but for fibre production drilling is always preferable as this gives more even maturity at pulling time.

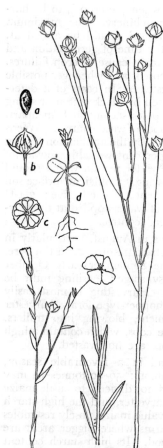

FIG. 45.—LINSEED (*Linum usitatissimum*).

a, seed. *b*, seed capsule or boll. *c*, boll in transverse section. *d*, seedling.

The seed should not be buried deeply, 10–30 mm being ample in most cases, followed by a light harrowing and rolling.

Harvesting. When grown for seed the crop should be cut when the seed in the ripest capsules is plump and a shiny pale-brown colour. Most crops show very uneven ripening and if left until all the seed heads are ripe a good deal of shedding is likely to take place. The sheaves should be small to facilitate drying and should not be carted until perfectly dry. Careful adjustment of the threshing drum is essential to secure clean working. The straw is useless for feeding or for bedding, but the chaff is a very valuable food for making linseed jelly or feeding direct.

Yield. Seed crops usually average 1·5 tonnes/ha, with exceptional crops in the region of 3 tonnes. An average yield of flax is about 5–6 tonnes/ha with no more than 0·5–0·6 tonne of seed.

GROWING ROOT CROPS FOR SEED

In recent years the production of seed of root and other crops has developed very considerably, and this country is no longer so dependent upon imported seed as it used to be. The outstanding example is the sugar beet crop. Before 1939 most of the sugar beet seed sown here came from abroad, but by 1945 home production of seed was sufficient to supply all requirements and leave a small excess for export.

Mangel and Sugar Beet. The production of seed from these two crops is so similar that they may be considered together. The plants are biennial under normal methods of cultivation, growth during two seasons (or more accurately, exposure to low temperature) being necessary for the formation of flowers and seed. Formerly seed was sown from July to mid-August in rows 200–300 mm apart in seed-beds. One ha of seed-bed, sown with 30 kg of seed, provides enough stecklings (or young plants) for planting out 10 ha. Stecklings can be transplanted from the seed-bed in autumn or in early spring.

However, growing seed crops from stecklings brought from isolation areas was expensive and transplanting proved risky; stecklings transplanted in the late autumn were susceptible to frost damage and attack from pests during the winter and those transplanted in the spring sometimes failed because of drought.

A new practice of sowing stecklings under barley to grow on as seed crops *in situ* resulted in effective control of virus yellows provided that the barley was harvested after aphid numbers had declined, and the stecklings were sprayed with an organo-phosphorus insecticide after the barley was removed. This system of undersowing is now the usual practice.

Suitable isolation must be provided for these seed crops because mangels, sugar beet, garden beet, sea-kale beet and spinach beet will all cross-pollinate. At least 1 km should separate seed crops of this group. In the chief seed-producing areas of the eastern counties special zoning schemes are in operation to facilitate isolation.

Inter-row cultivations and hand hoeing where necessary are needed to keep down weeds. The leading shoot is topped, or cut off with a knife, usually when the plants are about 450 mm high, some 80–120 mm being removed. Topping causes the plant to bush out and produce several flower-bearing stems. Harvesting takes place when about half the seed has turned brown; it is inadvisable to delay harvesting until all the seed has ripened because of the losses in shedding. The crop is cut with hooks, tied into sheaves and stooked.

Sheets are used freely when carrying to the rick to collect shed seed. Ricks should be narrow and well ventilated. The average yield is 2·5 tonnes/ha of seed.

Brassica Crops. Swedes, turnips, kale and rape are all grown for seed in very much the same way. They are all biennial, requiring growth in two seasons to complete the life cycle. Isolation must be very carefully considered, and crops should be separated from other brassica crops by at least 1 km.

There are two methods of obtaining plants, by direct drilling or by transplanting ; usually only swedes are transplanted. Swede seed is drilled in the field during mid-July to mid-August at the rate of 2–4 kg/ha. Turnip seed is drilled from mid-August to mid-September at 2–4 kg/ha. Kale and rape are drilled in June and early July at 4 kg/ha. The width apart of the rows varies with the equipment available, and may be from 450 to 750 mm. In spring the plants are thinned out to 250–360 mm, and an application made of 50–75 kg of nitrogen. Kale is frequently left unthinned. For transplanting, swede seed is sown in seed-beds, 1·5–2·5 kg producing enough plants for a ha. Transplanting takes place from November to early March, according to weather and labour conditions. Hand labour or a mechanical transplanter can be used, the stecklings being set out 250–450 mm apart. Tractor and hand hoeing are done as required, and a certain amount of earthing up may be carried out to support the plants.

Seed crops may be harvested by combine or binder. Careful handling at all stages is necessary, since the ripe pods shatter very easily. The seed must be carefully dried.

The average yield per ha of swede seed is 1 tonne, of turnips 1·5 tonnes, of kale 0·8 tonne, of rape 1·5 tonnes.

Rape for Oil Production. Various types of rape are grown increasingly in this country to produce seed from which oil is extracted, for a market that has expanded considerably since 1972. In 1976 about 40,000 ha were grown in Britain. Depending upon the variety, rape is sown in spring or late summer in rows 180–360 mm apart. The crop requires adequate lime and 150 kg of nitrogen in the seed-bed. It is harvested in July/August either from windrows or combined direct. The yield of seed is about 2–2·5 tonnes/ha and it contains about 45 per cent oil.

Chapter XI

THE CONSTITUENTS, ESTABLISHMENT AND MANAGEMENT OF GRASSLAND

IN Britain the flora of the cultivated grasslands, namely, meadows and pastures, is composed mainly of grasses and clovers. Although about one hundred species of these are commonly found, less than twenty can be considered of much agricultural importance. These for the most part are capable of growing under a wide range of both soil and climatic conditions, though few of them are found in much abundance outside the limits of enclosed land.

FACTORS INFLUENCING THE FLORA OF GRASSLAND

Soil. Some plants show a sufficiently marked preference for the conditions obtaining on, for instance, chalky or clay soils to make them the dominant species of the associations occurring in such situations, but this does not imply that they will not be found on gravelly or loamy soils. All soils, indeed, tend to carry their more or less distinctive grass flora when the fields remain down for a number of years. For instance, a very distinctive flora occurs on soils overlying the Chalk on which the smaller fescues and crested dogstail thrive ; and again on the sour soils of the Coal Measures and Millstone Grit formations where bent, Yorkshire fog, mat grass and wavy hair grass form the bulk of the vegetation. These natural associations have, generally speaking, a very low productive capacity, and the essence of good grassland husbandry is to cultivate the more productive species. The soils which are especially favourable for the development of good grassland are the well-drained loams, preferably those on the heavy side. Good second-class grassland occurs on the clay soils of the Midlands and Eastern Counties, whilst sandy and gravelly soils are characterised by producing a thick, low-yielding type of turf. On chalk and limestone soils the indigenous grassland is best fitted for grazing, the herbage being short and sweet and well liked by stock.

Management. Apart from the natural tendency to form a certain type of sward according to soil and climate, the flora depends upon the use made of grassland. If a field is mown year by year and only the aftermath grazed, the flora tends to differ from that of a field which is systematically grazed. The taller grasses which form the bulk of the hay crop shade the smaller species and the white clover sufficiently to check their growth, finally suppressing them. The herbage thus tends to consist of the stronger growing species such as perennial ryegrass and cocksfoot. Where grazing is practised these vigorous growers are kept under control and other grasses, together with white clover, have a better chance of developing. Because of this a common method of management is to mow and graze in alternate years in order to maintain some balance between the species. In many parts, however, meadows and pastures are kept separately and treated in the same manner year after year. The modern technique of grassland management stipulates either a rotational system of management to include both hay and grazing, or the provision of leys composed of mixtures of grasses and clovers specially compounded for these two methods of utilisation.

Climate. The productivity of grassland, whether meadow or pasture, depends more on climate than any other single factor. As a result of the dependence of the grasses on an abundant rainfall, the grasslands of Britain are concentrated in the west and north, being better suited to and more productive under the wetter climatic conditions prevailing in these parts compared with the drier eastern and southern areas. Deficiencies in the rainfall may, however, be offset by higher water tables which provide a continuously moist root zone. The plants then grow steadily throughout the spring, summer and autumn, and provide the grazier with an unusual uniformity of production. An example of this is to be found in the so-called "marshes" of East Anglia where rich grazings occur that continue growth even during a dry summer on account of readily available soil water.

Such conditions can also be secured by irrigation. This, in essence, was the system used in the water meadows of Wiltshire and Dorset, although the practice has by now fallen into disuse. Irrigation by artificial means is, however, used on an ever-increasing scale coupled with systems for combining irrigation with that of the disposal in a semi-liquid form of dung and urine from stock housing. Where these methods of providing pasture growth are applied the yield of herbage can be increased some 50–100 per cent during the growing season. The application of the water must however be done systematically, for water in excess of soil-holding capacity is detrimental to the growth of the better grasses. Under waterlogged conditions tussock grass, rushes and sedges are the dominant species in the sward.

The main bulk of the root system of grasses does not range deeply in the soil and the depth to which it penetrates is determined largely by the soil's physical condition. On badly aerated clay soils most of the roots will be found in the top 50 mm, whilst in loams and the lighter types of soil they may extend their range to a depth of 150 mm. This relatively shallow rooting habit is characteristic of all of the grass and clover species of agricultural importance, and none of them is capable of making much use of the moisture present in the deeper layers of the soil. Drought must always present the grassland farmer with serious feeding problems, but its effect can be offset to some degree by adopting husbandry techniques in terms of suitable seeds mixtures, fertiliser usage and, where practicable, irrigation.

Season. The rate of growth of herbage plants in British grasslands is dependent on the seasons. At the beginning of the farming year, in October, it slows down and, except in mild seasons, it practically ceases with the coming of winter. In severe weather the foliage may die off almost completely. There are differences, though, in the hardiness of the various grasses, but practically the only permanent grasses resistant to winter burning are perennial ryegrass, crested dogstail and rough-stalked meadow grass. Even with these, however, care must be taken not to allow excessive leaf growth to go through into the winter as this renders the plant more susceptible to frost damage. After lying dormant through the winter, growth begins again in the spring, the various grasses starting into growth at different periods. With the coming of warmer weather, and especially of an ample rainfall, the rate of growth accelerates, reaching a peak from mid-May to mid-June, according to season and district. At this stage most of the species, if ungrazed, begin to flower. Growth then slows down and the subsequent production of aftermath depends largely upon climatic conditions and the level of fertility. The variability of grassland from season to season constitutes the grazier's chief difficulty, for he must estimate at the beginning of the season the head of livestock that the pastures can carry, and so control the grazing that the plants are eaten when at their maximum nutritive value, which is when the leaves are some 100–150 mm in length and when sunlight and water are freely available to promote sugar formation and mineral uptake.

THE CHARACTERISTICS OF GRASSES

Two of the most important characteristics of grasses are first their fibrous, adventitious roots, and second their habit of tillering, or producing new shoots at about ground-level.

Tillering. The production of numerous fibrous roots means that grasses have greater powers of recovery after injury than have plants with a main tap root. If a tap root is severed or broken it may mean death to the plant, but grasses are often stimulated by root injuries to produce yet more fibrous roots. That is one of the reasons why severe harrowing of old pastures may have beneficial results.

The tillering capacity of grasses is obviously of great economic importance. A grass with a single shoot can hardly be as productive as a grass which forms numerous side shoots ; a grass which tillers profusely will cover the ground much more rapidly than a grass which tillers less freely. This is one of the reasons why some of the new varieties of perennial ryegrass such as Aberystwyth S.23 or Melle Pasture will form closer swards at a low rate of seeding than will pastures seeded with varieties having a poorer tiller production capacity. Some grasses which tiller freely tend to form tufts, for example cocksfoot, but if the tillers are kept grazed or mown this tuftiness is masked and the herbage in the sward remains even. In a grass such as tussock or tufted hair grass, however, the tillers are so unpalatable to grazing stock that they are neglected and huge tussocks standing well above ground may be developed. Some grasses, like smooth-stalked meadow grass, do not tiller in the accepted sense of the word, but spread by underground stems or rhizomes ; the habit of growth is matted rather than tufted. Any bud-bearing portion of a rhizome which is broken off the plant can start another independent plant. It is this characteristic which makes couch grass such a pest in arable land (Fig. 47).

General Structure. The general structure of a grass is shown in Fig. 46. The aerial parts of a grass consist of the leaves, stem and inflorescence or ear. The leaf is made up of a sheath, which clasps round the stem, and a blade. At the junction of a sheath and blade are two structures. First the ligule, a delicate, colourless membrane pressed close to the stem, and second the auricles, or claws. The claws have already been mentioned in describing the shoot structure of wheat (see p. 219), and they are of considerable value in recognising

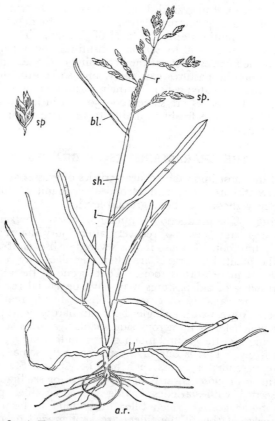

FIG. 46.—A TYPICAL GRASS—ANNUAL MEADOW GRASS (*Poa annua*).

a.r., adventitious root system. *l*, ligule. *sh.*, leaf sheath. *bl.*, leaf blade. *r.* rachis. *sp.*, spikelet.

the grasses. The shape of the blade and its appearance in cross-section are also useful aids to identification (Fig. 48).

The inflorescence, or ear, is of two kinds. Usually it is a branching ear, or panicle, with the individual spikelets on short or long branches, as in meadow fescue or rough-stalked meadow grass. Sometimes the spikelets are set direct upon the rachis or stem forming a spike as in perennial ryegrass or timothy. Each spikelet has two chaffy coverings called glumes within which is a short stalk, the rachilla, upon which are situated the florets. Each floret has its own chaffy coverings, the

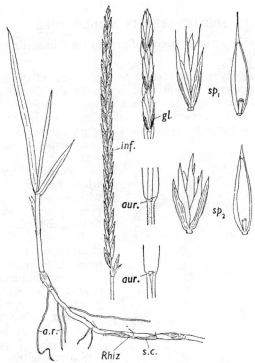

FIG. 47.—COUCH GRASS (*Agropyron repens*).

Rhi-, rhizome.   *a.r.*, adventitious roots.   *sc.*, scale leaves.   *inf.*, ear.   *gl.*, glumes.   *aur.*, aurieles.
sp₁, *sp₂*, spikelet and seed of the awned and awnless forms.

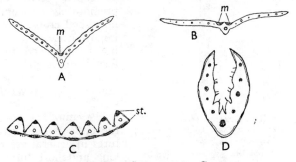

FIG. 48.—LEAF STRUCTURE IN GRASSES.

Transverse sections of A, Cocksfoot; B, Smooth-stalked Meadow Grass; C, Tussock Grass D, Sheep's Fescue.

m, motor cells.   *st.*, strengthening tissue.

paleae, or pales, enclosing the ovary and stamens. The so-called seed of grasses consists of the ripe ovary (grain or caryopsis) enclosed within the pales, and sometimes the glumes as well. This is a very fortunate arrangement because the grains of the grasses are all very similar, but the chaffy portions differ widely in appearance, and so enable us to recognise the commercial seed of grasses fairly easily.

SPECIES USED IN AGRICULTURE

The more important species of grasses used in seeds mixtures are described below.

Perennial Ryegrass (*Lolium perenne*) is one of the most widely distributed species in Britain and is usually the dominant grass in first-class pastures (Fig. 49). It is a lasting perennial on good soils, but on light land, especially in dry districts, it tends to die out after a few years. Perennial ryegrass has a long growing season, is palatable and responds to manurial treatment. It stands cutting and grazing very well, but when grown in association with other grasses it has a suppressive action on them and on the companion clovers.

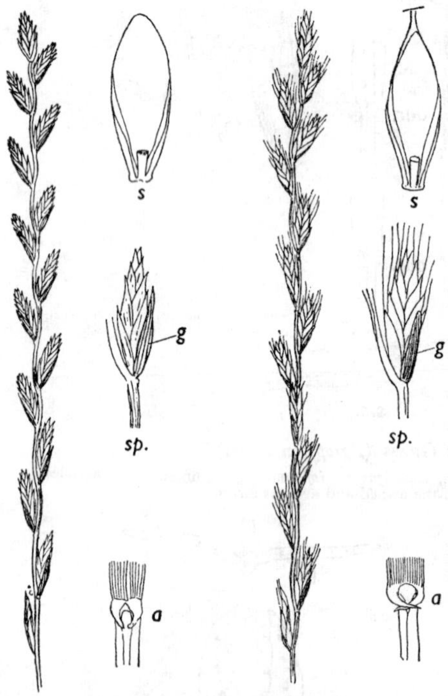

The seed is relatively cheap, easy to sow and its quick germination gives an early establishment of the sward. The germination is generally 85–95 per cent, 80 per cent being the lowest legal limit. A high hectolitre weight of 35 kg is an indication of the quality of the sample. Lower weight volume ratios mean less well-developed seeds harvested prematurely or from undernourished crops. Too much importance, however, must not be attached to a high hectolitre weight. It is no index of the value of the plant which grows from the seed.

Perennial ryegrass forms the basis of most seeds mixtures in general use for both long- and short-duration leys.

FIG. 49.—THE RYE GRASSES (*Lolium spp.*) :
A, PERENNIAL RYEGRASS (*Lolium perenne*) ;
B, ITALIAN RYEGRASS (*L. multiflorum*)

s, seed. *sp*, spikelet. *g*, glume. *a*, auricles.

The young plants grow rapidly, and under favourable conditions produce a good yield of herbage in their first year, but reach their maximum productivity by the second year. Most of the seed is produced in the UK with imports from the Continent, notably Denmark, and from New Zealand. The number of varieties has greatly increased in recent years following the enactment of Plant Breeders Rights. Broadly the varieties can be grouped according to their mean dates of ear emergence following the pattern of the early introduction of the " S " varieties by the Welsh

Plant Breeding Station. The early flowering group is typified by the variety S.24 whose characteristics are those of early seasonal production for grazing or silage or early cuts of hay, but which are not as high tillering and persistent as the medium late heading group typified by S.101. These serve the same purposes with a later seasonal pattern of growth. The late heading varieties typified by S.23 are the most durable as grazing swards, being able to withstand hard seasonal grazing by virtue of their very high tillering habit on a wide variety of soil and climatic situations. Within each group there are tetraploid varieties, obtained by a doubling of the chromosomes. These varieties are more palatable to stock, form less dense swards due to their lower tillering habit, and have higher water contents which with their thicker cell wall structure render them more difficult to conserve.

The full range of varieties appears in the National List of the *Plant Varieties and Seeds Gazette* published by the Ministry of Agriculture, Fisheries and Food, and by the Classified List of Herbage Varieties produced by the National Institute of Agriculture Botany who also produce an Annual Recommended List.

Italian Ryegrass (*Lolium multiflorum* var. *italicum*) is a biennial and on moist, fertile soils, especially those irrigated with sewage, it is capable of producing a greater bulk of herbage than any other species of grass which can be cultivated in this country. The yields under the more ordinary conditions of farming are also high, and in addition the grass is remarkably winter-green and makes good growth early in the spring (Fig. 49B). It is this characteristic that is now being increasingly exploited for the provision of " early bite " in the spring, particularly in the feeding of dairy cows. It is a first-class, short-duration grazing plant, which recovers rapidly after defoliation.

It is readily established from seed, and the plants rapidly reach their maximum productivity ; for this reason Italian ryegrass must be used sparingly in mixtures which include the slower-growing grasses and clovers because its vigorous growth tends to stifle them. Under these conditions the seed rate should seldom exceed 3·3–4·5 kg/ha. Reliable seed with a germination capacity of 85–95 per cent can always be obtained.

Westerwolds ryegrass (*L. multiflorum* var. *Westerwoldicum*) may be described as an annual form of Italian ryegrass. It establishes rapidly from seed, runs to head quickly and is stemmy in growth. Improved varieties are now commercially available which retain the rapid growth characteristics and annual form, but which are a little more leafy and persistent than the unimproved varieties. Westerwolds ryegrass can be useful in providing a hay crop in the year of seeding under conditions where for some reason or other a seeds mixture sown for this purpose the previous year has failed to become established.

The range of varieties available has been greatly extended, and now includes varieties of longer duration lasting some two to three years. In the drier areas these leys form the basis of grass silage production enabling several cuts to be taken during the season.

Hybrid Italian and perennial ryegrass varieties are also available, both as tetraploids and diploids. These varieties are intermediary in character, generally closer to that of the Italian parent.

Cocksfoot (*Dactylis glomerata*) is a grass capable of producing a good bulk of fodder, although possibly not of the highest feeding quality. It is widely distributed and of particular value on light, sandy soils in regions of low rainfall where it thrives better than other species. Its growth begins fairly early in the spring, depending upon the variety and the management in the previous autumn. The young shoots in the spring are very palatable and care must be taken not to overgraze at this time. Understocking when the plant is growing rapidly must, however, be avoided, for the foliage tends to become harsh and unpalatable and the plants, especially when widely spaced, grow into rough, compact tufts which may unprofitably occupy a large proportion of the grazing surface. Some varieties of cocksfoot " burn " badly in the winter and these should, wherever practicable, be grazed down cleanly in the autumn. Other varieties not quite so susceptible to "burn" are used for winter grass production on account of the capacity of cocksfoot to make good growth in the autumn. On thin, light land where ryegrass tends to die out after the second year, cocksfoot should be included in the seeds mixture, for it is under such conditions that its greatest value will be found.

There are some 22 varieties on the National List. Whilst much of the seed is produced in the UK, imported varieties come from Scandinavia, chiefly Denmark, and from New Zealand. There are some 3–4 weeks between the heading dates of the earliest and latest heading varieties. Generally the latest heading varieties are the most persistent, and the earliest heading varieties least persistent, the intermediate varieties being variable in their persistence rating.

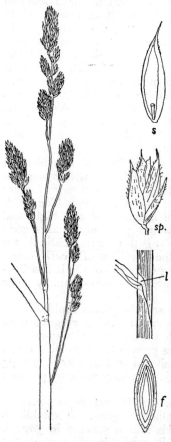

FIG. 50.—COCKSFOOT (*Dactylis glomerata*).

s, seed. *sp.*, spikelet. *l*, ligule. *f*, section of the flattened shoot.

The extremes within the total range are typified by the Aberystwyth S.345, which is the earliest heading variety used for its earliness as grazing and hay production, but which is in short supply due to its low seed yield, and by the Aberystwyth S.143 as an extreme pasture type used in reseeding the poorer uplands for the longest possible duration where it is able to withstand the hard seasonal grazing by sheep. The use of cocksfoot has greatly diminished in recent years due to the more widespread use of perennial ryegrass leys with higher levels of nitrogen fertiliser use.

Timothy (*Phleum pratense*) thrives best on deep fertile loams and clays. When sown on light soils in dry districts it may fail to establish or, if established, fails to persist. It reaches its maximum productivity by the first harvest year ; it is late in flowering and its characteristic broad, bluish-green leaves are produced rather late in the spring so that the palatable growth made at this time helps to supplement pasture growth as this declines in other grasses. Autumn growth does

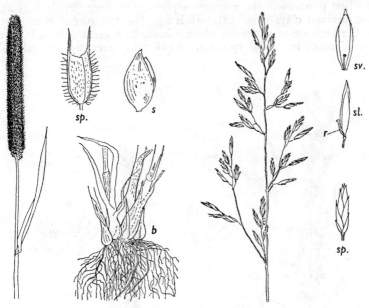

FIG. 51.—TIMOTHY (*Phleum pratense*).

s, seed. *sp*, spikelet. *b*, swollen base of stems.

FIG. 52.—MEADOW FESCUE (*Festuca pratensis*).

sv., seed (ventral view). *sl.*, seed (lateral view). *r*, rachilla. *sp.*, spikelet.

not produce any bulk of herbage, but the plant remains green into the winter (Fig. 51). As a hay plant timothy excels, and well-made timothy hay always commands a high price and a ready market, for it is especially valuable for horses. Under suitable soil and climatic conditions this grass may be sown alone to form a timothy meadow or, as is more general at the present time, it is sown with meadow fescue as these two grasses grow extremely well together. Timothy is one of the most palatable grasses and care must be taken to avoid it being overgrazed on this account.

Ample supplies of seed are available, coming mainly from Canada, Europe, New Zealand and as home grown. There is a wide range of varieties amounting to some 32 in number on the National List. Amongst the earliest heading is the Aberystwyth S.352 and selections from the Scots land race type which are more suited to hay production and which are more persistent than similar varieties derived from more extreme continental climates. These have a shorter season of growth. Those best suited for general purposes, both grazing and conservation, are of the S.48 type having a long season of growth and high persistency. At the extreme grazing range, though earlier than S.48 in its heading, is the sub-species *Phleum bertilonii* of which S.50 is an example. This has a limited market for the reseeding of poorer upland soils and for use under close mowing in top fruit orchards.

Meadow Fescue (*Festuca pratensis*) is often found in some of the best pastures in the country, especially on deep, fertile soils and well-drained clays (Fig. 52). It is an almost universal constituent of the herbage of water meadows where, to a great extent, it takes the place of perennial ryegrass. Until fairly recently the valuable

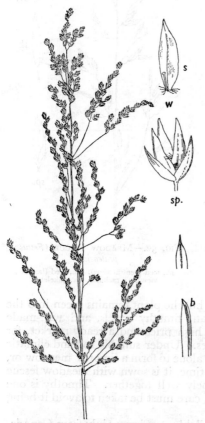

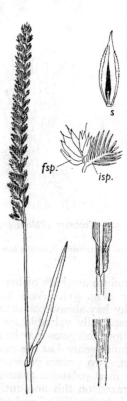

FIG. 53.—ROUGH-STALKED MEADOW
GRASS (*Poa trivialis*).

s, seed. *w*, web of hairs. *sp*, spikelet.
b, " boat-shaped " tip of leaf.

FIG. 54.—CRESTED DOGSTAIL
(*Cynosurus cristatus*).

s, seed. *fsp*, fertile spikelet. *isp*,
infertile spikelet. *l*, leaf.

potentialities of this grass were not appreciated, because it was usually sown in mixtures containing more aggressive grasses with which it was unable to compete. In consequence it was looked upon as a grass not readily established from seed ; sown by itself, however, or in company with a less aggressive grass such as timothy, it has been shown to be one of our most valuable agricultural grasses. It is extremely palatable and grows well in company with white clover as well as timothy. It has a mid-season habit of growth, making only a little autumn growth, and this in turn tends to " burn " if excess of leaf is left on the plant during the winter.

Commercial seed supplies are mainly from Denmark and Holland though a high proportion of that used in the UK is home produced. The amount used in the UK has greatly diminished in recent years, and amounts only to a small fraction of UK grass seed usage.

There are some 21 varieties on the National List. They are divided into early and late groups, typified by the Aberystwyth S.215 variety as early, and the Aberystwyth S.53 as a late variety.

Tall Fescue (*Festuca arundinacea*). In Britain this grass can be found growing under natural conditions ranging from moist soils in low-lying meadows to the drier soils of the chalk and sands. Its chief characteristic is its capacity to grow in the late autumn, winter and early spring. Until recently, however, this feature has not been exploited for grazing purposes on account of the coarse nature of the plant and its consequent lack of palatability. Improvements are being made in this respect, and varieties are being bred which will lead to this grass being more widely used, particularly for winter grazing purposes and for " early bite " in the spring. Its slow establishment from seed can be largely overcome by growing it in drills 250–530 mm apart, and under these conditions it will produce a good bulk of leafy fodder. Its main usage is for the grass drying industry in the lower rainfall areas.

Commercial supplies of seed are obtained from Europe and the USA with a small proportion home grown. There are 4 varieties on the National List, the best known of which is the Aberystwyth S.170 variety.

Rough-Stalked Meadow Grass (*Poa trivialis*) is the most valuable of all the " bottom " or close-growing grasses, its short runners and fine, close growth being ideal for the formation of a closely knit sward (Fig. 53). The foliage is fairly abundant, relished by all kinds of stock and produced over a long season. The grass is widely distributed, and though generally to be found in the greatest abundance under somewhat moist conditions of rich loams or water meadows, it does fairly well on poor soils provided that the rainfall is adequate. On light soils it is apt to burn out in dry weather and disappear completely. In such conditions its place should be taken by the rhizomatous Smooth-stalked Meadow Grass (*Poa pratensis*).

Both meadow grasses are essentially pasture plants and better suited for long leys than for temporary leys, since they grow comparatively slowly and take some three years to reach the stage of maximum development. Very little use is now made of the species in reseeding, but where used the seed rate should not exceed 1–2 kg/ha. Seed supplies are usually of Dutch, Danish or USA origin.

Crested Dogstail (*Cynosurus cristatus*) is a widely distributed grass thriving not only on fertile loams but also on heavy clays and on the light soils overlying the chalk (Fig. 54). Its yield of herbage is small compared with that of perennial ryegrass and because the seed heads are not consumed by stock it is a prolific self-seeder and may become the dominant species in a sward. For this reason its chief value is on the poorer classes of soil where the more productive species fail to attain their full output, and where its winter-green character can be put to good use by sheep.

THE CHARACTERISTICS OF CLOVERS

For farm purposes no sward can be considered complete unless it contains a proportion of clovers. There are two reasons for this, the more important one being the power that clovers possess of removing

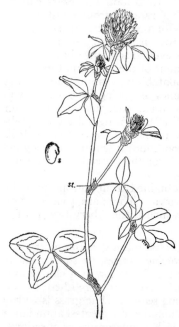

FIG. 55.—RED CLOVER (*Trifolium pratense*). *S. seed. st, stipule.*

nitrogen from the air via the root nodule bacteria (see p. 403). The results of this behaviour are first that the foliage of clovers is relatively richer in protein than that of grasses, and second that a crop of clovers leaves a residue of nitrogen in the soil in the same way as a crop of beans and peas enriches the soil. Another reason for including clovers in a ley is the relative richness of some members in mineral salts. White clover contains proportionately more calcium and magnesium, for example, than most grasses. This may be an important factor in preventing symptoms of mineral deficiency in the livestock grazing or feeding on such herbage.

Clovers are dicotyledonous plants whose seedlings are provided with tap roots and whose cotyledons, or seed leaves, are brought above ground. The first true leaf is always a simple one, but later leaves are trifoliate. The root nodules begin to develop at about the same time as the first leaves, and the nodules themselves vary in shape and position from species to species.

One useful structure for recognising clovers is the stipule, a small modified leaf found in pairs at the base of the leaf stalks. In alsike clover the tip of the stipule tapers gradually; in red clover it terminates abruptly in a sharp point (Figs. 55, 57). The growth habit of clovers is also variable. Red and alsike clovers are erect growing, with basal shoots or tillers. White clover spreads by stolons, a habit which is of great importance in turf formation.

Of the true clovers only three are used to any extent in the formation of leys, namely red, white and alsike clovers.

Red Clover (*Trifolium pratense*). The history of the origin of red clover, as of so many agricultural crops, is obscure, but there are good reasons for thinking that it is not a native of the British Isles. It was the subject of considerable notice in the agricultural literature of the 16th century, especially on the Continent, and there is a strong probability that it was introduced into England from there by Sir Richard Weston, or reached here as a result of his strong advocacy. Its value for stock feeding, and as a preparation for wheat, soon became recognised, and it has been used mainly for this purpose from then on. At the present time its use in this respect is declining, particularly with the advent of fertilisers containing nitrogen. It is almost certain that the original stocks contained a large number of varieties, and the crop as now grown is remarkable for a large number of localised types. These differ not so much botanically as in features such as persistency, resistance to disease, time and duration of growth and the bulk of the produce (Fig. 55).

Three types of red clover are of interest agriculturally : Early-flowering red clover, also called Broad red clover or Double-cut red clover ; Late-flowering or Single-cut red clover ; and the Extra-late flowering red clover.

Broad red clover is about a fortnight earlier in flowering than late-flowering red, and in its first harvest year—the year after seeding —it will give a cut of hay or silage followed by an aftermath for cutting or ploughing-in or for seed purposes. It makes little recovery growth after this and is therefore mainly used as the chief constituent of one-year leys, or as a part contribution in the first year to leys of longer duration. Commercial seed stocks are mainly of local varieties such as Essex, Drewitts and Berrys ; some may be imported from Canada or New Zealand ; somewhat later flowering and persistent Broad red types are Aberystwyth S.151, Dorset Marl and the new persistent tetraploids.

The late-flowering red clovers tiller more abundantly and as a type are more persistent and hardier than Broad red. In seasonal growth they give a good cut of hay but little aftermath, and they remain active in growth for two to three years. They are less susceptible to clover sickness (*Sclerotinia trifoliorum*) than the early-flowering types and are generally sown as part of a general seeds mixture of long duration. Local varieties are available, but considerable quantities of Canadian seed, particularly of a variety called Altaswede, have been brought in to Britain in recent years.

The extra-late-flowering red clovers are late in spring growth, have a high tillering capacity and are persistent. They are, therefore, used under conditions where the fields will be grazed fairly intensively, particularly by sheep in the winter and early spring, followed by a hay cut. This form of management may take place over a number of years. Seed is available as New Zealand Grasslands Turoa, and there are very limited supplies of home origin namely Montgomery,

Aberystwyth S.123 and Cornish Marl. Some 29 diploid varieties of red clovers are listed, and 14 tetraploids.

White Clover (*Trifolium repens*). It is probable that white clover, like red clover, found its way into England from the Low Countries. The plant is characterised by its creeping habit ; the runners or stolons root freely, and when sown with grasses in a seeds mixture they knit the grasses together to form a close sward. It is essentially a grazing plant (Fig. 56).

There are two distinct types of white clover at the extremes of range, the large-leaved white and the smaller leaved wild white. Ten varieties are listed as large-leaved of which Kersey and Blanca are examples. Four varieties are listed as small-leaved, the best known

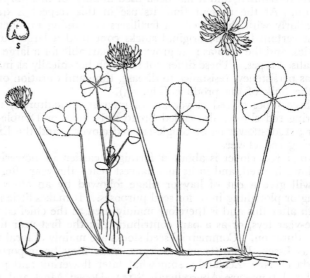

FIG. 56.—WHITE CLOVER (*Trifolium repens*).
s, seed. r, runner or stolon.

being Aberystwyth S.184 and Kent. Between this range are several varieties of various leaf size, examples of which are the well-known Aberystwyth S.100 and the New Zealand Grasslands Huia.

Plant size is related to leaf size, the larger leaved varieties are generally least persistent, whilst the wild white types are truly perennial, but much depends upon their susceptibility to clover rot disease (*Sclerotinia trifoliorum*). In this respect those varieties of continental origin in the large-leaved varieties generally possess some higher resistance. In the absence of the disease the large-leaved white clovers will last from two to five years. They are especially useful with single varieties of grass for high-quality cuts for silage where little nitrogen is used. As grazing plants their excessive growth is liable to cause bloat. The small-leaved wild white type clovers are more suited to leys of the longest duration in upland situations and where required for hard persistent grazing.

Although white clover has a strong tap root it also roots freely at the nodes of the prostrate runners. The plant responds very markedly to phosphatic fertilisers, particularly basic slag. As a turf former it has no equal. Some of the nitrogen which it accumulates from the air in its root nodules is released in the soil and stimulates the surrounding grasses.

Alsike Clover (*Trifolium hybridum*) is similar in habit of growth to late-flowering red clover although not so persistent (Fig. 57). It has the advantage of being a good deal hardier and less susceptible to soil acidity and clover sickness, and hence is generally included in mixtures for sowing on land where red clover is likely to fail. It is an excellent companion plant for timothy in hay mixtures. The supplies of seed are almost wholly Canadian.

Trefoil (*Medicago lupulina*) is an annual plant related to the true clovers, found in abundance on chalk and other soils containing much lime. It does not contribute greatly to the bulk of material obtained from seeds either for hay or grazing, but it is useful for providing an early bite for sheep on chalk land. It is a ready self-seeder and consequently almost functions as a perennial unless grazed before it sets seed. If sown too generously in mixtures it may have a smothering influence on the other species. Supplies of seed are chiefly English.

Miscellaneous Herbs. In association with the grasses and clovers, one frequently finds in a pasture or meadow, especially when these have been down for a number of years, miscellaneous plants like ribgrass or plantain, chicory, yarrow and burnet. The specific value of these plants is not fully known, but farmers in many grassland districts are firm believers in them, even to the point of including them in seeds mixtures. It is known that they have a high mineral content and are well relished by stock which will seek them out in a pasture. It is also believed, though not proved, that they counteract scouring in stock grazing a lush pasture. They are all deep rooting and capable of withstanding drought, and on thin upland soils ribgrass may be the only plant contributing to the keep of the stock during a prolonged period of dry weather. There are grounds for suggesting

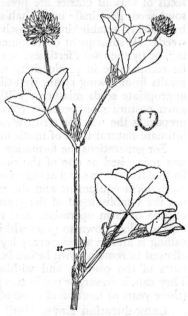

FIG. 57.—ALSIKE CLOVER (*Trifolium hybridum*).

s, seed. *st.*, stipule.

that such disorders as hypomagnesaemia in grazing stock may be reduced in incidence by the inclusion of herbs in the diet by way of pasturing or in the silage and hay feed. The chief drawback to the use of these plants is the amount of space they occupy in relation to the amount of keep they provide. For this reason they may be sown alone in a narrow " herb strip " across the field where the stock can help themselves, and where the herbs can have freedom for development yet are restricted to a reasonable area of land.

THE ESTABLISHMENT OF GRASSLAND

Arable land left to itself will in time grass itself down, often with a considerable degree of success which at first would appear to indicate that no special care is needed in the establishment of grassland. That this is far from the truth was amply demonstrated, prior to the agricultural revolution which took place with the outbreak of the second world war, by the derelict grassland resulting from the tumbledown methods of laying arable land down to grass following the 1914–18 war. When arable land tumbles down to grass the immediate flora is derived from the perennial weeds present on it and from the seeds present in the soil. Probably the grasses would consist principally of couch, soft brome and creeping bent. The next contribution to the flora is provided by wind-borne seeds, amongst which the minute seeds of various grasses are present in abundance. These naturally sown seeds germinate in the half shade of the existing vegetation, and by quickly establishing themselves tend to crowd out the annual weeds. If subsequent development is controlled by mowing or grazing, or by the use of fertilisers, a turf will result. Such turf can never be comparable in productivity with the well-established turf which results from sowing clean, well-tilled and well-fertilised land with an appropriate seeds mixture. Without a measure of control the self-sown pasture soon becomes colonised by brambles and thorn, thus preparing the way for scrub woodland and forest trees which is the ultimate natural flora of much of our agricultural land.

For generations the formation of permanent grassland from arable was recognised as one of the most difficult operations on the farm. Indeed it is safe to say that only since 1939, with the advent of improved mechanical equipment and the use of soil analysis on a wide scale, has the establishment of the grass sward become part of the normal routine of farm operations and rotations.

Land laid down to grass within the particular farm rotation prevailing is known as temporary ley. The length of time such leys are allowed to remain down before being ploughed out varies in different parts of the country and within the system of farming practised. They can, however, be put in two main groups, those of long duration (three years or more) and those of short duration (one to two years).

Long-duration Leys. Until comparatively recently the seeds mixtures sown for the purpose of establishing leys of a three-year duration or longer were made up of many different species of grasses and clovers which in total amounted to a seeding rate of some 45–55 kg/ha. It was not uncommon for the mixtures used to contain as many as, or more than, twenty different species ; the assumption being made that sward formation would be assured by the establishment of some, at least, of these grasses and clovers. As the sward developed so some of the species present would grow when drought conditions occurred ; some would be better when conditions were more moist ; some would be early in spring growth while others would develop in midseason or in the autumn. Since the time that this type of seeds mixture was in vogue a great many grassland management studies have been carried out, together with the breeding of herbage plants suitable for particular growth environmental conditions. Furthermore, the factors influencing competition between species and varieties are better understood and as a result the tendency has been for seeds mixtures to become much less complex in their make-up and for far less seed per ha to be sown.

Long-duration leys can therefore be divided into two main types ; those in which the main grass will be perennial ryegrass and those without ryegrass. The use of non-ryegrass mixtures is nowadays very limited. Their use is confined to special purposes, such as the grass drying industry, and areas of special difficulty such as the lighter sands and gravels in low rainfall areas, or as special-purpose leys as alternative grazing during midseason, where the use of nitrogen is limited. The non-ryegrass mixtures are generally of meadow fescue with timothy and white clover, and cocksfoot may be added on the lightest of soils. Cocksfoot with ryegrass is generally confined to stock-rearing farms where there is more limited use of nitrogen. Timothy is a common constituent of ryegrass mixtures.

In addition to the grasses there is the possibility of including one of the varieties of red clover, usually late-flowering red clover, particularly if the ley is to be taken for hay or silage. Normally if the ley is to be grazed the red clovers will be omitted, and only white clovers put in to the mixture.

Long-duration leys often use lucerne as the main herbage plant, and in some instances grasses are included in these leys. Examples of these are given on page 294.

It is in the lower rainfall areas of the east and south that lucerne leys are mainly grown. Approximately half the total of 16,000 hectares grown in Britain is to be found in East Anglia.

Typical examples of seeds mixtures that make use of bred varieties available on the commercial market, and of three to five years' duration are given in the table at the top of page 320.

The varieties that may be used in the compounding of seeds mixtures, such as those given on page 320, may be other than those bred by British plant breeders. It is important, however, that when compounding any mixture the main growth characteristics of the varieties used should be known.

In many cases simplification of seeds mixtures and the reduction in the amount of seed sown have gone even further than in the examples noted above, so that ultimately a seeds mixture may consist solely of one grass variety and one variety of white clover, with a total seed rate of around 4·5 kg/ha. In this way it is possible to make up seeds mixtures of different varieties of the same species or of different species and to sow them on separate areas. The potential growth character peculiar to any variety or species can then be fully expressed

PERENNIAL RYEGRASS-DOMINANT, LONG-DURATION LEY

	kg/ha
Perennial ryegrass S.24	9
Perennial ryegrass S.23	4·5
Cocksfoot S.37 ⎤	
or ⎬	6·7
Timothy S.48 ⎦	
Late-flowering red clover	2·2–3·3
White clover, S.100 or Kersey	1·7–2·2

MEADOW FESCUE-DOMINANT, LONG-DURATION LEY

	kg/ha
Meadow fescue S.215	11
Meadow fescue S.53	2·2
Timothy S.48	6·7
Cocksfoot S.37	2·2–3·3
Late-flowering red clover	1·7–2·2
White clover S.100 or Kersey	1·7–2·2

to provide herbage for pasturage or conservation at a more or less specified time during the growing season under the soil and climatic conditions prevailing.

These are called ultra-simple mixtures : examples which have given good results are :

		kg/ha
(a)	Perennial ryegrass S.23	6·7
	White clover S.100 or similar type	4·5
(b)	Cocksfoot S.37	6·7
	White clover S.100 or similar type	4·5

The ultra-simple seeds mixture, although successful in the hands of a farmer familiar with sowing and managing such mixtures, can be disappointing where attention cannot be paid to details. In the latter circumstance the general type of seeds mixture usually gives more satisfactory results. With the ultra-simple seeds mixture it is essential to use a high tillering variety of the grass selected. Light as these seedlings appear to be they have given good ground cover. It is important, however, to ensure maximum plant establishment by sowing them only on land adequately supplied with lime and phosphate. When sown under a cereal nurse crop they should be sown immediately after the corn crop, thus minimising the amount of competition between the two for light, air, moisture and plant food.

Short-duration Leys. Whilst there is a measure of uniformity in the seeds mixtures prescribed for long-duration leys—and in practice there is virtually no difference between a mixture for a three- to five-year ley and one used for sowing down to permanent grass—for leys of shorter duration there is little uniformity. Probably because the purpose of these short leys varies so greatly in different districts, and even between one farm and the next, standardisation of seeds mixtures is impracticable. In the case of the one-year ley the purpose varies from the primary one of replenishing soil fertility under a four- or five-course system of farming to that of providing the maximum amount of fodder or grazing. For instance, the requirement may be for a heavy cut of hay ; or two crops of hay ; or a cut of hay plus aftermath for ploughing in or taking a seed crop, or simply abundant pasturage for sheep or cattle. In the case of the two-year ley the requirement may be for hay in each year ; for hay in the first year and grazing in the second year or vice versa ; or for grazing in both years. In most cases the two-year ley will be a compromise between the one-year ley and the three-year ley.

In general, for one-year leys, seeds mixtures designed primarily for hay will contain early-flowering or Broad red clover only ; or a mixture of grasses, principally Italian ryegrass and/or perennial ryegrass with a few kilos of Broad red clover. The latter type of mixture may also be grazed, in which case some large-leaved white clover and trefoil may be added, or trefoil alone may be used for stubble grazing. Where soil conditions are moist, alsike may be included and where clover rot is likely to attack diploid Broad red clover, alsike may be used as a substitute.

For two-year leys either cocksfoot, timothy or meadow fescue may be introduced instead of, or as well as, the two ryegrasses, in which case the early varieties should be used. In addition 0·5–1 kg/ha of a large-leaved white clover, such as S.100 or Kersey, will be included to give both high-quality fodder and a mass of clover roots to plough down at the end of the second year.

Below are a number of examples of seeds mixtures for one- and two-year leys. These may be modified to suit specific requirements and particular conditions of soil and climate.

SEEDS MIXTURES FOR ONE–TWO YEAR LEYS IN kg/ha

	One-year leys			Two-year leys	
Italian ryegrass 22·5–28	6·7	6·7	6·7	3·3	
Perennial ryegrass S.24		4·5	4·5	9	
Cocksfoot S.37				9	
Timothy S.51					6·7
Meadow fescue S.215				9	
Alsike		3·3			
Trefoil			2·2		
Broad red clover	15·5	9	4·5	2·2	
Late-flowering red clover			4·5	2·2	6·7
White clover S.100 or Kersey			1	1	

A short-duration ley suitable for the purpose of providing " early bite " for dairy cows or ewes and lambs, or to provide successive cuts for silage, or silage and hay is one based solely on Italian ryegrass of a leafy persistent variety. This is sown at 22·5–28 kg/ha when undersown in cereal or at 33·5–39 kg/ha sown direct. Such a ley can be grazed intensively for 18–24 months provided it is well supplied with nitrogen.

SOWING SEEDS MIXTURES

The sowing of seeds mixtures used in the formation of temporary or permanent grassland requires more care and attention to details than the sowing of the seed of any other farm crop. Conditions of soil and weather which would result in a satisfactory plant of wheat or swedes may often prove unsuitable for the establishment of grasses and clovers.   In addition, because the seeds of these herbage plants are so very small, adverse factors such as a shortage of available plant food are likely to have much more serious consequences than in the case of the stronger seedlings arising from large, well-nourished seeds.

To ensure a successful " take " it is necessary to provide a clean, well-worked, compact seed-bed with a surface of 50–75 mm of finely crumbled soil.   This can be most easily obtained after a root crop or fallow.   If the sowing has to follow a cereal crop the recently shed seeds of annual weeds should, as far as possible, be forced to germinate by stirring the surface soil well before ploughing.   Then, in the spring, the weathered furrow slices should be worked down to secure a fine, deep tilth.   During the latter stages of the process a dressing of fertiliser supplying 62·5–87·5 kg/ha of phosphoric acid, 75 kg/ha of potash and 37·5 kg of nitrogen should be given.   When the seeds are undersown these fertilisers may be combine drilled along with the cereal nurse crop, in which case it may be expedient to use a phosphate/potash compound fertiliser for combine drilling and to apply the nitrogenous fertiliser as a top dressing.   In this way the latter will benefit the cereal crop as well as assist in the early establishment stages of the grass and clover seeds mixture.

As soon as the surface soil is in a fit condition the seed should be sown broadcast by hand, fiddle drill, or seeds barrow, or drilled with a suitable coulter drill.   In districts of average to high rainfall, broadcasting on the surface gives good results even on dry soils, but in the low rainfall areas it is necessary to place the seed in contact with the sub-surface moisture : this means using a coulter or disc drill and may involve placing the seed as deep as 37–50 mm below the surface, although shallower drilling is desirable.   Some drills sow at 90–100 mm spacing, but if a corn drill spacing of 180 mm is used, to secure uniform distribution and to avoid undue competition between the germinating grasses and clovers in the drill, half the quantity of seed may be sown in one direction and the rest at right angles to it.

The Nurse Crop.   Temporary leys are sown for the most part using a cereal as the " nurse " crop.   With a spring variety of wheat, barley or oats the cereal is drilled as soon as conditions are favourable in March or early April.   This crop is harrowed in and the land rolled before the grass and clover seeds mixture is sown ; it is advisable that the sowing of the small seeds should take place immediately following that of the cereal ; the small seeds in turn are harrowed in and rolled.   In some cases the cereal may be brairded before the seeds mixture is sown, but in these circumstances there is always a risk that the cereal, already growing vigorously, will smother out the small, tender grass and clover seedlings.

When autumn-sown wheat or oats is the nurse crop there is usually a strong leafage by the time the seeds mixture is to be sown.   To ensure that the small seeds will get a fair chance of establishment it is a good practice to remove the " flag " of the cereal by grazing it off with sheep or cattle, or even by mowing it.   Provided this is carried out in most seasons before the end of April, the cereal does not suffer unduly and the grazing can be a useful additional feed at a time of year when a fresh succulent feed is very scarce. The practice has the additional advantage of counteracting lodging of the cereal. To obviate as far as possible the danger of lodging and the smothering of the seeds mixture, it is important to drill the cereal at an under-average seed rate and to select a short, stiff-strawed variety.

Seeding Direct. When long leys are sown or it is intended to sow a field to permanent grass, the nurse crop is often dispensed with and the seeds mixture sown on bare ground from March to May, or July to August—these being in most districts the " safe " periods for sowing. In such conditions it is not uncommon to include in the seeds mixture about 3·3–4·5 kg/ha of Italian ryegrass or 2·2 kg/ha of rape which, germinating rapidly, provide grazing for sheep or cattle, the treading of the stock being beneficial in the establishment of the sward. It is important to ensure that neither the Italian ryegrass nor the rape is allowed to get out of hand as this will lead to suppression of the more permanent constituents of the seeds mixture.

Weed Control. It is also important, whether the leys are undersown or direct seeded, that measures be taken to control the growth of broad-leaved weeds that may appear during the early stages of establishment. As with cereals and other arable crops, the means of weed control during the early phases of ley establishment have progressed considerably in recent years. Selective herbicides have been developed which will control many broad-leaved weeds when applied at the correct stage of cereal development, when the seeds mixture is undersown, and when the grass and clover seedlings are at a particular developmental phase when the seeds are under-sown or direct sown. The selective herbicides used in this context are of the growth regulator type MCPB and 2,4-DB, or contact herbicides such as Dinoseb. The value of these lies in the fact that they may be used without harm to the clovers in the seeds mixture.

Grazing. If the season has been favourable a considerable amount of herbage will be produced in the autumn by the time the cereal crop has been harvested. Stock can then be turned in to graze it lightly if the soil is not so wet as to be injured by treading. When sowing is done without a nurse crop, grazing will be possible from 5 to 10 weeks after sowing the seed, depending upon weather conditions. The choice of stock for this grazing will be determined largely by the class of farm and the stock available ; some prefer sheep, others cattle. Cattle will graze the upright shoots of the grasses, whereas sheep prefer the side shoots : hence there is a danger that the sheep may graze too hard and injure the vitality of the plant. On the other hand, they tread more lightly. Perhaps the safest compromise is to select young cattle, but in any case care must be taken to avoid drastic grazing in the early stages whilst the sward is becoming established, and before a good sole or bottom has formed which is capable of carrying the stock.

In the case of a direct seeding without a nurse crop the establishment of the young seedlings can be assured by irrigation should facilities in terms of water and equipment be available for this purpose.

This is particularly so if drought conditions are experienced shortly after seeding.

To Hay or Graze. In the following season the decision must be made between taking a crop of hay or continuing to graze. This decision does not apply, of course, in the case of one-year hay or grazing leys where the purpose is clear. Here again diverse views are held, some averring that grazing for the first two years when establishing a sward is important, others contending that the best procedure is to take a hay crop. In the latter case it is essential to cut the crop early, when the grasses first come into flower, and graze the aftermath : with this management little difference in results between mowing and grazing is likely to be noticed. On the other hand, to leave the hay crop until late in the season to gain the maximum weight per ha is harmful to the future of the sward.

THE MANURING, MANAGEMENT AND UTILISATION OF GRASSLAND

Soil conditions, climatic conditions and the constituents of the seeds mixtures sown all play a part in the production and development of grassland. The degree to which such grassland is capable of further herbage production and its conversion into saleable animal products rests largely with the standard of subsequent manuring and with the general management of the sward and method of utilisation. The quantity and quality of herbage growth, whether converted into animal products by way of pasturing or by conservation practices, hay, silage and dried grass, will largely depend on the availability of plant nutrients, particularly nitrogen, phosphate, potash and lime. These must be present in adequate quantities and at the correct time in relation to plant requirements. It is necessary, therefore, to have some understanding of the role that these play in maintaining and increasing grassland output and of how they may be used to advantage. To this end attention must be given to the condition of the grassland receiving the fertilisers to ensure favourable conditions for growth by mechanical treatment and proper drainage, and to the conversion of the herbage produced to animal products by the grazing animal directly or indirectly through methods of herbage conservation.

Manuring of Grassland. Since 1856 the effect of different fertilisers on permanent grassland has been under investigation at Rothamsted, the grassland in this case being used for the production of hay which enables yield data to be recorded. During this time also large numbers of manurial trials on both meadows and pastures have been carried out all over the country. It is on the results of these experiments that the manuring of grassland is now based.

Nitrogen. The first effect of applying to grassland any dressing capable of acting as a plant food material is to upset the equilibrium arrived at between the species composing its flora. Even if the rate of growth only is affected, some species grow more vigorously than others, with the result that they tend to crowd out others from the beginning. Under a system of grazing this is counteracted, and in the case of meadows may be controlled by grazing in some years instead of leaving the herbage to grow to its maximum length for hay production. By the judicious blending of fertilisers allied to wise grazing management it is possible to maintain a well-balanced sward at a high level of productivity for a long period of years.

The initial result of applying nitrogenous fertilisers to grassland is a great increase in the growth of the plants. The new growth is characteristically dark green in colour and distinctly more succulent than that of plants grown without an additional supply of nitrogen. All the nitrogenous fertilisers have this stimulating action, and their value for increasing the yield of hay or producing an early bite in the spring is well known. The application of 25 kg/ha of nitrogen may be confidently expected to result in an increased yield of from 750 to 1,250 kg/ha of green herbage if the weather conditions are favourable. When the nitrogenous fertiliser is applied in mid-February to sheltered, well-drained fields, about a fortnight's earlier growth is obtained compared with fields not so treated. The value of this practice in connection with Italian ryegrass has already been noted and this is a system which is used more extensively as time goes on. Fields used for this purpose are top dressed with 50–75 kg/ha of nitrogen in the late winter. Fields put up for silage must also be top dressed with similar quantities of nitrogenous fertiliser to give both bulk and quality. Nitrogen also has an effect on the quality of the hay produced, for when applied from ten to fourteen days before the crop is cut for hay an appreciable portion of the inorganic nitrogen is taken up immediately and converted into organic nitrogen, with the result that the protein content of the hay is increased. Any nitrogen not taken up by the plant is utilised later for the production of aftermath. The continued use of nitrogenous fertilisers on pastures and hay fields will reduce the proportion of clover present, partly by the overshadowing of the clovers by the increased growth of grass stimulated, and partly by the toxic effect of the nitrogen on the clover plant itself. To reduce this suppressive effect of nitrogen, adequate levels of phosphate and lime must be maintained.

Phosphate. Numerous trials with various phosphatic fertilisers have shown that, with comparatively rare exceptions, the result of their application is an abundant growth of all leguminous plants, and especially of white clover. Fields so treated, on which only traces of the plant can be found after careful searching, may become an almost continuous white clover mat after a single dressing. No complete explanation of this extraordinary development is available yet, but it is generally assumed that it is in some way connected with the bacteria-containing nodules which are so distinctive a feature of leguminous root systems. This stimulation of the clovers is of indirect benefit to the grasses, since the former accumulate nitrogenous materials which in time find their way into the soil and become available as food material for the grasses.

Potash. Lack of potash results in a decline in the yield of hay and silage, a retardation of the flowering period of the grasses and, in some cases, poor development of the clovers. Not all soils respond to potassic dressings and probably the potash requirements of grass-land flora are amply met by the supply available in the soil, except in the case of those of an extremely sandy or gravelly nature, or where a soil analysis reveals a serious shortage of potash. With the increased use of nitrogenous fertilisers, particularly on temporary leys, attention must be given to additional potash requirement. Care, however, needs to be exercised in this respect as the excessive manuring of grassland with potash may lead to metabolic disturbance of the animals grazing thereon (see p. 679).

Lime. Although the grassland flora of soils overlying both chalk and limestone is usually rich in species the grasses, as a whole, are not calcareous plants, and most of those of agricultural value can thrive on soils with a low calcium content. Acid soils, however, are distinctly prejudicial to their growth, and very few species contrive to exist on really sour soils (see p. 31). Quite apart from the fact that the most productive herbage plants demand a non-acid soil, the maintenance of an adequate lime content in the soil ensures a high calcium content in the herbage, and this is acknowledged to be the best way in which to make sure that livestock have an ample supply of calcium in their diet. This is especially important for young growing animals, milking cows and all breeding stock.

Signs of lime deficiency in grassland are a preponderance of bent grasses, sheep's fescue and Yorkshire fog, together with tormentil (*Potentilla tormentilla*), heath bedstraw (*Galium saxatile*) and sour dock (*Rumex acetosa*). Under such conditions white clover will be present in traces only or it may be completely absent, and it is not uncommon for there to be considerable patches of bare earth. If the turf is lifted on these areas, the edges show sections of a skin of peat-like material consisting of the undecomposed remains of previous years' growth. This tends to increase in thickness year by year, for the processes of decay go on slowly under sour soil conditions. Rainwater falling on the surface is retained by the peat, with the result that after a dry season the soil below it may be dry for weeks after rain has fallen.

When heavily limed the peat mat gradually disintegrates, the process often going on so slowly that no effects are visible for some years. As it disappears the clovers re-establish themselves and the grazing improves considerably. But no pronounced changes occur in the flora, and the bent grasses still predominate. By tearing out the peaty material with heavily weighted spiked harrows prior to the application of lime, the decomposition of the mat is speeded up and the improvement greatly hastened. To be really effective, however, the harrowing should be drastic to the point of near destruction of the original sward. Phosphates, lime and some nitrogenous fertiliser are applied together with a simple grass seed mixture. In this way it is possible in time radically to improve the herbage, although the tough weed species will still persist and may ultimately regain dominance in the sward unless care is taken to nurture the sown species. This process is a form of pasture renovation discussed in more detail later, and is a method of reclamation advocated by some for certain conditions of terrain and soil type in place of the more drastic treatment necessitated by ploughing and reseeding.

To secure the best results from temporary leys, of long or short duration, it is essential to keep up the fertility of the soil. With long leys or permanent grassland the swards deteriorate as time goes on owing to the removal of food materials from the soil, either in the form of hay and silage or in the carcasses of livestock or in milk. This deterioration is characterised by a markedly increased weediness in the grass flora. The judicious use of fertilisers in association with farmyard manure and lime will do much to prevent this deterioration. So great, however, are the differences in grassland floras, in soils and climates, as well as in the requirements of individual farmers, that it is impossible to do more than give general indications on the best fertiliser treatment.

Manuring for Hay and Silage. On dairy farms where the whole supply of farmyard manure is not required for the arable land, the yield of grassland for hay or silage may be maintained by the application of farmyard manure at a rate varying from 25 to 50 tonnes/ha. The disposal of dung and urine and cowshed washings by collection in a suitable tank container and its application in a diluted form to grassland being conserved as hay or silage is being practised on more holdings as time goes on. In Switzerland this practice has been carried out for many years and is known as the " Gülle " system. Where this practice is continued year by year the tendency is towards the production of a bulky and coarse herbage which is often rich in Yorkshire fog and weeds like dandelion, and singularly deficient in the better grasses and clovers.

Where farmyard manure is not available nitrogenous fertilisers are essential for the continuous production of herbage for hay and silage. Nitrogenous fertilisers for this purpose are applied in March or early April at the rate of 50–75 kg/ha of nitrogen with the expectation, if weather conditions are favourable, of a yield of some 4·8 tonnes/ha of hay or 12–14 tonnes/ha of silage.

The use of farmyard manure and nitrogenous fertilisers alone is inadvisable as deterioration of the herbage is then inevitable : they should be supplemented at intervals with phosphatic fertilisers and lime. On light land an occasional application of a potassic fertiliser may also be called for.

A definite manurial rotation such as farmyard manure in the first year, phosphatic and nitrogenous fertilisers in the second, and nitrogenous fertilisers alone in the third and fourth year will do much towards keeping up the quality of the herbage as well as producing heavy crops for conservation. Fertiliser applications of this type may also be given as compound fertilisers containing the required nitrogen, phosphate and potash in suitable combination (see p. 189). An application of lime at intervals of four to seven years, depending upon the type of soil and the climate, will of course be necessary in most cases.

Manuring for Pasture. Leaving aside the value of an application of at least some nitrogenous fertiliser to grassland used as pasture, fertiliser requirements under these conditions will be mainly phosphates, particularly on the heavy and medium classes of soil, and of potash on the lighter sandy soils. Phosphates on permanent pasture may be applied in the form of basic slag, those containing a high content of phosphate being the most efficient. Increasing use is being made on temporary leys of the more soluble forms of phosphate, both superphosphate and triple superphosphate. Phosphate in compound form combined with nitrogen or potash may also be applied.

Recent work has shown that annual dressings of small amounts of phosphatic fertilisers to pastures give better returns in terms of herbage yield and quality than do heavier dressings at less frequent intervals.

Pastures on many different soil types benefit from liming; even those situated on chalk formations may need to be dressed with lime. Pastures which demand heavy and frequent applications are those situated on the Millstone Grit and Coal Measures soil formations, especially when these are situated in industrial areas where the acid fumes aggravate the natural lime deficiency of the soil. Dressings are now usually made as ground limestone which is easy to store and handle. Under average conditions an application of 3·7 tonnes/ha of ground limestone every four years should be adequate to meet the needs of the pastures and the grazing stock.

Where the system of pasturage practised is extremely intensive, as is the case on some dairy farms in Britain, the application of nitrogenous fertilisers in quantity is necessary. This usually takes the form of an application of some 75–100 kg/ha of nitrogen in late February or early March to provide early bite when the cows are first turned out to grass in the spring on to such grazings as Italian ryegrass. This may be followed by a similar further top dressing of 75–100 kg/ha of nitrogen in April to stimulate herbage growth for a second grazing, or for conservation as silage, or in preparation for taking a seed crop. Where it is possible to irrigate, this dressing can be repeated in June but, unless water is available for irrigation, mid-season dressings of nitrogenous fertiliser may be largely ineffective in giving growth response.

Pastures of a more permanent nature than those used for early bite are given a dressing of about 75 kg/ha of nitrogen in March or early April. This stimulates early growth, after which greater reliance is placed on maintaining the white clover in the sward to provide the nitrogen needed by the grass. Excess application of nitrogenous fertilisers to long-duration leys or permanent pastures will result in the suppression of the clovers.

Nitrogenous fertilisers may also be applied to stimulate autumn growth of pastures and so prolong the grazing season at that time of year. This autumn growth is being increasingly used to provide field grazing up to Christmas and the New Year.

Obviously, under an intensive system of grazing involving the use of heavy applications of nitrogenous fertiliser, it is vital to see that the soil is not depleted of the essential phosphate, potash and lime necessary to maintain an adequate and balanced supply of those plant nutrients. It is also important to ensure that the grazing stock are provided with mineral supplements when pastured in this manner.

THE CONTROL OF GRAZING

The intensification of pasture production by the breeding of improved varieties of grasses and clovers and the application of suitable fertilisers will lead to increased financial outlay. To recoup this it will be necessary to convert such increased production of herbage into an increase of saleable animal commodities. As little as possible of the herbage grown should be wasted and its utilisation by grazing livestock will need to be controlled. This is particularly so at certain times of the year when growth is rapid or when growth already made has to be rationed.

Utilisation by Cattle

Soilage or " Zero Grazing ". This practice, which in essence is that of cutting green fodder and feeding it to stock confined permanently in yards or other suitable feeding area, has recently been revived in this country. This is largely due to the availability of forage harvesters capable of cutting and collecting green fodder including grassland herbage, quickly and efficiently. In this way the main drawback of the system previously, the heavy demand on labour for cutting, collection and carting, has been eliminated.

The system offers considerable scope for maximum utilisation of any feed grown, and by its adoption cattle can be kept off the land so that the minimum of wastage due to soiling through stock excreta and treading takes place, although some damage may be caused by the cutting machinery. Fields at a distance from the homestead or milking unit, or badly placed for watering and fencing, can be laid down to temporary leys as part of the arable rotation, thus providing feed for the cattle from fields that are otherwise inaccessible to the grazing stock, while at the same time contributing towards the maintenance of soil fertility. Elaborate equipment for cutting and feeding is not necessary, and the use of self-feeding trailers for field collection and for feeding is a method of dispensing with the need for trough space and unloading of the trailers. The advantages of the system, however, have to be weighed against the fact that increased costs are involved in the daily cutting and carting of the fodder, as compared with allowing the stock to do their own harvesting by grazing.

Close Folding by Electric Fence. Apart from the particular circumstances prevailing when soilage is adopted as a method of utilisation control, field grazing will be the normal practice in this country. Intensive grassland husbandry for dairying or feeding for beef depends on the control of the grazing, partly to avoid wastage by treading and partly to ensure a high quality of feed over an extended period. This may be the case when grazing an early bite of Italian ryegrass in the spring or the feeding of " winter grass " on pastures laid up from the previous late summer or early autumn. The easiest system of control for this purpose is by close folding with the electric fence. This may entail one or two daily moves of a forward fence to allow sufficient feed to satisfy grazing requirements, and the bringing up of a back fence to prevent the continual grazing of the pasture already fed to the cattle.

Paddock Grazing. The basis of grazing control by paddock grazing is again to provide a high quality feed at a stage of growth to give the greatest return as animal products. The size of the paddocks is such as to provide grazing for the head of cattle carried for three to four days at a time. Alternatively the paddocks may be only of a size to provide the herd with one day's grazing, as in the case of close-folding with the electric fence. The fencing used can be of a temporary nature such as electric fencing, or of a more permanent type using post and wire. Ease of access to each paddock is essential. The rotation of grazing between a predetermined number of paddocks should allow a sufficient rest period for recovery of growth of young leafy herbage to take place by the next grazing. Any paddocks that grow beyond the grazing stage are omitted from the grazing rotation and cut for conservation.

Field Grazing. The extension of the paddock method of grazing control is field-by-field grazing. Here the cattle will be kept longer on a particular area, in all probability for more than ten days at a time. Attention should be given to the type of grass/clover mixture sown to ensure that the field comes for grazing at more or less specified times during the season. Field grazing can result in as high an output of cattle products as some of the methods of closer control, but the tendency is for field grazing to be lax, resulting in undue wastage of the pasturage.

Utilisation by Sheep. In parts of the country where there are extensive areas of grass for sheep grazing, as in the north and west, control of the grazing is largely based on a low stocking density, coupled with a not too rapid growth of grass. Where a higher stocking rate is possible, more especially of ewes and lambs, greater attention must be given to the provision of suitable grazing with regard to quality and type of feed and with a view to minimising the danger from intestinal parasites. The degree to which such control need be exercised will depend largely on the intensity of stocking envisaged. The greater the number of ewes and lambs carried per unit of area the greater the need for care and control of the grazing available.

Paddock or Field Grazing. Under a rotational system of paddock or field grazing, the carrying capacity of the grass will be 9–12 ewes and their lambs per ha during the spring and summer. Stock density should be such that sufficient high-quality leafy pasturage is available for the ewes for the first six to eight weeks after lambing to stimulate a sufficient flow of milk for the lambs. As the lambs get older such pasturage will allow them to make rapid gain in weight. The system involves either a permanent pasture or a temporary ley of four to six years' duration. Pasture management should be such as to allow, from the commencement of lambing, pasturage which grows with the ewes and lambs as these are transferred to it from the lambing field.

It is extremely important that in the grazing of these fields a year of grazing should alternate with a year in which the herbage is cut and conserved, or failing this, grazed with cattle. The grazing should allow for the maximum exposure of the base of the sward to sun and air, while at the same time it should not be so hard as to force the sheep to graze the basal herbage, containing the greater danger from parasite harbouring. This method of grazing control, coupled with anthelminthic dosing of ewes and lambs, will help to minimise the danger of parasitic infection, which to some degree is unavoidable, particularly as this affects the lambs. Essentially the grazing control should ensure as far as is possible that during its early life the lamb should graze as clean a pasture as possible, uncontaminated for at least a twelvemonth by previous sheep stocking and under conditions in which the ewe has but a low burden of intestinal parasites.

Creep Grazing. Where the carrying capacity is to be even higher than that under paddock or field grazing, that is 14–20 ewes and their lambs per ha from late March to late July, and where the maximum output of lamb is required, then still stricter control of both pasture quality and grazing method will be necessary. This can be provided by creep grazing which, in essence, is arable sheep folding transferred to an intensive grass regime. For this method of intensive grazing the field should be completely free from parasitic infestation likely to affect sheep. To ensure this, a field that has not been grazed by sheep for at least a twelvemonth, or preferably a new ley, must be provided. Superimposed on this will need to be a system of fencing of a temporary type by means of which pasturage can be made available to the ewes for three to four days at a grazing. In addition, the fencing must allow the lambs to have access to pasture not grazed by the ewes. This will be possible by forward creep grazing of the lambs where eventually the pasture will also be grazed by the ewes ; or by sideways creep, where the pasturage provided for the lambs need not be grazed by the ewes later on and so contaminated.

By the adoption of a creep grazing system immediately following lambing, the ewes can be given high-quality pasturage and the lambs will gain weight rapidly from the milk flow so stimulated. As the lambs grow older the milk flow of the ewes can be pulled down by the gradual reduction in the quantity and quality of the pasturage, while the lambs will continue to improve by the provision of a greater area of clean, high-quality grass.

Irrigation of Grassland. In the eastern and southern parts of Britain it is considered that grassland irrigation will give increased yields of herbage four years out of five. The amount of this increase will vary from year to year giving a 50 per cent increase in an average rainfall summer to over 100 per cent increase in dry summers. Normal pasture yields in Britain from temporary leys average some 6,725 kg/ha of dry matter. This may be increased to some 10,000–13,450 kg/ha dry matter yield by a combination of irrigation and nitro-genous fertiliser application. From 225 to 336 kg/ha increase of dry matter yield will be obtained per 25 mm/ha of water applied, and the best returns are to be had when the water deficit is not allowed to get beyond 12–25 mm. The financial returns from grassland irrigation must be carefully estimated before embarking on such a scheme. It is important to keep the cost of the supply of water and its means of distribution as low as possible to make the enterprise a worthwhile proposition (see p. 215).

Winter Grazing. The improvement in the winter greenness quality of grass varieties and the availability of quick-acting nitrogenous fertilisers has led to an extension of the period of growth of pastures in the late summer and autumn. In some areas where conditions of soil and climate are favourable it has been found possible to extend this period up to Christmas and even into the New Year. For this purpose the growth of suitable grasses is stimulated by the application of about 75 kg/ha of nitrogen in mid-August and this resultant growth grazed from late October onwards. Grasses that have given good results in this respect are cocksfoot and tall fescue, particularly when these are sown in drills 250–350 mm apart or in alternate drills with lucerne at the same spacing. The seeding rate of the grasses should be 5·6–6·7 kg/ha and of the lucerne 4·5–6·7 kg/ha. The grazing of the herbage grown in this way must be controlled by a rationing system. The subsequent summer production from these pastures is not reduced by the winter grazing. The sowing in drills prevents undue rotting of the herbage before it is grazed in the winter and the lucerne drills, consisting at this time of dormant plants, which are not harmed by the grazing, help to increase the amount of herbage produced in the following spring and summer.

DRAINAGE AND MECHANICAL TREATMENT OF GRASSLAND

Good drainage is a prerequisite of good grassland management. Soils that are badly drained restrict the rooting range of most herbage species. Where the water table is high it will result in the killing out of the deeper ranging roots so that if dry conditions occur later on, only the surface roots will be functioning and these will be unable to draw on the water reserves in the lower horizons and at the same time supply the minerals necessary for herbage growth.

Productive grassland should not be allowed to deteriorate into a matted condition. At least once in the season grasses and clovers benefit from a good harrowing with spiked harrows which cut into the turf and aerate the soil. Special turf-cutting blades can be fitted to spring tine harrows for this purpose, or one can purchase a special grass harrow of which there are many efficient types on the market. Combing to tear out dead grass can best be carried out during late autumn or winter. Before growth commences in the spring a light harrowing to spread mole-hills, droppings and the like is beneficial and, in most cases, grassland this time of the year benefits from a good rolling. This is especially useful on stony land which is to be mown for hay, since the stones are pressed into the soil out of the way of the knives of the mowing machine.

Intensive pasture utilisation often leads to unpalatable herbage developing in the vicinity of the dung and urine patches. The spreading of the dung pats is effected by winter and spring harrowing and the degree of unpalatability developing may be minimised by alternate grazing and cutting of the herbage either during the season or from year to year and by heavier top dressings of suitable fertilisers.

The mowing machine, of a reciprocating knife type or of the cylindrical or horizontal rotary knife type, should also be used as an aid to good pasture maintenance. Deterioration of a sward is rapid when persistent understocking allows the more vigorous grasses to become dominant and crowd out the finer grasses and clovers. To prevent this during periods of rapid growth the area allowed to the grazing animals should be restricted by using temporary fencing and cutting the surplus grass for hay or silage. An occasional " mow-over " is also useful to prevent tufts of grass from forming and to remove any seed heads which have formed, thereby encouraging the grasses to produce more leaf.

Weed Control. A good control of weeds in long-duration leys or permanent pastures can be effected by the use of the mowing machine. This control can be supplemented by the use of selective herbicides similar to those used in the control of broad-leaved weeds during the seedling phase of grassland establishment. Research work has shown that where weeds such as buttercups are eliminated from pastures heavily infested by these plants, increased production of herbage will only result following the application of suitable fertilisers in addition to the herbicide.

Pastures infested with Soft rush (*Juncus effusus*) can be improved by spraying the rushes with MCPA or 2,4-D. The spraying should take place before flowering when the rushes are growing vigorously and the rushes should then be cut some four weeks after treatment.

Gorse (*Ulex spp.*) in pastures may be cleared by the use of a 2,4,5-T spray and the underlying sward encouraged to improve by suitable renovation techniques.

GRASSLAND RENOVATION

The renovation of permanent grassland by mechanical treatment and the application of lime and fertilisers with the addition of a seeds mixture, has already been described briefly.

Sod-Seeding. Interest is being shown in what are described as the " sod-seeding " methods of improvement ; in some instances where these methods are applicable the grassland may be fairly open and renovating seeds can be drilled in to supplement the existing herbage ; in other situations the grasslands consist of a close turf and before seeds are drilled in a wedge of soil is removed by the coulter of a drill adapted for this purpose ; the seed and fertiliser are then sown into the ground exposed by the removal of the wedge. The drilling of seeds direct into a close turf would result in little improvement of the sward as the sown seeds have little chance of survival in competition with the already existing grasses. In many cases the improvements brought about by these treatments may be both slight in amount and slow in development : in some cases there may be no improvement and little return for the money and time expended.

Ploughing and Reseeding. Better and quicker returns can in many situations be obtained by ploughing and direct reseeding. Direct reseeding involves ploughing the old turf completely under and bringing to the surface sufficient soil to form a seed-bed. This is then treated with the appropriate amount of lime and phosphates (determined by soil analysis), these being worked into the surface soil during the seed-bed preparing operations, which are carried out in most cases by means of disc harrows. Potash fertilisers may be applied if necessary, and prior to sowing the seeds mixture 25–50 kg/ha of nitrogen should be applied to encourage rapid growth of the young seedlings. With favourable growing weather these come away rapidly and grazing is likely to be possible about two months after seeding. Careful costings have indicated that in most cases direct reseeding in this way is in the long run cheaper than renovation, although the initial cost may be considerably higher.

The Pioneer Crop. Under conditions of extreme poverty the sowing of a good seeds mixture in the first instance is seldom justified and, following ploughing and the application of the appropriate fertilisers, it is frequently advisable in such conditions to sow what is termed a pioneer mixture. A common mixture of this type is:

Italian ryegrass.	18 kg/ha
Rape	2·2 ,, ,,
Hardy green turnips	2·2 ,, ,,

This will provide abundant keep for sheep or cattle a few weeks after sowing. The crop is grazed down during the first year, and in the second a good disc harrowing may be given, after which the same seeds mixture can again be sown with the aid of 25 kg/ha of nitrogen. This crop is again stocked heavily, and the dung and urine passed into the soil encourage rapid bacterial activity which brings about the decomposition of the mat. Lime and phosphates become intimately mixed with the soil, and after this pre-treatment, the land can once more be ploughed, thoroughly cultivated, refertilised and a good seeds mixture sown with every hope of securing a really productive and lasting turf.

Chemical Destruction of the Sward. With the discovery of grass-killing chemicals such as dalapon, paraquat and glyphosate, a method of pasture renovation has been developed whereby the sward is first destroyed by chemical spraying. The method is particularly useful where conditions of terrain are unsuitable for ploughing or where it may be difficult to reconsolidate the soil once consolidation has been broken up by ploughing. After the killing of the turf the land is rotary cultivated as shallow as possible to produce a surface tilth ; fertilisers are applied, the seeds sown and the land harrowed and rolled. As with the ploughing method this type of renovation may be carried out in steps ; first by improving the condition of the soil with pioneer cropping and following this with the permanent seeds mixture when these initial improvements have taken place.

GRASSLAND POTENTIAL

In recent years the aim in grassland management has been to increase the production from grassland by growing greater quantities of herbage by more intensive utilisation of this herbage and by better and cheaper methods of conversion into the main products resulting from grazing stock. By paying attention to the environmental conditions under which grassland herbage is grown, by improving the varieties and seeds mixtures in use, by adequate fertiliser application and by weed control, much improvement has been effected. Management practices with regard to " early bite ", summer production and the extension of grazing into the autumn and winter have played their part coupled with grassland conservation as good-quality hay and silage. The development of other techniques where these are suitable, for instance irrigation, zero grazing and pasture renovation, helps to increase the potentiality of our grasslands in the feeding of the appropriate livestock over a greater part of the year. This in itself can also lead to a reduction in the amount of supplementary feed required from other sources. In a pastoral country, such as Britain, there is need for expansion along these lines and every endeavour should be made towards the adequate feeding of our livestock from all sources of home feed production.

GROWING GRASSES AND CLOVERS FOR SEED

In Britain up to the outbreak of the second world war the cultivation of grasses for their seed was a local occupation carried out in parts of Northern Ireland, Scotland, Kent and East Anglia. This was also the position with the growing of clovers for seed, and certain local varieties were well known by their district of origin. Examples were to be found in Essex, Suffolk, the Cotswolds, Hampshire and Montgomeryshire. As a result of the breeding of new varieties, particularly those from the Welsh Plant Breeding Station, the need arose for these varieties to be multiplied and supplied to the seed trade under conditions that would ensure complete authenticity of the varieties grown.

In consequence an organisation, under the direction of the National Institute of Agricultural Botany, Cambridge, known as the National Scheme for the Comprehensive Certification of Herbage Seeds, was set up. This scheme is now statutory (July 1975) and controls the production of all varieties used in Britain through the various stages of seed multiplication necessary between the breeder and the retailer. Some 19,000 ha of certified herbage seed were grown in Britain in 1973.

Most herbage seed crops are perennial, with the exception of such species as Italian ryegrass and Broad red clover which are virtually biennial. On arable farms the grass or clover seed crop takes the place, in the rotation, of leys of three to four years' duration. A well-managed herbage seed crop has the advantage of a cash crop of high potential financial return while at the same time providing a break in the sequence of arable cropping as well as improving soil condition.

On mixed, or predominantly grassland farms, herbage seed crops combine the function of providing some grazing for the stock with that of a cash crop, providing low-quality hay as a by-product.

The main herbage seed-producing areas in Britain are to be found in the east and south, where a relatively low rainfall allows harvesting to be carried out with reasonable chances of success. Other areas, for example parts of Northern Ireland and Scotland, also contribute to this section of the agricultural industry.

GRASS SEED PRODUCTION

The chief grasses grown for seed in Britain are perennial ryegrass, Italian ryegrass, cocksfoot, timothy, meadow fescue, tall fescue and red fescue. In choosing a field for the growing of a seed crop of any species of grass, and in particular of any of the bred varieties such as those from Aberystwyth, it is essential that the land be free from seeds of that species, and that the isolation of the crop shall be such as to prevent cross-fertilisation from neighbouring crops.

Clean Land. With regard to clean land it is obviously desirable to avoid complications from the usual annual and perennial weeds, some of which have seeds that are difficult to clean out of the seeds of the crop. For example, it is not easy to separate completely the seeds of the ryegrasses from cocksfoot, or of blackgrass (*Alopecurus myosuroides*) from the ryegrasses or from cocksfoot or meadow fescue. If a variety of any grass species is to be grown for seed it is necessary to make as certain as possible that seeds of any other variety of that species are not present in the soil, for they would germinate, grow and form seed along with the crop. To prevent this when growing grass seed under statutory control an interval of four years' satisfactory cropping must have taken place before sowing the seed crop. A longer period is necessary when it is intended to grow certain local varieties for seed. In a similar way contamination from stray seeds may arise through the feeding of hay upon leys on arable land.

Isolation. Since grasses are wind pollinated it is desirable that a seed crop should be as far away from potential sources of contamination—such as a hay crop—as possible. Usually a distance of not less than 50 m from a possible source of contamination is insisted upon. A closely grazed pasture is not regarded as a likely source, and it is sometimes possible to satisfy isolation requirements by mowing-over at the correct time a field of grass and clover seeds situated less than the required distance from the seed crop.

It is not sufficient that the field be clean and well isolated. It must also be in good heart, for the production of satisfactory seed crops depends very much upon a fertile condition of the soil. Yields of seed are greatly increased by judicious manuring, but it is possible also to over-manure and give rise to a laid and tangled condition which greatly hinders harvesting.

Drilling Seed. Cocksfoot, timothy, tall fescue and red fescue are usually grown in drills 450–600 mm apart, although both cocksfoot and timothy are sometimes sown broadcast. Perennial ryegrass, Italian ryegrass and meadow fescue are, however, grown as broadcast stands or in narrow drills up to 350 mm apart.

The wide-drill crops are best sown without a cover crop, while those sown broadcast suffer less when undersown. Should it be necessary to undersow the wide drill crops, care should be taken to avoid undue suppression by the cover crop by reducing its seeding rate or by blocking a coulter from sowing the cover crop and planting the grass seed in this vacant drill. Large seeded grasses such as the ryegrasses or cocksfoot may be sown 19–38 mm deep, but timothy with a small seed should be sown as shallow as possible, normally not more than 12 mm deep. Seeding rates should be as low as possible consistent with satisfactory establishment. For large seeded grasses in drills 4·5–6·7 kg/ha is ample, and 11–17 kg/ha when sown broadcast. The smaller seeded grasses are usually sown at 2·2–4·4 kg/ha in drills and 6·7–9 kg/ha broadcast.

Weed Control. Undue competition from weeds during the early stages of establishment must be avoided. As most grass seed crops do not have a companion clover a greater choice is possible in the use of selective herbicides for this purpose, without danger of damaging the grass seedlings. When the grasses are beginning to tiller, MCPA, 2,4-D, CMPP and dinoseb will give satisfactory results with the particular weeds which each of these effectively controls. Where clovers are present with the grasses only MCPB and 2,4-DB can be used for spraying. In the year when the grass seed crop is being put up for seed these chemicals can again be used to control broad-leaved weeds in the crop. Care must be taken not to spray earlier than four to five weeks before ear emergence is expected or later than the time of ear emergence.

Management. Cereal cover crops should be removed as soon as possible and the stooks or straw must not be left on the field longer than is absolutely necessary. Vigorous crops of ryegrass may be grazed lightly some six to eight weeks after planting if not undersown, but this grazing should not be prolonged and the crops should be allowed to recover before winter. Other grass seed crops are not normally grazed in the seeding year unless excessive leaf growth is present and there is a likelihood of frost damage if this is carried into the winter. In this case a light grazing to remove the excess leaf is advisable.

Direct sown crops are normally given a complete fertiliser dressing at seeding time, and these should not require more fertiliser before the following spring unless they show lack of growth. In the event of this a top dressing of 25–37·5 kg/ha of nitrogen should be given in the autumn. Undersown crops should be given a complete fertiliser dressing at this time of 25–50 kg/ha of nitrogen, 50 kg/ha of phosphoric acid, and 37·5–75 kg/ha of potash.

In the spring of the first harvest year most crops will need a top dressing of nitrogen in late February or early March. The quantity given will vary according to the grass seed crop, but usually ranges in amount from 50 to 100 kg/ha of nitrogen in the case of meadow fescue and ryegrass to 125–200 kg/ha for cocksfoot.

Harvesting. Grass seed crops may be harvested in a number of different ways ; by cutting with a binder, stooking and then threshing from the stook or from the stack using the combine harvester as a threshing drum ; by cutting with a mower into swathe and threshing from the swathe using a combine harvester with a suitable pick-up

attachment ; or by combining direct.   Threshing out of the swathe
with the combine harvester or direct combining usually necessitates
artificial drying of the seed under carefully controlled temperatures.

After harvest the grass crop stubble should be cleaned up by
grazing or burning or topping over with a mower and further fertiliser top dressing given.   For most crops this will entail a dressing
of 75 kg/ha of nitrogen, 62·5–75 kg/ha of phosphoric acid and 75–150
kg/ha of potash.   After autumn resting the stand may be grazed
during the winter ; the time and duration of the grazing will vary
with the different grass crops.   Cocksfoot, for example, may be grazed
up to the end of January and timothy to the end of February without
reduction in the yield of the following seed crop.

A spring dressing of nitrogen will again be given in the second
harvest year similar to that of the first harvest year.   Treatment in
subsequent harvest years will then follow this same general pattern.

CLOVER SEED PRODUCTION

The two clovers most widely grown for seed in Britain are red
clover and white clover ; some interest is still shown in sainfoin and
more recently in lucerne.   Except for certain areas which have
specialised for some time in producing seed of well-known local
varieties, the harvesting of red clover seed has been somewhat haphazard.   In times of seed scarcity farmers frequently allow the aftermath of Broad red clover to mature and thresh out the seed which
may be of very mixed origin.   Under EEC Regulations only authenti-
cated varieties which have qualified under various statutory tests for
the National List are permitted to be produced for sale.

The choice of a field for clover seed production will be governed
by the same general standards with regard to the cleanliness of the
field, previous cropping and isolation as for the grass seed crops.
Most soils are suitable except land likely to dry out badly in summer,
and very fertile land where excessive leaf may be produced thus
reducing the amount of flowering and making harvesting conditions
difficult.   Small fields are preferable to large ones as they are worked
more thoroughly by the pollinating insects.   This is particularly the
case with late flowering red clover where the main pollinator is the
humble bee.

Seeding.   Red and white clovers are both normally sown under
a cereal cover crop fairly early in the spring, by mid-April if possible in low rainfall areas or early May in the wetter districts.   Preferably they should be sown immediately following the sowing of the
cereal.

Seed rates of 6·7–9 kg/ha of red clover and 2·2–4·4 kg/ha of
white clover are customary, with a sowing depth of not more than
25 mm and 12 mm respectively.   White clover may be sown also with
3·3–6·7 kg/ha of meadow fescue or of a pasture variety of perennial
ryegrass, but seed yields are likely to be reduced.

Red and white clovers need adequate reserves of lime in the soil,
while both benefit from reserves of phosphate and potash or from the
application of these fertilisers prior to seeding : 62·5–75 kg/ha of

phosphoric acid and 75 kg/ha of potash are the standard applications at this time.

At harvest the cover crop and its attendant straw should be removed as early as possible. A light grazing at this time may be beneficial, more especially if the growth of the clovers is forward and likely to go into the winter in too proud a condition. This grazing must, however, not be too severe.

The control of weeds in the seed-bed in the seeding year is important. For this purpose, provided the cereal has reached its resistant growth stage and the clovers are in their first trifoliate leaf, MCPB and 2,4-DB are satisfactory. In the year of putting up to seed, carbetamide herbicide may be used to control grass and broad-leaved weeds in clover seed crops and MCPB and 2,4-DB may be used for crops of white clover. With the latter the spray must be applied immediately previous to the crop being put up for seed in mid to late May.

Under reasonable conditions of soil fertility, fertiliser application in the first harvest year will not be necessary.

Harvesting. The decision as to when a crop of clover should be closed up for seed is not easy to make, particularly when dealing with white clover. Broad red clover may be taken for silage or hay in late May or early June, while crops of late flowering red clover are best cut or grazed in early spring. This must be done not later than the end of May with the extra late varieties. White clover is generally closed up from the second to third week in May depending on the season, a dry season necessitating an earlier closing time.

Broad red clover is only taken for seed in the one year, while late flowering red clover varieties may be taken a second time although the yield is usually reduced in the second crop. White clover can be taken for two to three years.

Seed harvest of white clover is in late July and of red clover between late August and the end of September. In the past clover seed crops were cut and stacked and then threshed with a clover huller. Nowadays it is more usual to cut and swathe and to thresh out of the swathe using the combine harvester fitted with pick-up attachment. Red clover may be direct combined and to assist in this under moist weather conditions it is advisable to desiccate the clover leaf previous to harvesting. Suitable desiccants are sulphuric acid, diquat or DNOC.

Chapter XII

THE CONSERVATION OF FODDER CROPS

IN Britain, during winter, pasturage is scarce and of poor nutritive value. Therefore, in order adequately to feed cattle and to a less extent sheep, it is necessary to conserve for later use a proportion of what is grown in spring and summer. Fodder conservation is not considered to include the saving of roots, tubers or grains but only those crops which require processing in some way to prevent deterioration. The raw materials include a range of crops and by-products of crops, of which the grasses and clovers are the most important. The problem with grasses and clovers is the high content of water at the stage of growth when they are suitable for preservation.

Faced with this situation, the farmer has two alternatives. He can either dry the crops to the stage at which bacterial and fungal decomposition ceases (about 15 per cent moisture) : or he can make them into silage, which is preserved by organic acids formed during the fermentation process. Almost all the fodder crops are suitable for silage, but some are less suitable for hay making or artificial drying, the two common methods of removal of water.

ARTIFICIAL DRYING

Artificial drying is usually known as " grass drying ", although a large proportion of the green fodder dealt with in this way is composed of legumes. It differs fundamentally from hay and silage making, firstly in the relatively high expenditure on machinery and fuel, and secondly in the stage of crop growth at which it is conducted. Because of the expense of the process, crops must be young and nutritious so that the resultant product is a concentrated rather than a bulky food and therefore correspondingly valuable.

Suitable crops for grass drying are permanent pastures or leys in the wetter areas, and pure lucerne or lucerne/grass mixtures in the drier areas. These may be cut three or four times in a year at an early stage of growth and produce a total of 6-10 tonnes of dried grass per ha. The dried material may be made into wafers of various sizes or ground and made into pellets.

A tonne of young grass growing in a field may contain 175 kg of dry matter and 825 kg of water. During the drying process it is necessary to evaporate 813 kg of water to obtain 187 kg of dried grass of 6–7 per cent moisture content. Efficient use of fuel and labour is very important. Early grass driers were coke-fired but these have been superseded by oil-fired driers. Concurrently the earlier endless-conveyor driers have given way to rotary-drum driers and both capacity and cost have increased substantially. Designers have found it possible to obtain high thermal efficiency of fuel and economical use of labour only by combining these large driers with correspondingly large field machines. Consequently the process is no longer appropriate to the farm of medium size but requires the co-operative organisation of a large acreage of grassland.

In the drier air heated to about 150 °C is blown through the crop, and it passes first through grass that is nearly dry and secondly through grass that has only just begun to dry: this is in order to obtain a high drying efficiency. After drying, the crop is cooled before baling or milling.

Fuel cost per tonne of dried grass can be reduced if the crop is cut and left in a swathe to wilt for 24 hours before being collected. Nevertheless, apart from periods of high cattle food prices, artificial drying cannot often be justified economically on the ordinary farm. It is the most efficient method of conservation available if a comparison is made between the total food value of the growing crop and the total food value of the conserved product : losses usually only amount to about 5 per cent. The carotene content of dried grass is also high, varying from 200 to 400 mg/kg. On average, dried grass contains 6–10 per cent of moisture, and from 10 to 20 per cent of crude protein.

HAY-MAKING

Crops grown for hay include grasses, legumes, and occasionally cereals. They may be used as pure species or more usually as mixed swards. Taking the country as a whole, most of the hay comes from permanent pastures and this is especially the case in pastoral areas, but in the drier arable districts one-year and other short-term leys are predominant. The clovers are the commonest legumes, with white clover contributing to long leys and permanent grass and red clover to the short leys ; but lucerne, sanfoin, vetches and peas are also used.

Quality in Hay. One of the most important principles in all fodder conservation work is this : *a nutritious fodder can only come from a nutritious crop at the date of harvest.* If the harvest is too late nothing done subsequently can retrieve a position already lost. As pasture plants mature so the protein content and, to a less extent, the ash content falls. At the same time, the fibre content rises steadily. Leaves have a higher protein and ash content than stems and a lower fibre content. Thus, if there is any choice in the matter, hay is better harvested from leafy, early cut swards than from stemmy, late cut, ones.

The percentage of protein is a useful index of herbage quality but not the only one : the digestibility of the crop is most important. It has been found that grasses and clovers vary between different species, but in general they remain highly digestible until a certain stage of growth and then lose digestibility at an increasing rate thereafter. In this way, although the total quantity of protein and carbohydrate may continue to increase, the quantity of digestible nutrients declines. The actual date at which this factor becomes a dominant consideration varies geographically and from crop to crop. With first cuts in the south of England it is usually in the latter half of May : further north it will be later. The growth stage at which it occurs is just before the emergence of the majority of the flowering heads.

Although there are strong reasons for cutting hay at an early growth stage, before it has reached its maximum weight per hectare, it is unfortunate that the most suitable weather is not normally in the early part of the summer. Hay making requires maximum use to be made of dry weather and, on average, the chances for field drying improve from the middle of May to the middle of July and deteriorate from the middle of August to the middle of October. It is for this reason that so much hay is made too late to give it a chance of being of good quality, e.g. in July.

Another difficulty of early hay making, in addition to poor weather, is the fact that immature crops contain a higher proportion of water and dry less quickly than mature crops. On the other hand light crops dry more quickly than heavy crops.

Taking all the known principles together, their effects may be summarised by stating that although the risk is greater in making hay early from a young crop, the possible advantage of doing so is sufficient to make the risk worth taking, and additional skill in management can be used to redress the balance.

Methods of making Hay. In the description of methods below it is assumed that the crop dealt with is young and leafy. If it is old and stemmy the task is much easier and does not warrant detailed discussion.

The process involves nothing more than the evaporation of water to such a level that the material can be stored with safety. This simple statement is intended to dispel any impression that the operation involves any mysterious " curing " : hay making is often difficult but none the less simple in principle.

The crop is cut with a hay mower. Many farmers believe in starting this task after the morning dew has been dispelled, but there is experimental evidence that no practical advantage is gained from so doing. If fine weather is expected a start can be made as soon in the day as convenient.

Freshly cut grass has two important characteristics. Firstly, it is supple and, therefore, can be shaken or tossed vigorously by machinery without suffering any damage, whereas nearly dry hay is easily fragmented by rough handling and leafy parts are lost in the stubble. Secondly, newly-mown hay is not susceptible to damage from rain because the still turgid cells and waxy epidermis make the plants waterproof, whereas partly dried hay is liable to lose a considerable proportion of the water-soluble nutrients by leaching. Consideration of the latter factor leads to the practical conclusion that there are occasions in showery weather when cut hay suffers less damage if drying has been delayed : but there are more often compelling reasons why the drying process should be hastened in order to limit the period during which the crop remains in the field.

An unbroken swath, as left by the mower, exposes only a small surface area to sun and wind : it dries slowly and unevenly. If, on the other hand, the swath is fluffed up, the evaporation of water is encouraged by free circulation of air through it. The sooner a tedding machine is used after mowing and the more often it is used thereafter, the more quickly the hay will dry, so long as the weather remains fine. At this stage the hay should be treated in swaths which cover the whole field, rather than in windrows which only cover a part and thus limit evaporation. Some of our leading farmers use a modification of this principle whenever possible. They " ted " the crop in the swath during the day, and then windrow it in the evening to reduce the area of hay upon which the dew can fall and in addition the ground between the windrows is allowed to dry before the hay is spread out again the next day.

Conditioning Machines. Speed of drying is the essence of making high-quality hay and there are various machines which may be used to accelerate the process. In the ordinary way plant stems dry more slowly than the leaves, but " conditioning " machines aim to flatten or kink the stems without breaking them, so that stems and leaves dry at the same speed.

The Flail Mower. Although originally designed for cutting silage the flail mower is an important hay making tool because of its reliability and quick drying effects but it has to be used with understanding and skill. If the crop is young, heavy or leguminous there can be considerable fragmentation and loss of leaf but if it is light and fibrous the same operation is attended by little damage. Under all circumstances losses are reduced by maintaining as low a rotor speed as possible in relation to the particular ground speed of the machine. Flails can also be used as tedding machines but it is normally advisable to follow flail cutting with a conventional tedder and not flail the same crop twice.

The Crimper. Of the special purpose machines the " crimper " may be taken as an example. It picks up and passes the crop between two interlocking fluted rollers which crush or crimp the stems at short intervals ; it leaves behind it a well-ventilated swath which dries very quickly. The crimper is used only once, i.e. immediately after cutting : a tedder and side rake are still required for subsequent operations.

Field studies have shown that if a heavy crop is treated with a crimper it is likely to reach 25 per cent moisture content (suitable for baling) about two days earlier than a similar crop which has only been tedded.

The fear has often been expressed that hay treated in the way described may be damaged more by rain than hay treated with a tedder. Experience has shown such fears to be exaggerated if not groundless. In one detailed study, when an inch of rain fell on partly-dried hay that had been tedded and partly dried hay that had been crimped, the weight per hectare after collection of the latter was equal to the former and the nutritive value was equally high. It was also found that both before and after the wetting the crimped hay dried more quickly than the tedded.

A very large area of hay in total is made on very small holdings where the scythe and hand rake are more appropriate than an array of expensive machinery. The system is simple and laborious and not likely to be assisted by a discussion of fundamental principles. From these simple methods to the fully mechanised there are grada-

N

tions in methods of collection too numerous to mention and only one traditional system will be dealt with briefly. On many farms, hay that has been partly dried in the swath is collected together and built into cocks or pikes in the field and it is left in this form to dry further before carting. It is usual for the cocks to be small in the drier parts of the country and large (containing about 750 kg each) in the wetter areas. In the drier areas the cocks are allowed to dry completely and they are then built into large stacks, or baled. In the wetter areas field drying does not proceed so far, and the cocks are made into small stacks in the rickyard in the expectation that further drying will take place afterwards.

The Pick-up Baler. On most farms where hay was formerly made into cocks and carted loose, use of the pick-up baler has now taken over. The chief reason for this change is that the baler sub-stantially reduces the amount of manual labour involved by older methods, in collecting, transporting, stacking, and finally cutting out and feeding hay stacked loose. The quantity of hay in bales can also be estimated accurately and rationed feeding is easier.

The pick-up baler has often been blamed for producing mouldy hay. Like many another useful machine it only gives good results when operated intelligently and in suitable conditions. It requires that the hay should be dry enough for baling and the bales should not be neglected thereafter. The decision when to bale is helped by knowing what will happen to hay at various stages of dryness if it s baled :

Hay containing 35 per cent moisture will rot.
,,   ,,   30 ,, ,,   ,,   will mould and heat.
,,   ,,   25 ,, ,,   ,,   may mould unless it is given the opportunity to dry out further.
,,   ,,   20 ,, ,,   ,,   is safe to store in Dutch barns and covered stacks.

From the information above it is seen that hay baled at 25 per cent moisture, as it often is in practice, must be left to dry further in the field or some special arrangements must be made for it indoors, whereas hay a little drier may be baled, carted, and stacked, imme-diately. The indoor conditions in which further drying is possible include good ventilation and stacking with numerous air spaces between the bales. After a short period the bales can be moved and stacked more closely, thus reducing the space required.

If the baler invariably picked up and collected the whole of the crop lying in the field, it would nearly always be better to dry to a low moisture content before baling, so that handling thereafter is simplified ; but unfortunately the baler shatters the leaves of dry crops and much is left on the ground. On a hot day the quantity lost may be considerably more in the afternoon than in the morning, and considerations such as these make field experience essential in deciding when to bale and what to do afterwards. As a rough guide it may be stated that hay is not fit to bale at all if there is any extrusion of moisture when it is twisted with the hands. Bales that are fit to collect into large stacks are springy within the strings.

Bales made by a pick-up baler may be scattered in singles all over the field ; collected together in heaps by means of a sledge drawn behind the baler ; or delivered directly into a trailer drawn behind. The first-named method has the advantage of requiring only one man for baling but uses the most labour in the end. Loads of bales are carried out of the field either with trailers or on buck-rakes. The best method of storing baled hay is under a barn, and if this is not available, thatching or covering of stacks with waterproof sheeting is essential.

SPECIAL METHODS OF HAY-MAKING

The beginning of making a crop into hay is assisted by taking advantage of a special weather forecasting service run by the Meteorological Office, but long-range forecasting is subject to error. Therefore, any method which requires only short spells of fine weather has an advantage. Tripod hay-making is one such method and barn drying is another.

Tripod Hay. Hanging partly dried crops on tripods, wires hut-racks, etc., to keep them clear of the ground and to allow air to circulate is chiefly of historical interest for hay in this country but is still common on many small farms on the Continent.

If such appliances are loaded carefully and wet weather follows, a proportion of the rain runs off the outside of the crop rather than through it, and the amount of leaching is less than with hay on the ground. At least one of these methods, which uses hut-racks, can be partially mechanised and there is also a special machine for loading tripods, but in general much hand forking is required and the labour requirement is high.

The details of construction of " tripods ", " quadpods " (formed from four poles instead of three), racks, etc., vary, but they usually consist of poles or frames about 2·4 m long which are leaned and fixed together in such a way that they stand 2 m high. When they are covered with hay there is a pyramid or wedge-shaped empty space around which the hay is disposed and although the thickness of hay may be 0·6–0·7 m in some places, the opportunity for it to dry is favourable. Each unit will carry about 250 kg of dry hay, and it is usual to erect about twenty per hectare.

Barn Dried Hay. In the discussion of artificial drying it was explained that the main drawback is the expense of evaporating a large quantity of water with purchased fuel. In field hay-making the drawback is the risk of qualitative and quantitative losses during natural drying. Barn drying is a compromise between the two. The objective is to remove about three-quarters of the crop moisture in the field and one-quarter indoors. Because moisture evaporates much more quickly from freshly-cut hay than from nearly-dry hay, it is usually possible to complete the field drying operation in two days and this improves the prospect of making use of short-term weather forecasting services to decide when to cut the crop.

In this country the first experience of barn drying was obtained with long or loose hay. However it was quickly found that with suitably designed equipment the process can be operated with rectangular bales made by a pick-up baler and this has reduced handling costs.   It is the bale-drying method that is discussed below.

Field Drying.   The field-drying part of the operation is conducted in the usual way, but instead of baling at about 25 per cent moisture content it is usually done at 35–50 per cent moisture content.   At this stage full-sized bales weigh 30–45 kg each, and some farmers prefer to modify their balers so that they deliver half-sized bales which are easier to handle.   By contrast the " big baler " makes bales weighing about 0·5 tonne each.   These must not be too moist at baling.

Drying Methods.   Two common methods of barn drying are called " batch " drying and " deep storage " drying.   In the former each batch of hay is stacked in the drier, is dried and is removed elsewhere for storage.   In the latter the hay is introduced in layers, and after each layer is dried another is placed on top until the drier is full. " Tunnel drying " is a batch method in an unwalled area.

Batch drying is suitable for hay containing up to 50 per cent of moisture.   The size of batch to be dealt with at one time depends upon the floor area of the drier, each square metre of which will take about 250 kg of dried hay : thus one bay of a barn measuring 33·5 m² deals with a 9 tonnes batch.   The floor usually consists of strong steel mesh, 75 mm square, through which air is forced by a fan.   The floor is surrounded by air-proof walls rising to a height of at least 2 m and the bales of hay are stacked neatly into this space.   The bales are laid on edge and alternate courses have their long axes at right angles to each other.   So long as the total depth of hay does not exceed about 3 m (6 courses), drying should proceed satisfactorily provided the air flow is sufficient.

Ventilation.   It is important that the ventilating fan should be capable of forcing about 1·3–1·4 m³ of air per minute through every square metre of the floor.   At this rate the air flow passes through the bales in sufficient quantity to keep them cool and to remove moisture from them.

The moisture extraction capability of the ventilating air varies inversely with its relative humidity, and since the relative humidity of air can be reduced by applying heat it follows that heated air removes water more quickly than unheated air.   In the initial stages of drying, moisture can be removed more easily than in the later stages and, therefore, the need for supplying heated air increases as drying proceeds.   There are many variable factors of which to take account in drier management, e.g. relative humidity of atmospheric air, moisture content of bales, time available before the drier is required for the next batch, volume of air supplied by the fan.   Because of these factors it is advantageous to be able to alter air flow and heat supplied according to the conditions.   Some plants are designed to raise the temperature of the ventilating air by 14 °C, others by a less amount.   Although the time taken to dry each batch is particularly affected by the ingoing moisture content it is generally possible in about five days.

Deep Storage Drying.   Deep storage drying is most suitable for baled hay collected from the field at 35–40 per cent moisture content, although cases are on record of successful operation at higher moistures. It requires a ventilated floor, impervious sides and a copious flow of air in the same way as a batch drier but the air temperature rise is usually less and the air speed greater. The depth of loading is not more than 2 m at a time but as each layer is dried another one is placed on top until the final depth of hay may be about 6 m, i.e. up to the eaves of the building. The advantage of storage drying is the reduction of handling due to drying and storing in the same place. One disadvantage is the necessity of providing a large area of ventilated floor.

The total labour required for storage drying is less than for field drying and even batch drying compares favourably with field drying. The reduced time during which the hay is in the field decreases the amount of tedding and turning and makes up for the labour of stacking bales in the drier and transferring them from there to the storage place. On the other hand the capacity of the drier may limit the volume of crop which can be dealt with at one time and for this reason it is advisable to start hay-making earlier and finish later than normal. Mid-season and late-harvested crops need not be mature and fibrous if they are taken from fields which have been grazed at well chosen times in the spring.

The cost of fuel for barn drying varies within wide limits according to the initial moisture content of the hay and the thermal efficiency of the plant. Most driers are either heated electrically or with oil heaters, and of these oil is the cheaper fuel although oil-fired equipment may be dearer to instal and maintain. Some driers were heated by coke which is the cheapest fuel of all. They had to be designed so that sparks could not enter the air stream, as must the oil-fired type.

An important aspect of barn drying is the reduction of field losses as compared with conventional hay-making. It is usual to collect about 15 per cent more hay because less of the valuable leafy material is lost when baling at the appropriate moisture content. Feeding experiments have also shown that the feeding value of barn hay is superior to field-dried hay. The danger is that superior quantity and quality may be procured at too high a cost. In order to guard against this possibility it is necessary to ensure that the field machinery is well chosen, the plant well and cheaply designed and management directed towards extracting as much moisture as possible by wind and sun and as little as possible by artificial means.

The Economics of Barn Drying. The total cost of barn drying includes labour, fixed capital and fuel, all of which vary according to the design of the plant and method of operation.

Fuel cost is tremendously affected by management, and the basic reason is illustrated by the following figures given at the top of page 348 which show the amounts of water to be removed from the crop at various moisture levels.

It has been said earlier that some driers are capable of dealing with hay containing 50 per cent moisture. But because hay at 35 per cent moisture only requires half as much water to be evaporated from it, a good manager will ensure that hay is usually carted at 30–40

Moisture Content before barn drying (per cent).	Weight of water to be removed to produce 1 tonne of dried hay at 15 per cent moisture content (kg).
60	1,125
50	700
40	417
30	214

per cent moisture content rather than 50 per cent. By so doing he will halve the fuel cost and at the same time reduce the risk of mould formation which is apt to happen even in a barn hay drier when very damp hay is baled.

The art of the manager is to dry the crop far enough in the field to reduce drying costs, but not so far that leaf is lost before and during baling. In this latter respect the difference between 35 per cent and 25 per cent moisture content is critical. The last 10 per cent is not only slow to evaporate in the field but it is also accompanied by a substantial loss of crop.

On average, for every 1,000 kg of barn hay collected at 35–40 per cent moisture content there will be only 875 kg of field hay because of the final drying in the field and the shattering caused by baling of dried hay.

This extra yield potential of barn drying goes a long way towards paying for the hay drying plant. And needless to say, barn drying uses less labour in the field because less tedding and turning are required.

In this account of the process the moisture contents at various stages of dying have been mentioned for illustrative purposes, but it should not be thought that in practice such measurements are essential.

So long as a farmer already has a good grasp of the principles of quick hay making he can soon judge when the crop is fit for barn drying. It is not quite ready for baling under " ordinary" circum-stances and makes a bale which is firm without being compact. After drying the bales become distinctly springy but not so loose that they fall apart.

ENSILAGE

Silage can be made from a wider range of crops than dried grass or hay : any crop which is suitable for feeding to livestock in its natural state, and which contains sufficient moisture and sugars to produce the necessary fermentation, is usually suitable for making into silage.

Grass Silage. Grass silage is by far the most important and will be considered first. For the present purpose it includes pure grasses, pure legumes (such as lucerne) and mixtures of the two either in the form of temporary or permanent swards.

The main objectives are to collect the crop at a nutritious stage ; to deal with it in such a way that field losses and in-silo losses are minimised, and to achieve a type of fermentation which results in a palatable product of high feeding value. Silage is preserved by organic acids which are produced by fermentation of the sugars contained in the cell sap. If the sugars are plentiful a beneficial lactic fermentation takes places : if they are scarce, other undesirable types of fermentation may occur, the most common of which leads to the formation of butyric acid. Grasses contain more sugar than legumes and mature plants more than immature plants. Hence young and mainly leguminous crops are more difficult to make into good silage than older and mainly grass mixtures. A desirable type of fermentation is also encouraged, within limits, by a reduction of the water content of the crop and by chopping or laceration.

Wilting. One of the important skills necessary for making good quality silage is an ability to recognise the conditions liable to lead to a poor type of fermentation, together with knowledge of the appropriate corrective action. When the objective is to make nutritious silage early in the season the most common difficulties are too much water in the herbage and too low a concentration of sugars. In this case it is often only necessary to correct the former because a reduction of water content automatically increases the concentration of the cell sap. For example, a leguminous crop containing 20 per cent or less of dry matter is much improved for ensilage if it is wilted for 24 hours, during which the dry matter content may rise to 25 per cent or more. Experiments have shown that the quality of the silage is almost invariably improved and at the same time the in-silo losses are reduced.

Molasses. Should wilting not be possible, the necessary conditions may be obtained by the addition of a sugar solution in the form of molasses, which is a by-product of the sugar refining industry. The usual rate of application for leafy young grass is 9 litres molasses diluted with 9 litres water per tonne of grass ; half as much again should be used if the crop is a pure legume.

Preservatives. An alternative to providing sugar for fermentation is the addition of a chemical preservative to inhibit fermentation. Compounds such as formic acid or mixtures of formic or sulphuric acid with formalin are very effective so long as they are used at the recommended rates and distributed evenly through the mass of silage. They have the advantage over molasses of being easier to handle and forage harvesters can be equipped with mechanical " applicators " which mix them with the crop most efficiently.

Losses in Silage Making. There is always a substantial loss of food material between filling a silo and taking out the silage, the three chief sources being the fermentation energy dissipated as heat and carbon dioxide ; the waste products of putrefaction, some of which have to be discarded from the outsides of the silage before feeding ; and the effluent liquid which drains away as the herbage is consolidated.

Under good conditions in walled clamps the total loss may be about 25 per cent, of which 1 per cent may be due to effluent liquid, 5 per cent the tangible products of putrefaction, and the rest due to fermentation of various kinds. In practice total losses as high as 40 per cent and as low as 15 per cent have been measured, the former often associated with unwalled clamps out of doors and the latter with specially designed and roofed structures.

Temperature Control. When grass is deposited in a heap it continues to respire and generate heat so long as it has access to a supply of oxygen. The quantity of oxygen will be increased if it is dry and loose and decreased if it is wet and consolidated, so that the farmer has a ready method of temperature control. If the temperature is allowed to reach and remain above 50 °C for very long, the result will be a sweet smelling brown silage. This is palatable to stock but of low feeding value—partly because a large proportion of carbohydrate has been used up in the heating and also because the high temperature reduces he digestibility of the protein.

It has been recommended in the past that the objective should be a temperature of about 32–38 °C, but recent experimental work has shown that under favourable conditions the temperature should be kept as low as possible by quick filling and efficient consolidation. When this method is used for ensiling high quality young grass it should if at all possible be wilted to 25–30 per cent dry matter—and preservatives are not necessary. But if the dry matter is below 25 per cent a suitable preservative is required to prevent undesirable types of fermentation taking place.

Method of Cutting. A crop intended for silage may be cut either with a hay mower type of knife bar or the rotating flails of a forage harvester. In the former case it may be transported to the silo as *long* material either with or without a preliminary wilting period, or it may be picked up from the swath with a flail harvester or a crop chopper. Crops cut with a flail may be transported as *flailed* (lacerated) material or the primary machine may both flail and *chop* at the same time before delivery to the transport vehicle. From the foregoing methods it is obviously possible to select many different combinations, but the resultant products fall into three broad groups, i.e. " long ", " lacerated " and " chopped ". Because the last two groups involve more damage to the individual plants there is a smearing of the cell sap through the mass and this usually leads to a better fermentation.

Silos can be filled with chopped material by means of a tractor-driven blower but long and lacerated material cannot be dealt with in this way. It has to be moved with hand forks, elevators or buck-rakes, all of which involve more labour than a blower.

Type of Silo. Silage can be made in unwalled clamps or stacks. temporary or permanent walled clamps, tall box silos or tower silos, Stacks are usually situated in the open ; walled clamps may be outside or under cover : box or tower silos are always roofed. Efficiency of conservation usually increases in the order in which the silos are named above, efficiency being improved by exclusion of air from the mass and protection from the weather.

Stacks have gradually become less popular owing to their high labour requirement, the usual method being hand forking into an elevator and hand building. Draw-over clamps without walls, or bunker silos (walled clamps), are filled with long or lacerated grass and are the most numerous. If the clamps are large enough it is possible to draw trailer loads over, but such a practice is always difficult and often dangerous, especially on narrow clamps, and the task is better done with a buckrake. Even a buckrake can be dangerous unless the tractor is always driven backwards up the slope and is kept sufficiently far from the sides when the contents reach above the height of the retaining walls. Boxes and tower silos are particularly suited to chopped grass and are usually filled with a forage blower.

Covering the Silo. Two of the details that are frequently given insufficient attention on farms are the disposal of effluent liquid and the methods of covering the silage and protecting it from the weather.

Unless it can be guaranteed that the crops will always be dry enough to eliminate seepage (i.e. 30 per cent dry matter or over) it is essential to provide adequate drainage. Young herbage ensiled without wilting may easily produce liquid to the extent of 70 litres per tonne and account must be taken of the fact that if this is discharged into watercourses it may lead to serious pollution, for which the farmer may be called to account.

Finishing off a silo implies protection against leaching by rain and against oxidation by air. Weather protection may be achieved by a variety of obvious means, including coverings of soil, dung, plastic sheeting or roofing materials. The exclusion of air is assisted by covering with plastic sheeting which is applied as soon as possible after filling and held in place by materials such as ground limestone, sawdust, straw bales, etc. Air should also be prevented from filtering in at the sides by filling cracks in the walls. Such methods are worth while with indoor as well as outdoor silos except those that are gas-tight in construction.

Other Types of Silage. Mixtures of oats and vetches, or oats and peas, or beans, vetches and oats, are still grown specially for silage in some parts of the country, but the practice is dying out. Wheat, barley and oats can all be grown as forage to yield 7·5/10 tonnes of dry matter per hectare when ensiled.

Maize. The National Institute of Agricultural Botany has classified maize varieties according to their suitability for silage making in different parts of the country. It is recommended that the crop should be harvested when the seeds are at the " soft dough stage ". It is essential to chop or severely lacerate the crop in order to achieve the necessary consolidation, and maize harvesting attachments are available for forage harvesters : otherwise maize is treated in the same way as grass and very palatable silage usually results.

Pea Haulms. This valuable by-product is available when peas are harvested green for human consumption. The whole plants are collected and passed through the shelling machine or " viner " in which the peas are separated ; the residue consists of broken stems, leaves and empty pods. This material does not need molasses or other additions in order to make good silage and the procedure is the same as for grass silage.

Beet Tops. These must be clean, otherwise feeding the silage will result in digestive disturbances. The tops can be ensiled either whole or chopped and there is no necessity to add anything or to encourage heating. A large volume of effluent liquid must be allowed for and its disposal may present problems.

Wet Beet Pulp. All that is necessary is to make a heap and exclude air as soon as possible with a weighted plastic sheet or other suitable material such as a layer of soil.

Surplus Potatoes. If they are cooked first, potatoes may be treated in the same way as wet beet pulp : but if they are uncooked they cannot be ensiled by themselves. On the other hand they can be introduced in shallow layers into grass silage if they are available at the right time of the year : the heat engendered during fermentation of the grass partly cooks and preserves the potatoes.

Kale, Rape, Cabbage. These crops need to be chopped or lacerated. There is no need for additives or for heating to occur in the silo. Stock do not usually find kale silage as palatable as grass silage and it is not recommended that cruciferous crops should be grown specially for the purpose but only ensiled when they would otherwise go to waste.

Feeding Silage. Silage feeding methods come under three main headings, either (a) the cattle go to the silage and help themselves (self-feeding) or (b) the silage is taken to the cattle or sheep (rationed feeding), or (c) the silage is fed mechanically (mechanised feeding).

For self-feeding it is desirable that the quality of the silage should be uniform in order to avoid any tendency for relatively unpalatable layers to be refused. The settled depth of silage should be about 1·5–1·8 m so that animals can reach to the top but not over the top. A suitable feeding fence should be placed between the cattle and the silage and moved forward as the silage is consumed : it should be so constructed and managed that silage is not pulled to the floor and trodden underfoot.

The number of animals must be adjusted to the width of feeding face so that they do not need to bully each other in order to eat to appetite. The width per beast will depend upon the length of time allowed for feeding, and the minimum is about 150–200 mm in the case of animals with continuous access for 24 hours per day.

The floor of a self-feed silo should be of concrete and should slope away from the silage face. It should have provision for disposal of the slurry which is inevitable and copious. Attempts to soak up the slurry with straw usually fail and it is better to plan at the beginning to keep the area in front of the silage clean. If the water trough is at a distance from the silage face the slurry problem seems to be reduced a little.

One simple but very important principle is occasionally overlooked, i.e., the necessity of providing sufficient silage for animals which are allowed to help themselves : dairy cows of the larger breeds may eat from 40–50 kg per day.

When silage is carted to the stock the method of removal from the silo varies according to the conditions in which it has previously been ensiled. Long and lacerated silages need to be cut vertically before they can be forked out, and although there are various mechanical devices for the purpose few have any advantage over an ordinary hay knife. Chopped materials score in respect of the ease with which they can be removed mechanically. For example, silage that has been chopped very short will flow through mechanically driven augers or it can be removed with a tractor fork.

Feeding Value of Hay and Silage. In the past a great deal of emphasis has been placed on the value of silage making as a conservation process and it has often been compared with hay-making to the latter's detriment. But the fact is that the two methods are complementary rather than competitive and each is capable of producing first-class fodder by following the correct methods.

It is not usually possible to make hay in the field as early in the season as silage, but when hay is barn dried at the " silage stage " it appears to be at least as good as well-made silage and sometimes better. When comparisons have been made between barn-dried hay and silage, both of which were cut from the same field on the same day, it has been evident that both sheep and cattle found the hay more palatable—appetite being measured by the consumption of dry matter on *ad. lib.* feeding. The hay also had a lower fibre content in the dry matter and a higher digestibility. With bulky fodders palatability is an important factor, and just as cattle find barn hay more palatable than silage, so they find high dry matter silage more palatable than low dry matter silage.

If field-made hay could be collected at the same growth stage as silage there seems no reason why it should not be just as nutritious, and it has not been possible to distinguish in feeding trials between barn hay and field hay made in good weather. On the other hand, hay that has been badly weathered will certainly have a comparatively low feeding value.

All these considerations lead back to the original postulate that management is more important than method, and that bad management can spoil a good crop but good management can never improve a bad one : it can only prevent further deterioration.

This chapter should be read in conjunction with Chapter VI, which gives a more detailed description of the machinery used in reserving green crops.

Chapter XIII

FUNGUS AND VIRUS DISEASES OF CROPS

THERE is no clearly defined boundary between a healthy and an unhealthy condition in plants. In practice, yield is a common criterion of good health, for it is generally recognised that a large crop cannot be produced when the plants composing it are in an unhealthy condition. The causes of unhealthy crops and low yields are many and varied, but the more important are these : the land may be in poor heart and the standard of fertility low ; the soil may be sour or badly drained and in an unsatisfactory physical condition ; an ample supply of food materials may be lacking, or some essential plant nutrient may be absent, or present in a form in which it cannot be absorbed. The weather, too, may adversely affect plants, periods of drought or excessive rain may intervene and check plant growth.

Apart from all these considerations, however, the crop may be diseased. If the damage caused by the depredations of insect pests and eelworms is excluded, the main agencies of disease are fungi, bacteria and certain extremely minute disease agents called viruses.

THE NATURE OF FUNGI

The fungi are a lowly group of plants completely lacking in chlorophyll, the green colouring matter of other plants. As a consequence they cannot build up the food required for their growth from the simple plant nutrients in the soil and the carbon dioxide of the air. To live they require, therefore, a ready-made supply of complex organic foods, such as carbohydrates and proteins. The main sources from which they obtain these are either living plants or their dead remains.

Parasites and Saprophytes. The fungi thus fall into two groups namely, *parasites*, those dependent on a living host, and *saprophytes* those able to feed on dead plant tissues. It should be noted that this distinction, though useful, is not an absolute one, for some fungi bridge the gap by living saprophytically for a time and then becoming parasites. A fungus of this nature is *Botrytis*, species of which cause the well-known chocolate spot disease of field beans.

From the farmer's point of view the fungi parasitic on plants naturally seem to be the most important, but without the saprophytes there would be no agriculture, for they play an essential part in breaking down and returning to the soil and air the materials locked up by the growth of previous generations of plants. Their activity in this direction has its troublesome side when it is responsible for the decay of timber used in farm buildings, gateposts, stakes, etc. Much of this can be prevented, however, by the use of suitable wood preservatives.

354

The Mycelium. Structurally, the fungi are far simpler than all other plants, with the exception of the algæ. The fungus body is (with rare exceptions) built up of very minute, thin-walled, branching threads or *hyphæ*, collectively known as the *mycelium*. It may be so small that it is invisible to the unaided eye, or the individual strands may be matted together to form large conspicuous bodies, as in mushrooms, toadstools and puff-balls.

Spores. Fungi reproduce their kind by means of spores, but these are so small that they can be seen only with the aid of a microscope. They may be formed either vegetatively or by the simple abstriction of the tip of a fungus thread or by its division into a chain of spores, or again, as the result of a sexual process. Most fungi produce two or more forms of spores which are often so unlike one another that, before the complete life cycle of the fungus was discovered, the different stages of development were classified in totally different groups. These various forms of spores fall into two broad categories, namely those responsible for the rapid multiplication of the fungus, and those by means of which it tides over periods of adverse climatic conditions or contrives to persist in the absence of the host plant. Those of the former group, often popularly called " summer spores ", are usually short-lived and their life is very dependent upon weather conditions. For instance, a short spell of dry weather in the summer may result in the wholesale destruction of the spores of *Phytophthora infestans*, the fungus responsible for potato blight, whilst damp muggy conditions will favour their development. Spores of the latter group, sometimes called " winter spores ", usually result from some form of sexual process, and they often have a lengthy period of life even under conditions inimical to growth. Their walls are commonly thick and dark in colour, or, where this is not the case, they are enclosed in resistant spore cases. Some fungi, however, form *sclerotia*, bodies which consist of hard compact masses of mycelium which are capable of surviving in a dormant state for many years. A fungus of this kind is *Sclerotinia trifoliorum*, the cause of clover and bean rot. The spores of the different kinds of fungi vary in size and shape, but they are all very minute and are produced in enormous numbers. The spores of the Bunt fungus are comparatively large, but a single bunted ear of wheat may contain 250,000,000 of them.

The dispersal of fungus spores is effected mainly by air currents, and at some seasons of the year, except after heavy falls of rain, the air is full of them. In addition to this mode of distribution, the spores of some fungi adhere to the coats of grain and seeds, and it is these which cause the so-called " seed-borne " diseases, such as the smuts of cereals. It should be noted, however, that not all the smuts are caused in this way for some, as will be seen later, are due to the presence of mycelium deeply embedded within the grain. Spores are also dispersed by rain, and by insect agency, and examples of these will be given in due course.

Method of Attack. Parasitic fungi gain an entrance to the tissues of their host plants in various ways. Some are only capable of attacking them at the seedling stage, so that once this is passed there is no likelihood of the plant becoming diseased. More generally, though, the spores germinate on the surface of a leaf, stem or young twig, and give rise to fungus threads which either penetrate between the guard cells of the stomata or bore directly through the cuticle. Other fungi, however, are incapable of attacking sound and healthy tissues, and to gain an effective entry into their hosts the spores must germinate on wounds (Fig. 58c). These " wound parasites " are often very troublesome to the fruit grower for they cause such diseases as apple and pear canker (*Nectria galligena*), and the silver leaf disease (*Stereum purpureum*). Once an entry has been secured the

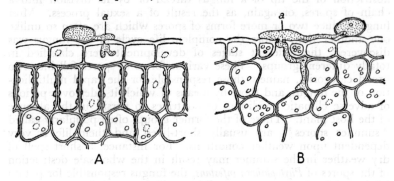

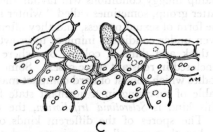

FIG. 58.—FUNGUS THREADS ATTACKING PLANT TISSUE.

A, germinating spore penetrating an epidermal cell.
B, germinating spore penetrating stoma.
C, spores germinating in wound.

a, appressorium.

hyphæ spread either between the cells of their host plant, so forming an *intercellular* mycelium, or by boring directly through them, that is, growing *intracellularly*. In the former case the actual contact of the fungus with its host's food supply is effected by means of button-like or branching suckers, or *haustoria* (Fig. 59). The penetration of the tissues does not necessarily result in immediate and obvious damage to the plant. It may, indeed, have no visible effect for some time as, for instance, in the case of Bunt of wheat, where infection which occurred at the seedling stage does not become apparent until harvest time.

The irritation set up by fungus organisms penetrating into plant tissue often acts as a stimulus to the further growth of the infected tissues, and galls or tumour-like outgrowths are formed. This occurs in wart disease of potatoes (*Synchytrium endobioticum*) and club root of turnips and swedes (*Plasmodiophora brassicæ*). Many fungi, however, do not gradually drain their host plants of foodstuffs, but bring about the immediate death of the cells with which they come into contact. The track of their mycelium is then marked by brown dead tissues, as it is in the case of barley infected with the leaf stripe fungus, *Pyrenophora graminea*.

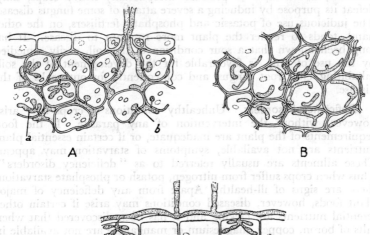

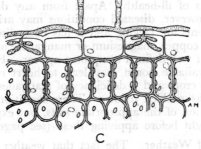

FIG. 59.—DIFFERENT TYPES OF HYPHÆ.

A, intercellular hyphæ showing penetration of haustoria into cells : (a) branched, (b) knob-like haustoria.
B, intracellular hyphæ.
C, superficial hyphæ with haustoria penetrating epidermal cells.

Specialisation. The fungi show great specialisation and are generally only capable of attacking either a particular host plant or a limited range of closely allied plants. Some of them, especially those responsible for the " rust " diseases of cereals and grasses, are very specialised indeed, for they exist in many strains with sharply limited infective capacities. This extreme physiologic specialisation prevents their indiscriminate spread amongst cereal crops and wild and cultivated grasses. Where the specialisation is less marked the parasites are generally confined to the limits of a particular natural order. Thus the fungus responsible for the club root disease of swedes and turnips is capable of attacking many other cruciferous plants, or again, blight of the potato may occur on other members of the Solanaceæ, but neither can infect plants outside of these two groups.

The extent to which fungi attack their host plants, or the intensity of an outbreak of disease, is largely dependent on the environment. The conditions both of the soil and of the atmosphere play an important part. As a rule crops growing in the most fertile soils are those which are most badly attacked. When growth is forced, and the leaves and stems are luxuriant and sappy, plants are usually more susceptible to parasites. Thus a too liberal use of nitrogenous manures, in attempts to produce exceptionally heavy crops, may, defeat its purpose by inducing a severe attack of some fungus disease. The judicious use of potassic and phosphatic fertilisers, on the other hand, tends to render the plant more resistant to disease. It has long been known that a sour condition of the soil, easily remedied by the use of lime, is favourable for the development of the soil-inhabiting club root fungus, and consequently for outbreaks of the disease.

Deficiency Disorders. Unhealthy conditions of crops may arise however, without the intervention of any parasite. If the food requirements of the plant are inadequate, or if certain essential plant nutrients are not available, symptoms of starvation may appear. These ailments are usually referred to as " deficiency disorders ". Thus when crops suffer from nitrogen, potash or phosphate starvation, there are signs of ill-health. Apart from any deficiency of major plant foods, however, diseased conditions may arise if certain other essential nutrients are lacking and it has been discovered that when salts of boron, copper, magnesium or manganese are not available in the soil certain well-defined symptoms appear. For instance, a soil deficient in available boron may lead to heart rot and dry rot in the sugar beet crop, and a deficiency of manganese to a debilitating disease known as speckled yellows. These conditions can be remedied by an application of the appropriate salts, but expert advice should always be sought before applying these (see pages 22-4).

Influence of Weather. The fact that weather conditions largely determine the extent to which diseases affect crops has been recognised from the earliest historical times. Indeed it was not until the fact had been established that fungi could be the direct cause of plant diseases that the view was abandoned that the weather was solely responsible for epidemics of disease. Parasitic fungi do not all require precisely the same conditions for optimum development. Moist weather is favourable for the spread of most of the serious plant diseases occurring in this country, but some fungi, such as the powdery mildews, thrive best in dry seasons. Although moisture is necessary for the germination of fungus spores, the process does not require long periods of wetness, and a film of dew, lasting for a few hours only, is usually sufficient for most spores to start growth and gain an entry into the tissues of their hosts. Temperature, too, is important, partly because high temperatures may lead to the desiccation and death of thin-walled spores and partly because their germination takes place only within certain limits. Much progress has been made in reducing the enormous losses caused by plant disease, and this is largely because the life cycles of many important fungus parasites have now been worked out in detail, and information gained on the environmental conditions which are suitable or unsuitable for their development.

THE CONTROL OF PLANT DISEASES

The methods in use for controlling fungus diseases of plants are for the most part preventive rather than curative. Hitherto, once the mycelium established itself within the tissues of the host it was out of reach of fungicides. With the advent of fungicides with systemic properties, however, the possibility is now presented whereby fungicides applied to one part of the plant can be translocated to other parts and so confer protection or, in some cases, eradicate already-established infection. A detailed knowledge of the life histories of parasitic fungi is particularly valuable when attempts are being made to devise methods for preventing the diseases which they cause, as it usually reveals the stage of growth at which measures can be applied with the best chances of success.

Chemical Seed Treatment. If it is known that a disease is caused by some specific parasite carried externally on the seed coat, then the organism can be killed by treating the seed with a suitable fungicide before sowing. Such fungicides are widely used and many important diseases of crops, notably on cereals, are prevented. To the unaided eye, however, apparent cleanliness of the seed can be deceptive as the contaminating spores are minute. With microscope examination spores of fungi causing the following diseases may well be found although in these days but rarely. Wheat may be contaminated with bunt (*Tilletia caries*), and barley affected with covered smut (*Ustilago hordei*) or leaf stripe and net blotch (*Pyrenophora graminea* and *P. teres*). As for oats, these may carry the spores of covered smut (*Ustilago hordei*) and loose smut (*Ustilago avenæ*) and in addition leaf spot (*Pyrenophora avenæ*). On rye the stripe smut fungus (*Urocystis occulata*) and bunt (*Tilletia caries*) may be present. These are all seed-borne diseases and the enormous losses which they used to cause are now prevented by seed disinfection. Substantial reduction is obtained also of seedling damage due to seed infection with *Septoria nodorum* and *Fusarium nivale*. In these examples, however, the fungicide is acting as a disinfectant. With the advent of fungicides with systemic action it is now possible to effect control over a number of seed-borne diseases not hitherto controlled by seed disinfectants ; among these are loose smut of wheat (*Ustilago tritici*) and loose smut of barley (*Ustilago nuda*) in which infection is deep seated within the embryo. It is also possible by seed treatment with some systemic fungicides to confer protection against foliar diseases resulting from airborne infection, e.g. cereal powdery mildews.

Infection from Soil. Although seed treatment can confer on seed a degree of protection from invasion by some parasitic fungi surviving in soil, its effectiveness is limited. In many instances the organisms must be starved from the soil by ceasing to grow susceptible host plants for a period or, better still, a build-up can be avoided by refraining from too frequent cropping with such host plants. The interval will depend on the level of infective material in the soil and on the nature of the parasite involved ; in some cases a year will suffice, but in others a much longer period will be required. In these circumstances, therefore, it is clear that the sequence of cropping must be modified, and susceptible crops omitted. Weeds, too must be suppressed because some of these may harbour the parasite and enable it to linger on from year to year. " Groundkeepers " or self-sown plants are an important means of carry-over of foliar and other diseases. For diseases that can only survive on the living host they are the so-called " green bridge ". It is, unfortunately, still not generally realised that if the same crops are grown too frequently on the same land disease is likely to appear, yet this is the case. When red clover is grown too frequently on the same field, clover rot (*Sclerotinia trifoliorum*) gains a hold; similarly when land is corned too heavily, take-all and whiteheads (*Gaumannomyces graminis*) usually appear.

Foliar Application of Chemicals. Although major developments have occurred in recent years in which some diseases spread by airborne spores can be controlled by the use of systemic fungicides either as soil drenches or seed treatment, foliar application of spray chemicals is still the major defence against diseases arising from airborne infection—supplemented by the use of disease resistance. In horticulture the direct prevention of pests and disease with chemicals applied as sprays, dusts or smokes has long been practised as also has control of potato blight (*Phytophthora infestans*) among agricultural crops. In recent years, however, the development of new fungicides and insecticides, many of them systemic in action, and new methods of applying them, have received considerable impetus. The recognition of the economic significance of many of the diseases of agricultural crops, notably the foliar diseases of cereals, has also made the use of chemical methods of control commercially and economically acceptable.

Resistant Plants. In addition to chemical methods of attacking plant parasites, another means of avoiding the losses they cause is by raising plants capable of resisting, either partially or completely, those species which normally attack them. The possibility of this is dependent upon the fact that most of the crop plants, when exposed to infection under uniform conditions of growth, show very different degrees of susceptibility, some even proving immune. Much of this capacity of resisting the disease is known to be inheritable in a simple Mendelian manner, and thus the building up of new resistant varieties has become practicable.

Prohibition of Imports. One other aspect of the problem of controlling the fungus diseases of plants has received much attention of late years. The fact has been recognised that widespread as many of the important ones already are, all of them are by no means universally distributed. Many countries, for instance, though growing potatoes, know nothing of wart disease or even of blight. It may be that local conditions are unsuitable for the growth of the particular fungus or, on the other hand, its absence may be due to the fact that it has, so far, failed to reach that country.

There is clearly much to be said for trying to prevent the introduction of new diseases, especially when it is realised that in their first appearance they are often very destructive. This is possible by regulating the movement of plants or seeds likely to carry them. Most countries now have legislation designed for this purpose, and require the inspection and certification for freedom from pests and diseases of plants intended for export and import, and the provision of health certificates. Each country has its individual problems and requirements, and these may on occasion involve complete prohibition or seasonal prohibition of certain plants or produce according to the countries of origin. In this country the Destructive Insects and Pests Acts have from time to time prohibited the importation of a comparatively small number of specified plants. They also prescribe the measures to be used in checking the spread of a number of new or newly introduced pests.

CLASSES OF FUNGI

There are four main classes of fungi :

Phycomycetes,	Basidiomycetes,
Ascomycetes,	Deuteromycetes.

This classification is based on the method of spore formation. In most Phycomycetes spores are formed in *sporangia* at some stage in the life cycle, in the Ascomycetes in *asci*, and in the Basidiomycetes on *basidia*. In the Deuteromycetes or Fungi Imperfecti, there are no sporangia, asci or basidia in the life cycle so far as is known, and only asexual spores, chiefly conidia, are formed. Another class of organisms, the Actinomycetes, is not grouped with the fungi because of its closer relationship to the bacteria, but mention of it is made, as it contains *Streptomyces scabies*, the organism causing common scab of potato.

In the following section the life cycles of a number of parasitic fungi are described, together with the symptoms of the diseases they cause and the methods used in controlling their spread. Many important species have been omitted, and those dealt with have been chosen largely because they provide examples of the various methods in use for the control of plant diseases. Before proceeding with fungi, however, it would be appropriate to give an account of a disease caused by an Actinomycete.

Common Scab of Potato (*Streptomyces scabies*). Probably the commonest of all potato diseases, common scab is seen on the tubers at lifting time as unsightly scabs resulting from the production of loose corky tissue by tubers in resisting invasion by the parasite. Infection is usually superficial, but various forms of scabbing may occur, including the deeper pitting of the tubers. This may be because of the existence of different strains or species of *Streptomyces*. Although usually most noticeable on the tubers, all other below-ground organs are susceptible to attack. The internal flesh of the tubers, however, is not affected.

The causal organism, *Streptomyces scabies*, possesses very slender, sparsely septate, simple or much-branched filaments, and from these arise aerial hyphal branches that terminate in a characteristic spiral twist. It is here that sporulation takes place by fragmentation of the filament progressively from the tip backwards. Infection occurs through lenticels and small wounds, and may take place at the earliest stages in the development of the tuber and continue as long as growth of the tuber is still taking place. The organism is soil-borne, and especially prevalent on light, gravelly soils low in organic matter, and on freshly broken land. Although scab may sometimes be very severe after newly ploughed up grass, the level of attack often becomes negligible after several years of arable cultivation. Severe attacks often follow hot, dry summers, and may arise on land that has received heavy applications of lime, ashes or other alkaline substances.

Control Measures. Potato varieties differ in their susceptibility to common scab : Arran Pilot, Maris Peer and Pentland Crown are fairly resistant ; Home Guard and Craig's Alliance are less so, and Majestic, Maris Piper and Désirée are very susceptible.

Irrigation reduces the incidence of scab, especially on the lighter soils. In a dry season it may be of advantage to irrigate rather earlier than usual—at the beginning of tuber initiation.

The disease is often reduced following a making up of a deficiency of organic matter in the soil by the ploughing in of a green crop.

Scabbed seed may well produce a clean crop where conditions are not conducive to scab. With very badly scabbed seed, however, if the eyes have been injured, a gappy crop will result. Chitting the seed will guard against this, as any tubers failing to sprout through scab or any other disease can be discarded before planting.

DISEASES CAUSED BY PHYCOMYCETES

Club Root of Swedes and Turnips (*Plasmodiophora brassicæ*) Until recently it was usual to classify the organism causing club root, or finger-and-toe as it is sometimes called, with the slime fungi or Myxomycetes, but it is now included in the Phycomycetes. It is completely lacking in the mycelium characteristic of most fungi and consists instead of masses (plasmodia) of much-nucleated protoplasm with no defining walls. The plasmodium breaks up into small spherical spores, which on germinating produce amœba-like bodies capable of changing their shape and travelling by a slow creeping movement. *Plasmodiophora brassicæ* attacks only those plants belonging to the Cruciferæ. Swedes, turnips, rape, cabbage, brussels sprouts and mustard are the crops most commonly affected, but weeds such as charlock, wild radish and shepherd's purse are also attacked (Fig. 60), as also are wallflowers.

The symptoms of club root are irregular masses of nodule-like swellings on the root system. These remain firm during the summer months, but as the autumn advances they decay and form a foul-smelling brown mass. The plasmodium of club root cannot be seen by the unaided eye, but it can be demonstrated by the use of appropriate staining methods and the aid of a microscope. It occupies abnormally large cells, the contents of which, in unstained prepara-tions, appear to be normal. Towards the end of the summer the plasmodium in the " giant cells " gives rise to small rounded spores. When the diseased tissue decays these are set free and contaminate the soil. On germination the spores give rise to minute bodies which are able to move about in the soil water. These spores and the slow-moving amœboid-like bodies into which they subsequently

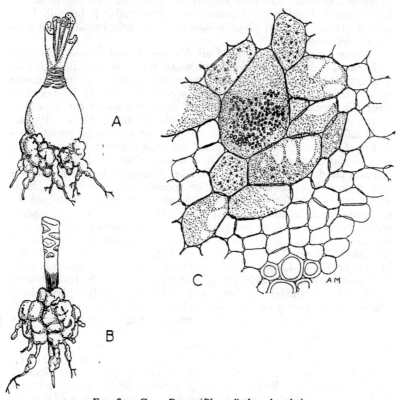

FIG. 60.—CLUB ROOT (*Plasmodiophora brassicæ*).

A, swede, B, cabbage, affected with club root.
C, enlarged section of diseased root showing giant cells and stages of spore formation.

develop bring about the infection of the root hairs of susceptible plants. Once inside the root the fungus increases rapidly in bulk, and it may subdivide to form more than one plasmodium in a cell. As fresh cells are penetrated they are stimulated to further develop-ment, forming giant cells and the mass of gall-like growths.

Control Measures. Club root is very prevalent on soils which are deficient in lime, especially if these are waterlogged or badly drained. Its occurrence is usually as reliable an indicator of lime shortage as the presence of spurrey or sheep's sorrel. The only direct approach towards controlling the disease is by the improvement of drainage and the application of lime to the soil. Freshly slaked burnt lime, hydrated lime or ground quick lime are best. Ground limestone, chalk and other forms of lime are not so good. It should be given as soon as the crop is cleared, but it must be applied in a finely divided state and well incorporated into the soil. Where club root has been troublesome, however, it is wise to seek advice before applying lime because, if it is applied in excess, it may pave the way to other troubles. On soils subject to club root the use of super-phosphate and sulphate of ammonia is inadvisable ; instead, basic slag or ground mineral phosphate, and nitro chalk or nitrate of soda should be used.

Where possible the roots of all diseased plants should be burnt. This can be done in gardens and small holdings, but it is usually impracticable where susceptible crops are grown on an extensive scale. In farm practice, therefore, precautions should be taken to limit the spread of the parasite, and then to remedy the condition of the affected field. Great care should be taken in the disposal of an infected crop. If the roots are used for feeding in yards, frag-ments of diseased material may fall into the litter and so find their way to the manure heap, and in this way clean land may become contaminated. If the roots are fed on grassland, however, there is less chance that the disease will be spread. All cruciferous plants which are bought for transplanting should be carefully examined for the presence of club root, and any which show signs of disease should be burnt. None but perfectly healthy plants should be used for transplanting. It should be noted in this connection that under the Sale of Diseased Plants Orders, 1927–52 and the Sale of Diseased Plants (Scotland) Orders, 1936–53, plants substantially affected with club root may not be sold for planting purposes. On transplanted crops, a good control can be obtained by dipping the plants in a mixture of 50–100 g of pure calomel to 1 litre of water. On land where the disease has appeared in severe form, as long a period as possible should elapse before a susceptible crop is again grown. Dazomet can be used as a soil sterilant on small areas.

So far, none of the commonly grown cruciferous crops has been found to be immune to club root, although the yellow-fleshed, purple-top turnip called the Bruce and grown in Scotland for many years possesses a high degree of resistance. Some varieties of swedes, too, including the purple swedes Coxton Crofter and Chignecto and the green swedes Wilhelmsburger and Wilhelmsburger Prima have shown a high degree of resistance in trials in this country. The constant selection inherent in seed multiplication of swede varieties, however, can result in changes in detailed varietal characters and the degree of resistance can, in consequence, vary according to the seed source. It is known too that physiological races of *Plasmodiophora brassicae* exist, and these might conceivably result in differential re-actions according to the variety of the host. Of the brassicas used for forage crops, kale is probably the least seriously affected.

Powdery Scab of Potato (*Spongospora subterranea*). All below-ground portions of the plant are subject to attack, but there is little adverse effect except on the tubers, on which the disease occurs most commonly as circular scabs. These originate as raised spots or pimples from which the skin later ruptures and breaks away leaving a frill round the edge and exposing a brown, powdery mass of spore balls beneath. When the spore balls are mostly shed, it is often difficult to distinguish between these scabs and those of common scab without microscopic examination. Unlike common scab, the organism is favoured by relatively low soil temperatures and high moisture content, and under conditions especially favourable to attack the disease may assume a much more serious canker form, in which the tubers may be malformed and considerable portions of the flesh become involved. Occasionally too, a raised scab may continue to grow and produce a tumour-like outgrowth that may readily be confused with the abnormal growths of wart disease. Pentland Crown commonly shows this symptom.

Control Measures. The disease is always worst in rainy seasons and in cold wet soils, and where the land is cropped with potatoes too frequently. A suitable rotation should be adopted and measures taken to improve drainage. Potatoes substantially affected with the disease should not be used for seed, although if infection is only slight, unsprouted seed tubers can be disinfected before planting by steeping them for 3 hours in a solution of formaldehyde containing 0·5 l of formalin (40 per cent formaldehyde) in 120 l water. Under the Sale of Diseased Plants Orders, 1927–52, potatoes substantially affected with powdery scab may not be sold for planting.

Wart Disease of Potatoes (*Synchytrium endobioticum*). Wart disease was first discovered in this country in 1900 in Cheshire and Lancashire. From these counties it has spread mainly in a northerly and southerly direction and, at present, it is confined to approximately the western and the midland parts of the country.

The haulm and foliage of infected plants have a normal appearance and the disease only becomes evident at lifting time, when rough-surfaced, rounded or irregular, tumour-like masses are found on the tubers (Fig. 61). These warts vary in size : on tubers slightly attacked they are little more than small, rough swellings situated invariably on the eyes, but in bad attacks the whole surface of the tuber may be covered with them. In the early stages of growth they are not unlike pieces of the curd of a cauliflower, but as they age their colour changes to a deep brown or black. At this stage they contain many thick-walled, dark-brown sporangia, which are able to survive in the soil for several years. When conditions are suitable, they liberate many motile spores (zoospores), and when these come into contact with the eyes of young tubers infection results. This is followed by the repeated divisions of the neighbouring cells, resulting in the formation of a mass of succulent tissue which in turn is liable to fresh infections. In this large numbers of sporangia, but no signs of hyphæ, develop. These sporangia give rise to motile spores which either re-infect neighbouring cells or tubers near them, or act as gametes and fuse in pairs. The fusion-products, on infecting fresh tissue, give rise to the thick-walled resting sporangia which are set free in the soil when the warts decay.

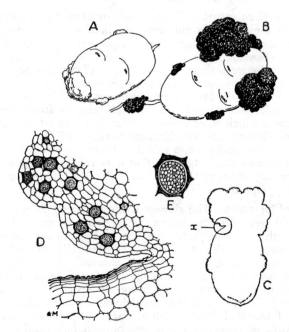

Fig. 61.—Wart Disease (*Synchytrium endobioticum*).

A, immature warts on tuber.
B, mature warts on tuber.
C, tuber cut lengthwise.
D, section taken at (*x*) showing diseased tissue containing resting sporangia.
E, resting sporangium.

Control Measures. Fortunately, early in the study of this disease it was found that some varieties, even when grown on heavily contaminated soil, escaped attack, and so a method of checking the spread of this dangerous disease was available and advantage of it was promptly taken. The disease was scheduled under the Destructive Insects and Pests Acts. Provision was thus made for compulsory notification of all outbreaks, and with this was coupled the requirement that, on land on which the disease had occurred at any time, planting of potatoes was restricted to approved immune varieties. These and other measures are incorporated in the Wart Disease Order, 1958. Under this Order, moreover, Protected Areas may be defined in which potatoes sold for planting must be certified officially as having been grown on land free from wart disease or, alternatively, be uncertified once-grown seed raised within the Area. An area round the Wash is so far the only Area so defined. Among other provisions of the Order are that neither potatoes visibly affected with wart disease nor tubers from a crop in which the disease has been found may be sold or offered for sale for planting.

Immune Varieties. The following list gives the names of the more commonly grown immune varieties of potato approved by the Ministry of Agriculture, Fisheries and Food under this Order.

Early: Arran Pilot, Home Guard, Ulster Chieftain, Ulster Prince.
Second Early: Craig's Alliance, Craig's Royal, Maris Page, Maris Peer.
Maincrop: Majestic, Maris Piper, Pentland Dell, Record, Désirée, Kerr's Pink, Pentland Crown.

Some of the better-known older varieties susceptible to wart disease are : Arran Chief, British Queen, Eclipse, Duke of York, Epicure, King Edward, Ninetyfold, Royal Kidney, Sharpe's Express and Up-to-Date. (See also Chapter X.)

The problem of finally controlling this disease rests with the plant breeder. The mode of inheritance of immunity and susceptibility has now been investigated and found to take place on simple Mendelian lines, and thus the raising of varieties which the fungus cannot attack is a relatively simple matter. The difficulty, however, is to raise sorts which, whilst possessing immunity, are equal in every respect to the best known. The breeder has to produce large numbers in the hope of finding one or two worthy of general cultivation. All new varieties of potatoes are now submitted for an official test in the early stages of their multiplication, and if a variety does not prove immune from wart disease there is little prospect of its further propagation or introduction into commerce. The situation is complicated, however, by the presence in other countries of strains of the fungus to which varieties of potatoes resistant to wart disease in this country may be susceptible.

Potato Blight (*Phytophthora infestans*). The potato crop, from the time of its introduction in Elizabethan days until about the year 1845, does not appear to have suffered from this well-known disease. In that year it broke out in the Isle of Wight about the middle of August and, in the course of a few weeks, spread throughout the British Isles. Contemporary accounts state that it destroyed between one-quarter and one-half of the crop. In the following year an epidemic started earlier, and was even more destructive. In Ireland it was so severe that nine-tenths of the crop rotted. The Irish famine of 1846 was the direct result of this outbreak.

In recent years the plant breeder has progressed considerably in the production of potato varieties resistant to blight. Varieties differ in their relative susceptibility, however, and do not always combine both foliage and tuber resistance. King Edward is very susceptible in both tubers and leaves, Majestic is less susceptible in the leaves and still less so in the tubers. Whilst Stormont Enterprise possesses considerable resistance in both foliage and tubers the reverse is largely true for Ulster Beacon. As with other diseases, the plant breeder has to contend with physiologic races of the pathogen, and a number of races of *P. infestans* is now known to exist. Varieties at first resistant may therefore later succumb, and the same variety of potato may also not always respond consistently in different parts of the country, depending upon the geographical distribution of the races.

The first symptom of blight is the appearance of dark green, water-soaked patches on the tips and margins of the leaves. These increase in size and number from day to day, turning, as they age, to a dark brown colour (Fig. 62). The disease generally spreads

with extraordinary rapidity, and it is not unusual for every plant in
a field to be infected a fortnight after its first appearance. Later,
the tubers may become infected, in which case discoloured and slightly
depressed areas appear on the surface. These, when cut across,
show rusty-brown stains extending inwards from the skin. Blight
does not grow down through the haulm to the tubers; infection
occurs from spores that have been washed down through cracks in

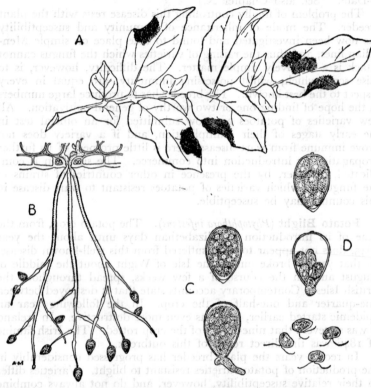

FIG. 62.—POTATO BLIGHT (*Phytophthora infestans*).

A, symptoms on leaf.
B, branches of fungus emerging from stoma.
C, spore germinating directly without producing zoospores.
D, sporangium liberating zoospores.

the soil. If blighted tubers are clamped they decay rapidly. The
rotting is usually the work of other organisms, mainly bacteria, which
invade the infected tissue.

When the under surface of an infected leaf is examined more
closely, a fine white fringe is seen towards the margin of the infected
areas. This consists of branching hyphæ which bear lemon-shaped
spores. If these fall on to foliage wetted by rain or dew they give
rise to a number of minute motile zoospores, each of which is capable
of causing infection. A hypha is sent out from the germinating
zoospore and pushes its way through the epidermis or the breathing

pores in the leaf. A mycelium then develops within the leaf which discolours and kills the cells. If the spores are washed down to the tubers these also become infected, and their cells are killed and discoloured in the same way.

In some fungi, such as *Pythium* and *Peronospora* which resemble *Phytophthora infestans* very closely, thick-walled resting spores (oospores) capable of surviving through the winter are commonly formed. These germinate in the following spring, or even a year or more later, and thus ensure survival of the fungus, which would otherwise be eliminated by the death of the readily perishable conidia and zoospores. Similar resting spores have been obtained in laboratory cultures of the blight fungus, but their occurrence has only recently been satisfactorily demonstrated in the field and this only in Mexico. It would seem, therefore, that the fungus is usually perpetuated from year to year solely by mycelium in blighted tubers. These may be left in the field the previous autumn or discarded when the crop is graded. The disease may also have its origin in blighted tubers which are inadvertently planted with the healthy ones. The first outbreaks of blight are frequently found on " volunteer " plants on or near the site of old potato dumps. Where there is self-sown rubbish of this sort on a farm there is usually disease.

It has long been recognised that weather conditions mainly determine the severity of potato blight, and that warm, damp, muggy conditions, especially during the latter part of July, favour severe epidemics. The motility of the spores is clearly dependent upon a film of moisture on the foliage and stems and, the more persistent it is, the better are the chances for germination and subsequent infection. Rain, too, splashes the spores further and further afield. Indeed it is now known that from June onwards an outbreak of potato blight may be forecast with reasonable accuracy if, over a continuous period of 48 hours, the relative humidity of the atmosphere exceeds 75 per cent and the minimum temperature does not fall below 10 °C. These infection periods are known as Beaumont periods after the British plant pathologist who defined them. Temperature and humidity, moreover, determine whether the fungus will spread rapidly once it has made an appearance. Hot dry weather makes a severe epidemic improbable and prevents the spread of any infections that have already become established.

Control Measures. As self-sown tubers may be a source not only of blight infection but of virus diseases as well, they should not be left on the fields at harvest time. At the end of harvesting the field should be harrowed and all exposed tubers collected and stored separately. Blighted tubers rejected at the time of grading the crop from the store or clamp should never be left around to serve as a source from which blighted plants can arise the following season. Such plants act as primary infectors. In the absence of wet rots, such tubers when boiled can be used for feeding, otherwise they should be destroyed. Neglected potato dumps are a fertile breeding-ground for disease ; they should be sprayed with a suitable weedkiller to destroy " volunteers ". Wherever possible seed should be boxed, periodically examined and diseased tubers destroyed.

Spraying. In addition to taking precautions for the reduction of sources of infection, it is necessary in many areas to protect the growing crops. This can be done by spraying the haulm with a suitable fungicide which will kill the spores when they come into contact with it. Two copper fungicides that were at one time used extensively are Bordeaux and Burgundy mixtures. A good Bordeaux mixture for potatoes is one containing 10 kg of powdered or granular copper sulphate (bluestone) and 12½ kg of best hydrated lime in 1,000 l of water. The bluestone should be at least 98 per cent pure, and the hydrated lime (bought in paper bags) should be purchased fresh each year. Burgundy mixture is prepared in the same way, but contains 12½ kg of washing soda instead of the hydrated lime. These now historic mixtures have been replaced by proprietary fungicidal preparations sold as powders or pastes for ready mixing with water. Copper has largely been superseded by organic fungicides such as the dithiocarbamates and organo-tin compounds which, although they are less persistent, are less liable to scorch or check the growth of the potato foliage. For some years low-volume spraying against potato blight, either with ground machines or from the air, has been practised successfully in many parts of the country. Only about 225 l or less of spray liquid are used per ha. In south and south-west England, in Wales, and other districts where the disease appears first and is most severe, spraying should be regarded as a routine summer operation. In these districts, however, the crop unit may be very small indeed, and measures more in keeping with the farm economy would be timely planting, good earthing up, and delayed lifting until after the haulm is completely dead.

The date at which the crop should be sprayed varies with the season and the district, but it should be done before, or as soon as, blight spots are seen on the foliage. In this country, blight warnings based on the widespread occurrence of Beaumont periods are reported in the Press and by sound radio and television, together with records of first outbreaks. The information is useful even if it may not immediately be applicable locally, for it calls attention to the advisability of having everything in readiness. On susceptible main crop varieties and in the larger potato growing areas, however, it is customary to apply a first " insurance " spray in the first week of July or just before the crop meets across the rows, whichever is the earlier. Subsequent applications are made according to weather conditions. In these later applications the use of one of the tin-containing fungicides affords some degree of protection from tuber infection. Earlies usually do not require spraying. In some districts, where there is difficulty in carting water, dusting with proprietary copper powders is practised. Four, five, or more applications at fortnightly intervals may be necessary, beginning before the plants are earthed up.

Fine weather is essential for spraying operations, and no opportunities should be missed of carrying out the work on a still, windless day, when the spray mist will settle down uniformly. Once the coating has dried thoroughly on the foliage it will adhere for a long time, but if rain falls before this happens much of it will be washed away. If dusting is practised it should be carried out early in the morning whilst dew is still on the leaves. Spraying or dusting with copper fungicides should not be done in the immediate vicinity of large industrial areas, as injury to the foliage may result owing to an interaction of the spray and acid fumes in the atmosphere.

The result of protecting the crop from infection is that the haulm continues to grow for a longer period than it otherwise would and, as a consequence, the crop from a field which has been thoroughly sprayed is larger than that from a field in which the foliage has been prematurely killed by blight. Many farmers, however, do not carry out this spraying or dusting operation themselves, but engage the services of spraying contractors.

Protection of the Tubers. The better the earthing-up of the rows, the less chance blight spores have of reaching the tubers during the growing season. A great deal of tuber infection may take place at lifting time, however, during the period that the potatoes lie exposed on top of the soil. If lifting can be delayed until the tops have been quite dead for at least a fortnight, the risk of this happening is much reduced, as most of the spores will have perished. If the tops remain green very late, however, and blight is present, they should be " burned off " about 10 days before lifting. One of the most efficient substances for potato haulm destruction is sulphuric acid, at the rate of 150 l B.O.V. in 1,000 l/ha of water : it is still occasionally used but requires acid-resistant machines and considerable care in its use. Newer chemicals, more convenient to apply and at the same time non-corrosive have been developed and have come into general use as destroyers of potato haulm—and of other vegetation as well. Of these substances dinoseb and diquat have been found very effective. Mechanical methods of destroying potato haulm have been introduced in recent years, and these have also been shown to reduce tuber infection at lifting. An associated and very valuable advantage of burning off potato haulm is the destruction of weeds, which greatly facilitates the digging and lifting operations.

Two other Phycomycetes, related to the potato blight fungus and belonging to the sub-class Peronosporales are *Peronospora schachtii* causing downy mildew of sugar beet ; and *Peronospora parasitica*, various physiologic strains of which are found causing downy mildew on wild and cultivated members of the Cruciferae, including brassicas and wallflowers.

Downy Mildew of Sugar Beet (*Peronospora schachtii*). This fungus attacks sugar beet, beetroot and mangels and can be serious on sugar beet crops in the midland and eastern counties of England, especially where seed and root crops are grown in close proximity to one another. Seedlings are occasionally attacked, but symptoms are more often seen after singling, when the growing point and central rosette of leaves may become affected. The young leaves become pale-green, thickened, curled and brittle, and on the under surface the fungus may be seen as a lilac-grey, downy growth which later changes to buff-grey. Spore production is encouraged by mild weather and high humidity and, under conditions favourable to infection, the older leaves too may be attacked, but mostly towards the basal parts of the leaf blade. With the death of the heart leaves, confusion can often arise through the similarity of symptoms with those of heart rot resulting from boron deficiency. On the seed crop the symptoms at first resemble those on the root crop, and many of the stecklings may be killed. Infected plants, moreover, are more readily killed by frost. Infected plants that survive are stunted, with suppressed axillary floral shoots, and later infections cause direct damage to the seed-bearing stalks with resultant reduction in yield. Oospores may be formed in the leaves and flower parts, and it is thought that the disease might be seed-borne, although there is little indication of this being a serious factor in this country. The greater danger lies in the continuity throughout the year of sources of infection from air-borne conidia. This occurs where seed crops and root crops are grown in close proximity.

Control Measures. Stecklings raised for the seed crop should be grown in isolation and planted as far removed as possible from root crops or other sources of infection. Individual plants found later to be affected should be lifted immediately and buried *in situ*, to minimise dissemination of spores through the crop. Conversely, root crops should not be sown in the vicinity of crops being raised for seed. Control by the application of fungicides is scarcely practical or effective on root crops, although some benefit might be derived from protecting the more restricted steckling beds by the periodic application of a suitable fungicide in the early stages of growth. Spraying, however, should be discontinued when there is the likelihood of frost occurring.

Downy Mildew of Brassicae (*Peronospora parasitica*). Many cruciferous plants are attacked, including most of the cultivated species. Infection is commonly found on the older leaves of kale, cabbage, brussels sprouts and cauliflower, where it is usually of little consequence. The fungus occurs as a thin, greyish-white, downy growth in scattered, angular patches bounded by the veins on the underside of the leaves. The corresponding areas on the upper surface are often pale and chlorotic.

Control Measures. There are no practical measures that can be adopted for controlling the disease in field crops, and its incidence there is not usually of sufficient economic significance to warrant any special concern. The disease is favoured by high humidity, however and very severe infection can arise in seedlings such as cauliflowers growing under glass, especially when the seedlings are dry at the roots and atmospheric humidity is high. In these circumstances abundant ventilation should be given, and it is helpful if the seedlings are not too crowded. Protective spraying at regular intervals with zineb or a fungicide containing copper will sometimes hold the disease in check but, to be effective, applications must commence before the disease becomes established. If copper is used, care should be taken to avoid it reaching the root as in some circumstances it may cause damage.

DISEASES CAUSED BY ASCOMYCETES

The chief characteristic of the Ascomycetes is the formation at some stage in the life-cycle of an *ascus* (a sac-like cell inside which a definite number of spores, usually eight, is produced). This class, which includes many important parasites, is widely distributed, and some of the more important types are now described.

An important family is the *Erysiphaceæ*, for its members are responsible for the powdery mildews which occur on a wide range of host plants. These include clovers, swedes, apples, roses, gooseberries, hops, cereals and grasses.

The Powdery Mildews. The mycelium is superficial and spreads over the foliage, the young shoots, and occasionally the young fruits, in the form of a delicate cobweb-like layer. This is white in colour at first, but becomes grey or brown as growth proceeds. The hyphæ are fixed to the host plant by haustoria driven through the cuticle of the epidermal cells, and it is by means of these suckers that the fungus absorbs food. Though the invasion of the host by the parasite is extremely slight, the infected plants often die prematurely, and consequently these mildews are responsible for much damage. Later, from the superficial weft of mycelium, upright branches arise, and transverse walls form to divide them up into chains of elliptical conidia, which are abstricted one after another. At this stage the surface has a characteristic mealy appearance, but after a time the mycelium becomes denser and often changes in colour. In the thickened weft small yellow, and finally brown or black, dots develop. These are the cleistocarps. They are completely closed, globular structures containing one or more asci, within each of which are the ascospores. The wall of the cleistocarp is formed of a layer of dark, close-fitting cells, from some of which simple or branched appendages grow. The various genera comprising this family are identified by the shape of these appendages and, also, by the number of asci within the cleistocarps; it is almost impossible to distinguish them by the conidial chains, which are very similar to one another in general appearance.

The spread of powdery mildews during the growing season of their hosts is brought about by conidia, or " summer spores " as they are sometimes called. These are readily caught up in air currents and are scattered far and wide. When one of them reaches the surface of a leaf or shoot of a suitable host plant it anchors itself by the formation of a sucker-like disc (appressorium); a hypha is then formed from which an haustorium is soon driven into an epidermal cell. It is the ascospores within the cleistocarps, however, which provide the over-wintering stage of the fungus. They are set free explosively in the spring and early summer, and on germination give rise to mycelium similar to that developed from a germinating conidium. The life history of *Erysiphe graminis*, which is responsible for the widely distributed powdery mildew of cereals and grasses, is typical of most powdery mildews.

Powdery Mildew of Cereals (*Erysiphe graminis*). This fungus

(Fig. 63A) affects wheat, barley, oats, rye and the common grasses, and it exists in a number of specialised strains, each usually specific to its own host. The disease is most noticeable in crops which have been thickly seeded and liberally manured ; it is also often severe on late-sown spring corn, especially when this has been checked by drought. Lodging of cereals is sometimes attributed to mildew, but there is no evidence for this, although in most seasons it reduces the yield and quality of grain. Cereal varieties vary in their susceptibility

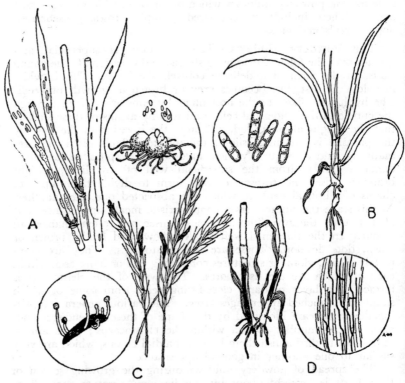

FIG. 63.—FOUR DISEASES CAUSED BY ASCOMYCETES.

A, Powdery mildew (*Erysiphe graminis*). *Inset :* cleistocarps, asci and ascospores.
B, Barley leaf stripe (*Pyrenophora graminea*). *Inset :* conidia.
C, Ergot (*Claviceps purpurea*). *Inset :* ergot germinating.
D, Take-all and whiteheads (*Gaumannomyces graminis*) *Inset :* Runner hyphæ on rootlet.

to mildew and at one time control depended almost solely on the skill of the plant breeder in producing numerous varieties. Such new varieties soon succumbed to mildew consequent upon a rapid expansion in area devoted to them and the resultant selection pressure giving rise to build-up of mildew races pathogenic to the new varieties. Knowledge of varietal susceptibility is now supplemented by the use of fungicides applied either as a seed treatment or as foliar sprays, combined with careful husbandry to reduce carry-over of infection on volunteer plants. Much is also known now of conditions favouring infection thus enabling more precise timing for the application of foliar sprays.

An important sub-class of the Ascomycetes is the **Pyrenomycetes,** or Flask Fungi, examples of which are the fungi causing the well-known ergot of cereals and grasses (see Fig. 63c), and choke of cocksfoot and other grasses.

Ergot of Cereals and Grasses (*Claviceps purpurea*). The hard, purplish-black, horn-like ergots of *Claviceps purpurea* are commonly seen protruding from the florets of grasses, especially ryegrass. The fungus exists in a number of physiologic strains affecting cereals and a wide range of grasses. Of the cereals, rye is most susceptible and wheat and barley much less so. The now rarely seen Rivet, and varieties derived from Rivet, used to be varieties of wheat more commonly affected. Oats are seldom infected, and only once has it been recorded on oats in this country.

Only the inflorescence is affected, and infection by means of ascospores takes place through the open florets. The fungus invades the ovaries to convert them into a white, compact mass of mycelium ; the early stage in the production of the sclerotia or ergots. On the tip of the developing ergot, conidia are abstricted and a sugary fluid or " honey-dew " is secreted. This is the sphacelia stage, when rapid dissemination of the fungus takes place from flower to flower by means of insects or rain splashes. The ergots gradually lengthen, and become hard and black, so that by harvest time they are mature. They then fall to the ground or they may be threshed out and sown with the grain.

If not buried too deeply in soil, the ergots germinate in early summer to give rise to small, club-shaped stalks that swell at the tips to produce stromata in which are embedded numerous flask-shaped perithecia. It is within these perithecia that the asci, each containing eight thread-like ascospores, are produced, at about the time of flowering of the host plant. The ascospores are then forcibly ejected and carried by the wind to cause fresh infections in newly opened florets.

The ergots of *Claviceps purpurea* have long been known for the toxic and medicinal properties of the alkaloids that they contain ; gangrenous and nervous disorders in both man and animals can arise if food containing quantities of ergot as a contaminant are consumed. The alkaloidal content of ergots, however, can vary very widely. In his book *Ergot and Ergotism* (1931), George Barger describes very fully the symptoms and history of human ergotism.

Control Measures. Rotation of crops, deep ploughing to bury fallen ergots, and the use of clean seed will remove sources of infection. Ergots buried deeply in the soil do not germinate, and they are relatively short-lived. Grasses in pastures, especially ryegrass, are sometimes heavily infested, but this should not occur if the grasses are prevented from flowering by close grazing or mechanical topping. With contaminated grain samples, much of the ergot is removed in the cleaning operations, but seed can also be freed from contamination by pre-soaking in water and floating off the ergots in a saturated salt solution.

o

Choke of Cocksfoot (*Epichloe typhina*). Many kinds of hedgerow and pasture grasses besides cocksfoot are affected. The disease is conspicuous because of the bright-yellow stromata of the fungus, *Epichloe typhina*, which sheathes and suppresses the flowering tillers. The mycelium of the fungus is systemic within the tissues, and perennates in the rhizomes and vegetative organs of the host, only appearing on the surface of the plant during the summer months. In early spring the mycelium emerges as a weft encompassing the youngest leaf of the developing flowering tiller. As the mycelial growth increases in density to form a stroma, the inflorescence becomes trapped and degenerates, a feature from which the name choke has been derived. At this stage the stromata are white, and from them are abstricted numerous conidia. As the season progresses the colour of the stroma changes to bright yellow, and immersed in it are numerous perithecia within which are produced asci, each containing eight thread-like ascospores. The ascospores when mature are forcibly discharged into the air and carried away by wind. The role of conidia and ascospores is obscure, and little is known about the control of the disease. Fortunately the disease appears to be of little consequence in pastures, and its economic importance lies more in the damage sometimes caused to crops raised for seed production and in the shortening of their period of productivity.

Other troublesome fungi in this sub-class of the Ascomycetes are those responsible for apple and pear canker (*Nectria galligena*), apple scab (*Venturia inaequalis*), pear scab (*V. pirina*), and certain species of *Pyrenophora* (conidial stage = *Drechslera*) which attack barley and oats. For example, leaf stripe of barley (*Pyrenophora graminea*), and leaf spot of oats (*P. avenæ*) are seed-borne parasites that are often responsible for much damage. A very similar disease, net blotch of barley (*P. teres*), is also widely distributed, but it is not so important. The causal agents of these three diseases are often referred to as *Drechslera graminea*, *D. avenæ* and *D. teres*, since their perfect stage is rare and has not been definitely recorded in Britain.

Leaf Stripe of Barley (*Pyrenophora graminea*). The characteristic symptoms of leaf stripe are long and narrow brown stripes on the leaves. The fungus causing the disease is present in a dormant state on the surface of the grain or in the seed coat, but it cannot be detected by the unaided eye. When infected seed is sown the fungus resumes growth and attacks the young shoot as it emerges from the seed. The course of the disease is usually progressive ; the first leaf becomes infected, then the second, the third, and so on, and finally the ear. This may be so severely affected that a condition of " blindness " results, in which the grains fail to develop normally. When adverse weather conditions prevail (low soil temperatures favour the disease) the young shoots may be killed before they pierce through the soil, and affected seedlings that do emerge may be so weakened that they eventually succumb. Leaf stripe, therefore, may lead to thin stands, reduced yields, and grain of low quality. In the past this disease caused much loss, but it can be prevented by treating the seed with an Approved organo-mercury or other seed disinfectant.

The related disease, net blotch of barley (*Pyrenophora teres*), is controlled in the same way.

Leaf Spot of Oats (*Pyrenophora avenæ*). Leaf spot of oats is very similar to the related leaf stripe disease of barley. The fungus is carried on the surface of the grain or in the husks and seed coat, and the young shoot is infected as it emerges from the seed. In cold wet weather many of the infected seedlings die ; others are severely crippled, and their leaves bear reddish-brown streaks. For a period after tillering the symptoms of the disease are not very noticeable, but later brown spots appear on the maturing upper leaves, and it is from these that spores are spread which later infect the grain. The disease can usually be avoided by treating the seed with an Approved organo-mercury seed dressing. Most reputable seed merchants now treat cereal seed as a routine measure so that leaf stripe and leaf spot are much less common than in the past.

Take-all (*Gaumannomyces graminis*). Take-all and whiteheads of wheat, barley and oats is another important disease, for it often reduces the yield considerably, especially when cereal crops are grown too frequently on the same field. The disease is due to a soil-inhabiting fungus (*Gaumannomyces graminis*) that infects the roots and the base of the stem (Fig. 63D). There are at least two strains of the fungus: an oat strain that can affect all three cereals and is confined mainly to the wetter parts of the country, and a wheat and barley strain that does not affect oats. Affected plants ripen prematurely and produce bleached ears containing little or no grain, and it is this feature that has led to the use of the term " whiteheads ". Somewhat similar symptoms, however, may arise from other causes, but when the condition is due to *Gaumannomyces* there is a black discoloration at the base of the stem and the roots are black and rotten. As crops other than cereals and grasses are not attacked, the fungus can be starved from the soil by avoiding susceptible crops and by keeping the land free from weed grasses. It is for this reason that, in an arable rotation with a high proportion of cereals, non-cereal break crops are introduced with the two-fold purpose of cleaning the land of weed grasses and reducing take-all level. In practice it has been found possible to crop continuously with spring barley and, subsequently, with winter wheat. Indeed there occurs a decline in severe take-all with an increase in slight and less damaging infection with the take-all fungus as the succession of susceptible cereals lengthens after the fifth or sixth crop. The phenomenon is referred to as take-all decline.

The effect of the disease can be minimised by early cultivation of infected stubble, spring sowing, reduced rates of seeding and the maintenance of good drainage and a high level of fertility.

Another important sub-class of the Ascomycetes are the **Discomycetes** or Cup Fungi, characterised by asci borne on open, frequently cup-shaped, fruiting bodies called *apothecia*. It includes the important genus *Sclerotinia*, members of which give rise to many destructive diseases including clover rot, and brown rot of apples and plums.

Clover Rot (*Sclerotinia trifoliorum*). When red clover is grown too frequently on the same land a condition known as " clover sick-ness " often appears.   This is sometimes due to the presence in the soil of a parasitic eelworm, but more often it is caused by the clover rot fungus.   The disease is likely to be most severe during a mild winter following a wet autumn, but when dry frosty conditions prevail the rot is checked, and spells of such weather may lead to a partial crop recovery.

The first symptoms appear in the autumn months on plants

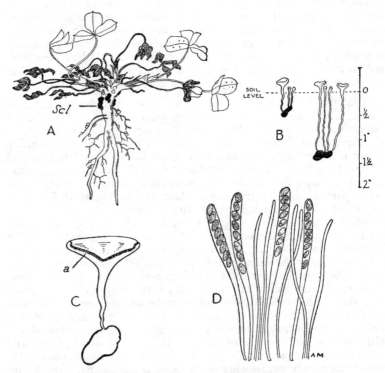

FIG. 64.—CLOVER ROT (*Sclerotinia trifoliorum*).
A, affected plant with sclerotia (*Scl.*) on root.
B, germinating sclerotia.
C, sclerotium and apothecium cut lengthwise showing position of asci (*a*).
D, asci containing ascospores.

arising from sowings made in the previous spring.   The early stages of the disease, characterised by a " peppering " of the foliage with small brown spots, are not generally noticed, and it is only in the late winter and early spring, when widening patches of dead plants are seen, that the extent of the trouble is realised.   On the upper part of the root and the neck of such plants black bodies develop which vary in size from a pin's head to a small pea (Fig. 64).   These are hard compact masses of mycelium known as sclerotia, and, as they are able to withstand desiccation and freezing, they function as a resting stage for the fungus.   It is these bodies which contaminate the soil and are responsible for subsequent attacks of clover rot. They can also sometimes be carried over as contaminants in seed and thus play a part in spreading the disease. They remain dormant in the soil during the summer, but, in the early autumn, those within 25 mm or so of the surface germinate and produce flesh-coloured, saucer-shaped, stalked bodies. These are known as apothecia and they bear asci which discharge ascospores. When these come into contact with susceptible plants, and the surfaces of the leaves are wet, the spores immediately germinate and infection results. Sclerotia more deeply buried may decay, but some remain in a living state for several years. When cultivations bring them near the surface, however, they too may germinate.

The clover rot fungus only attacks clover and related plants, and they vary in their susceptibility to the disease. Common or broad red clover is very susceptible, but late-flowering red clover (single-cut cow grass) may also be severely attacked. Trefoil, although often affected, withstands the disease better. Alsike and white clover are less susceptible. In their first year sainfoin and lucerne may be attacked, but failures seldom occur and they show increasing resistance after one season's growth. Vetches, peas and spring-sown field beans appear to escape the disease, but this is not so with winter-sown beans for they are sometimes severely affected with bean rot, a disease caused by a varietal form of the same fungus.

Control Measures. In fields where clover and bean rot has occurred the only method of avoiding further losses is to starve the fungus from the soil by growing only those crops which are resistant to the disease or less susceptible to it. An interval of eight years should pass before common red clover and late-flowering red clover are taken again. During this period sainfoin or lucerne can be sown if the soil conditions are suitable, or alsike or white clover may be substituted, either alone or with Italian ryegrass. As peas and vetches escape the disease they too may be grown, but it is unwise to sow winter beans.

DISEASES CAUSED BY BASIDIOMYCETES

The Basidiomycetes include such well-known saprophytes as mushrooms, toadstools and puff-balls. At some stage in the life-cycle a more or less club-shaped body called a *basidium* is formed, and this normally bears four spores. For example, the gills of mushrooms bear basidia, and these also line the tubes of the large fruiting bodies of the bracket-shaped fungi that cause disease in forest trees and hedgerow timber. The basidia may be septate as in the Sub-class Heterobasidiomycetes, or without septa as in the Sub-class Homobasidiomycetes. Three Orders contained within the Sub-class Heterobasidiomycetes are the Tremellales, the Uredinales or rust fungi, and the Ustilaginales or smut fungi, and one of the Orders within the Heterobasidiomycetes is the Agaricales. Some typical examples are now described.

The majority of the members of the Order Tremellales are saprophytes, although some are parasitic on mosses, and the Order includes the Genus *Helicobasidium*, of which the species *H. purpurea* is parasitic, and produces violet root rot in a wide range of cultivated crops.

Violet Root Rot (*Helicobasidium purpureum*). Of the cultivated crops, sugar-beet, carrots and potatoes can at times be affected seriously and the disease may especially be troublesome in asparagus beds if the seedling plants have been raised on infested land. Other susceptible crops include lucerne, clover, beans, swedes and celery, and the fungus has been found on such diverse hosts as cucumber and carnation, besides a wide variety of weeds. The fungus is soil-inhabiting and is more common in light, alkaline soils. It spreads slowly, and usually occurs in sharply-defined patches in the wetter parts of the field to produce stunted, unthrifty-looking plants. Soil tends to cling to such infected plants on lifting, but when washed off a characteristic violet-coloured weft enveloping the affected parts can be seen. On severely attacked plants the below-ground parts are pervaded by a brown rot, often followed by secondary invaders to cause a slimy bacterial breakdown.

Just visible to the unaided eye, but clearly distinguishable with a hand lens, numerous dark-coloured raised spots can be seen distributed along the strands forming the purple web of mycelium. These infection cushions, or " corps miliaires " as they are sometimes called, enable the fungus to penetrate and parasitise the host tissue. Besides these, the mycelium may produce small rounded, reddish-violet sclerotia up to 6 mm in length. These resting bodies are shed into the soil when the roots rot and are a means of ensuring survival of the fungus in the absence of a host plant.

The spring stage in which the basidia are formed is rarely observed, but it may be seen in early summer as a white " bloom " on a felt of purple mycelium on stems of affected plants just above ground level and spreading over the surface of adjacent soil.

Control Measures. Although light, dry soil conditions are said to favour attack by the fungus, survival of the mycelium and sclerotia is greatest under wet conditions. Land should be kept in good heart, at a high level of fertility and well drained, for under these conditions severe attacks are rare. Cereals are not known to carry infection, and their place in the rotation provides a suitable break from susceptible crops, but groundkeepers and weed hosts must be kept under control. As far as possible, infected material should not be ploughed back into the land.

The Uredinales or Rusts. All the members of this large and complicated Order are obligate parasites which attack a wide range of host plants, including some of the most important crops.

The life history of the rusts differs markedly from that of all other kinds of fungi for, in those in which the cycle is complete, there are three well-defined stages, the *uredospore*, the *teleutospore* and the *æcidia* (Fig. 65). The uredo stage consists of thin-walled, one-celled uredospores, brown or rusty-red in colour. They germinate rapidly when conditions are favourable and quickly cause infection. It is this stage which serves for the rapid dispersal of the fungus. It is followed by the teleuto stage, consisting of thick-walled, dark-coloured teleutospores which in some genera are single celled, in others two, three or many celled. These usually function as resting spores. On germination they give rise to a basidium with cross walls, each section

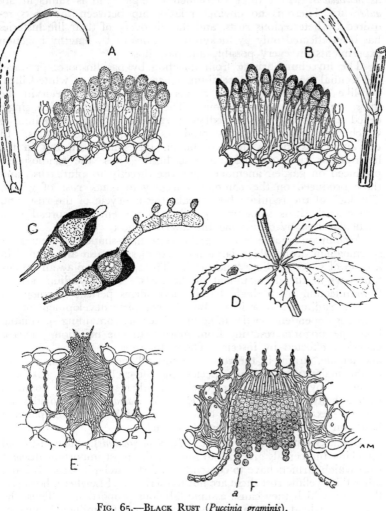

FIG. 65.—BLACK RUST (*Puccinia graminis*).

A, uredospore pustules on leaf, and single pustule enlarged.
B, teleutospore pustules on stem, and single pustule enlarged.
C, germinating teleutospores.
D, cluster cups on under surface of Barberry leaves.
E, section of spermogonium.
F, section of æcidium showing æcidiospores (*a*).

of which produces a single basidiospore on a slender stalk. These spores infect their appropriate host, and this may be one of the same kind as that which produced the teleutospore or it may be totally different. For example, the teleutospores of plum rust lead to the infection of anemones, and the teleutospores of black rust of wheat to a disease of the common barberry. Rusts requiring two different kinds of host plants to complete their life-cycle are described as *heterœcious*, whilst those for which a single kind is sufficient are *autœcious*. There is no obvious relationship between the hosts required by heterœcious rusts, and the discovery of their life histories has been effected only by lucky observations or by lengthy infection experiments on every possible alternate host.

The mycelium arising from infection by basidiospores produces the æcidial stage. Each æcidium, when mature, is a white-frilled, cup-like structure containing chains of spores which, seen in the mass, are a bright orange colour. The spores are thin walled, single celled and capable of immediate germination. The mycelium to which they give rise quickly produces a crop of uredospores. In the autœcious species this will be on the same host, but in the heterœcious rusts on the alternate host. Thus, the æcidiospores produced on garden anemones give rise directly to plum rust, and those produced on the common barberry to black rust of wheat. The lack of the requisite host breaks the life-cycle of the rust, and consequently it is sometimes possible to prevent the spread of a particular rust by destroying the alternate host.

Accompanying the æcidia are minute flask-shaped structures or spermogonia, from which large numbers of spermatia are exuded in a drop of sweet-smelling " nectar ". The spermatia play an important part in the development of the rusts. The mycelium arising from infection by a single basidiospore forms perfect spermogonia and the rudiments of æcidia. But the complete development of the æcidia is dependent on the mingling of nectar, containing spermatia, from spermogonia resulting from more than one infection. This is generally effected by insects. There are thus + and −, or male and female, strains in the rusts. Whilst all these spore stages occur in the majority of the rust species, the life-cycle of some has been simplified by the omission of one or another stage. The common rust of hollyhocks and mallows (*Puccinia malvacearum*) provides an extreme example, for the teleutospore, with its accompanying basidium stage, is the only one produced.

Black Stem Rust of Wheat (*Puccinia graminis*). The rust fungi attacking the cereals give rise to some of the most important diseases with which farmers have to contend, but fortunately, in Great Britain, other than yellow rust these are not too serious. Elsewhere, however, the losses which they cause assume fabulous proportions. Thus, the Canadian wheat crop in 1916 was only some $6\frac{1}{2}$ million tonnes as compared with the previous year's crop of 11 million tonnes, the difference being almost entirely due to an intense outbreak of black rust (*Puccinia graminis*).

Although black rust attacks wheat, oats, barley and rye, as well as various grasses, it cannot spread indiscriminately from one cereal to another, or from the grasses to the cereals. It exists in a number of varieties, each of which is more or less rigidly confined to its own special host or group of host plants. Thus, although it is not possible, even by the aid of a microscope, to distinguish black rust on wheat from black rust on oats, or rye, the fungus on wheat cannot readily attack these other hosts, and vice versa.

Physiologic Races. Indeed, it is even more specialised than this, for within a given variety of black rust, such as *Puccinia graminis tritici* (the variety attacking wheat), there is a large number of physiologic races. When two similar series of wheat varieties are inoculated with uredospores of *P. graminis tritici* obtained from two distinct sources, the results are often different. Thus the rust from one source may fail to infect some of the varieties of wheat, whilst that from the second source attacks them readily, or vice versa. Or again, the rust from one source may produce a heavy infection on one wheat and that from the other only a slight one. Many thousands of infection experiments, carried out on a special standard series of wheats, have fully established this fact and led to the recognition of many hundreds of physiologic races. The demonstration of the complex nature of black rust has lead to an explanation of the puzzling phenomenon that wheats immune to its attacks in one country are often susceptible when grown in others. Their power of resistance has not broken down under the new conditions, as was often assumed, but they have been exposed and have succumbed to the attacks of different races of the parasite.

Control Measures. The control of black rust has proved to be difficult. The extermination of the alternate host, the barberry, was recommended as a means of control a century and a half before the life-cycle of the fungus was known. It was enforced by legislation as early as 1756 in Massachusetts. In some of the American states its destruction has been carried out on a large scale for a long period. Unfortunately, the rust can survive without completing its life-cycle on the barberry provided that plants are available for the production of uredospores, or as long as these are not killed by adverse weather conditions. This occurs in Australia, where the barberry is unknown, and in the southern United States. In the Canadian prairies barberries are also absent, and in these areas rust starts from wind-borne uredospores from infected crops in the United States. In England Ogilvie and Thorpe have shown that the barberry plays little part in the initiation of epidemics and that the severe attacks that do occur in some seasons, mostly in the south-west and along the south coast, follow widespread and early infections by uredospores carried northwards from North Africa and Spain in high-altitude air streams.

The possibility of raising wheats which are resistant to the attacks of black sust has been patiently investigated for many years. A pronounced immunity has been found amongst the members of two groups, namely, the macaroni wheats (*Triticum durum*) and the emmer wheats (*Triticum dicoccum*). Certain varieties of these have been found to come practically unscathed through the severest of epidemics. After many attempts their rust-resisting power has now been built by plant breeders into typical bread wheats, and the foundations have been laid for securing new types suitable for cultivation under various conditions. Although serious in some parts of the world, black rust is rarely of economic significance in Britain although the disease can still be found on occasion, mainly in the south-west and south of England.

Yellow Rust (*Puccinia striiformis*) is the common rust of wheat and barley in England, and it can readily be distinguished from other cereal rusts by the deep cadmium yellow colour of the uredospore pustules which are closely crowded together in parallel lines. The severity of the disease varies greatly from year to year, and in some years can cause very appreciable losses. Weather and soil conditions play an important part in determining the intensity of yellow rust attacks. Cool moist weather favours its development, but in hot dry seasons it makes little progress. It is most severe on rich, well-farmed land, especially when nitrogenous manures have been used in an attempt to secure a heavy crop. Carry over from year to year is dependent upon infection surviving on volunteer plants.

There are two stages, the uredospore and the teleutospore, but, as far as is known, no æcidial stage. The starting-point of epidemics in this country is provided by the over-wintering of the uredospores. These remain viable for a long period under normal climatic conditions, though a spell of hot dry weather in the early autumn may bring about their wholesale destruction. Further, an almost continuous supply of host plants is available owing to the necessity for sowing wheat during the autumn months. Yellow rust, however, makes little progress during the cold weather, and it is only exceptionally that heavy infection occurs by the end of March. More generally the epidemic starts to become severe towards the end of May, the uredospore stage continuing to develop as long as any green foliage is available.

Control Measures. Losses through attacks of yellow rust can be minimised by growing varieties which are not markedly susceptible to its attacks, or on susceptible varieties by being prepared to protect the crop by foliar applications of rust fungicides. The common varieties vary in their resistance to rust, but it does not necessarily follow that because a variety is not attacked that it is immune to disease. Yellow rust also exists in specialised races, and these may not be uniformly distributed geographically. It is only by observations made over several years that the relative resistance of varieties to rust can be assessed. Atou, Bouquet and Flinor are among the more resistant varieties in cultivation at present. Maris Ranger, Champlein and Templar on the other hand are much more susceptible. Between these extremes, and differing in their degree of susceptibility, are the remainder of the commonly grown varieties. In the choice of a cereal variety account must be taken of its intended position in the sequence of other cereals, its relative susceptibility to diseases including yellow rust and its suitability for the soil type and standard of fertility. Some diversification in varieties is necessary to spread the risk, with preparedness to resort to fungicide control where other overriding circumstances make necessary the choice of a variety susceptible to rust diseases.

Other Rusts. There is a third rust which occurs on wheat in Britain, brown rust (*Puccinia recondita*). Although the scattered brown

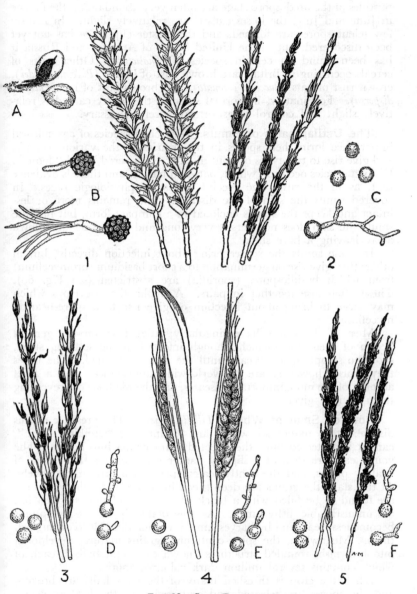

FIG. 66.—SMUT DISEASES.

1, Bunt of wheat (*Tilletia caries*). A, bunt balls, B, germinating spores.
2, Loose smut of wheat (*Ustilago nuda*). C, germinating spores.
3, Smut of oats (*Ustilago* spp.). D, germinating spores.
4, Covered smut of barley (*Ustilago hordei*). E, germinating spores.
5, Loose smut of barley (*Ustilago nuda*). F, germinating spores.

pustules of the uredospore stage are often very abundant on the foliage in June and July, the losses caused are relatively slight. In Britain few teleutospores are formed, and the cluster cup stage has not yet been discovered, but in the United States of America and Russia it has been found on certain species of *Thalictrum*. Other rusts of cereals occurring in Britain are brown rust of barley (*Puccinia hordei*), crown rust of oats (*Puccinia coronata*), and brown rust of rye (*Puccinia dispersa*). Fortunately, in normal seasons, the injury caused is relatively slight and control measures are rarely necessary.

The Ustilaginales or Smuts. The chief species of agricultural importance form their spores in the ovaries of the various cereals, and give rise to two types of smut diseases—" covered " and " loose ". Distinct species occur on wheat, oats and barley, and there is evidence that, as in the rusts, the species consist of physiologic races. In covered smuts the skin of the diseased grain remains more or less intact and keeps the spores enclosed in a compact form, but in loose smuts the skin does not persist very long and the spores are blown away leaving a bare stalk.

In some smuts the spores bring about infection directly, but in others they give rise on germination to a short basidium (promycelium) from which basidiospores (sporidia) are abstricted (see Fig. 66). These often fuse together in pairs. A hypha then develops which may serve to bring about infection or give rise to a spore-bearing mycelium.

Most of the smuts have singularly little effect on the general health of their hosts which, though infected in the seedling stage, show no symptoms of disease until the grain is formed. The losses they cause, however, are of world-wide importance, and a brief account of the more important diseases, and the methods of preventing them, is now given.

Covered Smut of Wheat (*Tilletia caries*). The presence of this disease, other names for which are " bunt " or " stinking smut ", cannot be detected until the affected plant comes into flower. The young ears are then of a slightly darker green colour than those of healthy plants, and the chaffy scales open out a little more widely. At this stage the grains are deep green in colour and, when crushed, are found to be filled with a black greasy mass of spores which has an unmistakable fishy smell. As the plant ripens the symptoms become less clear, and by harvest time it is often difficult to distinguish them. Meanwhile, the spore-containing grains mature, developing into brownish, rounded structures known as " bunt balls ", each of which contains several million dark coloured spores.

When the crop is threshed many of the bunt balls are broken, and the spores are released and scattered over the healthy grain. If this is sown without treatment the spores germinate, giving rise to basidia which bear terminal clusters of narrow, sickle-shaped basidiospores which frequently fuse together in pairs. As the grain is germinating, hyphæ from the basidiospores penetrate into the young shoot and work their way into the tissues at the tip of the stem. As the stem elongates the growth of the fungus keeps pace with it, and the mycelium persists in the terminal growing point. When the flowers begin to form the fungus strands push their way into the tissue which normally gives rise to the grain, and then divide into millions of spores. Thus instead of a normal wheat grain developing a " bunt ball " is formed.

Control Measures. Covered smut of wheat can be prevented by treating the seed before sowing with a suitable seed disinfectant. At one time weak solutions of copper sulphate or formalin were used, but both these wet treatments have certain disadvantages and have now been entirely superseded by modern methods of seed disinfection using dust or liquid formulations of proprietary organo-mercury seed dressings. These products are manufactured by several different firms, but most of them bear the official approval mark (Fig. 67) of the Ministry of Agriculture, Fisheries and Food and the Department of Agriculture and Fisheries for Scotland.

AGRICULTURAL CHEMICALS APPROVAL SCHEME

[*Ministry of Agriculture, Fisheries and Food.*]

Fig. 67.—The Approval Mark for Pesticides.

Combined seed disinfectants are also available containing an organo-mercury or other fungicide for the control of seed-borne diseases and an insecticide, such as gamma BHC for wireworm and wheat-bulb fly control, or carbophenothion or chlorfenvinphos for the control of wheat-bulb fly on autumn-sown seed to be sown before the end of December. (See p. 434.) Seed already treated with one or other of these dressings can be obtained from most seed merchants.

Provided the grain is kept dry after treatment, and the dressing is correctly used, there is no risk of injury to germination.

Loose Smut of Wheat (*Ustilago nuda*). The symptoms of loose smut differ from those of covered smut of wheat inasmuch as the soot-like mass of spores is set free by the breaking down of the ovary walls. This occurs when the healthy plants are in full flower, and by the time their grains have started to swell all that remains of the ears of infected plants is a blackened stalk. The spores, therefore, are not harvested with the crop but drift about in the air, some of them falling on the feathery stigmas of the healthy wheat flowers. When this happens they germinate directly, if conditions are favourable, giving rise to a mycelium which travels downwards into the ovary and then into the developing embryo. There is no visible effect of this on the seed, which ripens in the normal way. Nevertheless, although the grains appear to be healthy, they are infected internally, and when sown give rise to diseased ears, the mycelium growing upwards with the tips of the main and lateral shoots as in covered smut, and again forming spores at the flowering stage.

Control Measures. As the fungus is not on the surface of the grain but inside it, control is difficult because the mycelium within the seed must be killed without injuring the vitality of the grain itself.

At one time the only way of doing this was by means of a warm-water treatment, known after the name of its discoverer as Jensen's method. The first operation is to place the seed wheat in a sack, incompletely filled to allow for the subsequent swelling of the grain, and to immerse it in cold water for four hours. It is then taken out, drained for a few minutes, and submerged for ten minutes in a tank of warm water. During this period the temperature of the warm water must be kept between 52 and 54 °C which is the critical temperature for the fungus. The margin of safety is very narrow, for if the temperature rises above 54 °C the vitality of the grain will be seriously affected, and if it falls below 52 °C the fungus will not be killed. After the sack has been taken from the warm water the grain is spread out in a thin layer on the barn floor to dry, and when dry is sown. Although this process is effective in preventing loose smut it requires very careful supervision otherwise the risk of injuring the grain is great, and for this reason it is not a practical treatment for the average farmer to carry out. The best way of avoiding loose smut is to select seed corn from crops which have not shown the disease. With the systemic fungicide carboxin it is now possible to control the disease with a seed disinfectant. Such treatment, however, is at present economically worthwhile only on seed intended for multiplication.

Covered Smut of Barley (*Ustilago hordei*). The life cycle of this fungus, and the course which the disease takes, is very similar to that of covered smut of wheat, with the exception that the spores produce a basidium with cross walls on which are borne small rounded basidiospores.

The disease can be prevented by treating the seed with an Approved organo-mercury seed disinfectant. This not only prevents covered smut but also leaf stripe, *Pyrenophora graminea*, an important seed-borne disease which may kill the seedlings or cause partial or complete blindness of the ears (see p. 376).

Loose Smut of Barley (*Ustilago nuda*). Control by the warm-water method is very similar to that for loose smut of wheat but, after the four hours' soaking in cold water, the grain is steeped for 5 minutes in water at 49 °C and then for 10 minutes in water kept as close as possible to 51 °C. If the temperature is allowed to rise higher than this the grain may be damaged. Here again the best practical means of avoiding loose smut is to obtain seed from crops in which the disease has not been present. Among spring varieties, Mazurka and Maris Mink are very susceptible and Proctor relatively resistant. Here again a seed disinfectant containing carboxin will control the disease.

The Oat Smuts. The oat crop is also affected by two smut diseases, but it is often very difficult to distinguish them apart. Covered smut is caused by *Ustilago hordei*, and loose smut, the more common of the two, by *Ustilago avenæ*. Unlike the fungi causing the other loose smuts, however, *U. hordei* and *U. avenæ* do not penetrate deeply into the grain. In both diseases active infection occurs from spores (or dormant mycelium) on the grain, and consequently they are amenable to chemical methods of seed disinfection.

Control Measures. Both the smut diseases of oats can be much reduced by disinfecting the seed with an Approved organo-mercury seed dressing, and this treatment has the added advantage that it also prevents leaf spot and seedling blight, a seed-borne disease caused by the fungus *Pyrenophora avenæ*. Treating seed oats with a solution containing 0·5 l of commercial (40 per cent) formaldehyde in 160 l of water is a very effective treatment, but it has the serious disadvantage that it does not prevent leaf spot, and it is a treatment unlikely to be used except in special circumstances.

There are few farmers nowadays who do not disinfect their seed wheat to prevent disease, but there may still be some, especially with home-saved seed, who fail to adopt this simple precaution for barley and oats. Since the introduction of the combined seed dressings that are effective also against insect attack, however, it is probable that seed of all three cereals supplied through seed merchants is mostly treated.

The **Agaricales** is another Order in the Basidiomycetes (Subclass Homobasidiomycetes), and included in it are families such as the *Thelephoraceæ*, the *Polyporaceæ* and the *Agaricaceæ*. The fruiting bodies are mostly large and each basidium usually bears four basidiospores at its apex. A brief description of certain fungi in these families is now given.

Black Scurf and Stem Canker (*Corticium solani*). This is a disease of potato caused by a member of the family, *Thelephoraceæ*. Black scurf is descriptive of the appearance of the sclerotia of the fungus on the skin of affected tubers. These black, carbonaceous-like aggregates of mycelium are most conspicuous on moist tubers from which the soil has been rinsed, but on drying they lighten in colour and become less noticeable. They are readily detached with the finger nail and, as the mycelium does not penetrate the skin, the flesh of the tuber is unaffected. Such infected tubers used for seed can be a potential source of trouble to the resultant crop, especially in the early stages depending on seasonal and soil conditions. Slow growth resulting from cold wet soil or, on some of the lighter soils, prolonged dryness, can often predispose a crop to severe attack.

The growing tips of the young sprouts are readily attacked by the mycelium and become blackened and dead. The resultant lateral shoots may later be attacked similarly and, at worst, the tuber may fail to produce a plant. Shoots that manage to emerge are seriously weakened and the leaves may assume an appearance simulating primary symptoms of leaf roll, and can be mistaken for such unless the plant is pulled up and the stem canker condition observed on the below-ground parts. Under less severe conditions, although the mycelium may ramify over the below-ground parts of the plant to produce numerous brownish, slightly collapsed areas on otherwise relatively white roots, stolons and shoots, the plant survives and only the occasional shoot completely girdled by a stem canker succumbs. It is on such plants later in the season that the perfect stage is not infrequently seen as a thin, skin-like white felt enveloping the surface of the stem and spreading upwards for an inch or more above soil level. This is the "collar fungus" stage consisting of a hymenial layer bearing basidia of the fungus. The vegetative mycelium is branched and colourless at first, but later amber coloured. The branches arise at right angles and are constricted where they originate from the main hyphae. These are very characteristic features of the fungus as seen under the microscope.

Control Measures. A counsel of perfection would be to avoid planting tubers bearing black scurf. In Scotland and Northern Ireland it has been shown that potential infection from contaminated tubers can be reduced by disinfecting them with organo-mercury compounds, but the thicker sclerotia make fully effective control difficult. Boxing up and chitting the seed will enable weakly-sprouted tubers to be discarded before planting and at the same time afford greater latitude in the choice of favourable planting conditions.

Dry Rot (*Merulius lachrymans*). In the *Polyporaceæ* the fruiting bodies are fleshy, leathery or woody in texture, and they usually grow in the form of bracket-like structures standing out at right angles from tree-trunks. Their lower surfaces bear a hymenium or spore-bearing surface composed of tubes, the cavities of which are lined with basidia bearing basidiospores. The mycelium is confined to the wood, which in time is usually reduced by its action to powdery touchwood. Included in this family are species of *Polyporus* and *Fomes* (many of them wound parasites which are responsible for serious diseases of forest trees and hedgerow timber), and *Merulius lachrymans*, a most destructive fungus which causes dry rot of timber.

Dry rot is comparatively rare in the open but it often occurs in farmhouses and farm buildings, where its destructive effects are well known. In any moist situation such as in cellars or in the space between the soil and a floor, it develops a vigorous mycelium in the form of a dense, white, felt-like sheet which can readily be stripped from the joists or floorboards. As this grows, thick strands of hyphæ, packed with nutrient material and capable of transporting water, are formed. These often grow to great lengths, not only on wood but also on the surface of brick walls, etc., with the result that wood far removed from the original source of infection may become attacked. An outbreak at soil level, therefore, may lead to the destruction of roof timbers. The surfaces of the mature fruiting bodies, which are rusty-brown, flat incrustations 150 mm or so across, are covered with folds and wrinkles forming a series of shallow pits, the surfaces of which are lined with basidia. These produce such enormous quantities of spores that the first symptom of the presence of the fungus, apart from its characteristic mouldy smell, is the accumulation of a rust-coloured dust on floors, shelves and furniture.

When the spore-bearing stage has been reached the wood is usually so badly affected that the next symptom is the sagging of beams or the collapse of the floorboards. The spores germinate readily on moist wood and quickly give rise to the mycelial sheets. The hyphæ of these invade the wood, partially destroy the cell walls and reduce it to a spongy, brownish mass which readily absorbs and retains water. When dry, the decayed portions shrink and crack, and the fissures forming at right angles to one another divide it into cubical blocks which can easily be rubbed down into a powder.

Control Measures. Where an outbreak occurs it should be dealt with immediately. Every piece of wood which has been reached by the mycelium should be cut out and replaced with sound timber which has been thoroughly brushed over with a suitable wood preservative. Precautions should then be taken to prevent a further attack. As the growth of the fungus is dependent on a supply of moisture, an endeavour should be made to trace this to its source and prevent its further access. One of the commonest points of entry, especially in old buildings where damp courses are non-existent or inefficient, are joists let into walls at or below the soil level. An external surface drain will do much to keep such a wall dry, and the liberal use of creosote, both on it and on the end of the joists inserted in it, should stop any fresh infection. Provision should also be made for a thorough ventilation of the space enclosed between the ground and the floorboards.

The Honey Fungus (*Armillaria mellea*). This is one of the *Agaricaceæ*, or gill fungi, such as mushrooms and toadstools, the basidia being formed on gill-like plates radiating from a common centre. Most of the gill fungi are saprophytes, but *Armillaria mellea*, the honey or shoe-string fungus, is parasitic, causing a serious root rot of trees

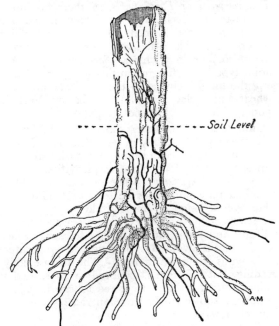

FIG. 68.—ARMILLARIA ROOT ROT (*Armillaria mellea*).
Rhizomorphs on root.

The mushroom-shaped fruiting bodies are formed in large clusters at the base of infected trees or on stumps which have been left in the ground. They are honey-yellow in colour and the top of the cap bears many dark brown, fibrous scales. If the bark of an infected tree is stripped off, a vein-like network of strands, or rhizomorphs, not unlike black leather bootlaces, will be found on the surface of the wood. These pass over in their younger portions to delicate white sheets of mycelium, the hyphæ of which penetrate and destroy the tissues of the wood (Fig. 68). They also spread through the soil, travelling for considerable distances at a depth of about 100 mm. On coming into contact with a sound root, their hyphæ penetrate between the crevices of its bark and develop rapidly between it and the wood. Once the roots have become infected and the mycelium has worked its way round the base of the tree, it dies suddenly.

Control. The disease is difficult to control. Using reserves of carbohydrates in stumps and large pieces of root as a base, the fungus can invade neighbouring roots. Removal of stumps and as much as possible of the root system reduces the ability of the fungus to colonise new roots. Creating a barrier by deep cultivations or insertion of thick polythene sheet in a slit trench between infection sites and living healthy trees will afford some protection.

DISEASES CAUSED BY DEUTEROMYCETES

In the *Deuteromycetes* or *Fungi Imperfecti* there are no sporangia, asci or basidia in the life-cycle so far as is known, and only asexual spores, chiefly conidia, are formed. These are usually borne on hyphæ or groups of hyphæ (*Moniliales*), on cushion-like masses of hyphæ (*Melanconiales*) or in pycnidia (*Sphæropsidales*). An example of the pycnidial type is the fungus causing gangrene of potato. Examples among the fungi included in the *Moniliales* are those causing eyespot of wheat and barley, leaf blotch of barley, Verticillium wilt of lucerne and dry rot and skin spot of potatoes.

Potato Gangrene. This tuber disease has become of serious significance in recent years, mainly in seed potatoes. The characteristic symptom is the occurrence of thumb-like dark depressions accompanied by a progressive dry form of rot with cavities. It is caused by the soil-borne fungus *Phoma exigua*. The pinhead-sized black fruiting bodies or pycnidia may sometimes be seen protruding through the blackened skin over a lesion or lining a cavity.

Gangrene is especially prevalent in some seasons, associated with weather at lifting and temperatures during storage. The fungus is a wound parasite and is favoured by low storage temperatures during which tuber wounds are slow to heal. The disease may increase following grading, but infection is believed to be related mainly to injuries sustained during lifting. In stored tubers the disease is most prevalent in pockets subject to lower temperatures, e.g. badly fitting doors, broken windows, etc.

Slightly affected tubers in which the eyes are still viable should still produce satisfactory plants but severely affected tubers should be discarded. Chitting before planting is the most practical means of reducing the risk of gappy crops. Good control can sometimes be obtained by immediate post lifting chemical treatment (see p. 396).

Glume Blotch of Wheat. This disease caused by *Septoria nodorum* has become increasingly serious in recent years, especially in the wetter parts of the country. Early symptoms of the disease occur as a leaf spot followed later by chocolate-brown blotches on the glumes of the ripening ear and by brown discoloration around the nodes. In severe attacks there is considerable reduction in yield and an increase in shrivelled grain and tail corn. The fungus is seed-borne and can carry over from one season to the next on infected crop debris. Seed treatment with an organo-mercury seed disinfectant will reduce seedling damage and foliar sprays with benomyl and other fungicides before heading have given some control and substantial yield increases. Autumn cultivations reduce carry-over on debris. The fungus can also infect rye and rough-stalked meadow grass.

A related fungus, *Septoria tritici*, can also cause a leaf spot on wheat but rarely produces glume blotch symptoms.

Leaf Blotch of Barley (*Rhynchosporium secalis*). Although familiar for many years and observed from time to time to be severe on the occasional barley crop, it is only since the late 1960's that the disease has assumed any serious significance. With changing varieties, an increase in the barley area and the cultivation of barley on a greater scale in the wetter parts of the country, leaf blotch has become a factor to be reckoned with. The causal fungus is seed-borne and it can be carried over from season to season on infected crop debris and on volunteer plants. There is no indication that any control is achieved by seed disinfection. The disease affects both barley and rye, causing greyish blotches with dark brown, often zoned, margins on any part of the leaf, but often at the junction with the stem. The blotches may coalesce to involve the greater part of the leaf.

Varieties differ widely in their susceptibility, and in districts in which severe attacks may be expected this largely determines the choice of variety. Of the winter barleys Senta and Astrix are very resistant while most spring barleys are relatively susceptible, Maris Mink being especially so, with only Armelle relatively resistant.

Control is by choice of variety, stubble cultivations and latterly by use of foliar fungicide sprays as a preventive early in the season when cool wet weather favours infection.

Eyespot of Wheat and Barley. Eyespot is caused by the fungus *Cercosporella herpotrichoides*, which attacks all commercial varieties of wheat and barley, especially autumn-sown crops. Oval or eye-shaped areas with brown borders, and small black dots in the middle of the areas, appear in spring on the leaf sheaths just above soil level. The fungus weakens the individual straws, which tend to fall over in all directions : this straggling is quite different from the lodging caused by storms or excess of nitrogen, in which the crop usually collapses in one direction. Infection is spread by spores spread in rain splashes.

The symptoms can be confused with those of sharp eyespot caused by *Corticium solani*. The latter produces somewhat asymmetrical lesions with more clearly defined margins.

Control Measures.   A sound rotation in which wheat and barley do not occur too frequently is the best way of preventing eyespot, but a two-year break will usually reduce infection substantially. Oats and ryegrass are relatively non-susceptible and can be used in the rotation to reduce eyespot.   A thin seeding, together with balanced manuring, also helps to reduce losses from this disease.   Varieties differ in susceptibility and choice of variety should take into account shortness of straw, standing power and resistance to eyespot.   Most of the recommended winter wheats are now resistant to eyespot. Promising results are also being obtained from the use of fungicide sprays applied early in the growth of the plant.

Verticillium Wilt of Lucerne (*Verticillium albo-atrum* and *V. dahliæ*).   Observed in this country in recent years, this disease is now commonly found wherever lucerne is grown.   Field symptoms occur from June to September and consist at first of the flagging of the upper leaves of infected plants during warm, bright periods, often with recovery in the cool of the evening.   As infection proceeds, the lower leaves wilt permanently, turn yellow and later become bleached and dried up.   They are then readily shed, causing defoliation. These symptoms intensify progressively from the base of the shoots upwards, and the stem bases become enveloped in a greyish growth of sporophores.   Internally the xylem tissues are stained, and in completely wilted plants this dark-brown discoloration may extend throughout the vascular system of the plant.   As the stems die the bases become blackened due to the formation in the cortex and epidermis of resting mycelium or, with *V. dahliæ*, micro-sclerotia.

The disease can be introduced with the seed on infected plant fragments to produce scattered infections in the resultant crop.   Subsequent spread can be rapid by means of air-borne spores and fragments of diseased plants, both of which are spread widely by implements such as grass-cutters and forage harvesters.   Severe infection localised near gateways and round the headlands is often indicative of introduction of the disease from another crop by contaminated implements.

Control Measures.   Organo-mercury and thiram seed dressings have shown promise in controlling introduction of the disease with seed, and on the farm some measure of success in localising outbreaks may attend disinfection of machinery and footwear with formalin before moving from infected fields.   If possible, infected areas should be cut last.

Importation of seed from Verticillium-free areas offers a means of avoiding introduction with the seed, and so far there has been no record of this disease in America.   Promising, too, is the occurrence of varietal differences in susceptibility which may be exploited by the plant breeder to produce resistant varieties.

Dry Rot of Potatoes (*Fusarium spp.*).   One of the most destructive diseases of stored potatoes in this country, especially of early varieties, is dry rot caused by species of *Fusarium*, of which the most important is *F. caeruleum*.   Although the fungus is soil-borne, the disease does not affect the growing plant but occurs on mature tubers after storage, infection having taken place from adhering soil particles. Susceptibility to infection increases with maturity of the tuber and the disease develops most rapidly at about 16 °F. in conditions of high humidity. Early stages of infection appear on the surface as brown patches overlying a brownish rot somewhat reminiscent of blight. Internal shrinkage of the rotting tissues, however, soon results in the very characteristic concentric wrinkling of the overlying surface and the development of cavities within the rotted tissue. At this stage pink, white or bluish pustules of the fungus begin to break through the skin to produce large numbers of spores. Microscopically, these are characteristically sickle-shaped and multi-septate. In some spores, individual cells may become thick-walled resting spores and enable the fungus to survive unfavourable conditions. Where infection is severe in large bulks of potatoes, soft-rotting bacteria often supervene resulting in overheating and extensive losses.

The fungus is essentially a wound parasite and enters through cuts and abrasions. This may occur at lifting time, but as susceptibility increases with tuber maturity, mechanical injuries sustained late in storage are especially subject to infection. Most early varieties are very susceptible, Arran Pilot especially so, and, of the main crop varieties, the once widely grown Doon Star is also very susceptible.

Control Measures. Mechanical damage to the tubers at lifting time and subsequently should be kept to a minimum, and all unnecessary handling during storage should be avoided. For tubers intended for seed, early grading and boxing up is advisable, and subsequent " picking over " should be avoided until as near planting time as possible. Satisfactory control can be obtained by chemical means if treatment takes place very soon after lifting (see p. 396). Tubers affected with dry rot should not be planted as they usually rot in the soil and result in a gappy crop.

Skin Spot of Potatoes. This disease affecting the underground parts of the potato plant is caused by the fungus *Oospora pustulans*. Familiar tuber symptoms are the presence of pimple-like spots 1–2 mm across. These become noticeable eight or more weeks after lifting when pockets of infected tissue, sealed off from the remainder of the tuber by a protective layer of cork, begin to die. Further infection can take place through minute wounds sustained at lifting. The most serious effect of planting infected seed is the gappiness in the crop due to failure to grow of tubers on which all the eyes have been killed. Less seriously affected tubers may produce weak sprouts, slow to emerge.

There are varietal differences in response to infection. Majestic, relatively resistant to skin spot, shows little eye damage, while King Edward, a susceptible variety, will sustain considerably more eye damage than might be indicated by skin spot symptoms. Seed tubers found to be infected on arrival at the farm should be boxed and chitted. Tubers that fail to sprout can thus be discarded before planting.

In early work on chemical control of skin spot and other tuber diseases, good control was sometimes obtained by washing and then dipping the seed tubers in an organo-mercury solution immediately after lifting. More recently, fumigation of tubers with 2-amino-butane has been found to be effective, and treating tubers with a finely-divided spray containing thiabendazole has given promising results, provided always that chemical treatment takes place soon after lifting.

VIRUS DISEASES

In addition to the diseases caused in agricultural crops by the attacks of such microscopic organisms as fungi and bacteria, there is a large group of diseases which falls into quite a different category. These are known as *virus diseases* and they must be considered as a class apart because they differ fundamentally both in appearance and treatment from disorders due to bacteria or fungi.

The word "virus" comes from the Latin and means literally a poison, but in its modern sense it refers only to a particular kind of disease agent. Viruses are not only of extreme importance from the practical point of view since they cause disease in every kind of living organism from bacteria to man, but they are of great scientific interest. They may be regarded as a kind of link between the "living" organism and the "non-living" chemical substance, and they possess certain properties of both organisms and chemicals.

It is necessary to emphasise at this point a few of the outstanding characteristics of viruses in order to explain more clearly how they differ from other disease agents. Viruses are often referred to as "filterable" or "ultramicroscopic", and this denotes one of their chief characteristics—their extremely small size. Most viruses, and all those which attack plants, are too small to be seen other than with an electron microscope, and this property also allows them to pass through filters which hold back even the smallest bacteria. Another characteristic of viruses is the fact that unlike so many microscopic organisms, they cannot be cultivated in an artificial nutrient medium such as broth or agar : viruses must have a living cell in which to multiply.

The method of spread of viruses is another important character-istic because it may govern to a large extent the control measures to be applied to the disease. Viruses depend in many cases upon an insect, eelworm or other organism to transmit them from the diseased to the healthy individual. This is true not only of plant viruses but of some animal viruses also, yellow fever and the mosquito for example. Of the insects concerned with the spread of plant virus diseases, aphids or greenfly are perhaps the most important, although thrips, weevils and flea beetles are vectors of a number of viruses of economic importance. Free-living eelworms (*Xiphinema* and *Longi-dorus* spp.) in the soil too have been shown to carry soil-borne viruses that affect strawberries, potatoes and a number of other hosts. In some virus diseases, infection may be carried on or in the seed ; of these, lettuce mosaic is an outstanding example.

Viruses are the cause of disease in all types of organisms. Yellow fever, small-pox and influenza in man ; foot-and-mouth disease in cattle ; dog distemper ; fowl pox, are all virus diseases : a large number of separate and distinct viruses cause disease in plants of every kind. From the agricultural point of view there are certain viruses which are of first-rate importance, such as those which attack sugar beet, potatoes and cruciferous crops, particularly cauliflowers, broccoli and other brassicæ. In this section attention will be mainly confined to the virus diseases of these crops : *barley yellow dwarf virus*, which attacks cereals and grasses, is not discussed.

VIRUS DISEASES OF SUGAR BEET AND MANGELS

Two of the virus complexes affecting sugar beet and mangels in the British Isles are those causing *mosaic* and *virus yellows* respectively, the latter being the more important disease of the two. In the first disease the symptoms consist of a mottling of light and dark green on the central leaves somewhat resembling a mosaic pattern ; this mottling is brightest on the youngest leaves. On the whole, the effects of the disease on the growth of the plant are not very severe.

Virus Yellows. Two viruses are involved in the beet yellow, complex ; *beet mild yellow virus* (BMYV) and *beet yellow virus* (BYV). Symptoms may occur as early as June or July resulting in considerable losses in yield with some reduction in sugar content. A crop heavily infected in July may lose 50 per cent of its sugar. Yellowing of the outer and middle leaves from the tips and upper margins characterises the disease and the leaves become thickened and very brittle and crackle when handled. BMYV infection results in the older leaves becoming bright orange-yellow and relatively upright. Its effect on yield is much less than with BYV. With the latter, leaf coloration is more a dull yellowing or bronzing accompanied by necrotic brown or reddish spots between the veins. Both viruses are frequently present in the same plant.

It is important to realise that, once a sugar beet or mangel plant is infected with these virus diseases, it is too late to do anything about it. The plants cannot be cured by any kind of chemical treatment. Efforts must be made, therefore, to prevent the plants becoming infected, and the first step in this direction is to understand how the viruses are brought to the plants. As mentioned earlier, many plant viruses depend entirely on insects for their spread and the insect vectors mostly associated with the spread of viruses affecting agricultural crops are aphids or greenfly. More than one species of aphid may be concerned in the spread of a particular virus but not any other kind of insect. In the field, mosaic and yellows of sugar beet are spread by two species of aphid, the small green potato and peach aphid (*Myzus persicæ*, Sulz.) and the common black bean aphid (*Aphis fabæ*, Linn.), and not by any other means. So far as is known at present the virus is not transmitted through the seeds.

Control Measures. Control measures must therefore be developed to prevent infected aphids gaining access to the sugar beet and mangel crops, or to kill any aphids arriving on the crop by spraying with a systemic insecticide. Although this may be too late to prevent an infective aphid from transmitting the virus to the plant on which it has fed, build-up of aphid infestation is prevented and transmission of virus from plant to plant within the crop is reduced substantially by a timely spraying.

As an aid to farmers, spray warning cards are sent by the factories to sugar beet growers based on aphid levels assessed in widespread field observations by British Sugar Corporation fieldsmen. Any measures which help to reduce the sources of infection are useful because an aphid must first feed on an infected mangel or sugar beet before it can infect a healthy plant. Volunteer beets or mangels which are often heavily infected are a common source of virus infection. The sprouts of clamped mangels and fodder beet held late into the season can be very dangerous in this respect. Another important measure is the isolation of the steckling beds, which can be heavily infected with virus, as far as possible from the root crops of sugar beet and mangels. Seed used for sowing stecklings should be treated with systemic insecticide. Spraying the stecklings with a systemic insecticide at the four true leaf stage and subsequently at fortnightly intervals, will prevent aphid infestation. A last application at the end of October will safeguard against any that might otherwise overwinter. Direct-sown seed crops are grown in the south and centre of England away from the main sugar beet areas.

Put very briefly, the following are the recommendations for the control of the virus diseases of sugar beet and mangels :

(1) Isolation of the steckling beds as far as possible from root and seed crops of sugar beet and mangels.

(2) Treating seed for stecklings with a systemic aphicide.

(3) Spraying of the steckling beds with a systemic insecticide to reduce aphid infestation.

(4) Destruction of the steckling beds if heavily infected with virus.

(5) Early sowing of the sugar beet crop, and spraying with systemic insecticides as soon as aphids begin to build up on the plants.

(6) Destruction of sources of infection such as volunteer beets and mangels, and the early removal of mangel clamps.

VIRUS DISEASES OF THE POTATO

The potato plant is subject to infection with a number of different viruses and their study is difficult and complicated. From the viewpoint of the practical grower the virus diseases of the potato may be divided into three categories, (1) " *leaf-roll* " disease due to infection with one specific virus, (2) the various virus diseases which make up " *potato mosaic* ", (3) soil-borne viruses such as *spraing* and *mop top*

Leaf-roll. In potato *leaf-roll*, the outstanding symptom is the rolling of the leaves. The rolling takes place upwards and inwards, and the leaves themselves are thickened and yellowish in colour. In some potato varieties there may be a purple coloration of the leaves together with the formation of small aerial tubers. The whole plant is dwarfed and stunted and the yield may be reduced by 70–90 per cent. Some potato varieties, e.g. Record, are relatively susceptible to leaf-roll, others such as Pentland Crown and Majestic are less so.

The virus of potato leaf-roll is carried from diseased to healthy plants in the field by the agency of aphids. The aphid chiefly responsible is the potato and peach aphid, *Myzus persicæ*, though one or two other species play a minor part in the spread of the virus.

Potato Mosaic. In this disease several viruses are concerned, the most important being severe mosaic, sometimes called *leaf-drop streak*. The virus causing this disease is known as potato virus Y and is spread by the same potato aphid transmitting leaf-roll. The first signs of infection appear as dark streaks on the veins of the under sides of the leaf. These streaks increase in size and coalesce, causing the whole leaf to shrivel and hang down ; it is from this character that the name " leaf-drop " is derived. These leaves usually remain attached to the stem, on which there frequently develops a longitudinal streak. The symptoms described above are those of a current season infection ; in subsequent years there is less leaf-drop but the plant is stunted with mottled, distorted leaves.

Potato virus A is another aphid-transmitted virus similar in some ways to potato virus Y.

Another common mosaic disease of the potato is known as *potato mottle* or *mild mosaic*, the chief symptom being a mild mottling of the leaves with light green or paler spots and patches. On some varieties, however, such as Epicure, King Edward and Ninetyfold, the disease takes a different form, and instead of a mottling of the leaves the growing point and young leaves are killed. This type of disease is known as *top necrosis*, and plants reacting in this manner are considered to be field immune to the virus since it does not spread from such plants. The causative virus of potato mottle is known as potato virus X and it differs in many ways from potato virus Y, chiefly in that so far as is known it is not insect-transmitted, but spreads by contact of the leaves of diseased and healthy potato plants. Potato virus X is exceedingly common, sometimes latent, in many varieties of potatoes and is present in much of the Scotch Stock Seed. Some varieties, such as Kerr's Pink, are almost universally infected.

Rugose mosaic is a composite disease and is due to the two viruses A and X acting in unison ; a somewhat similar disease is caused by the two viruses Y and X. The symptoms are severe ; the plant is bushy and dwarfed and the leaves show a pronounced puckering and downward curling. Diffused yellowish areas occur all over the foliage, which is brittle and easily injured.

Spraing of Potato. A form of internal reddish-brown discoloration in potato tubers which shows as arcs in the flesh on cutting has been familiar for many years. The trouble is most pronounced on the lighter soils and some varieties are more affected than others. The condition, termed spraing, is now known to be due to infection with *tobacco rattle virus* (TRV) transmitted by the free-living nematode, *Trichodorus* sp. The virus has a wide host range but is not well adapted to potatoes, and only limited invasion of the plant occurs Transmission in seed tubers is not common. In varieties in which the virus is more systemic such as in King Edward, a stem mottle symptom may appear in the foliage and there is less evidence of tuber symptoms. Severe tuber symptoms can occur in Ulster Prince, Arran Comet and Pentland Dell. Other varieties may show only slight symptoms and Arran Pilot appears to be resistant to attack.

Light sandy soils known to carry the virus should not be cropped with varieties in which the most severe symptoms are produced. Efficient weed control will reduce alternate hosts.

Potato Mop Top Virus (PMTV). Symptoms somewhat similar to those of spraing occur in tubers infected with PMTV. In first-year infections the internal reddish-brown arcs are similar, but their presence can often be detected externally as raised, wheal-like, concentric rings. A variety of foliar symptoms can also be present including golden-yellow areas on the leaves, often in the form of chevrons, and a mop-like habit of growth due to shortening of the internodes. Chlorotic markings can be especially pronounced in Red Craig's Royal.

Symptoms in the tubers in the second year take the form of malformation with cracking, often minute cracking, forming a reticulate pattern. Pentland Crown and Ulster Premier can be severely affected.

Unlike TRV, infection with PMTV is more prevalent on medium to heavy loams. The vector is the powdery scab fungus *Spongospora subterranea*.

Control of Potato Virus Diseases. In a crop like the sugar beet or mangel discussed above, it is easy to start with a clean crop because beet is grown from seed and few viruses are seed-transmitted. With the potato crop, however, the state of affairs is different. The " seed " potato is a method of vegetative propagation, and all vegetative parts of a virus-infected plant give rise to a plant similarly diseased. For this reason the virus problem is vitally important in all crops which are raised from cuttings, tubers, runners, etc., and this includes such crops as potatoes, strawberries, raspberries, hops, dahlias and all flower bulbs. The reason why the English potato grower obtains fresh seed potatoes from special seed-producing areas every year lies in their comparative freedom from insect-borne and other virus infection. The first essential, then, is to start with as clean a crop as possible. Suitable seed potatoes for ware production should be first-quality commercial seed with an " A.A. " certificate or healthy commercial seed with a " C.C. " certificate (see p. 264).

Where the intention is to produce a crop for seed then it is essential to start with as clean a crop as possible. In these circumstances the seed used should have an " F.S. " certificate (foundation seed). Seed bearing a "V.T.S.C." certificate (virus tested stocks propagated from stem cuttings—a grade not issued in England and Wales) is basic seed intended primarily for further multiplication as seed. It is also important to select sites least vulnerable to invasion by the insect vector that spreads the viruses. Aphids will not fly if there is a wind blowing at more than 6 km.p.h., if the temperature is lower than 18 °C and if relative humidity is more than 75 per cent. These conditions hold in the Scottish and Irish districts where the best seed potatoes are grown and in many of the hill areas in western parts of England and Wales at altitudes above 120 m. It is also well to remember that

Myzus persicæ overwinters as an egg on peach trees, that it breeds in glasshouses on various plants all the year round and that it may also spend a mild winter out of doors on cabbages or brussels sprouts. All these facts should be borne in mind when growing potatoes, especially crops destined for seed. A certain amount of control of potato virus diseases is possible by carefully roguing out and burning infected plants and any tubers that may be on them. Such roguing should be done when the plants are small, firstly because they are easier to remove and will not have infected their neighbours with virus X by contact of their foliage ; and secondly because they will not have formed tubers which might be left behind in the soil.

In recent years it has been shown that the useful life of seed stocks in the ware-growing districts can be extended if either the ware crop from which seed is to be saved, or a special area planted for seed, and isolated from the ware crop, is sprayed with a systemic insecticide to prevent the build up of aphids. Where an area is planted specifically for seed production it is of benefit to apply a systemic aphicide as granules with the seed at planting time followed later in the season by a foliar spray. This, combined with careful roguing, makes possible the production of home-saved seed even in ware-growing areas. Unfortunately the treatment will not prevent the introduction of virus from outside sources. The project requires great care to ensure success.

VIRUS DISEASES OF CRUCIFEROUS CROPS

Of late years growers of cauliflowers, broccoli, brussels sprouts and turnips have been perturbed at the damage done to these crops by virus infections. There are several viruses concerned but only the more important ones are dealt with here.

Cabbage Black Ringspot attacks cabbages, cauliflowers and turnips, besides various ornamental plants belonging to the same family, such as wallflowers, stocks, arabis, etc. On cabbages, cauliflowers and brussels sprouts the symptoms of the disease consist of black, ring-like spots on the leaves which frequently coalesce, forming patches of dead tissue. As the disease progresses the veins and midribs become blackened and the leaves turn yellow. Diseased plants are stunted and useless for market. Affected turnips are badly crinkled and distorted and make little growth.

Cauliflower Mosaic. Here the first sign of infection is a yellow clearing or intensification of the colour of the veins of the youngest leaves, a symptom which is absent in the black ringspot disease. This is followed by a veinbanding, consisting of narrow dark-green areas parallel to and adjoining the midrib and lateral veins. The effects of this disease are severe and the head or curd of cauliflowers is greatly reduced in size.

Both these viruses are aphid-transmitted, the two most important aphid species being the mealy cabbage aphid, *Brevicoryne brassicæ*, and the peach and potato aphid, *Myzus persicæ*. No very efficient control methods for these diseases are known, but any measures which reduce aphid numbers are helpful since the diseases cannot spread in the absence of these insects. Brassica plants are frequently infected in the seed-bed, so that effort should be made to protect the seedlings from aphid infestation. Screening the young plants by planting barrier crops round the seed beds, and spraying with a systemic insecticide, are possibilities.

Self-sown plants and plants remaining after the harvesting of a brassica crop should be destroyed as soon as possible, as these form a serious reservoir of inoculum from which virus can be spread.

Turnip Yellow Mosaic has in recent years become prevalent in brassica crops in the north-east of England. The virus is transmitted by the common turnip flea beetle and this is unusual since the beetle is a biting insect. Apparently only insects with biting mouthparts can carry this virus. The first sign of infection is the appearance of yellowing along the veins of the young leaves, and this is followed by an intense yellow mottling which resembles a variegation. The virus attacks turnips, swedes, kale and radishes, and can also affect the weed, shepherd's purse. Cabbages and cauliflowers are not easily, if at all, infected.

BACTERIA

The bacteria are a group of unicellular organisms resembling the fungi in their mode of nutrition but differing from them in their method of growth. They are exceedingly minute, the largest of them seldom being more than ten thousandths of a millimetre in length. Their shape may be that of spheres, long or short rods, commas or corkscrews. Multiplication is effected by the simple division of the individuals into halves, the new ones so formed either becoming separate or remaining united in chains, sheets or solid masses when the divisions take place in one, two or three planes respectively. The successive splittings follow with great rapidity, and a single cell may become two within twenty minutes after its formation if the conditions for growth are favourable. Reproduction may also be effected by means of spores formed either internally or by the thickening of the cell wall. The spores are highly resistant to extremes of temperature and to drought, and consequently they provide the bacteria capable of forming them with a resting stage. Many species are motile owing to the presence of one or more cilia.

It is difficult to realise the great importance of the group to mankind. They are mainly responsible for the decomposition of the organic matter built up by plants and animals. To counteract this, in the case of materials required as food, various processes of preservation, such as cold storage, deep freezing, freeze drying, canning, bottling, desiccation, pickling, and so on, have had to be devised. They play a great part in the manufacture of various food products, in which their activity is often brought under control by sterilising, with the object of destroying unwanted species. In the soil and manure heaps their development determines the production, or in some cases the destruction, of essential nitrogenous plant food materials, as well as the reduction of cellulose-containing tissues to humus.

They are, however, better known generally as the causative organisms of many diseases, such as typhoid, cholera, tuberculosis and diphtheria. Plants are also attacked by them and one of the diseases they can cause, fire blight of pears and apples, has been most disastrous in its effect on pears growing in this country since its discovery in Kent in 1957. The disease is now also common in the southeast of England on ornamental species including *Pyracantha* and *Pyrus* spp. A number of other bacterial diseases, some of considerable importance, affect agricultural crops. Examples are blackleg of potatoes (*Pectobacterium carotovorum* var. *atrosepticum*), soft rot of turnips (*Pectobacterium carotovorum*) and crown gall of sugar beet (*Agrobacterium tumefaciens*). On the other hand, the symbiotic association of bacteria and leguminous plants is a matter of great economic significance.

The Nodules of Leguminous Plants. The clovers and leguminous plants in general have been described as " nitrogen-accumulators " because their tissues contain, mostly in the form of proteins, an unusually large percentage of nitrogen.

These plants can build up large quantities of protein when growing on soil deficient in nitrogenous food materials. In fact, crops such as lupins can be grown satisfactory on sandy soils incapable of carrying any non-leguminous crop, and their growth actually enriches the soil sufficiently for it to be used as a method for bringing such land under general cultivation. The value, too, of the nitrogen-collecting clover as a preparation for the wheat crop has been realised for several centuries.

Occasionally a soil is found to be lacking in a nodule-forming organism essential to the healthy growth of a particular crop. Lucerne is the commonest example. Inoculation of the seed, and consequently of the soil, is easily carried out by mixing with the seed a commercial culture of the required organism.

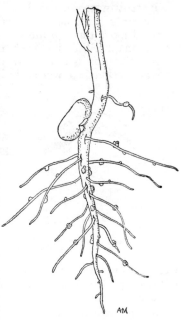

FIG. 69.—NODULES ON THE ROOTS OF FIELD BEAN.

Chapter XIV

INSECT AND OTHER PESTS OF CROPS AND STOCK

INSECTS AND ALLIED ANIMALS

ANIMALS from four great groups are important on the farm either as pests or benefactors :

ANNELIDA (or segmented worms)—Earthworms ;
NEMATODA—Round-worms and Eelworms ;
MOLLUSCA—Slugs and Snails ;
ARTHROPODA (or animals with jointed limbs)—Insects, Woodlice, Spiders, Mites, Millipedes and Centipedes.

Earthworms have a long cylindrical body consisting of many similar rings or segments (Fig. 70). They have neither antennæ nor legs and move through the soil with the help of stout bristles. They are hermaphrodite, i.e. each individual combines both sexes, but cross-fertilisation is usual. Worm eggs are enclosed in a " cocoon " and newly hatched worms have the same form as adults. Earth-

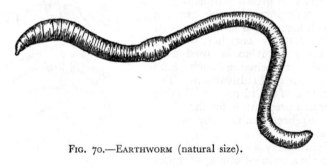

FIG. 70.—EARTHWORM (natural size).

worms should be present in all agricultural soils since they assist in breaking down organic matter and their burrows loosen and aerate the soil.

Round-worms and Eelworms have the same shape as earthworms but their bodies are smooth and unsegmented. They have a stout, muscular body-wall which gives them great flexibility (Fig. 85). Males and females are separate individuals and eggs are produced by the females after fertilisation. Many round-worms and eelworms are parasitic in stock and crops.

Slugs and Snails have a soft unsegmented body which is withdrawn under a fold of the body-wall, the *mantle*, when the animal is at rest. The conspicuous spiral shells characteristic of snails are secreted by the mantle, as are also the small shells hidden under the mantle of slugs. In slugs (Fig. 71) and snails the head bears two pairs of retractable tentacles, the upper pair carrying the eyes. The mouth has a strap-like tongue, the *radula*, with transverse rows of teeth. The ventral surface of the body is flattened into the broad muscular foot which expands and contracts as the animal moves. The whole body is thickly coated with slime which gives some protection from the effects of adverse weather, from the action

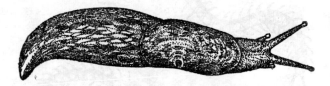

FIG. 71.—GREY FIELD SLUG : × 2.

of irritants present in soil and also from the attacks of enemies. Slugs and snails are hermaphrodite but reproduce by cross-fertilisation, each individual acting as both male and female. Slugs are major pests of farm and garden crops, and a small green fresh-water snail, *Limnæa*, the mud snail, is the host of liver-fluke, a parasite of sheep.

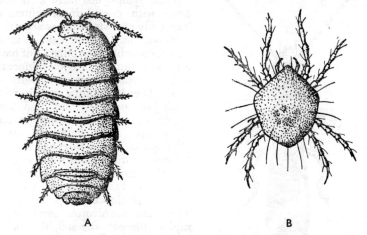

A   B

FIG. 72.—A. COMMON WOODLOUSE (*Armadillidium*) : × 6.
B. RED SPIDER MITE : × 25.

From *Insect Pests of Glasshouse Crops*, by H. W. & M. Miles (Crosby Lockwood).

Arthropoda. The fourth group of animals, the Arthropoda or animals with jointed limbs, consist of four main groups :

1. CRUSTACEA—Woodlice, Shrimps, Crabs and Lobsters.
2. ARACHNIDA—Spiders, Mites, Ticks and Scorpions.
3. MYRIAPODA—Millipedes, Centipedes and Symphylids.
4. HEXAPODA or INSECTA—Insects.

1. Woodlice (Fig. 72) and their allies have five or more pairs of

jointed legs. The head and thorax are united into a *cephalo-thorax* which has two pairs of antennæ.

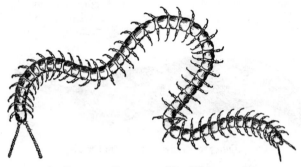

FIG. 73.—CENTIPEDE (*Geophilus*) : × 4.

2. Spiders, mites (Fig. 72), ticks and scorpions have 4 pairs of legs. The head and thorax are united to form a cephalo-thorax which has no antennæ.

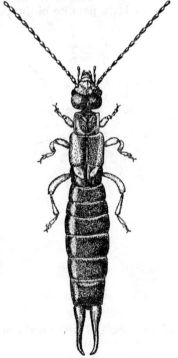

FIG. 74.—EARWIG ♀, : × 3.
Note the three main divisions of the body and the presence of three pairs of legs on the thorax.

3. Millepedes, centipedes (Fig. 73) and symphylids have a distinct head with one pair of antennæ. The body consists of many segments each with legs. Millepedes to some extent resemble caterpillars but have a greater number of body segments, most of which bear two pairs of legs. They have biting mouths with weak jaws or mandibles, and feed mainly on decaying organic matter. They are often found in crops already attacked by slugs and insects. Centipedes differ from millepedes in that they have only one pair of legs on each body segment and the first pair are developed into poison fangs. They are carnivorous and move rapidly through the soil, devouring small slugs, snails, insects and other creatures.

Symphylids are fragile white creatures about 6 mm long, with long, slender antennæ, twelve pairs of legs and a pair of tail-feelers. They usually abound in soil rich in organic matter and feed on soil organisms and on plant roots. They are often troublesome in market garden greenhouses where they damage roots of lettuce and tomatoes.

4. The **Hexapoda** or six-footed animals form the largest and most varied group of **Arthropoda**. They are generally called **Insecta**, or insects, because the body is divided into three distinct sections, namely head, thorax and abdomen. The head carries one pair of antennæ, the thorax carries three pairs of legs and, in winged insects, one or two pairs of wings, and the abdomen is without true (jointed) legs (Fig. 74).

THE FORM AND STRUCTURE OF INSECTS

The Head and Mouthparts. The head is thought to have developed from the union of several segments each of which carried a pair of limbs, and in the course of evolution the limbs have become antennæ and mouthparts. Insects have two kinds of eyes : compound or faceted eyes, and simple eyes or *ocelli*. In most insects there are three ocelli arranged in a triangle near the top of the head ; these are missing in earwig (Fig. 75) but are seen in frit fly (Fig. 77). The

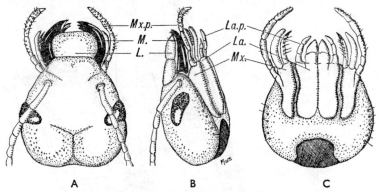

FIG. 75.—HEAD OF EARWIG : × 12.

A, head of earwig from above, showing L., labrum ; M., mandible   Mx.p., palp of maxilla
B, side view to show relationship of mouthparts.
C, head from below, showing La., labium ; Mx., maxilla (labrum and mandibles removed).

compound eyes are often of a greenish or reddish colour, and may occupy most of the top and sides of the head. Insect mouthparts are complicated structures and two distinct types occur, the biting or mandibulate mouth and the piercing and sucking mouth.

Mandibulate Mouthparts. In beetles, wasps, earwigs, grasshoppers and cockroaches the mouth structures conform to a more or less typical and easily recognisable pattern (Fig. 75). The mouth is closed on the upper side by a tough, horny flap, the *labrum* or upper lip. Beneath it are the mouthparts (Fig. 75) which consist of three pairs of jaws overlying one another and working in a horizontal plane : and the upper jaws or *mandibles*, the middle jaws or first pair of *maxillæ*, and a *labium* or lower lip formed by the fusion of the second, or lower, pair of maxillæ. The mandibles are thick and strong and generally consist of a single segment. They may be sickle-shaped as in carnivorous ground beetles (Carabidæ) and glow-

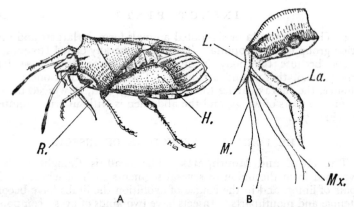

FIG. 76.—SHIELD BUG (*Acanthosoma*) : × 3½.

A, shield bug : R., rostrum ; H., hemi-elytra.
B, head enlarged to show L., labrum ; M., mandibular stylets ; Mx., maxillary stylets; La., labium.

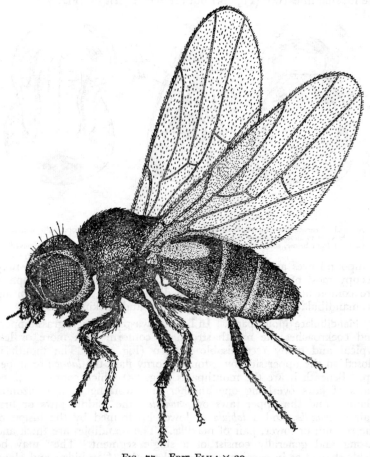

FIG. 77.—FRIT FLY : × 30.

worms (Lampyridæ), or broad and ridged or toothed for tearing and grinding plant tissue, as in caterpillars and wireworms. The middle pair of jaws, or first pair of maxillæ, are delicate complicated structures often fringed with spines and bearing a pair of mouth-feelers, the *maxillary palps*. The maxillæ assist in the mastication of food. The lower lip or labium consists of the second pair of maxillæ fused to form a single structure with a pair of feelers, the *labial palps*. It closes the mouth from below and may have attached to it the open end of the salivary duct. In caterpillars the salivary duct is highly developed and the labium is modified to form a *spinneret* from which saliva passes like a continuous thread of silk. The salivary strand of some moths, as for example *Lasiocampa* and the silk-worm moth (*Bombyx mori* L.), is used to make silk.

Sucking Mouthparts. In bees, flies, butterflies, mosquitoes, aphids and bugs (Fig. 76) the mouthparts have undergone much change of form and arrangement to enable the insects to feed on liquids. Houseflies and allied flies have a short broad sucking proboscis for lapping superficial liquid. The honey bee has a long sucking proboscis for gathering nectar and a pair of stout mandibles for removing wax or other obstruction. In butterflies and moths there is a long flexible proboscis which, when not in use, lies curled under the head. Mosquitoes, gadflies, bugs and aphids have mouthparts adapted for piercing the animal or plant host and sucking liquid from below the surface (Fig. 91).

The character of the mouthparts and the manner of feeding have greatly influenced the measures taken to check insect damage to crops and stock. Injurious insects with biting jaws devour leaves, stems and fruits, and may be killed by coating the food plants or bait with stomach poisons like lead arsenate and Paris green. They may also be killed with contact poisons like nicotine, rotenone (derris), DDT (dichlor-diphenyl-trichloroethane) and benzene hexachloride. Biting insects are not usually carriers of plant and animal diseases, and their presence in small numbers on established plants is generally of little importance. Injurious insects with piercing and sucking mouths are always a serious menace. Their feeding causes distortion and stunting of plants, and swelling and irritation in stock, and they transmit virus diseases in plants and fevers in man and animals. They draw food from below the surface and consequently are not affected by stomach poisons spread over their hosts. Contact or systemic insecticides must be used against sucking insects.

The Thorax and Limbs. The midbody or thorax occupies the region between the head and the hind body, but its junction with the hind body may be so complicated that it is difficult to see where one ends and the other begins. The thorax consists of three segments, the pro- (or fore) thorax, the meso- (or middle) thorax and the meta- (or hind) thorax. It carries the legs and wings, and is usually stout and strong in order to support the powerful wing and leg muscles which almost fill the interior. Each thoracic segment has a pair of legs, and when wings are present these are attached to the middle and hind thoracic segments.

The Legs. Insect legs have a common basic pattern of segments though the size and shape of the various segments show remarkable variation. They consist of five parts : coxa, trochanter, femur,

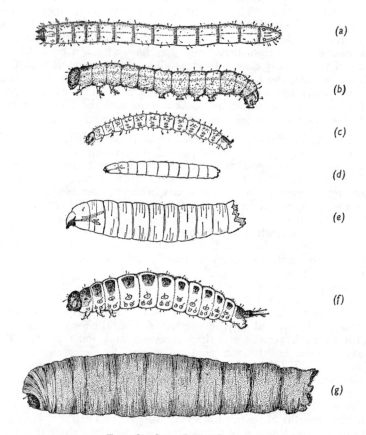

(a)

(b)

(c)

(d)

(e)

(f)

(g)

FIG. 78.—SOME INSECT LARVÆ.

(a) wireworm, grub of beetle (dorsal view) has 3 pairs of true legs, and a false foot on tail segment (*Agriotes*, Coleoptera) : × 5.
(b) caterpillar of moth with 3 pairs of true legs, false legs on segments 6, 7, 8, 9, 13. A grass stem-borer (*Apamea*, Lepidoptera) : × 4.
(c) flea beetle grub with 3 pairs of true legs, and false foot on the tail segment (*Psylliodes*, Coleoptera) : × 10.
(d) maggot of frit fly with reduced head, mouth hooks instead of mandibles, no legs (*Oscinis*, Diptera) : × 8.
(e) maggot of wheat bulb fly (*Hylemyia*, Diptera) : × 8.
(f) grub of wheat mud beetle with 3 pairs of true legs, and a false foot on tail segment (*Empleurus*, Coleoptera) : × 8.
(g) leatherjacket, maggot of daddy-long-legs or crane fly, with small head withdrawn into body mandibles, no legs (*Tipula*, Diptera) : × 8.

tibia and tarsus. The *coxa* is a short, broad, flattened segment closely joined to the body. The *trochanter* is generally a small triangular strut-like segment at the articulation of coxa and femur. The *femur* or thigh is a long strong segment greatly enlarged for jumping in grasshoppers and flea beetles. The *tibia* is usually long and slender and often has characteristic spines and hairs. The *tarsus* consists of 1–5 small tarsal segments tipped with sensitive hairs, and a pair of tarsal claws.

The Wings. The number and character of the wings are used in the classification of insects. Butterflies (Lepidoptera) have two pairs of wings thickly clothed with scales. Ants, bees, wasps and sawflies (Hymenoptera) have two pairs of membranous wings, each with a stigma or small thickened area near the middle of the front edge. Beetles (Coleoptera) have a pair of leathery fore wings or *elytra* that form a protective cover for the membranous hind wings. Some of the plant bugs (Hemiptera or Rhynchota) have *hemi-elytra* : that is the fore wings are thickened at the base and membranous at the extremities, and the hind wings (Fig. 76) are membranous. The flies (Fig. 89) (Diptera) have a pair of membranous fore wings and a pair of knob-like *halteres* or balancers which are thought to be degenerate hind wings.

The Abdomen. The abdomen or hind body of an insect consists of as many as 10 rings or segments, some of which may be telescoped within the body. In adult insects the abdomen is legless, but caterpillars of butterflies, moths and sawflies have several pairs of abdominal false feet or *pseudopods* (Fig. 78, *b*). In some insects the tip of the abdomen bears a pair of tail-feelers or *cerci* (Fig. 78, *f*). The bristle-tails (silver-fish and other Thysanura) have a pair of long slender cerci, sometimes also a long median tail, the *telson*. In

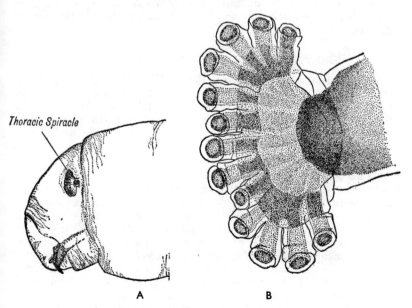

Thoracic Spiracle

A   B

FIG. 79.—STRUCTURE OF A SPIRACLE.

A, head of cabbage root fly maggot, showing mouth hooks and thoracic spiracle : × 16. B, spiracle much enlarged.

earwigs the cerci are conspicuous pincers (Fig. 74). The abdomen contains the reproductive system, and the egg-laying mechanism which projects from the tip of the body in grasshoppers, cockroaches, ichneumon wasps and sawflies.

The Respiratory System. Most insects breathe by means of a system of air-tubes or *tracheæ*, whose branches extend throughout the body and limbs. The main trunk air-tubes lie internally on each side of the body and air enters from pores or *spiracles* in the sides. In most insects spiracles occur on the mid and hind thoracic segments and on the first eight abdominal segments, but in the maggots of blowfly, frit fly, cabbage root fly (Fig. 79) and their allies there are only two pairs of spiracles, one pair in the prothorax close to the head and another at the tip of the tail. The air tubes are strengthened and kept open by a spiral thickening, and the inflow of air is regulated by the opening and closing of the spiracles. Much of the carbon dioxide passes out of the body through the skin.

The Exoskeleton. The form and character of insects depends on the outer case or *exoskeleton* which encloses the body. It is a continuous covering or *integument* with thickened horny plates or *sclerites* over the segments, and with thin intersegmental membrane folded under the sclerites. The horny sclerites give shape and protection to limbs and vital organs, and the membrane between the sclerites gives the insect the flexibility necessary for movement and permits some expansion of the body. The skin of insects is largely composed of *chitin*, a complex substance related chemically to the cartilage of vertebrate animals. Chitin is insoluble in water, ether and other organic solvents, and in dilute acids. It may be softened in dilute and concentrated alkalis.

GROWTH AND CHANGE OF FORM IN INSECTS

Most insects hatch from eggs (Plates XIX and XX). Insect eggs show remarkable variation in colour, form and texture, from the tall smooth elliptical bright yellow eggs of ladybird beetles to the flat greenish or drab scale-like eggs of many small moths. The period of incubation also varies from about three days for the summer eggs of cabbage root fly to several months for the overwintering eggs of the vapourer moth (*Orgyia antiqua* L.) and aphids.

The Moult or Ecdysis. Insects begin to feed soon after they hatch but their tough inelastic skin will not permit of a gradual increase in size. As feeding takes place the old skin becomes distended and finally splits and is shed. The moult, or *ecdysis*, is immediately followed by a considerable expansion of all parts of the body. When an insect is about to moult it rests until the contents of the alimentary canal are absorbed or voided. Moisture is secreted between the old and new skins so that the two are completely separated, and as a result of strains and pressure the old skin ruptures, usually near the head. The insect then withdraws its head and gradually pushes the old skin backwards off the body. Attached to the old skin are the lining of the alimentary canal and main air-tubes (spiracles and tracheae), and the hard outer shell of jaws, feelers, spines, legs and claws. The insect rests until the new skin toughens and assumes its normal colour. Before the new skin finally hardens and oxidises considerable expansion takes place, and many insects are approximately one-quarter larger after the moult than they were before. The number of moults varies in different insects and in the sexes, many caterpillars having 6-7 moults while maggots of flies may reach maturity after 2-3 moults.

Metamorphosis. Many insects undergo great change of form in the course of growth while others retain the same form through life. Simple insects like bristle-tails (Thysanura) and spring-tails

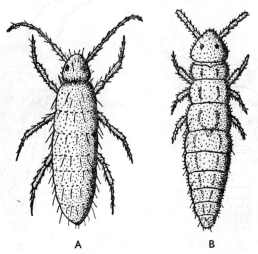

A   B

FIG. 80.—SPRINGTAILS.

A, longhorned springtail (*Tomocerus*) : × 22.
B, shorthorned springtail (*Hypogastrura*) : × 30.
    Examples of the Apterygota, insects wingless at all stages, with no change of form from hatching to maturity. From *Insect Pests of Glasshouse Crops*, by H. W. and M. Miles (Crosby Lockwood).

(Collembola) (Fig. 80) emerge from eggs in a form similar to that of the adults. They grow larger at each moult but undergo no change of form or metamorphosis.

Insects which retain the same form throughout life are called *AMETABOLA* (without metamorphosis) ; they are also known as *APTERYGOTA* (without wings) because the adults are wingless.

A second group of insects undergoes a partial change of form in the course of development : they are known as *HEMIMETABOLA*. Grasshoppers, earwigs, cockroaches, dragonflies, aphids, thrips and plant bugs are hemimetabolous. The young emerge from the eggs in a form resembling that of the adults. Mouthparts, antennæ, legs and general body-shape are similar in young and adults but the young are wingless and are known as *nymphs*. Wings develop externally in pouches on the back and at each moult the wing pouches grow larger. At the last moult the pouches are shed with the skin and the wings assume the size, shape and colour of adult wings. Insects whose wings develop externally in pouches or sacs on the back are also known as *EXOPTERYGOTA* (external wings) (Fig. 81).

The third main insect group is the *HOLOMETABOLA*, insects that undergo complete transformation during their development. This group includes beetles, ants, bees, wasps, sawflies, flies, moths, butterflies, caddis flies, weevils and lacewings. Beetles begin their active lives as grubs with three pairs of legs (Fig. 78, *c* and *f*) ; weevil grubs and the grubs of bees and wasps are legless. Butterflies,

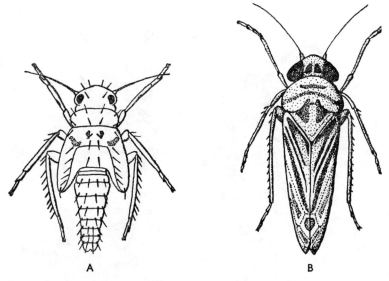

A   B

FIG. 81.—NYMPH AND ADULT OF LEAF HOPPER (*Erythroneura*).

A, nymph with well-grown wing-buds.
B, adult with fully developed wings: × 18.
  An example of Exopterygota, insects with external wing development.   From *Insect Pests of Glasshouse Crops*, by H. W. and M. Miles (Crosby Lockwood).

moths, caddis flies and sawflies pass the first part of their active life as caterpillars (Fig. 78, *b*) feeding on leaves, stems, roots, fruits and seeds or, in the case of caddis flies, mostly on small aquatic animals. The immature stages of flies are legless maggots, worm-like in the stilleto fly (*Thereva*), and stout and wrinkled in the crane fly and wheat bulb fly (Fig. 78, *e* and *g*).

Grubs, caterpillars and maggots are still in an embryonic stage. The eggs from which they hatch contain insufficient food to permit them to complete their development and they emerge as *larvæ* and spend their time feeding. When a sufficient store of food has been accumulated adult tissue and limbs begin to grow within the larval body and when the last larval skin is shed the insects become inactive chrysalids or *pupæ*. During the pupal stage the larval body is broken down and remade into adult tissue, and adult limbs and organs acquire their characteristic form. When the transformation of larva into adult insect is complete the pupal case is shed and the adult or *imago* is revealed. Insects that pass through the larval and pupal stages before reaching maturity are classed as *ENDOPTERYGOTA* (internal wings) because early wing growth takes place within the larval body.

The chief characters of the three groups of insects can be summarised as follows :

AMETABOLA OR APTERYGOTA. Young resemble adults in form and mode of life ; no marked change or transformation occurs during growth, and adults are wingless. (Silver-fish, bristle-tails, springtails.)

HEMIMETABOLA OR EXOPTERYGOTA. Young resemble adults in form and, except in dragonflies and mayflies, have a similar mode of life. Wings develop externally in pouches on the back (Fig. 84). There is no pupal stage and insects pass immediately from nymph to adult. (Cockroaches, earwigs, grasshoppers, dragonflies, mayflies, aphids, thrips, bugs.)

HOLOMETABOLA OR ENDOPTERYGOTA. Insects of this group undergo complete transformation. There is a larval stage (as caterpillar, grub or maggot) concerned wholly with feeding, a resting stage as a pupa or chrysalid (Fig. 87) which is concerned with internal development, and an adult or imaginal stage concerned mainly with reproduction. Adults are winged and wing development takes place inside the body. (Butterflies, moths, beetles, weevils, ants, bees, wasps and flies.)

IMPORTANT ORDERS OF INSECTS

The main insect groups just described are divided into twenty-three Orders, ten of which are of interest and importance in agriculture and horticulture.

APTERYGOTA

Thysanura (Bristle-tails, Silver-fish) are slender active insects about 12 mm long, with the body soft and covered with scales and having at the tip two or three long cerci. They are wingless and have no metamorphosis, and are thought to resemble the ancestral type from which all insects have developed. Silver-fish (*Lepisma saccharina* L.) are common in old farmhouses where they feed on dust and dirt about kitchens and larders and on the paste of wallpaper and of books.

Collembola (Springtails) (Fig. 80) are tiny wingless insects some of which have a forked tail-structure, the *furcula*, held in a catch or *hamula* on the underside of the body. The release of the furcula from the hamula propels the insects forward and earns them the name " springtails ". There are two types of springtails : those with the body cylindrical and elongate and the segmentation well defined (*Anurida*, *Tomocerus*), and those with a rather globular body with obscure segmentation (Sminthuridæ). The former are often found about the soil and under stones, the latter about foliage and among seedlings. Springtails have weak mandibles and feed on soft plant tissue. They frequent decaying organic matter, mushrooms and other fungi, and sometimes attack seedlings and sickly plants. They are often found in large numbers on swedes and mangolds damaged by hoeing or broken off by " strangles ".

EXOPTERYGOTA

Orthoptera (Cockroaches, Crickets, Grasshoppers and Locusts) have strong biting mouths, two pairs of straight stiff wings, and legs adapted for running (cockroaches) and jumping (grasshoppers, bush-crickets and locusts). The fore-wings or *tegmina* are long, narrow and thickened to form a leathery cover for the broad membranous hind wings that fold fan-wise when at rest. Metamorphosis is incomplete and wings develop externally. Cockroaches and crickets are omnivorous and occur in boiler-houses and in warm food stores and preparation-rooms. Bush-crickets are largely carnivorous and devour other insects. Grasshoppers and locusts are plant-feeders and in some parts of the world they occur in devastating swarms.

Dermaptera (Earwigs) (Fig. 74) are closely related to Orthoptera but have short leathery fore-wings and semi-circular membranous hind-wings that fold into a double fan beneath the forewings. The cerci or forceps are sickle-shaped in males and straight in females. Earwigs feed on decaying organic matter. They are mainly scavengers but in late summer and autumn they devour leaves and flowers of cultivated plants. Earwigs hibernate as adults in soil, in hollow stems and under bark. Eggs are laid in clusters in the soil in spring and the female tends her eggs during the incubation period and looks after the young for a short time after they hatch.

Hemiptera or Rhynchota (Bugs, Aphids, Scale insects, Mealy Bugs and Whiteflies) have mouthparts adapted for piercing and sucking liquid from their hosts (Fig. 76). The lower lip is modified into a beak or *rostrum* and the two pairs of jaws, the mandibles and maxillæ, are modified into slender piercing stylets. Metamorphosis is incomplete and the wings develop externally. There are usually two pairs of wings and the character of the first pair is used to separate Hemiptera into two sub-orders, Heteroptera and Homoptera.

HETEROPTERA (Plant bugs, Shield-bugs, Capsids) (Plate XXIX) have the fore-wings modified into *hemi-elytra*, that is the base is tough and leathery and the tip is membranous. The hind-wings are entirely membranous and both pairs fold closely over the body. The head is usually small and triangular, and the beak arises from its front margin. The shield bugs (*Pentatomidæ*) (Fig. 76) have long 5-segmented antennæ and shield-shaped bodies and when disturbed or handled they give off an unpleasant odour from large pores on the underside of the thorax. Many British shield bugs are predaceous. Capsid bugs (*Lygus* and *Plesiocoris* spp.) are slender, oval and greenish. They are common about herbaceous plants, trees and bushes and feed by puncturing young stems and leaves and sucking the sap.

Their feeding causes such severe distortion that they rank as major pests of fruit trees and bushes, strawberries, and also of many flower crops.

HOMOPTERA (Aphids, Frog-hoppers, Leaf-hoppers) have wings that are entirely membranous or uniformly thickened. The beak appears to arise from the hind margin of the head and point backwards. The leaf-hoppers (Jassidæ) and frog-hoppers (Cercopidæ) have the fore-wing slightly thickened, and the antenna short and tipped with a long slender bristle. Leaf-hoppers are common about foliage and jump when disturbed, and their feeding-sites show as

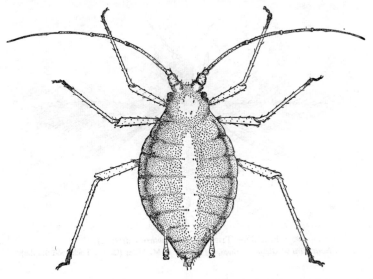

FIG. 82.—WINGLESS FEMALE OF PEACH-POTATO APHID (*Myzus persicæ*) : × 22.
From *Insect Pests of Glasshouse Crops*, by H. W. & M. Miles (Crosby Lockwood Staples).

small pale or bleached spots on the leaves. The nymphs of frog hoppers cover themselves with froth and are called " cuckoo-spit " insects.

Aphids or plant-lice (Aphididæ) have long slender antennæ and a pair of tubular siphunculi on the back near the tip of the body (Fig. 82). Winged aphids have two pairs of membranous wings ; the fore-wings are larger than the hind-wings and have a characteristic arrangement of veins. The typical annual life cycle among aphids includes overwintering eggs ; wingless females of the spring generation that give rise to colonies of wingless aphids on the spring host ; winged summer migrants that fly to the summer host and start the summer colonies of wingless aphids ; and winged autumn migrants that return to the winter host and give rise to males and females that together produce overwintering eggs. During spring and summer reproduction is parthenogenetic, that is, without the act of mating, and aphid colonies are produced rapidly in this way. The mealy cabbage aphid (Plate XXIV) (*Brevicoryne brassicæ* L.) passes the winter in the egg stage on brussels sprouts, kale and cabbage, but in mild seasons parthenogenetic reproduction may continue throughout the winter. The black aphid of beans, sugar beet and mangels (*Aphis fabæ*) overwinters in the egg stage on spindle (*Euonymus*), a common hedgerow shrub in many districts, and the number of eggs found on spindle in winter enables observers to forecast the intensity of attack by black fly on susceptible crops in the following season. Aphids are important carriers or vectors of virus diseases in plants. The control of mosaic and leaf-roll diseases of potatoes is largely a matter

Fig. 83.—Thrips (*Parthenothrips dracenæ*).
From *Insect Pests of Glasshouse Crops*, by H. W. & M. Miles (Crosby Lockwood Staples).

of preventing aphid infestation of the crop. The method selected is one of prevention rather than cure, stocks of seed potatoes being usually grown in districts where aphids are absent or rare. Protective spraying of seed crops against aphids is a recent development.

Thysanoptera (Thrips) are tiny slender insects with a rather triangular head and short stout antennæ. The mouthparts are peculiar in that one mandible is modified into a piercing stylet, and the insect sucks sap after piercing the tissue. Thrips have two pairs of narrow wings fringed with long hairs (Fig. 83). Metamorphosis is incomplete, but in some species the last nymphal stage is quiescent in the surface soil for some time before the insect emerges as an adult. Thrips are common about crops, and at harvest they settle on faces and arms of the workers. One species carries spotted wilt, a virus disease of glasshouse plants ; other species occasionally attack seedling brassicas and may infest leeks and onions on market gardens.

ENDOPTERYGOTA

Lepidoptera (Butterflies and Moths) have two pairs of wings thickly covered with coloured scales arranged in characteristic patterns. The antennæ are long and in males they are often conspicuously plumed. The second pair of jaws (maxillæ) form a long proboscis for sucking nectar and capable of being coiled beneath the head when not in use. Lepidoptera undergo complete metamorphosis, passing through larva and pupal stages before reaching maturity. Larvæ of Lepidoptera are caterpillars with stout biting mouthparts, three pairs of thoracic legs and usually with five pairs of false legs (prolegs or pseudopods), tipped with small hooks or crotchets on abdominal segments 3, 4, 5, 6 and 10. Caterpillars of winter moths (Geometridæ) have false legs only on the sixth and last abdominal segments and progress in a looping manner that has earned them the name of " loopers ".

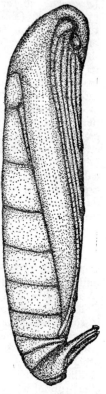

Most caterpillars are plant feeders but their habits, form and colour show great differences. Caterpillars of tortoiseshell, peacock and large cabbage white (Plate XXVII) butterflies and lackey and small ermine moths feed in colonies often in a web of silk. Many caterpillars are green and inconspicuous among the foliage while others like those of the peppered moth (*Biston betularius* L.) have a twig-like appearance. Caterpillars of tortrix moths make a feeding shelter by spinning together the tips of shoots ; others like the lilac leaf miner (*Gracillaria syringella* F.) feed between the upper and lower leaf surfaces of privet and lilac, and many, such as the leopard moth, goat moth, and apple pith moth, feed in the woody interior of branches and twigs of trees. The woolly bear caterpillars of the garden tiger moth are protected by their long irritating hairs, and caterpillars of puss moth (*Cerura vinula* L.) and the lobster moth (*Stauropus fagi* L.) by their menacing appearance.

Fig. 84.—Chrysalis of Apple Pith Moth (*Blastodacna*) showing outlines of developing adult : × 16.

Chrysalids or pupæ of butterflies tend to be spindle-shaped with the wings projecting on the underside of the body, and those of moths are elongate oval with the wings folded close to the sides (Fig. 84). Chrysalids of butterflies are often suspended by the tail and supported by a girdle round the thorax. Caterpillars of the drinker moth and its relatives (Lasiocampidæ) construct large silken cocoons and many caterpillars enter the soil and make earthen cells.

Many common pests belong to the order Lepidoptera. Swift moths (Hepialidæ) have very short antennæ and fly rapidly at dusk

during May and June. The whitish caterpillars live in the earth
and feed at the roots of grass and in the tap-roots of grassland herbs,
and are occasionally injurious in strawberry plantations and market
gardens. Clearwing moths (Sesiidæ) resemble bees and wasps in
appearance ; the caterpillars live in the woody twigs and branches
of currants, fruit trees, willows and poplars. The large caterpillars
of hawk moths (Sphingidæ) feed on foliage, those of the death's head
(*Acherontia atropos* L.) devouring potato haulms and those of the
eyed hawk (*Smerinthus ocellatus* L.) devouring the leaves of fruit
trees. The looper caterpillars of the large family Geometridæ feed
on the leaves of forest trees, fruit trees and ornamental trees and
shrubs. Those of the winter moth (*Operophtera brumata* L.) and
its allies are of special interest to fruit-growers who use various
insecticides against them at green cluster stage of fruit bud development.

The owlet moths (Noctuidæ) which fly at night and shelter in
herbage during the day include the large yellow underwing (*Noctua
pronuba* L.), turnip moth (*A. segetum* Schiff.), heart and dart (*A.
exclamationis* L.), antler moth (*Cerapteryx graminis* L.), light arches
(*Apamea lithoxylea* F.), tomato moth (*Lacanobia oleracea* L.), and
cabbage moth (*M. brassicæ* L.). Caterpillars of some of these moths
are cutworms and surface caterpillars. They shelter in the soil by
day and feed by night at or near ground level. They then pupate
just below the soil surface and the chrysalids are chestnut brown,
smooth and polished.

Small injurious moths (Tineina) include the diamond-back
(*Plutella xylostella* Curt.) which feeds on cultivated brassicas ; the
small ermine (*Yponomeuta padella* L.) whose caterpillars feed in colonies
under webs or tents among the leaves of apples; the house moths
(*Borkhausenia* and *Endrosis*) living on feeding stuffs in granaries.
Caterpillars of the tortrix moths (Tortricidæ) generally live concealed
in rolled leaves or in the tips of shoots. The oak leaf-roller (*Tortrix
viridana* L.) sometimes defoliates oak forests, and apple tortrix
(*Archips podana* Sc.) devours leaves and damages fruit of apple and
other fruit trees. Caterpillars of some tortrix moths are seed-feeders.
Codling moth caterpillars (*Cydia pomonella* L.), notorious the world
over, tunnel into apples and eed on the seeds ; and those of the pea
moth (*C. nigricana* F.) feed on peas within the pods. Amongst the
butterflies the cabbage whites (*Pieris brassicae* L., *P. rapae* L., *P. napi*
L.) are troublesome on brassica crops.

Coleoptera (Beetles). These are characterised by the hard or
leathery fore-wings or elytra that meet in a straight line along the
middle of the back, and usually cover the abdomen (Plate XIX). The
hind-wings are membranous and when not in use are folded under
the elytra. The legs are either long, for running, in ground beetles
(Carabidae) ; broad and stout, for digging, in dung beetles (*Aphodius*);
or modified to form swimming paddles in whirligig beetles (*Gyrinus*).
The antennæ are long and slender in long-horn beetles (Longicornia);
clubbed in burying beetles (*Necrophorus*) ; and with a series of leaf-like
segments or *lamellæ* in the cockchafer (*Melolontha melolontha* L.). Beetles
have biting mouths equipped with well-developed mandibles, very

large and antler-like in the male stag-beetle (*Lucanus cervus* L.), and sharp and sickle-shaped in tiger beetles (*Cicindela*).

Beetles and weevils undergo complete metamorphosis. Beetle grubs vary greatly in size and shape from the slender wireworms (*Agriotes*) (Plate XIX) to the large fleshy grubs of cockchafer and dor beetle (*Geotrupes*) but all have well-developed heads and three pairs of true legs on the thoracic segments. The grubs of weevils are wrinkled, fleshy and without legs (Fig. 86). Pupation takes place in the soil or in the feeding-sites. The pupæ are enclosed in a thin membrane with the limbs free.

Mode of life and habits show great variety among beetles. Ground beetles and their grubs are active carnivorous insects which prey upon slugs and other insects and are therefore beneficial. Ladybird beetles (Coccinellidæ) and their grubs feed on aphids. Rove beetles (Staphylinidæ) are readily recognised by the short elytra that cover only about a third of the body and by the habit of turning the tip of the body upwards in a menacing fashion. Both adults and grubs are carnivorous on soil creatures, small rove beetles (*Aleochara*) being parasitic on maggots and pupæ of cabbage root fly and its allies. Chafer beetles (*Melolontha, Amphimallus, Phyllopertha*) feed on the leaves and blossoms of fruit trees and shrubs and the grubs feed on the roots of grass, shrubs and fruit trees and bushes. Click beetles (Plate XIX) are elongate oval, and when placed on their backs they aright themselves by springing upwards with a sharp clicking sound. The turnip flea beetles (Plate XXVII), mustard beetles and willow beetles feed on foliage. They belong to a large group of beetles, the Phytophaga, and have the head directed downwards, long slender antennæ and shining greenish, bluish and bronze elytra. The Colorado beetle (*Leptinotarsa decemlineata* Say.) which feeds on potato haulm also belongs to this group. The wood-boring beetles (*Anobium* and *Lyctus*) are major pests of timber ; the longhorn beetles (Longicornia) are also wood-borers, and the typographer beetles (Scolytidæ) are bark-borers whose feeding-tunnels appear as curious engravings on fencing-posts and rails after the bark has been removed.

Weevils (Rhynchophora) are beetles with a long snout and elbowed antennæ clubbed at the tip (Plate XXVII). Pea and bean weevils (*Sitona*) have short stout beaks and feed on the leaves of peas and beans. The blossom weevils of the apple and strawberry (*Anthonomus*) have a slender beak of medium length with which they bite a hole in the blossom bud and deposit an egg in the hole. The nut weevils (*Balaninus*) have slender gracefully curved beaks, longer than the body. Grain weevils (*Calandra*) and a large number of associated beetles (*Ptinus, Niptus, Dermestes* and *Sitodrepa*) feed on grain and other stored products.

Hymenoptera (Bees, Wasps, Ants, Ichneumon wasps (Plate XXIII) and Sawflies) have two pairs of iridescent membranous wings. The antennæ are usually long and conspicuous and may be elbowed, clubbed, simple or branched. Mandibles are present but in some groups the maxillæ and labium are modified to form a sucking proboscis. In bees, wasps, ants, ichneumon wasps and other parasitic Hymenoptera the body has a slender waist-like constriction near the junction of thorax and abdomen, but in the sawflies the abdomen is broadly joined to the thorax. Throughout the Order the females have well-developed organs for egg-laying called *ovipositors* (Plate XXIII). The saws of sawflies, the stings of bees and wasps and the piercing organ of parasitic species, which in some ichneumon wasps is about as long as the body, are all modified forms of ovipositor.

In the Hymenoptera there are two types of larvæ, the caterpillar-like larvæ of sawflies and the legless grubs of ants, bees, wasps and ichneumon wasps. Sawfly caterpillars can be distinguished from those of butterflies and moths by their large conspicuous eyes since caterpillars of Lepidoptera have clusters of tiny ocelli that cannot be seen without a lens, and by the absence of hooks or crotchets on the abdominal false feet or pseudopods. Like the caterpillars of Lepidoptera, sawfly caterpillars are green and inconspicuous, or brightly coloured, spotted or striped or clothed with large branched spines. The fleshy, white, legless grubs of ants, bees, wasps and ichneumon wasps spend protected, inactive lives surrounded by food.

Caterpillars and grubs of Hymenoptera usually pupate within silken or parchment-like cocoons. The pupæ are pale and soft and enclosed in a thin membrane which leaves wings and limbs free. The pupal period in summer generations of bees, wasps and ants may last about a week, or may extend over two winters in wheat stem sawfly and apple sawfly.

Reproduction by parthenogenesis, that is by unfertilised females, is common in Hymenoptera, and in some species (the honey bee and the pale spotted gooseberry sawfly (*Pteronidea leucotrocha* Hart.)) unmated females produce only males, while in others (the small green gooseberry sawfly (*Pristiphora pallipes* Lep.) and the greenhouse white fly parasite (*Encarsia formosa* Gahan)) females are produced and the species reproduces indefinitely without the intervention of males (Plate XXIV).

Social life reaches its highest development in bees, wasps and ants. In bumble bees (Bombidæ) and wasps (Vespidæ) the colonies start each year from fertilised females which shelter throughout the winter about buildings, hedges and evergreen trees. The colony depends on workers which are under-developed and barren females, and males are generally produced towards the end of the summer to ensure the fertilisation of overwintering females. Ant colonies persist year after year, but towards the end of the summer males and females are produced and, after fertilisation, the females lose their wings and settle in a small soil cell to lay eggs and start new colonies. The hive bee (*Apis mellifera* L.) has been carefully studied and its value as a pollinator in fruit plantations is widely recognised. A queen bee (fertilised female) may produce eggs for several seasons, and eggs, larvæ and pupæ are tended by workers. In an artificial colony or hive there may be over 30,000 workers engaged in cell-building, foraging and all operations necessary for the care and maintenance of the colony. New colonies are established by swarming,

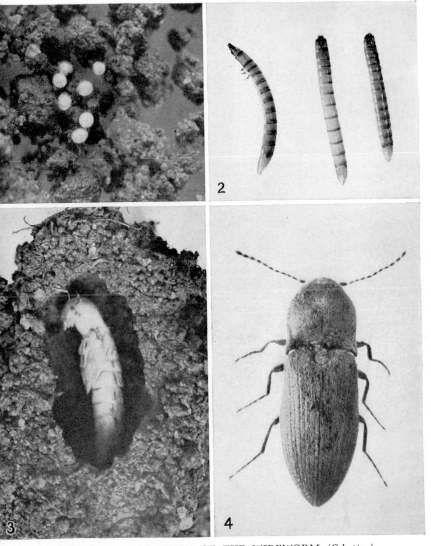

PLATE XIX. LIFE CYCLE OF THE WIREWORM (*Coleoptera*)

, eggs (× 10) ; 2, wireworms, the larval stage (× 2) ; 3, the chrysalis in soil cell (× 5) ;
4, the click beetle, the adult stage, *Agriotes obscurus* L. (× 6)

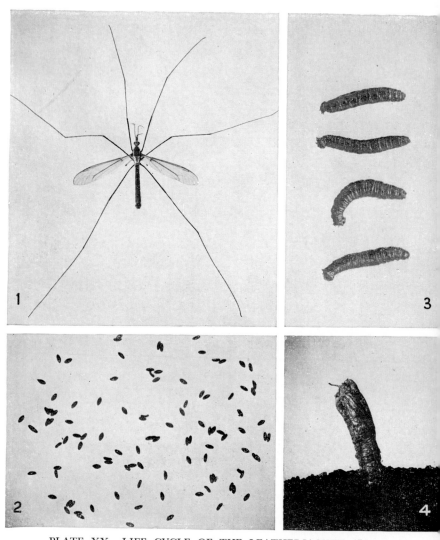

PLATE XX. LIFE CYCLE OF THE LEATHERJACKET (*Diptera*)

1, the crane fly, the adult stage, *Tipula paludosa* Meig. ♂ ; 2, eggs (× 3); 3, leather-jackets, the larval stage (nat. size) ; 4, chrysalis protruding from soil and ready for the emergence of the adult (× 2)

A

B

PLATE XXI. TWO EXAMPLES OF TRUE FLIES (*Diptera*)

    A. Celery fly (*Philophylla heraclei* L.) ♀ (× 8)

    B. Gad fly (*Tabanus autumnalis* L.) ♀ (× 4)

PLATE XXII. EXAMPLES OF INSECT EGGS

A.   Cluster of eggs of vapourer moth (*Orgyia antiqua* L.) laid by the wingless female upon the cocoon from which she emerged (× 2)

B.   Part of base of apple twig in winter, showing black shining eggs of aphids ; narrow, pointed (light-coloured) eggs of apple sucker ; and tiny spherical eggs of fruit tree red spider mite, in rows in the scars on the left at the junction a swarm consisting of a large number of worker bees and a queen. Swarming in bees is related to the honey flow, and new colonies are formed only at times when food is sufficiently plentiful to allow them to become successfully established.

The Hymenoptera are divided into two main groups : the Symphyta or sawflies with a sessile abdomen, and the Apocrita which includes all the rest and is characterised by a waisted or petiolate abdomen.

SYMPHYTA are vegetable feeders and include many pests of fruit trees and bushes, forest trees, ornamental shrubs and flowers. The woodwasps or horn-tails (Siricidæ) are large insects with a stout ovipositor for piercing timber. The larvæ are wood-borers in dying or dead conifers. The stem-boring sawflies (Cephidæ) are slender insects with slightly clubbed antennæ whose larvæ live in the stems of grasses and cereals. The wheat stem sawfly (*Cephus pygmaeus* L.) attacks wheat in Europe and North America. The leaf-eating saw-flies (Tenthredinidæ) often feed gregariously along the edges of leaves and may cause severe defoliation. Certain species are associated exclusively with particular host plants, and gooseberry, willow, larch, poplar, strawberry, hazel, spruce and pine have specific sawfly pests. Some sawflies are gall-formers, *Pontania* causing the red bean galls on the leaves of willows and *Euura* causing woody galls on the shoots. Fruit-eating sawflies feed in apples, pears and plums and are serious pests in fruit plantations.

APOCRITA, including ants, bees, wasps, ichneumon wasps, and other parasitic Hymenoptera, are mainly beneficial insects. Ichneu-mon wasps (Ichneumonidæ) are external (*ectoparasitic*) or internal (*endoparasitic*) parasites on other insects. *Rhyssa* is the largest British ichneumon and its long ovipositor is used to deposit eggs deep in timber in the feeding-tunnels of wood-wasp grubs. *Aphidius* feeds in the bodies of aphids. The fairy flies (Mymaridæ) are so minute that they reach their full development within the eggs of other insects, particularly those of leaf-hoppers and frog-hoppers. The gall wasps (Cynipidæ) cause galls on leaves and stems and the larvæ live in the galls. Many species are found on oak. The social wasps (Vespidæ) are well known for their attacks on ripening fruit, but it should not be overlooked that when they are rearing their brood they capture and destroy large numbers of injurious caterpillars. Solitary wasps (Eumenidæ) construct curious nests of mud. They lay eggs in the nest and fill it with caterpillars on which the wasp grub feeds. Bumblebees (Bombidæ) frequent flowers and are useful pollinators of farm and garden crops, but they often bite holes in the corolla tube of flowers of beans in order to reach the nectar easily. Ants (Formicidæ) have nests in the soil, in grass-grown humps above soil or in loose heaps upon the soil. Many species are predatory and feed on other insects which they dismember and carry to their nests. Ants are a nuisance in dwelling-houses and greenhouses, and ants' nests in dry meadows and pastures are associated with deterioration of the herbage.

Diptera (Flies). The true flies have one pair of membranous wings (Plates XX, XXI), the fore-wings, which are sometimes barred

Q

or mottled, and a pair of small knob-like structures, halteres or balancers, in place of hind-wings. The antennæ may be long and delicate as in gall midges and mosquitoes, short and slender as in crane flies, or short and thick and with a feather bristle as in the blue-bottle fly, cabbage root fly and the hover flies. The eyes are large and conspicuous and in the males they occupy most of the top and sides of the head. The mouthparts show several types of adaptation for taking liquid food : mosquitoes have slender stylets for piercing and sucking, gadflies and clegs have stout lancets, and houseflies and root flies have a stout trunk with a lapping surface at the tip.

In Diptera metamorphosis is complete. The larvæ are legless maggots with fleshy retractile false feet and rows of spines to help them to move about. Maggots of St. Mark's flies, false leather-jackets (*Bibio* spp.) which live gregariously at the roots of grass and cereals, and the slender maggots of fungus gnats which live in mush-rooms and in decaying plants, have distinct black heads with biting jaws. Leatherjackets, the maggots of crane flies (Tipulidæ), have biting jaws, but the head is small and incomplete and withdrawn into the body (Fig. 78, *g*). Maggots of root-feeding flies and blow-flies are practically headless (Fig. 78, *e*) ; the jaws are reduced to a pair of strong, downward-curving hooks (Fig. 79) which lie side by side and tear at food material as they are pushed forward and retracted again in an almost continuous rhythmic motion. Most of these degenerate maggots have delicate skins and live surrounded by moisture among decomposing plant and animal tissue. Some are injurious to crops and stock but many are scavengers and play a useful part in the disintegration of plant and animal remains.

There are two distinct types of pupæ among the Diptera. In crane flies, mosquitoes, gnats and midges the fully grown maggot casts the larval skin and reveals a pupa which shows the outlines of wings and limbs (Plate XX). In houseflies, rootflies, frit flies and warble flies the last larval skin hardens to form a barrel-shaped chrysalis called a *puparium*. The pupa is sheltered and protected within the puparium, which bursts open at one end when the fly is ready to emerge.

Flies usually reproduce by laying eggs. Some flesh flies (Sarco-phagidæ) and parasitic flies (Tachinidæ) retain the eggs within the body of the female until hatching takes place, and in consequence they are said to be *larviparous*, that is, they deposit living larvæ. Another group of flies is known as Pupipara from their habit of nourishing the larvæ within the bodies of the females. The pupiparous flies are external parasites on the bodies of animals and birds, the forest fly (*Hippobosca equina* L.) feeding on horses and cattle, and the sheep ked (*Melophagus ovinus* L.) (Fig. 92) feeding on sheep. They deposit their fully grown maggots on the ground, or in places fre-quented by their hosts, and pupation takes place at once. In some of the gall midges (Cecidomyidæ) reproduction by egg-laying adults is supplemented at certain times of the year by larval reproduc-tion or *pædogenesis*, that is, the maggots give rise to other maggots. Pædogenesis may continue as long as temperature and food supply are favourable but, after a few pædogenetic generations of larvæ, development becomes normal again, pupæ are produced, and egg-laying flies reappear in the cycle.

The Diptera contains many families of flies important to man and his crops and stock. Crane flies or daddy-long-legs (Tipulidæ) (Plate XX) form an extensive family of long-legged flies, some of which are small and fragile like mosquitoes, while others are nearly 50 mm long with hard black and yellow bodies and mottled wings. Their maggots are called leatherjackets and are often injurious to crops in poorly drained fields and weedy gardens. Mosquitoes (Culicidæ) are notorious the world over as carriers of malaria, yellow fever and ague. They lay eggs in water, either singly or in small floating masses. The larvæ are aquatic and live on minute aquatic creatures. They obtain air through pores (spiracles) at the tip of a tubular projection arising near the tail. Pupation takes place in the water ; the pupæ retain the power of movement and obtain air through tubes in the thorax.

Gall midges (Cecidomyidæ) cause various deformities to stems, leaves, flowers and fruit. The maggots are red, pink, orange, yellow or almost colourless, and usually have a small horny, slightly forked, breast bone (anchor process) on the underside. Injurious gall midges include the hessian fly (*Mayetiola destructor* Say.), an important pest of wheat in the United States ; blossom midges (*Contarinia, Sitodiplosis*) which feed on developing wheat grains ; swede midge (*Contarinia nasturtii* Kieff) which feeds in the terminal bud of swedes and other brassicas ; pea midge (*Contarinia pisi* Winn.) which feeds in the pods of peas, and the pear midge (*Contarinia pyrivora* Riley) which feeds in the young fruits of pear. Fungus gnats (Mycetophilidæ) feed in fungi and decaying plants, and their maggots have been found in rotting swedes, mangels, potatoes, bulbs and mushrooms.

Many of the short-horned flies (Brachycera) are beneficial insects. Both larval and adult stages of down-looker flies (Leptidæ), robber flies (Asilidæ) and stiletto flies (Therevidæ) are carnivorous. Larvæ of bee-flies (Bombyliidæ) are parasitic. The horse flies, gadflies and clegs (Tabanidæ) (Plate XXI) are large stout-bodied flies whose females are vicious biters of both man and stock. They are strong active fliers and settle on their hosts with such lightness and precision that the sharp prick of the piercing proboscis is often the first indication of their presence. The males are harmless and feed on nectar, and the larvæ live in damp meadows and ditches along the edges of woodland and feed on small earthworms, woodlice and insects.

The hover flies (Syrphidæ) are rather large handsome flies, often with black-and-yellow-barred bodies that give them a wasp-like appearance. They abound in woods, fields and gardens, and the high-pitched hum as they hover is characteristic. The flies feed on nectar. They lay eggs near aphid colonies and the rugged green, pinkish or yellowish maggots feed entirely on aphids. The chrysalids are pear-shaped and are found attached to leaves, stones, twigs and bark of trees. Maggots of bee-like hover flies called drone flies (*Eristalis*) live in liquid manure tanks and in water in holes in trees, and their long breathing tubes have earned them the name of " rattailed maggots ". The narcissus flies (*Merodon equestris* F., and

Eumerus spp.) are among the few injurious hover flies ; their larvæ feed in the bulbs of narcissus and related plants.

In a large group of flies, the Schizophora, the adult at emergence forces open the puparium by inflating the intersegmental membrane (*ptilinum*) of the face. After emergence the ptilinum is gradually deflated and withdrawn into a crescent-shaped scar (*frontal suture* or *lunule*) lying above the antennæ. Many important fly pests are members of this group. Frit flies and gout flies (Oscinidæ) attack the shoots and flower heads of cereals all over the world. The carrot flies (Psilidæ) are small shining blue-black flies whose maggots feed on the tap roots of carrots, parsnips and celery and may render them unfit for market ; closely related species infest the root stocks of chrysanthemums and the tap roots of lettuce. Leaf-mining flies (Trypetidæ) with characteristic mottled wings (Plate XXI) feed in the leaves of celery, parsnip, parsley and chrysanthemums in irregular blotch mines, while other leaf-miners (Agromyzidæ) make long winding tunnels in chrysanthemums, clover and cereals. Dung flies (Scatomyzidæ) are the common yellowish and brownish flies often seen in great numbers about fresh cow dung. The flies are predacious upon small insects and the maggots live in excrement.

The bot flies and warble flies (Œstridæ) are large bee-like flies whose larvæ are parasites of horses, cattle and sheep. The horse bot flies (*Gasterophilus*) are found in the stomach of the horse ; warble flies (*Hypoderma*) live in the backs of cattle, and maggots of the sheep nostril fly (*Oestrus ovis* L.) inhabit the nasal cavities of sheep. Tachina flies (Tachinidæ) are stout, medium-sized bristly flies, usually grey and black, which lay eggs or larvæ upon caterpillars, grubs and other insects. Tachinid maggots are internal parasites ; they completely devour their hosts and often form chrysalids within the host's body. The blowflies, greenbottles and bluebottles (Muscidæ) frequent dead animals, decaying organic matter and excrement, and their maggots live in similar situations. Many are scavengers and consequently are beneficial, though they are carriers of bacteria to food and milk in houses, shops and dairies. Greenbottle flies and some bluebottles at times lay their eggs about the tails of sheep and lambs and the maggots may penetrate the flesh and cause the death of infested animals. There are several generations of blowflies and their relatives during the year and they pass the winter in the soil as larvæ or pupæ ; adults hibernate in the shelter of buildings, shrubs and evergreen trees. The root flies (Anthomyidæ) frequent decaying vegetation and freshly turned soil. The cabbage root fly, wheat bulb fly, mangel fly and bean seed fly are major pests of farm and garden crops, but many related species are scavengers upon decaying tissue or are predators that seek a living among the soil fauna.

PRINCIPLES OF INSECT CONTROL

The control of insect pests can be attempted in a number of ways. Before describing these reference must be made to the natural factors which adversely affect insect increase, namely, the action of unfavourable weather, parasites, predators and, in the final result, exhaustion of food supply. The interaction of these factors produces fluctuations in insect populations from season to season so that a particular pest is now relatively scarce, now in epidemic proportions. The study of these events with a view to ultimate prediction is a main concern of the entomologist. By these influences insect numbers are kept within bounds over a period, but it is small comfort to a grower whose crops are at the moment being destroyed to know that next year, or the year after, his plants will escape attack. He must have at hand weapons he can use the moment his crops appear seriously threatened.

These " artificial " control methods may be conveniently classified into the indirect and the direct. *Indirect* control measures consist mainly of the modification of ordinary farm practices, giving them a twist, as it were, so that they act against the pest or pests. Consequently they are inexpensive and examples are the arrangement of ploughing and sowing dates to the best advantage (frit fly, leatherjacket, wheat bulb fly, slugs) ; correct rotations (eelworm) ; and the growing of resistant varieties. *Direct* control refers mainly to the use of specially prepared chemicals (insecticides), and since these are not inexpensive, economics become important. Therefore, for the most part, chemical methods tend to be used in the horticultural and fruit industries with their higher turnover, though the development of the more potent insecticides and the low-volume sprayer have in fact considerably extended the possibilities of use in general agriculture.

Chemical Control. The spate of new chemicals available for trial as insecticides since the war is a new phenomenon and shows little sign of abating. Not only are new and better substances available but diverse formulations (granules, dusts, wettable powders, emulsions, aerosols, smokes, etc.) designed for different purposes still further increase their effectiveness. Machinery is now available for applying the insecticides and there has emerged the spraying contractor. The position in general agriculture, therefore, is a very different one from that prevailing not so very long ago when the poison bait was about the only chemical control that could be prescribed and when, indeed, if a spray was suggested there was no means to apply it.

In the fruit world, however, there may be said to have been a tendency to the reverse. Between the wars a complicated annual routine (not to say ritual), the " winter washing programme ", had developed, using tar oils, DNC, petroleum oils and other compounds in their variety. But more recently the view has gained ground that the logical process is to spray according to the needs of the orchard rather than according to the calendar. The new insecticides have facilitated this, and it is now possible to get reasonable control of orchard pests by omitting winter washing altogether.

The New Insecticides. It used to be convenient to classify the old insecticides according to their mode of action. Thus certain pests like aphids were best controlled by sprays aimed directly at them and acting by *contact* via the cuticle or the spiracles : examples are nicotine, pyrethrum and derris. Other pests, those with biting mouthparts, could best be poisoned by spraying the foliage of the attacked plant with a *stomach poison*, the best example being arsenate of lead. Finally, there were other pests, notably the carrot fly and the cabbage root fly, against which neither type of insecticide was effective ; here the attempt was made to repel the insect, or perhaps to disguise the food-plant, by using strong-smelling substances or *repellents* like naphthalene and creosote. So, the first essential, the detailed working out of life history of the pest, having been observed, the most suitable of these methods would be applied.

The new insecticides have improved the position very considerably. Their main characteristics are their extreme potency, their ability to act both as stomach and contact insecticide so that the old distinction between these becomes somewhat blurred ; and most important of all, their persistence. This means that instead of repeated spraying being necessary, a single spraying will remain effective for days or, in some cases, weeks. This is a great advantage to the grower. Some examples of new types of formulations have already been referred to, but special mention must be made of the seed dressings now available for use against wireworms, flea beetles and the carrot fly. These developments, if their success can be sustained, are of the first importance.

One other advance must be mentioned, and that is the emergence of the *systemic* insecticide, a chemical that is actually taken up by the plant by absorption through leaf or root like a nutrient solution and which is moved in the sap stream through the whole system of the plant. An insect sucking up such sap, even after a lapse of several days, or in some cases weeks, is poisoned.

Before referring in more detail to the main types of insecticides, two drawbacks resulting from their properties of toxicity and persistence should be pointed out. First, spray residue problems become important and this has resulted in the laying down of minimum intervals between application and harvest to protect the consumer. Second, and this applies particularly to the persistent organochlorines, they affect wild life, particularly the predators at the end of food chains such as birds of prey, and pollute the environment.

The organochlorines. The first two and best known are DDT and BHC. There are now a further ten of these chemicals on the market. Of these, five are insecticides and four are acaricides. The remaining one is a nematicide called D-D and is used mainly under glass.

The most toxic and persistent organochlorines have been the subject of two Government Reviews (Cooke 1964 and Wilson 1969) which have restricted the use of aldrin, dieldrin, endrin, endosulfan, DDT and TDE. As less persistent alternatives become available the existing uses of these organochlorines are likely to be further restricted and they will almost certainly eventually disappear from the market.

BHC is still freely available but has the disadvantage that it taints vegetable crops, particularly potatoes and carrots. BHC and dieldrin are both used as vegetable seed dressings for protection against flea beetles, carrot fly, bean seed fly and onion fly. Dieldrin and aldrin seed dressings for use on winter wheat against wheat bulb fly have recently been replaced.

Resistance to some of these chemicals is now widespread in cabbage root fly, carrot fly, bean seed fly and onion fly populations.

Some of the earlier acaricides have become obsolete because fruit tree red spider mite has developed resistance to them. Several are still available including chlorbenzide and chlorfenson which are summer ovicides, and dicofol, tetradifon and tetrasul. Resistance seems to develop after about five years of continuous use.

The organophosphorus compounds. There are now about twenty-five materials in this group on the market in this country. Two of the earliest introductions, HETP (TEPP) and schradan, are now becoming obsolete and are little used.

Discovered first by Schrader in Germany during the war this series of compounds is a very powerful one. The first to be introduced was hexa-ethyl tetraphosphate (HETP), the effective constituent of which was later found to be tetra-ethyl pyrophosphate (TEPP). This, although extremely poisonous, breaks down soon after application. The next compound was parathion, which has the ability of penetrating leaf tissue thus killing aphids in curled-up leaves. The next one was the first true systemic insecticide called schradan, and in the early years it was used extensively against aphids on Brussels sprouts and hops. Other related systemics followed, notably demeton-methyl and dimethoate, which because of their lower mammalian toxicity could be sprayed without the operators wearing protective clothing. This in effect made the ordinary farmer independent of the spray contractor as he could deal with his aphid problems himself.

Many other organophosphorus compounds have since been introduced and a number of these are systemic. Malathion is one of the least toxic, although non-systemic, and is widely used in the private garden. It is impossible to mention all of the newer materials. Mevinphos is systemic and active against aphids and caterpillars and has a short life. Other systemics such as menazon, disulfoton and phorate are active for several weeks, the latter two being available in granular form only. Phorate gives some control of wireworms in potato crops when put in with the seed at planting time.

Others worth mentioning are chlorfenvinphos and carbophenothion, used as seed dressings against wheat bulb fly, and azinphosmethyl and trichlorphon used against various caterpillar pests.

Biological Control. Biological control seeks to make use of natural enemies for the control of pests. It has been successfully developed in America where the fluted scale (*Icerya purchasi* Mask.), a pest of citrus trees, had been introduced without its natural enemy, a ladybird beetle, *Novius cardinalis* Muls. The introduction and establishment of the ladybird in citrus-growing states effectively checked the ravages of the scale insects. Under the stable agricultural conditions prevailing in Britain natural control is an important but not a spectacular method of pest control. Almost every species of insect has its parasites and predators and the examination of an aphid colony will give some indication of the extent of their activities. It is usual to find the grubs of ladybird beetles, the maggots of hover flies, numbers of small black ichneumon wasps and the grubs of lacewings all congregated about the colony and feeding upon the aphids. In this country effective use is made of the parasite introduced from the Far East—*Encarsia formosa* Gahan—to control the greenhouse white fly. The mite, *Phytoseiulus riegeli*, from Chile is being used against glasshouse red spider mite, *Tetranychus urticae*. In the orchard, *Aphelinus mali* Hald. was introduced many years ago from the USA to control woolly aphid.

Legislative Control. Legislative measures are occasionally adopted to prevent the introduction and establishment of new pests and to check the increase of those already established. The Colorado beetle (*Leptinotarsa decemlineata* Say.) is a notifiable pest and, when its presence is reported to the Ministry of Agriculture, steps are taken to secure the complete destruction of the pest by supervised spraying and soil treatment. Other pests that have been the subject of legislation include the fluted scale (*Icerya purchasi* Mask.), chrysanthemum gall midge (*Diarthronomyia chrysanthemi* Ahl.), mediterranean fruit fly (*Ceratitis capitata* Wied.), potato moth (*Phthorimaea operculella* Zell), vine phylloxera (*Phylloxera vastatrix* Planch.), and amongst stock pests, the ox warble fly and the sheep scab mites.

A successful experiment in legislative control has been the enforcement of an adequate rotation in fields in the east of England where beet eelworm (*Heterodera schachtii* Schmidt) is known to occur. This measure has maintained yields of beet and prevented waste of land and labour in efforts to produce sugar beet where the eelworm population is so high that it constitutes a serious risk to the crop.

Indirect Pest Control. Most agricultural crops are mass cultures of plants with a comparatively low individual value, and farmers generally adopt indirect methods of pest control which involve little outlay as part of the routine of crop production.

Cultivations. These often take the form of advancing or delaying the time of routine cultural operations in order to create unfavourable conditions for particular pests. It is an advantage to plough ley or pasture before corn harvest in order to prevent frit fly oviposition on ryegrass in the ley. When leys are ploughed late, frit fly maggots feeding in the shoots of ryegrass are able to survive ploughing operations and they finish their development in the seedlings of the new corn crop. In east and central England where wheat bulb fly (*Hylemyia coarctata* Fall.) is prevalent, the aim should be to cover the ground in late summer with a forage crop like rape or common turnips or a green manure like mustard so that there is no bare soil to attract the egg-laying flies. Cultivation to keep down weeds is also an important factor in pest control. Weed grasses like couch and bent harbour pests of cereals, and charlock and knot-grass harbour pests of brassicas and beet. Weeds encourage egg-laying by owlet moths and afford food and shelter for the young cutworms.

Time of Sowing. Time of sowing affects the incidence of such pests as frit fly, flea beetles and carrot fly. If oats are sown very late in the spring, damage by frit fly is liable to occur. Crops sown during February and March are usually well established before frit flies are on the wing in great numbers and infestation is negligible, but crops sown in late April and May often suffer severely from frit fly attack. In districts where flea beetles cause heavy loss farmers try to avoid attack by advancing or retarding the date of sowing brassica crops so that the emergence of the seedlings does not coincide with hot dry weather and maximum beetle activity. Late sowing is commonly practised to avoid loss to carrots by carrot fly. The new insecticide treatments have, however, made these practices less necessary.

Time of Harvesting. Damage to potato tubers by wireworms and slugs can be limited by early harvesting. The autumn feeding period of wireworms usually begins in September, and wireworm injury to potato tubers increases steadily during autumn if they are left in the ground after the middle of September.

Crop Rotation. Pest control by crop rotation is of the greatest value against specific pests like cyst eelworms which have an exceedingly low capacity for migration. Losses from attack by potato cyst eelworm (*Heterodera rostochiensis* Woll.), beet eelworm (*H. schachtii* Schmidt), cereal cyst eelworm (*H. avenæ* Woll.) and pea eelworm (*H. göttingiana* Lieb.) can be reduced by increasing the interval between host crops. It is hardly an exaggeration to say that these eelworm pests, especially the cyst eelworms, are a greater threat to agriculture even than insects. Since rotation is often the only effective means of preventing eelworm attacks, and the only economic means of dealing with them should they occur, the importance of rotation in modern agriculture can scarcely be stressed enough. It should be remembered that once a cyst population has been built up to danger level, the soil is likely to remain always infested.

On the other hand, neglect of crop rotation and the cultivation of a limited range of crops within an area are followed by severe infestation by specific pests capable of migrating from field to field. Intensive carrot growing in certain districts has built up such high local populations of carrot fly that the production of clean carrots is almost impossible. Similarly, many market-growing areas have high populations of cabbage stem weevil and turnip gall weevil (*Ceuthorrhynchus* spp.), flea beetles and cabbage root flies. Overcropping with cereals in parts of the south of England during the inter-war years built up high populations of gout fly (*Chlorops taeniopus* Meig.) and wheat stem sawfly (*Cephus pygmæus* L.) which were locally destructive during the war years when cereal growing was compulsory.

Manuring. The use of manures and fertilisers has little direct influence on the prevalence of insect pests. The value of manures lies in their effects on the plants. They serve to promote plant growth and vigour, and produce healthy robust plants generally able to withstand insect attack. The early establishment of cereals in fertile soil is of prime importance in preventing loss of crop through wireworm attack, and sugar beet, mangel and barley seedlings are especially susceptible to soil acidity. Combine drilling fertiliser with the oats may circumvent frit fly attack.

Farm Hygiene. Proper attention to the disposal of crop remains makes an appreciable contribution to pest control. Mustard beetles, wheat stem sawfly, pea and bean weevils and cutworms shelter in the stubbles of various crops. Burning hedgesides and the stubble of cocksfoot seed crops in autumn and early winter does much to check cocksfoot moth. The chrysalids of this pest hibernate in the

old flowering stems. Early and deep ploughing of stubbles buries
hibernating insects and may prevent their survival and emergence
in the following spring. Bulky crop residues should be buried deeply,
preferably after an application of a nitrogenous fertiliser such as
ammonium nitrate to promote rotting.

The storage of crops should be done carefully to avoid loss from
rotting and from the attacks of secondary pests like woodlice and
millepedes. When the clamped crops of potatoes and other roots are
finally removed it may be advisable to treat the site with soil insecticides to destroy slugs, wireworms, eelworms and other pests that have
left the crop and entered the soil below the clamp to complete their
development. Mangel clamp hygiene is a matter of urgency to avoid
the spread of virus yellows. Corn bins, granaries and forage stores
should be thoroughly cleaned and disinfected before new crops are
stored : this has assumed special importance since the introduction
of the combine harvester and the consequent storage problems. Walls
should be faced with cement to prevent insects sheltering in the
crevices and a light dusting of a suitable insecticide such as DDT,
BHC, malathion or pirimiphos-ethyl on walls and floor, or more
conveniently the same substances in the form of " smokes ", will
destroy many wandering insects without affecting the palatability of
stores in sacks and bins.

Soil and crop debris in sacks and other containers should be
burnt. This is especially important in the case of sacks used for
seed potato tubers since they may sometimes contain cysts of potato
cyst eelworm.

Clean Seed and Healthy Plants. The purchase of clean seed of
good quality is always worth while. Important pests such as the stem
eelworm (*Ditylenchus dipsaci* Kühn) are carried on the seeds of onion
and clover, while the ear cockle eelworms (*Anguina tritici* Steinb.) are
carried in wheat seed. It is also advisable to know something of the
history of fields from which brassica seedlings, strawberry runners and
onion sets are purchased for the cysts of root eelworms and the infective stage of stem eelworm are often carried to new sites in soil
at the roots of purchased plants.

Birds and Animals as Agents in Pest Control. Plovers
(peewits or lapwings) feed almost exclusively on insects ; they destroy
wireworms, leatherjackets, chafers, slugs and snails in grassland, and
these and other insects in arable land. Rooks, jackdaws, magpies
and starlings also feed on grassland insects. It has been estimated
that about a tonne of insects is distributed over each 2 ha of grassland
and each 8 ha of arable land, and the insectivorous birds help to
prevent an excessive increase in the insect population. Blackbirds,
thrushes, warblers and chats are active in destroying insects about
crops and fruit plantations, especially during the breeding season.
Mice, voles, shrews and hedgehogs, together with reptiles, lizards,
frogs and toads also play a part in keeping down the numbers of
insects.

MAJOR PESTS OF FARM CROPS AND STOCK

In the following account of some of the more important insect and allied pests of farm crops and stock attention is focused on life history and habit, and for details of structure the student should refer to the appropriate paragraph in the section on the classification of insects on pages 415–26.

PESTS OF CEREALS

Wireworms (Plate XIX, Fig. 78, *a*). Wireworms are the grubs of click beetles (Elateridæ). Their natural home is grassland, and because cereals are usually the first crop after ley or grass, wireworm injury is generally associated with cereal crops though it can be equally severe on sugar beet, potatoes, brassicas, market garden crops and strawberries. Wireworm feeding is greatest in **autumn** during September and October, and in spring from March to May. Autumn-sown cereals are subject to attack at both periods and spring-sown cereals are attacked before they are established. Attack by wireworms consists of hollowing out the grains, severing young plants from the seeds and primary roots, and tunnelling in the crown of the established plants so that shoots and tillers are destroyed. When wireworm attack occurs simultaneously with low fertility and adverse weather conditions the crop may die over large patches and re-drilling may be necessary.

Wireworms are smooth, slender, yellow grubs, 18–25 mm long when fully grown. The head has strong biting jaws and the body consists of twelve segments, the first three of which bear the true or thoracic legs and the last segment bears a single false foot and has two small brown oval pits on its upper surface. Wireworms hatch from eggs laid in the soil in early summer. They feed intermittently and grow so slowly that they require at least four years to reach maturity. During the fourth summer they make earthen cells in which they become soft white chrysalids, and about a month later the beetles emerge. Newly formed beetles may shelter in the pupal cell throughout the autumn and winter, or they may leave the cells and shelter in the soil, under clods or in thick herbage.

The common click beetle of grassland (*Agriotes obscurus* L.) is about 8 mm long, elongate-oval in shape and brownish in colour. The antennæ are slender and about one-third of the length of the body ; legs and antennæ may be held so close to the body that the beetle is not easily seen in the soil. Click beetles feed on the leaves of grass, cereals and clover but do little harm. In spring they are often found in pieces of turf on recently broken grassland.

Insects with a long larval feeding period occur under stable conditions, and wireworms are found most abundantly in permanent grassland where the surface has been undisturbed for a long time and fluctuations of temperature and humidity are at a minimum. Wireworm populations vary greatly with the geographical position, altitude, soil, drainage, etc., but now, even in old grass it is unusual to find more than $1\frac{1}{4}$ million per ha. As long as fields are under grass the wireworm population is of negligible importance, but when grassland is ploughed for arable cultivation wireworms are a menace for two or three years. Ploughing creates unfavourable conditions for egg-laying and the survival of young wireworms, and in consequence the wireworm population gradually diminishes when grassland is cultivated. Approximately a quarter of the wireworms reach maturity each year and, since the numbers are not appreciably replenished, the low wireworm populations characteristic of arable land are usually reached by the fourth year of cultivation.

Wireworms are not generally a serious pest of agricultural crops on established arable land, but when wireworm injury coincides with lack of fertility, water-logging, adverse weather or other conditions unfavourable to plant growth, it has an exaggerated influence on the backward crop.

Control Measures. Loss from wireworm attack on new arable land may be limited by growing resistant crops, by increasing the seeding rate of cereals even up to 50 per cent or by cross drilling, and by ensuring for the crop a firm seed bed in fertile soil. Good ploughing and consolidation and, where necessary, the correction of lime and phosphatic shortages likely to be present on old grassland are essential. In recent years, however, the wireworm problem has been completely changed by practical methods of direct chemical control. The first successful step in this was the introduction of a crude low-content benzene hexachloride dust which was applied direct to the seed bed. Crude BHC consists of a mixture of four or five isomers of which only one, the gamma isomer, is effective as an insecticide. The separation of this isomer on the commercial scale was the next step, making possible the production of a low-content dust in which taint hazards were considerably lessened. Perhaps the most important result, however, was the development of concentrated high-content gamma-BHC seed dressings which, combined with the already established mercurial fungicidal dressings, provided a dual protection for cereals and beet in all but the heaviest infestations— at any rate with spring crops. Aldrin is equally successful against wireworm and although the practice is not generally recommended it is still permissible to use aldrin on a potato crop planted in ploughed-up old grassland which is known to be heavily infested with wireworm. Phorate may be used to protect a potato crop against low wireworm populations.

Leatherjackets (Plate XX, Fig. 78, g) are the tough-skinned, grey-brown, legless grubs of the daddy-long-legs or crane fly (*Tipula* spp.). They constitute one of several pests—notably the slug, frit fly, wheat shoot beetle, rustic moth—any or all of which are liable to attack cereals after ley when ploughing and sowing closely follow one another. In recent years much attention has been paid to leatherjacket numbers : by a fairly simple technique, populations can be estimated and the general level of attack can be forecast with some accuracy and the appropriate warnings issued to farmers in good time. There were population peaks in 1952, 1963 and 1969.

Crane flies lay their shining black eggs in the herbage of damp fields in late summer and early autumn. The eggs hatch in about a fortnight, and the grubs feed during the winter. In some years serious damage may be done by the turn of the year. In spring they vary in length from 12 mm to 25 mm, and can speedily cause damage to germinating and seedling cereals, root and other crops. Whole fields may be destroyed. Pupation takes place in the soil in summer, and the crane fly emerges, leaving the empty pupal case protruding from the soil. There is one generation per year.

Control Measures. Fortunately, control is fairly cheap and satisfactory by use of the well-known poison bait, 1 kg Paris green to 30 kg bran broadcast per ha. Recently it has been found that a somewhat cheaper, more convenient, and more effective substitute for the Paris green is 1 kg of 50 per cent wettable DDT or BHC spray powder. The bait should be only slightly dampened so that the flakes do not adhere, and the broadcasting should take place towards evening, for the leatherjackets come to the surface at night to feed. Sprays of DDT or the new material fenitrothion are effective. Aldrin may be used only on DDT-susceptible varieties of barley.

Frit Fly (*Oscinella frit* L.) (Fig. 77). Frit flies are small shiny black flies about 2 mm long which can be found from spring to early autumn. There are three main phases to their attack. First, in May eggs are laid on the seedling oats and the maggots attack and kill the centre shoot, which typically turns yellow. Flies resulting from this attack lay their eggs in the glumes of the developing panicles and the maggots eat out the young grain. As much as 50 per cent grain attack has been recorded. Flies from this generation frequently cover the outsides of stooks and stacks and the insides of barns. Eggs are then laid in the grasses in leys, ryegrasses especially, on volunteer corn or early sown cereals, producing the autumn shoot attack. Often when a ley is turned in and corn sown immediately, the larvae leave the buried rye-grass and attack the young cereal seedlings. These autumn larvae live through the winter and give rise to the flies of the May attack already mentioned. The generations may overlap somewhat, and a late shoot attack in June may result in damage to the developing ears, producing a sort of blindness which is, however, quite distinct from the more common " blast ". Frit fly attack is mainly associated with oats, but wheat is also affected.

Control Measures. Frit fly attack causes annual loss at one level or another both in the shoot and in the grain. The latter is more easily estimated and records show that a 10 per cent grain attack is a fair average figure : but as has already been mentioned, in some fields a figure of 50 per cent has been recorded. The acreage under oats has greatly decreased in recent years and consequently fewer complaints of frit fly damage arise. The use of insecticides is not an economic proposition.

The only really practical measure of control is early sowing against the early summer shoot attack. Dates vary somewhat, but February and March sown crops are usually safe, while May and later sown crops can be almost complete failures. An exaggerated example of the latter is where oats are attempted as a catch silage crop after early potatoes. They may never get beyond the " grass " stage.

Wheat Bulb Fly (*Hylemyia coarctata* Fall.). This insect used to be known as a more or less localised pest of wheat following land that had been fallowed in August. Recently, it appears to be equally dangerous after potatoes, both early and late, and root crops. The epidemic of 1953 on the east side of England when thousands of hectares of wheat were destroyed will be long remembered. Fortunately since then the damage has been on the decline.

Eggs are laid from late July to the end of August in bare soil. Hatching takes place the following February onward and the grubs mine into and kill the central shoot in much the same way as with the frit fly. If the wheat is well tillered heavy attacks can be withstood, but untillered wheat rapidly succumbs.

The most successful control measure is early October sowing in wheat bulb fly areas. The 1953 epidemic stimulated much research resulting in the development of suitable seed dressings of BHC, aldrin, and dieldrin. In view of the danger of the last two of these to wild life, it is officially recommended that they should no longer be used. Two organophosphorus materials, chlorfenvinphos and carbophenothion are effective alternative seed dressings but should be used only in areas at risk to the pest. Emergency spray treatments applied in spring give variable results.

Corn and Grass Aphids. Three species of aphids may in epidemic years do appreciable damage to corn and grass and are also capable of spreading a virus disease, barley (cereal) yellow dwarf. The three species are *Rhopalosiphum padi* L. *Metopolophium festucae* Theob. and the grain aphid *Sitobion (Macrosiphum) avenae* F.

Slugs are persistent pests of young cereals. During autumn and winter, seed grains may be hollowed out and young seedlings damaged so that the leaves are frayed in a very characteristic manner. Slugs are more numerous in the heavier, damper soils. They live in the soil and vary slightly in their feeding habits according to species. The grey field slug (*Agriolimax reticulatus* Müll.), for example, feeds mainly at the soil surface in damp mild weather, hiding in the ground during daylight or cold weather. The underground or keeled slugs, which infest gardens and horticultural land, feed especially on the underground parts of crop plants, and because they spend most of their time beneath the surface are correspondingly more difficult to control with surface baits. Eggs are laid in clusters in the soil and rotting vegetation. They are spherical and translucent, about 2 mm in diameter. The young slugs hatch in about a month.

Control Measures. Slugs flourish in damp soils rich in organic matter and the opposite conditions tend to discourage them—the use of fertilisers instead of farmyard manure, for example. On a small scale, clearing of plant debris is helpful. Direct chemical control consists mostly of the use of metaldehyde, and this remarkable substance has considerably eased the position. Bait, made of 1 kg metaldehyde to 30 kg bran, is the cheapest form : but in the darker months when the sun has less power, a mixture of 1 kg Paris green, ½ kg metaldehyde, 30 kg bran is more effective (it is also useful against leatherjackets). Recently the contact action of metaldehyde has been exploited and wettable powders for spraying have been marketed, but the cost of these limits them to special work. A new material, methiocarb, has now become available and is proving better than metaldehyde.

Slugs have many natural enemies of which the duck is probably the best known. Pheasants, partridges, wood-pigeons and other birds devour them, and ground beetles feed on young slugs.

Eelworms. Two main types are concerned, the stem eelworm (*Ditylenchus dipsaci*) which distorts the shoot and causes " tulip root " in oats (Plate XXVI), and the possibly more serious cyst eelworm (*Heterodera avenae* Woll.) which, like its better known relation the potato cyst eelworm, disorganises the root system in which it forms

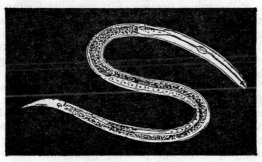

FIG. 85.—STEM EELWORM (*Ditylenchus dipsaci*), ADULT ♀ : × 160, showing position of mouth spine.

cysts. These cysts are persistent and may keep the soil infected for many years. In both cases the trouble is brought about by unwise rotations, i.e. the too frequent cropping of oats. But whereas in the case of the stem eelworm loss may be minimised by growing other cereals or by growing varieties of oats which are to some extent resistant (Maris Quest, Peniarth, Early Miller and Manod), in the case of cyst eelworms there is as yet no effectively resistant oat, and other cereals act as hosts. But, because of the decline in the area of oats and the associated increase in the area devoted to barley, the cyst eelworm is of far less importance than it was a decade or so ago : it can now be regarded as a pest of minor importance only.

PESTS OF SUGAR BEET AND MANGELS

Mangel Fly (*Pegomyia betæ* Curtis). The mangel fly, like the root flies, is brownish grey and about the same size as, but more slender than, the housefly. The eggs are white, elongated and sculptured, and are laid usually in clusters on the undersides of the leaves of mangels and beet. They are easily visible with the unaided eye and should be recognised by the farmer since control measures depend on this. In a few days the young larvæ hatch and enter the leaves, forming bladder-like tunnels which, if the attack takes place early, can destroy all the assimilating tissue. When full fed the maggots desert the leaves and pupate in the soil. Later a second generation occurs which is of little importance.

Control Measures. Mangel fly attack always looks much worse than it really is, for even when defoliation has been so severe that a burnt-up appearance is produced, recovery is usually complete and loss in ultimate yield is small. However, in very severe cases there is undoubtedly some delay and slight loss : now that a fairly cheap control is within the reach of the farmer who possesses a low-volume sprayer the economics of the situation are rather different. A convenient material to use is dimethoate which also controls aphids. Alternatively trichlorphon may be used. If spraying has not been done a top dressing of a nitrogenous fertiliser is advantageous.

Black or Bean Aphid (*Aphis fabæ* Scop.). This is perhaps the best known of the aphids because of its partiality for the broad bean. Its life history is one of classic interest. During the summer months it feeds mainly on the undersides of the leaves, causing them to curl and in bad cases to wither. Towards autumn a migration takes place to the spindle tree where mating is accomplished and eggs are laid. It is possible from a numerical estimate of these eggs to form a fair forecast of the potential outbreak the following season, although the weather of spring and early summer has the final say. The eggs hatch in early spring and large colonies of aphids are produced which begin to migrate from April onwards. Many go to beans, but early beet may also be infested and in such cases serious damage is possible in June and July from this primary infestation. Often, however, and this is especially true of the western areas, the main migration is a secondary one, largely from the bean crop, in July. The capacity for damage is then much smaller, because by August the colonies usually begin to die away, largely from fungus attack. It is the survivors that return to the spindle tree.

Control can be readily achieved with one of the systemics ; in the case of sugar beet an effective system of spray warnings has been developed and is carried out in conjunction with the factories.

Peach-Potato Aphid (*Myzus persicæ* Sulz.) (Fig. 82). This small, greenish aphid is really a greater threat to the beet crop than is the bean aphid, for though it is much less spectacular and will probably not be noticed by the average farmer, it introduces and spreads the well-known yield-and-sugar-reducing virus, sugar beet yellows (see p. 397). The bean aphid does take some small part in this spread but it cannot introduce the virus. The damage done depends on the time of infection. If it goes ahead early in the season, say in June or early July, the loss may be serious. If it only gets going in August the loss is considerably smaller.

It is important to know where the aphid gets the virus, and how the virus is kept going from year to year. In seed-growing areas the beet, because it is biennial, provides its own source : but in other areas it is believed that the aphids and the virus are carted into the clamps along with mangels and overwinter there. In the spring, from late March onwards, should any mangels be left, the aphids

PLATE XXIII. BENEFICIAL INSECTS

, Ichneumon wasps (*Hymenoptera*), parasitic on small caterpillars, e.g. diamond back moth ; female (with long ovipositor) above, male below (× 5) ; 2, the ladybird grub (*Coleoptera*) is a predator ; it devours aphids (× 5) ; 3, chrysalids of a parasitic wasp (*Paracodrus apterogynus* Hal.) emerging from the dead body of a wireworm in which they have developed (× 8) ; 4, maggot of hover fly (*Lasiopticus pyrastri* L.) (*Diptera*) feeding on an aphis colony (× 2)

PLATE XXIV. EXAMPLES OF SUCKING INSECTS

A. Mealy cabbage aphis, developing colonies on young cabbage (nat. size)

B. Greenhouse whitefly on the back of a tomato leaf; the white scales are the immature nymphs. The black scales have been destroyed by, and now contain, the chalcid parasite *Encarsia formosa* Gahan. (× 10)

PLATE XXV. EELWORM DAMAGE

Above: Typical appearance of a potato crop infested with potato cyst eelworm (*Heterodera rostochiensis* Woll.)

Below: Reduction in oat crop due to cereal cyst eelworm (*H. avenæ* Woll.). The crop at A is the second oat crop in 5 years. At B, the land has carried 3 oat crops in 5 years, the rotation being oats : roots : oats : roots : oats.

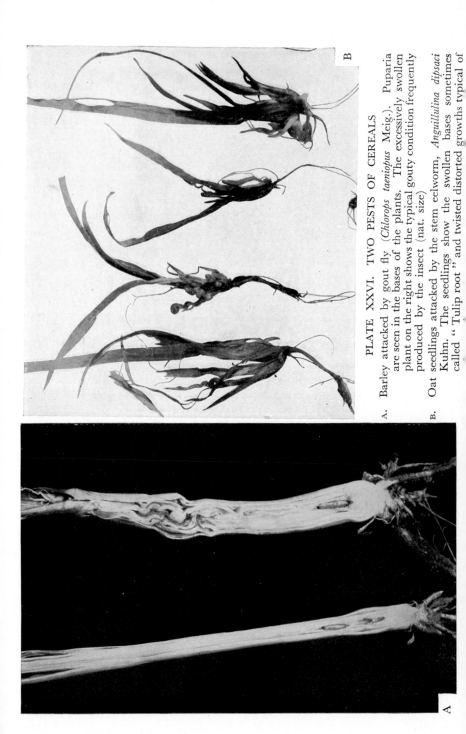

PLATE XXVI. TWO PESTS OF CEREALS

A. Barley attacked by gout fly (*Chlorops taeniopus* Meig.). Puparia
   are seen in the bases of the plants. The excessively swollen
   plant on the right shows the typical gouty condition frequently
   produced by the insect (nat. size)

B. Oat seedlings attacked by the stem eelworm, *Anguillulina dipsaci*
   Kuhn. The seedlings show the swollen bases sometimes
   called "Tulip root" and twisted distorted growths typical of

1. Attack by turnip gall weevil on swede. The galls on the left contain the grubs. Similar galls also occur on cabbage, cauliflower and other market garden brassicas (nat. size)

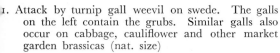

PLATE XXVII. PESTS OF BRASSICA CROPS

2. Turnip gall weevil (*Ceutorrhynchus pleurostigma* Marsh.) adult (× 10)
3. Turnip flea beetle (*Phyllotreta nemorum* L.) adult (× 10)
4. Swedes damaged by caterpillars of cabbage white butterfly (*Pieris brassicae* L.)

PLATE XXVIII. PESTS OF ROOT CROPS

A. On left, two sugar beet plants showing the condition known as strangles; on right, two plants damaged by rabbits

B. Attack by carrot fly (*Psila rosæ* F.) Note that the maggots have attacked the tips and the sides of the roots. Growth and normal development have been severely checked and the roots are unfit for market (nat. size)

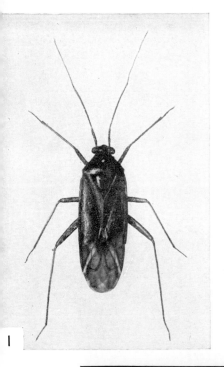

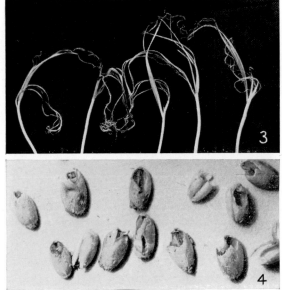

PLATE XXIX. SOME PLANT PESTS

1. Capsid Bug (*Plesiocoris rugicollis* Fall.) (× 8)
2. Colorado Beetle (*Leptinotarsa decemlineata* Say.) (× 6)
3. Shoots of cereal damaged by slugs, showing the characteristic
"ribboning" of the leaves
4. Seed grain damaged by slugs

PLATE XXX. FRIT FLY ATTACK ON OATS

1. Left: whole grains and kernels from oats attacked by frit flies. Right: normal grains and kernels. Centre: frit flies to show relative size
2. Oat plant attacked by maggots of frit fly showing stunted tillers
3. Pupae of frit fly in the grains of oat

breed upon the infected shoots and carry the virus into the beet fields. Consequently this should be prevented by using an efficient aphicide, and if success is to be achieved against virus yellows this precaution must be universally taken. A useful measure where mangels are usually clamped with their tops on, is to clamp topped mangels only in one end of the clamp and to use these last, or such portion may be sprayed with a systemic insecticide before clamping.

Beet Eelworm (*Heterodera schachtii* Schmidt). As with all the other species of cyst eelworms, this pest builds up its numbers during too close rotations not only of sugar beet and mangels but also of brassicas, for these can act as hosts to this eelworm. Rotational precautions are provided for in the sugar-beet factory contracts that the farmer signs, and restrictions on brassica growing have been recently tightened. Beet eelworm was first discovered in the east of England in the 1930's, and with the knowledge of what happened in Germany and also of the unrestricted spread of the potato cyst eelworm, legislation was at once enacted to prevent, or at least confine, the spread of the pest. All infected fields are scheduled under the Act, and the growing of plants of both the beet family and the brassica family is controlled.

Other Pests. All the general soil pests, wireworm, leatherjackets, cutworms, and slugs will destroy young beet. The mangel flea beetle (*Chætocnema concinna* Marsh) in some years seriously delays the growth of the young seedlings, pitting the leaves in the characteristic flea beetle manner. Normally its host plants are in the dock family (*Polygonaceæ*). This flea beetle is not usually a pest of any consequence. Beet carrion beetle (*Acalypea opaca* L.) occasionally will destroy the foliage of young beet but recovery usually takes place. The larvæ are black and are reminiscent of wood lice. The pygmy mangel beetle (*Atomaria linearis* Steph.), formerly a destroyer of the germinating stages, is little heard of since the rotation clauses (preventing beet after beet or mangels) were put in the sugar beet growers' contracts.

PESTS OF BRASSICA CROPS

Brassicas in great variety occupy an important place among farm and market garden crops. They are attacked by flea beetles, stem weevil, gall weevil (Plates XXVII and XXVIII), swede midge, root flies, slugs, cutworms, caterpillars of cabbage butterflies and moths, mealy aphid (Plate XXIV), whitefly and eelworms : the seed pods are attacked by pod weevil and pod midge.

Flea Beetles (*Phyllotreta* spp.) (Plate XXVII) are a serious pest of spring-sown seedlings of cabbage, kale, turnips and swedes. They are small dark inconspicuous beetles, in colour black, blue-black or bronze-black, sometimes with a yellow stripe down each wing case. Their hind-legs have thighs, or femora, enlarged for jumping. The beetles assemble about the seed rows and shelter in the soil when the weather is dull. They are active in sunshine and feed on the seed leaves, often completely devouring them or so crippling them that the plants fail to make normal growth. In a period of dry weather flea beetles may kill the seedlings of several successive sowings unless protective measures are taken.

Flea beetles lay yellow eggs in the soil near the seedlings, and the grubs devour root and stem tissue or burrow into the stems of lower leaves, the feeding-site and habits depending on the species of flea beetle. Pupation takes place in the soil, and beetles emerge during the summer and feed on the leaves of established brassica crops and cruciferous weeds. Towards autumn they seek winter shelter in rough herbage, stack bottoms, hedge sides, and ditch sides, and in spring they leave these sites to feed on seedlings. Flea beetle injury is usually worst during the germinating stage. Beetles fly freely in bright weather and can travel some distance to food plants.

Control Measures. It is now possible to dress the seed with a concentrated gamma-BHC seed dressing sold for the purpose, which should protect the early stages from attack. Should it be necessary a low-volume spray of DDT, or a dusting with powder containing DDT or BHC, will give complete control if properly applied. The dusts can be applied by means of insecticide dusters, powder blowers, adapted manure or seed drills and improvised shakers such as loosely woven bags. These dusts give good protection to the seedlings and a routine application will usually ensure a stand of seedlings from the first drilling. Turnip " fly " is consequently much less important than it used to be.

Cutworms. Cutworms are the caterpillars of the turnip moth (*Agrotis segetum* Schiff.), heart and dart moth (*A. exclamationis* L.), and garden dart moth (*Euxoa nigricans* L.). The moths are dull, light or dark brown, with obscurely patterned wings and are often seen at haytime when grass is cut and turned. They are active only at night and hide during the day. They lay grey eggs in flat clusters on leaves, and after about a fortnight the caterpillars emerge to feed on foliage. As they get larger they become scattered and take to sheltering by day in the soil and coming to the surface at night to feed.

Cutworms are liable to attack the roots and stems of root crops and pretty well any other row crop, feeding at ground level and biting completely through the younger plants and excavating the root of the older ones. They will also eat into potato tubers. An examination at ground level of a wilting plant will usually discover one or more of the waxy-looking greyish caterpillars curved into half circles. There are two main phases of attack. The overwintering caterpillars feed on into late spring, when they attain a length of about 37 mm. The new generation occurs in June and early July and may do serious damage. Attacks often occur on land that has been weedy.

Control Measures. Cutworms can often be controlled by the Paris green-bran poison bait as recommended for leatherjackets (p. 435). This will have to be applied to the base of the plant when an attack is in progress. Quicker results can be obtained by the use of a DDT emulsion watered in at the base of the plant : or if land is known to be infested before planting, by working in a dust of DDT or BHC.

Mealy Cabbage Aphid (*Brevicoryne brassicae* L.). This grey aphid is probably most familiar to the farmer as a pest of swedes and these can be quickly ruined in epidemic years. The attack may be accompanied by mildew : the two troubles are often confused and it is important to distinguish them. All brassicas are attacked and Brussels sprouts are particularly vulnerable in that the aphids may get into the sprouts and are then difficult to reach even with a systemic spray. Attacks on swedes are easily controlled.

The aphid overwinters in the egg stage and in mild winter in the adult stage also : it builds up during spring on its host plants from which migration occurs from May onwards. Care should be taken that all field brassicae whose useful life is over, such as old Brussels sprout crops, are well buried by this time. As summer proceeds, especially in a hot season, brassica crops should be watched for the beginnings of a build-up and a systemic insecticide applied as soon as the pest becomes threatening.

Cabbage Caterpillars. These are larvæ of the large white butterfly (*Pieris brassicæ* L.), the small white butterfly (*P. napi* L.) and the cabbage moth (*Mamestra brassicæ* L.).

Large white caterpillars cause defoliation and occasionally in epidemic years whole fields are affected. Usually, however, the damage is confined to patches. There are two generations, early summer and late summer–early autumn : it is this second generation that does most damage. This generation is usually heavily parasitised by the Braconid *Apanteles glomeratus*, and groups of the deep yellow cocoons are a common sight in the autumn attached to the shrunken empty bodies of the caterpillars.

The small cabbage white butterfly lays single eggs which hatch into greenish caterpillars. DDT as dust and spray is quickly effective: mevinphos, trichlorphon or derris can also be used if preferred. The cabbage moth lives in the heart of the cabbage and is more difficult to get at and needs early treatment.

PESTS OF PEAS AND BEANS

Black or Bean Aphid (*Aphis fabae* Scop.). This is one of the most serious pests of field and broad beans and has already been referred to as a pest of beet and mangels. A systemic spray can be used in the early stages, and since attack often begins around the headlands special attention should be paid to these.

Pea and Bean Weevils (*Sitona* spp.). Pea and bean weevils attack seedling peas and beans in early spring by eating semi-circular notches out of the leaves and stems. They may also be a serious bar to the establishment of lucerne and sometimes of clover. The commonest weevil (*Sitona lineata* L.) is brownish yellow or greyish yellow and covered with tiny scales forming light and dark bands down the elytra. The head is short and stout, the elbowed antennæ are brownish red, the eyes are dark and prominent and between them is a deep groove that extends to the short snout. The weevils may be seen pairing among pea and bean seedlings, but when disturbed they fall from the plants and lie motionless on the soil.

The eggs are laid in the soil. They are at first cream-coloured but become darker after two or three days : they hatch in about three weeks. The grub (Fig. 86) is white, wrinkled, and legless, with a brown head and strong biting jaws, and with fine bristles that assist movement through the soil. The grubs feed through the summer on roots, stem bases and root nodules of peas and beans, and their feeding shortens the life of the crop. In about two months the grubs cease feeding and burrow into firm soil to form earthen cells for pupation. The pupa is enclosed in a transparent membrane which leaves the limbs free. It is white at first, but it gradually

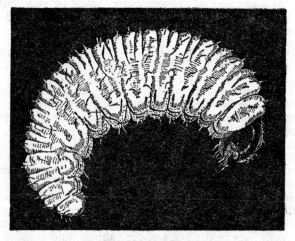

FIG. 86.—GRUB OF PEA AND BEAN WEEVIL (*Sitona*) : × 15.

darkens as it develops, and in two to three weeks the weevil is ready to escape. The new generation of weevils feed in late summer on the leaves of leguminous plants, and in autumn they seek shelter in stubbles, hollow stems and in dry rubbish.

Control Measures. Control of pea and bean weevils in market gardens can be readily secured by dusting in spring with DDT or BHC dusts at the rate of approximately 33–44 kg/ha. The dust should have a heavy base so that it falls quickly on to the seedlings and does not blow about in a cloud. The protection afforded by a suitable insecticidal dust is soon apparent, and in the course of a few days dusted plants produce new leaves that remain uninjured. Field beans sown in autumn may be heavily attacked by the weevils before they seek winter shelter, but such crops are usually well established by spring and the stimulation to growth given by hoeing and cleaning is sufficient to enable them to withstand weevil injury.

Pea and Bean Beetles. Seeds of field beans, broad beans and occasionally peas, often have circular holes and cavities in the cotyledons. The holes are exit holes of small squat brownish grey beetles (*Bruchus* spp.) having short wing cases that leave the grey tip of the body exposed. The beetle, which is about 4 mm long, lays its eggs on the young bean pods. The grubs penetrate the pods and finally burrow into the developing seeds. They pupate inside the beans and the beetles are ready to escape about the time the beans are harvested. Some hibernate within the beans and others shelter in crevices about walls and floors of buildings and in the bark of trees. In spring the beetles frequent flowers and fruit blossom, and gradually assemble in bean fields for feeding and egg-laying.

Control Measures. Since injury by bean beetles seldom affects the embryonic region, the germinating capacity of bean seed is not seriously impaired by attack, but damaged seeds often rot in the ground and may be subject to infestation by secondary pests like millepedes. Modern methods of seed drying and seed fumigation are usually effective in destroying beetles hibernating in the beans. A light dusting with DDT or with BHC dust over the sacks, and over the floor and walls of barn or storehouse, will destroy beetles emerging in autumn from harvested beans.

Pea Moth. Late-sown peas and peas allowed to ripen before harvesting are often attacked by pea moth (*Cydia nigricana* F.). The moth has a wing expanse of about 12 mm ; the fore-wings are dark brownish black with a satiny sheen and distinct white streaks on the front margin ; the hind-wings are lighter brown with a pale fringe. The flight period lasts from June to August and eggs are laid on leaves, stems and flowers of peas. The caterpillars tunnel into the pods and feed on the seeds, filling the pods with silk and excrement. When mature they are whitish and about half an inch long. They leave the pods and enter the soil to hibernate in cocoons of silk with adhering soil particles. Pupation takes place in spring and the moths emerge in summer. There is one generation a year.

Control Measures. The number of moths appears to fluctuate greatly, and the pest may be prevalent for several seasons and then become comparatively scarce. Crop rotation is a valuable method of limiting attack by pea moth on field peas, and carefully timed applications of azinphos-methyl or carbaryl to destroy the young caterpillars before they penetrate the pods can be very effective : they keep down attacks by thrips as well.

Pea Eelworm (*Heterodera göttingiana* Lieb.) is an important pest of peas in market gardens. It feeds in the roots and forms cysts. Attack causes patches of dwarfed yellow plants which fail to set pods. Host plants of pea eelworm include peas, beans, clover and vetches, and the pest is associated with close cropping with these leguminous crops. Remedial measures consist of widening the rotation and improving the fertility of the soil so that susceptible crops are enabled to withstand some degree of attack.

PESTS OF POTATOES

Potato Cyst Eelworm (*Heterodera rostochiensis* Woll.) (Plate XXV) is associated with patchy failure in potato crops and occurs wherever close cropping with potatoes has been practised. The eelworms can be seen as white or yellow spherical cysts protruding from the roots and stolons (Fig. 87). Mature cysts are brown and may contain several hundred eggs and young eelworms. In the absence of their host plants the cysts remain viable in the soil for many years, but the presence of growing potatoes stimulates the hatching of eggs and the emergence of young eelworms. The eelworms invade the young

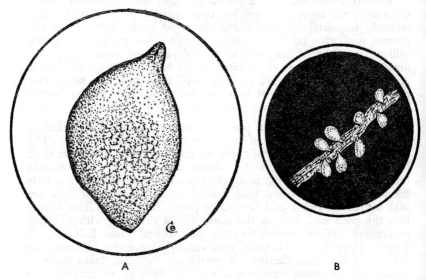

A   B

FIG. 87.—CYST EELWORM (*Heterodera*).
A, cyst of beet eelworm : × 75.
B, cysts of pea eelworm protruding from root : × 8.

roots and destroy the tips, and in consequence the plants suffer from malnutrition. Attacked plants are dwarfed and yellow and produce few tubers.

Once potato cyst eelworm is established in fields intensively cropped with potatoes its multiplication is rapid and yields of tubers are greatly reduced. The greatest source of infestation is the soil adhering to seed potato tubers and in the bottom of sacks and containers. A wide rotation is very desirable. Resistant varieties of potato are now available and will be of most use in the Eastern Counties where pathotype "A" is predominant. Soil sterilants or fumigants like aldicarb, oxamyl, dazomet and D-D mixture are of value.

Colorado Beetle (*Leptinotarsa decemlineata* Say.). Colorado beetle was introduced into Europe from America late in the 19th century. It has gained entry to Britain several times but owing to a good system of inspection and the use of effective control measures, every outbreak has been promptly eradicated. The beetle is rather rounded, about 10 mm long. In colour it is yellow with four back stripes extending lengthwise along each wing cover, and there are black markings on the thorax. The beetles hibernate in the soil, often at a depth of 250–300 mm, and emerge in early summer. They walk or fly to potatoes where they feed on the foliage and lay clusters of upright yellow or orange-yellow eggs. The eggs hatch in ten to fourteen days and the little dark-coloured grubs resemble the grubs of ladybird beetles. At first the grubs feed in colonies on the leaves but as they get older they become dispersed over the plant. Colorado beetle grubs are fully fed in about three weeks and by that time they have eaten the leaflets to the midrib and have fouled many leaves with excrement. Mature grubs are about 12 mm long, rather humped and swollen, and reddish with head and legs black and two rows of black spots along the sides. They penetrate to a depth of 50–75 mm in the soil for pupation, and in less than a fortnight, in late July or early August, the new generation of beetles emerges. These beetles feed locally on potato foliage and may migrate by flying to another crop before returning to the soil to hibernate. In recent years there have been no outbreaks and this is partly due to precautionary measures taken on the nearest Continental coasts.

Control Measures. In Britain control of Colorado beetle is the responsibility of the Ministry of Agriculture, to whom any outbreak must be promptly reported. Inspection of potato crops in suspected areas is now a routine practice, and when outbreaks occur infested crops are sprayed with Approved insecticides and the soil is fumigated with carbon bisulphide under strict official supervision.

Potato Aphids. At least four aphids commonly infest the potato, but though they can sometimes cause much direct harm their importance lies in the fact that one of their number, the peach-potato aphid (*Myzus persicæ* Sulz.), is mainly responsible for spreading the virus of leaf-roll in potato stocks. Because of this, new seed has continually to be imported into the ware growing districts from the relatively aphid-free areas, thus adding to the expense of growing the crop.

PESTS OF GRASS AND CLOVER

Grasses are related to cereals and are subject to attacks by many of the pests of cereals. Ryegrass is attacked by maggots of frit fly and the grass aphids may occasionally devastate a seed crop or ley. Midges infest grasses and the flower heads are often distorted through their attacks ; the foxtail midges prevent the formation of seed in meadow foxtail grass. The cocksfoot moth (*Glyphipterix simpliciella* Stephens) feeds on the seed of cocksfoot grass.

Leatherjackets and chafer grubs are pests of old grassland. Leatherjackets tend to occur in damp, chafer grubs in drier, situations, e.g. on the thin soils of hill and mountain slopes, and in the dry and cold sandy pastures of the Midlands and East Anglia.

Chafer Beetles. Three species of chafer beetles are common in Britain. The cockchafer (*Melolontha melolontha* F.) is about 30 mm long with the head black, the prothorax covered with long greyish hairs, and the elytra light chestnut brown and distinctly ridged. Antennæ and legs are brownish ; the apex of the abdomen is exposed and tapers to a broad down-turned tip. The garden chafer (*Phyllopertha horticola* L.) is about 12 mm long with the prothorax a dark metallic green or blue and the elytra brown ; the long legs and the tips of the antennæ are black. The summer chafer (*Amphimallus solstitialis* Latr.) is about 18 mm long ; the prothorax and elytra are yellowish brown and the elytra have indistinct ridges that suggest stripes. The cockchafer flies at night in May and June and rests by day among the foliage and blossoms of trees. The garden chafer flies in bright sunshine in May and June and hides in grass and vegetation at night and in dull weather. The summer chafer flies in late evening in the latter part of June and the beetles often swarm about high trees in grass country. The grubs of all three species live in the soil. The garden chafer reaches maturity in one year and the cockchafer takes three to four years to complete its life cycle. The summer chafer and the garden chafer are largely grassland insects but the cockchafer is most abundant in areas of mingled woodland and pasture.

Remedial measures against chafers and leatherjackets are similar. They consist of breaking the old matted turf, correcting any mineral or lime deficiency, and re-seeding with a suitable grass mixture. Birds are fond of chafer grubs and feed regularly in infested fields.

Young clovers are attacked by pests of peas and beans (see p. 441). Pea and bean weevils (*Sitona* spp.) devour stems and leaves and are sometimes responsible for the poor take of clover in young leys. Clover leaves are also attacked by several species of leaf-eating sawflies including *Pteronidea myosotidis* F. Clover for seed is often infested by seed weevils (*Apion* spp.), small black pear-shaped weevils whose grubs feed in the seed heads and reduce the yield and quality of the crop.

Stem and Bulb Eelworm (*Ditylenchus dipsaci* Kuhn) is an important pest of clover on heavy land. It contributes to " clover sickness ", a name for the repeated failure of red clovers on old arable land. The eelworms (Fig. 88) are present in the soil when the seed mixture is sown and they penetrate the seedlings soon after germination. Infested seedlings become swollen and distorted and many are killed. Some broad red clovers are very susceptible, others are resistant. Some late-flowering types are resistant to attack but others are killed out. Alsike is fairly resistant, and white clovers and hoptrefoil are almost immune. In arable districts where red clovers regularly die out of the leys, seed mixtures should contain the resistant types with ryegrass, cocksfoot and timothy. Stem eelworm is spread in hay from infested pastures, in fragments of infested plant tissue, in seeds from infested crops and in the soil on implements and on the hoofs of livestock. The diagnosis of eelworm disease is difficult and technical help should always be sought when the presence of eelworms is suspected. It may be confused with the disease brought about by the clover rot fungus described on page 377.

PESTS OF VEGETABLES

Vegetable crops suffer from attack by numerous pests including cabbage white butterflies, cabbage moth, garden pebble moth, angle shades moth, asparagus beetle, celery leaf miner, turnip gall weevil and many aphids. Two fly pests are of special importance, cabbage root fly and carrot fly.

Cabbage Root Fly (*Erioischia brassicæ* Bouché) is an Anthomyid fly somewhat resembling a housefly. The flies emerge from over-wintering pupæ in April and early May and lay white eggs in the surface soil close to the stems of seedling brassicas and transplanted cauliflowers and cabbage. The eggs hatch in four to five days and the maggots feed on the surface tissue of the rootstock and lower part of the stem. Rotting frequently accompanies attack by root maggots and the plants may be killed or their growth may be severely checked. The maggots reach maturity in about three weeks and become puparia in the soil in the vicinity of the plants. The puparia are reddish-brown, barrel-shaped and about the size of a grain of wheat. Flies emerge from most of the summer puparia in two to five weeks, but occasionally emergence is delayed for several months. There are two or three generations a year. No cruciferous crop is free from attack by root maggots ; wallflowers, radishes, kales, swedes and young Brussels sprouts may be infested.

Control Measures Although aldrin and dieldrin may still be used in the form of dips in which the roots of brassica plants are immersed before planting, the cabbage root fly has in recent years developed marked resistance to these materials in a number of important brassica-growing areas. In these districts chlorfenvinphos and diazinon are widely used as spot treatments. Seed-bed protection is also possible by spraying.

The Carrot Fly (*Psila rosæ* F.) appears early in May, and after feeding in flowers of wild Umbelliferæ it flies to young carrots for egg-laying. The eggs are laid in the soil near carrots and the slender white maggots burrow into the tap roots near the tips. Attack results in a check to growth and the foliage flags and loses its fresh green colour. In some plants the root tips are severely tunnelled or destroyed, and the tap roots may become forked or distorted (Plate XXVIII). Maggots of carrot fly reach maturity in about four weeks and become puparia in the soil. In July and August there is a second generation of flies. These give rise to maggots that continue feeding during autumn in maincrop carrots, celery, parsnips and parsley. Flies often congregate in May in hedgerows and were formerly attacked there with a 0·5 per cent DDT spray. This method was superseded by BHC seed dressing which provided a complete answer to first generation attack. Granule applications of phorate, diazinon, disulfoton or chlorfenvinphos are now in use, the last two on mineral soils only.

PESTS OF GRAIN AND FEEDING-STUFFS

Grain for seed or for feeding, peas and beans, flour, meal and cattle cake that have to be stored for some time often deteriorate through infestation with insects and mites. Beetles, weevils, cater-pillars and mites are common about farm granaries and food stores, and care must be exercised to restrict their numbers and to protect the stored food.

Granary Weevil (*Calandra granaria* L.). This beetle is about 3 mm long and dark brown, with elbowed antennæ that originate near the base of the snout, a long coarsely punctured thorax, and elytra with longitudinal grooves and ridges. The eggs are laid in holes bitten in the grain and the soft yellowish-white, legless grubs feed inside the grain. The pupa is formed within the grain and the adults emerge in six to nine days. The life cycle is completed in four to six weeks and, when temperature and humidity are suitable, generations of weevils follow each other quickly and grain soon becomes heavily infested.

Moths. Several species of moths infest farm granaries but the most common are the brown house moth (*Borkhausenia pseudospretella* Staint.) and the white-shouldered house moth (*Endrosis lactella* Schiff.). The brown house moth has a wing expanse of about 18 mm ; the head and thorax are light brown, and the fore-wings light or dark brown with three distinct dark areas that look like three pairs of black spots when the moth is at rest ; the hind wings are light grey. The white-shouldered house moth has the head and thorax conspicuously white, the fore-wings yellowish-grey speckled with black and white, and the hind-wings pearly grey with a deep fringe of light hairs. Both species lay eggs on almost any kind of stored food and the caterpillars spin silk in which frass, or pellets of excrement, become entangled. When fully grown the caterpillars are about 12 mm long, white with the head brown, and with conspicuous brown plates on the thorax just behind the head ; stoutish bristles are sparsely scattered over the body. The larvæ spin tough silken cocoons in cracks and crevices in bins, about the walls, floors and beams of granaries and store-rooms, and in sacks and sacking. There are several generations a year. Infested food in bins and sacks becomes fouled by the silk and frass of the caterpillars and the wastage is considerable.

Flour Mites. Several species of flour mites (Tyroglyphidæ) infest food stores. They are minute creatures, pale grey or whitish in colour, and when numerous they resemble flour dust. Examination with a lens shows that mites have eight legs, a short pointed head and a rounded oval body bearing a number of rather long hairs or setæ. Mites are able to eat holes in cereal grains, and infested corn soon develops an unpleasant musty smell. The life cycle of mites is completed in about three weeks and breeding goes on practically all the year round.

Other pests about farm buildings and food stores include the larder beetle (*Dermestes lardarius* L.), the cheese and bacon fly (*Piophila casei* L.) and the meal worm (*Tenebrio molitor* L.).

Control Measures. Control of pests of stored foods is largely a matter of granary hygiene. Walls, floors and roofs should be properly cleaned before fresh stores are brought in. Walls and stone floors should be faced with cement to reduce shelter and wooden floors should be sealed with a bitumastic preparation. Bins and other receptacles should be kept clean and empty sacks should be removed and sterilised before using them again. Light applications of insecticides like BHC, DDT and malathion to walls and floors of store-rooms before taking in fresh consignments of food are also helpful in keeping down pests, as are insecticidal " smokes ".

PESTS OF LIVESTOCK

Insect and allied pests of livestock cause appreciable losses to British farmers. Biting flies make dairy herds restless and irritable during milking ; the proximity of blood-sucking and warble flies causes gadding and stampeding and the tormented herds soon lose condition ; infestation by lice, fleas and ticks checks the growth of young animals and may infect them with diseases and worm parasites ; and the attacks of parasitic flies on sheep and cattle may bring about internal disorders that result in death. Much research has been carried out on the pests of livestock during recent years, a wide range of chemicals have been investigated and some are now available which give better control. In the following pages the present know-ledge of the life histories and habits of common pests of farm livestock and the means of controlling them is summarised ; but future ex-tensions of the livestock industry are likely to be followed by further developments in the protection of flocks and herds, and the student should read farming papers for new and up-to-date information on insecticides and methods of using them.

Warble Flies. Cattle are attacked by two species of warble flies : *Hypoderma bovis* de G. and *H. lineatum* Vill. They are on the wing in spring and summer and lay eggs on the under parts and legs of the animals. The larger species, *H. bovis*, lays eggs singly and the smaller, *H. lineatum*, lays them in rows on hairs low down on the legs. Warble fly eggs are yellow and are attached to the hair by short clasp-like extensions of the base (Fig. 88, B). They hatch in three to six days, and the maggots burrow into the skin at the bases of the hairs and cause swelling and inflammation. Maggots of *H. bovis* travel through the body tissue of the host arriving in the back in late spring, some may enter the spinal canal. Maggots of *H. lineatum* burrow in the connective tissue to the gullet, which they reach when they are about half-grown, and from the gullet they migrate to the back. These maggots reach the back from February onwards ; they perforate the hide in order to breathe and remain stationary in pus-filled cavities. The swellings made by the maggots are about the size of a walnut and can be easily felt by running the hand along the back from withers to tail. The maggots attain a length of 16–26 mm, and are dull yellow or cream-coloured (Fig. 88, C) and much wrinkled. By April they begin to reach maturity. The fully developed maggots enlarge the breathing pores, escape through the hide and fall to the ground to pupate. The pupal stage lasts five to six weeks.

The emergence of the bee-like adults completes a life cycle lasting about a year, of which ten to eleven months are spent feeding within the bodies of the hosts.

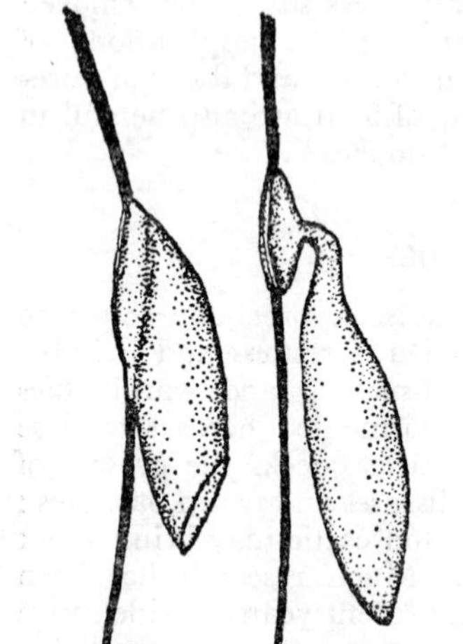

FIG. 88.—EGGS AND MAGGOTS OF WARBLE FLY AND BOT FLY.
B, egg ; C, maggot of Ox Warble.
A, egg ; D, maggot of Horse Bot.
(Eggs : × 26 ; maggots : × 2.)

Warble flies on the wing make a buzzing sound that causes the cattle to run wildly about the fields, and in consequence they lose condition rapidly. The passage of the maggots through the body tissues, the gullet and the spinal canal, and their establishment in the back, causes great irritation and inflammation which further contribute to loss of health and vigour in infested animals. Attack by warbles also reduces the value of the hides because perforations made by the maggots persist in the leather and entail much waste when it is used.

Control Measures. After detailed studies of the distribution of warble flies in Britain the Ministry of Agriculture and Fisheries introduced the Warble Fly (Dressing of Cattle) Order in 1936, which made it the responsibility of stock owners to destroy the maggots or warbles in cattle that were visibly affected. Approved dressings contained rotenone (the active principle in derris root) and usually consisted of a mixture of derris and soft soap in water. The dressing had to be applied to the backs of infested animals at monthly intervals from 15 March to 30 June. An alternative treatment was the removal of warbles, by squeezing or other means, at ten-day intervals during the same period.

Since cattle are the only hosts of the ox warble flies the enforcement of the order requiring the destruction of warbles in infested cattle had secured an appreciable reduction in the severity of warble fly attack by 1940, when lack of derris led to its suspension. Dressing of visibly affected animals was again made compulsory in 1948 but the order was revoked in 1964. Insecticidal dressings were applied with a stiff brush that exposed the warble pores, or they were swabbed on the backs of the animals with a cloth in such a way that they penetrated the breathing pores and reached the warbles. In recent years the use of organophosphorus materials such as trichlorphon and fenthion as " pour on " treatments to the backs of cattle has been easier and more effective. Animals are best treated in the autumn from October to mid-November. They must not be treated in December, January or February but can be from mid-March onwards.

Sheep Nostril Fly. The sheep nostril fly, *Œstrus ovis* L., is a yellowish-brown, bee-like fly related to the warble flies. It is active in hot summer weather, and deposits young maggots in the nostrils of sheep. The maggots are spiny and easily work their way through the nostrils into the nasal cavities, and finally into the frontal sinuses of the head where they attach themselves to the mucous membrane. Their feeding causes irritation and inflammation in the head, and when the brain becomes affected the sheep appear dazed, a condition called " false gid " or " false staggers " to distinguish it from a similar condition produced by the larval stage of the tapeworm, *Tænia multiceps* Leske. The maggots of the nostril fly remain in the frontal sinuses for up to ten months and attain a length of about 25 mm. When mature they detach themselves from the mucous membrane and are sneezed out on the soil, into which they burrow for pupation. The life cycle is completed in about a year.

Control Measures. Sheep nostril fly is difficult to control, but the provision of shade for the animals and the use of repellent smears containing tar or creosote were recommended for reducing attack. In recent years many new chemicals have been tested for controlling this pest and some have shown promise but no simple method of treatment is available for farmers to apply. Examination of slaughtered sheep has shown that the presence of a few maggots of sheep nostril fly may not seriously affect the condition of the sheep, but when a higher number of maggots occur the sheep are visibly affected. When sheep are obviously distressed and attack by sheep nostril fly is suspected, the advice of a veterinary surgeon should be sought.

Horse Bot Fly. The horse bot fly, *Gasterophilus intestinalis* de G., another relative of the warble flies, is a large, brown, hairy fly somewhat like a honey bee in appearance. It lays elongate yellowish-white eggs (Fig. 88, A) and attaches them to hairs, usually on the fore-legs within reach of the horse's tongue, but occasionally on the shoulders, parts of the belly and the hind-legs. It appears to be an advantage to the species that the eggs are laid on parts of the body that can be licked. Friction from the lips helps the eggs to hatch, and the young maggots are licked into the mouth. They burrow under the tongue and in due course make their way through the gullet to the stomach, where they attach themselves by their mouth hooks and remain for nine to ten months obtaining nourishment from the stomach contents. When mature they are over 12 mm long, rather barrel-shaped, and with girdles of short, chitinous, backwardly directed spines on the body segments (Fig. 88, D). They leave the host with the fæces and enter the soil for pupation. The pupal stage lasts about a month and the flies emerge in the later part of summer.

Control Measures. Effective control of horse bot flies is difficult. Clipping the hair on the legs and under parts and rubbing down the animals with mild disinfectants, or with a little derris extract on a damp cloth, help to prevent infestation. The provision of shade permits the animals to escape from the attentions of these sun-loving flies during the egg-laying period. Various chemicals have been used against bots with some success, and where there is reason to suspect a heavy infestation a veterinary surgeon should be consulted as to the best and safest treatment to evacuate the maggots.

Gadflies and Clegs.   Gadflies and clegs belong to a family of flies (Tabanidæ) in which the females suck blood to nourish their eggs. In Britain three groups are troublesome to farm animals, and they can be distinguished by their colour and wing patterns. Gadflies (*Tabanus*) are large dark flies with clear wings (Plate XXI, B). Breeze flies (*Chrysops*) are dark with conspicuous triangular yellow marks on the upper surface of the abdomen, and the wings have a

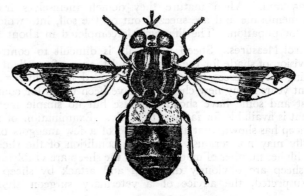

Fig. 89.—Breeze Fly (*Chrysops cæcutiens* ♀) : × 3.

broad dark band across the middle and a less distinct band near the tip (Fig. 89). Clegs (*Hæmatopota*) are ashy-grey with the entire surface of the wings finely mottled. Flies in all the groups have large brilliant eyes, often banded or spotted, and short, stout head-feelers, and they are able to travel long distances in search of a meal of blood. They are on the wing in bright sunshine from May to September, and animals in open low-lying situations and about the edges of woodlands are especially liable to attack.

Females of these blood-sucking flies lay eggs in masses on the leaves and stems of water plants and on sticks and posts projecting from ponds, pools, ditches, lakes and shallow sluggish water courses. The newly hatched maggots fall into the water and burrow in the mud at the bottom, where for two years they feed on small insects, worms and other soft-bodied animals. When they have finished feeding they seek drier soil beyond the swampy margin of the water, and pupate at a depth of about 25 mm. The pupæ have a thin brown skin with rows of bristles on the segments, and the developing fly is visible within. At emergence the pupal case is left behind in the soil and the fly crawls to the surface to expand and dry its wings.

Control Measures. No methods of controlling gadflies, clegs and breeze flies have yet been devised. The breeding grounds in swamps and the bottoms of shallow lakes, ponds and ditches are difficult to treat, but draining and reclaiming damp low-lying land and keeping water courses clean and open limits the extent of suitable feeding-sites for the maggots. Since blood-sucking flies are most active in bright sunshine in the middle of the day the provision of dense shade by means of trees or sheds gives some protection from attack. Horses at work were protected by coarse cord netting slung over the neck and back; the constant movement of the net disturbed the flies and reduced their opportunities for biting. Repellent dressings containing carbolic acid helped keep the flies away, and horses were rubbed down lightly with a pad or cloth damped in a solution of 1 part carbolic to 5 parts water.

Biting Midges and Black Flies. Females of biting midges (*Culicoides*) and black flies (*Simulium*) suck the blood of man and animals, and are especially annoying to farm animals when they bite the tender tissue about the eyes, mouth, nose and ears. Biting midges are exceedingly common over most of Britain. They are tiny, greyish, two-winged flies that appear in swarms at dawn and dusk and on dull damp days in summer. Their bites cause inflammation. *Culicoides impunctatus* Meig. is a common and troublesome species and *Culicoides nubeculosus* Meig. transmits parasitic worms to horses. Biting midges breed in damp soil, in shallow standing water and in moisture in holes in trees, and are active from spring until autumn.

Black flies are larger than the biting midges and inflict severe and painful bites. They are usually about 3 mm long, dark coloured, stoutly built and somewhat hump-backed; one of the larger species, *Simulium ornatum* Meig., the ornate black fly, is black with light markings on the thorax, abdomen and legs. Black flies are less well known than biting midges, but they can be recognised by their curious direct and darting flight. They occur in open and wooded country but are seldom found in stables and cow sheds. They breed in swift-flowing rivers and streams, and the maggots anchor themselves to stones and rocks by means of sucker-like discs and hooks at the tip of the body, and feed on animalculæ that are caught in the head fans. Pupation takes place under water, and at emergence the fly floats in an air bubble to the surface from which it takes flight immediately. The life cycle takes ten to twelve weeks in summer and there are two or more generations a year. The flies are sometimes found about fruit blossom, and are on the wing periodically during summer and autumn.

Control Measures. Control of biting midges and black flies presents a serious regional problem for which there is no effective and safe solution. It is not possible to treat the major breeding sites such as the edges of ponds or streams, since serious contamination of water courses could occur. Sheep may have some protection from chemicals used for dipping against maggot fly. Cattle may have some protection from pybuthrin or other safe suitable cattle fly sprays which are available. If it is possible, move the animals for some grazing time to pastures with lower infestations. Repellents based on di-methyl-phthalate (DMP) are available for humans.

Sheep Maggot Fly. Sheep maggot flies or greenbottle flies of the species *Lucilia serricata* Meig. are the most important external parasites of sheep in Britain, especially in sheltered pastures. Eggs are deposited in the fleece and the resulting maggots feed on the skin and flesh giving rise to a condition known as " strike ". The bluebottle flies or secondary maggot flies are attracted to " struck " sheep and oviposit in the infested area. When neglected, infested sheep rapidly lose condition and may die.

Maggot flies have a short life cycle, and several generations can occur during spring, summer and autumn. The eggs hatch in a few hours after being laid and maggots are fully grown in 2 to 3 days. Pupation takes place in the soil. The flies emerge in 2 to 3 weeks and are active in sunshine, especially when the atmosphere is damp and sultry.

The condition of fleece has an important influence on the occurrence of " strike ". The flies are attracted to wool contaminated with faeces and urine, and consequently strike can be more frequent on the hind quarters'. In warm, showery weather when the fleece is kept moist by rain and wet vegetation the risks of sheep being fly-struck are far greater than in dry weather conditions.

In clean dry sheep, conditions in the fleece are unattractive for egg laying and unsuitable for maggots to survive, but any factors altering these conditions favour the development of strike.

Breeds of sheep with open or coarse wool fleeces are less likely to be struck than sheep with close, fine wool fleeces since they remain moist far longer after wetting. Sheep remain relatively free from strike for a few weeks following shearing provided they remain free from contamination by urine or faeces.

Control Measures. The use of dips containing dieldrin greatly reduced the incidence of strike and made shepherding much easier in summer. A great advantage which this material possessed over the arsenic-sulphur dips was its persistence in the fleece. Unfortunately, dieldrin persisted in the flesh as well as the wool and was a possible source of dieldrin contamination for the " meat " consumer. Organophosphorus chemicals such as bromophos, butacarb, carbophenothion, chlorfenvinphos, coumaphos, diazinon, dichlorfenthion, fenchlorphos, iodofenphos, phosalone and a few other materials have now replaced the " banned " dieldrin. The new chemicals remain in the fleece to give effective protection but break down rapidly in the flesh. To obtain a fairly long period of protection care must be taken to maintain the dip wash at the correct strength—most chemicals are " stripped " from the dip wash by the fleece as the sheep pass through, leaving a much weaker dip wash. Sheep should have wool at least 25 mm long and should be kept in a dip wash for 30 seconds.

Showering, carefully done, can be a substitute for plunge dipping, but spraying in spray races, unless repeated at short intervals, is much less effective.

Crutching helps to reduce susceptibility to " strike ". There are dressings containing insecticides which kill maggots and assist in healing the wounds on " struck " sheep.

Stable Fly. The stable fly, *Stomoxys calcitrans* L. (Fig. 90), is a greyish fly about the same size as a house fly, and its stiff projecting

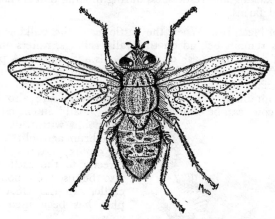

FIG. 90.—STABLE FLY (*Stomoxys calcitrans* ♀) : × 5.

proboscis (Fig. 91), divergent wings and abdomen with dark spots make it easy to recognise. Like the house fly it breeds in rotting vegetation and dung, and thrives best in the drier parts of manure heaps containing cow dung. The flies frequent cowsheds, stables and

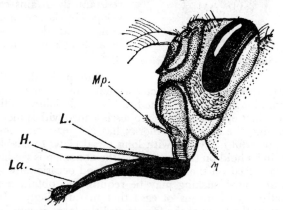

FIG. 91.—HEAD OF STABLE FLY TO SHOW PIERCING MECHANISM.
Mp., maxilliary palp ; L., labrum ; H., hypopharynx ; La., labium. × 20

gate-posts of pastures and trunks of trees near drinking-places. The females require several meals of blood before the eggs are mature, and they obtain the blood from cattle, horses, pigs and man. Each female is capable of laying upwards of 500 eggs in small batches in the moist crevices of manure heaps. The life cycle takes about six

R

weeks and there are several generations during the year. The flies are on the wing from spring to autumn and are most abundant in July, August and September. They are a special nuisance to dairy herds. They follow the cows into the byres and milking-sheds, and by their persistent "biting" of the legs of the animals they cause the stamping, kicking and restlessness during milking that is characteristic on infested farms.

Control Measures. When the cattle leave the milking sheds the stable flies remain behind on the walls, ceilings, rafters and pillars, and out of doors they bask on sunny walls and doors. A proportion of such resting places may be sprayed with bromophos or malathion but the surface should first be rendered non-absorbent. Special insecticide-containing paints and synthetic resins are also available. Incoming cows can be cleared of their fly " load " in enclosed collecting points with water sprays or pyrethrum aerosols. After milking crotoxyphos may be applied directly to the cattle; it may also be used on non-lactating cattle. In the USA fenchlorphos has been used on cattle rubs.

FIG. 92.—SHEEP KED (*Melophagus ovinus* ♀) : × 8.

Finally, much can be done to prevent or reduce breeding in manure heaps by general hygiene and the use of insecticides. It should be remembered that resistant fly strains can be developed by the constant use of the same insecticide and there should be a change about from time to time.

Sheep Ked. The sheep ked, *Melophagus ovinus* L., is a curious flat hairy, brown, wingless fly about 6 mm long (Fig. 92). The head is closely joined to the thorax, and has distinct eyes and a long proboscis armed with spines that assist in penetrating the skin; the thorax has neither wings nor halteres but has powerful hairy legs with long tarsal claws to enable the ked to maintain its hold on the sheep, and the abdomen is sac-like. Keds of both sexes feed on the blood of sheep and occasionally goats, and twenty-four hours' sucking may be required to obtain a full meal. Reproduction is by means of eggs that mature singly. Incubation takes place in the body of the female and the larva is nourished there until it is fully grown. Gestation lasts about eleven days and each female may produce 10–15 larvae. Each larva sticks to the wool, becomes a pupa without feeding, and the adult emerges in three to four weeks. Females of the new generation produce their first fully developed larva about two weeks after they emerge, and though each female produces only 10–15 larvæ the progeny of one pair may total some hundreds in the course of the season.

The movement and feeding of keds on infested sheep causes constant irritation, and when the parasites are numerous the sheep become unthrifty. Keds migrate from sheep to sheep, and from sheep to lambs, during the flocking operations incidental to management, but dips as used for the control of sheep maggot flies (p. 454) give satisfactory control of the pest.

Fleas. Fleas (Aphaniptera) are small brown wingless insects, characteristically flattened laterally ; they have long hind-legs adapted for jumping, and piercing mouth structures with which they puncture the skin of their warm-blooded hosts and suck the blood. The narrow body clothed with backward-pointing bristles, and the long powerful hind legs, enable fleas to move rapidly among the hair or feathers of their hosts, and give them some protection against rubbing and scratching. Fleas occur on most small rodents, dogs, cats, rabbits, birds and poultry. Rat fleas are of special importance because they frequently attack other animals and man and spread a fatal disease called bubonic plague. Other species of fleas transmit tapeworms in dogs and children. The irritation caused by the feeding of numbers of fleas soon lowers the condition of animals and makes them restless and unthrifty.

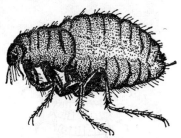

FIG. 93.—DOG FLEA (*Ctenocephalus canis* ♀) : × 15.

Two species of fleas common about farms are the dog flea, *Ctenocephalus canis* Curt., and the hen flea, *Ceratophyllus gallineæ* Schrank. The adults stay on the hosts only to feed but breed in places used by the hosts. They feed frequently and draw relatively large quantities of blood, some of which is voided in a partly-digested form that can be found as small dry pellets on the hosts, about nest boxes and in places where dogs sleep. The females of some species lay eggs in the feathers of their hosts and the eggs fall or are shaken out into the bedding or nest-box material. They hatch into active white maggots with brown heads, and the maggots move freely by means of bristles : they bear a close resemblance to the grubs of small carabid beetles. They feed upon waste skin cells, feather base scales and organic particles in dust, and upon the dry blood-containing excrement of the adults. They reach maturity in two to three weeks and pupate in cocoon-like cases. New adults emerge in a few days and the complete life cycle from egg to adult requires only about three weeks. Under some conditions the adults remain dormant in the cocoons for a long period and emerge when there is some marked change in the environment ; this phenomenon sometimes accounts for the sudden emergence of large numbers of fleas in unexpected places. Frequently cocoons and fully fed maggots fall through floor crevices and accumulate in the debris under kennels, hen houses and rabbit hutches.

Control Measures. It is obvious from a knowledge of the life history and habits of fleas that hygiene plays an important part in their control. Clean bedding frequently changed, the treatment of bedding, nest boxes and dust baths with pyrethrum, derris, lonchocarpus, sevin or malathion, and the occasional application of these insecticides as dusts on animals and poultry keep the flea population at a minimum. Floors and walls of hen houses and kennels may be sprayed or dressed with light creosote, or the interior of hen houses may be treated with malathion or sevin.

Biting Lice and Sucking Lice. Biting lice and sucking lice belong to the Order Anoplura, and are wingless degenerate Exopterygota (see p. 410) that spend their whole lives on their hosts and soon die if they are removed from the warmth and shelter of the hosts' bodies. The biting lice (Mallophaga) have broad heads and mandibulate mouths, and are found on most farm animals, *Damalinia ovis* on sheep, and *Damalina bovis* on cattle. The common biting louse of the dog is *Trichodectes canis* Deg., a broad yellowish species that multiplies rapidly on neglected animals and is known to transmit tapeworms. A common and injurious louse of poultry is *Menopon*

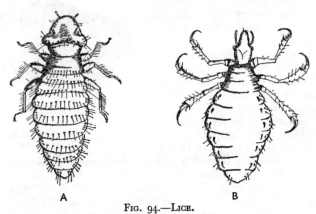

A   B

FIG. 94.—LICE.

A, biting louse of poultry (*Menopon pallidum*) : × 15.
B, sucking louse of pig (*Hæmatopinus suis*) : × 10.

gallineae. (Fig. 94, A), an active insect with four-segmented antennæ, that crawls readily on the hands and arms of persons handling infested birds. Biting lice feed on fragments of dead skin, hair, fur and feathers, which they bite off with their sharp mandibles. The sucking lice (Siphunculata) have narrow heads and mouth parts adapted for piercing and sucking. A number of species infest farm animals. The pig louse, *Hæmatopinus suis* L. (Fig. 94, B), attacks pigs about the neck and jowl, the bases and insides of the ears, the belly and the inner sides of the legs. A smaller species of *Hæmatopinus* feeds on cattle about the shoulders, neck, forehead, escutcheon and base of the tail, and other species attack horses and sheep. The sucking lice have large curved tarsal claws with which they maintain their hold on the hosts. *Linognathus ovillus* (Neuman) feeds on sheep.

The eggs of lice are laid singly and stuck to hair, fur or feathers of the hosts ; they are easily seen with the unaided eye, and are well known as " nits ". They hatch in four to five days and the empty egg shells remain attached to the host for long periods. The young lice resemble the adults in appearance and feed in the same manner. Lice are spread when their hosts are in contact with each other, and the biting and sucking and scratching with their long sharp claws as they wander about the hosts causes great discomfort and irritation. Infested birds and animals take little food or rest and their health soon deteriorates. Lice transmit disease, the body louse of man being notorious as the carrier of typhus fever.

Control Measures. Animals infested with lice do much rubbing and scratching and the irritation is followed by dullness of coat, falling hair and bare patches. The presence of numbers of lice among farm stock and poultry necessitates the thorough cleaning and disinfecting of their quarters : sties, byres, sheds, loose boxes and poultry houses. Woodwork may be treated with insecticidal wash, but the main effort should naturally be concentrated on the animals themselves. Dusting powders containing derris and pyrethrum have proved useful against lice, but BHC and malathion are more effective. Chemicals used for maggot fly control lice.

Sheep Ticks. Sheep ticks, *Ixodes ricinus* L. (Fig. 95), are troublesome pests in parts of Wales, north England and Scotland, where the hilly country provides extensive sheep and cattle ranges. They infest cattle and sheep from April onwards through the summer and autumn, and feed about the neck and ears and on the inner sides of the fore-legs and inguinal regions. Ticks are eight-legged animals, members of the class Arachnida, Order Acarina (p. 405). The head is extended into a " blunt beak " ; the central part of this is called the *hypostome* and has chelicerae and side palps (Fig. 95, C). Ticks

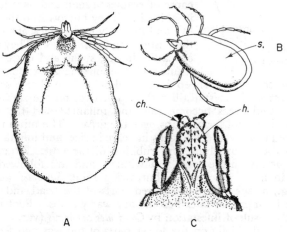

FIG. 95.—SHEEP TICK (*Ixodes ricinus*).

A, engorged female × 6: B, male : × 6 ; C, front of head to show mouthparts : × 36, *s*, shield; *p*, palp; *h*, hypostome; *ch*, chelicerae.

insert their mouth parts into the host and suck blood. Females are flattened dorso- ventrally with a distinct black shield on the thorax ; after feeding the abdomen is distended with blood, and engorged adults may attain a length of 8 mm (Fig. 95, A). When they finish feeding they fall to the ground where they lay eggs. The eggs hatch in autumn and the young ticks, which hibernate without feeding, have only three pairs of legs. After the first meal in spring they leave their hosts and moult and the fourth pair of legs is revealed. After each moult they seek a new host. Male ticks (Fig. 95, B) are small, black and completely covered by the dorsal shield. The feeding of ticks causes irritation and inflammation and they also transmit tick-borne fever and louping ill in sheep and red-water fever and tick-borne fever in cattle.

Control Measures.   Ticks require abundant moisture and a high degree of humidity in summer for their survival, and they are only numerous in districts where rank vegetation covers water-holding soil. This suggested that the improvement of hill pastures by draining, liming and the application of basic slag, and the removal of gorse and bracken cover by cutting and burning where this is possible would do much to create un-favourable conditions for ticks ; and where such measures have been carried out, trouble from ticks has abated. The use of the new sheep dips can reduce the tick problem.

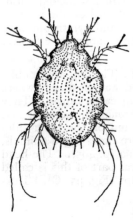

FIG. 96.—PSOROPTIC MITE OF SHEEP (*Psoroptes communis ovis*) origerous female : × 50.

Itch Mites, Scab Mites and Mange Mites.   Itch mites, scab mites and mange mites, microscopic members of Acari, are the cause of scabies in man and parasitic mange in animals. They burrow into the skin and live and multiply under the scabs that cover the pustulating feeding sites. They are found on cattle, horses, swine and dogs, and con-stitute such a serious menace to sheep-rearing that their control on sheep is compulsory. The sheep-scab mite, *Psoroptes communis (ovis)* Raillet (Fig. 96), lives on the surface of the skin. Infested areas become raw and inflamed and the yellowish discharge dries and forms scabs over the areas. The mites under the scabs breed rapidly and the scabs increase in size, and infestation is in-dicated by the scratching and rubbing of the host. Attacks occur most frequently on the back and withers, and are followed by loss of wool in large patches. A burrowing mite, *Sarcoptes scabei* (var. *ovis*) Meg., attacks the hair-covered parts of the head and produces small hard lumps that later become dry warty scabs. Foot mange of sheep is the result of infestation by *Chorioptes communis* (var. *ovis*) Verh., which is found chiefly on the lower parts of the legs and feet. Foot mange does not spread rapidly through the flock but it causes great discomfort and pain to the victims.

Control Measures. Animals infested by sheep scab mites frequently rub themselves against posts, palings and walls in their efforts to relieve the irritation. In this way scabs become detached and mites may spread to other individuals in the flock. Indeed, so contagious is parasitic mange that the sarcoptic and psoroptic forms in both sheep and horses must be notified to the Ministry of Agriculture. Until 1948, arsenic-sulphur dips were used against sheep scab but were replaced by BHC dips which helped eradicate the disease in England, Wales and Scotland by 1952. BHC (HCH) was also used in controlling the outbreaks which occurred in England and Wales in the early seventies.

Poultry Mites. There are four mites troublesome to poultry— Red mite *Dermanyssus gallinae* Deg., the Northern fowl mite *Liponyssus sylviarum* C. & L., the Depluming mite *Cnemidocoptes gallinae* Raill. and the Scaly Leg mite *Cnemidocoptes mutans* Robin. Other mites may also be found associated either with food or litter, and though they may sometimes give rise to alarm they are not dangerous and can usually be cleared up by good hygiene.

The red mite is the most important of the four and attacks most species of poultry, especially hens and turkeys. It may live temporarily on domestic animals. Feeding takes place at night while the birds are on their perches and during the day the mites return to crevices in walls and perches. The engorged mite is red, but as digestion goes on it becomes grey in colour. Eggs are laid in crevices and in hot weather the life cycle may be completed in a week, hence build-up may be rapid. Infestations may be detected by the reddish black excremental marks most easily seen on the eggs. The fowls may suffer severely from blood loss if control measures are not quickly taken. These consist of the thorough cleansing and creosoting of the house followed by malathion, carbaryl, dichlorvos and bromophos all having been found to be very effective.

The northern fowl mite closely resembles the red mite but is permanently attached to its host ; hence it has to be dealt with by use on the birds of suitable and safe insecticides. Scaly leg mite, which burrows into the tissues, is not common and can be treated by carefully dipping the affected areas only into paraffin ; care should be taken to dip only the scaly part of the leg since paraffin irritates skin.

The depluming mites cause intense irritation and this results in the birds pulling out their feathers. When infestations are heavy the mites spread very rapidly. Control is difficult, infested birds must be isolated, and house disinfestation is the first measure to take.

Chapter XV

PLANT BREEDING

PLANT breeding applied to agricultural plants is a method of crop improvement, and as such it seeks to add to the efficiency and productivity of crop husbandry. There are many ways by which plant breeding can make positive contributions to crop husbandry, and the methods used vary according to the crop in question and the aims of the breeder, who must, if he is to contribute anything of value, study and appreciate the problems and requirements of the agriculturist. The great potentialities of plant breeding lie in the possibilities of offering the grower something new which either possesses advantages not found in the types already in cultivation, or which offers scope for cultivation under conditions which have hitherto been impossible or unprofitable for the particular crop. The greatest value of plant breeding is, therefore, its creative powers in the controlled production of desirable types, but it may also play an essential part in maintaining the standard of the varieties, strains and stocks already in cultivation. This latter function, however, important as it is for intensive and efficient crop husbandry, does not bring into play the essential feature of crop improvement which is the true role of plant breeding.

THE SCIENTIFIC BASIS OF BREEDING

Ever since plants were first taken into cultivation they have been subjected to unconscious or conscious improvement by growers, but it is only within recent historical times that plant breeding as a method of crop improvement has been developed intensively and organised on a rational basis. It was not until the discovery of sexuality in plants, and the understanding that came subsequently of the reproductive processes and life-cycle phenomena, that attention was focused on the causes of variability in plants and the possibilities of improving cultivated forms, particularly by creating new types. In the eighteenth century it was still a matter for debate as to whether all new varieties of cultivated plants arose from the variability of the soil in which they grew, or whether such things as natural hybridisation played a part ; and it was not until artificial hybridisation was successfully accomplished and the progeny studied that the causes of variability in plants were more clearly appreciated and the possibilities of controlled breeding were visualised.

Towards the end of the 18th century the idea of plant characters being inherited first attracted serious attention from those interested in crop improvement ; early in the 19th century systematic hybridisation, with the study of the inheritance of individual characters, was a matter for investigation among a few interested workers. During this period, also, certain investigators became sufficiently knowledgeable and critical to realise that agricultural crops often consisted of a mixture of types, and that, by selection, new and improved varieties could be obtained. It also became apparent that the mixed nature of varieties as then cultivated was due partly to natural hybridisation, and that the only way to get rid of these " impurities " was to select from a single plant, or even a single grain. In this way the foundations of modern plant breeding were laid by the gradual accumulation of knowledge on reproduction, the causes of certain kinds of variability, the occurrence of natural hybridisation, the heritability of characters and the powers of artificial selection.

With this knowledge a great deal of valuable improvement of crop plants was effected, and agriculturists became very " variety conscious ". The means of improvement were showing themselves, but the methods still lacked certain essential knowledge to make them intelligible to the investigator and to show the full possibilities. This essential knowledge, which has done so much to enlighten the plant breeder and give him a better control of his methods and technique, as well as indicating the limitations and scope of plant breeding, has only come in the present century with the development of the sciences of genetics and cytology, and also with improved and more exact information on plant behaviour and physiology. The biological principles which have gradually been unfolded have done a great deal to replace in varying degrees the hit-or-miss methods of a plant breeding art by the more logical and controlled methods of scientific plant breeding which is using all the available proven knowledge and facts. That there are still serious limitations to the practical application of this knowledge to breeding will become apparent later in this account, but first it is necessary to consider the established principles and accepted theories on which modern methods of breeding are based. It will then be possible to discuss the application of these breeding methods to the improvement of the different kinds of crop plants and thereby to assess the value of the scientific knowledge available.

METHODS OF REPRODUCTION

Sexual reproduction is the normal method of reproduction in higher plants, but some species have developed asexual or vegetative methods in addition to, or in some rare cases more or less in place of, sexual reproduction. Asexual, or vegetative, reproduction may be used as an artificial means of propagating certain cultivated plants, such as apples, plums and cherries, which are naturally reproduced sexually, but in the case of the potato the natural vegetative means is also the plant breeder's and cultivator's method of propagation.

The higher plants differ from the higher animals in that the former are commonly hermaphrodite, or bisexual, while the latter are invariably unisexual. This means that in most plant species both male and female mature germ-cells, or reproductive cells (which are termed the *gametes* in sexual physiology and genetics), are produced on one plant. This consequently makes it possible for bisexual plants to be self-fertilised ; but self-fertilisation is by no means the invariable rule even in plants which bear bisexual flowers, and in hermaphrodite plants with unisexual flowers cross-fertilisation is common. In those plant species in which individual plants are unisexual, cross-fertilisation is a necessity for seed setting, and only the female plants bear seed and fruit. Thus reproduction in plants shows conditions varying from almost complete self-fertilisation to absolute cross-fertilisation, depending in the first place on the distribution of the sexes in the flowers and on the plants.

These fertilisation relationships are sometimes complicated by various kinds of *sterility* and *incompatibility*, which interfere with normal sexual relationships in certain plant species. Sterility may be *morphological*, due to the abortion or suppression of the stamens or ovaries ; or it may be due to a failure in the normal development of pollen, ovules, or early seed development, in which case it is described as *generational*. Sexual incompatibility, which may be partial or total, is not caused by any functional failure of the sexual parts or post-fertilisation development, but is due to partial or complete failure of pollen to reach the ovule and bring about fertilisation, in spite of there being sufficient pollen on the flower stigmas. It is possible to speak of self- and cross-fertility, and of self- and cross-incompatibility, according to whether one or more varieties, strains or species are involved.

It is obvious that all phenomena concerned with the method and mechanism of plant reproduction are the basis of methods of plant breeding, because it is only through the plant's reproductive capacity that propagation can be effected and breeding methods and technique can be devised. The breeder must know all the facts concerned, and in particular he must appreciate the distinction between asexually and sexually reproduced types, the amount of self- and cross-fertilisation, and the incidence of sterility and incompatibility.

HERITABLE AND NON-HERITABLE VARIATIONS

It is characteristic of all living organisms to show variability, and no two individuals are exactly alike. The visual and non-visual characters and characteristics of any individual at all stages of growth and development are the product of its hereditary constitution and of its environment, the latter modifying the former and causing *non-heritable variations*, *modifications* or *fluctuations*, which are not handed on to the progeny. It is well known that even identical twins are not identical in every respect, and the relative effects of " nature " and " nurture " have long been a problem of discussion and study and have considerable practical significance.

It was the recognition of the fact that selection within the progeny of pure breeding individuals cannot lead to any inherited improvement, because the variations are not heritable, that led to the concept of " pure lines ". This concept has had very important repercussions on plant breeding methods in relation to methods of selection and the basis for selection technique. It should be pointed out, however, that many so-called " pure lines " are in effect nothing more than " single plant selections " and are not necessarily " pure " in the genetic sense.

The recognition of the distinct nature of heritable and non-heritable variations was a great step forward for breeders of plants and animals. Early investigators were much hampered by the lack of any clear understanding of this distinction, although it was recognised that certain characters were transmitted from parent to offspring. On the other hand, although " like tends to beget like " through sexual reproduction, and in asexual reproduction the vegetatively produced individuals are all possessed of the same hereditary characters, variability is one of the essential characteristics of living organisms. This is really only another way of saying that it is not characters as such which are inherited, but merely the ability and tendency to develop and manifest in a particular way under different environments ; and the circumstances of the environment may not only modify the expression of individual characters, but may actually determine whether or not certain characters ever express themselves at all.

Heredity, through sexual reproduction, is a potent cause of variation by bringing about the recombination of parental characters and the production of new types in the offspring. Hereditary variation brought about by hybridisation is one means by which evolution has taken place through natural selection, and it is also the hope of the breeder who is striving to create new and improved forms. The early hybridists were bewildered by the prolific production of new types by hybridisation, and most of them failed to profit by this method of breeding because they had no knowledge of the mechanism of inheritance and consequently did not understand how to handle hybrid progenies.

MENDELISM

The first successful scientific attempt to explain the mechanism of inheritance resulted from Mendel's experiments on hybridisation, as a result of which he formulated certain " laws ". Mendelism is often loosely thought of as synonymous with heredity and the modern science of genetics, but this is not so, although Mendel was the first investigator to analyse satisfactorily the mode of inheritance of certain characters. As a result of his experiments Mendel was able to establish, not only certain principles governing a particular type of inheritance (Mendelian inheritance), but also to indicate a mathematical precision in inheritance and a method of handling progenies which has provided the basis of breeding techniques.

Mendel worked with the garden pea, which is a self-pollinating plant, and he studied the inheritance of green and yellow seed leaves (cotyledons), grey and brown seed coat, round and wrinkled seeds, green and yellow pod, purple and white flowers, distribution of the flowers, and tall and dwarf plants. The significant features of Mendel's work were that he studied the inheritance of each character separately ; that he observed critically the characters in the first hybrid generation ; that he counted the proportions of the plants with the different characters in the second generation ; and that he followed the hereditary behaviour into the third generation. During this work he appreciated the necessity for keeping the progeny of individual plants separate ; in this way he was able to study the breeding behaviour of single plants with regard to the particular characters, and he used this information to interpret the inheritance of the individual characters in the whole progeny. Not content with this evidence, Mendel tested and verified his hypothesis concerning the mechanism of the inheritance of these characters by a full analysis of the second-generation progenies, and thereby introduced the important technique of progeny testing which is now regarded as essential for all work on heredity and breeding.

Mendel's contribution to the study of inheritance is of outstanding importance because he established three important concepts and a fundamental technique. In addition, he demonstrated that the inheritance of certain characters, as expressed in the second-generation progeny, is mathematically so precise that it can be represented in absolute ratios according to the numerical representation of the characters in the individuals of the progeny.

The three Mendelian concepts are first, that the characters he studied behaved as separate units in inheritance. Second, that in the second generation the characters are independently assorted and that they separate or " segregate " at random among the individuals of the progeny. Third, that a particular character expression can dominate an alternative character expression so that, although both expressions may be carried by one individual, only the " dominant " character expression can show itself.

In addition to observing the way in which the characters he studied were inherited, Mendel provided the explanation in terms of the hereditary constitution of the male and female gametes, and in this way for the first time associated hereditary behaviour with the processes involved in sexual reproduction.

Mendel had observed that constant forms appear in the progeny of hybrids, and that this happened with regard to all combinations of the associated characters. He confirmed that " constant progeny can only be formed when the egg cells and fertilising pollen are of like character, so that both are provided with the material for creating quite similar individuals " and he amplified this by stating, " we must therefore regard it as certain that exactly similar factors must be at work also in the production of the constant forms in the hybrid plants. Since the various constant forms are produced in one plant, or even in one flower of a plant, the conclusion appears logical that in the ovaries of the hybrids there are formed as many sorts of egg cells, and in the others as many sorts of pollen cells, as there are possible constant combination forms, and that these egg and pollen cells agree in their internal composition with those of the separate forms. In point of fact, it is possible to demonstrate theoretically that this hypothesis would fully suffice to account for the development of the hybrids in the separate generations, if we might at the same time assume that the various kinds of egg and pollen cells were formed in the hybrids on the average in equal numbers."

Mendel confirmed these assumptions also by carefully planned experiments and thus completed the evidence on which his principles are based, and from which the science of genetics has been developed.

GENETICS AND CYTOLOGY

Although Mendel's account of his work was published in 1866 it attracted little attention until 1900, when three investigators (de Vries, Correns and Tschermak) gave independent accounts of their own experiments in which they were able to confirm Mendel's principles in plants. Two years later, Bateson and Cuernot showed that the inheritance of certain characters in animals could also be explained on Mendelian principles. With the accurate accounts which had just previously been given on the mechanism of fertilisation and the recognition of the importance of chromosomal activity in cell division and fertilisation, the essential basic facts for the development of genetics and cytology were realised and appreciated.

Such fundamental facts as the constancy of chromosome numbers, and the splitting of chromosomes in cell division in the body cells, were known before the re-discovery of Mendel's paper, and the idea of the cell nucleus as being the basis of heredity, with the later but still pre-1900 discovery of the reduction of the number of chromosomes in the reproductive cells, cleared the way for the rapid developments of the 20th century. This study of the cell and its contents, particularly the nucleus, with the cellular interpretation of the phenomenon of sexual reproduction, quickly suggested a working hypothesis to explain what was then known of inheritance and in 1902 the close association of chromosomal behaviour and Mendelian segregation was first recognised.

Chromosomes and Heredity. The parallel development of knowledge in cytology and heredity rapidly led to the formulation of the chromosomal theory of inheritance, which received further confirmation from the demonstration that there were definite sex chromosomes in higher plants and animals and that sex is an inherited character which is determined by chromosomal behaviour. It is now generally accepted that the chromosomes are the bearers of the factors which determine most hereditary tendencies and manifestations. This explanation gives a constructive basis for studying the principles of inheritance, although there are undoubtedly exceptions, particularly demonstrated in plants, of *extra-chromosomal* inheritance and maternal effects which are due to influences outside the chromosomal mechanism.

Genes. The chromosomal theory of inheritance is based on the concept of the chromosomes bearing, in a linear arrangement, *genes* or factors which determine the development of the characters of the particular organism, and also which control the manifestation of each character within the limits set by the particular environmental conditions. Each chromosome may carry many genes, and a particular character expression may be decided by one or more genes, so that the ultimate characteristics of any organism are decided not only by the presence of particular genes as they affect a particular character expression, but also by the interaction of these genes and the modifying effects of the environment. The simplest forms of inheritance are found when a character difference depends on a single gene difference, as Mendel found in the pea characters with which he worked.   But it is commonly the case that more than one gene is involved, and when this is the case the mode of inheritance may become difficult to interpret, particularly when the environment has a strong modifying effect on the expression of the character.

Meiosis and Mitosis.   If the chromosomes are the principal determiners of the hereditary constitution and behaviour of living organisms, it is obvious that their stability, behaviour and structure are of fundamental importance.   Every organism is characterised by a particular chromosomal complement, by which is meant a definite chromosome number composed of individual chromosomes with a characteristic and specific structure.   Of the total number of chromosomes present in the body cells of any individual, half are contributed by each parent when the two reproductive cells or *gametes* come together to form the *zygote* on fertilisation.   It is clear that if the chromosome number is to remain the same from generation to generation, and not to double with each successive fertilisation, there must be a special mechanism to provide for this.   The means by which this is brought about is that in the formation of the reproductive cells the number of chromosomes is halved by a *reduction division* during the nuclear divisions (Meiosis) which take place in the formation of gametes in the reproductive organs.   All other nuclear divisions which take place when the zygote develops after fertilisation are characterised by a longitudinal splitting of each chromosome, so that each body cell has represented in it the hereditary material of each and every chromosome.   This type of nuclear division is termed Mitosis, and by this means every body cell of the individual carries an identical chromosomal complement.   (See Plates XXXI, XXXII.)

During mitotic division the nucleus of the cell passes through a number of distinct phases which take place in an orderly and invariable sequence in normal divisions.   The characteristic appearance of the " resting " nucleus shows a dense granular structure, spherical in shape, more or less in the centre of the cell (Plate XXXI, Fig. 1).

PL. XXXI.   STAGES IN THE DIVISION OF A CELL
NUCLEUS : MITOSIS

1. Cell with nucleus in resting stage.
2. Chromosomal threads showing double structure (prophase).
3. Later stage showing contraction and thickening of chromosomal threads (prophase).   The individual chromosomes are still too long for separate identification.
4. Final contraction of chromosomes and arrangement on equatorial plate : side view (metaphase).
5. Daughter chromosomes moving to opposite poles of spindle : side view (anaphase).
6. Two daughter cells formed from the division, with the dividing cell wall evident.
7. The ten chromosomes (two sets of five) of a species of *Trillium*.   These show the structural differences between different homologous pairs of the complement.
8. The forty-eight chromosomes of the domestic plum.
9. The twenty-four chromosomes of the Sitka spruce.

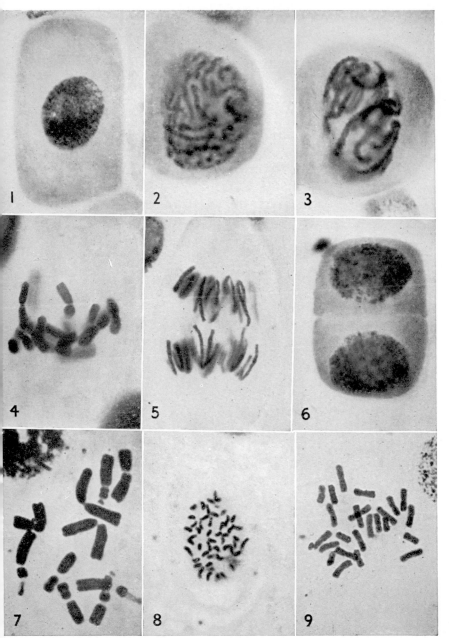

PLATE XXXI. STAGES IN THE DIVISION OF A CELL NUCLEUS—MITOSIS
Full legend on opposite page

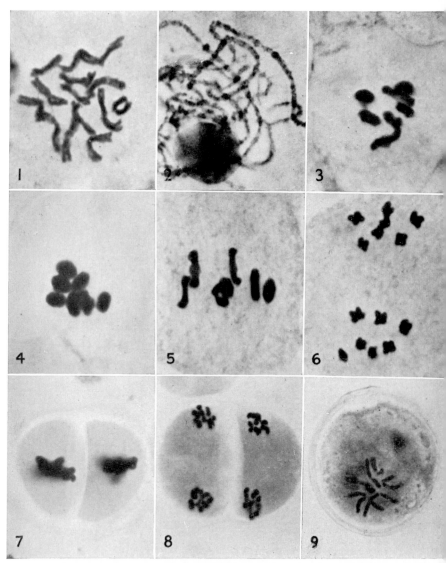

PLATE XXXII. STAGES IN THE DIVISION OF A CELL NUCLEUS—MEIOSIS
Full legend on opposite page

During the first phase (*prophase*) this granular structure resolves itself into a number of double threads ; these are the chromosomes, which have divided longitudinally at an earlier stage (Figs. 2 and 3). Subsequent continued coiling results in contraction and thickening of the chromosomal thread. When the chromosomes have reached their maximum state of contraction, the membrane round the nucleus disappears, the double threads become arranged on a flat plate across the cell (*metaphase*), and the so-called spindle is formed, the equator of which is the plate (Fig. 4). The " daughter " chromosomes formed by the splitting of each individual chromosome now separate, and one of each twin moves to the opposite end of the cell where the poles of the spindle are situated (Fig. 5). This is termed the *anaphase*, and the net result is that two identical chromosome groups are formed which are also exact replicas of the original. The chromosomes then uncoil and the original granular nature of the nucleus is restored (*telophase*) in the two daughter nuclei between which a new cell wall is formed (Fig. 6).

The characteristic structure of chromosomes in individual species of plants can best be seen by examining nuclei from the polar view, when the number and special features of the chromosomes can be seen as the latter lie on the spindle plate before separation to the poles (Figs. 7, 8, 9). At this stage, also, the double nature of the chromosomes formed by splitting can commonly be distinguished

PL. XXXII. STAGES IN THE DIVISION OF A CELL NUCLEUS : MEIOSIS IN *LOLIUM* (RYE GRASS)

(*Only Figs. 2–8 show stages of Meiosis. Figs. 1 and 9 illustrate mitosis, Fig. 1 to show diploid chromosome number for comparison, and Fig. 9 to show later development of pollen grain.*)

1. The fourteen chromosomes (two sets of seven) from vegetative (somatic) tissue to show the diploid chromosome number.
2. The pairing of homologous chromosomes, which at this stage are long, thin strands (first division prophase).
3. Paired chromosomes (seven pairs) after contraction. The pairs are called bivalents.
4. The seven bivalents on the equatorial plate : end or polar view (first division metaphase).
5. The seven bivalents on the equatorial plate : side view (first division metaphase as in Fig. 4 but slightly flattened).
6. The seven daughter bivalents formed at each end of the cell as the result of reduction division (first division late anaphase).
7. The daughter bivalents arranged on the two plates preparatory to the second (mitotic) division (metaphase, side view).
8. The resulting four cells (the tetrad) each with half the somatic chromosome number. Each chromosome shows a middle constriction (second division telophase).
9. One pollen grain with the half chromosome number (seven). Each pollen grain will usually have a different combination of genes due to exchanges of partner within the bivalent and also due to the random orientation of the bivalents during division. Mitosis is taking place in this pollen grain, with the production of the generative and tube nuclei.

very clearly, while the distinct structure of individual chromosomes within the complement of the plant is sometimes very striking (Fig. 7).

Linkage and Crossing Over. It is by means of reduction division that segregation of genes, and consequently of the hereditary characters that they control, takes place. The random assortment of the half, or *haploid*, number of maternal and paternal chromosomes which follows, means that reproductive cells of many different genetic constitutions are formed according to the particular chromosomes that they carry. But there cannot be a complete random assortment of all the hereditary characters at this time because the genes which are carried on any one chromosome tend to be *linked*, and the closer together they are on the chromosome, the closer will this linkage be.

However, in addition to the random assortment of individual chromosomes at meiosis, there is an additional source of variability in the kinds of reproductive cells formed by reason of the peculiar behaviour of the chromosomes. During meiosis, maternal and paternal chromosomes of similar structure (*homologous pairs*) pair and become physically connected, and during the close pairing transverse breaks may occur in the chromosomes, and when they join up different pieces may come together to form a new chromosome. By this means, *crossing over* of chromosomal pieces can take place, and the pairs of chromosomes come out of the pairing with one or more parts of their lengths interchanged. In this way there is a recombination of parts of similar chromosomes, and structural changes in the chromosomes result (see Plate XXXII).

Meiosis shows a number of distinct phases which are characteristic of this form of nuclear division and which normally take place in an orderly and invariable sequence as is the case in mitosis. Nuclear division begins before the chromosomes have split in the nuclei of cells which are going to form gametes, the homologous pairs come together and pair side by side (Plate XXXII, Fig. 2). At this stage the chromosomes are long thin strands which are differentiated along their lengths, and they then divide to give four strands, thus providing opportunity for exchanges of partner within the paired chromosomes by means of breakage and reunion of strands as mentioned in the previous paragraph. The chromosomes then contract and thicken, the pairs still lying together, after which they become arranged on the spindle plate (Figs. 3 and 4); then one chromosome (still showing its double structure through splitting) moves to opposite ends of the cell (Fig. 5). This results in half the original chromosome numbers aggregating at each end (Fig. 6), and then by a subsequent nuclear division, during which the chromosomes split and separate, each daughter nucleus forms two nuclei (Fig. 7). The two divisions thus result in four daughter nuclei, each with half the original chromosome number (Fig. 8), and each representing a potential gamete as in the case of the pollen grain illustrated in Fig. 9. It may be seen, therefore, that meiosis consists of two divisions of the nucleus with only one division of the chromosomes.

Homozygosity and Heterozygosity. The mode of inheritance of characters is studied by analysing the progeny in the first, second, third and possibly subsequent generations after hybridisation. Genetics is therefore essentially progeny testing and analysis based on quantitative data and capable of mathematical expression in terms of numbers and ratios. In order to study critically the mode of inheritance of any character it is necessary to choose two parental types that are true breeding for the contrasting character expressions, as Mendel did with his purple- and white-flowered peas. Such pure breeding individuals are said to be *homozygous* for the particular character, while individuals which are not true breeding are *heterozygous*. For example, the first generation hybrid is heterozygous for all the characters in which the two parents differ, and such hybrids will not breed true for those characters. But in the second generation and subsequently, a certain proportion of true breeding forms will be formed, the number depending on the number of genes which control the expression of the character.

Genotype and Phenotype. The genetic constitution, or *genotype*, of any individual cannot be determined by its appearance, or *phenotype*, because it may not be true breeding for one or more of the characters expressed in its body or by its behaviour, and it may be carrying *recessive* genes which cannot express themselves because of the presence of *dominant* genes. It may also happen that there may be an *inhibiting* gene, or that the expression of a character depends on *complementary* genes or complex *gene interaction* and the presence of *modifiers*. All these possibilities make it necessary to concentrate on genotypic constitution as shown by progeny testing and breeding behaviour when studying inheritance, and they are absolutely essential to bear in mind when applying genetic knowledge to practical breeding.

Polyploidy. Further complications can arise in genetic behaviour through more than two sets of chromosomes being present in any individual. This condition is known as *polyploidy*, and organisms having more than the normal *diploid* chromosomal complement are called *polyploids*. Polyploidy is rare among animals, being absent in higher animals, and occurring regularly only in certain lower animals that are hermaphroditic or which reproduce asexually (partheno-genetically). Polyploid forms are, however, common among higher plants, certain families and genera tending to be prolific in their production, and the condition is particularly characteristic of certain groups of cultivated plants. In the Gramineae, for example, the genus *Triticum* (wheat) has diploid species as well as species with four sets of chromosomes (tetraploids) and species with six sets of chromosomes (hexaploids). There are other genera, however, which show a wider range of polyploidy than does wheat, and very high multiples occur in such crops as sugar cane. Polyploidy has undoubtedly played a most important part in the evolution of some cultivated plants, and it has a very significant effect on plant breeding. Not only does it influence hereditary behaviour, but it also introduces complications in relation to the crossability of forms possessing different chromosome numbers, while the artificial production of polyploids is a means of overcoming hybrid sterility and a method of creating new forms.

THE APPLICATION OF GENETICS AND CYTOLOGY

In the early days of Mendelian and genetic experimentation it was thought that the problems of plant, and to a lesser extent of animal, breeding methods were solved, and it only remained for breeding material to be subjected to a complete genetical investigation for the way to be made clear for the creation of new desired types at will. But breeding has never become merely a matter of applied genetics, although with cytology, genetical science has supplied an invaluable means for obtaining a greater measure of control over breeding material, while the genetical and cytological approach to some breeding problems has yielded astonishingly profitable results in practical plant breeding. Methods of breeding are to a large extent based on the knowledge which has accumulated in genetical research, while cytological phenomena have not only provided the explanation of certain genetic principles, but they have also suggested the means for the development of new techniques in breeding methods. But it is well to remember that the methods of the geneticist and the cytologist on the one hand, and of the practical breeder on the other, are quite distinct ; for whereas in the former case the object is interpretation and analysis, in the latter the objective is selective and creative with the production of new types of practical worth.

Naturally a knowledge of the fundamental concepts and the most up-to-date developments in genetics and cytology is a virtual necessity for the breeder, although the extent to which both the knowledge and the developments can be applied to the actual breeding methods and technique must vary considerably with the plant or animal in question, the problems involved, and the objectives. In some agricultural and horticultural plants the whole basis of the breeding rests on exact genetical and cytological knowledge, and the limitations and technical complications of the breeding technique are firmly grounded on a genetical and cytological approach. At the other extreme the breeder has little to guide him other than his general knowledge and experience of the crop, combined with his skill in the art of plant breeding.

BOTANICAL RELATIONSHIPS

The breeding of any crop requires more than a thorough knowledge of the agricultural varieties already in existence and cultivation. The breeder needs to be familiar with all the botanical variations available in the species and genus with which he is concerned, as well as understanding the relationships with allied groups of forms in different species and genera. It is only through such knowledge that the possibility or limitations of transferring valuable characters from one group of plants to another can be decided, and the possibilities of improvement by " wide-crossing " or " distant hybridisation " be assessed. If such knowledge is not available, and such possibilities are not taken into consideration, the breeder is immediately imposing limitations on his methods and the scope of the improvements which he can visualise.

PHYSIOLOGICAL AND DEVELOPMENTAL PHENOMENA

The study of the hereditary behaviour of plant characters has often tended to obscure a proper knowledge and understanding of the characters themselves. Genetic knowledge is quite inadequate without a fundamental knowledge of the characters, particularly when these characters are of a complex physiological nature affecting growth, development, yield and the various important attributes included in " quality ". Great advances have been made in the knowledge of some of these plant characters and attributes during recent years, and their inter-relationship or incompatibility are matters of prime importance to the breeder. It is seldom that the plant breeder is not concerned with physiological problems sooner or later, and in particular is he concerned with the plant and its environment and the fundamental processes governing growth at all stages. The morphological manifestations of physiological activity, such as ear size and weight, and grain characters in cereals, or root conformation, weight and sugar content in sugar beet, are equally important as subjects of study and investigation, and in all cases must there be as complete an understanding as possible of all the plant characters.

GROWER AND CONSUMER REQUIREMENTS

Scientific knowledge combined with skill as a breeder are essential for success in plant breeding, but by themselves they are insufficient. The improvement of agricultural plants requires a close and intimate association with agriculture, so that the problems of crop production can be readily appreciated. To be most effective the breeder must set out to provide the grower with new varieties or strains of crop plants which contribute to the efficiency of production either by the improvement of field characters, or by meeting the requirements of the consumer more satisfactorily by improvement of the quality of the product. It is by either or both of these means that plant breeding can bring about progressive improvements in crop husbandry by supplying the grower with better material for cultivation.

METHODS OF PLANT BREEDING

The methods employed in the breeding of improved varieties and strains of agricultural plants are determined in the first place by the method or methods of reproduction, the fertility and incompatibility relationships, and the life cycle phenomena of the plant or plants with which the breeder is working. In addition to these biological considerations, the breeder will have to plan his methods in relation to the kind of improvement he has in mind and the material with which he is going to work, or which is available to serve as a basis for his investigations.

The following types of sexually reproduced plants as affecting the methods of breeding can be distinguished :—

    1. Normally self-pollinated, and showing no intra-specific self-sterility complications, e.g. wheat, barley, oats.

2. Normally cross-pollinated, and showing no self-sterility complications, e.g. maize.
3. Normally cross-pollinated, but showing varying degrees of self-sterility complications, e.g. sugar beet, grasses and clovers.
4. Mostly self-pollinated, but with a high degree of cross-pollination, e.g. beans.
5. Invariably cross-pollinated due to unisexual individuals, e.g. hops.

There are, in addition, those species which are normally reproduced vegetatively and which show complete or partial sexual sterility according to the variety, as in the domestic potato ; while in rare cases, as for example in smooth-stalked meadow grass, seed is produced without normal sexual fertilisation (i.e. apomictically).

These reproductive phenomena determine not only the planning of breeding programmes, the material that can be used and the methods to be adopted, but also the finer points of handling the breeding material and the technical procedure during the carrying out of the programme once it has been planned.

But, in spite of these complications (which after all are primarily concerned with deciding and modifying the technique of breeding), there are only two important fundamental methods of breeding crop plants, and these two methods are applicable to all the common crops. Although it is convenient to consider the two methods separately, it is seldom that either is used exclusively for any particular crop ; one method may serve as the basis of improvement, while at the same time incorporating the other. We may, therefore, consider breeding methods and their application to common farm crops under the major headings of Selection and Hybridisation. In the course of the following discussion it will become apparent that these methods are not mutually exclusive, but that most breeding involves the use of both.

SELECTION

It may be said that some form of selection is the basis of all plant breeding ; but, for the sake of clarity, selection may here be considered without the complications of hybridisation or other methods of creating new types and greater variability on which to practise selection. For the moment, therefore, we may consider selection as a means of improvement by the exploitation of already existing variability in mixed populations. The various kinds of selection applicable to different crops are as follows :—

Single Plant Selection. This may be used in mixed populations of self-pollinating crops, and is the first and most obvious means for seeking improvement when starting the breeding of such crops. Before intensive breeding of such crops as the cereals was organised extensively in this country, a large number of improved varieties was produced by selecting the outstanding plants within commercial varieties, and multiplying new stocks by propagating from single plants. By this means improved stocks of wheat, oats and barley were put on the market ; in more recent times Victory oats, Squarehead's Master 13/4 wheat, the various kinds of Archer barley and the early selection of Spratt-Archer barley named Earl were developed by this type of selection.

The scope for this type of improvement naturally depends on the amount of variability present in the stocks that are used for selection. In some cases, where stocks are obviously composed of easily discernible mixtures, improvement may result from " purifying " these stocks, or by picking out particularly outstanding plants similar to the bulk of the stock. Where stocks are apparently " pure " it may be necessary to select apparently similar types, and test the progenies of the single plants with great care to see whether any improvement can be obtained in such characters as yield, quality or perhaps resistance to lodging. The ease with which improvement can be obtained, and the type of improvement possible, will therefore depend on the variation in the stock before selection starts ; obviously the chances of improvement become less as the stock is progressively purified. The stocks of cereals in this country at the moment offer little scope for any great improvement in this way because all the standard varieties have been subjected to intensive selection.

In the past, great improvements in certain cereals have been effected by the chance selection of isolated plants of unknown origin, as in the case of Potato oat, Chevalier and Goldthorpe barley, and possibly also Squarehead's Master wheat, although there is some doubt of the origin of this wheat variety. The possibility of bringing about radical improvement by such chance selections is naturally becoming less and less as the commercial stocks and the standard of the cultivated varieties improve, but the history of our cereal varieties shows the important part played by this means.

Mass Selection. This may be practised in both self-pollinating and cross-pollinating crops. In the self-pollinating cereals (wheat, barley and oats), the best plants typical of the variety may be selected and bulked together to form a new nucleus stock for multiplication. Such selection rarely leads to any radical improvement, but the method may be used to maintain the stock at a satisfactory level.

In cross-pollinated crops, such as rye and sugar beet, a modified form of mass selection is a standard method for improvement. The individual plants are normally grown on and the value of each selection is judged by the behaviour of the progeny, i.e. " progeny testing ". The lines, or progenies, thus established may be inbred for a varying number of generations to test their performance, and this is normally accompanied by a loss in vigour varying in amount with different crops and within the lines of any one variety of a crop. When the lines have eventually been chosen, they are all grown on in one isolation plot and allowed to interpollinate to produce the new stock of selected seed.

Group Selection. In herbage plants the basis of strain (i.e. variety) building may be the selection of small groups representing a desired type, e.g. hay or grazing type, that it is desired to establish. Single plants showing the desired characters are grown together and their type is tested and judged by multiplying the individual plants vegetatively to establish *clones* which offer a better opportunity for selection of the right type.   The selections may be inbred for a few generations to obtain a better idea of their genetic constitutions before allowing interpollination between the selections ; or alternatively the most satisfactory clones may be allowed to interpollinate to produce the stock seed of the new strain for trial.

In both mass and group selection breeding, the value of the individual lines for serving as the basis of new stocks may be tested by hybridising the lines in pairs.   In this way the breeding value of the lines can be judged in various combinations, and the expression of any " hybrid vigour " can be observed.   This is a most important procedure in the breeding of all cross-pollinated crops, and it has been found that not only does the compatibility between different lines vary, but also the hybrid vigour expression.   On this basis the best combinations of individual lines for cross-pollination can be utilised for establishing a new strain.

It is obvious that the basis of selection, and the technique of handling the material, will depend on the life cycle of the crop plant under investigation and the nature of the character or characters that are being studied.   In the case of self-pollinated annuals such as wheat, barley and oats, selection on all the important characters may proceed during each growing season in an uninterrupted sequence. But where the crop is a biennial, as in sugar beet, mangels and certain Brassicas, selection for the important field characters is made in one year, and then the selections have to be grown for seed the following year before the material can be tested again.   Further complications may be experienced in such biennials where the actual seeding habit and seed characters themselves are important for selection study in addition to the actual crop product itself.   This is the case in sugar beet, where not only must selection be based on the root, but also on the type of seeding plant and the morphology and physiology of the " clusters " (i.e. the commercial seed).

In perennial herbage plants it is necessary to continue selection of the same material over a number of years before the true value of the material can be judged, particularly when the breeder's judgment and tests have to be based on sustained productivity over a number of years, and on persistence under different systems of management.   In these circumstances it may take years before even initial selection of basic material for breeding can be satisfactorily achieved, and progress is necessarily slow.   There is, however, the advantage in herbage plant breeding of there being a vast reserve of naturally occurring variation on which the breeder can practise selection without the necessity of creating new variability by controlled hybridisation.

It is seldom that any breeding material or any form of selection for direct commercial improvement allows the breeder to concentrate on one character only.   Such a luxury can only be experienced when the material is stable for all the important field and commercial characters which are likely to affect the value of the selection product, other than the one on which the breeder is selecting.   In most cases, however, attention has to be paid to all the economic characters during the period of selection to prevent selection of one character being made at the expense of losing the best expression of other characters.

HYBRIDISATION

The use of hybridisation as a method of crop improvement is commonly directed to three main aims—the transference of one or more desirable characters ; the increase of the range of variability by introducing various character recombinations ; and the direct and controlled exploitation of hybrid vigour. In the first two cases the main objective sought is increased variability by combining characters of more than one individual in a single plant ; while in the third case the aim is either to restore vigour which has been lost by inbreeding, or to improve the level of productivity by concentrating the desired genetic factors.

As a method of crop improvement, hybridisation is effective only in so far as it brings about the kind of character recombination or expression of greater productivity and vigour that is desired ; and it must be accompanied by efficient selection. Haphazard and random hybridisation as a mere speculation is seldom effective, and all breeding based on hybridisation should be made with a clear objective. To plan improvement in this way, therefore, it is first necessary to select the parental material that is most likely to give the type of improvement required, and before this can be done the full range of potential parents must be known intimately. With reliable knowledge of this kind, and with a definite object in view, selection in the hybrid generations which follow can be most efficiently carried out. The success of any hybridisation experiment is determined by the thoroughness and efficiency of this subsequent selection, provided always that the choice of the parents was a happy one, and that the objective was practicable.

According to the relationship between the parents chosen, hybridisation may be planned under the following categories :

Intra-Varietal (and Intra-Strain) Hybridisation. It has been claimed that hybridising selected individuals within one variety of a self-pollinated crop can bring about improvement, particularly by enhancing yield. It seems clear that such improvement is only possible if the so-called variety is in reality a mixture of different genotypes which on hybridisation produces new combinations showing better combinations of economic characters. No effective improvement by this means has been recorded in this country where the standard of purity of reliable stocks of commercial varieties is high.

In cross-pollinated crops, however, such as sugar beet, grasses and clovers, where the varieties are genetically heterogeneous, controlled hybridisation within the varieties is used to maintain and improve individual varieties. The possibilities of improvement by this method can be readily appreciated by what has already been said of the genetic variability of these cross-pollinated populations ; the scope of the improvement possible depending on this genetic variability, the effectiveness of the selection of lines for hybridisation, and the extent to which compatible combinations are obtained which show increased vigour after selection of the lines by inbreeding. Intra-strain hybridisation is a standard and recognised method of breeding certain types of cross-pollinated crops, and the so-called "hybrid maize" developed in the USA is an outstanding example of this type of improvement.

Inter-Varietal Hybridisation. The basis of the improvement of most self-pollinating crops is at present inter-varietal hybridisation, and by far the greatest number of new varieties at present in cultivation has been produced in this way. In most cases, also, successive improvements have been built up by an orderly system of planned breeding whereby one improvement has been used as the means of the next step forward by intensive schemes of crossbreeding within related forms. The most important progress has also resulted where breeders engaged in particular areas and engaged in special lines of improvement, have concentrated on the use of specially valuable parental varieties which have proved their worth in breeding. These features are illustrated in the following cereal pedigree charts:

WINTER WHEAT

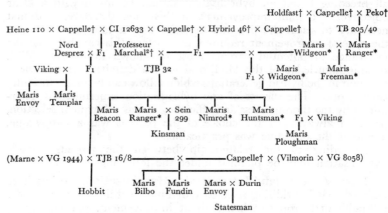

SPRING WHEAT

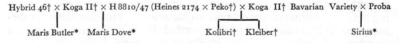

SPRING BARLEY

WINTER BARLEY

Proctor* × Pioneer†	259711 × (Ares × Hatif De Grignon)	Tria × (820 × 989)	Weihenstephaner Stamm × Dea
Maris Otter*	Astrix*	Sonja*	Senta†

(Hybrid 456 × Feebar) × Hatif De Grignon

Hoppel*

Complex Hybrid
(Emir, M. Concord†, C 205, M. Otter,* Carstens, France 7, M. Puma†)

Maris Trojan*

WINTER OATS

Blenda† × S. 172†——— × (S. 147† × 01747/10/7)		Manod† × Peniarth†
Maris Quest*	Peniarth*	Maris Osprey*

SPRING OATS

05443 × Condor† × Cebeco 725	Cc 4146/4 × Condor†	Astor† × Manod†
Mostyn*	Leanda*	Maris Tabard*	Maris Oberon*

Complex Cross of Five Varieties	Palu × Saxo
Nelson*	Selma*

* Currently recommended varieties by the National Institute of Agricultural Botany.
† Previously recommended varieties by the National Institute of Agricultural Botany.

These pedigrees show how some of the important varieties cultivated in Britain have been derived. Foreign bred varieties have been introduced into this country on a large scale, and some have achieved considerable success in agriculture. These varieties, and other foreign varieties possessing special characters of particular value, have been used in breeding programmes in this country, while current breeding work is tending to exploit a wider range of varieties.

Such methods must be accompanied by intensive selection after hybridisation. It is usual to make single plant selections in the second generation and to assess the progenies in the third generation purely by observation, at which time it is normal to select single plants again. Observation and selection of this type, accompanied by testing for disease resistance, may continue for several years—up to the sixth or seventh hybrid generation—until the desired type is " fixed " for all the observable characters. This stage is then followed by small scale yield trials, testing for quality and perhaps also for field characters when judged in bulk, for three years, by which time the value of the selections should be known with some precision if all the tests have been accurately conducted. It may be noted that the early generations of selection are judged very largely on the eye judgment of the breeder, and there is little else of a practicable nature to take the place of this judgment except in certain cases where controlled tests as for disease resistance can be used. Although the mode of inheritance of some or all of the characters with which the breeder is working may be known, this will help very little in the actual selection of the desired types in the field.   Moreover, it is more commonly the case that the complex agricultural and economic characters with which the breeder works are difficult to analyse genetically and are much affected in their expression by the growing conditions.

An alternative to the above method of successive single-plant selections in the early generations is to bulk the produce and take a sample for sowing in each of the first four or five generations, and only start to make single-plant selections subsequently.   This method allows the plants to become increasingly homozygous before selection, and also has certain advantages in allowing natural selection to act on the mixed population before " artificial " selection starts.   There is, however, the great disadvantage of not having the material under close selection control in the early generations, and the method is not generally popular.

With vegetatively propagated plants like the potato there is no need to continue selection until the hybrids are pure breeding or approaching the homozygous condition.   Indeed, homozygosity may be considered undesirable because the plants when heterozygous can be expected to show some hybrid vigour.   Accordingly the first generation hybrids are each bulked up and the so-called " seedlings " are judged and selected in relation to the particular characters for which the worker is breeding, and there is in consequence selection in one generation only.   This means, of course, that any new variety which may be put on the market in this way is a selected clone derived from a heterozygous first generation hybrid which remains stable only because it is vegetatively propagated.

Although this method of breeding has certain advantages for the breeder it has obvious disadvantages in that vast numbers of hybrid seedlings have to be raised in the search for improved types.   Potatoes, however, show a high degree of sexual sterility of one form or another in the great majority of varieties, and one of the great handicaps in potato breeding has always been the impossibility of using some of the best varieties for hybridising purposes, and the consequent restriction of the parental material as far as commercial varieties are concerned, to those which show sexual fertility.   This difficulty has been lessened in recent years by the raising of sexually fertile seedlings which can be used for further breeding.

Inter-Specific Hybridisation.   In spite of the great range of genetic variability that is available in the varieties of many cultivated species, the use of inter-varietal hybridisation may offer definite limitations for the improvement of valuable characters.   It may also be the case that the desired expression of certain characters is not found in one species, and further improvement is held up through lack of the required parental types, or by a restricted range of variability which has been tried in various combinations by hybridisation within the species.   When these conditions are met it may be necessary to turn to allied species for some desired character or characters, and to try to incorporate the required character, or induce a wider range of variability, by resorting to inter-specific hybridisation.

This method of breeding has been used with success in several crops including wheat and oats, and has been the basis of the attempts to breed potatoes resistant to late blight (*Phytophthora infestans*). In both wheat and potatoes there is a range of species related to the important cultivated species (in which are included the most widely grown varieties) showing a degree of disease resistance not found in common commercial varieties. In many cases of inter-specific hybridisation there are difficulties concerned with sterility often associated with different chromosome numbers of the species concerned. Special techniques are often required for the most effective use of inter-specific hybridisation, and in the case of wheat, new species have been synthesised by hybridisation and artificially doubling the chromosomes to overcome sterility and make further hybridisation possible.

The development of the technique of inter-specific hybridisation adds considerably to the possibilities of artificially creating new types which would be quite impossible under natural conditions. The collection of a wide range of forms and species that do not exist together naturally, or in any one area of cultivation, is now a fundamentally important part of plant breeding. The large number of species allied to the cultivated potato offers a great range of genetic variation, and certain characters would be of great value if they could be introduced to cultivated varieties suitable for this country. Several expeditions have visited the centres of the occurrence of these wild and cultivated forms of *Solanum*, which occur in Central and South America, to collect potentially useful species for breeding. Other expeditions have visited the Iberian Peninsular and North Africa to collect grass species such as tall fescue and cocksfoot, while the centres of variation of the principal cereals, and of tropical crops such as bananas, have been similarly closely studied.

Inter-Generic Hybridisation. In the same way that inter-specific hybridisation offers a means of effectively combining new characters normally outside the scope of hybridisation within one species, so does inter-generic hybridisation present the possibility of even wider combinations. There is, however, less scope for inducing generic combinations because of the difficulties in the actual technique of hybridising. The method is, however, being used in the related genera which include wheat, rye, and the couch grasses (the genera *Triticum, Secale* and *Agropyron*). These hybrids have been produced with two primary objects—the breeding of wheats with the capacity to survive extreme climatic and soil conditions (*Triticum* x *Secale*, and *Triticum* x *Agropyron*), and perennial wheats and forage grasses suitable to land reclamation work (*Triticum* x *Agropyron*). Much hybridisation has also been effected between *Triticum* and the closely related genus *Aegilops* (Goat grasses), and it has been possible to transfer rust resistance from this genus to wheat by the use of X-radiation. Special techniques have been developed for the transfer of whole chromosomes or parts of chromosomes of both *Secale* and *Aegilops* to cultivated wheat, and this synthesis of what are called chromosome substitution, and chromosome addition lines, offers particularly for substitutions an effective inter-generic hybrid technique in wheat breeding.

Back-Crossing. " Straight " hybridisation, involving the application of only one parental crossing, is in certain cases ineffective in bringing about the type of character recombination that is desired, and it may be necessary to resort to back-crossing to one or other of the parents, or even to more than one different parental type. The back-cross method is used particularly, for example, when the object is to transfer a single character like disease resistance from an otherwise undesirable form to good commercial stocks and it has been applied successfully to breeding disease resistance into cereals and potatoes. The method is applicable to inter-varietal crosses, as has been done in transferring resistance to mildew (*Erysiphe graminis*) from one barley variety to another, or to inter-specific crossing as in the case of late-blight in potatoes. In the latter case, where domestic varieties are hybridised with a wild species which possesses blight resistance (*Solanum demissum*) a succession of back-crosses to different commercial varieties may be practised to restore the desirable domestic characters to hybrid seedlings which have been shown by test to be resistant to blight as a result of the first cross.

Multiple or Poli-Crossing. In both self- and cross-pollinated crops crossing involving more than two parents may be used. The material may be handled in various ways, but a common way in self-pollinated crops is to make a number of "straight" crosses and then hybridise the first generation hybrids of these crosses in various combinations. The object in all cases is to introduce the "blood" of as many forms as is considered practicable in order to provide character combination for selections which would be impossible by a single cross.

Limitations of Hybridisation. Although hybridisation offers the most obvious and practicable means of creating new variability and new character recombinations, it is important to realise its limitations as a method of improvement, particularly with regard to bringing about character recombinations. In attempting to mould new plant types showing enhanced economic characters, the breeder often finds that it is impossible to select from the hybrid populations the particular combination for which he is looking. The reason for this may be genetic linkages or repulsions, or it may be regarded as due to a fundamental physiological incompatibility of characters which make it impossible to accumulate in one individual the highest expression of a combination of certain characters. An example of this may be seen in a character complex like yield which is dependent for its expression on a number of morphological characters and certain physiological characteristics. The particular yielding capacity of different commercial varieties depends on certain combinations of all the yield characters, and these combinations are not necessarily the same in all varieties—indeed they rarely are. When two varieties showing different yield characters are hybridised, it does not follow that it will be possible to select in the hybrid generations individuals showing a combination of all the highest expressions of the yield characters. In fact, it has been shown that this is impossible, and all the breeder can do is to select the best combinations available.

This example shows that hybridisation is not simply a means of mechanical re-shuffling and recombination of characters, and that the breeder cannot produce new types at will. The living plant is a delicately balanced organism and there are limits to what can be effected in attempts to concentrate desirable economic characters in one individual. It was a lack of appreciation of this fact that led the early Mendelian hybridists to express such extreme optimism on the future of plant breeding by the application of "Mendelian methods" of breeding, an optimism which led some investigators to state that the problems of breeding in animals and plants had been solved. Nevertheless, hybridisation followed by judicious selection and testing has been the means of effecting far-reaching and fundamental improvements in plants, and when properly applied offers the most effective means of organised and systematic progress in crop husbandry through crop improvement.

Chromosome Doubling. The artificial production of polyploids by treating individual plants with the drug colchicine is an attempt to bring about artificially what is known to have occurred under natural conditions in the evolution of cultivated plants. Straightforward doubling of chromosomes, i.e. the production of *autotetraploids*, has been achieved in many cultivated plants by this method, and useful new varieties have resulted with certain crops grown for their vegetative parts. Such autotetraploids often show " gigantism " in certain characters, but this may have no practical value to the crop. In sugar beet and fodder beets the most promising results at present appear to be in hybridising the artificial tetraploid with the normal diploid. The resulting triploids have been found in certain cases to produce higher yields of dry matter than either the diploid or the tetraploid. In sugar beet ' polyploid ' varieties are now available, these being mixtures of tetraploids, diploids and the triploid hybrids between them. Tetraploid forms of some herbage plant species have also been produced in recent years, and polyploid commercial varieties of perennial ryegrass, Italian ryegrass, tall fescue and broad red clover are available.

The alternative use of chromosome doubling in plant breeding is in wide crossing where the hybrid is sterile. By doubling the chromosome number a fertile *allopolyploid* or *amphiploid* is produced and such forms can be used for further experimental work. These amphiploids have been produced in considerable numbers in crosses between different species of *Triticum*, and in crosses of *Triticum* with *Aegilops* and with *Secale*, and they can be used as ' bridging ' forms to hybridise with cultivated wheats or as the starting point for the production of the chromosome substitution and addition lines already mentioned.

Mutations or " Sports ". Many well-known varieties of certain cultivated plants, for example potatoes and some fruits, have arisen by spontaneous changes in the chromosomes and genes resulting in the development of new characters of commercial value. Attempts have been made in many plants to accelerate the rate of mutation, or to induce mutations artificially, by exposing plants to X-rays, extremes of temperature, chemicals and similar sudden shocks likely to cause chromosomal irregularities or aberrations. Mutations have been obtained in this way, but with few practical results as far as useful or improved types is concerned. This technique for the breeding of agricultural crops, as distinct from the production of novelties in horticultural ornamentals, needs considerably more research before it could compete with standard hybridisation methods. It will be necessary to devise methods of increasing the rate of mutation and also to obtain a better control over the means of directing the type of mutation before the technique is really economic. However, it is obviously possible to visualise circumstances in which ' mutation breeding ' could be effective in solving a special problem in improvement, but more research is needed for a better definition of the most appropriate techniques.

THE FUTURE OF PLANT BREEDING

During the many thousands of years during which plants have been cultivated, improvements have consisted, until recent times, in the cultivator taking advantage of new types which have arisen naturally. The mere act of cultivation, with the growing of plants in large numbers and their careful tending with the selection of good types for seed, are contributory causes to the origin and survival of new forms. The growing of different varieties, or mixtures, adjacent to one another offers the means of natural hybridisation with the production of new types, while any new form arising by natural mutation is likely to be perpetuated artificially if it has any value.

But the development of organised plant breeding on an intensive scale seeks to accelerate the natural means by which diversity and new types arise, although the breeder has necessarily to use these natural means, as in artificially controlled hybridisation, selection, polyploid induction and mutational changes. The greater scope in planned breeding is due to the use of a wider range of material, the directed hybridisation and selection, and the use of scientific aids to effecting combinations that would not occur naturally, as well as in using selection methods and testing with greater degrees of precision. In addition to better-controlled and better-directed methods of breeding in themselves, the opportunity of producing improved types is enhanced by a greater understanding of the utilisation of the product by the consumer. The scientific basis of quality, the precise requirements in processing agricultural products, and the underlying principles of the feeding value of plant products gives greater scope to breeding and enlarges the whole aspect of plant improvement. But in the final analysis, it is the interests and requirements of the grower that must receive the lion's share of the breeder's attention, and it is towards plant breeding that attention can be focused for important contributions to the efficiency of crop husbandry.

Chapter XVI

ANIMAL BREEDING

THE general aims of animal breeding are the same as those of plant breeding, to improve the efficiency and productivity of the stock and, if possible, to ensure that the level of performance in the next generation is at least as good as, and preferably better than, that shown by the present generation. The concept of bringing about improved performance in the offspring as compared with their parents is fundamental to the whole idea of constructive livestock husbandry ; and in so far as performance is affected by genetic conditions, the same basic principles as those applied in plant breeding are relevant to animal breeding. Also, as has been pointed out in the case of crop plants, more complete knowledge of the physiology and development of the characteristics which build up performance is needed, as well as of the qualities of the various animal products which influence consumer demands, in order that the directions along which improvement is required can be more clearly defined.

Some of the early " improvers " of farm livestock tried to assess demands so that they could discover the ways in which their existing stock fell short of requirements and thus how and where improvement was needed ; for instance, it is on record that Ellman, to whom the development of the improved Southdown sheep is due, studied closely the joints and their relative weights and prices in butchers' shops as a means of deciding the most desirable carcase conformation and weight at which to aim. The aim determined, the early breeders developed, largely by trial and error, selection and breeding methods which they thought would achieve the desired progress. The discoveries of the principles influencing the processes of reproduction and inheritance have explained the results of the early improvers' efforts but they have not yet caused any spectacular or radical change in the methods used. Only in recent years have deliberate applications of the scientific principles begun to make much impact on practical procedures in animal breeding and that mainly in the direction of making selection methods more accurate and breeding methods more precise.

LIMITATIONS ON ANIMAL BREEDING METHODS

Some of the major differences between plants and animals in terms of reproductive and genetical processes have been indicated in Chapter XV. These differences affect the selection and breeding systems which can be used. Farm livestock are unisexual, so that self-fertilisation or the development of " pure lines " (pp. 473, 476) are impossible. Sexual reproduction continually sets the stage for genetic variation to be expressed ; mating is between animals of the same species (except in such cases as mule breeding) so that sterility and incompatibility problems in the previous sense do not arise ; nor

485

does hybridisation, in the narrow sense of crosses between species, occur to provide variations on which natural or other selection can work. Yet deliberate production of crosses between breeds of the same species to give hybrid stock which are not intended for subsequent further breeding has become a widespread commercial practice, e.g. in the poultry industry. This development is more appropriately discussed later as a form of " crossbreeding ". Polyploidy also does not occur in farm stock, and since there are large numbers of chromosomes involved in each species the chances are remote of linkage and crossing-over being exploited in recombinations of genes ; only very few cases of linkage in livestock have as yet been discovered.

A major significant difference between plants and animals which severely restricts the scope of operations in animal breeding lies in the forms of populations of individuals or communities. Very large numbers of plants are usually distributed on relatively small areas, whereas animals are more widely spread in fewer numbers in smaller breeding units. Even a numerically large " breed ", say of sheep, can total many fewer individuals than one particular field of crop plants. Except in the case of numerically very small " breeds ", the head of stock of the " breed " owned by, or available to, any one breeder amounts to only a small fraction of the total of the " breed ". Moreover the breeding units—herds or flocks—are usually dispersed over a variety of management systems and environments. The reproductive rate is low, except in poultry and pigs ; parents have very few offspring, while in plants a new population can be built up very rapidly from a small amount of parental material.

These considerations imply that :

(1) in animals, more parents have to be kept to provide enough offspring to keep the " breed " going, whereas in plants whole populations, e.g. varieties, can be rapidly replaced by the descendants of a small amount of parental material ;

(2) selection of parents in plants can be much more rigorous ;

(3) there is less uniformity of aims in animal breeding and less rationalisation or consistency of methods and objectives ;

(4) progress in any direction is slower and " improved " breeding material is less rapidly spread into commercial practice.

It takes a considerable time for the individual animal to develop sufficiently to give a return to the breeder. During this time the growing animal is exposed to differing environmental conditions. For example, a lamb ready for slaughter at four months old has spent more than half its life as a foetus ; for about 3 or 4 weeks after birth it has been entirely dependent on its dam's milk for its nutriment ; the rest of its life has been spent on a diet in which milk becomes progressively less important and grazing, and perhaps supplementary feed, more significant. In dairy cattle, the earliest actual measure of milk performance that can be made is that of the first lactation ; if a heifer is mature enough for first mating at 18 months old, then she will be over twice this age before her first standard lactation is ended. During this time, the varying and complex series of feeding, management and environmental conditions under which she has been reared and kept have affected her physiological development and the degree to which her genetic potential for milk production is expressed. The animal breeder has continually to try to control, or to allow for, the phenotypic differences due to environmental influences before he can distinguish the genotypes.

BREEDING MATERIAL : VARIATIONS

As in plants, the breeder has to make use, for the most part, of the variations already available in his stock as a basis for improvement, or to bring about recombinations of genes to provide further variation on which he can work. The reservoir of breeding material—and genes—from which he can draw is made up of " breeds ", varieties, or types of the species. Unfortunately, none of these terms, especially " breed ", has a clear and specific meaning, except in so far as it may denote a group of animals—a sample of the whole range of variation of the species or race—which possess certain highest common factors of characteristics so that they can in general, or on average, be distinguished more or less clearly from other groups.

All breeds, say of cattle, have some characters in common ; all produce milk, all put on some flesh and lay down some fat at some stage or other : but the yields of milk and its butter fat content vary from animal to animal, and from one herd to another, within the breed as well as from breed to breed. None the less, some breeds have characteristically high yields or low yields, high versus low butter fats and so on. No breed is uniform in itself in respect of the performance characters which typically distinguish it from others, the ranges of performance levels overlap to some extent. While it is proper to say that the Jersey is a high butter-fat breed of relatively low milk yields as compared with the Friesian, this does not mean that all Jersey milk is invariably high in butter fat content or that all Friesians give relatively high milk yields ; it means that the gene combinations favouring one or other main character are differently distributed throughout the two breeds. Moreover, these two breeds are distinguished by differences in other characteristics such as coat colour, and body conformation or type ; the defined black and white colour pattern of the Friesian does not occur in the smaller framed, more extreme " dairy type " Jersey. Yet colour and pattern are not uniform within either breed.

Breeds are often distinguished by characters which are inherited in a fairly simple Mendelian manner with the controlling genes distributed throughout the breed. For example, the Aberdeen Angus is typified by the simple dominant polled and black coat characters (as opposed to the recessive horned and basic red characters) while the Hereford breed is marked by its dominant white head pattern on its recessive red body colour.

Where colour patterns, as in Hereford cattle or in Saddleback pigs, can be regarded as " breed labels ", the precise extent or distribution of the pattern may vary between individuals because of the action of modifiers, which may be subject to selection : in such cases the conformity to breed standard or trueness to type of a strain may be interpreted as indicating the consistency or " purity " of its selection and breeding.

In contrast to such simple genetic situations, the inheritance of most performance or productive characters is more complicated. There is overwhelming evidence that physiological characters such as growth rate, fleshing, milk yield, etc., depend upon the action of large numbers of complementary genes, many genes with additive effects, or complex gene interactions, so that they are expressed as levels of " quantitative " rather than " qualitative " inheritance. Variations in performance tend to be continuous, grading smoothly throughout the range and not being exhibited in marked steps or jumps from one category to another so that the effects of a single gene or of a few genes cannot be detected. During the development of a breed these quantitative characters have been built up to give the breed its characteristic performance levels ; and where a definite breed label is involved this has sometimes led to the idea that the label gene itself may have some direct or appreciable effect on per-formance type. For example, in the Aberdeen Angus breed its polled black labels are now associated with the conformation, growth, and fleshing abilities which have been developed in it ; in the Hereford, the white-faced pattern is associated with the other breed characters of beef conformation and growth. Yet in other breeds, the general beef performance characters and conformation have been built up under other labels, e.g. in the Sussex and Devon with their whole red coat colours, or under no specific coat colour pattern as in the Beef Shorthorn, where it ranges from white through the various roans to red.

Where the breed labels are of dominant characters, such associa-tions are used in the " colour marking " of crosses between breeds. In the cross between the Hereford and the Welsh Black, the offspring are black with white faces. Along with the dominant gene for white face some of the genes affecting the physiological development of beef characters are also brought in from the Hereford.

The offspring of a cross between Aberdeen Angus and, say, Ayr-shires are basically black and polled, with more beef potential than that found in the Ayrshire, but with less milk yield potential ; similarly, Aberdeen Angus × Friesian animals exhibit the effect of the polled gene as well as those of the genes influencing earlier fleshing.

Again, in sheep, the speckled brown faces of lambs out of white faced ewes such as Cheviot, Scots Half-bred, or Welsh Mountain indicate not only that their sires were of a dark faced breed such as the Suffolk or Hampshire but also that they must have some measure of improved genetic potential for growth and development for lamb production. If, however, such crosses (breed hybrids) were to be interbred, segregation of the genes affecting all characteristics would be expected, resulting in a wide variation in the second crossbred generation.

Another aspect of these breed label situations must be noted. Blacks do not occur in the basically red breeds, but recessive reds may appear in the black breeds if and when their parents are hetero-zygotes ; the appearance of a red calf at once incriminates both its parents as heterozygous for the black-red pair of genes and the frequency of recessives indicates the frequency of heterozygotes in the breed or strains within the breed. (The same applies in the case of inherited defects.) But so long as the contrasting qualitative genes have no major effects on the quantitative characters, there is no other difference in genetic potential for performance between the animals showing the recessive character and those which conform to the breed standard ; thus, in every other respect red and white are similar to black and white Friesians or red to black Aberdeen Angus.

In general, the simple gene situations are not affected by differences in environment, whereas the complex quantitative characters respond to varying degrees to external influences. This means that among the total character variation with which the breeder can deal, some of it is largely due to inheritance, while some is attributable to environmental conditions. By the use of modern statistical methods it is possible to analyse the relative amounts of phenotypic variation to genotype, to genotype-environment interactions, and to environment : that is, to estimate the " heritability " of the various performance characteristics and thus gain some idea of the procedures to be used by the breeder to manipulate the genetic material in the direction of improved or more reliable and efficient performance.

The estimates of heritability so far made do not give a complete coverage of all the characters which contribute to the economic worth of farm stock ; also, the estimates vary among themselves for the same character within any one type of stock.

Without attaching too much importance to any particular estimate, it is possible, however, to compare grades of heritability in respect of different characters, and to distinguish between those where environmental effects are so important that they can override or mask the genetic variation and those in which the genetic effects are more significant. The former groups are of low, the latter of high heritability.

Among dairy cattle, milk yield is of lower heritability than butter fat percentage ; birth weight in beef types and the rate and economy of gain of live weight after weaning are more highly inherited than conformation grade at weaning, and the dressing-out percentage at slaughter higher than carcase grade.

In pigs, the carcase characteristics of length, area of eye muscle, backfat thickness, and percentages of cuts in the carcase weight are all of very high heritability, with body and leg lengths in the live pig less heritable though still high, and weight at five to six months relatively high, as compared with live conformation score, numbers farrowed and weaned, and weight at weaning which are generally of low heritability.

In sheep, there are indications that the characters in the relatively low heritability group include type or body conformation and multiple births, while wool staple length and fibre diameter and face covering are in the high group ; body weights at birth, weaning and twelve months, fleece weight and milk yield are of intermediate or moderate levels of heritability.

SELECTION—TESTING—MATING SYSTEMS

Selection. As in plant breeding, selection of the parents of the next generation is the main tool available to the breeder. Except for some simple recessive gene combinations an animal's appearance or performance is the result of the genes inherited from its parents and the environment to which it has been exposed, so selection can only be made on the basis of an estimate of true breeding merit. The ways differ in the accuracy of the estimate they supply but are—on the basis of (a) the individual's own merit, (b) the merit of its ancestors or relations, and (c) the average merit of its progeny.

Own Merit. An animal's appearance or performance on a test estimates its breeding merit directly ; the best are selected for breeding (or for further testing if necessary). The method is particularly useful and cheap for those traits which can be measured on the individual whilst it is being grown to breeding size and age, e.g. growth rate, feed efficiency and carcass quality when this can be measured indirectly. The major advantages of the performance test are that (a) the estimate of breeding merit is usually available before the animal is used for breeding, (b) it relates to the animal concerned though it can be supplemented by information on close relatives, (c) it is cheap, because only one " animal place " is required to evaluate a potential parent. Except for characters which are moderately to highly heritable it is usually less accurate than a progeny test.

Merit of Ancestors or Relations. Selection on the merit of relations is selection on pedigree ; it is applicable to sex-limited traits such as milk production where full or half sibs provide a certain amount of information and when the heritability of the characteristic is low. Performance information and also type classification scores are included in breed society pedigree certificates but these are difficult to interpret for comparative purposes unless the feeding and management levels are known.

Average Performance of Offspring. Selection on the average performance of a large number of offspring (a progeny test) comes near to assessing the *true* breeding merit of the animal ; it is appropriate when (a) the heritability of the character is low, (b) the character is sex-limited, e.g. litter size, and (c) if it is necessary to slaughter animals to assess carcass quality.

There are two drawbacks to a progeny test, (a) the cost is usually high because a large number of progeny have to be tested and (b) because the generation interval (the time between generations) is increased. A progeny test is unlikely to be justified, technically or economically, unless sufficient sires are tested to allow selection of the top 25 per cent of sires after test. The number of progeny needed to give a particular accuracy depends on the heritability of the characters but the accuracy has to be a balance between the number of sires that should be tested, the number of " tester " females available and the cost of the programme. In the dairy bull test programmes, run by artificial insemination organisations, the aim is to record the first lactations of 40–50 daughters, but where females are more prolific, e.g. in pigs and poultry, fewer dams are needed to supply the necessary number of progeny. Although a performance test can be a very valuable way of screening for superior merit, it is accepted that a progeny test should be undertaken before a sire is used as widely as is technically possible today; thus progeny testing schemes require large breeding units or co-operative undertakings before sufficient numbers and resources are available.

Testing Stations. Both performance and progeny tests may be carried out at testing stations. The advantages are that (*a*) individual potential sires or samples of the sire's progeny are measured under standard conditions so that individuals from different herds and sire progeny groups can be compared more accurately for a given number of progeny, and (*b*) more measurements can be taken because specialist recording staff and equipment can be justified. Fundamental questions in any central testing programme are (*a*) the choice of management régime and (*b*) the health risks from the collection of stock from many sources. Test stations are used for boar and beef bull performance tests; the latter because beef breeding herds are generally small and provide few effective or accurate comparisons but there is little organised performance testing of sheep in UK. Test stations are used for progeny tests of beef bulls and boars prior to their widespread use in artificial insemination but not for progeny tests of dairy bulls.

Field Testing. Field testing is used for assessing dairy bulls in most countries as it enables groups of offspring of a sire to be recorded under farm conditions and in comparison with the progeny of other sires at the same time and kept under the same wide range of environments. The basic assumption is that contemporaries are of similar age and treated similarly within each herd. " On farm " performance tests of pigs, beef bulls and to a lesser extent sheep are sponsored, officially; sophisticated recording, analysis and evaluation procedures are readily available to farmers. Field testing is cheaper and with less health risks than station testing, but fewer detailed or complex records can usually be made on individual animals.

Contemporary Comparisons. In both station and field testing contemporary comparisons are used instead of absolute performance figures; the results are given as plus or minus differences from the averages of the contemporaries. Sometimes the differences for several characteristics are combined through a selection index which " weights " the pieces of information according to their economic values and their relationships with other characters to provide a " total score " of breeding merit. Selection indexes are used for pig testing but they can be applied to cattle and sheep also because recording schemes are available. In dairy cattle, because it is difficult to measure conformation objectively and to give it an economic value, conformation is assessed by eye; the information is published separately from that for milk yield and quality. In Great Britain the first lactation yields and quality (corrected for age and season of calving) of a bull's daughters are compared, within each herd and within each defined calving season, with those of contemporary daughters of other bulls. A computer is necessary to produce the " *improved contemporary* *comparison* " for milk yield and fat and protein yields and percentages. This sophisticated calculation takes into account the genetic merit of the sires of contemporary heifers ; the published figure gives the *best estimate available* of each bull's breeding merit for the production traits.

Mating Systems. The ways in which parents are mated together affect the distribution and recombination of genes ; mating systems are usually distinguished by the degrees of genetic relationship between the parents. All members of a breed have some genes in common so mating within a breed or line is called " pure breeding " (or within the registered section of the breed type " pedigree breeding "). There are two generally recognised forms, first, the mating of animals or genotypes more closely related than the average of the breed is called " inbreeding ". Matings of brother to sister, father to daughter or son to mother is " close inbreeding " whilst in " linebreeding " the animals are kept as closely related as possible to some illustrious ancestor. Inbreeding increases homozygosity of many gene combinations and leads to greater uniformity, but undesirable gene combinations also accumulate leading to deterioration particularly for reproductive characters. The second form, " outbreeding " or " outcrossing ", is the crossing of different strains or families to get all the advantages of mating relatively unrelated animals of the same pure breed.

By contrast, when parents are from different breeds they are unrelated or only very distantly related ; this is " crossbreeding " and has similar effects to those described under hybridisation of plants. Crossbreeding is widely used in commercial pig, sheep and beef cattle to exploit " hybrid vigour " or " heterosis " (the boost above the average performance of the parental types in the crossbred) which is usually seen in the characters of low heritability like conception-rate, litter size, etc.

Two forms of crossbreeding are recognised. The first is where the new generations of crossbreds are produced each generation from the two parent breeds involved ; this is called " discontinuous ", for example the Blue-Grey suckler cows ; other examples occur in sheep and pigs. Such first cross females are considered to give the maximum amount of hybrid vigour, but both pure breeds must be maintained in order to propagate them. There is little planned discontinuous crossbreeding in dairy cattle, mainly because the resulting heterosis is not sufficient, economically, to compensate for the loss of milk in the cross compared with that of the better parent and the low reproductive rate in cattle.

The second form is " continuous " crossbreeding or " cyclical breeding " where two or more breeds of sire are used in rotation ; so that all the females and progeny are always crossbred. When two breeds of sire are used alternately the system is called " criss-crossing "; the animals will, after a few generations, have two-thirds of their genes from the sire's breed and one-third from the other breed. Using three or more sires in rotation is a more complicated variant of the two-breed system. Cyclically bred females may not exhibit as much " hybrid vigour " for reproductive characters as first crosses between two breeds but performance traits like growth, feed efficiency and carcass quality need not suffer because they depend entirely on the merit of the sires used in each generation. The main problem which limits its widespread use in pigs and dairy cattle is to find a succession of suitable sires from widely different breeds.

ANIMAL BREEDING IN THE FUTURE

Most of the theory concerning the inheritance of quantitative or metric characters (those which vary in degree, e.g. growth rate) and the procedures for statistical analysis of breeding data are already available. But much has yet to be implemented possibly because the sophisticated techniques require large populations of animals in co-operative or company programmes either of which may be multinational. Efficient use of the theory and analysis procedures presupposes precision in the evaluation of the breeding merit of candidates for selection. In the future there will be improvement in methods of conventional recording and also through the study of blood groups, hormone levels and mineral metabolism. Knowledge of biochemical and physiological processes may well lead to earlier assessment of breeding merit for litter size and maternal qualities than is possible today ; it may also permit the assessment of, for example, a bull's merit for milk yield and quality directly on him when young rather than through the laborious and expensive progeny test methods carried out today.

Techniques whereby fertilised ova are transplanted from donor to foster females will have some effect on improving the rate of genetic gain, but when the techniques are economically viable they can rapidly increase the speed at which one breed can be substituted for another on commercial farms. Thus, transplantation can do for ova from cows of high merit what artificial insemination does for semen from superior bulls—it seems doubtful, however, whether the former will be as effective as the latter in numbers of progeny from one parent.

The free interchange of breeding stock between countries will mean that breeders have access to a wider range of genetic merit, and much work will concentrate on understanding and quantifying the differences between breeds to identify their particular roles in purebreeding and crossbreeding and permit efficient planning of particular breed or strain crosses to meet specific commercial objectives. Selection as the tool to manipulate gene combinations will, therefore, remain of paramount importance.

Chapter XVII

THE STRUCTURE AND FUNCTIONS
OF THE ANIMAL BODY

CATTLE, sheep and pigs, which are the principal large farm animals, are all mammals. The mammals are the highest of the five classes constituting the *Vertebrata* or animals with backbones. The other classes in descending order are the birds (which include the domestic fowl, another important farm animal), reptiles, amphibians and fishes.

Mammals are warm-blooded, hairy animals which suckle their young and which, with only a few exceptions, undergo a period of embryonic development within the womb of the mother. Two of the exceptions are the duck-billed platypus and the spiny ant-eater from Australia, and these form the lowest division of the mammals ; they can be described as egg-laying mammals. A covering of hair is characteristic of mammals and distinguishes them from other vertebrates, but in some species the hair covering is scant. Most of the mammals are quadrupeds, but in some aquatic mammals, e.g. seals, whales and dolphins, the hind-limbs are modified, reduced or absent.

The birds are also warm-blooded, but the characteristic hair covering of mammals is replaced by feathers. The hind-limbs are well developed, but the fore-limbs have become adapted as wings used for aerial flying : in the domestic fowl this power of flight is greatly reduced. The birds are egg-laying animals and the embryonic development of the chick takes place within the egg.

The study of the structure and form of the body is known as *anatomy* ; it may represent a study of parts visible to the naked eye but may also involve more detailed microscopic studies (*histology*). It is quite as important to study the uses and functions, or *physiology*, of the various organs of the body. Further examination of the actual nature of the chemical reactions occurring as the body functions proceed can be described as the study of *biochemistry*.

The development of an animal can be divided into two parts, an early phase of development within the body of the mother (mammals) or in the egg (birds), and a phase after birth when the animal is free-living. The former can be referred to as the pre-natal and the latter as the post-natal phase.

THE CELL

The body of any higher animal such as a man, an ox or a fowl, is composed of a number of microscopic structural units or cells built up in various ways to produce tissues and organs. The cell is the smallest unit of living matter capable of leading an independent existence, and has the power to divide and produce other cells of its own kind. Some of the simplest animals, e.g. *Amœba*, consist of a single cell, and bacteria, which are looked upon as plants, are probably the smallest

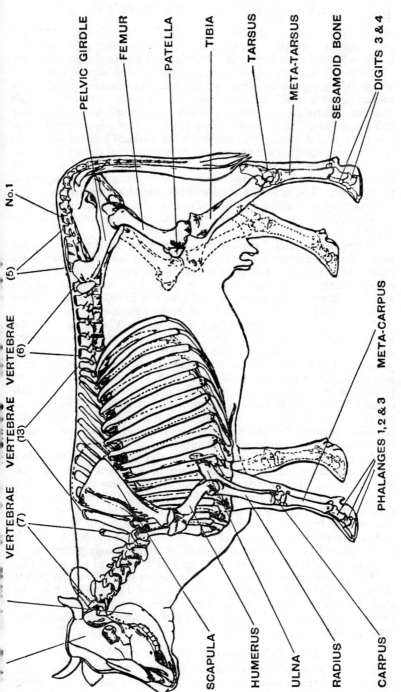

VERTEBRAE (7)   VERTEBRAE (13)   VERTEBRAE (6)   (5)   No. 1   PELVIC GIRDLE   FEMUR   PATELLA   TIBIA   TARSUS   META-TARSUS   SESAMOID BONE   DIGITS 3 & 4

SCAPULA   HUMERUS   ULNA   RADIUS   CARPUS   META-CARPUS   PHALANGES 1, 2 & 3

FIG. 97.—SKELETON OF THE COW.

495

of all single-celled (unicellular) organisms. The viruses, which are even smaller than bacteria, can only reproduce when they are inside living cells and therefore cannot themselves be regarded as true cells.

Each cell has two main parts. In the central area is the *nucleus* which is the organising centre of the cell and contains the chromosomes (see Plate XXXI). Surrounding the nucleus is a variable amount of *cytoplasm*, and the whole cell is contained within a delicate cell-membrane. The cellular material is collectively referred to as *protoplasm* and this complex material is the physical basis of life. Life itself is dependant on the continuous bio-chemical activities (metabolism) taking place within the protoplasm, and the protoplasm is not merely a receptacle in which these activities occur but is a dynamic participant. The activities consist of innumerable chemical reactions and, as they take place, the actual nature of the protoplasm changes ceaselessly. Upon these diverse activities the phenomena of life depend.

When normal cell-division occurs there is an exact division of the material of the nucleus so that each cell has a nucleus which is an exact replica of that in the cell from which it was derived. Also, during division there is a sharing of the cytoplasm between the parent cell and its offspring. However, not all cells are identical and different types of cells become adapted for different functions in the body.

The body of a higher animal develops from a single cell, the fertilised egg or *ovum* which is derived from the union of two specialised cells called *gametes*, one from the male and one from the female (see p. 468). The fertilised ovum is the parent cell of the future offspring and divides first into two cells, then each of these divides again to produce four cells and so on, until the many millions of cells which constitute the animal body are produced. It is not strictly true to say that the body is completely composed of cells, for there is material deposited between the cells (inter-cellular material) which is important in maintaining the structure and functions of the body.

An examination of different cells in the body reveals that they are by no means all identical as might be supposed from the description of cell-division above. During development the animal has the power to produce different families of cells of different shapes and sizes which are adapted to their functions in the body. An examination of cells from nervous tissue, bone and muscle will demonstrate just how great these differences can be. The mechanism by which these different types of cells are produced from a single parent cell is obviously complex and begins at a very early stage of pre-natal life.

The process of cell-multiplication continues throughout the period of growth of the animal and to a lesser extent during adult life. In adult life cell-multiplication is largely concerned with the repair of tissues as cells age and die, and as this process slows down the animal itself eventually ages and dies.

THE ORGANISATION OF THE ANIMAL

In a simple unicellular animal like *Amœba* the single cell is extremely versatile in that it carries out all the functions of the body. Living in a watery environment, the cell can absorb oxygen and take in food

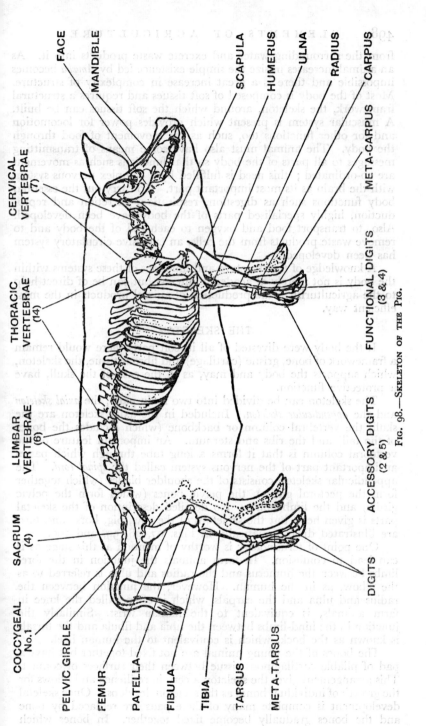

FIG. 98.—SKELETON OF THE PIG.

COCCYGEAL No. 1

SACRUM (4)

LUMBAR VERTEBRAE (6)

THORACIC VERTEBRAE (14)

CERVICAL VERTEBRAE (7)

FACE

MANDIBLE

SCAPULA

HUMERUS

ULNA

RADIUS

CARPUS

META-CARPUS

FUNCTIONAL DIGITS (3 & 4)

ACCESSORY DIGITS (2 & 5)

DIGITS

META-TARSUS

TARSUS

TIBIA

FIBULA

PATELLA

FEMUR

PELVIC GIRDLE

from the surrounding water and excrete waste products into it.   As an animal increases in size the simple existence led by *Amœba* becomes impossible and there is a great increase in complexity of structure. Most of the body is composed of soft tissues and requires a structural framework, the skeleton, around which the soft tissues can be built. A muscular system is present which provides power for locomotion and for other functions too, such as the movement of food through the body.   The animal must also have some means of transmitting messages to all parts of the body so that functions such as movement are co-ordinated ; this need is fulfilled by a complex nervous system with the brain as its most important part.   To carry out the essential body functions such as digestion, respiration, excretion and repro-duction, highly specialised parts of the body have been developed. Also, to transport food and oxygen to each cell of the body and to remove waste products from the cells, an extensive circulatory system has been developed.

A knowledge of the structure and functions of these systems within the body is not merely of academic interest but can be of direct help to the agriculturist in the production of animal products in the most efficient way.

THE SKELETON

If the body were divested of all its soft parts there would remain a framework of bone, gristle (cartilage) and fibrous tissue, the skeleton, which supports the body and may, as in the case of the skull, have a protective function.

The skeleton can be divided into two main parts, the *axial skeleton* and the *appendicular skeleton*.   Included in the axial skeleton are the skull, the vertebral column or backbone (which includes the bones of the tail) and the ribs and sternum.   An important feature of the vertebral column is that it forms a long tube through which passes an important part of the nervous system called the *spinal cord*.   The appendicular skeleton consists of the shoulder blades (which together form the pectoral girdle), the pelvic bones (which form the pelvic girdle) and the limb bones.   No detailed description of the skeletal parts is given here, but the skeletons of the cow, pig, horse and bird are illustrated diagrammatically in Figs. 97, 98, 99, and 100.

One point of terminology is worthy of mention at this stage as it can lead to confusion.   In farm animals the junction in the fore-limbs between the humerus and the radius and ulna is referred to as the elbow, as in the human.   However, the junction between the radius and ulna and the carpals, which is often called the knee in farm animals, is equivalent to the human wrist.   Similarly the junction in the hind-limbs between the tibia and fibula and the tarsals is known as the hock, which is equivalent to the human heel.

The bones of the young animal are not fixed together but have a pad of pliable cartilaginous tissue between their surfaces of contact. This arrangement gives the skeleton a certain resilience and allows for the growth of individual bones as the skeleton develops.   Once skeletal development is complete many of these pads are replaced by bone and the bones gradually become fixed together.   In bones which

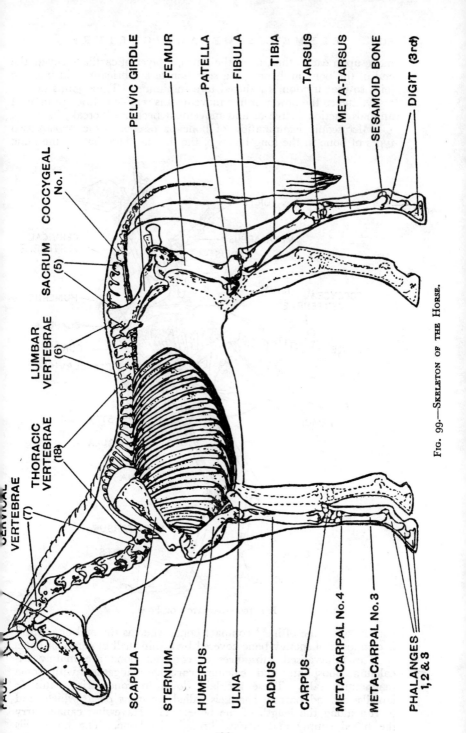

CERVICAL VERTEBRAE

PELVIC GIRDLE

FEMUR

PATELLA

FIBULA

TIBIA

TARSUS

META-TARSUS

SESAMOID BONE

DIGIT (3rd)

COCCYGEAL No. 1

SACRUM (5)

LUMBAR VERTEBRAE (6)

THORACIC VERTEBRAE (18)

CERVICAL VERTEBRAE (7)

SCAPULA

STERNUM

HUMERUS

ULNA

RADIUS

CARPUS

META-CARPAL No. 4

META-CARPAL No. 3

PHALANGES 1, 2 & 3

Fig. 99.—Skeleton of the Horse.

499

move upon each other, as in the limbs, a layer of cartilage covers the end of the bone and there is a secretion of a lubricant fluid into the joint so that friction is reduced to a minimum.   These joints do not ossify, and so the power of free movement is retained although in aged animals friction increases and movement becomes " creaky ".

Microscopic examination of bone or osseous tissue reveals two types of bone in the long bones of the limbs.   The shaft of the bone

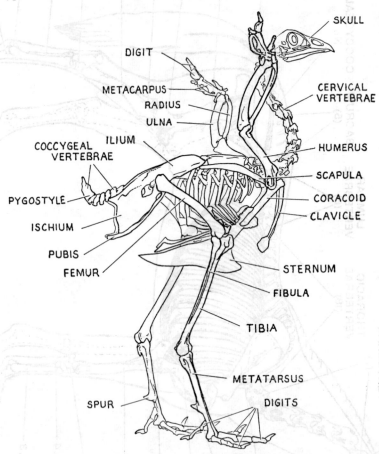

FIG. 100.—SKELETON OF FOWL.

consists of a tube of hard compact bone, whereas the end or *epiphysis* is composed of spongy bone covered by a thin shell of compact bone.

The mineralised substances of compact bone are mainly tricalcium phosphate and calcium carbonate together with some magnesium salts.   These are laid down in concentric plates or lamellæ around narrow channels called *Haversian canals* which tend to run along the length of the bone.   The Haversian canals carry the blood supply and nerves through the bone.   The bone cells (osteocytes) lie in small cavities called lacunæ between the lamellæ. The hollow central part of the bone shaft is filled with a yellow fatty material called yellow marrow.

Spongy bone is made up of numerous slender rods of osseous tissue called traberculæ which interlace to give a sponge-like structure. The spaces between the traberculæ are filled with red marrow which is responsible for the manufacture of red blood cells. Spongy bone with its associated red marrow is found, not only in the ends of the long bones, but also in the ribs, vertebræ and skull.

DENTITION

A study of the teeth is of special interest as variations in the number and condition of the teeth are commonly used to make approximate estimates of age. The teeth are hard structures set firmly in spaces in the jaw-bones and are used both in the intake of food and in its subsequent mastication ; they may also serve as weapons. Domestic animals have two sets of teeth during their life-time. At birth, or soon after, the young animal has a set of temporary teeth (often called " milk teeth "), but these are gradually replaced during growth by a set of permanent teeth.

Teeth can be classified into three groups :

1. Incisor teeth situated at the front of the mouth and used for seizing food.
2. Canine or eye teeth (of special importance to carnivorous animals) used for tearing and cutting food.
3. Cheek teeth which include the premolar and molar teeth. The premolars are situated in front of the molars and the molars are only represented in the permanent teeth. Both types are used in the grinding and mastication of food before swallowing.

Dental Formula. The dentition of an animal can be conveniently expressed as a dental formula, e.g. for the dog :

$$\text{Incisors } \frac{3-3}{3-3} \quad \text{Canines } \frac{1-1}{1-1} \quad \text{Premolars } \frac{4-4}{4-4} \quad \text{Molars } \frac{2-2}{3-3}$$

This indicates that there are three incisors on each side of the mouth in both the upper and lower jaws, and so on.

Dentition of the Ox. The dental formula of the ox is as follows :

$$\text{Incisors } \frac{0-0}{4-4} \quad \text{Canines } \frac{0-0}{0-0} \quad \text{Premolars } \frac{3-3}{3-3} \quad \text{Molars } \frac{3-3}{3-3}$$

The molars and premolars bear crescentic ridges which are characteristic of the ruminant. There are no incisors or canines in the upper jaw, these being replaced by a cartilaginous dental pad against which the incisors bite. In grazing the ox does not sever grass from the pasture with its teeth but harvests it with a long prehensile tongue. The permanent incisors are very much larger than the " milk teeth " and are known as " broad teeth ". Animals slaughtered when young may have no " broad teeth ".

The approximate arrangement of the teeth at different ages is :

Up to one month—temporary incisors up.
Eighteen months—central incisors up (two " broad teeth ").
Two and a half years—central and middle incisors up (four " broad teeth ").
Two and three quarter years—three incisors up (six " broad teeth ").
Three and a half years—all incisors up (eight " broad teeth ").

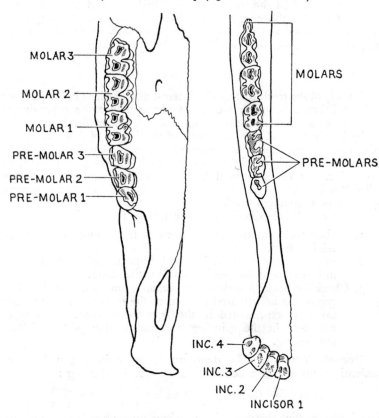

UPPER JAW   LOWER JAW

FIG. 101.—DENTITION OF OX.

Dentition of the Sheep. The sheep has the same dental formula as the ox. Its method of grazing is somewhat different because grass is actually severed from the pasture by the teeth. This process is assisted by a " split " upper lip which enables the sheep to graze much closer to the ground than the cow.

The approximate arrangement of the teeth at different ages is :

Birth to one month—temporary incisors up.
12 to 18 months—two central permanent incisors.
18 to 24 months—two middle permanent incisors.
30 to 36 months—two lateral permanent incisors.
3 to 4 years—two corner incisors or canines (eight " broad teeth " or a "full mouth ").

By the time the sheep is six to eight years old the permanent teeth develop gaps between them and begin to fall out ; the sheep is then referred to as being " broken-mouthed ". As sheep are often rejected from the flock when they become broken-mouthed, it is important to investigate the factors causing wear in sheeps' teeth.

Dentition of the Pig. The pig has the following dental formula :

$$\text{Incisors } \frac{3-3}{3-3} \quad \text{Canines } \frac{1-1}{1-1} \quad \text{Premolars } \frac{4-4}{4-4} \quad \text{Molars } \frac{3-3}{3-3}$$

The pig differs from ruminants in having incisors and canines present in the upper jaw. The canines are enlarged to form the tusks, which are exceptionally large in the boar. The molars and premolars are provided with knobs or bosses in place of the crescentic ridges of the ruminant. The order of development of the teeth is as follows :

Birth—two temporary teeth on each side of the jaw (incisors and canines).
Three months—full mouth of temporary teeth.
Nine months—two permanent teeth on each side (incisors and canines but the latter only just above the gums).
Thirteen months—a pair of permanent central incisors in each jaw.
Eighteen months—a pair of lateral incisors in each jaw (full mouth).

THE MUSCULAR SYSTEM

The most important property of muscle is its power of contraction which enables it to effect movement. Thus muscles are found at any point in the body where movement is required including not only the muscles of the limbs, trunk and neck, but also those of the heart, the blood vessels, the digestive organs and many other parts of the body.

There are different types of muscle which can be classified in various ways. For example, microscopic examination of muscle shows that it is composed of thread-like fibres. If the fibres exhibit cross striations or stripes the muscle is known as striated muscle. If, however, there are no cross striations, the muscle is called smooth or unstriated muscle.

Muscles can also be classified on the basis of the way they function. If a muscle is under voluntary control, e.g. the muscles of the limbs, it is called a *voluntary* muscle. If the action of a muscle is outside the control of the will, e.g. movements of the muscles of the digestive system, it is called an *involuntary* muscle. Generally, striated muscle is voluntary and smooth muscle is involuntary. The muscle of the heart is exceptional as it is partially striated but involuntary, and for this reason is usually classified separately as *cardiac muscle*.

A further name is commonly given to striated muscle ; as it is almost invariably attached to the skeleton it is known as skeletal muscle.

Skeletal Muscle. This is responsible for the movement of the limbs, trunk and head. It is adapted to respond rapidly to a stimulus and to work hard for limited periods. Thus violent exercise soon leads to muscular fatigue and a rest period is required.

Muscles are attached to bones, and movement is brought about in the following manner : one of the bones to which a muscle is attached is held in a relatively fixed position by the action of other muscles and then the muscle which makes the movement (the prime mover) has a fixed point against which to contract. This point of attachment of a muscle to a bone is called its origin. The point of attachment of the other end of a muscle is known as its insertion. At its origin the muscle is attached more or less directly on to the bone, but at its insertion blends into a tough fibrous tendon which actually unites with the bone.

Individual muscles can be classified according to the types of movement which they cause. For example, muscles which bend a limb are called flexors, and those which straighten a limb, extensors.

An interesting point about the structure of skeletal muscle is that the number of fibres in a muscle does not change after birth. Therefore increases of muscular size during growth must be due to an increase in the size of individual fibres and the incorporation of non-muscular material, e.g. fat into the muscle.

To the farmer, muscles are important not only because they effect movement, but also because they form the most important final product of the meat animal—the lean meat. The quality of meat is largely determined by the lean meat. The muscle must be tender, not too strong in taste and of a bright pinky-red colour. This type of meat can be produced from the well-fed, fast growing animal slaughtered at a young age.

Smooth Muscle. This type of muscle responds slowly to a stimulus but is capable of working for long periods at a slow steady pace, without tiring. It is found in the walls of the intestines and blood vessels and in all the organs of the body cavity where sustained movement is required. The activity of this type of muscle is outside the control of the will.

Microscopic examination shows that smooth muscle fibres are shorter than those of skeletal muscle and are rather spindle-shaped. They do not exhibit cross striations. Also, unlike skeletal muscle fibres, the fibres of smooth muscle can divide by mitosis to produce more fibres.

Cardiac Muscle. The heart has very special needs, to which cardiac muscle is adapted. Like smooth muscle it can work steadily day after day without tiring but, like skeletal muscle, it is capable of periods of intense activity when the dem nds of the heart are great.

THE NERVOUS SYSTEM

In a higher animal with its complex structure there must be some means of co-ordinating the activities of the body to avoid any conflict of purpose. This is provided by the nervous system.

The basic structural unit of the nervous system is the nerve cell or *neuron* (Fig. 102). The body of the nerve cell has several spidery processes called *dendrites* and a single, long conductile thread called the *axon*. These processes of the nerve cell can be referred to as nerve fibres. Nerve cells are linked together by the axon of one cell making contact (but not structural union) with the dendrite of another cell.

Contacts of this type (synapses) can be found in the central nervous system which is described below.

Some nerve fibres transmit messages from the extremities of the body arising from touch, sound, light, heat, etc., and give rise to the sensations ; they have a sensory function. Other fibres carry messages to the muscles causing movement and these are called motor nerves. Usually the nerve fibres are bound together in bundles to form what are generally called nerves.

The nervous system can be divided up into two parts, made up of the central nervous system and the peripheral nervous system.

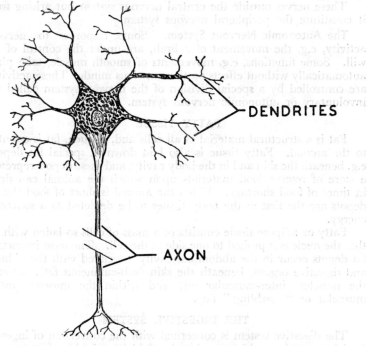

FIG. 102.—A NERVE CELL OR NEURON.

The Central Nervous System. This includes the brain and the spinal cord. The brain is the headquarters of the nervous system and is well protected against damage by the bony structure of the skull. It receives messages and impressions from all parts of the body, and gives rapid and decisive orders for appropriate action by the body. The importance of the brain is so great that even minor brain-injuries may seriously affect the ability of the animal to co-ordinate its activities.

The spinal cord arises from the rear part of the brain and runs the entire length of the vertebral column lying in the well protected neural canal.

Both the brain and the spinal cord are composed of white and grey matter. The reason for this colour differentiation is that the white matter is composed only of nerve fibres, whereas the grey matter contains the cell bodies of the neurons and their shorter processes.

The Peripheral Nervous System. At the point where the spinal cord and brain join, there arises a series of nerves which supply the head and certain other parts of the body. These are known as cranial nerves as they originate from within the cranium. The spinal cord also gives rise to nerves at intervals along its length and these pass to all parts of the trunk and limbs. Injuries to the spinal cord generally result in a paralysis of most parts of the body posterior to the point of injury.

These nerves outside the central nervous system but arising from it constitute the peripheral nervous system.

The Autonomic Nervous System. Some responses to nervous activity, e.g. the movement of a limb, are under the control of the will. Some functions, e.g. movements of smooth muscle, take place automatically without effects on the conscious mind. These activities are controlled by a special division of the nervous system called the involuntary or autonomic nervous system.

FATTY TISSUE

Fat is a structural material of all cells and, as such, fat is essential to the animal. Fatty tissue is also laid down in special fat depots, e.g. beneath the skin and in the body cavity, and these depots represent a store of reserve food materials upon which the animal can draw in times of food shortage. When the animal is short of food the fat depots are the first of the body tissues to be depleted as a source of energy.

Fatty or adipose tissue consists of a mass of cells so laden with fat that the nucleus is pushed to one side of the cell. The most important fat depots occur in the abdominal cavity associated with the kidneys and digestive organs, beneath the skin (sub-cutaneous fat), between the muscles (inter-muscular fat) and within the muscles (intra-muscular or " marbling " fat).

THE DIGESTIVE SYSTEM

The digestive system is concerned with the conversion of ingested food into portions which can be absorbed into the blood stream, by which nutrients are transported around the body. The principal chemical constituents of the food are carbohydrates, proteins and fats to which must be added certain essential minerals, vitamins and water. These various substances are considered in Chapter XVIII. Although these essential food substances are required by all animals the nature of the diet from which they are obtained differs markedly from species to species. The digestive systems of different species have become adapted to deal with the different types of food eaten. The digestive systems of farm animals can be divided into three groups :

1. The simple stomached or monogastric type, e.g. the pig.
2. The ruminant type, e.g. cattle and sheep.
3. The avian type, e.g. the domestic hen.

The Monogastric Digestive System. The digestive tract consists essentially of a long coiled tube, the alimentary canal or gut, which traverses the entire length of the body. The main regions of the alimentary canal are the mouth, pharynx, œsophagus or gullet, stomach, small intestine and large intestine (Fig. 103).

In the mouth, food is ground by the molar teeth, and as this process continues the food is moistened by the secretion of saliva from the salivary glands of the mouth. When chewing is complete small boluses of food are swallowed, passing through the pharynx and into a long distensible tube, the *œsophagus*. The œsophagus extends through the thoracic cavity and passes through the diaphragm (a musculo-membranous partition separating the thorax and abdominal

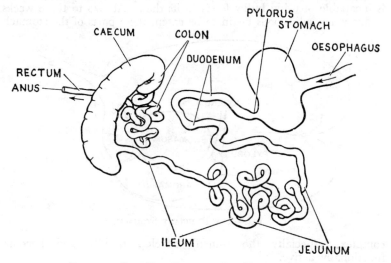

FIG. 103.—DIGESTIVE TRACT OF PIG (DIAGRAMMATIC).

cavity) into the abdominal cavity. In the abdominal cavity the gut enlarges to form a sac, the stomach. The junction between the œsophagus and the stomach is guarded by a muscular valve, the *cardiac sphincter*, and the exit from the stomach is guarded by a similar valve called the *pyloric sphincter*.

From the stomach leads off an extremely long coiled tube with thin walls called the *small intestine*. Near the stomach the small intestine forms a loop called the *duodenum*, with which is associated a secretory organ known as the *pancreas*. From the duodenum onwards the small intestine is attached to the dorsal wall of the abdomen by a membrane called the *mesentery*, which carries blood vessels and nerves to the intestinal wall.

The small intestine leads into a much wider tube, the large intestine, which can be divided into three regions, (*a*) a blind sac, the *cæcum*, (*b*) the *colon* and (*c*) leading from the colon, the *rectum* which opens externally by the anus.

The **Ruminant Digestive System.** Cattle, sheep and goats eat large quantities of bulky foods such as grass, and the digestive tract is modified for the storage and digestion of this large bulk. It is popularly believed that the ruminant has four stomachs, but in fact there is only one true stomach. The other three stomach-like compartments are really enlargements of the posterior end of the œsophagus (Fig. 104). The names of the four compartments are, (*a*) the *rumen* or paunch, (*b*) the *reticulum* or honeycomb stomach, (*c*) the *omasum* or book and (*d*) the *abomasum,* reed or rennet stomach which is the true stomach.

In the newborn calf or lamb the rumen, reticulum and omasum are relatively poorly developed and are non-functional. At this stage, which can be called the pre-ruminant phase, the young animal is incapable of using bulky foods in its diet. At two to three weeks of age when solid foods begin to be eaten, these parts of the stomach

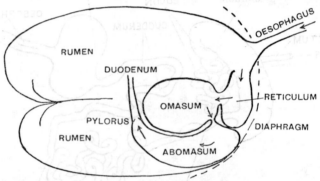

FIG. 104.—THE RUMINANT STOMACH.

complex (especially the rumen) develop rapidly and become functionally active.

The capacity of the rumen in adult cattle is enormous, amounting to some 225 to 270 litres, and the rumen therefore fills the greater part of the abdominal cavity. The remaining compartments are all much smaller. The reticulum has an internal lining, the structure of which resembles a honeycomb ; it is in this compartment that heavy objects such as pebbles eaten accidentally tend to accumulate. Internally the omasum has many leaves of tissue resembling a book, and is an organ concerned with the absorption of water. The abomasum is a relatively long tubular sac and is the only part of the stomach complex which secretes digestive juices (see p. 510). The abomasum is sometimes called the rennet stomach.

The large quantities of vegetable material eaten by the ruminant are stored in the rumen. Subsequently this material is regurgitated into the mouth, where it is mixed with the copious secretion of the salivary glands and ground into fine particles by the molar teeth. This activity is known as *cudding* or *rumination.* When the material is finely ground it is swallowed, water is absorbed by the omasum and the fine particles of food pass into the abomasum. From this point the processes of digestion are very similar to those of the monogastric animal.

It would be wrong to assume that the rumen is merely a place where food is stored, for active digestion of some food products takes place in this organ. This is possible because the rumen contains an enormous population of micro-organisms including bacteria and protozoa. These have several important functions. They are responsible for the digestion of the fibrous cellulose found in vegetable foods which cannot be utilised by the normal digestive processes : the cellulose is broken down by bacterial activity to produce short-chain fatty acids (acetic, butyric and propionic acids) which are

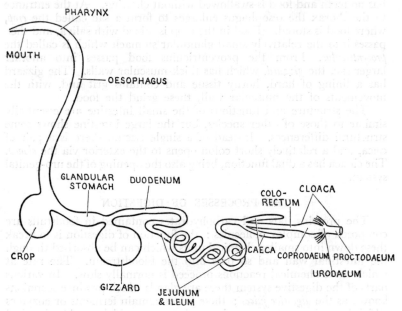

FIG. 105.—DIGESTIVE SYSTEM OF FOWL (DIAGRAMMATIC).

absorbed through the lining of the rumen, pass into the blood stream and are used as a source of energy by the animal.

A second important function of the rumen micro-organisms is that they take plant proteins and certain other nitrogenous compounds and build them up into microbial protein. When the micro-organisms die, they pass into the abomasum and along the digestive tract where they are digested, thus forming a valuable protein food. Not the least important activity of the micro-organisms is that they have the power to synthesise certain vitamins essential to the animal, including many of the B vitamins and vitamin C. The rumen provides an example of a relationship between a higher animal and micro-organisms which operates to give mutual benefits.

An important result of microbial activity is that large volumes of gas are produced in the rumen ; normally this gas is removed by belching or eructation, which is a normal phenomenon in the healthy ruminant.   If the belching mechanism is inhibited gas accumulates and causes distention of the rumen, giving rise to a condition known as *bloat* or *hoven* (see p. 679).

The horse, which also eats large quantities of vegetable food, has a rather different arrangement of the digestive organs.   The horse is a monogastric animal, but microbial digestion of cellulose takes place in the large intestine which is much enlarged in this species.   The process appears to be less efficient than in the ruminant.

The Avian Digestive System.   The general arrangement of the digestive organs of the hen is illustrated in Fig. 105.   The hen has no teeth and food is swallowed without chewing.   At the entrance to the thorax the œsophagus enlarges to form a sac called the *crop*, where food is stored.   Food in the crop is mixed with saliva, and then passes into the relatively small glandular stomach which is called the *proventriculus*.   From the proventriculus food passes into another larger sac, the *gizzard*, which has thick muscular walls.   The gizzard has a lining of hard, horny tissue and contains grit and, with the movements of the muscular wall, these grind the food.

The structure and functions of the small intestine are essentially similar to those of other species, but the large intestine shows some structural differences.   Instead of a single cæcum there is a pair of cæca, and a relatively short colon opens to the exterior via the *cloaca*. The cloaca has a dual function, being also the opening of the uro-genital system.

THE PROCESSES OF DIGESTION

The proteins, fats and carbohydrates contained in foodstuffs are composed of complex molecules : the function of digestion is to break these down into smaller chemical units which can be absorbed through the intestinal wall and so pass into the bloodstream.   The rate at which these chemical reactions proceed is normally slow.   In various parts of the digestive system there are glands which produce secretions known as the *digestive juices* ; these juices contain ferments or enzymes which have the power to accelerate the rate at which specific chemical reactions occur.   As the food passes through the digestive system different digestive juices are progressively secreted on to the food.

The passage of food through the alimentary canal is accomplished because the walls of the digestive organs contain smooth muscle fibres. Under the control of the autonomic nervous system these muscles contract rhythmically and gradually force food along the alimentary canal, continually churning the food and mixing it with the digestive juices as they do so.   This process is called *peristalsis*.   Occasionally food may be moved in the opposite direction, and this is referred to as anti-peristalsis.

The Digestive Juices.   It has already been stated that the digestive juices contain enzymes which bring about chemical changes in food.   The most important of the juices are :

Saliva.   This is produced in the salivary glands which are situated at the base and sides of the mouth.   Saliva has **the** important function of lubricating the food during mastication and so rendering it easy to swallow. It also contains an enzyme, *ptyalin*, which aids the conversion of starch into simple sugars. Enzymes which act on starch in this way are called amylases. There is no evidence of the presence of ptyalin in the saliva of cattle, sheep, horses or goats.

The secretion of saliva is under neural control and frequently the sight or thought of food induces a flow of saliva into the mouth.

The Gastric Juice. The gastric or stomach juice contains (*a*) *rennin* which curdles milk, (*b*) *pepsin* which converts insoluble proteins into more soluble peptones, (*c*) *lipase* which has a weak action on very finely divided (emulsified) fat droplets such as those of cream, and (*d*) hydrochloric acid which provides the acid conditions under which the enzymes of the gastric juice work best.

Rennin coagulates milk, producing a flocculent mass called a curd, and a clear fluid called whey. This reaction has been used from ancient times in the manufacture of cheese, employing rennin extracted from the stomach of the calf. Because of its action on proteins, pepsin is known as a proteolytic or protein-splitting enzyme.

As digestion proceeds the acidic contents of the stomach are gradually released into the small intestine where they are subjected to the action of the bile, the pancreatic juice and the intestinal juice.

Bile. The bile is produced in the liver, and is then stored in a small pear-shaped body called the gall-bladder, which is attached to the liver. There is no gall bladder in the horse, and it may be absent from individuals of other species without adverse effects. From the gall bladder the bile flows along the bile duct, which enters the small intestine in the region of the duodenum.

The bile itself contains no digestive enzymes ; it is largely composed of alkaline salts which have two important functions. Firstly, the bile salts cause fats to emulsify, i.e. to be reduced to minute droplets, in which state they can be more readily digested. Secondly, bile neutralises the acidity of material passing from the stomach and provides the slightly alkaline conditions in the small intestine necessary for the efficient activity of enzymes secreted by the pancreas and small intestine.

The Pancreatic Juice. This is also an alkaline secretion and includes three important enzymes, (*a*) *trypsin* which is proteolytic, (*b*) *amylopsin* which is an amylase, and (*c*) *steapsin* which is lipolytic or fat-splitting.

Trypsin acts upon proteins and peptones, breaking them down into smaller units called peptides. Amylopsin continues the breakdown of starch into sugars and steapsin splits fats into their component fatty acids and glycerol.

The Intestinal Juice. An alternative name for the intestinal juice is the *succus entericus*. It contains (*a*) peptidases which effect the final breakdown of peptides into amino acids which are the basic structural units of proteins, (*b*) Sucrase, maltase and lactase which act upon the sugars sucrose, maltose and lactose, breaking them down into simple sugars (largely glucose), and (*c*) lipase, a lipolytic enzyme which continues the digestion of fat.

The lower parts of the digestive tract, the cæcum, colon and rectum, secrete no digestive juices.

The Absorption of Digested Materials. Digestion produces simple sugars, amino acids, fatty acids and glycerol which can be absorbed into the circulatory system. The indigestible residue is voided in the fæces.

Almost the whole absorption of food products takes place in the small intestine, though some absorption of water and minerals (and possibly glucose) takes place in the stomach ; the absorption of water takes place in the large intestine and, in the ruminant, absorption of fatty acids occurs in the rumen. The small intestine is specially adapted for the absorption of food. This is due to the formation of millions of minute finger-like projections or *villi* (Fig. 106) which develop on the internal surface of the small intestine and so increase the absorptive area. Each villus is covered by a layer of delicate epithelial cells surrounding a network of blood capillaries which are fed by a small artery and drained by a small vein. Within this network is a lymph capillary or lacteal (see p. 517) opening into a small lymph vessel which passes away from the villus.

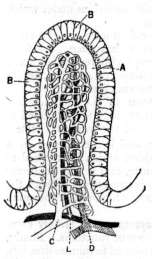

FIG. 106.—A VILLUS OF THE SMALL INTESTINE.

A, body of the villus. B, external covering of epithelium cells. C, the small artery entering the villus and breaking up into capillaries, which reunite to form—D, the small vein which leaves the villus. L, the lacteal radicle which occupies the middle of the villus.

The epithelial covering of the villus allows the passage of smaller molecules such as glucose which enter the blood capillary and are thence transported to all parts of the body. Absorption is not merely a passive process ; the epithelial cells play an active part in absorption and may exercise a selective effect in absorbing one product before another.

The absorption of fat is rather different from that of other food products. It has been suggested that fat is broken down by lipolytic enzymes into fatty acids and glycerol. This is partly true, but a large proportion of fat is not subjected to the activity of lipolytic enzymes, and remains in the small intestine in emulsified form. This emulsified fat can pass directly through the epithelial cells into the villus. Fat is not transported by the blood vessels but enters the lacteal and is transported by the lymphatic system.

THE CIRCULATORY SYSTEM

The circulatory system provides a means of internal transport in the body. Essentially it consists of a system of tubes through which flow the blood and lymph which act as media for exchange. The blood system which carries the blood, and the lymphatic system which carries the lymph, are best considered separately.

The Blood System. This consists of a closed system of tubes through which flows the blood.

The Blood. Blood is composed of a fluid medium, the *plasma*, in which are carried innumerable cells or *corpuscles*. There are two types of corpuscles, red and white.

The red corpuscles are by far the more numerous, and each consists of a biconcave disc ; unlike most cells they have no nucleus. Individually the cells are yellowish-red in colour but, in a mass, give the blood its characteristic red colour. An important function of the blood is to transport oxygen to the tissues, and this is a property of the red corpuscles. The colour of the red corpuscles is due to a substance called *hæmoglobin* which has a remarkable ability to absorb oxygen. This takes place in the lungs, where oxygen forms a loose compound with hæmoglobin, called oxyhæmoglobin. In this state the blood is said to be oxygenated and has a bright scarlet colour. Blood passing through the tissues rapidly gives up its oxygen to the tissues, and the hæmoglobin is now known as reduced hæmoglobin. In this state the blood is said to be deoxygenated and has a dark purplish-red colour. Carbon dioxide, produced as a waste product of tissue activity, is removed in blood passing away from the tissues, part being attached to the hæmoglobin and part being carried in the plasma.

The white cells, or *leucocytes*, are larger and less numerous than the red corpuscles and have nuclei. These cells constitute an important defence mechanism of the body against invading micro-organisms. Leucocytes are capable of passing through the thin walls of the smallest blood vessels and wander through the tissues, tending to congregate in sites of bacterial infection. Here they engulf and digest pathogenic bacteria, so protecting the body from disease. The pus which exudes from a festering wound is largely composed of white cells.

It is important that, should a blood vessel be damaged, all the blood does not leak out of the blood system. In fact this does not occur (unless the damage is extensive) because the blood has the power to form a jelly-like clot. This clot acts as a block to further blood and the bleeding ceases.

The blood system consists of the heart, the arteries, the veins and the capillaries.

The Heart. The heart is a hollow muscular organ and is the central force pump of the blood system (Fig. 107). It is composed of two halves, the right and left halves, which are each divided into a thin-walled anterior compartment, the *auricle*, and a thick-walled posterior compartment, the *ventricle*. The heart is enveloped in a delicate membrane called the *pericardium* and has a lining membrane called the *endocardium*. Between the auricle and ventricle of each side is a communicating passage guarded by a system of valves. The exits from the ventricles are also guarded by a system of valves. These valves are so arranged that blood can pass from the auricles to the ventricles and from the ventricles outwards, but not in the opposite direction. There is no direct communication between the two halves of the heart.

The Arteries, Capillaries and Veins. The arteries carry blood from the heart towards the tissues, and the two largest arteries are the *aorta*, which is the sole outlet of blood from the left ventricle, and the *pulmonary artery*, which receives the contents of the right ventricle. Traced outwards from the heart, the arteries branch and subdivide like the branches of a tree, becoming progressively smaller until, in the tissues, they give rise to very small *arterioles*. The walls of the arteries and arterioles are well supplied with bands of smooth muscle innervated by the autonomic nervous system. Contraction and relaxation of the walls of the arterioles alters the internal diameter of these vessels and so regulates the amount of blood passing to any part of the body.

In the tissues, each arteriole breaks up into an extremely fine network of narrow tubes, the *capillaries*, which are just large enough to permit the passage of blood corpuscles. The wall of each capillary tube is composed of a delicate membrane only one cell thick. This wall holds the blood in the capillaries but allows some fluids, food products and oxygen to pass into the tissues, and waste products of tissue activity to pass from the tissues into the bloodstream. Thus it is in this region that the essential exchange of products between the blood and the tissues takes place. The walls of blood vessels larger than the capillaries are impervious to both gases and fluids.

The capillary network gradually re-forms into larger tubes through which blood begins its journey back towards the heart. The small vessels collecting blood from the capillaries can be called venules and these become confluent into larger veins which ultimately discharge their contents into the auricles of the heart. The walls of the veins are much thinner and less muscular than those of the arteries.

Circulation of Blood. The general lay-out of the blood system is illustrated diagrammatically in Fig. 107. Contraction of the left ventricle drives scarlet, oxygenated, blood into the aorta. The aorta sends branches to the head and fore-limbs, and then passes in a posterior direction, lying beneath the vertebral column. From this part of the aorta, branches supply blood to the liver, kidneys, digestive system, hind limbs, etc.

Blood in the aorta is under pressure, but as blood passes into the smaller arteries pressure falls, and in the arterioles and capillaries blood pressure is low. This fall in pressure is essential because otherwise the thin-walled capillaries would burst.

Blood leaving the capillaries is returned to the auricles of the heart through the venous system. Blood pressure in the veins is even lower than in the capillaries, and it has been noted that the veins are only weakly muscularised. How then does blood return to the heart? This is due to several factors, the most important being (*a*) the left ventricle is continually pumping blood into the arteries, thus forcing blood through the capillaries and into the venous system, (*b*) the thorax is always at a slightly negative pressure (see p. 518) which causes blood to be drawn towards the heart by suction, (*c*) muscular contraction presses against the veins and helps to force blood along them—animals may stretch after a period of sleep to aid this effect,

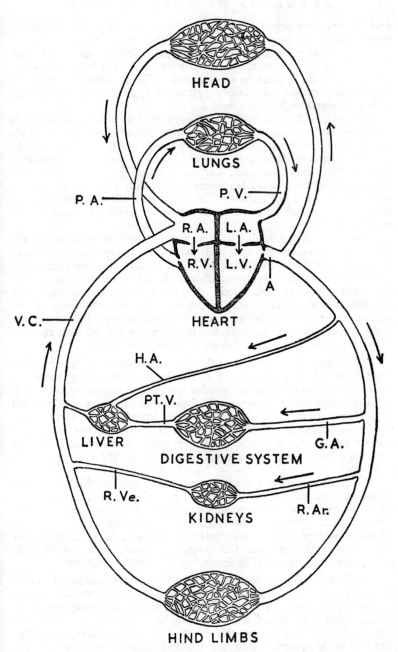

FIG. 107.—DIAGRAM OF THE CIRCULATION OF BLOOD.

R.A., right auricle ; R.V., right ventricle ; L.A., left auricle ; L.V., left ventricle ; P.A. pulmonary artery ; P.V., pulmonary vein ; A., aorta ; H.A. hepatic artery ; G.A., artery supplying intestines; R.Ar., renal artery   R.Ve., renal vein ; PT.V., portal vein ; V.C., vena cava.

and (d) veins are provided with flap-like valves which allow blood to pass towards the heart but not in the reverse direction.

Dark, venous blood enters the right auricle of the heart, but, before being passed to the left ventricle to start a further journey round the body, it must be oxygenated. This is achieved through a special loop of the circulatory system, the pulmonary circulation, which carries blood to the lungs.

Venous blood in the right auricle is forced into the right ventricle by auricular contraction. Then contraction of the right ventricle forces blood into the pulmonary artery which passes to the lungs. In the lungs the capillaries line the air spaces where carbon dioxide is released from the blood and oxygen is taken up from inhaled air. Oxygenated scarlet blood is returned to the heart by the pulmonary vein, entering the left auricle. Contraction of the left auricle forces blood into the left ventricle and the whole cycle of circulation is recommenced.

The pulmonary circulation gives rise to a unique situation in the animal ; it provides the only case where an artery (the pulmonary artery) carries dark deoxygenated blood and a vein (the pulmonary vein) carries scarlet oxygenated blood. In all other cases veins carry dark, and arteries scarlet, blood. The arrangement of the pulmonary circulation also explains why there are two separate sides of the heart ; the right side is concerned with collecting blood from the body and pumping it through the lungs, while the left side receives blood from the lungs and pumps it round the body.

The Pulse. The working of the heart is so co-ordinated that first the two ventricles contract and then relax as the auricles contract ; then the auricles relax as ventricular contraction occurs again. This cycle of operations is known as the beating of the heart. As the left ventricle contracts, the rush of blood being forced into the aorta causes throbbing of the larger arteries. This is known as the pulse and can be felt in places where an artery passes over the surface of a superficial bone, e.g. in the human wrist or on the underside of a cow's tail.

The pulse rate gives a measurement of the rate at which the heart is beating, and the normal number of pulsations per minute is 55 in the adult ox, 75 in the sheep and pig and 312 in the domestic fowl. These figures cannot be accepted rigidly because pulse rates are subject to wide variations, increasing with exercise or excitement and being at a minimum when the animal is at rest. Abnormally high pulse rates are generally indicative of a feverish condition.

The Spleen. The spleen is a purplish-red organ situated in the abdomen near the stomach, and it plays an important part in the circulation of blood. Internally the spleen contains many spaces or sinuses filled with blood and these act as a reservoir ; when the demand for blood increases the spleen contracts, releasing its blood into the blood system. The spleen also has the important function of destroying wornout red blood corpuscles, and is a site where certain leucocytes are manufactured.

The Lymphatic System. It has been mentioned that fluids,

food products and oxygen pass out of the capillaries into the tissues ; here the fluid material is known as the tissue fluid. When the exchange of materials between these fluids and the tissues is complete, part of the fluid passes back into the venous end of the capillaries, but the remainder is drained from the tissues into the lymphatic system. The fluid drains into small capillary vessels of this system called *lymphatics*. Lymphatics present in the villi of the small intestine are given the special name of *lacteals*. The lymphatics become confluent into larger lymph vessels which are similar in structure to the veins. These ultimately open into two main ducts (*a*) the thoracic duct lying beneath the vertebral column, which drains the greater part of the body, and (*b*) the right lymphatic duct, which drains some regions of the head, fore-limbs and thorax, and part of the liver. These two main ducts open into a large vein in the neck region and thus all the fluids are finally returned to the blood system.

Tissue fluids entering the lymphatic system become known as the *lymph*, which is almost identical in composition with the tissue fluids and, in addition, contains large numbers of leucocytes. Normally no red blood corpuscles are present in the lymph. The ability of leuco- cytes to engulf and destroy bacteria has already been mentioned and their defensive activity is aided by the formation of small round bodies called lymph-nodes at strategic points along the lymph vessels. These filter out bacteria from the lymph and hold them in a position where they can be more easily attacked and destroyed by leucocytes ; in this way many bacteria are prevented from entering the blood stream. The lymph-nodes are also a site for the manufacture of leucocytes.

Movement of lymph is caused by muscles pressing against the lymph vessels during muscular contraction. Valves situated in the lymph vessels (which are similar to those of veins) ensure that lymph flows in the correct direction.

THE RESPIRATORY SYSTEM

The vital exchange of oxygen between atmospheric air and the blood takes place in the lungs, and air is drawn into the lungs through a series of air passages. These air passages are the nasal cavities, pharynx, larynx, trachea or wind-pipe, bronchi, bronchioles and air sacs (Fig. 108).

From the nose or mouth air passes through the pharynx and larynx and then enters the trachea which extends from the larynx into the thorax. The *trachea* is composed of a series of resilient rings of cartilage which give it its characteristic appearance. In the thorax the trachea divides into two branches, the right and left *bronchi*. Each bronchus enters the corresponding lung where it divides and subdivides into smaller branches, the terminal twigs being known as *bronchioles*. These open into irregularly shaped air sacs which are lined by the blood capillaries of the pulmonary circulation. It is at this point that air is brought into close contact with the blood for the purpose of gaseous exchange.

The lungs themselves consist mainly of air spaces surrounded by blood vessels and a pale-coloured elastic tissue. The lungs are each invested in a double membrane called the *pleura*, which enable the lungs to glide smoothly over the internal surface of the thorax as the respiratory movements take place.

During the process of gaseous exchange carbon dioxide passes into the air sacs and oxygen is taken up by the blood, reduced hæmoglobin being converted to oxyhæmoglobin. At the same time air in the lungs becomes warm and takes up some water vapour; this means that the lungs can play a part in the excretion of water and in the regulation of body temperature—points which will be discussed later. As a result of the gaseous exchange the air in the lungs becomes richer in carbon dioxide and water vapour, and poorer in oxygen than atmospheric air. Clearly, before any further exchange can take place, air must be expelled from the lungs and be replaced by fresh atmospheric air. This is accomplished by a series of breathing movements.

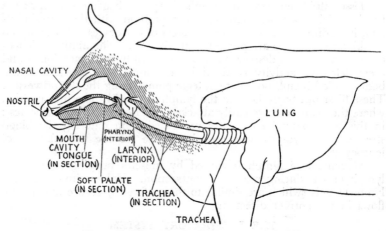

FIG. 108.—DIAGRAM OF RESPIRATORY SYSTEM.

The Breathing Movements. The thorax is bounded by the ribs, the sternum, the backbone and the diaphragm, and is almost completely filled by the lungs and heart. The thorax is at a pressure slightly below that of the atmosphere due to the elasticity of the lungs.

The drawing of air into the lungs is known as inspiration or inhalation, and its expulsion as expiration or exhalation. Inspiration and expiration alternate rhythmically and occur at a rate of 26–30 per minute in the resting cow, 10–20 in the resting sheep or pig and 15–48 in the domestic fowl. Respiration rate is more rapid in young animals, after exercise and under conditions where the animal is hot or feverish.

Essentially the breathing movements are similar to the action of bellows where the nozzle of the bellows is equivalent to the trachea. At inspiration the volume of the thorax is increased, the lungs expand and air is drawn into the larger air tubes. This increase in the volume of the thorax is brought about by movements of the ribs and diaphragm. The intercostal muscles between the ribs contract and cause movements of the ribs and sternum which increase the volume of the thorax. At the same time the muscles of the diaphragm (a dome-shaped sheet of muscle which separates the thorax from the abdomen) contract ; as contraction occurs the dome becomes flattened and again the volume of the thorax is increased.

In expiration the intercostal muscles relax and the ribs and sternum return passively to their original position. At the same time the muscles of the diaphragm relax and it rises towards the thorax, this movement being aided by the negative pressure existing in the thorax.

In all these movements the elastic lungs play a purely passive role, expanding and drawing air into the lungs as the volume of the thorax increases during inspiration, and contracting as the volume of the thorax diminishes during expiration so that air is expelled from the lungs. It is important to note that the respiratory movements do not carry air directly into the air sacs ; the movement of gases to and from the air sacs through the bronchioles takes place purely by gaseous diffusion.

The breathing movements are normally under autonomic control, i.e. at the sub-conscious level, but can be controlled at the conscious level. Breathing is inhibited during swallowing to prevent food materials passing into the trachea. Coughing represents a rather violent modification of the breathing movements in response to irritation of the lungs.

THE EXCRETORY SYSTEM

The body possesses specialised excretory organs, the kidneys, but these are by no means the only organs concerned with the excretion of waste products. Some waste products may be excreted in the saliva, while others are excreted into the alimentary tract and pass out of the body with the fæces. It has been noted that carbon dioxide and some water vapour are eliminated from the lungs, and in this respect the lungs must be regarded as excretory organs. The liver also has an excretory function, for the bile, as well as aiding in the digestive processes, contains products of excretion. The skin, too, has an excretory function, for the perspiration carries excretory products. The kidneys have a most important function, as excretory organs, and are essential for the continuance of life.

The Kidneys. The two kidneys have a characteristic shape and are situated in the abdomen in a well-protected site beneath, and on either side of, the back-bone. Each kidney is embedded in a protective cushion of fat.

The excretory units of the kidney are microscopic nephrons (Fig. 109). Each *nephron* consists of two parts, a cup-shaped unit (Bowman's capsule), and leading from this capsule a long, looped tube, the *renal tubule*. The kidneys receive their blood supply from the renal artery and, in the kidneys, arterioles pass to the nephrons where they form a special capillary network, called the *glomerulus*, within Bowman's capsule. The vessel leaving the glomerulus passes

T

on and forms a second capillary network laced round the renal tubule. Blood is drained from the kidneys by the renal vein.

Fluid materials leave the bloodstream from the glomerulus, pass into Bowman's capsule and then down the renal tubule. The fluid passing into Bowman's capsule contains not only waste products, but also substances (such as glucose) which are required by the body. These latter substances are selectively reabsorbed in the renal tubule, and pass back into the bloodstream through the capillary surrounding the tubule. The remaining fluid, the *urine*, passes on down the renal

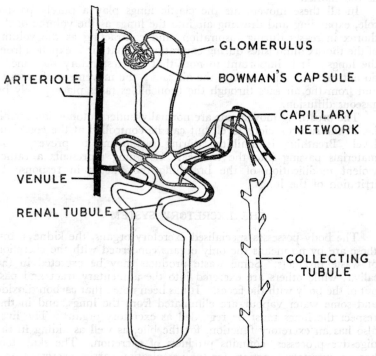

FIG. 109.—A KIDNEY UNIT OR NEPHRON (DIAGRAMMATIC).

tubule. Gradually the tubules become confluent into larger tubes and urine leaves the kidney by a single tube, the *ureter*. In mammals the ureters from each kidney pass to a storage organ called the bladder, situated in the pelvic region of the abdomen. From the bladder a single tube, the *urethra*, conveys urine to the uro-genital organs, from which it passes to the exterior. In birds there is no bladder and the ureters open into the *cloaca*.

Urine. The urine is composed largely of water, but contains several salts (including sodium chloride) and some nitrogenous compounds. The main nitrogenous compound produced by mammals is urea and by birds uric acid. Because of the presence of these materials, urine has a high fertilising value.

THE SKIN

The skin, in addition to serving as a protective covering for the body, is of considerable functional significance, being concerned both with excretion and with the regulation of body temperature.

Microscopic examination of the skin shows it to be composed of two layers, an upper *epidermis* and beneath this the *dermis*. The epidermis has mainly a protective function and is composed of several layers of cells, the outer layers being formed of dead scaly cells. The hairs also originate from the epidermis, though their roots are found in the dermis. The dermis is a thicker layer which contains the blood vessels and nerves of the skin : there is much adipose tissue in the lower layers of the dermis (the sub-cutaneous fat). From the dermis arise many coiled sweat glands, which pass through the epidermis and discharge their contents on to its surface. Also the roots of the hair follicles are found in the dermis, and each follicle has associated with it a sebaceous gland which secretes a lubricating fluid on to the hairs. The fowl has no sweat or sebaceous glands but the feathers are similar in origin to the hairs. Both hairs and feathers each have associated with them a small muscle which causes their erection ; the erection of hairs is most evident in frightened or enraged animals though it may also be observed when the weather is very cold.

Pigs have a rather sparse covering of hair, while cattle have a thick cover over most of the body. In both species the colour and pattern of hair distribution is distinctive of breed. In sheep the main cover is formed of a modified type of hair known as wool though true hairs are also present.

Temperature Control and Excretion. The normal average body temperature is 38 °C in the cow, 39 °C in the sheep, 39 °C in the pig and 41 °C in the fowl. Normally, body temperature is controlled within narrow limits and the control mechanism is associated with the skin. Heat is lost from the skin surface to the generally cooler air surrounding it. In temperature regulation the blood acts as a carrier of heat to a part of the body from which it can be lost. The blood vessels of the skin are under neural control and the magnitude of heat loss from the skin can be increased or decreased through variations in the quantity of blood passing through the skin. For example, if body temperature tends to rise (as on a hot day) the blood vessels of the skin dilate, more blood is brought from the interior of the body to the skin, and heat loss increases accordingly. Conversely, under cold conditions the blood vessels of the skin become constricted so that blood flow is reduced and heat is retained in the body.

If, under hot conditions, the increased blood flow to the skin fails to maintain temperature within normal limits, a nervous impulse from the brain stimulates the sweat glands to secrete sweat. As the sweat evaporates from the skin surface it has a cooling effect on the body. Under these conditions the animal is said to sweat or perspire.

Some excretory products are lost from the body in the sweat, but in the healthy animal the skin is relatively unimportant as an excretory organ.

Under extremely cold conditions body temperature may tend to fall in spite of constriction of the blood vessels ; if this occurs spasmodic muscular contractions are initiated and the animal is said to shiver. The heat produced as a by-product of this muscular activity helps to bring body temperature back to within normal limits.

There are, of course, limits within which normal body temperatures can be maintained. Small or moderate fluctuations can be quickly rectified but, under conditions of extreme heat or cold, the temperature-regulating mechanism may break down with serious consequences. The new-born animal is extremely sensitive to temperature changes and may quickly die under cold conditions.

THE ENDOCRINE SYSTEM

The way in which enzymes can accelerate the rate at which particular chemical reactions occur has already been described in relation to the digestive processes. Other substances produced in the body have the power to regulate whole body processes rather than single reactions ; these are known as *hormones* and are secreted by the glands of internal secretion or *endocrine glands*. These hormones are secreted into the bloodstream and are thus rapidly conveyed to the organ or organs upon which they exert their effect.

The most important endocrine glands are the thyroid glands, the parathyroid glands, the adrenal glands, the pancreas and the pituitary gland.

Thyroid Glands. The thyroids are a pair of glands situated in the neck on either side of the trachea. Their secretion, *thyroxine*, is a hormone which regulates the general level of activity of the body. If thyroxine secretion is inadequate the activities of the body become very sluggish and, conversely, if the rate of secretion is excessive the body becomes over-active with a rapid loss of weight and an excitable nervous condition develops.

In the region of the thyroid glands can be found two smaller glands called the parathyroids. They are concerned with the storage and utilisation of calcium and phosphorus : milk fever of dairy cows (see p. 678) is a symptom of temporary parathyroid inactivity.

Adrenal Glands. The adrenal glands are situated in the abdominal cavity, one being near each of the kidneys. Each gland consists of two distinct parts, an internal medulla and, enclosing this, the cortex. The adrenal medulla secretes *adrenalin* in times of excitement, fear or intense muscular activity. It causes the heart rate to increase and the arteries of the skin are constricted so that the blood flow to the skeletal muscles is increased. At the same time there is a rapid release of glucose from the liver which is conveyed to the muscles, where it is used as a fuel for muscular work. A further effect of adrenalin is to cause contraction of the smooth muscles attached to hairs or feathers, and this explains why they are elevated during fear or excitement. The adrenal cortex secretes many hormones, which are referred to collectively as corticoids. These have effects on a whole range of body functions and are intimately concerned with the control of water and salt levels in the body.

The Pancreas. The pancreas has been described already as an organ which elaborates digestive enzymes, but it also has an endocrine function. Its secretion, *insulin*, is concerned with the regulation of blood sugar levels ; insulin insufficiency results in high levels of sugar in the blood and the appearance of glucose in the urine. This condition is referred to as diabetes : human diabetics can lead normal healthy lives as long as they receive regular injections of insulin.

Pituitary Gland. One of the most important endocrine glands is the pituitary gland because many of its secretions regulate the activities of other endocrine glands. The pituitary gland is situated in a well-protected site in the floor of the skull and is divided into two main regions called the anterior and posterior lobes. The posterior lobe secretes *oxytocin*, a hormone which causes smooth muscle fibres to contract and which, amongst other phenomena, is responsible for milk " let down " (see p. 623). The anterior lobe produces six hormones, (*a*) growth hormone which stimulates growth (particularly of bone and muscle), (*b*) thyrotrophic hormone which stimulates the thyroid glands to secrete thyroxine, (*c*) adrenocorticotrophic hormone which stimulates the adrenal cortex to secrete corticoids, (*d*) and (*e*) two gonadotrophic hormones which control the activities of the gonads (the ovary and testis), and (*f*) prolactin which stimulates milk secretion.

The gonads may also be regarded as endocrine glands, but are described separately below.

REPRODUCTION IN MAMMALS

The Female Reproductive Organs. That part of the reproductive tract visible from the outside of the body is the *vulva* surrounding the uro-genital opening. Associated with the vulva is a small erectile organ which is analogous to the male penis and is called the *clitoris*. From the vulva leads a broad tube called the *vagina*, some inches along which is a narrow neck called the *cervix* which opens into the *uterus* beyond it. The main body of the uterus divides into two uterine horns which are suspended from the body wall by a ligament. Each uterine horn leads into a long narrow tube called the *fallopian tube* or *oviduct*, which opens into a membranous funnel. This funnel is open to the body cavity but partly envelops the *ovary*, which is a small ovoid structure (Fig. 110).

Stimulation of the ovaries by gonadotrophic hormones results in the production of the female gametes, *ova* or eggs and the female sex hormone called *œstrogen*. The ova are produced in small spherical follicles which project slightly from the surface of the ovary. During formation of the ovum there is a meiotic or reduction division of the nucleus (see p. 468) so that the gamete is haploid. Each ovum is a microscopic spherical body consisting of a nucleus surrounded by a series of delicate cell membranes. Œstrogen is produced by the cells lining the ovarian follicle and is responsible for, (*a*) the female characteristics, (*b*) heat or œstrus (the period during which the female is receptive to the male), and (*c*) the development of the mammary gland or udder.

As the ovum develops with the follicle, the follicle itself increases in size and finally ruptures in the act of ovulation, or setting the ovum free. The ovum falls into the funnel at the head of the oviduct and passes into the oviduct where fertilisation occurs. After fertilisation the ovum passes into the right or left horn of the uterus to the wall of which it becomes attached and where development of the fœtus occurs.

After ovulation the cells lining the ovarian follicle proliferate rapidly and form a knob-like projection on the surface of the ovary known as the *corpus luteum*. This is an endocrine gland and secretes *progesterone*, a hormone which stimulates the uterus to nourish the developing fertilised ovum and which, like œstrogen, is concerned with the development of the mammary gland.

The Male Reproductive System. The essential male organs are two *testes* which lie in the *scrotum*, a sac outside the body cavity.

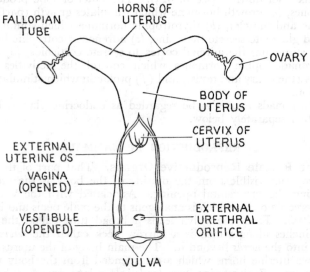

FIG. 110.—DIAGRAM OF FEMALE GENITALIA.

The reason for this apparently exposed position of the testes is that they function most efficiently at temperatures below those of the body cavity. Gonadotrophic hormones stimulate the testes to produce the male gametes or *spermatozoa* and the male sex hormone called *androgen*.

Androgen is responsible for the male characteristics and sexual desire. Castration (removal of the testes) inhibits the development of these characters and is commonly practised where animals are being reared for meat production, for it has a quietening effect on the animal and improves the quality of the meat produced.

Each minute spermatozoon has an ovoid head region, a short cylindrical body and a long thin tail which by whip-like movements is capable of causing movements of the spermatozoon. Like the ovum, each spermatozoon has a haploid nucleus. After production spermatozoa are stored in the *epididymis*, a tube situated within the scrotum. From the epididymis a duct leads back into the abdominal cavity, where it joins with the urethra in the region of the bladder. The urethra is therefore a common uro-genital duct and passes to the exterior through the *penis* (Fig. 111).

Fluids are secreted into the urethra from a number of accessory glands including the seminal vesicles, the prostate gland and the bulbo-urethral or Cowper's gland; the mixture of these fluids and spermatozoa is known as *semen*, and each drop of semen may contain many thousands of spermatozoa.

The male copulatory organ or penis is formed of erectile tissue, and at copulation it becomes erect due to the accumulation of blood within this tissue. In copulation the penis is inserted into the female vagina, and neural stimuli cause contractions of the epididymis and vas deferens thus forcing semen out through the penis and into the

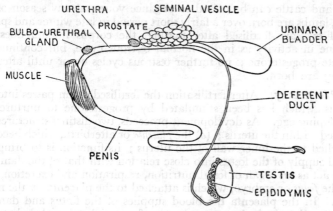

FIG. 111.—DIAGRAM OF MALE GENITALIA (BULL).

upper end of the vagina; this is called ejaculation. The quantity of semen produced at ejaculation shows considerable species variation, amounting to 1–2 cc in the ram, 5 cc in the bull and up to 500 cc in the boar. Once deposited in the vagina, spermatozoa are transported to the oviducts, partly due to movements of the spermatozoa themselves and partly by rhythmic contractions of the uterine wall.

In fertilisation a single spermatozoon enters the ovum; there is a fusion of nuclear material between the two cells, and the resulting fertilised ovum is the parent cell of the future offspring.

The Œstrous Cycle. The lives of both the ovum and spermatozoon are strictly limited, and for successful fertilisation it is therefore essential that mating should occur near the time of ovulation. This is ensured because, although the male will mate at almost any time, the female is only receptive to the male when she is in " heat ", or œstrus. The phenomenon of heat is due to the action of œstrogen produced by the ovarian follicle, and ovulation usually occurs towards or at the end of heat. This arrangement ensures that, after mating, spermatozoa will be present in the oviducts as ova pass from the ovaries.

The average duration of heat is eighteen hours in the cow, thirty-six hours in the ewe and two to three days in the sow : heat normally occurs at regular intervals. The interval between heats is known as the œstrous cycle. The period of sexual quiescence between heats corresponds to the time when the corpus luteum is actively secreting progesterone. As the activity of the corpus luteum declines, œstrogen is again secreted by the developing ovarian follicle and a further heat occurs.

In the non-pregnant cow and sow, heat occurs at intervals of about twenty-one days throughout the year.

The sheep, however, is a seasonal breeding species and in the ewes of most breeds breeding activity is confined to the autumn and winter months. During this period heat occurs at intervals of about sixteen days. In the spring and summer months the ewe is unreceptive to the ram, and this is known as the anœstrous period. Thus, although pigs and cattle can be mated to produce young at any season of the year, lambs are born over a fairly short period in late winter and spring.

If ova are fertilised after mating, the corpus luteum does not decline in activity as in the normal œstrous cycle, but continues to secrete progesterone ; no further œstrous cycles occur until after the young are born.

Pregnancy. After fertilisation the fertilised ovum passes into the uterus, which has been stimulated by progesterone to nurture the developing egg. As development proceeds, two distinct structures are formed within the uterus ; (a) the *placenta* or afterbirth, which becomes attached to the inner wall of the uterus ; its function is to bring the blood supply of the fœtus into close relation with that of the dam and so to act as an organ of fœtal nutrition, respiration and excretion, and (b) the *fœtus* or embryo, which is attached to the placenta by the navel cord. In the placenta the blood supplies of the fœtus and dam do not actually mingle, but an exchange of materials takes place between the capillaries of the two blood systems.

When fœtal development is complete, *parturition* (the act of giving birth) occurs. At this time the muscular walls of the uterus contract rhythmically with great force, and in this they are aided by muscular contractions of the abdominal wall. These contractions break the connection between the fœtus and placenta and the young are expelled through the vagina. The placenta is subsequently voided from the uterus in the same way.

The average duration of pregnancy or gestation is 116 days in the sow, 147 days in the ewe and 283 days in the cow. The pig has a relatively short gestation period, and is the only farm mammal capable of regularly producing two litters each year. Although the ewe is potentially capable of producing two crops of lambs per year, the limited duration of the breeding season precludes this in most cases.

Fertility. High levels of fertility are the basis of economic success in any form of livestock production. For the different classes of stock different standards are proposed. A sow should produce two litters per year, with nine or ten pigs reared from each litter. The cow should produce a calf each calendar year. The lambing percentage in sheep, i.e. the number of lambs reared expressed as a percentage of the number of ewes mated, should be 150 on lowland farms.

Fertility is controlled by several factors :

1. Age at sexual maturity, which determines when the animal can be first used for breeding. Gilts are normally mated at 8 to 12 months old, ewe lambs at 6 to 10 months (on lowland farms) and heifers at 15 to 20 months.

2. Ovulation rate (the number of ova shed from the ovaries at ovulation). Cattle normally ovulate one ovum, but double ovulations sometimes occur. Ovulation rates in sheep fall within the range of 1–3 depending upon breed and nutrition ; generally the higher the ovulation rate the more lambs will be born. The sow may ovulate as many as 30 ova but, again, breed and nutrition are important.

3. The number of ova fertilised. Failure of fertilisation may imply either that the male or female is infertile, or that the timing of mating was incorrect. Incorrect timing may be a cause of poor fertilisation rates where artificial insemination is used.

4. The number of fertilised ova which develop successfully to produce viable young. Sometimes diseases may result in the death of the fœtus, e.g. contagious abortion of cattle caused by *Brucella abortus*, but some fœtuses die and are reabsorbed by the dam when no disease organism can be identified ; up to 40 per cent of fertilised ova may be lost in this way.

5. The length of useful breeding life.

Artificial Insemination (A.I.). A.I. involves the collection o semen from the male and its insertion into the female reproductive tract during heat or œstrus. For the collection of semen the male is generally allowed to serve into an artificial vagina from which the semen can be recovered. Insemination of the female is carried out by trained inseminators, who insert semen into the female tract (either the uterus or vagina), using special syringes.

The large-scale use of A.I. in this country began in cattle in 1942 and has become particularly important in dairy cattle. The majority of bulls used in A.I. are stationed at centres controlled by the Milk Marketing Board. Artificial insemination of cattle results in conception rates which are similar to those achieved by natural service, and it is possible to dilute semen (with semen diluents such as egg yolk-phosphate, egg yolk-citrate or milk preparations) by 50 to 100 times without any appreciable loss of fertility. This is important because bulls of proven ability can be used to inseminate large numbers of cows. Further, in 1952 a technique was developed whereby semen could be stored for long periods at low temperatures by the addition of glycerine to the semen and freezing to $-78°$ C. Semen stored by this " deep freeze " technique has shown only a slow decline in fertility even after several years' storage. The technique has made it possible to preserve semen from outstanding bulls for use after the bull itself has died.

Now that bulls can be used on such an extensive scale it has become necessary to carry out extensive progeny testing (see p. 490) so that bulls of outstanding genetic merit can be discovered.

A.I. in sheep has not been used to any great extent in this country, though it has proved successful on a large scale in Russia and has been used in trials in Australia and South America. The development of a successful technique for pigs has proved difficult, but recent successes suggest that A.I. may be widely used for pigs in the near future.

Lactation. In early post-natal life the young animal is entirely dependant for its nutrition on milk produced in the mammary gland. The development of the mammary gland is under the dual control of œstrogen and progesterone, and is synchronised with pregnancy so that a supply of milk becomes available at parturition.

The udder is composed of secretory tissue associated with some fatty and connective tissue, and milk leaves the udder through teats situated on its ventral surface. In the sheep and cow the udder is suspended from the posterior region of the ventral surface of the abdomen, but in the pig it may extend over most of the ventral surface of the abdomen.

Although the sow and ewe produce only sufficient milk for their young, scientific breeding and feeding of dairy cows have resulted in a level of milk production far in excess of the needs of the calf.

REPRODUCTION IN BIRDS

The reproductive organs and the processes of reproduction in birds are different from those of the mammal. In the male the testes are retained within the abdominal cavity : the male copulatory organ consists of a pair of papillæ projecting into the cloaca which enter the female cloaca during mating. Another important difference is that semen retains the power of fertilisation for several days after mating. (Fig. 112).

Castration of the cockerel produces a heavy, fat bird known as a capon which is suitable for the table, but castration by surgical means is not usually practised. Normally cockerels are " castrated " by implanting pellets of female sex hormone beneath the skin of the neck ; these inhibit androgen production from the testes and produce the same effects as removal of the testes.

In the female only the left ovary normally develops. At ovulation the whole of the ovarian follicle passes into the oviduct, the egg is coated with a thick layer of albumen (the white of the egg) and a hard chalky shell is formed around it. There is no œstrous cycle comparable with that of the mammal, and eggs are laid at roughly daily intervals.

Development of the fœtus takes place within the egg, all the nutrients necessary for fœtal development being present in the egg when it is laid.

THE PATTERN OF DEVELOPMENT OF ANIMALS

Comparisons of young and adult animals of the same species reveal that the young animal is not merely a smaller edition of the adult but has a different form. These differences occur because, during development, the different tissues, organs and regions of the body develop at different rates. For example, there is an early phase of rapid development of nervous tissue, followed by bone, then muscle and finally fat. Also different regions of the body show an ordered development. A primary wave of growth passes from the skull anteriorly to the facial parts and then posteriorly to the loin (the region of the lumbar vertebræ). Secondary waves pass from the extremities of the limbs towards the body and ultimately converge on

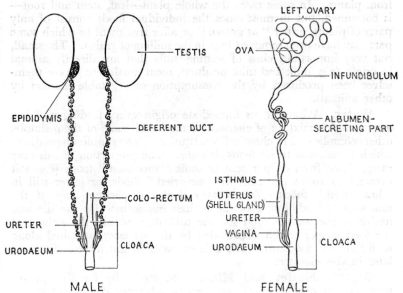

FIG. 112.—REPRODUCTIVE SYSTEM OF FOWL (DIAGRAMMATIC).

the loin. Finally, if the rate of development of different organs is compared, it is found that organs which are required to function early in life, e.g. the brain, eyes, heart, liver and kidneys, develop earlier than an organ, such as the rumen, which does not function until some time after the animal is born.

These developmental effects become evident early in fœtal life and result in an animal which, at birth, has a relatively large head and long limbs (with the extremities as their best developed parts) but a rather narrow, shallow trunk with a poorly developed loin region. In post-natal life the regions which are poorly developed at birth develop rapidly until the animal assumes its adult form.

The pattern of growth and development may be modified by various factors. Firstly, some breeds develop more rapidly to maturity than others, and this is of prime importance in animals being used for meat production. Secondly there are marked differences in size and appearance between males and females of the same species. And finally, nutrition has fundamental effects on the rate at which animals grow and develop.

Chapter XVIII

FOODS AND FEEDING

THE GENERAL CHEMICAL CHARACTER OF FOOD

ALL materials used as food for animals come directly or indirectly from plants. In some cases the whole plant—leaf, stem and root—is consumed, but in most cases the individual foods consist of only parts of plants, either " as grown " or after treatment in which some part may have been removed (e.g. in the milling of grain). The small, but very important, class of feeding-stuffs that are directly animal in origin (e.g. milk and milk products, meat meals, etc.) have themselves been produced by the consumption of vegetable matters by other animals.

Water. Whatever its immediate origin every feeding-stuff is a very complex mixture of chemical compounds, some of them simple, others complicated in chemical constitution. The simplest ingredient, which is common to all foods, is *water*. The proportion of this may range down from 95 per cent in some " succulent " foods (e.g. soft turnips) to 10–15 per cent in " air-dried " foods, or lower still in " heat-dried " foods. Water itself does not perform any of the functions of food, but is of the highest importance for the life and health of the animal, and for the utilisation within the animal of the food proper which is provided by the *dry matter* supplied along with it in the foodstuff. The role of water in nutrition is discussed later in this chapter.

Organic Matter and Mineral Matter. By far the greater part (90 per cent or more) of the dry substance consists of *organic matter* (i.e. material which can be burnt away, or oxidised), the rest consisting of a mixture of *mineral matters* (ash). The basic explanation of the differences in nutritive characteristics and value of different feeding materials lies in the nature and proportions of the various chemical compounds of which these two main sub-divisions of the dry matter are composed.

Mineral elements are found in every tissue and fluid of the animal body and its products, some being essential to the proper functioning of the body metabolism, whilst others are apparently not essential, though not necessarily useless if present. Apart from the need in appreciable quantities of certain mineral elements (notably calcium and phosphorus) for the structural needs of the animal, the functions of the mineral matters resemble those of the water, in that they do not so much function as " food " but as activators and regulators of the food utilisation processes and of the efficient working of the bodily organs.

The food proper is thus confined entirely to the organic matter, and a knowledge of the nature and functions of this part of the animal's dietary is consequently necessary for an understanding of the nutritive value of a feeding-stuff or ration, and an efficient control of its use.

The Analysis of Feeding-stuffs. In terms of the number of different chemical compounds present in even the simplest-looking feeding-stuff, the organic matter is so complex that any attempt to ascertain the proportion of each separate chemical compound present in it would be a matter of prolonged research for the chemist.

For the practical purposes of the farmer and the feeding-stuffs trader, however, it is essential that the chemist shall be able to give guidance fairly quickly as to the composition of the different food materials submitted to him. For these purposes, therefore, a simplified scheme of analysis is used which, though based upon assumptions which are never entirely accurate, gives data which have been proved by long experience to be reasonably adequate for the practical purposes of the farm. The most important of these assumptions are (1) that all nitrogenous compounds in foods are proteins containing 16 per cent nitrogen ; (2) that all ether-soluble matter in foods are fats. Because of these uncertainties scientists prefer in setting out the results of analysis the modified terminology for the items that are shown to the right of the example given on page 532.

This analytical scheme takes advantage of the fact that almost the whole of the chemical compounds which make up the organic matter of foods belong to the three great chemical groups or " families " of *carbohydrates, fats* and *proteins*. Each group is sharply differentiated from the others ; within each group there is sufficient resemblance between the individual members in composition and nutritive characteristics to make it possible for practical purposes within wide limits to treat each group as a nutritional unit. Visualised along these lines the composition of the food can be set out as indicated in the schedule below.

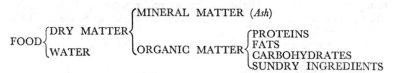

The scheme of analysis in common use in all countries is based upon the foregoing schematic conception of the composition of food, with the modification that the carbohydrate item is sub-divided into *soluble carbohydrates* and *crude fibre*. Furthermore, since the particular scheme of analysis adopted involves the embodying of the " sundry ingredients " item in the " soluble carbohydrates " figure, the latter is commonly returned as "soluble carbohydrates, etc.". A further extension sometimes made is to divide the " protein " figure (described as " crude protein ") into " true protein " and " non-protein nitrogenous matter " or " amides ". Another refinement is to indicate the proportion of sand, etc. (dirt), present in the ash of the food.

A typical analysis of wheat grain carried out according to this scheme in full might thus be set out as follows :

ANALYSIS OF WHEAT GRAIN

Per cent.

Water	12·8	(= Loss of weight at 100 °C)
Crude protein . . .	11·9*	$\left(= \text{Total N} \times \dfrac{100}{16}\right)$
Fats	1·9	(= Ether extract)
Sol. carbohydrates, etc. . .	69·6	(= Nitrogen-free extractives)
Crude fibre	2·0	(= Fibre)
Mineral matter (ash) .	1·8†	(= Ash)
Total . . .	**100·0**	

* Including True protein . . . . 11·1 $\left(= \text{Protein N} \times \dfrac{100}{16}\right)$

"Amides" or non-protein nitrog. subs. 0·8 $\left(= \text{Non-protein N} \times \dfrac{100}{16}\right)$

† Including sand, silicious matter, etc. 0·2.

In carrying out the analysis, direct determinations are made of five of the six items, viz. water, crude protein, fats, crude fibre and mineral matters ; the remaining item—soluble carbohydrates, etc.— is then arrived at by difference, i.e. by subtracting the sum of the five directly ascertained percentage data from 100. Carbohydrates form by far the most abundant group of food ingredients in the great majority of our feeding-stuffs, and it is unfortunate therefore that a direct determination of their amount cannot readily be made. The difference figure must necessarily carry the full weight of the combined errors in the results obtained in the five direct determinations made. These errors may or may not cancel each other out, but in any case the practical effect will be serious only in the case of foods poor in carbohydrates.

Attention has already been drawn to the further effect of the difference method in embodying in the carbohydrate figure any non-carbohydrate ingredients of the organic matter of the food other than proteins, fats and fibre. The vitamins of the food, for example, will mostly be included in this figure.

For a more detailed critique of the current scheme of analysis here outlined reference must be made to textbooks of agricultural chemistry. This applies also to the detailed exposition of the chemical characteristics and metabolism of the various classes of ingredients present in foods. In this chapter it is only necessary to outline the main features upon which rest the practical utilisation and value of foods.

THE CONSTITUENTS OF FEEDING-STUFFS

Carbohydrates. The carbohydrate group comprises a very large number of individual members, all composed of the three elements carbon, hydrogen and oxygen, present in proportions corresponding to the general chemical formula, $C_x(H_2O)_y$, where x and y may be the same or different numbers. For the purposes of animal nutrition the only class of carbohydrates of any appreciable value are those in which x is either 6 or some multiple thereof (hexosans). Apart from these, the only other class of carbohydrate present in foods in more than very small proportions, if at all, is the five-carbon class (pentosans), of which coarse fodders, husks and seed coatings may contain appreciable quantities. They are mainly contained in the fibre fraction of the food, but beyond the heat arising from their oxidation it is doubtful whether they contribute much to the needs of the body.

Of the hexosans the most valuable are the sugars, dextrins, starches, hemi-celluloses and celluloses. Of these the sugars are the simplest in chemical constitution and fall mainly into the two classes of the *glucose* group ($C_6H_{12}O_6$—glucose, fructose, galactose) and the *sucrose* group ($C_{12}H_{22}O_{11}$—sucrose, maltose, lactose). They are all soluble in water and of high nutritive value, but being readily attacked by bacteria are apt to suffer some loss in the digestive tract before they can be assimilated.

The carbohydrates of the highest nutritive value are those of the starch class [$(C_6H_{10}O_5)_n$] since they are rather more concentrated in energy-yielding elements (C, H) than the sugars and being less soluble suffer rather less waste by bacterial action.

The digestive juices contain enzymes which quickly digest the sugars and starches, but they do not contain any enzyme capable of digesting the more complex celluloses and hemi-celluloses. In so far as the latter can furnish useful nutritive material, therefore, it can only be through the agency of bacteria, and can only occur to any considerable extent when there is abundant opportunity for vigorous and prolonged bacterial activity in the digestive tract, as in the rumen of cattle and sheep. Under such conditions some part of the cellulose is converted into volatile organic acids that can be assimilated and made useful for nutritive purposes. It is in the form of such products that the useful material digested from the carbohydrates, whether sugar, starch or cellulose, enters the blood-stream and is applied to meet the needs of the body.

Owing to the great preponderance of carbohydrates in the food-supply, even in diets of relatively low carbohydrate content, and the readiness with which they liberate their energy in the metabolic processes as heat and other forms of energy, the carbohydrate group makes by far the greatest individual contribution to the needs of the animal for both maintenance and production purposes, whether the production takes the form of growth-, milk-, work- or egg-production. Even in the fattening animal the greater part of the fat laid down originates from the carbohydrates digested from the food.

Fats resemble the carbohydrates in being made up entirely of carbon, hydrogen and oxygen, but differ sharply from them in being far more concentrated in carbon and hydrogen, and correspondingly poorer in oxygen, and consequently being more concentrated in stored-up energy which can be placed at the disposal of the animal. Measured in terms of the heat given out on complete oxidation, fats are nearly $2\frac{1}{2}$ times as concentrated in energy as carbohydrates. Economic and dietetic considerations, however, usually restrict greatly the proportion of fat that can be effectively included in the food and, except in the milk-feeding of young animals, it is unusual to find more than about 5 per cent of fat in the rations used in farm practice. Within the limits thus imposed upon their use the fats can discharge the same functions in energy supply and fat production as the carbohydrates.

Fats are chemical combinations of glycerol with organic acids. A pure fat may contain one, two or three acids, but most natural fats are mixtures of fats and therefore may contain many more acids in their make-up. These acids are of two general chemical types— *saturated* and *unsaturated* ; those of the former type found in fats are mostly solid within the ordinary range of temperatures of the temperate zones, whereas the unsaturated acids are liquid down to much lower temperatures. This difference is reflected also in the texture of the fats containing them in combined form, the fat tending to be softer the higher the proportion of unsaturated to saturated acids present in it. Whether by direct transference or indirectly, the food fat tends to impress its character in this respect upon the animal fat, whether body-fat, milk-fat or egg-fat. It thus has a hardening or softening effect which may need to be taken in account in feeding practice when fat texture is an important factor in the assessment of the quality of the product, as in the production of bacon and butter. This influence of the food fat seems to be more directly exercised through the unsaturated than the saturated acids and, generally speaking, it is easier to produce softening than hardening effects.

In the fats group as a whole by far the most abundant fatty acids are the two saturated acids, stearic and palmitic acids, and the unsaturated acid, oleic acid, the former tending to preponderate in the solid fats and the latter in the soft and liquid fats (" oils ").

Proteins. Although from the point of view of chemical constitution the proteins are the most complicated of the three great groups of nutrients, they are little more complex than the carbohydrates and fats in terms of the chemical elements from which their complex structure is built up. The differences, however, are very significant and important, especially the presence, along with the carbon, hydrogen and oxygen, of some 15–18 per cent of nitrogen in the make-up of the protein molecule. In some cases the mineral elements sulphur (up to 2 per cent) and phosphorus (up to $1\frac{1}{2}$ per cent) are also present. Clearly therefore, in so far as nitrogen enters into the composition of the animal body and its products, the supply of it must be a special charge upon the food-supply, and in particular upon the proteins present in it, since there is no other class of ingredient present in it in more than small proportions that contains nitrogen. It is hardly an exaggeration, therefore, to make the generalisation that nitrogenous substance in the body can only come from proteins in the food. On a full diet consisting only of fats and carbohydrates, an animal would assuredly die of nitrogen starvation.

Amino-acids. In the plant the protein, starting with carbon dioxide, water and simple nitrogenous compounds (ammonia, nitrates), is built up in progressive stages into more and more complicated compounds until eventually the perfected protein is produced. At one stage in this building-up process the various products then formed belong mainly to the two chemical classes of amino-acids and amides. When the protein is consumed by the animal the protein-digesting enzymes or proteases break it down mainly to this particular stage and no further ; the amino-acids and amides are then assimilated into the blood-stream and there rebuilt into animal proteins. These intermediate constituents of the proteins, of which more than 20 have been identified, are the protein-building units upon which the animal depends. In their value to the animal the importance of the individual units varies, some being absolutely essential, whilst others, though possibly useful if present, are not essential to the protein purposes of the animal.

Protein " Quality ". The amount and kind of the different amino-acids, etc., present in different proteins varies widely, and one or more of them may be completely missing from some proteins. If the missing item chances to belong to the essential class, then that protein by itself clearly cannot serve efficiently as the only source of protein, but must be blended with another protein whose digestion products will supply the missing item. This is one of the many reasons why in feeding practice it is safer to include a variety of foods in the ration rather than depend too much upon one or two particular foods. Variety as a safeguard of protein efficiency is determined by the origins rather than the commercial descriptions of the foods. Thus a mixture of maize, maize germ meal and maize gluten feed does not ensure protein variety, whereas a mixture of maize, palm kernel meal and bean meal does.

Even this mixture is not nutritionally complete because recent research shows that proteins of animal origin (e.g. in fish meal, milk, dried blood) contain a factor (the Animal Protein Factor or A.P.F.) which is essential for the normal growth and well-being of the young animal, especially for non-ruminants like pigs and poultry.

Intensive work on this factor resulted in 1948 in the discovery of a pure chemical substance containing the element cobalt which was extracted from liver tissue. Not only is this compound extremely potent in promoting growth but it is also specific against pernicious anæmia in man. On this account it is now classed with the vitamins and is known as vitamin B_{12}. Vitamin B_{12} is not the only component of the A.P.F. but it is certainly a very important one.

Vitamins. All foods contain organic ingredients which are not properly classified as proteins, fats or carbohydrates, and which, as explained above, are mainly embodied in the " soluble carbohydrate, etc." item given by the conventional food analysis. These ingredients are mostly present only in very small proportions, and until the opening years of the present century their presence was largely ignored in the consideration of nutritional problems. Modern research is steadily revealing, however, that some of these items have a profound bearing upon, and are essential to, the efficient metabolism of the major nutrients. Details of the nature and nutritional significance of these items will be found in textbooks of agricultural chemistry, and it must suffice here to outline the salient points concerning one particular group of ingredients which has aroused more widespread interest than any other—the *vitamins.*

Evidence had long been accumulating that the supply of proteins, fats, carbohydrates and minerals, although certainly the major part in terms of quantity, did not cover the whole of the requirements of the animal from its food, and in particular that certain specific forms of ill-health, such as rickets and scurvy, were attributable to deficiencies in the make-up of the diet. Conclusive experimental evidence of the correctness of this view was not obtained, however, until the early years of the century, since when it has been abundantly justified and amplified. In default of knowledge as to the precise chemical nature of these essential ingredients, the term " accessory factor " was at first applied to them, but this was soon replaced by the simpler term " vitamin ", which has now become the class name for the numerous substances of this type that have since been discovered. It should be noted, however, that, unlike the proteins, etc., the vitamins are not grouped together on grounds of chemical similarity, but on the basis of the kind of effect that their absence or shortage produces in the animal.

At first, until the precise chemical nature of the individual vitamins could be established, chemists agreed to name them simply by letter, as vitamin A, vitamin B, etc. When further investigation showed that what were being regarded as single ingredients were really mixtures of vitamins, the A, B, C, etc., became group denominations, and the different items within each group were distinguished by numbers, e.g. A_1, A_2, A_3, etc. ; B_1, B_2, B_3, etc. This nomenclature is still widely used, but with the growing knowledge of the chemical nature of the individual items, the present tendency is to replace it by specific names, often indicative of the chemical nature of the vitamin.

On the farm the risks of vitamin shortage of any serious consequence vary with the class of animal, the system of management and the season of the year. Broadly speaking, they are less with ruminants than non-ruminants, with outdoor than indoor systems, and in summer than in mid-winter. The groups of vitamins of which our present knowledge is the most extensive are vitamin A (growth-promoting) ; B (anti-neuritic and growth-promoting) ; C (anti-scurvy) ; D (anti-rickets), and E (anti-sterility). The vitamins A, D and E are fat-soluble ; B and C, water-soluble.

In ordinary farm practice trouble due to deficiency of vitamins of the C and E groups is rarely met with, but in some " intensive ", largely indoor, systems of management the supply of the A, B and D groups needs to be watched, whilst in poultry feeding, especially in chick-rearing, the supply of other vitamins not listed above may need attention. A brief outline of points of practical interest in relation to the A, B and D vitamin groups follows.

Vitamin A is essential to life and to the production and repair of tissues that are essential to growth and maintenance. It is thus of special importance in the feeding of growing animals and in the milk designed for their early support. Its lack or deficiency is claimed to lead to increased susceptibility of the animal to bacterial infection, and therefore to disease, but this is due rather to its general effect upon the health of the animal than to any specific anti-infective action.

It can be produced in the animal from the yellow pigment, carotene, which occurs in green plants along with the chlorophyll ; such materials containing carotene thus serve as important sources of vitamin A for the animal. In the form of the vitamin it occurs only in animal products and in association with the fats, especially liver oils (e.g. cod liver oil). Whether supplied as carotene or pre-formed vitamin, if the supply is adequate some part of it passes into the animal products, such as milk and eggs, and a reserve supply is accumulated in the liver. In the case of the pregnant female a similar reserve is stored in the liver of the fœtus.

It is fairly resistant to heat but gradually loses its value when exposed for long to the air. For this reason the pre-mixing of vitamin A or carotene supplements with food to be consumed without much delay.

A characteristic effect of vitamin A deficiency is a tendency to night blindness and other eye defects, together with the development of sores in the throat and mouth. In the young animal growth is retarded and in severe cases may cease entirely.

Vitamin B. The vitamin B complex comprises at least six water soluble vitamins. Vitamin B_1 (thiamine), the antineuritic vitamin, is not stored up in the body and hence must be entirely supplied through the food, except in ruminants, since these can derive their B_1 supply from bacterial synthesis in the rumen. It is present in a wide variety of foods, particularly brewers' yeast and wheat germ. Cereal grains, young fodder, liver, kidney and egg-yolk are well supplied with it.

Symptoms of its absence from or deficiency in animals are loss of appetite, and retardation or cessation of growth, followed by the development of paralysis and early death. Lack of it may lead to sterility in both sexes.

Of the other vitamins comprised in the vitamin B complex the best known is *riboflavin* (or B_2, or in America, vitamin G). Like B_1 it can be synthesised by bacterial action in the rumen, but must be supplied pre-formed in the food of non-ruminants. It is present in appreciable amounts in green leafy fodders, oil meals, meat meals and yeast. Cereals and their by-products are less well provided with the B_2 than with the B_1 vitamin. Dried yeast and milk, especially separated milk and whey, are relatively rich in this vitamin.

The symptoms of B_2 deficiency vary widely in different animal species, apart from the general tendency common to many vitamins of inducing lower growth rates and unsatisfactory consumption and utilisation of food. In chicks the most characteristic symptom is " curled-toe paralysis ", often accompanied by diarrhœa. In laying birds the most obvious symptoms are low egg-production and poor hatchability. In pigs, a B_2 deficiency often causes stiffness of the joints and skin eruptions.

Vitamin B_{12}, already referred to as the major constituent of the so-called " Animal Protein Factor " (A.P.F.), is essential for optimum growth in the young animal, and is likely to be required as a special supplement where the diet consists almost entirely of vegetable foods. Good sources are fish and meat meals, fish solubles, dried blood and cow's milk. Commercial vitamin B_{12} is usually produced by fermentation with a specific mould organism such as *Streptomyces griseus*. Incidentally the products of decomposition of organic matter by fermentation have now assumed major importance in nutrition as well as therapeutics and are known as *Antibiotics* (see p. 544). Vitamin B_{12} is synthesised by bacteria in the paunch of ruminant animals and it is more than likely that the available supply of cobalt in the herbage consumed limits the production of this important substance and hence the ability of the animal to digest its food.

Vitamin D, which like vitamin A is fat-soluble, was found, when separated out from the original vitamin A, to have specific effects in warding off the disease of rickets. It does this through its influence upon the deposition of calcium and phosphorus in the bones. Apart from this connection with bone growth the D vitamin is in other ways essential to health. Later research has shown that there are at least four different forms of it, of which the most important in practice are D_2 and D_3, the latter being the form especially required by growing chicks.

The amount required to ensure healthy bone development varies with the mineral composition of the diet and also with the species of animal. Broadly speaking the amount of vitamin D required is lowest when the ratio of calcium to phosphorus lies between 2 : 1 and 1 : 2, but no amount of vitamin will compensate for actual deficiencies of either mineral. The role of vitamin D is much less important in the adult than in the growing animal, except for the female during periods of reproduction and lactation.

The long-known therapeutic value of sunlight in the treatment of rickets found an explanation, perhaps only partial, in the discovery that one or more of the D vitamins can be produced by exposure of the skin of the animal, or of many foods, to certain of the ultra-violet radiations from the sun or from special types of artificial light that are rich in these rays. The vitamin is produced in this way from substances of the sterol class that are present in small proportions in the skin or the foods irradiated. From this possibility one may rightly infer that in the tropical and temperate zones animals that live wholly, or in substantial part outdoors, are unlikely to be short of vitamin D, except possibly under winter conditions. In the case of animals kept indoors, even in well-lighted quarters, the effects of irradiation will not be obtained unless the windows are glazed with special glass (" vita-glass ") and kept clean, or preferably so constructed that they can be thrown wide open in fair weather.

The symptoms of rickets in the young are the familiar deformation of the leg bones, accompanied by swellings and tenderness of the joints. In adults, if the deficiency is prolonged, the bones become brittle and easily broken, and in laying fowls the thickness of the eggshell is reduced. In very severe cases the effects may develop to the stage of paralysis.

The best natural sources of vitamin D are animal fats, liver, and certain fish liver oils (e.g. cod liver oil), whilst, as indicated above, many foods (notably yeast) may be " reinforced " with the vitamin by irradiation. Growing greenstuffs do not contain the vitamin, but after cutting and exposure, as in haymaking, a certain amount of it may be produced in them.

Vitamin E has been found to be essential to reproductive activity under experimental conditions, but information is still very scanty as to the extent to which deficiency of vitamin E may be a causative factor of sterility in farm animals kept under farm conditions.

It is a fat-soluble vitamin, and is widely distributed in the common foods, especially in green fodders and other leafy materials. It is also present in seeds, especially in the germ. Many vegetable oils contain it, wheat germ oil being regarded as the most concentrated source.

Chemically, vitamin E acts as an antioxidant and as such may suffer destruction in foods under conditions which favour oxidative processes.

The condition known as muscular dystrophy in calves, and also that recognised as crazy chick disease in chickens, are associated with a deficiency of vitamin E.

Mineral Matter. The animal body contains a considerable variety of mineral elements, many of which have been shown to be essential to its efficient working. Until recent years the list of essential mineral elements was thought to be confined to the elements sodium, potassium, calcium, magnesium, iron, phosphorus, chlorine, sulphur and iodine ; but to these must now be added manganese, copper, cobalt, zinc, fluorine, and possibly others may still have to be included. These additions to the old list share with the vitamins one feature in common, in that they are required only in extremely small proportions, and may have detrimental effects if supplied in excess.

Our knowledge of the precise function of each element and of the amounts needed by the animal is in most cases still very inadequate. Of the mineral elements that are required in (for minerals) relatively large amounts the lead is taken by potassium, calcium and phosphorus, followed by sodium and chlorine, then by iron, magnesium and sulphur, and finally by the rest, most of which are only needed in such minute amounts that they are commonly referred to as " trace minerals ".

Some of the mineral elements are present in both the food and in the body both as mineral salts and combined in organic forms. Thus the sulphur, which appears in the ash wholly as sulphates, is present in the original plant or animal matter almost entirely in organic forms (e.g. proteins) ; the same applies also in the case of phosphorus (except in ash from bones).

Foods used on the farm vary greatly in their mineral make-up. As a broad general rule the vegetative parts of plants are richer in mineral elements than the seeds ; the dry matter of young greenstuffs is richer in useful minerals than that of maturer vegetation ; that of " roots " is relatively rich in soluble minerals. Animal products used as food, unless they contain bone, show mineral contents, expressed as ash, ranging roughly from 3 per cent (dried blood) to

7 per cent (dried milks) ; but in these cases all the ingredients are useful and readily assimilated, whereas in foods of vegetable origin the ash usually includes a proportion of useless ingredients.

The general functions in the body of the essential mineral elements are varied and complex.   Some of them are essential to the structural soundness of the bones and teeth.   The soluble salts serve a variety of functions in the blood and other body fluids, such as the main-tenance of normal pressure relations and the desirable balance of acidic and basic elements.   They also play a part in regulating the activities of muscles and nerves.   Apart from such general functions discharged by groups of minerals, each mineral element has specific roles, such as those outlined below.

Calcium and Phosphorus, which together make up more than 70 per cent of the ash of the body, are so closely associated in meta-bolism that it is convenient to consider them together.   Reference has already been made (p. 538) to the necessity for a suitable balance between the supplies of calcium and phosphorus for sound bone formation, as well as the co-operation of vitamin D.   They are particularly abundant in the bones and teeth, which together contain about 99 per cent of the calcium and 80 per cent of the phosphorus present in the body.   This high proportion of calcium phosphate in the bones not only assists in conferring strength and rigidity, but also serves as a reservoir of calcium and phosphorus out of which tem-porary shortages of these elements in the food may be made good.

Although the proportions in the soft tissues and fluids are thus only small, they are widely distributed and play parts of fundamental importance in the body metabolism.   Reference has already been made to the fact that the phosphorus is mainly present in organic combinations, such as some of the proteins ; this applies also to some extent to the calcium.

The concentration of calcium in the blood, though low, is a matter of importance, and is controlled by a special agent or hormone pro-duced by the parathyroid glands.   Low blood calcium leads to tetany and may be a contributory cause of milk fever and rickets ; high calcium leads to unduly rapid drain of calcium from the reserve in the bones.

Foodstuffs vary widely in their content of these two elements. Broadly speaking, all seeds and products derived from them are poor in calcium in relation to animal needs, but relatively rich in phos-phorus.   The content of grass and roughages varies with the soil, manuring and conditions of growth, leguminous produce being gener-ally richer in calcium, but less regularly so in phosphorus, than non-leguminous.   Outstanding in high content of both elements are naturally the animal by-products containing bone (meat meal, fish meal).   Dried separated milk also ranks as a valuable source of the two elements.

In view of the relatively large amounts of calcium and phosphorus required by the animal—especially where growth, milk-production and egg-production are concerned—and the consequent frequency with which deficiencies of supply of these minerals are liable to arise in practice, the addition of mineral supplements to rations has become common. For this purpose fine bone meal and precipitated bone flour are useful in supplying both elements. If mineral phosphate is used for the purpose, care should be taken to avoid material containing more than about 0·5 per cent of fluorine, since this element, commonly present in mineral phosphates, though essential to the animal in minute proportions, exercises a cumulative toxic effect at higher levels of intake.

Magnesium occurs in the body in much smaller proportions than calcium and phosphorus, but is closely associated with them (especially phosphorus) in its metabolic action. It is essential for the production of sound teeth, and may play a part in connection with rickets which may be helpful or harmful according to the amount fed and the relations of the calcium and phosphorus present in the diet. The possible effect of magnesium deficiency which has attracted most attention is its apparent association with certain types of tetany, such as " grass staggers " (grass tetany), to which grazing animals are sometimes subject and which is often accompanied by subnormal magnesium content of the blood. Other dietary factors probably enter also into the causation of this type of trouble.

Potassium, Sodium and Chlorine form another group of elements which, severally and collectively, play important parts in the body functions. They occur almost entirely in the soft tissues and fluids, potassium predominating over sodium in the tissues, and the reverse in the fluids, with the outstanding exception of milk. Together, the three elements exercise a very important influence upon the movements and utilisation of water in the body. In chemical character sodium and potassium are both powerful bases (alkalis) and in some of their actions are to a certain extent mutually replaceable.

Potassium exercises an influence upon the metabolism of carbohydrates, and a marked, but as yet little understood, influence upon muscular activity. Since it is relatively abundant in all plant products it is probably only rarely that potassium deficiency will cause a practical feeding problem on the farm, except possibly where high milk production coincides with deficient potash manuring.

Sodium, as above mentioned, is the dominant basic element of the body fluids and is thus largely responsible for the maintenance and regulation of a suitable base–acid ratio in them. It is not stored in the body, and any excess of supply is rapidly removed mainly through the kidneys, but also to some extent in perspiration.

Chlorine, an acidic element, is present in the body partly in organic combination, but mainly in combination with bases as chlorides, principally sodium chloride (salt). The fact may also be recalled that the gastric juice contains chlorine both as a small percentage of hydrochloric acid and as chlorides (see p. 511).

The importance of common salt in the human dietary has been recognised from ancient times, and the practice of making it available to farm animals, especially to milch cows, is also of very long standing. The body has the power of adjusting itself within wide limits to the intake of salt, and any serious effects of an actual deficiency may be revealed only very slowly in unthriftiness, loss of weight, and fall in milk yield. The moderate condimental use of salt tends to stimulate the secretion of saliva and to promote the action of the starch-digesting enzymes, whilst its excessive use leads to the familiar development of thirst, owing to the large amounts of water then needed to remove the excess of salt. Owing to the character of the basic foods consumed by them, cattle have a relatively greater need for salt than pigs or poultry, but because of the great preponderance of salt-poor cereals and cereal products in the common dietaries of the latter classes of stock, the addition of a little salt is often necessary to ensure optimum feeding results.

Iron, although present in the body to the extent of only about 4 parts per 100,000, is of vital importance through being a constituent element in the make-up of the red colouring matter of the blood, hæmoglobin, by means of which oxygen is absorbed from the air inspired by breathing, and so carried to all parts of the body (see p. 513). Small amounts of iron in other active forms are also present in the body, whilst a small amount is stored in the liver and kidneys.

Deficiency of iron is the main cause of the type of anæmia termed " nutritional anæmia ", which may occur at any time of life but arises most frequently in quick-growing species of animals during the period of suckling, since milks are very low in iron. On the farm this trouble is common only in suckling pigs, and can be averted or remedied by administration of a little iron (plus a trace of copper) direct to the young animal. The iron content of the mother's milk cannot be raised by addition of iron to her food. When the young animal begins to eat other food, the risk of anæmia quickly passes away since most of the farm foods are liberally supplied with iron in proportion to the needs of the body.

Copper is required in minute amounts because without it the hæmoglobin does not function effectively, though the precise role of the copper is not yet clearly understood.

In farm practice copper-deficiency only arises in areas where the natural herbage is so poor in copper that even the very small needs of the grazing stock are not met. Such areas have been identified in several parts of the world, especially in north-west Europe and Australia, and parts of the hill-regions of northern England and Scotland (see p. 642).

Cobalt is now regarded as an essential element of food because it is a component of the important vitamin B_{12}.

Though the actual amount of cobalt required by animals is extremely small, both soil and herbage may be deficient in the element. Cobalt deficiencies are known to occur in certain areas throughout the world, and are associated with wasting diseases, such as pining (see p. 642).

Iodine is an important essential mineral element, because although the body as a whole may contain no more than 4 parts per 10,000,000, more than one-half of this is in one particular gland, the thyroid gland, which plays an essential part in the regulation of some of the most important vital processes, such as gestation, growth, egg-production and milk-production. This regulation is effected by means of the hormone, thyroxine, which is present in the internal secretion produced by the gland and contains a large proportion of iodine.

A marked deficiency of iodine causes the gland to swell up and the serious condition known as goitre develops. This trouble is particularly liable to develop in regions where the soils, crops and water-supplies are practically devoid of iodine, such as many districts remote from the sea ; but it is also met with in other districts where the soil iodine content is not particularly low, so that other factors are apparently also involved in the causation of goitre.

This serious risk involved in iodine deficiency has led to an exaggerated and indiscriminate tendency to add iodine supplements to diets regardless of any evidence of the need for them. Whether iodine deficiency in farm rations is sufficiently widespread and general-ised to warrant this practice is doubtful, and the evidence from experiments is as yet conflicting, but on balance suggests that it is not common under ordinary conditions of management. Where an iodine supplement is thought to be necessary, it can be provided in the form of a few ounces of potassium iodide per ton of food.

Sulphur is a constituent of some of the animal proteins, and the amounts provided by the common varied diets appear to be adequate for the sulphur needs of the body. Sulphur-containing proteins (" keratins ") are notably present in horn, hoof, claws, feathers, skin and wool. Some of the essential amino-acids (e.g. methionine) contain sulphur.

Manganese is required in minute amounts for normal reproduction, and it has an influence on the rate of growth. In poultry its absence also gives rise to a specific malformation of the leg bones of growing chicks described as " slipped tendon " or perosis. Certain other mineral and organic factors are probably also concerned in the causation of this trouble, e.g. the supplies of calcium and phosphorus.

Zinc has been found to be an essential trace element in laboratory experiments with rats, and for the prevention of an unhealthy con-dition of the skin, known as " parakeratosis " in pigs. It is also an essential element for the growing chick (40 ppm), the laying hen (especially breeders) (60 ppm) where a deficiency may adversely affect hatchability, and the turkey poult (70 ppm).

Fluorine is another example of a mineral element that is apparently essential in minute amounts, but seriously toxic at excessive, though still very low, levels of supply.

Aluminium and Silicon. Apart from the mineral elements listed as essential, foods will generally contain other mineral elements taken in from the soil, but exercising, so far as present knowledge goes, no essential function in the food. Outstanding in this respect are the elements aluminium and silicon, which are universally present.

In the case of silicon no essential function has been traced, although it enters extensively into the mineral metabolism ; nor does any harmful effect seem to develop from its presence in more than average proportions. Whether the presence of aluminium in foods may be harmful or not is a matter of current controversy in relation to foods used for human consumption.

Toxic Elements. Certain other elements which may be present in foods grown in special areas are definitely toxic in their effects, e.g. arsenic, molybdenum, selenium, etc. Thus the presence of molybdenum in the herbage explains the scouring trouble experienced on the so-called " teart " lands of Somerset and Warwickshire (see p. 668). Recently selenium has been shown to be an essential element.

Antibiotics. These are recently discovered chemical substances prepared from the products formed during the growth of certain moulds. Penicillin is the best known example, but others that have been found useful include aureomycin, streptomycin, terramycin, bacitracin and chloromycetin. Their original, and still principal, use was to control and dispel diseases of microbial origin—to act so to speak as disinfectants. It was found later, however, more or less by accident, that certain antibiotics possess growth stimulating powers, probably by reason of the control exercised over intestinal bacteria, although this may not be the complete explanation. Experiments with pigs and poultry in which very small amounts of selected antibiotic substances have been incorporated in the food (14 g per tonne of food) have repeatedly given enhanced growth rates in young stock, usually of the order of about 10 per cent, while the medicinal value of these drugs (at a heavier rate than the above) has been demonstrated particularly with a fairly high proportion of runt pigs.

Additives. The major components of the mixed concentrate rations of all classes of livestock are concerned with meeting the energy and protein requirements of the animal. However the rations of today may include a number of minor factors at less than a 5 per cent level and very often below 1 per cent, and these exert an appreciable effect on the health and performance of the animal. Some of these minor factors are direct nutrients, such as sources of minerals and vitamins ; others have beneficial effects in widely different ways. These factors are usually referred to as " additives ", although it is virtually impossible to give a comprehensive definition of the term.

Amongst " additives " we may include such nutrients as vitamins, minerals and synthetic amino-acids, the latter enhancing the value of the particular proteins of the food for certain species. Other additives may protect labile factors against loss through atmospheric oxidation and rancidity and are called *anti-oxidants* : these include " natural " substances such as vitamin E, or synthetic substances such as butylated hydroxyanisole (BHA). Some additives may have direct growth-promoting properties ; these include the antibiotics (at low level usage) and copper sulphate (for pigs) at specified levels. Other additives are used for directly controlling or preventing disease, such as the *coccidiostats* and the anti-blackhead group of drugs, and thus enable the animals fed on them to remain in sound health and make the fullest use of the food offered. Growth-promoting additives also include synthetic anti-hormone substances (e.g. thiouracil), synthetic hormones (e.g. hexoestrol), tranquillisers, surfactants, arsanilic acid derivatives and so on.

It is evident, then, that the term " additive " includes a wide and ever-increasing range of substances that broadly are used to promote health and well-being and to stimulate various forms of production.

They should not be regarded as unmixed blessings, for many of them have a narrow margin between the limits of safe usage and the production of toxic effects ; the greatest care must be exercised in their uniform incorporation in the food. It is impossible in this short reference adequately to cover all aspects of additives from the point of view of toxicology, undesirable side effects, margins of safety, residues in animal products and carcases, legislation and so on. In recent years a vast amount of research has gone into the production and use of additives, and it is evident that we are only at the beginning of an era of additive supplementation of foodstuffs. There are tremendous possibilities in the use of medicaments added to food for the control of disease in animals, for health promotion, and for stimulating productivity. At the same time it must be appreciated that additives can be powerful forces for evil as well as good, but without doubt, their uses in the future will continue to expand.

Water. At every stage in the development of the animal, water is the most abundant ingredient in the make-up of its body, ranging down from 80 to 90 per cent in the fertilised ovum to 70–80 per cent at birth and then to a fairly steady 50–60 per cent in the full-grown, lean animal. Even in a prime fat animal there will still be some 40 per cent of water, so that a 500 kg fat bullock will contain 200–250 kg or litres of water in its body substance ; despite the steady loss of water in the excreta and by evaporation the supply must be kept up at this level if the animal is to continue healthy and active.

In computing water requirements the water contained in the food must be taken into account, and this in the case of succulent, may cover a considerable part, or even the whole, of the animal's water needs. These needs vary widely with changing conditionss and are best met by the provision of free access to water. Excessive consumption of water, with ultimate deleterious effects, is only likely to arise where the animal is forced to it, e.g. by heavy root feeding. The water requirements of farm livestock are set out on page 101.

In adult animals with free access to drinking water there is usually a rough correlation between the amount of water consumed in food and drink and the total dry matter consumed in food. For most farm animals the ratio is usually in the region of 3–4 kg of water to 1 kg of dry matter, the proportion of water tending to rise with the fibre content (and also the protein content) of the food. The greater water consumption in hot than in cold weather is also familiar.

In young growing animals the water-requirement in proportion to live-weight steadily falls as growth progresses. Thus the suckling calf requires about twice as much water per kg of live-weight as the full-grown steer. Shortage of water is thus especially serious with the young animal, and even a moderate shortage may cause serious retardation in the case of the quicker-growing species, such as pigs and chickens. In milk-production and egg-production, where large amounts of watery products are removed from the body, the provision of an adequate water-supply is obviously one of the master keys to high production.

THE FEEDING-STUFFS

For the successful and economical construction of rations for farm animals, some knowledge is required of the composition and digestibilities of feeding-stuffs and of their suitability for each class of animal. The reader will find data for the average composition, digestibility and nutritive value of feeding-stuffs in readily available tables. Here it is proposed to discuss only the general nutritive and dietetic qualities of feeding-stuffs and to indicate the various factors responsible for variations in composition and digestibility.

THE CLASSIFICATION OF FEEDING-STUFFS

Comparison and tabulation are, of course, facilitated by classification and it so happens that animal feeding-stuffs fall conveniently into four main groups based on the characteristic qualities of (1) succulence, (2) richness in fibre or (3) high concentration of digestible nutrients. This method of grouping brings together feeding-stuffs with similar dietetic properties, and has the practical advantage of facilitating the selection and replacement of one food by another in animal rations. Sub-division of these major groups can be made in a variety of ways in accordance with some other property, such as the origin of the feeding-stuff or a particular nutritive quality like high protein content. No rigid classification is possible since some feeding-stuffs may have characteristics which would justify their inclusion in more than one group. The particular choice of group is, then, just a matter of opinion.

I. Dry Roughages or Coarse Fodders

    (1) Hay—grass, cereal and legume hays.
    (2) Straw—cereal straws, pea and bean straws.

II. Succulent Foods

    Roots and tubers—mangels, swedes, turnips, carrots, potatoes, etc.

III. Green Forages

    Immature crops of grass, clovers, lucerne, sainfoin, vetches, kale, cabbage, rape, maize ; silage, beet tops, etc.

IV. Concentrated Foods

    (1) Cereal grains and their by-products—oats, barley, wheat, maize, rye : weatings, bran, etc.
    (2) Leguminous seeds—beans and peas.
    (3) Oil seeds—linseed, sunflower seed.
    (4) Oil cakes and meals—ground-nut cake, linseed cake, soya-bean meal, etc.
    (5) Animal products—meat and bone meal, fish meal, etc.

THE DRY ROUGHAGES

The dry roughages are characterised by a high content of crude fibre (23–43 per cent) and by the physical quality of bulkiness. On these accounts, the dry roughages are best suited to form part of the maintenance rations of cattle, sheep, goats and horses ; they are useless for feeding to pigs and poultry.

Hay. Hay, made by the sun-drying of green, but nearly mature, grasses, clovers and cereals, is the traditional form in which these crops are conserved for winter feeding. Unlike roots and concentrates, the hays are usually good sources of calcium, and they contain important amounts of potassium, magnesium and other minerals. On the other hand, unless obtained in good, green condition, the hays tend to be rather poor and frequently inadequate sources of carotene and vitamin E, although they may contain significant amounts of vitamin D for animals kept indoors.

The hays, and to a less extent the straws, are extremely variable in composition and digestibility. Even where concentrates are carefully rationed, the influence of hay quality on the milk output from a herd of dairy cows can become sharply manifest when the winter milk yield following a good hay harvest is compared with that following a poor haymaking season. The crude protein content of meadow hay varies from 4 to 13 per cent, the fibre content from 23 to 35 per cent and the starch equivalent from 18 to 40 per cent.

The poorest hays are certainly no better than oat straw, and it is scarcely possible for a cow to eat enough poor hay to maintain body-weight and condition. On the other hand, the best hays may be fed, if it is desired, as the sole food for maintenance and the production of 10 litres of milk. The difference between poor and good hay is actually greater than it would appear from their digestibilities, for it seems that the digestible nutrients and minerals in good hay can be utilised more efficiently than those in poor hay. Moreover, the difference in palatability is such that cows will eat a ration of good hay in less than three-quarters of the time taken to consume the same weight of poor hay. Most well-made meadow hays fall within somewhat narrower limits of composition than those given above, the crude protein content ranging from 8 to 10 per cent, the digestible protein content from 3 to 6 per cent and the starch equivalent from 28 to 35 per cent. The differences in protein content are important. At the lower level of protein, the ratio of starch equivalent to digestible protein is about 10 to 1 and there is no surplus of protein relative to energy requirements for maintenance. At the higher level of 10 per cent of crude protein, the starch equivalent-digestible protein ratio is about 7 to 1 and there is enough protein in a full maintenance ration of hay to make it necessary to feed only cereals or even mangels for the first 5 litres of milk.

Factors Influencing the Quality of Hay. Recognition of this variability in composition of even " well-gotten " meadow hay should guard the reader against a common tendency to overestimate the feeding value of such hay. It has to be remembered that the getting of hay in good weather satisfies only one essential for the production of hay of high feeding value (see p. 341). Four factors are mainly responsible for variations in the nutritive quality of hay.

(1) The **Stage of Growth** or the degree of maturity of the herbage at cutting. This is a most important factor. For convenience in the making of hay, grasses must be allowed to reach the flowering stage, but as cutting is delayed beyond this point, there is an increasing deterioration in the protein content and digestibility of the crop.

Lignification of the stems increases rapidly and the leaves die as their nutrients are transferred to the seeds. The dropping out of the seeds means the loss of material approaching oats in feeding value.

(2) **The Botanical Composition of the Herbage.** The superior grasses, e.g. perennial ryegrass, cocksfoot and timothy, are of higher feeding value than Yorkshire fog, crested dogstail and the bromes and bent grasses of acid meadows. The protein and calcium contents of the hay tend to increase with the proportion of clovers.

(3) **The Weather Conditions at the Time of Making.** Rain and dull cool weather are unfavourable to haymaking. Loss of digestible carbohydrate through respiration is increased by slow drying ; moulding is very apt to occur and rain leaches out soluble constituents from partially made hay. All these losses fall on the more digestible nutrients. Excessively hot and sunny weather can also be detrimental since it is then easy to overcure hay which becomes too dry, harsh and brittle, while loss of carotene may be almost complete.

(4) **The Method of Making.** Leaves dry out more readily than the stems. This is particularly true of the clovers, and when curing takes place in the swath, the leaves tend to become overdried, brittle and easily broken off during handling. Since the leaves are richer in protein and other nutrients than the stems, the loss of leaf inevitably means loss of feeding value. Drying in windrow and cock reduces the difference in the rates of drying out of leaves and stems and permits a leafier hay to be made. At the same time, the crop is less exposed to the bleaching action of the sun and the leaching action of rain. Hay which has been " barn-dried " (see p. 345) is usually of better quality than hay made by traditional methods.

Types of Hay. Hay taken from short-term leys is characterised by consisting mainly of the superior grasses and clovers and of far fewer species of herbage plants than meadow hay. It appears to be equally variable in composition and the average sample is no higher in feeding value. It is usually a harder hay.

" Seeds " hay is made from a one-year ley mixture of grasses and clovers grown on arable land, with ryegrass and red clover usually predominating. Potentially, it is capable of being much richer in protein and calcium than meadow hay, since a high proportion of clover is sown and intended to be present in the mixture. In practice, however, the proportion of clover is very variable. Moreover, much clover leaf is lost in making and " seeds " hay is therefore seldom very much richer in protein than meadow hay. Its digestibility and starch equivalent tend to be slightly lower than those of meadow hay owing to the presence of a higher proportion of clover stem.

In spite of their greater content of protein, clover, lucerne and sainfoin hays tend to have energy values or starch equivalents which are on the average appreciably lower than those of meadow hay. This is because lignification and the development of woody stems proceeds very rapidly when flowering commences in these crops. Hence, if good-quality hay is to be obtained, cutting should preferably start as flowering begins and it should not be delayed beyond the point where most of the crop has come into flower. Since the great value of these crops lies in the leaves, which may contain 25 per cent or more of protein in the dry matter, it is worth while spending extra care in the making. The differential rate of drying between leaves and stems is far greater in the case of clover, lucerne and sainfoin than it is for the grasses. The dry leaves are very easily shattered. The crop should therefore not be left to lie long after cutting but should be put into windrow or cock or perhaps preferably made on tripods.

Heavy yields of hay may be obtained from oat and pea or oat and tare mixtures grown on arable land (see p. 292). The protein content and digestibility, like those of other hays, largely depend on the proportion of legume and the stage of growth at which the crop is cut. Mowing should not be delayed beyond the point where the pea or vetch pods have just formed.

Straw. Compared with the hays, the straws are markedly inferior in feeding value since they are more fibrous, less digestible and considerably poorer in protein, minerals and vitamins. In fact, no kind of straw from which ripened seed has been threshed is capable of supporting any class of animal without some other food. The full-grown bullock or barren cow may subsist for a comparatively long time, if need be, on good oat straw because it has no urgent nutrient requirements for growth and production, but it cannot maintain body-weight and condition indefinitely on such food.

Straw can therefore be fed most usefully to strong stores, beef breeding cattle, slowly fattening bullocks and other cattle in need of roughage to fill capacious digestive tracts. It can replace only strictly limited amounts of hay in the rations of more productive animals such as young growing cattle, milking cows and cattle in forward condition of fattening.

Such replacement of hay by straw is possible on the understanding that there is an increased need for concentrates or other food richer in protein and minerals. This fact is commonly overlooked when straw is fed during the late winter months to bulling heifers.

The feeding value of straw varies with the species and variety of cereal or grain crop ; with the degree of maturity of the crop and the weather conditions at harvest ; and with the soil and climatic conditions under which the crop is grown. It is advisable to feed straw in the long state, allowing the animal to select the more palatable and digestible portions and to reject the woody butt ends. Chaffing and grinding may save some energy of mastication, but the animal is forced at the same time to consume more non-digestible material and to reduce correspondingly its intake of useful food.

Oat straw is undoubtedly the softest and most digestible of the straws. This may be attributed in part at least to the fact that the oat crop is usually cut before it is fully ripe. This is particularly the practice in Scotland, northern England and on hill farms where the varieties of oats grown are best suited to the shorter season and the cooler and moister climatic conditions, and where there is every inducement to harvest the crop at the earliest possible moment. In some districts of Scotland and northern England it is consequently possible to fatten cattle on oat straw and turnips, whereas this is impossible in the drier and warmer south where the oat crop is harvested in a more mature state.

None of the straws is really suitable for feeding to sheep, but oat straw is the best for sheep to pick over when hay and other winter food is scarce.   Horses digest straw much less efficiently than cattle, and only good oat straw is suitable for feeding to these animals.   Long oat straw may be fed to horses which are not working hard, and it is the custom to mix oat straw chaff (1–2 kg) with oats and other concentrates to ensure adequate mastication.

Barley straw is somewhat inferior to oat straw because the barley crop is almost always allowed to stand until it is fully mature before being cut and harvested.   The feeding value of the straw, however, is often enhanced by the presence of clover from an undersown " seeds " crop, and the mixture may then be as good as, or even better than, straight oat straw.   Barley straw is best fed low down in order to avoid the danger of the awns entering and lacerating the eyes of cattle.

Wheat and rye straws are undoubtedly the least digestible of the cereal straws.   They are more fibrous, more lignified and more silicious.   (The ash of straws consists very largely of silica.)   Very seldom is the attempt made to feed rye straw to farm animals and the main use for both straws is for bedding, thatching and packing. If wheat straw has to be fed, it is best reserved for the older and low-producing cattle.   It is useless for horses, which can digest only about 20 per cent of the organic matter and actually spend more energy during the digestion process than is available from the straw.

Pea and bean straws are richer in protein and calcium than cereal straws but the stems are less digestible : bean straw, in particular, can be very fibrous, woody, unpalatable and costive.   It is not easy to harvest these straws in good condition, and they are apt to become mouldy during storage and therefore unsuitable for feeding.   Good pea and bean straws may be fed to store cattle, bullocks and low-yielding cows, but the allowance should preferably not exceed one-third of the total dry fodder intake.   It is often the practice to chaff these straws, but they are better fed in the long condition so that cattle can pick out the pods and the more nutritious tops and leaves.

Linseed, vetch, rape and clover straws have little or no feeding value.

Chaff and cavings are separated during the threshing of grain. Their feeding value can be higher than that of the corresponding straw.   Cavings consist of pales from the ears, broken stems and leaves and other material from the sheaves.   A high proportion of clover leaf can enhance the feeding value but this may be counteracted by the presence of soil, mould and unpalatable weed seeds.   Chaff consists of glumes, broken awns and leaves and the feeding value largely depends upon the proportion of leaf.   Both chaff and cavings are light, dry materials which are best fed by mixing and moistening with pulped roots.   Barley chaff may contain useful amounts of clover leaf, but the presence of broken awns reduces the value of the chaff, at least in the dry state, since the awns are liable to penetrate the mucous membrane of the mouth and thereby facilitate the entry of the actinomycosis organism.

Linseed chaff consists of husk from the broken seed bolls, broken leaves and a variable amount of unseparated seeds, particularly small, unripe seeds. The feeding value of the chaff depends almost entirely on the proportion of leaf and seed present. At its best, linseed chaff is a very fibrous material and, if it must be fed, then it should be given only in small amounts.

Oat Chaff. It has long been the custom in some localities to chaff oat sheaves without threshing out the grain. The product is therefore a mixture of oat grains and oat straw chaff. Its feeding value depends upon the proportion of oats to straw as well as upon the factors affecting the quality of straw and grain. The proportion of straw may run from about 50–70 per cent or more. Allowing for the chaffed state of the straw, the starch equivalent of the mixture may be taken to be somewhere between 35 and 45 ; the energy value of chaffed oat sheaves is therefore about 20–40 per cent higher than that of hay fed in the long state, but hardly better if the hay is also chaffed. The digestible protein may be estimated to be between 3 and 4 per cent, which is about the range for poor to medium hay. From the nutritional standpoint, the feeding of chaffed oat sheaves is to be deprecated since the animal is given no opportunity to reject the less nutritious butt ends of the straw. On the other hand, the cost of chaffing is considerably less than that of threshing, and where large amounts of oat straw have to be fed, the practice may have some economic justification.

SUCCULENT FOODS

The succulent foods are characterised by a high content of water, ranging from about 70 to 93 per cent. They possess mildly laxative properties and, usually, they are highly digestible. These qualities make the succulent foods nutritionally and dietetically valuable for all animals, particularly when on winter diets of rather costive, dry fodders : they are especially valuable for high-producing or pregnant animals or for those that are ill. Apart from their feeding value, roots and green forages are usefully grown as cleaning or weed-smothering crops.

All the succulent foods have a low content of oil, protein and fibre, and negligible amounts of carotene (except carrots) and calcium.

Wet sugar beet pulp, wet brewers' and distillers' grains, wet maize gluten meal, swill, liquid skimmed milk and whey are also succulent foods, but these by-products have not the same dietetic properties as the natural succulent foods.

Roots and Tubers. The roots and tubers are succulent carbo-hydrate foods.

At similar dry matter contents, mangels and turnips have about three-quarters of the starch equivalent of oats and three-fifths of the starch equivalent of barley. In spite of being highly digestible (90 per cent), the carbohydrates in roots seem to be utilised less efficiently than the predominantly starchy carbohydrates of cereals. Soluble sugars, sucrose, fructose and glucose form about one-half to two-thirds of the carbohydrates in roots, the rest being easily fermentable pectins and cellulose.   On the other hand, the main carbohydrate in potatoes is starch, which is easily digested by all animals without the aid of bacteria.   This may largely explain why potatoes are a better carbohydrate food than roots for pigs and poultry.   The higher dry-matter content is also a factor, but potatoes are still a better food for cattle than sugar beet roots with almost the same dry-matter content.

The starch equivalent of roots increases with the dry-matter content and, for each kind of root, the dry-matter content is known to vary with the variety, soil, size of root, manurial treatment and season. Red soils seem to favour the production of roots with higher dry-matter contents than those grown on dark soils.   Large roots tend to be more watery than small roots.   The crude protein fraction in roots and tubers consists largely of " amides " and of only a small proportion of " true " protein.   The digestible protein contents are therefore very low.   The ratio of starch equivalent to digestible crude protein in mangels is close to the required ratio of 10 to 1 for maintenance foods.   On the other hand, sugar beet pulp, kohl rabi, carrots, and especially potatoes and artichokes, are relatively deficient in digestible protein even for maintenance rations.

Mangels.   The dry-matter content of mangels ranges from about 9 to 14 per cent.   No variety or strain of mangel can be safely fed until after Christmas, or sometimes not until the end of January. Unripe mangels contain nitrates and these and possibly other constituents can cause severe scouring in cattle and sheep.   The mangel crop is therefore lifted and stored in clamps to ripen as well as to keep until needed in late winter.   During the ripening process the nitrates are converted to " amides ", and the latter in turn are partly converted to true protein.   Mangels do not impart a taint to milk and, on this account, they are more useful than swedes and turnips for feeding to milking cows.

In the past, huge quantities of 50 kg or more daily of mangels and other roots were fed to dairy and fattening cattle.   It was then necessary to pulp and feed the roots with chaff.   Nowadays, cows are given more reasonable amounts of 10–20 kg of mangels per head per day.   This change of practice has been brought about by several factors, but mainly the high cost of growing and feeding roots, possibly assisted by the powerfully made suggestion that roots are too bulky for dairy cows.   Only cleaning and no pulping or slicing are necessary when feeding the above quantities of mangels.   About 15 kg are the maximum amount that can be comfortably consumed in one meal by Shorthorn cows.   Sheep are not folded on mangels as they can be on swedes and turnips.   Mangels are usually carted from the clamp and fed to sheep on grass or in folds on arable land in late winter, the maximum amount per adult sheep being about 7 kg per day.   Mangels may be fed in small amounts to calves over three months of age, to hard-working as well as idle horses, to sows and even to fattening pigs.   For the latter, the maximum amount should be about 2·5 kg of fingered roots per day, replacing about 0·25 kg of barley meal.   Double this quantity may be fed to pregnant sows.   Split mangels are also usefully fed to fowls kept in fold units or intensive houses. The roots help to keep the birds feeding and also provide useful bulky food normally supplied by grass when birds are on free range.

Swedes are an excellent succulent food for all classes of stock. The dry-matter content ranges from about 10 to 13 per cent according to variety and soil conditions. Newly pulled swedes can have severe scouring properties, but their laxative nature becomes milder after a period in the clamp. In some southern districts, swedes may be left in the ground until late winter and then folded with sheep.

White-fleshed turnips have the lowest dry-matter content (7–8 per cent) of all roots. The yellow-fleshed varieties have dry-matter contents ranging from 8 to 10·5 per cent. Turnips are quickly grown and are useful for folded sheep in the autumn. They are sometimes sown in mixture with rape as a pioneer crop on very poor ploughed-out grassland before it is sown down with better grasses.

Kohl rabi can be used as a feeding crop from August to March. The roots may be fed whole to cattle, sheep and horses. Since kohl rabi does not taint milk, it can provide a valuable succulent food for dairy cows during late summer droughts. The crop can be folded by breeding ewes and fattening lambs.

Carrots are grown primarily for human consumption but unmarketable and surplus carrots can be fed to, and are relished by, all classes of stock. The carotene content of the roots is high and carrots are therefore most valuable when the rest of the diet is poor in carotene. Hence they are especially useful for helping to maintain health and good coat condition in horses kept continuously on dry rations in town stables and elsewhere.

Potatoes are an extremely useful carbohydrate food for farm livestock. They must, however, be reasonably clean and free from sprouts which are apt to contain toxic amounts of the alkaloid solanine. The dry-matter content of potatoes averages about 24 per cent and it consists predominantly of the highly digestible carbohydrate, starch. For pigs and poultry, potatoes should be steamed or boiled in order to obtain most efficient and economical utilisation, although a few raw potatoes may be fed to all but young pigs. Cooking not only improves palatability, but it bursts the starch grains and renders the starch more accessible to the digestive juices. About 4 kg of potatoes may be taken, after cooking, as equal to 1 kg of barley meal in energy value. Pigs may be given cooked potatoes in quantities running up to a maximum of 7 kg per head per day according to the age and the purpose for which the pigs are kept. For poultry, cooked potatoes may comfortably form 50 per cent of the total ration, but the proportion may be increased to 80 per cent with proper choice of supplementary foods. Mature cows can be safely given 10 kg of raw tubers per head per day without causing scouring or blowing. About 1 kg of raw potatoes is equal to 2 kg of mangels. Mature fattening cattle can be given 20 kg and even 36 kg of raw potatoes per head per day provided care is taken to introduce them gradually into the ration. Furthermore, greedy feeders in yards must not be allowed to get too large a share at any one meal owing to the risk of bloat. Raw potatoes are unsuitable for young stock or pregnant animals.

Fodder Beet. The high dry matter (18–21 per cent) types are almost indistinguishable from sugar beets, while those with 15–18 per cent dry matter may resemble mangels. Fodder beet, sliced or fingered, can be fed to fattening pigs up to a maximum of about 9 kg daily for a pig of 100–105 kg live weight, each 4–6 kg of fodder beet (18 per cent dry matter) replacing 1 kg of cereals. About 25–30 kg of medium dry matter fodder beet can be fed daily to dairy cows.

Sugar Beet Pulp. Both wet and dried sugar beet pulp are valuable feeding-stuffs in the sugar beet growing areas where the acreage of other roots is necessarily reduced. Transport costs limit the use of the wet pulp mainly to farms within reasonable distance from the factory. Furthermore, the product is very liable to go sour when exposed to air, but it can be safely ensiled in pits until required for feeding. It can be fed in about the same amounts as mangels. Dried sugar beet pulp may be fed as a substitute for roots, or as part of the concentrate ration. About 1 kg of the dried pulp is equal to about 8 kg of mangels or 1 kg of oats. The maximum amount for dairy cows is 4 kg per day, but fattening cattle may be given up to 6 kg per head per day.

Dried sugar beet pulp swells considerably when moistened with water and so the larger quantities for cattle are often soaked before feeding. This treatment is specially necessary when sugar beet pulp is given to horses, since it would be dangerous to allow the swelling of the pulp to take place in a full stomach. Half the corn ration of horses has been replaced by a mixture of equal parts of dried beet pulp and molassed beet pulp with apparently satisfactory results, but beet pulp is a rather bulky food for hard-working horses.

Dried beet pulp can be fed in limited amounts to pigs. A considerable proportion of the fibre in beet pulp is broken down in the large intestine and cæcum of the pig, but the product is still too bulky and its digestible carbohydrates rather inefficiently utilised to permit sugar beet pulp to be freely fed to pigs. About 1 kg per head per day is the maximum amount for non-lactating sows and about 0·5 kg per head per day for fattening pigs. The food is best omitted from the rations of very young pigs and it is unsuitable for feeding to poultry. Molassed beet pulp is more palatable, but otherwise its feeding value is very similar to that of the plain dried pulp.

Beet molasses is the syrupy residue left after crystallisation of the sugar from beet juice. Its water content varies from about 20 to 30 per cent: at an average moisture content of 25 per cent, the syrup contains 66 per cent of soluble sugars, about 5 per cent of ash and 3·5 per cent of crude protein which is of very little feeding value, the main nitrogenous constituent being betaine. The ash consists mainly of potash, which is partly responsible for the laxative properties of molasses. The product is useful for enhancing the palatability of feeding-stuffs for cattle. It is largely fed to animals for this reason rather than as a straight food. It is a useful laxative and is used in the treatment of cases of acetonæmia in dairy cows. Its sticky binding properties make molasses a useful ingredient of compound cattle " cubes " and pig " nuts ".

GREEN FORAGES

Under the heading of green forages must be considered the grasses and young cereals, certain legumes and such brassicas as kale, rape and cabbage which are customarily grazed or cut for feeding. In contrast with the succulent fodders, the green forages are characterised by relatively high protein contents, particularly in the early stages of growth, and the fact that they are all good sources of β carotene, the precursor of Vitamin A.

Grass and Clover. The grass crop is pre-eminent among green forages on farms of the British Isles. Although pure stands of grass are grown, it is more usual to find mixtures of grass and clover or other legumes forming the basis of our grassland farming.

The major seasonal changes in the nutritive value of grassland herbage are those associated with maturation of the plant. As physiological age increases, there is a rapid decline in protein content, due mainly to a fall in the proportion of leaf present, and a marked fall in energy value caused by increased development of fibre. It is this change in the proportion of leaf to stem as the grass plant develops which holds the key to its nutritive value.

Leguminous plants also decline in feeding value with age. However, with these plants, the protein content does not fall so markedly with the onset of maturity as in the grasses, although the fibre tends to rise more quickly, with an accompanying fall in their digestibility and energy value.

Although maturation in plants is a continuous process, it is possible by dividing it into three arbitrary stages, to estimate the feeding value of the herbage from visual observation.

(*a*) *Leafy or early bud stage.* In the grasses the leafy stage is marked by a complete absence of flower heads or of the stem associated with the flowering parts of the plant. The corresponding stage in leguminous plants is the early bud stage, when there are a few flower buds forming on the plants but none is in full flower. At this stage of growth, the plant is rich in protein and energy, with a close ratio of digestible protein to starch equivalent (about 1 to 5) making it particularly suitable as a productive food for milk.

(*b*) *Early flowering stage.* At this phase of development of the plant, the early grasses are in ear, pollen is blowing about and the stem is very noticeable. In leguminous plants the flowers are forming on the lower parts of the plant though no seed should be present in the early flowers. The protein content and the digestibility have both fallen by this time, and the ratio of digestible protein to starch equivalent for mixed herbage is now wider (about 1 to 7 or 8). With pure leguminous crops the ratio is more narrow because the relatively high proportion of protein and greater development of fibre give a more marked fall in the starch equivalent value.

(*c*) *Full flower stage.* By this time the grass plant has almost completed its growth cycle except for the formation of seed. Lignification is now proceeding fast, nutrients are being transferred to the seed or roots, and there is an obvious increase in the proportion of stem to leaf, resulting in a steep fall in both the digestibility and the feeding value of the plant. There is a similar loss of value in leguminous plants. The ratio of digestible protein to starch equivalent value has widened still further, and in all except pure leguminous crops, will now be about 1 to 10, making it only suitable as a maintenance food.

A guide to the relationship between stage of growth, chemical composition and nutritive value of both grass and clover is given in the table below.

TABLE 5

RELATIONSHIP BETWEEN STAGE OF GROWTH, COMPOSITION AND NUTRITIVE VALUE OF GRASSES AND CLOVERS

Crop.	Per cent in the dry matter.				
				Calculated.	
	Crude Protein.	Crude Fibre.	Total Ash.	Starch Equivalent.	Digestible Crude Protein.
Grass					
Very leafy	23·5	17·0	10·0	75	18·9
Leafy	15·5	18·8	8·2	74	11·2
Early flowering . .	13·5	23·1	8·1	66	9·3
Full flower . . .	9·2	30·4	7·2	53	5·2
Red Clover					
Pre-bud	25·3	18·2	10·6	63	21·2
Early bud	20·7	23·9	8·8	58	16·6
Early flowering . .	18·0	27·5	8·5	54	13·9

The starch equivalent and digestible crude protein values in the above tables were calculated from work carried out in the Netherlands.

The mineral content of a pasture is influenced by the species present; generally speaking, the herbs and legumes are better sources of mineral matter, particularly calcium, than the grasses. At the young and leafy stage of growth, the mineral content of most species is relatively high, but falls with the onset of maturity. In mixed pastures, this decline with maturation may be overshadowed by a change in the botanical composition of the sward, giving an apparent increase in mineral matter. This is particularly so where the proportion of legumes in the sward increases as the sward matures; and is commonly found in practice where the aftermath cuts have a greater preponderance of clover than the earlier ones.

The practice of accumulating pasture growth during the late summer and autumn to provide grazing (foggage) for cattle and sheep in the winter, and hence extending the effective grazing season, is of long standing in certain parts of the British Isles. Recent developments in grassland management, together with the use of suitably improved strains of grasses and adequate dressings of compound fertilisers, have improved the quality of the herbage available for grazing in the winter period. Suitable grass strains for this purpose should exhibit a high resistance both to frost damage and also to rotting under the cold wet climatic conditions often prevailing in early winter. Such pastures, which have been rested from late July until early December, should have a good proportion of green leaf present in the mid-winter period ; the feeding value of this herbage may resemble that of summer grass at the flowering stage of growth.

Cereals. The sowing of autumn rye to provide early spring grazing for cattle and sheep has become another popular way of extending the grazing season in this country. This crop can advance the start of the grazing season by as much as 4 to 6 weeks, and since it is utilised at the young and leafy stage of growth, its feeding value will be similar to that of grass at a corresponding age.

In the arable districts advantage is sometimes taken to graze off the excess growth of winter-proud cereals during the early spring period. This, too, provides a highly nutritious food similar to that of young and leafy grass, but the grazing is only an incidental by-product of grain production.

Leguminous Plants. Apart from the clovers which have been already mentioned, a number of other leguminous plants require mention.

Lucerne. Lucerne has been found to have a relatively lower nutritive value, under conditions in Great Britain, than properly managed grassland herbage. This is due mainly to the more rapid lignification in the stem of the lucerne plant, causing a fall in both the digestibility and the starch equivalent value, when compared with grass at a similar stage of growth.

Sainfoin. Sainfoin is a crop confined to a few areas in this country. It is a typical legume in respect of its richness in protein and calcium, but does not tend to become unduly fibrous. It compares very favourably with lucerne in nutritive value. For certain areas where potash deficiency is likely to have a serious effect on the growth of lucerne, it may be preferred as a drought-resistant leguminous crop.

Maize. Maize is always used at a relatively advanced stage of growth, and hence the product is normally low in protein. The leaves and stem change relatively little during the later stages of growth, while the developing ear, containing a high proportion of easily digested carbohydrates, rapidly increases in feeding value. Thus, unlike most fodder crops, as the maize crop matures the increasing contribution made by the ears results in improved feeding value. The optimum stage of growth for harvesting is when the grains in the ear are in the soft dough stage. Maize, at this stage of growth, has been found to have a starch equivalent between 45 and 50, and a digestible protein of about 5 per cent, in the dry matter.

Brassicas. Certain species of Brassicae are very important indeed as green forage crops : their cultivation has been described in Chapter X.

Kale. There are two main groups of kale, namely the Marrow-stem and the Thousand-head. The Marrow-stem variety has a fleshy stem of varying thickness, and is less frost resistant than the harder stemmed Thousand-head variety. In both there is a significant difference between the composition of the leaf and the stem and hence the proportion of each present is very much reflected in the composition of the whole plant. Kale leaves, like those of the legumes, are rich in protein and minerals, particularly calcium, but the stems, while being higher in fibre than the leaves, are less fibrous than the stems of leguminous plants. This means that kale not only has the high protein and mineral status as found in legume plants, but, also, a relatively high energy value. The difference between the feeding value of the Marrow-stem variety and the Thousand-head variety is small when the two are grown under similar conditions. Normally both varieties have a starch equivalent of 60–5 in the dry material and a digestible protein averaging 10 per cent; but the digestible protein content of any particular crop is very dependent on the proportion of leaf to stem present.

When kale is grazed, the leaf and the top third of the stem are almost invariably eaten, but the remainder of the stem is rejected in varying degrees. This lower part of the stem has a higher content of dry matter and crude fibre but a lower content of protein, calcium and phosphorus. Thus when utilisation is incomplete, the least nutritious portion of the plant is rejected. Since the rejected portion has a high dry matter content relative to the whole plant, yields which are quoted for the crop without due regard to the probable degree of utilisation may give an exaggerated impression of the quantity of nutrients which it supplies.

Rape. Rape may be classified into one of three main groups, namely the giant rape, broad-leaved Essex rape, and the rape-type kale. The differences between these groups lie mainly in their height of growth and size of leaf. As with kale, there is no real difference in feeding value between the groups grown under similar conditions but variations in the protein content of plants are brought about by changes in the proportion of leaf to stem. Recent findings suggest that the feeding values of kale and rape are similar.

Cattle Cabbage. Cattle cabbage is grown on a limited scale as an alternative to kale, particularly where the crop cannot be grazed and has to be cut and carted. Since the stem in this crop makes up only a small proportion of the total plant, the composition of cabbage is less affected by the proportions of leaf and stem present. There are, however, marked differences between the inner and outer leaves which explain the differences between the drum-head and the open-leaved varieties. The former variety, which contains a high proportion of heart leaves, has a higher content of protein and soluble carbohydrates in its dry matter but a lower content of fibre and calcium than the open-leaved varieties. These differences are overshadowed in practice by the significantly higher dry matter content of the open-leaved varieties. In general, the composition and feeding value of the open-leaved varieties of cabbage closely resemble kale, whilst that of the drum-head varieties have only about 60–70 per cent of the feeding value of kale.

Sugar Beet Tops. Sugar beet tops consist of the leaves of the plant and also the upper part, or crown, of the root. They thus combine the properties both of a green fodder and of succulents. In the fresh state the tops have a composition and feeding value slightly inferior to those of kale.

CONCENTRATED FOODS

Under this heading it is necessary to consider the cereal grains and cereal by-products, leguminous seeds, oil cakes and extracted meals animal and marine by-products and one or two miscellaneous items.

Cereal Grains and their By-products. The cereals are chiefly carbohydrate with 8–15 per cent protein and a poor Ca/P ratio.

Wheat. Wheat may be fed to pigs and poultry, although there is always a danger that if ground too finely it will form a pasty mass in the mouth. Consequently, if wheat is fed to pigs (or to cattle) it should be crushed or only coarsely ground, and mixed with other meals. Wheat is said to be liable to cause skin troubles and laminitis when fed to horses.

Before considering wheat by-products, it will be as well to look at the structure of the wheat grain. This consists of an endosperm containing starch and gluten : around this comes the aleurone layer, which is rich in protein : enclosing the grain is the outer coat or pericarp, consisting chiefly of fibre and ash. In addition, there is also the wheat germ, which amounts to approximately 2 per cent of the grain. The process of milling is primarily designed to obtain the endosperm which we know as flour ; various types of by-product, or " offals ", arise, which consist of the pericarp and aleurone layer with varying degrees of endosperm. These offals consist of *bran*, and various other products which used to be grouped broadly into three main categories : (1) pollards ; (2) coarse middlings or sharps ; (3) fine middlings or thirds. Wheat offals, other than bran, are now marketed as *weatings*, with a guarantee of not more than 6 per cent fibre. The differences between the various grades in fibre content and minerals are shown in the published tables.

Bran is a palatable feeding-stuff, somewhat laxative and capable of " opening up " a heavy ration ; because of its content of protein and fibre it is much more suitable for inclusion in rations for breeding, growing, milking or working animals than it is for fattening stock. Like wheat itself, and the other by-products, it has a low Ca/P ratio. Weatings are very popular for both pigs and poultry and may also be given to cows.

Barley. Barley is essentially a fattening food, and if used for the production of milk and eggs must be balanced with appropriate amounts of protein and minerals. Barley meal has been recognised for a long time as a splendid feeding-stuff and it is normally used extensively in rations for fattening pigs.

In the brewing of beer, barley yields a number of useful by-products. After the grain has germinated during the process of malting, the root and shoot are removed, and this material is known as *malt culms*. Malt culms should always be soaked before feeding and can be fed to dairy cows, horses and sheep. The malt is steeped in water to extract the sugar, and after the sugary solution—the so-called wort—has been run off, *wet brewers' grains* remain. These consist chiefly of protein and fibre, the starch having been lost during the malting and steeping processes. Wet brewers' grains may be fed to dairy cows, but excess is liable to cause scouring. This feeding-stuff must be used immediately or else ensiled with salt in pits, since it turns sour very readily on keeping. It is too expensive to transport very far, but *dried brewers' grains* are also available ; these are useful for milk production and can also be used for fattening cattle and sheep. They are too fibrous for pigs. The distilleries also provide by-products of a similar nature to those just mentioned and the pearl barley industry gives rise to small quantities of *barley feed* which is useful for pigs.

Oats. A good sample of oats should be bright in appearance, and the higher the bushel weight the lower will be the percentage of husk. Oats have a better Ca/P ratio than other cereals, but are also richer in fibre. They are the favourite cereal for horses, to which they can be fed whole; it is usual to crush them when feeding to other stock. Dairy cows, fattening bullocks and sheep can all be given oats, and despite their fibre content they can be used fairly extensively in pig rations, although the husks are sometimes sieved out before feeding to young pigs. Oats are also given to poultry but *Sussex ground oats* are more popular for this class of stock. In oatmeal milling districts such by-products as *meal seeds, scree dust* and *oat-dust* are also available for pigs.

Rye. The previously mentioned cereals are extensively grown in these islands, but rye is a very minor crop, although it is a major cereal of northern Europe. It is not popular as a feeding-stuff with farmers, neither is it relished by stock, but if used it can be considered as being roughly equivalent to wheat.

Maize. This cereal is not grown for grain in this country except on a very minor scale, but large quantities are imported. Its composition makes it suitable for fattening rather than for other productive purposes, because it has a high carbohydrate and low fibre content while the protein is poor in quantity and quality and the grain is deficient in minerals. The yellow varieties of maize contain crypto-xanthine, a precursor of vitamin A, and are given to laying hens ; the white varieties containing no carotenoid pigment are sometimes fed to fattening birds in order to maintain a white flesh. Maize should be cracked or kibbled for dairy cows, but very often it is ground to give maize meal.

Maize meal is of two types, the one having the maize germ left in, the other having this removed and hence being lower in oil and less likely to go rancid. By treating maize with steam and then rolling and drying, *flaked maize* is produced, which, because of its lower oil content, is favoured by pig feeders, although the production of a soft oily carcass should not occur if maize is used with discretion. Flaked maize is also favoured by other stock feeders.

During the manufacture of cornflour and starch from maize a variety of by-products arise, the chief of which are maize gluten feed, maize gluten meal and maize germ cake. *Maize gluten feed* differs from *maize gluten meal* in that it contains the husk of the maize and hence is richer in fibre and poorer in protein ; the two are, however, excellent foods and the gluten feed can be given to all classes of stock. *Maize germ cake* or meal is the residue left after removal of the oil from the germ and is mostly used for dairy cows.

Rice. In the East, rice is the foremost cereal and is fed along with pulses to livestock ; it is deficient in protein, fat and minerals. When available here, it is fed to poultry.

For human consumption it is necessary to remove the hard husk from the grain, and this process, known as polishing, results in a product containing husk and portions of the kernel and germ and known as *rice meal*. This rice meal is quite a useful food provided that it does not contain more than 12 per cent fibre, but it is important to distinguish it from inferior rice grains ground up and given the same name. The latter is almost pure carbohydrate.

Other cereals which normally reach this country are *millet* and *sorghum*, both of which are used for poultry : *buckwheat* is usually classed under the same heading, although not a cereal.

Leguminous Seeds. Seeds of leguminous plants are richer in protein than the cereal grains and correspondingly lower in carbohydrate. They are often referred to as pulse grains.

Beans. These seeds, usually home grown, can be fed to all livestock but it is not advisable to feed when new. More often beans are ground to give bean meal, a product which, unfortunately, is very easily adulterated and does not keep well. Bean meal is given to dairy cows, though in excess it will produce a hard white butter ; to fattening cattle, especially to finish them off ; to pigs, in which it gives firm carcases ; and to horses. This feeding-stuff can also be used in milk substitutes for calves.

Peas. Peas are not as popular as beans, and pea meal is less extensively used. In comparison, the meal is somewhat lower in protein but it can be used for similar purposes.

Other leguminous seeds are : *vetch* and *lupins*, each of which needs to be soaked and steamed to remove the bitter principle before feeding, although " sweet " lupins with a low alkaloid content are now available ; *lentils*, used for dairy cows and poultry ; and *chick peas*, which can be fed to most classes of stock. It should also be mentioned that imported leguminous seeds may contain poisonous seeds of considerable danger to stock.

Oil Seeds. Only two oil seeds need be considered here, linseed and sunflower seed, both of which have been used, as seeds, only to a very small extent in this country for feeding purposes.

Linseed. Linseed is rich in oil and protein, and the whole seed may be fed either after crushing or after scalding or cooking. It is specially useful for rearing calves, and also for sick animals. Linseed absorbs water and forms a jelly if soaked in warm water, but may become dangerous if allowed to remain in a moist condition because of the formation of small quantities of poisonous prussic acid from a cyanogenetic glucoside present in the seed in small quantities.

Sunflower. Sunflower seeds contain from 32 to 45 per cent of edible oil which is of great value for margarine, cooking and medicinal purposes. Consequently the seed, as grown, is too valuable to use as a stock food, but the seed residues, after the oil has been extracted and the husk removed, form a useful feeding-cake. Sunflower seed may be fed to poultry at the rate of from 30 to 60g per day, and is said to be specially valuable during the moult.

Oil Cakes and Meals. Numerous seeds, chiefly grown in tropical and sub-tropical countries, are rich in oil : the oil is of considerable value in the manufacture of cooking fats, soap and other products, and the residues have proved to be excellent protein-rich feeding-stuffs for farm stock. The oil is removed by pressure or by solvent extraction Formerly hydraulic pressure was used but this is now obsolete as it left too much oil (8–10 per cent) in the residue and was expensive in labour. Modern methods use either expeller or extraction procedures. In the first process the oil seeds are forced by a worm screw along a tube of gradually diminishing bore, thus squeezing out the oil and leaving a residue containing usually less than 5 per cent oil. In the extraction process the coarsely crushed oil seeds are subject to the action of an organic solvent such as hexane which dissolves out most of the oil, leaving a residue with usually less than 1 per cent oil. The volatile solvent is recovered and used again, and the residue is referred to as an extracted meal.

Some oil seeds have a fibrous seed coat which is removed by the process called decortication : the residue from decorticated seed will have a much lower fibre content than the residue from undecorticated seed, and the other constituents will vary accordingly. It is important to realise that oil seed cakes and cake meals are different forms of the same thing ; that extracted oil seed meals are clearly named : and that oil seed meals, e.g. linseed meal, are meals from which the oil has not been removed.

The chemical composition of all the important oil cakes is given in published Tables, and the figures should be consulted in conjunction with the following notes upon the individual materials.

Ground Nut Cake. This cake and the corresponding meal are available both in the undecorticated and the decorticated forms : the latter is particularly rich in protein. It is made from the earth nut or ground nut, which is the underground fruit of a tropical or sub-tropical leguminous plant. Both forms are used for feeding to dairy cows, fattening cattle, sheep, pigs and poultry, although the extracted decorticated ground nut cake is more useful in pig rations since it does not contain sufficient oil to affect the carcase quality.

Cottonseed Cake. This also is available in the decorticated and the undecorticated form and, in addition, the source of origin somewhat modifies the composition of the cakes. The undecorticated cake is very useful for cattle out at grass because of its " binding " action. Both cakes can be fed to cattle and sheep, but should be fed sparingly to pigs and young stock because of their high fibre content. Cottonseed products are poor in calcium, the protein quality is low, and cottonseed poisoning has caused trouble, particularly in America.

Palm Kernel Cake. This is made from the kernel residues of the African oil palm fruit, and it is somewhat gritty and absorbs water only slowly. In consequence cattle do not readily consume this cake, but if it is gradually introduced into the ration, possibly along with molasses, they will soon eat it. It is not as rich in protein as the other cakes and therefore can be used in larger quantities. Palm kernel cake is balanced for milk production and is usually recommended for the " steaming up " of dairy cows. It is also given to fattening cattle and small quantities may be fed to pigs.

Coconut Cake. This cake is sometimes soaked before feeding, and it is a very safe food for dairy cows, fattening cattle and pigs. It has the merit of producing a harder fat than do most of the oil seed products and hence can be used to offset the effects of these.

Linseed Cake. For many years this cake has had the reputation of putting a " bloom " on fattening cattle, an effect which may be due in part to its mild laxative action which, in turn, is dependent upon the mucilage which it contains. Because of this reputation linseed cake always fetches a high price, and, to economise in feeding, it is possible for the farmer to feed linseed cake in the later stages of fattening only, using cheaper cakes earlier on. Linseed cake or the cake meal will produce a soft oily fat if fed in excess and the biological value of the protein has been shown to be low for milk production ; nevertheless, it is an excellent feeding-stuff. Linseed cake meal is quite distinct from linseed meal, which consists of the ground whole seed ; this latter material still contains all the oil and is used in small quantities for sick animals. Whole unground linseed should never be fed unless it has been boiled or well soaked in water. As already mentioned linseed products may sometimes cause poisoning in young stock if due care is not taken in feeding them.

Soya Bean Cake is rich in protein and contains an oil which is somewhat laxative. Extracted soya bean meal on the other hand has a low oil content. Dairy cows, fattening cattle, pigs and poultry can all be fed on it. It is laxative but quite safe in normal amounts, and is used to balance the cereals in a ration.

Miscellaneous Oil Cakes. Amongst the lesser-known cakes and meals are *Rape, Sunflower, Safflower* and *Sesame.* The former is not always safe because it may cause irritation of the alimentary tract, but the others are satisfactory when used in the right way.

Animal Products. These are mainly by-products from slaughter-houses and the fishing industry.

Meat Meal and Meat-and-Bone Meal. These products are made from trimmed-off portions of animal carcases shortly after slaughter : they are subjected to cooking in steam-jacketed cauldrons until the moisture content is reduced to less than about 10 per cent. Surplus fat is drained off and the residue is then pressed in an expeller to remove the greater part of the fat. The quality of the material is dependent upon the proportion of bone included, the higher the proportion of bone the lower will be the protein content. Meat meals must contain by law at least 55 per cent protein ; meat-and-bone meals between 40 and 55 per cent protein ; neither type of product may contain more than 4 per cent salt. These legal definitions limit the amounts of fat and bone in these products which can be usefully employed in the rations of pigs and poultry.

Whale Meat Meal. This is a product of a very high protein content and consists very largely of the flesh of the whale with a minimum amount of included bone. The processing is similar to that of meat meal.

Blood Meal. The coagulate from slaughterhouse blood is pressed to remove as much serum as possible and it is then dried under vacuum in steam heated cylinders to a moisture content of under 10 per cent. The product contains some 80 per cent protein, but it is normally a low source of minerals. It may be used up to a level of 5 per cent in the rations of pigs.

White Fish Meal. This consists of the dried and ground flesh and bones of white (i.e. non-oily) fish, the drying being effected at low temperatures to avoid damage to the nutritive value of the proteins. By the Fertilisers and Feeding Stuffs Regulations (1973) the product must not exceed contents of 6 per cent oil and 4 per cent salt. These legal requirements ensure a very high standard type of article that is particularly valuable for pigs and poultry. It normally contains at least 60 per cent protein ; its judicious use involves virtually no risk of tainting carcases or eggs.

Other Fish Meals (Herring, etc.). These are products of oily fish such as herring, pilchards and so on. After being partially cooked they are subjected to hydraulic pressure to remove as much oil as possible, and then the residue is processed similarly to that of white fish meal. These meals usually contain 7–10 per cent oil and are often richer in protein than white fish meal, and correspondingly lower in bone content, as reflected in their figures for calcium and phosphorus. The presence of this oil involves a slight risk of tainting edible products from the animals to which these types of fish meal are fed, but withdrawal of the fish meal some three weeks before slaughter will ensure that no taints are present in the carcase meat of pigs.

THE DIGESTIBILITY OF FEEDING STUFFS

The crude or analytical composition of a feeding stuff gives a very limited indication of its nutritive value to animals, since the latter can only utilise those portions of the food that are capable of being digested. Consequently in order to determine the digestibility of a food we have to use the living animal.

Digestion trials can be carried out with any class of animal, but the results obtained can be applied strictly only to that class. Again, the choice of animal depends upon a number of physical considerations such as available space, the quantities of food given, the weight of excreta obtained, and associated problems of sampling accurately the food and the excreta. Anatomical considerations may involve modifications of the technique employed ; in poultry, for example, the droppings represent a combined faecal and urinary excretion, and digestibility trials in consequence are more difficult to carry out.

Digestibility Trials. Most digestibility work on ruminants is conducted with sheep. Although the results so obtained are not identical with those obtained for the same food using cattle, the differences are not so great as to prevent those results from being applied to cattle. On the other hand, it would be incorrect to apply information derived from the use of sheep to pigs or to poultry. The castrated male sheep (wether) is usually employed for work on ruminants, because of the convenience of its size, the quantity of food consumed daily under laboratory conditions, and the firmness and ease of sampling of the faeces.

The assumptions made in carrying out a digestibility trial include that :

(1) the difference between what is eaten and what is excreted in the faeces is that which is digested ;
(2) the degree of digestibility of a food is independent of the animal's inability to take exercise whilst the trial is in progress.
(3) the degree of digestibility is independent of the plane of feeding.
(4) the degree of digestibility is independent of the breed, strain and individuality of the experimental animals.

The main sources of error probably lie in the first assumption.

The *faeces* of an animal comprise (a) undigested food residues, possibly altered in form by bacterial action in the large intestine, (b) the metabolic debris, resulting from the sloughing off of epithelial tissue from the walls of the alimentary tract during the passage of the food, (c) the unabsorbed residues from digestive fluids and juices added to the food in order to carry out the various phases of digestion and (d) certain mineral matter which is brought into solution and absorbed from the small intestine and re-excreted from the blood stream into the faeces via the large intestine. Of these, (b) and (c) are inherent in the act of digestion so that really we determine the degree of apparent digestibility of the food. Factor (d) may appear at first sight to be an insuperable difficulty, but since the digestibility of the mineral matter is not involved in any assessment of the energy content of the food, trials are carried out in the knowledge that whereas the returned mineral matter affects the weight and composition of the faeces, these are self-compensating and do not affect the total weight of organic matter voided.

The essential point to grasp is that a digestion trial is based on quantitative data of weights of individual food fractions given and voided. We supply the animal with daily known weights of food of composition determined by analysis ; we measure the daily weights of faeces voided and analyse this excreta just as we analyse a food. The feeding of 1,000 g hay containing 10 per cent crude protein involves supplying the animal with 100 g of protein in the hay ; an average excretion of 2,000 g wet faeces containing 2 per cent crude protein means that in the faeces were 40 g protein. So we argue, 60 g protein from the hay have been digested from the 100 g supplied, a *percentage digestibility* of 60 per cent ; or, putting it another way, the *digestion coefficient* is 60 per cent. The original hay contained 10 per cent protein of which 60 per cent is digestible and 60 per cent of 10 is 6. Hence this hay contains 6 per cent digestible protein.

Similar calculations and reasoning can be applied to the organic matter, the ether extract, the nitrogen-free extractives and the crude fibre of the hay. Indeed, it can be applied to the mineral matter as well, but we know in advance that we shall obtain a falsely low result since part of the mineral matter has been digested but has been re-excreted into the faeces.

No consideration is given to the urine in this reasoning; this is because what is excreted in the urine represents, amongst other things, the end-products of the metabolism by the body of the digested food. A digestion trial is primarily concerned with the fate of food in its passage through the alimentary canal. It is usual to collect, measure and analyse the urine at the same time since a great deal of additional information on the balance of certain elements in the body can be obtained for relatively little extra analytical work; but essentially the urine plays no part in digestibility trials.

It is not necessary to go into full details of the technique involved, except to mention that although a constant weight of food may be fed daily there are considerable irregularities in the daily weights of faeces excreted. Consequently, trials of less than one week in duration can give very misleading results, and it is usual to work for periods of 10 days once the animal has become accustomed to the restraint of the metabolism stall and the necessary harness to ensure complete collection of faeces and the avoidance of contamination with the urine.

A modification of the technique is required where the digestibility of concentrated foods is being studied. It is essential for a ruminant animal to ruminate in order to be normal, and rumination is adversely affected by an all-concentrate diet. The modified technique involves an initial period of feeding with a roughage such as hay during which its digestibility is determined. This is followed by a reduction in the amount of hay fed and the introduction of the concentrate food. The combined digestibility of both foods is then determined, whence by calculation the digestion coefficients of the concentrate food can be derived. Such a technique is very time-consuming, often involving the animal in a month's sojourn in the metabolism crate, but it appears to be the only reliable method of obtaining the necessary data.

FACTORS INFLUENCING DIGESTIBILITY

It has already been stated that results of the determination of the digestibility of a feeding stuff by one species of animal do not necessarily apply to another species. The differences in results are related to the ability of the animal to deal with the fibrous portion of the food. Ruminants have the greatest capacity for digesting fibre, although horses and rabbits, in both of which there is a relatively large caecum, can also cope reasonably efficiently with fibre. The major difference is that in the ruminant the complex biological changes that occur in the rumen takes place *before* normal gastric and intestinal digestion, whereas in other species the main digestion of fibre occurs *after* normal alimentary digestion. This peculiar attribute of the ruminant may allow of better subsequent digestion of non-fibrous constituents, because the cellulose cell-walls of the plant have been ruptured, making possible a better attack by the normal gastric and intestinal digestive juices on the contents of the cells of the food material. Hence fibrous foods are generally better digested by ruminants than by other classes of animals.

The physical size of the food particles is also of importance. Whole grains, unless thoroughly chewed may, in part, tend to pass through the alimentary tract unchanged and appear intact in the dung ; the same grains, even coarsely ground, will be better digested. On the other hand, with ruminants fine grinding to a meal of coarse fodders may lead to less efficient digestion, because in fine meal form the fodder may pass through the rumen and reticulum too quickly, allowing too little time for the microbial population there to do its work of splitting the fibre.

The level of feeding may exert some influence on digestibility, and there is evidence that animals fed to appetite with a particular food digest this less efficiently than at a lower plane of feeding. Certain foods, e.g. raw potatoes, are less efficiently digested by pigs and poultry than when fed cooked, although there seems little justification for steaming or cooking foods for most forms of livestock.

Foods that are naturally laxative and pass rapidly through the alimentary canal are less efficiently digested than when their rate of passage is slowed down. Spring grass fed in conjunction with some long fibrous food such as hay, straw or undecorticated cotton cake, is more efficiently digested than when fed alone. There is little scientific evidence to support the use of condiments as an aid to digestion ; the fact that better results are sometimes obtained with spiced foods is probably more likely due to increased food consumption than to a higher level of digestibility of the food itself.

Digestibility Coefficients only a Guide. Finally, although one has to accept tacitly published figures for digestibility coefficients of the various components of the food, it has to be remembered that these have usually been obtained with a very limited number of animals. In the absence of more detailed information these probably represent the best figures available, but they are really no more than a guide and should certainly not be accepted as absolute in any sense. For practical purposes they have to be used, but the reader should always be aware of the many assumptions that have to be made in this class of work and should accept the figures as indicative rather than absolute.

THE UTILISATION OF FOOD

In the course of digestion the soluble products of the food pass through the walls of the intestine into the circulating blood stream and thence to all parts of the body (see p. 510). In the various cells of the body, these products undergo a large number of very complex bio-chemical reactions. The net effect of these reactions is that the digested products of the food are either broken down ultimately to very simple substances, subsequently excreted from the body in the breath or the urine, with the liberation of energy, especially heat ; or they are built up into substances that form part of the body such as muscle and body fat, and secreted products such as eggs and milk. We refer to the breaking down reactions as *catabolism*, and the building-up processes as *anabolism*. The overall term *metabolism* covers both types of process.

For agricultural purposes food has two main functions. It provides for the *maintenance* of the animal, covering as it were the requirements for existence and normal health, and for the *production*, the normal economic reason for keeping the animal. Production may assume the form of growth and development, fattening, milk production, egg production, work, growth of wool or fur, and reproduction. Maintenance is, normally, a first charge on the food, and production begins only after maintenance needs have been met. In some circumstances the animal may draw on its bodily reserves in an effort to sustain production, even if the diet is insufficient, but obviously this cannot go on indefinitely.

MAINTENANCE

Maintenance is normally considered to be the state of the non-producing animal, in normal health and keeping its liveweight constant. Domestic animals are warm-blooded, with a constant temperature usually well above that of the environment in which they are placed. Consequently they are regularly losing heat to that environment. They breathe and move about, their hearts are pumping blood around the body, and even when physically at rest there is a great deal of activity going on within the body. Moreover, there is a certain amount of wear-and-tear that affects every cell and every part of the body. Thus the food has the job of providing the energy for the vital processes of the body, and materials to make good this normal wear-and-tear. Normally the energy needed is provided from digested carbohydrates, fibre, fats and oils, while the digested proteins cover most of the replacement needs. When the vital needs of the animal are met by the food without any gain or loss of weight, then the animal can be said to be receiving a maintenance ration.

This definition involves the assumption that liveweight changes necessarily reflect changes in the energy, protein and also the mineral status of the body. Over a reasonably long period of time this is broadly true, but over short periods liveweight changes can prove very unreliable. The major constituent of the animal body is water, and even a 1 per cent change in the water status of a 500 kg bullock amounts to 2–3 kg.

But the most important factor causing variations in liveweight of an animal from day to day is that of " fill ", that is the weight of the contents of the digestive tract. A 500 kg bullock at grazing may well consume 80 kg fresh grass each day, and may drink another 25 kg of water. Variations in the weight of dung and of urine voided each day can be very considerable, so that constancy of liveweight by itself can, over short periods of time, be a most unreliable index of the adequacy of a maintenance ration.

The daily output of heat from an animal can also vary. It depends to some extent on the environmental temperature, although the animal has some control of the degree to which it loses heat to its surroundings. The amount of energy involved in exercise is also variable, although this can be reduced by keeping the animal under stall conditions. Even so there is a greater output of heat when the animal is standing compared with lying down.

The concept of an adequate maintenance ration based on a steady liveweight figure can hardly be regarded as scientific except over a reasonably long period of time ; for fundamental investigational work, as we shall see later, a somewhat different definition has had to be made. Nevertheless, for general agricultural purposes a maintenance ration can be looked upon as supplying the nutrients necessary to maintain an animal in health without appreciable change in its liveweight.

PRODUCTION

The economic purpose of keeping farm animals lies in the very varied forms of production. Production may consist of straightforward growth, development and fattening ; it may involve materials removable from the body such as milk, wool, eggs ; it may be manifest in draught power or work, or it can involve the processes of reproduction, both with male and female animals. Different types of nutrients are involved in the various forms of production, and ultimately these derive from the digested portions of the food. Requirements of nutrients, both absolute and relative to the different species and types of production, are variable, and it is the aim of nutritional science to attempt to express in measurable form the nutrients necessary for various purposes.

There is considerable variation in productive requirements, not only between species and in relation to different forms of production but also within species and between individuals. The nutritionist attempts to define within determinable limits what are the specific requirements for various forms of production for what might be termed the average animal under average conditions. Indeed, some of the present-day doubts about figures for production requirements of the past are merely giving recognition to the reality of biological variation. Because of this variation nutrition must be looked upon both as an art and a science, and *scientific feeding should be regarded more as an intelligent guide than a dogmatic assertion of the exact needs of the animal.*

Nevertheless, the student of animal nutrition has to be aware of the commonly accepted feeding standards provided by modern science. The standards are expressed in many different ways, and may involve a large number of different factors. The major needs of any animal for both maintenance and production relate to an adequate supply of energy in the food and the provision of adequate levels of digestible protein of appropriate biological value ; minor needs are for adequate levels of appropriate vitamins and minerals in relation to the type of production called for. Practically all these needs are met by the food, although some minor factors may derive from the circumstances of an out-door life. It is now necessary to consider some of the more essential characteristics of feeding stuffs needed to provide for the energy requirements of animals for both maintenance and production, and with these the provision of adequate protein as well.

ENERGY CONCEPTS OF FOOD

The principal sources of energy derivable from food are carbohydrates, fats and oils ; no other component can take the place of proteins in relation to their specific functions, but supplies of protein, surplus to the animal's immediate needs, can serve as sources of energy after the nitrogenous portions of the surplus proteins have been split off, metabolised, and excreted into the urine. For the present, therefore, we shall consider protein to be also a source of energy, although strictly speaking this concept applies only to proteins surplus to the animal's needs.

The living animal is able by the various processes of digestion and metabolism to transform the energy of food into other forms of energy. In nutritional science energy is measured after it has been transformed into liberated heat, which in turn is measured by the physical process of calorimetry and expressed in terms of heat units. The standard unit of heat is the kilo-calorie[1] (kcal) which is the amount of heat required to raise the temperature of 1 kg water through 1 ° C (from 14·5 to 15·5 °C).

GROSS ENERGY VALUES

When a food is burnt in an atmosphere of oxygen in a bomb-calorimeter the heat liberated can be measured in terms of kcal per gram of food, and this is referred to as the *Gross Energy Value* of the food. The animal is not as efficient in its combustion of the food as the bomb, and part of the total energy of the food is lost from the body in the form of undigested solid matter (the faeces), the incompletely oxidised nitrogenous parts of the digested food appearing in the urine, and in the production of combustible gases produced by bacteria action on the food in its passage through the alimentary tract. This latter fraction is quite appreciable with ruminant animals which produce and belch out considerable amounts of the combustible gas methane. These losses, therefore, reduce the potential of the Gross Energy Value of the food by the combustible value of the excretion products, and the residual part of the original energy is termed the *Metabolisable Energy* or *Physiological Heat Value* of the food.

Even this reduced Metabolisable Energy Value of the food is not all available to the animal as was shown by the classical work of Armsby and his co-workers in the USA, for on placing a steer in an apparatus (the animal calorimeter) and feeding it on different planes of nutrition with the same food he was able conclusively to demonstrate a greater evolution of heat with increasing levels of feeding. This wasteful and uneconomic transformation of some of the energy of the Metabolisable Energy fraction of the food into heat which was dissipated from the animal body was termed Thermic Energy (or, when related to each 100 kg of the food, the Increment of Heat Production) so that only what was left was actually available to the animal for its physiological needs.

This available energy is called the *Net Energy Value* of the food. Hence Net Energy Value of 100 kg food = Gross Energy Value of 100 kg *less* the energy lost in the faecal, urinary and gaseous excreta

---

[1] The use of the term "calorie" is gradually giving place to "joule" for scientific reasons, 4·184 joules being equivalent to 1 calorie.

derived from 100 kg of food, *less* the Thermic Energy per 100 kg (or the Increment of Heat Production).

For large-scale work the kilo-calorie is too small a working unit of heat and the Therm,[1] equal to 1000 kcal, is employed.

A comparison of results obtained from a low-energy, poorly digested food such as wheat straw, with a high-energy, well-digested food like maize meal, gave the following results.

Therms per 100 kg-Dry Matter in Food

	Gross Energy.	Energy losses in Excreta.	Metabolisable Energy.	Heat Increment.	Net Energy.
Wheat Straw . .	444·3	306·1	138·2	113·7	24·5
Maize Meal . . .	444·1	111·5	332·6	128·5	204·1

The heat of combustion (Gross Energy Value) of the straw and the maize were almost identical. The lower degree of digestibility of the straw is reflected in the much greater loss of energy in the excreta (especially the faeces) with the result that the Metabolisable Energy value of the maize was nearly 2½ times that of the straw. The output of heat, reflecting the work of digestion and the plethora of nutrients in the tissues of the body, is larger for the maize than the straw, but with the latter it is such a high proportion of the Metabolisable Energy that only about 5 per cent of the original energy of the food is available for productive purposes compared with about 47 per cent of that in the maize meal.

Useful figures to remember are those for starch which has a Gross Energy Value of 410·4 Therms per 100 kg, a Metabolisable Energy value of 376·4, and a Net Energy Value of 235·6 (i.e. about 58 per cent of the Gross Energy), as applied to ruminant animals.

The concept of Net Energy Values is perhaps the most fundamental one in assessing the nutritive value of fodders, but it is more complicated and difficult to follow than the essentially similar but more practical concept due to Kellner—the Starch Equivalent System.

STARCH EQUIVALENT VALUES

Kellner argued that if an adult animal were on a true maintenance ration and additional food were then given, the extra energy from this food would be stored in the animal body in the form of fat, since the carbohydrate content of the body amounts to only about 1 per cent of the weight of the body. Any slight storage of protein could be converted into its energy equivalent of fat, on the basis that 5 g stored body protein are equivalent, in *terms of energy*, to 3 g body fat.

Kellner placed a bullock in a closed chamber—the respiration chamber—so that he could feed it, collect the faeces and urine, and determine the gaseous output of carbon dioxide and of methane and other combustible gases arising from bacterial action in the rumen. He adjusted the food intake so that the animal was approximately in

[1] The " Therm " used here must not be confused with the " Therm " used as a unit of gas supply.

carbon and nitrogen equilibrium. This meant that the intake of carbon in the food exactly balanced the output of carbon in the faeces, urine, the " brushings " from the coat of the bullock and the gaseous excreta, and that the intake of nitrogen was equal to that put out in the faeces, urine and the coat " brushings ". This provided the animal with a true maintenance ration.

Next, in addition to this maintenance ration, he gave known weights of pure nutrients such as starch, sugars, proteins, cellulose, and fats and oils, which were all converted into fat in the body of the bullock, and again measured the intake and output of carbon and nitrogen in the solid, liquid and gaseous excreta. From a knowledge of the carbon and nitrogen content of body protein and of the carbon content of body fat, he was able to calculate the storage of carbon (and any slight storage of nitrogen) in the body and hence the actual weight of fat stored for each unit of weight of the pure nutrients fed. For example, Kellner found that 1 kg starch led to the storage of almost 0·25 kg fat in the body in the conditions described, and 1 kg protein to the storage of 0·235 kg fat. By dividing the various weights of fat stored per kg nutrient by the figure obtained for starch (actually 0·248 kg), he was able to calculate the relative fat-forming capacities of different pure foods. These are shown below :

Relative Fat-forming Values of Different Foods

1. Pure Starch	1·00
2. Fibre (Cellulose)	1·00
3. Sugars	0·76
4. Protein	0·94
5. Fats (a) from oil seeds	2·41
(b) from cereals and legumes . . .	2·12
(c) from roots, and green and coarse fodders .	1·91

Work with bullocks on a carefully adjusted maintenance ration in a respiration chamber is very time consuming, exacting and laborious. Kellner argued that if the various digested components of a mixed feed behaved, from the point of view of their fat-forming values, like pure nutrients, he should obtain the same results as with the use of the respiration chamber. To test this, he again put his bullock on a true maintenance ration, and fed various concentrated foods like decorticated oil seed cakes, cereals and so on. To his delight he found that the fat-forming capacities of these foods, relative to starch, were very closely similar to those calculated from the digested components employing the factors outlined above ; within the limits of experimental error they were generally slightly less.

The " ideal " or " theoretical " Starch Equivalent value of a food was calculated by adding together the digestible nitrogen-free extractives, the digestible crude fibre, the digestible protein × 0·94 and the digestible oil × 2·41, or 2·12 or 1·91, according to its origin, as shown above. The " actual " starch equivalent value was obtained from the actual weight of fat stored divided by 0·248 (the fat-forming value per unit weight of starch). The ratio between the two values was expressed as a Value Number (or " V ") thus

$$V = \frac{\text{actual weight of fat produced per unit weight of food}}{\text{weight of fat per unit of food predictable from digestible composition}}$$

" V " numbers for concentrated foods usually lay between 0·95 and 0·98.

When Kellner employed less concentrated, mainly more fibrous, foods, the discrepancy measured by " V " was larger, i.e. the " V " values became smaller. For example, " V " for bran was shown to be 0·77, for swedes 0·85, for good meadow hay only 0·67 ; Kellner quickly saw that the more fibrous the food the smaller was the " V " value, and that each 1 per cent crude fibre in the food lowered the predicted Starch Equivalent (S.E.) value of the food by 0·58 unit.

For coarse fodders he arrived at the relationship

Actual S.E. value = Theoretical S.E. minus 0·58 × % crude fibre.

For green fodders the figure for the fibre correction factor varied from 0·29 to 0·58, perhaps a rather unsatisfactory and arbitrary method of applying a correction factor.

It is now possible to define the Starch Equivalent value of a food as *the number of kg of pure starch, which when fed in addition to a true maintenance ration, will lay on in the adult ruminant animal the same weight of fat as* 100 *kg of the given food.*

Each kg of starch or starch equivalent fed in addition to a maintenance ration to a ruminant causes the storage of 0·248 kg fat or 2,356 kcal Net Energy, in the body.

Unfortunately there are still many difficulties in assessing the starch equivalent values of roughages and other fibrous foods. After Armsby had developed the use of his animal calorimeter, he found that whereas his values for energy stored from concentrated foods agreed reasonably well with those of Kellner, his values for coarse fodders were usually appreciably higher. The practice in this country now is to increase arbitrarily the Kellner S.E. values for coarse fodders by 20 per cent, following a suggestion by the late Professor T. B. Wood at Cambridge. An example will make this clear in respect of meadow hay.

STARCH EQUIVALENT OF MEADOW HAY

Crude Composition	(per cent)	Digestible Composition (per cent)	Factor	" Theoretical " S.E.
Moisture	15·0	—	×	
Crude Protein	9·0	4·6	(0·94)	4·32
Ether Extract	1·9	1·0	(1·91)	1·91
Crude Fibre	26·7	12·8	(1·00)	12·8
Nitrogen-free Extractives	42·0	24·0	(1·00)	24·0
Total Ash	5·4	—		43·03

Fibre Correction = 0·58 × 26·7 = 15·50
Kellner S.E. (Unadjusted) = 43·03 − 15·50 = 27·53
Adjusted S.E. (Wood's 20 per cent increase) = 27·53 × 1·2 = 33

The starch equivalent system is essentially a measure, under specified conditions, of the fat-forming capacity of a food ; it will be noted in the calculations involved that protein serves essentially as a source of energy since the protein requirements of the adult animal have been met by the true maintenance ration.   Whereas 1 kg S.E. for fat forming purposes involves the storage of 2,360 kcal, for milk production purposes (due to the secretion of protein into the milk) 1 kg S.E. corresponds to the storage of about 3,000 kcal for milk production.

Notwithstanding the limitations and assumptions involved in the S.E. system of food evaluation, it does form a useful guide to the scientific rationing of ruminants.   The factors were worked out with ruminant animals, and hence it is wrong to use the term S.E. for evaluation of foods for pigs and poultry.

FOOD OR FODDER UNITS

Fundamentally, the starch equivalent system is another method of expressing net energy values of foods.   In northern Europe a further modification has been involved in what is termed the Food (or Fodder) Unit system.   It is empirical and, whilst based on Kellner's results, has been modified to take into account the results of practical group-feeding trials.   The food unit is 1 kg of average barley and corresponds to 0·71 kg S.E.

A further modification of the system has been evolved for pigs and poultry, and is known as Total Digestible Nutrients (T.D.N.).   This is an expression summing the digestible nitrogen-free extractives, the digestible crude fibre, the digestible protein, and 2·3 times the digestible ether extract.   Broadly, it corresponds to the uncorrected " theoretical " starch equivalent values of Kellner, and for concentrated foods the T.D.N. values are numerically very close to those for S.E.   For less concentrated foods, and in particular the more fibrous foods, the T.D.N. system of evaluation gives excessively high values. If we take the figures cited on page 573 for the digestible composition of the specified sample of meadow hay, the T.D.N. value works out at 43·7, much higher than the uncorrected Kellner S.E. value of 27·5 or the Wood-corrected value of 33.   Whatever may be the inaccuracies of the various Kellner correction factors, they at least help to avoid the patent over-evaluation of the more fibrous foods inherent in the T.D.N. system.

MAINTENANCE STANDARDS

Kellner worked with cattle in store condition to try and assess the amounts of coarse fodders required to maintain them in carbon and nitrogen equilibrium, and found that the 500 kg animal (0·32 kg) required 3·0 kg S.E., *including* 0·32 kg digestible protein, to achieve this.   It seems logical to assume that larger animals would have higher requirements and smaller ones lower requirements than these. The maintenance need of food by an animal should balance the amount of heat evolved in a 24-hour day, and this loss of heat is more closely related to the surface area of the animal than to its body weight.   In general, surface area is not directly proportional to live-weight but to a power of that weight.   Rubner's Surface Law implied that the rate of metabolism was a function of the surface area and that

$$\frac{\text{Total Heat evolution (kcal)}}{\text{Surface Area}}$$ was approximately constant.

Surface area was formerly assumed to be proportional to the two-thirds power of the weight, though more recent biometric work has suggested the power of 0·73 : in other words maintenance requirements vary as (Live Weight)$^{0.73}$. If we follow the " two-thirds " power law, M is proportional to $W^{\frac{2}{3}}$ or $M = kW^{\frac{2}{3}}$. The value for M at 500 kg liveweight is 3·0 kg S.E. $\therefore k = 3/(500)^{\frac{2}{3}}$

The maintenance requirement of a 600 kg animal (assuming the validity of the law) is therefore

$$\frac{3 \cdot 00 \times (600)^{\frac{2}{3}}}{(500)^{\frac{2}{3}}} = 3 \cdot 00 \times (1 \cdot 2)^{\frac{2}{3}} \text{ or } 3 \cdot 39 \text{ kg S.E.}$$

Similarly, calculations can be made for other liveweights.

At all adult liveweights the digestible protein requirement is between 0·27 and 0·36 kg. The maintenance requirements of young animals cannot be truly calculated from the formula since experience suggests that they are appreciably higher than predicted. It is difficult to define under practical conditions exactly what is meant by maintenance, since apart from obvious heat losses from the surface some allowance has to be made for a normal amount of movement, of standing up and lying down.

For practical purposes the maintenance requirements of the stall-fed or yarded animal over the liveweight range 350 to 600 kg can be considered to vary directly with the liveweight, and to change by about 0·22 kg S.E. daily per 50 kg liveweight up or down from 500 kg within this range. This puts the normal maintenance requirement for the 350 kg animal at 2·3 kg S.E., for the 500 kg animal at 3·0 kg S.E., and for the 600 kg animal at 3·42 kg. S.E.

Again, for practical purposes, there is little to be lost by the very empirical assumption that the daily digestible protein requirement is one-tenth of the S.E. in kg plus 0·02 kg. This puts the maintenance digestible protein requirement of the 350 kg animal at 0·25 kg, the 500 kg animal at 0·32 kg and the 600 kg animal at 0·36 kg. The figures are perhaps slightly on the generous side, but there are bound to be individual variations, and since maintenance in practical terms is virtually impossible to define there is probably no serious fault in being a little generous.

PRODUCTION STANDARDS

In the adult ruminant the principal forms of production are live-weight increase, largely in the nature of laying on of fat, and milk production. It is with the latter that the classical conception of production has been subject to most criticism of recent times, owing to the very wide range of production levels attainable from the modern dairy cow. The farmer is faced with extremes of production—from under 5 kg of milk to something of the order of 50 kg or more. The classical conception of production was that if x units of nutrients were required for one unit of production (say 5 kg of milk or 1 kg liveweight gain) an expectancy of n units of production would require nx units of nutrients. If m units of nutrients were required for maintenance the total nutrient requirement of the animal would be $m + nx$ units.

Such a formula necessarily assumes that food requirements for production are the same at all levels of production : that if 0·4 kg concentrates is required for 1 kg of milk, 16 kg would be required for 40 kg. Various authorities have pointed out the law of diminishing returns, to the effect that with increasing levels of nutrient intake the response tends to diminish.

Most controversy has centred around the modern dairy cow with her in-bred capacity to produce high levels of milk. In her efforts to respond to this genetic capacity she may well, and usually does, draw upon her bodily reserves in order to sustain high levels of production. As against this, the energy value of 5 kg of milk at high levels of production may well be lower than normal, reflected usually in a lower-than-average butterfat level. There seems little doubt that the relationship between nutrient intake and milk output is *not* a straight line as the equation $m + nx$ would necessarily imply, but is curvilinear. On the other hand experience suggests that at lower levels of milk production, say, under 25 kg, the deviation of the curve from a straight line is for practical purposes negligible, and that the ability of the cow to milk off her back tends to make up for the deficiency in nutrient intake between requirement and a constant rate of food intake based on so many kg per kg.

PRACTICAL APPLICATIONS

Feeding is unlikely ever to be a mathematically exact science, and for practical purposes it should be remembered that best results are obtained when the art of feeding is allied to the science. No two animals ever respond in identical manner, despite absolute identity of treatment, but the fact remains that most animals conform within reasonable limits of variation to standards that have been scientifically worked out.

In the discussions on requirements and standards that follow some dogmatic, unqualified assertions have necessarily to be employed. The practically-minded reader should accept these assertions on the understanding that they are guides to the art of feeding and may require slight modification in the light of the individuality of the animals and the circumstances in which they are being kept.

Appetite. One most important factor limiting an animal's productivity is its capacity to take in food. Simple experience will show that a bullock or cow can eat a greater weight of mangels than of hay ; it is not a question so much of palatability or acceptability of the food as the fact that an animal eats to satisfy its need for " fill ". The simplest conception of daily appetite capacity is bound up with the actual intake of food dry matter, and experimentally this normally (for cattle) is between about 2·5 and 3·5 kg dry matter per 100 kg liveweight. If for illustration we take the 500 kg cow or bullock, we can reasonably predict that its intake of dry matter daily will lie between 12·5 and 17·5 kg. Foods vary enormously in their dry matter levels, ranging from about 10 per cent for succulent foods like mangels to 85 per cent or more for dry foods like cereals, oil cakes and average hay. If we tried to supply 15 kg dry matter in the form of mangels our 500 kg animal would theoretically have to consume about 150 kg daily : 15 kg dry matter would be contained in 17·5 kg hay. Now if hay or mangels were the sole food of such an animal, it is highly unlikely that even with the food before it for 24 hours the animal would eat anything approaching these quantities. Balanced dairy cake is usually a very acceptable food, but there are probably few animals that would eat day after day the 16 or 17 kg of such cake as would supply 15 kg dry matter. But a daily ration of 6·5 kg hay (= about 5·5 kg dry matter), 27 kg mangels (= 2·7 kg dry matter) and 6·5 kg dairy cake (= about 5·7 kg dry matter) could quite comfortably be eaten by a dairy cow giving 18 litres of milk.

The imperfections of dry matter intake as an absolute measure of appetite are well recognised, but with a mixed and varied type of diet such an intake acts as a useful guide to daily food assessments.

FOOD REQUIREMENTS OF ADULT CATTLE

The following table sets out in condensed and simple form a useful practical guide to food and nutrient requirements of adult cattle.

TABLE 6

FOOD REQUIREMENTS OF ADULT CATTLE

Appetite

Allow 2½–3 kg dry matter intake per 100 kg liveweight. For dairy cows the upper limit of food intake may be of the order of 3–5 kg D.M. per 100 kg.

Maintenance

Requirements are as follows :

Live Wt. (kg).	Starch Equivalent (kg).	Digestible Crude Protein (kg).
350	2·3	0·25
400	2·5	0·27
450	2·75	0·30
500	3·0	0·32
550	3·2	0·34
600	3·4	0·36

Milk Production

Requirements per kg

	Starch Equivalent (kg).	Digestible Crude Protein (kg).
(1) Non-Channel Island breeds	0·25	0·05
(2) Channel Island breeds (average butterfat levels of 4 per cent or more)	0·30	0·07

Fattening

Requirements per each kg daily liveweight gain

	Starch Equivalent (kg).
Early fattening (store)	2·25
(fresh)	2·5
Half fat	3·0
Fat	4·0

The daily protein requirement at all levels of fattening to include both maintenance and production is between 0·6 and 0·7 kg digestible crude protein. The higher standard is suggested where (as in modern practice) younger animals are growing and fattening simultaneously.

CONSTRUCTION OF A DAIRY RATION

Reference to the standards for ordinary (i.e. non-Channel Island) milk production shows a requirement per 5 kg of 1·25 kg S.E. and 0·05 kg digestible crude protein. This gives an S.E. : D.C.P. ratio of 1·25 : 0·5, or 5 : 1. The foods used in dairy production rations are usually the concentrated foods, cereals, oil seed cakes and so on. If the S.E. : D.C.P. ratios of the commoner of these foods are grouped according to whether these ratios are wider or narrower than about 5 : 1 we have three possibilities.

(1) " *Cereal* " *types* (Ratio wider than 5 : 1)

	S.E. (per cent).	D.C.P. (per cent).
Barley	71	6·5
Oats	60	7·5
Maize . . .	77	7·7
Dried Sugar Beet Pulp .	60	5·0
Maize Germ Meal . .	84	10·0

(2) " *Balanced* " *types* (Ratio about 5 : 1)

	S.E. (per cent)	D.C.P. (per cent).
Wheat Bran . . .	45	10
Palm Kernel Cake .	73	17
Coconut Cake meal . .	74	15
Maize gluten feed . .	76	19

(3) " *Protein* " *types* (Ratio much narrower than 5 : 1)

	S.E. (per cent).	D.C.P. (per cent).
Bean Meal . . .	66	20
Dec. Ground Nut Cake .	70	39
Soya Bean Meal . .	64	38
Linseed Cake . . .	74	25

Obviously, then, to construct a balanced dairy ration we need appropriate mixtures of Groups (1) and (3) to which we may add in reasonable proportions foods from Group 2. What must always be appreciated is that each type of food makes some contribution to both the S.E. and D.C.P. components.

We have also to consider the relative degree of " concentratedness " of our final mixture. A ration of which 0·4 kg contribute some 0·25 kg S.E. and 0·05 kg D.C.P. must have an S.E. value of $\frac{0·25}{0·4} \times 100$ or about 63 per cent and a D.C.P. value of $\frac{0·05}{0·4} \times 100$ or 12·5 per cent. Similarly a 0·35 kg/kg mixture needs an S.E. of about 71·5 per cent and a D.C.P. of 14·3 per cent ; a 0·45 kg/kg mixture an S.E. of about 61 per cent and a D.C.P. of about 12 per cent. Evidently, if we are aiming for a 0·35 kg/kg mixture, our usage of relatively low S.E. foods such as oats, beet pulp and soya must be limited unless we compensate with really high S.E. foods such as maize and its by-products.

Simple Binary Mixtures. Working out binary mixtures, or mixtures made up of two components only, is fairly simple if one keeps an eye on the S.E. : D.C.P. ratio. Let us balance oats with beans :

1 part oats supplies	60 units S.E. and	7·5 units D.C.P.	
1 part beans supplies	66 ,, ,, ,, 20	,, ,,	
2 parts mixture supply	126 ,, ,, ,, 27·5	,, ,,	

Now $\dfrac{126}{27·5} = 4·6$, approximately, and the mixture is relatively a little rich in protein.

2 parts oats supply	120 units S.E. and	15 units D.C.P.	
1 part beans supplies	66 ,, ,, ,, 20	,, ,,	
3 parts mixture supply	186 ,, ,, ,, 35	,, ,,	

The ratio $\dfrac{186}{35}$ is about 5·3 and hence is a little low in protein.

3 parts oats supply	180 units S.E. and	22·5 units D.C.P.	
2 parts beans ,,	132 ,, ,, ,, 40	,, ,,	
5 parts mixture ,,	312 ,, ,, ,, 62·5	,, ,,	

The ratio $\dfrac{312}{62·5}$ is practically 5 and hence is balanced for milk production. As the arithmetic shows, the mixture will have an S.E. value of $\dfrac{312}{5}$ or 62·4 per cent and a D.C.P. value of $\dfrac{62·5}{5}$ or 12·5 per cent.

The requisite number of kg of the mixture required per kg of milk can readily be worked out, thus :

$$62·5 \text{ kg S.E. are supplied by 100 kg mixture}$$
$$0·25 \text{ ,, ,, ,, ,, ,, } \frac{0·25 \times 100}{62·5} = 0·4 \text{ kg}$$

Hence one example of a simple binary milk production mixture is

3 parts of oats to 2 parts beans (a)

By a similar process of trial and error we can work out the values for barley (S.E. 71, D.C.P. 6·5) and ground nut (S.E. 70, D.C.P. 39) and show that

$3\frac{1}{4}$ parts barley will balance 1 part dec. groundnut (b)

Maize and soya on the figures quoted earlier can be balanced at

$3\frac{1}{4}$ parts maize meal and 1 part soya bean meal (c)

Since (a), (b) and (c) are each balanced, it follows that
(1) a blend of all three mixtures will give a balanced ration, i.e. 3 parts oats, $3\frac{1}{4}$ parts barley, $3\frac{1}{4}$ parts maize, 2 parts beans, 1 part groundnut and 1 part soya bean meal.

(2) a balanced blend will still result by taking integral parts of each mixture, provided the ratio of the separate ingredients of the mixture is maintained. As examples :

oats : beans	at	3 : 2	
	or	6 : 4	
	or	12 : 8	
barley : groundnut	at	3¼ : 1	
	or	6½ : 2	
	or	13 : 4	
maize : soya	at	3¼ : 1	
	or	6½ : 2	
	or	13 : 4	

Thus we still have balanced mixtures if we use, say, 1 part of group (*a*), 2 parts of group (*b*) and 4 parts of group (*c*) giving the final ration as

3 parts oats
2 parts beans
6½ parts barley

2 parts ground nut
13 parts maize
4 parts soya

For illustration we have selected cereal and protein foods in pairs, oats and beans, barley and groundnut, maize and soya. But we are not compelled to adopt this particular pairing arrangement. We could equally pair any of the protein foods with any of the cereals and by the appropriate calculations arrive at the necessary balances.

It follows from this that we have an almost infinite number of possibilities of making balanced rations from these six ingredients alone. The great advantage of thinking and working in terms of S.E. : D.C.P. ratios is that rations can be adjusted within a very wide range of conditions, paying particular care to both availability and prices.

Balanced Foods. We have still to mention the use of the " balanced " type of food. This is easy, because these foods can be added to any of the mixtures within reasonable limits and proportions without upsetting the balance appreciably. As an example, 3 parts of oats balance 2 parts of beans : if we add to this mixture one or two parts of palm kernel cake or bran we do not appreciably change the balance. The value of a mixture of 6 parts oats, 4 parts beans, 1 part palm kernel and 1 of bran can be calculated thus :

	S.E.	D.C.P.	
6 oats	360	45	
4 beans	264	80	
1 palm kernel . . .	73	17	
1 bran	45	10	
12 parts mixture . . .	742	152	
1 part mixture . . .	61·8	12·7	(Ratio about 4·87)

This mixture fed at slightly over 0·4 kg/kg will meet the requirements of milk production.

The only other factor to be considered is the need for dairy minerals, which are usually added at a level of 2½ to 3 per cent. (It is a useful practical tip to remember that a 2½ per cent usage amounts to 25 kg in a tonne mixture.)

MAINTENANCE NEEDS

We have already seen that maintenance needs vary with liveweight. Such needs in cattle are normally met wholly or in part with " bulky " foods, using this term in the sense of relatively low nutritive value per unit weight. Bulky foods as thus defined may include the coarse fodders, such as hays and straws, characterised by fairly high fibre contents ; and the succulent foods, which although digestible have a low nutritive value because of their high moisture content. It is uncommon for a single bulky food to be made responsible for maintenance needs and usually bulky foods are used in combinations of at least two, e.g. hay and roots, silage and straw, and so on.

Half-maintenance Units. From the practical point of view there is much to be said for regarding maintenance foods in terms of *half-maintenance* units, since usually at least two foods are used to provide the maintenance part of the ration. For the average adult animal weighing between 450 and 550 kg, the S.E. requirements for maintenance lie between 2·75 and 3·2 kg daily. Consequently, half-maintenance foods supply between 1·4 and 1·6 kg S.E. and the following provide some typical examples. There is also a maintenance need for protein, and the table indicates whether the allowances of foods suggested are adequate for that purpose.

TABLE 7

HALF-MAINTENANCE RATIONS

Quantity and Type of Food.	Approx. amount of Dry Matter (kg).	Protein in excess of, or deficient for, ½M.
4·6 kg medium hay	3·9	o
16 kg kale	2·3	+ +
18 kg beet tops	2·7	+
23 kg mangels	2·3	— —
18–23 kg low protein silage . .	3·6–4·6	+
14–18 kg high protein silage . .	2·7–3·6	+ to + +
7·5–9 kg oat or barley straw . .	6·6–7·8	— —
2·3–2·7 kg oats	2·0–2·3	o
2·0–2·3 kg barley	1·8–2·0	o
2·3–2·7 kg sugar beet pulp . .	2·0–2·5	—
3·1 kg bran	2·8	+

It is inadvisable to use together two foods that are both deficient in protein, e.g. mangels and straw. Protein surpluses do not constitute a problem, but the point is worth making that where protein surpluses exist with the dairy cow it may be possible to obtain the first 5 kg of milk by the use of cereals only. For example a ration of 4·5 kg hay and 16 kg kale with 2 kg oats or barley can provide adequately for the needs of maintenance and the first 5 kg of milk.

The comparative figures for the amount of dry matter provided by these various half-maintenance rations are of value when high yielding cows are being rationed. Such cows have naturally a large need of concentrates, and it is important that the maintenance rations shall not be too bulky if all the necessary food is to be consumed. A low-yielding cow might derive all its maintenance needs from 10 kg medium hay, amounting to 8·6 kg dry matter intake. Reduction in the bulk of the maintenance foods to 6 kg dry matter is obtained from 4·5 kg hay and 16 kg kale *or* 23 kg mangels. A further reduction can be obtained by using 2·5 kg hay (M/4) and 35 kg mangels (3M/4) giving a daily dry matter intake of under 6 kg. The maintenance ration could be further concentrated by using 3·5 kg beet pulp (3M/4) and 9 kg beet tops (M/4) (4·6 kg dry matter intake), though normally it is inadvisable to reduce the long roughage intake below 2·5 kg per day, because of the risk of low butterfat content in the milk.

These rations provide examples of how to set about rationing problems, although it is always advisable to make full calculations to ensure that the stated requirements of animals are met. Bulky foods are seldom individually rationed but are given on a herd basis, so that there is every possibility that the greedy feeder may get more than its fair share. Intelligent rationing must go hand-in-hand with milk recording and careful attention to the production behaviour of each individual cow. Whereas some degree of overfeeding may not significantly raise the production of a naturally low-yielding cow, the effects of underfeeding are soon seen in the production levels of the higher yielders. Once again it is important to stress that scientific rationing must go hand in hand with the art of the stockfeeder, who quickly learns the individual habits of each cow.

The maintenance needs of cattle have been worked out under conditions of stall feeding in reasonably equitable environments. Where animals are exposed to cold, wet and unpleasant conditions, maintenance requirements must necessarily increase to an extent that is scientifically unpredictable. Maintenance needs must also increase when animals are turned out to grazing, because the extra energy of locomotion has also to be considered. Arbitrary assessment of those needs have been made to the extent of an extra 0·45 kg S.E. per day for maintenance even under conditions of good grazing, and of 10r1·4 kg extra when grazing is sparse and scanty. Unsatisfactory levels of milk production in cows calving in the autumn and lying out by night under chilly conditions can often be attributed to increased maintenance requirements in such circumstances.

RATIONS FOR FATTENING

Broadly the same principles apply to the feeding of fattening animals as for milk production, or indeed any other form of production. The maintenance needs of the animal are the first charge upon the food and the surplus nutrients in the food are available for fattening. Formerly little attempt was made towards serious fattening until a bullock was two or more years of age and after it had undergone a " store " period of either remaining stationary in liveweight or else gaining very slowly. The average bullock was usually about 450 kg before fattening in earnest was carried out and the greater part of its muscle and bone growth had then been achieved. Fattening consisted essentially of laying on fat both internally and within and around the muscles with the achievement of a final liveweight of upwards of 600 kg.

With changes in the public taste favouring smaller and leaner joints, lower liveweights were achieved with animals that were still growing, in the sense of laying down muscle and bone, as well as fattening. With this came the restriction or cutting down of the long store periods formerly given ; in effect this means that the animal intended for fattening needs either a fairly high, or a highly nutritious, diet throughout its much abbreviated life.

Existing scientific standards for growth and fattening for animals under about 300 kg are far from being accurately based, but present-day trends obviously demand higher levels of protein and minerals for the growing, fattening animal than were necessary when the animal was much more mature before intensive fattening began. At one time it was thought that the scientific requirements for protein in fattening bullocks were unduly generous, but with modern trends for earlier finished animals it would seem unwise to attempt any reduction of the protein. Accordingly, in attempting to lay down requirements, the feeder should ensure that the total daily intake of digestible protein for maintenance and fattening does not fall below at least 0·6 kg daily per animal.

The amount of starch equivalent required per kg of liveweight increase is of the same order as for 4·5 kg of average milk, about 1·1 kg, and for practical purposes this seems the best figure to work to. Further, the range of possible liveweight increase is much less than that of the range of milk production with the modern dairy cow, whose yields may vary from 5 to 50 kg at different stages of lactation. A liveweight increase of about 0·5 kg per day (i.e. 15 kg per month) is virtually a " store " standard of performance, and in present circumstances 0·9 kg per day can be regarded as a minimum ; 1·4 kg per day is fairly intensive fattening, and figures of 1·6 or 1·7 kg can be looked upon as very intensive.

The fattening animal of today is relatively small and it has a correspondingly smaller appetite ; this means that its diet will have to contain a higher proportion of concentrates and a lower proportion of the more indigestible bulky foods such as straw that were formerly fed to fattening animals. In effect this means some concentration of the maintenance ration as well.

Calculating a Fattening Ration. If we take as an example the needs of a 400 kg bullock putting on about 1·1 kg liveweight daily we have to supply, within an appetite capacity of about 9 to 11 kg dry matter, a diet providing

For maintenance	2·50 kg S.E.
For 1·1 kg L.W.I. (1·1 × 2·5) . .	2·75 kg S.E.

or about 5·3 kg S.E. which should include at least 0·6 kg digestible protein.

Let us suppose we have available limited quantities of medium hay, a good supply of kale to be followed by mangels, plenty of barley, some sugar beet pulp ; and that we can purchase some oil seed cake.

A ration of about 1·8 kg hay (quarter-maintenance) and 23 kg kale

($\frac{3}{4}$ M) will provide about 2·6 kg S.E. and 0·37 kg digestible protein within a 4·9 kg dry matter intake.

This leaves the more concentrated part of the ration to provide just over 2·7 kg S.E. and another 0·23 kg digestible protein within the compass of a further 4 kg or so of dry matter. A mixture of 2·8 kg crushed barley and 1·4 kg sugar beet pulp will supply 2·8 kg S.E. and about 0·25 kg digestible protein in a bulk of 3·7 kg dry matter. Thus the requirements are met and any necessary balance of appetite can be supplied by offering some barley straw for the bullock to pick over. No recourse has had to be made to additional purchases of protein, but the kale has contributed more than half the total protein needs.

When the kale is finished and has to be replaced by mangels, it would take over 31 kg of the latter to furnish the same amount of S.E. as the kale, and the joint contribution of hay and mangels to the protein needs is now only about 0·18 kg digestible protein. The barley and sugar beet pulp as before supply the necessary extra S.E. but only 0·25 kg digestible protein, leaving at least a further 0·17 kg protein in deficit. The replacement of 0·45 kg barley by 0·45 kg of ground nut cake would remedy the deficiency without any change in the supply of S.E.

In practice one would probably make up a mixture of 250 kg barley, 150 kg pulp, 50 kg groundnut cake and 10 kg dairy minerals, and feed 4·5 kg per head daily. This would guard against the possibility that individual animals fed the kale or mangels might fail to get their full share of the bulky food when a yard of bullocks is being group-fed.

Rationing with Self-feed Silage. A second example sets a different type of problem. The same types of bullocks are to be self-fed on a medium quality grass silage through the winter. Cereals, sugar beet pulp and oil cake are available as before.

First of all we have to assess the probable intake of silage. As a trial, we assume an average daily consumption of 23 kg silage (20 per cent D.M., 9 per cent S.E. and 1·2 per cent digestible protein). This would account for 4·6 dry matter intake with 2·1 kg S.E. and 0·28 kg digestible protein, leaving the concentrated food to supply 3·3 kg S.E. and 0·30 kg digestible protein.

The mixture of 5 parts barley, 3 parts pulp and 1 part ground nut (plus minerals) has an S.E. value of 67 per cent with about 9·5 per cent digestible protein: 5 kg of this supply 3·35 kg S.E. and just over 0·47 kg digestible protein in slightly under 4·5 kg dry matter. On the assumption that the bullocks would eat 23 kg silage in addition to this level of mixed concentrates, the needs for a 1·1 kg liveweight increase are fully met, with a generous surplus of digestible protein. Omission of the ground nut from the mixture and its replacement by barley would lower the intake of digestible protein from the 5 kg concentrates from 0·48 kg to 0·30 kg, giving a total protein intake from silage and concentrates combined of about 0·58 kg, a marginally adequate figure. Where such a high level performance is being demanded of the bullocks it is probably not worth the risk of being marginal in respect of protein, and in practical terms it would therefore seem best to adhere to the original 5 : 3 : 1 mixture suggested above.

Rations for Other Livestock. Further information about the feeding of cattle, sheep, pigs and other livestock is given in Chapters XIX, XX and XXI.

THE METABOLISABLE ENERGY SYSTEM FOR RUMINANTS

The Starch Equivalent system for expressing the energy value of feeds and the energy requirements of ruminants was introduced to Britain just prior to the first world war. As a system it has proved extremely useful and has contributed tremendously to the practical feeding of ruminants ; nevertheless, from the scientific angle, it has considerable limitations and imperfections. In the circumstances of today, when animal productivity has so greatly increased and systems of animal husbandry and feeding are so much more intensive, the Starch Equivalent system has become less reliable as a guide to accurate prediction of performance.

For these reasons, especially since the second world war, much scientific effort has been put into development of the concept of Metabolisable Energy (see p. 570) as a better unit for the evaluation of energy requirements of ruminants. This concept has been strongly recommended by the Agricultural Research Council and has been taken up by the Nutrition Chemists of the Ministry of Agriculture's Agricultural Development and Advisory Service for practical advisory purposes. For scientific reasons the Royal Society has recommended that the basal energy unit of the calorie (p. 570) should be replaced by the S.I. unit, the joule. The joule equivalent of the thermo-chemical calorie is 4·184 joules and for practical purposes the mega-joule (MJ), or a million joules, is the unit that will subsequently be employed.

This official change to a Metabolisable Energy system has been timed to coincide with the intended introduction within the United Kingdom of metrication and consequent replacement of Imperial units of measurement of mass, length, area and volume. In this way the confusing effects of the double change—Starch Equivalent to Metabolisable Energy, Imperial to Metric—can be minimised, with the advantage that all calculations relating to the use of the new systems are in the decimal notation and require a minimum of arithmetical manipulations.

The major weaknesses of the old Starch Equivalent system were that it ascribed to individual feeds a constant nutritive value which was independent of other ingredients of the diet, the amount of the diet given and of the type of production the diet was designed to support. These assumptions cannot be justified in the light of modern scientific knowledge. Although in many practical circumstances the Starch Equivalent system has served its purpose well without involving serious errors, it has proved to be less reliable than the Metabolisable Energy approach for the high levels of productivity demanded of livestock in today's economic circumstances. Moreover, the new system is much better adapted to the " whole dietary approach " so essential today. Foods are rarely supplied in isolation of one another in practical husbandry circumstances and are certainly not digested and utilised in an independent manner, as the Starch Equivalent system implicitly suggests. Performance by animals is governed not only by the intrinsic merits of the individual feed components but also by the capacity of the animal to consume feed. By being based on the whole dietary approach and the overall concentration of metabolisable energy in the dry matter of the diet and a knowledge of the efficiency with which that metabolisable energy is utilised for the various functions of the animal (maintenance, growth, pregnancy, lactation, fattening), the Metabolisable Energy system offers a more scientific and practical conception of the function of feed in relation to animal performance.

It is impossible, in the context of this section as a whole, to give a detailed description of the use of the system and of the calculations involved in its application. For this the reader is referred to official publications now in being or in course of preparation ; for those who are receiving or will receive in future more specific instruction in academic institutions the new concepts should be as readily grasped as were those in the past based on the Starch Equivalent system. The weaknesses and areas of doubt in the latter system were often lightly glossed over but, in view of the higher levels of production for which our present day livestock have the genetic potential, there is need to ensure that our recommended systems of feeding are capable of achieving this potential. This, then, is the major reason for the proposed change to the Metabolisable Energy system for ruminant animals.

Chapter XIX

CATTLE

THE total cattle population in Great Britain now exceeds 11½ million (June 1972 Census) with 9½ million in England and Wales and over 2 million in Scotland, compared with a population of some 7¼ million at the beginning of this century. Most of this increase in the cattle population has been in respect of cattle kept mainly for the production of milk, so that today the value of milk sold off farms is 35 per cent greater than the value of meat (as beef and veal). This trend towards increasing production of milk has now slowed down and more attention is being given to the expansion of beef production from dairy herds, involving new systems of breeding, rearing and feeding.

Of the 18 major breeds of cattle in Great Britain, we can distinguish three main groups ; *Dairy Cattle* kept mainly for milk production ; *Beef Cattle* kept mainly for beef production ; and *Dual Purpose Cattle* which have the characteristics of giving reasonable milk yields allied with fattening propensities. This classification is by no means rigid as strains within the different breeds can exhibit considerable variation.

CLASSIFICATION OF CATTLE BREEDS

Beef Breeds

Beef Shorthorn	Devon	Galloway
Welsh Black	South Devon	Lincoln Red
Hereford	Sussex	Charolais
Aberdeen Angus	Highland	

Dual Purpose Breeds

Dairy Shorthorn	Dexter
Red Poll	British Friesian

Dairy Breeds

Ayrshire	Guernsey
British Friesian	Jersey

CHANGES IN BREED DISTRIBUTION

The first census of cattle classified by breed was carried out in 1908 by the Board of Agriculture. The changes in breed distribution since that date are illustrated in the following diagram, which refers to England and Wales : the diagram is based upon the most recent census of the cattle populations by breed carried out by the English Milk Marketing Board in 1965.

In Scotland dual-purpose cattle have never attained the popularity they have enjoyed in England and Wales and, over the same period, the relative distribution between beef and dairy breeds in Scotland has remained relatively uniform with dairy cattle now 32 per cent of the total cattle population. The relatively greater importance of beef cattle, 68 per cent in Scotland compared with 17½ per cent in

England and Wales, reflects the difference in the cattle economy north and south of the Border.   In Scotland, with its much higher proportion of hill and upland grazings, beef cattle are the dominant

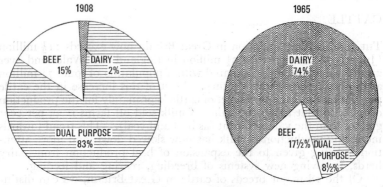

FIG. 113.—Changes in Breed Distribution, 1908–1965.

cattle enterprise. The more productive lowlands are devoted to milk production, or arable farming with cattle fattening, whereas hill farms are predominantly stock rearing farms—to a less extent this is also true of Wales and south-west England

BEEF BREEDS

The Beef Shorthorn.   The Shorthorn originated in the north-east of England : it owes its principal development to the Colling brothers who lived near Darlington, and whose herd was established from the best of the earliest foundation stock in 1784.   Robert Culley, who was a close friend of Robert Bakewell of Dishley, made the Collings familiar with the methods of livestock improvement that in Bakewell's hands had given such good results in the development of the Leicester breed of sheep and the Longhorn breed of cattle.   The Collings started their herds from existing good stocks of cattle.   The bull, " Hubback 319 ", is regarded as the father of the improved Shorthorn and one of his most famous descendants was the bull " Favorite 252 ", himself a product of inbreeding.   The influence of the Collings as breeders was to produce an animal of moderate size, shapely and early maturing, with well-sprung ribs, short legs and good thick hide covered with thick mossy hair.   Though the cattle were thick fleshed, by modern standards they tended to be patchy in their fleshing.

Two Yorkshire men, Thomas Bates and Thomas Booth then influenced the development of the improved Shorthorn in the early 19th century, and this breed dominated the cattle of the country for nearly a century.   Booth concentrated on the development of the beef type, whereas Bates was more concerned with the development of the dual-purpose characteristics and the dairying capacity of the breed.   At this time the improved Shorthorn quickly established itself in Scotland, and in Aberdeenshire in particular where its development was largely the work of Amos Cruickshank of Sittyton. As in the case of the Colling brothers, Cruickshank was equally dependent upon the use of the prepotent sire, the greatest of which in the Sittyton herd was a bull called " Champion of England 17526 ", calved in 1859.

The Shorthorn was developed subsequently to serve the needs of beef production, particularly in the northern region of England and Scotland. A polled strain of Shorthorns has been developed, mainly from polled Shorthorns imported from Australia.

Throughout the world, the Beef Shorthorn is highly regarded, for its type is that of a blocky animal, standing on short legs and possessing the capacity to produce a good carcase of beef. The distinguishing colours are red, white and roan (Plate XXXIIIA). The Shorthorn Society was formed in 1875.

The Welsh Black. These cattle constitute one of the most ancient breeds, and the modern breed is the result of merging several local types, but chiefly the breeds of South and North Wales, combining both milk and beef qualities. Because of hill land and lime-deficient soils, it does not attain a high standard of uniformity of milk production—yields are about 2,700 litres with butterfat of 4 per cent. The most valuable characteristics of this breed are the hardiness and ability of the cows to produce good yields of milk under poor and hard conditions (Plate XLIA).

The Hereford. This is a purely beef breed, for long associated with its native county. It is probable that this breed shares a common origin with the Devon and Sussex breeds, but the white face and characteristic markings originated from a cross with white-faced Flemish cattle imported to the county before 1671. The early improvement in this breed was due to the efforts of Benjamin Tomkins and his son between the years 1742 and 1815. The Tomkins selected for beef and were good judges of cattle. Thereafter the Hereford was skilfully bred and developed and was very successfully exhibited at the Smithfield Show in the first half of the 19th century. It is largely a grass beef breed, with the capacity to fatten readily on grass alone. Animals are bred and reared under natural conditions, the cows running on grass throughout the year and receiving extra treatment in the form of hay and straw during severe weather and at calving time. As a result the breed has been remarkably free from tuberculosis and the cows are good mothers, rearing their calves well.

Herefords are popular in North and South America and have proved themselves to be one of the best breeds for ranching purposes : they are able to resist better than most breeds the effects of shortage of water and fodder. Polled Hereford cattle are now being bred in this country, mainly from polled parent stock imported from Australia, USA and Canada, and a separate Poll Hereford breed Society is now in being. The parent Hereford Society was founded in 1878.

In size, weight and maturity they compare favourably with the Shorthorn. Their colour is distinctive ; the face, throat, chest, legs, lower portion of the body, crest and tip of the tail are white, with the rest of the body red (Plate XXXIIIB). The white face is a dominant characteristic of the Hereford breed : when this breed is crossed with other breeds, the white face is a valuable quality in the scheme for colour-marking calves where such animals are destined to be reared for beef purposes.

The Aberdeen Angus. This black-coloured, hornless breed is native to the north-eastern counties of Scotland and in particular the counties of Forfar and Aberdeen. In its modern form it descends from two similar types in these counties from which it derives its name, and there is evidence of its existence in 1523. The pioneer breeder was Hugh Watson of Keillor in Forfarshire. The Keillor herd was established in 1808 and the breeding policy was moulded on the methods adopted by the early breeders of improved Shorthorns. Watson's foundation cattle were the best that he could buy and he was a good judge of cattle and a skilful breeder and feeder ; his mating policy, though not entirely favouring close-breeding, nevertheless permitted its use when circumstances justified the practice. His herd was dispersed in 1865. The second notable influence on the breed was the work of William McCombie of Tillyfour in Aberdeenshire, whose herd was founded in 1830 and dispersed in 1880. The breed's international reputation was created largely as a result of McCombie's exhibits at the International Exhibition in Paris in 1878, when the Aberdeen Angus breed won the prize for the best group of beef-producing animals. The third pioneer was Sir George McPherson Grant, whose herd at Ballindalloch in Aberdeenshire played an important part after the dispersal of the Tillyfour herd.

As a producer of high-quality beef the breed is almost perfect. It does well under confinement, keeps its form in the fattening process without any tendency to patchiness, and the finished beast is a solid mass of firm meat. The proportion of dead- to live-weight is unusually high. The breed is early maturing and produced the first two-year-old champion at the Smithfield Show. There has been a tendency for the breed to lose its size as a result of a show-yard craze for small animals.

The bulls of the breed are used extensively for crossing purposes for the production of quick-maturing high-class steers. These grow into good beef animals and inherit the black colour and polled character of the sire (Plate XXXIVA). The Aberdeen Angus Society was founded in 1879.

The Devon. The Devon, or North Devon as it is frequently called to distinguish it from the South Devon, is native to North Devon and Somerset and, by contrast with the three preceding breeds, is more local in its distribution. The breed derived originally from the common stock which also produced the Hereford and Sussex, and the recent development cannot be attributed to any outstanding pioneers, though the Quartly family of Molland in North Devon exercised a great influence on the breed from the beginning of the 18th century to the middle of the 19th century.

These cattle are distinguished by their rich ruby-red colour, but they are larger than the preceding breeds. They are characterised by their extreme neatness, fineness of bone and evenness of fleshing.

For its size the Devon is remarkably thick-fleshed and produces meat of high quality that compares favourably with that of the Aberdeen Angus. Devons make excellent grazing cattle, particularly on pastures that would not be suitable for larger breeds and their maturity is average (Plate XXXIVb). In spite of the Devon's high quality as a beef breed, only a few herds are kept exclusively for beef production, a certain amount of milk being drawn during the spring and summer. In this respect the breed is more notable for the quality rather than the quantity of produce. The Devon Cattle Society was founded in 1884.

The South Devon. The South Devons are the country's heaviest cattle ; cows of the breed often weigh 750–800 kg and mature bulls up to 1,500 kg (Plate XLb). It is native to South Devon, and the Guernsey is thought to have played a part in the development of the breed.

The breed is pale red in colour, inclined to be strong-boned but with a good reputation as a dual purpose animal. The milk yield is not as high as the Dairy Shorthorn by approximately 10 per cent, but is very rich, often averaging 4·5 per cent butterfat.

The Sussex. This breed, which in external appearance bears a similarity to the Devon except that it is larger, is native to the county of its name, but has a popularity on the Wealden clays and the marsh-lands of Sussex, Kent and Surrey. Until the present century the breed was bred for draught purposes. It is one of the hardiest breeds, able to thrive under the most unfavourable circumstances and, unlike most cattle, it does not pick and choose when grazing, but will graze right through a pasture from one side to the other, taking the rough with the smooth. It will also keep itself as a good store on the poorest of pasture. Its maturity is good if well looked after and the quality of beef is equal to that of the Devon (Plate XXXVa). The Sussex Cattle Society was founded in 1878.

The Highland. This breed is probably descended from the aboriginal wild cattle of the country and its principal breeding area comprises the western Highlands of Scotland and the islands off the west coast of Scotland. No cattle in the United Kingdom have pre-served in greater uniformity their characteristic breed points than the Highlanders or " Kyloes " have done. This of itself points to their antiquity and purity of breeding. Indeed, attempts to improve them in their native environment by the introduction of alien blood have failed.

The Highland is most distinguished in its appearance, carrying a well-formed but low-set body, covered with a thick skin that produces a long growth of wavy hair. The short and well-balanced head carries horns of good length. The colour is variable, with yellow, light dun, brindle, red and black being common (Plate XXXVb).

The outstanding characteristic of the breed is its hardiness ; it can withstand exposure to wet and cold and can exist on scanty and poor pastures. This is the smallest of the beef breeds, apt to be some-what nervous, and rather slow in maturing. The quality of meat is excellent and they are popular for finishing on good pastures.

The Galloway. This breed takes its name from the province of Galloway, comprising the counties of Kirkcudbright and Wigtown in the south-west of Scotland. The breed has also been long associated with the counties of Dumfries and Cumberland. Their characteristics and qualities suggest that they are of considerable antiquity. Compared with the Aberdeen Angus they have a flatter poll and carry a greater length of hair on the skin, and are slower in maturing. Their colour is variable, but two pedigree types have been evolved, namely black and belted strains (Plate XXXVIA).

The breed is notable for its great hardiness and is only surpassed in this by Highland cattle. In its native habitat the cattle are wintered in the open. They are excellent grazers in rough pastures but also finish well on rich land. The quality of their meat is regarded along with the Highland, as the best of all beef, having a high proportion of lean to fat, with a fine grain and rich flavour.

They are highly valued for crossing purposes. The cows are frequently mated with white, beef Shorthorn bulls, the result being the famous polled " Blue Greys ". These are very popular cattle with both feeder and butcher, for they inherit the most valuable characteristics of both the Galloway and Shorthorn parents. The Galloway Cattle Society was founded in 1877.

Lincoln Red. This is a strain of the Shorthorn which has been selected for its cherry-red colour and it has been bred in its native county for over a century. The name that is associated with the early development of this type of Shorthorn is that of Thomas Turnell, of Wragby in Lincolnshire, who selected and bred only red-coloured Shorthorns. Slightly larger and coarser in the bone than the Dairy Shorthorn, they grow to a big size on the large arable farms of their native district and they have a high reputation with graziers in the east Midlands.

Recently Lincoln Red bulls have become popular for crossing with some dairy breeds to produce beef-type calves. A polled strain has been developed by Mr. E. L. C. Pentecost to meet the increasing demand for polled beef cattle which fatten more readily, require less yard space per beast and suffer less damage during transit (Plate XXXVIB).

Charolais. This breed comes from the grasslands of central France where it evolved over a period of many years, particularly in the region around Nevers and Charolles. The Herd Book was founded in France nearly a century ago. In 1961 the first importation of 26 bulls to A.I. stations, in order to test the merit of the breed, was approved and this was followed in 1965 by the importation of 27 bulls and 178 females. Since this date its value as a beef breed, particularly for crossing purposes, has been clearly recognised. A large animal, its ability for both rapid growth and the production of flesh without fat is particularly useful because these characteristics are stamped even on smaller, finer-boned cattle, such as the Ayrshire and Channel Island breeds. The British Charolais Cattle Society was founded in 1962 (Plate XXXVIIB).

DUAL PURPOSE BREEDS

The Dairy Shorthorn. The history and characteristics of the Shorthorn breed have been covered in the section on beef cattle, but the bulk of Shorthorn cattle in the United Kingdom are essentially dual-purpose animals. At the end of the last century there was a definite danger that the dairy qualities of this breed would be lost as a result of high prices offered for beef-bulls, so the Dairy Shorthorn Association was formed in 1925. The Northern Dairy Shorthorn Association, with a separate herd book, was formed in 1945.

Although the dual-purpose ideal implies the production of animals with the capacity to milk and fatten, the modern emphasis with both the Dairy Shorthorn and the Northern Dairy Shorthorn, is to develop still further the milking properties. With regard to milk yield, Dairy Shorthorns are about 810 litres below the Friesian and 270 litres below the Ayrshire level. The butterfat at 3·59 per cent has remained practically unchanged for the last 30 years.

In recent years numbers have declined but a good market prevails for calves for rearing for beef, and this is an important asset to the breed (Plate XLA).

The Red Poll. This breed originated in East Anglia from a cross between the Suffolk cattle with good milking qualities and the Norfolk beef animal, and became a distinctive type by the year 1846. It is still found in its greatest numbers in East Anglia but has spread to almost all counties.

The Red Poll is a medium-sized, active breed with the advantage of being naturally polled. They are probably more suitable for fattening early than at a later age because of their lack of size when mature.

Their milking properties are very reliable and the yield of milk and fat is very similar to that of the Dairy Shorthorn. Their chief weakness in the dairy sense is the lack of shapeliness of udder, but improvements have been made in this respect. Their type is also subject to variation in the absence of careful breeding (Plate XXXVIIA). The Red Poll Cattle Society was founded in 1888.

The Dexter. The smallest of our breeds, weighing on an average 325 kg (Plate XLIB).

It is coloured either whole black or whole red and has good dual-purpose qualities. This breed provides one of the most interesting examples of the operation of a lethal factor. Thus when pure bred about one-quarter of the calves are monstrosities with a very large head and usually do not live very long. This condition can be avoided by mating Dexters to Kerrys.

The British Friesian. Before 1880 when an embargo was placed on all importations of cattle for breeding purposes into this country, there had been constant importations of Dutch cattle into eastern England. In 1909 the British Friesian Society was formed to foster the interests of the breed, and in 1914, 50 specially selected animals were imported from Holland. Later there were further importations from South Africa and Canada. The 1936 and 1950 importations from Holland carried marked dual-purpose qualities in this breed. In the last 30 years the breed has forged ahead and now exceeds in numbers all the other dairy breeds put together.

Friesians are strongly built animals outweighing the Ayrshire and Shorthorn with the mature animal reaching 600–650 kg, with abundant bone and large barrel, indicative of the capacity to consume an abundance of roughage. The colour of this breed is black and white in sharply defined patches and the legs should be white immediately above the hooves (Plate XXXVIIIA). Occasionally dun- and red-coloured animals occur, but these are not eligible for registration in the main Society Herd Book. A separate Red and White Friesian Society was established in 1951 and by 1969 there were seventy members.

As a breed the Friesian is unchallenged on milk yield, but the milk tends to be lower in butterfat compared with other dairy breeds, while the fat globules are small and the cream line in bottles is not well emphasised. Breeders have done much to improve the quality of milk in this breed, and the percentage of butterfat is steadily increasing. Allied with the capacity for high yields is the development of large udders, though these are not so shapely as those of the Ayrshire.

The breed has good powers of adaptability, but is best suited to good conditions of housing and feeding, and responds well to generous treatment. The calves are large, often 40–45 kg at birth which gives them a marked advantage over the other dairy breeds for sale as veal. The change in the type of Friesian has also given the breed much prominence as a producer of rapidly growing steers. The Friesian cow when fat is a much better butcher's beast than any of the other single-purpose dairy breeds.

Another strain of Friesian cattle originating from Canadian importations, known as the British-Canadian-Holstein-Friesian with a separate Herd Book has been established in recent years principally in East Anglia, with characteristics broadly similar to the British Friesian but carrying less flesh.

DAIRY BREEDS

The Ayrshire. This is the only native breed of pure dairy cattle which has originated in the United Kingdom and is native to the county of Ayrshire in the south-west of Scotland. An early improver of this breed was John Dunlop, of Dunlop in Ayrshire, but the breed was developed along its modern lines during the 19th century.

During the past 35 years the breed has established itself in England and Wales where 15 per cent of dairy cattle are now of the Ayrshire type.

The Ayrshire colouring is varying shades of red or brown, and sometimes black, patches on a white background. Particularly characteristic is the shapeliness and quality of the cow's udder, a type of udder which breeders would like to see more widely distributed. Calving at an average age of about 2 years 8 months at about 450 kg or even less in commercial herds, the Ayrshire reaches 500 kg as a second calver and about 550 kg as a mature cow. Studies made of recorded animals of the breed show that the lifetime yield is not adversely affected by calving at a very young age, between 24 and 26 months, so that under good management the breed responds to early calving.

Specially selected and bred for its hardiness and its ability to make good use of its food for milk production purposes, the Ayrshire's milk yield performance is about half-way between the Dairy Shorthorn and the Friesian but the fat content is higher. Their milk is pale in colour and the fat globules are small, making the milk very suitable for feeding to infants, and for cheesemaking.

Since 1950 the expansion of this breed has slowed down owing possibly to the more recent encouragement given to beef by the Government's price policy and by their subsidy schemes for calves, for the breed is of a more extreme dairy type than the Friesian and in this respect approaches the Channel Islands breeds (Plate XXXIXb). The Ayrshire Cattle Society was founded in 1877.

The Guernsey. The islands of Guernsey, Alderney and Sark are the native home of the Guernsey cattle which are kept pure by the fact that since 1763 the importation of live cattle to the islands has been banned. Since 1820 there have been shipments of these cattle to Cornwall and the breed has established itself in the southern half of England and particularly on the Isle of Wight.

They are similar in many of their qualities to the Jerseys, but are slightly larger in size and give greater yields of milk of somewhat lower fat yield than the Jerseys, 4·57 per cent butterfat.

In colour they are various shades of fawn, usually associated with white markings (Plate XXXVIIIb). The milk produced is an even deeper colour than the Jersey milk and may earn a price premium. The Guernsey Cattle Society was founded in 1884.

The Jersey. This breed is again based initially on island bred cattle—the small island of Jersey contains less than 16,200 ha. As for Guernseys, restrictions on importations of cattle to the Island have been in force since 1763. Admission to the herd book entails inspection as well as registered parentage, and no heifer is inspected until she has calved ; this means that any animal in the Island herd book is not only pure bred, but also reaches a certain standard of merit.

The Jersey is one of the smallest of the dairy breeds but is very graceful (Plate XXXIXa). The colour may be light and dark fawn, brown, silver grey and mulberry, with a characteristic " mealy " ring round the muzzle.

In proportion to their size they have large barrels, an indication of their ability to consume bulky foods. This breed calves at an earlier age than any of the other breeds (2 years 4 months) and at even a younger age it shows no decline in performance.

They produce the richest of all milks produced by various breeds of cattle, average butterfat percentage is well over 5 per cent. The milk is very deeply coloured and large fat globules give a well-marked cream line in the bottle, and this milk may command a premium over the price paid for ordinary milk.

The Jersey has good powers of adaptability, and refinement of bone and frame are not necessarily a sign of constitutional weakness.

The breed is usually found where conditions and climate do not require the cattle to be subject to exposure to cold weather. It is used for mixing with both Ayrshire and Friesians to improve the cream line and the colour of the milk. The Jersey Cattle Society was founded in 1878.

LIVESTOCK JUDGING

From the description given of the various breeds of cattle in Great Britain it will be realised that each breed is characterised by certain breed "points" or trade marks typical of the breed. Such characteristics as coat colour, the presence or absence of horns, or the degree of fleshing as between a beef and a dairy animal are all identifiable by inspection and are, in conjunction with the general appearance of the animal, the basis of inspection judging as frequently carried out at Agricultural Shows.

Of recent years, however, increasing attention has been given, not only to inspection points, but to economic merit ; beauty with utility is the modern breeders' aim. Few of the inspection points are closely correlated to utility, so that to establish economic merit in cattle we have to resort to methods of measuring actual performance. To date, in the sphere of cattle production, the greatest progress has been made in measuring milking capacity by recording milk yields. Such recording is now possible for all dairy breeds under National Milk Records, administered by the Milk Marketing Boards (see p. 625). In beef cattle a similar trend is now appearing, whereby periodic weighings of beef cattle will give an indication of rate of live-weight gain : if these records are followed up to carcass assessment, they will give similar information to the beef producer as to the quality of his end product (beef) and is now available to the dairy farmer in respect of the fat percentage of milk.

TYPE IN CATTLE

Livestock breeders speak of cattle as being of a good or bad type, by which they mean how closely they conform to accepted standards of excellence, based on appearance. Since large numbers of cattle change hands every year without any performance records being available, external appearance or type still has a definite market value.

Type in Beef Cattle. The general conformation of a well-fleshed beef animal is illustrated by the photographs of the representatives of the beef breeds given in this book. A deep level body is set on short legs, with top line and under line running parallel, with very full development of the hindquarters where the most valuable cuts in the beef carcass are to be found. This is illustrated in the diagram opposite of a side of beef showing the various butcher's cuts, and their relative value. The top quality joints are used for roasting or grilling ; the second quality for roasting and stewing ; and the third and fourth quality for stewing only. It will be noted how the top quality joints are found along the back and in the hindquarters ; our modern breeds of beef cattle show outstanding development of these parts of the body as compared with native or unimproved cattle.

The beef breeder looks therefore for indications of high carcass quality, as determined by bodily conformation in beef cattle, at all stages of development. Conformation is, however, determined by the level of feeding or plane of nutrition on which cattle are reared, as well as by their heredity or inborn characteristics. Poor nutrition does, for example, retard the development of muscular tissue relatively more so than in the case of bone development, so that beef cattle

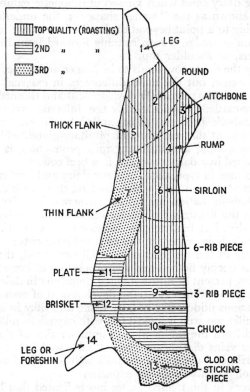

FIG. 114.—JOINTS AND THEIR VALUE IN A SIDE OF BEEF (after Wood and Newman 1928).

reared on a relatively low plane of nutrition grow long in the leg, narrow in the hindquarters and with disproportionately large heads. Such cattle tend, when eventually fattened, to produce carcasses deficient in muscle (lean meat) and with an unduly high proportion of bone. Modern taste is for lean meat, and the trend in beef production today is towards a sustained high level of nutrition throughout life with no lengthy " store " period, giving a high proportion of lean meat where the growth rate potential of the stock is high (see p. 583). Should, however, cattle of a high potential be subject to a period of underfeeding (or " store " period), this suitability for high-class beef production is impaired because conformation suffers. The buyer of store cattle for fattening consequently needs to be a sound judge of beef type to decide whether any faults evident in conformation are due to the breeding of the cattle or to faulty, or too severe restriction in, feeding.

Type in Dairy Cattle. In contrast to the deep fleshed " blocky " conformation of beef cattle, the dairy breeds exhibit a much more " angular " appearance, particularly in the cows. All high-yielding dairy cows when viewed in silhouette exhibit what has come to be known as the " wedge shape " ; the underline and top line converging to a point beyond the head of the cow ; viewed from behind, a similar wedge is seen running forward from the hip bones to the withers, or shoulder top.

Skeletally the dairy cow tends to be longer and flatter in the bone than the beef cow, but the main differences in external appearance are due to the greater mammary development and thinner covering of flesh in dairy cattle. Physiologically the difference arises from their differing metabolic processes as controlled by the endocrine glands, and it is significant that the anterior pituitary gland, which is so important in controlling and initiating milk production, is much more highly developed in a dairy cow than in a beef cow.

The difference in type, then, between dairy and beef cattle largely arises as a result of the use to which the food they eat is devoted. In beef cattle flesh formation has priority, whereas in dairy cattle milk production is the dominant physiological activity. Dairy cattle will fatten when dry, but tend to give inferior carcases to the pure beef cattle, particularly in respect of the most valuable cuts ; for example, the eye muscle in the rib joint is less well developed, the rump and thighs are less deeply fleshed and the proportion of bone is higher.

In assessing present or future milking properties in dairy cattle, the most reliable guide is the extent and development of mammary tissue. A large capacious udder, free of fatty tissue and silky in skin texture, and which milks out well is looked for, with capacious milk veins running forward from the udder along the under side of the abdomen. Each of these veins disappears into a " milk well " in the region of the 6th–8th rib, and the size of the milk well is as good an indication as we have of probable milking capacity.

Recent work in the United States has indicated that by handling the developing udder of heifers at about seven to eight months of age, an experienced operator can judge, with some degree of success, future milking ability. Apart from secretory capacity, however, the modern dairy cow requires an udder well attached to the body and which does not become unduly pendulous with age ; above all it must have teats of suitable size, well placed for machine milking, and with teat orifices large enough to give a full flow of milk. Quick milking cows are highly desirable with modern milking equipment.

Further characteristics which are important in judging cattle, whether of beef or dairy type, are ability to walk well (sound feet and legs are essential in long-lived breeding stock) and ability to convert bulky foods, such as hay, silage and grass, economically into meat or milk. Since cattle are ruminants, ample digestive capacity needs to be linked to high productivity, so that judges look for well-developed

PLATE XXXIII

A. Beef Shorthorn bull
B. Hercford bull

PLATE XXXIV

A. Aberdeen Angus heifer

B. Devon bull

PLATE XXXV
A. Sussex bull
B. West Highland bull

A

B

PLATE XXXVI

A. Galloway bull
B. Lincoln Red heifer

PLATE XXXVII
A. Red Poll cow
B. Charolais bull

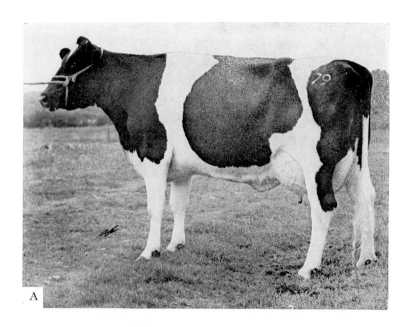

PLATE XXXVIII
A. British Friesian cow
B. Guernsey cow

PLATE XXXIX
A. Jersey cow
B. Ayrshire cow

PLATE XL
A. Dairy Shorthorn cow
B. South Devon cow

PLATE XLI
A. Welsh Black cow
B. Dexter cow

abdominal organs, or " barrel " capacity. Finally, temperament is of some economic importance ; cattle of an equable disposition are easier to handle, and expend less energy, thus reducing maintenance feed requirements.

ASSESSMENT OF MERIT IN CATTLE

It will be realised from the above discussion, that livestock judging by inspection still has its value but is by no means the final measure of an animal's merit whether for meat or milk production. We can summarise the place of livestock judging in assessing merit in our cattle in the following scheme :

I. Judging by Inspection establishes :

(i) Conformity to breed " points "—colour markings, udder shape and attachment, etc.

(ii) General appearance—beef or dairy type, levelness of top line, soundness of feet and legs, etc.

(iii) Temperament and state of health.

2. Productive ability is established by :

(i) Milk recording in dairy cows (see p. 625).

(ii) Recording of growth rate in beef cattle or *Performance Testing.*

3. Fertility is measured by :

(i) The conception rate with bulls, or percentage of conceptions to services ; 70 per cent is good ; 65 per cent is average ; below 60 per cent is poor.

(ii) Calving sequence with cows ; the *calving index* or period which elapses between successive calvings is a common index of fertility, though by no means rigid—an index of 365 days is often used as a standard. A good index is an important economic factor.

4. Breeding merit is assessed by :

(i) Testing the performance of progeny ; this is of the greatest value in judging the merit of bulls, particularly dairy bulls ; most livestock men today hold the view that this is the only logical basis of assessing the merit of bulls. For a further discussion on *Progeny Testing* see Chapter XVI.

No livestock judge is competent to assess cattle in order of merit in respect of categories 2, 3 and 4, unless he has the necessary recorded data available. At livestock shows in the future, it can be confidently predicted that such information will be much more freely available, to give real distinction to the awards of merit made. The breeder of cattle also needs to follow the same path in the selection of his breeding stock.

CALF REARING

The health and future well-being of the calf is laid by good feeding and management from birth to six months of age.

Methods of rearing vary, but common to all methods is the need to ensure that the young calf receives an adequate supply of colostrum which has the property of building up resistance to those infectious conditions of the bowels often referred to as " scours ".

Colostrum is more easily digested than ordinary milk, is rich in vitamin A and has laxative properties. If cows are milked before calving, their colostrum should be saved by refrigeration for the calves at birth. Where colostrum is not available a substitute can be used, and a calf should also be given an injection of blood serum as soon as possible after birth, thus transferring the protective antibodies from the blood of the cow to the calf, as normally happens where true colostrum is fed.

Adequate warmth is also important as young calves are easily chilled, and up to three to four weeks of age lack the heat-producing mechanism of the cud-chewing adult ; they should always be kept in warm dry quarters.

Single Suckling. The simplest and easiest method of calf-rearing is the single-suckling system, each cow rearing her own calf. This is most common in beef production, it is the method most expensive in overheads but it is justified where first-class suckled calves are the aim and where the cows can be outwintered and cheaply fed. Calving in such suckler herds is usually timed for March/April so that the calves are old enough to take the flush of milk at the beginning of the grazing season. Later calvings may mean too much milk for the calf and consequent severe scouring unless the cow is kept on a bare pasture for a time.

Multiple Suckling. This method is followed where better food supplies and buildings are available : each cow can be expected to rear seven or eight or more calves by a continuous process of multiple suckling. At her peak of lactation a good nurse cow may suckle four calves for, say, three months, then three more calves, and then two— making nine in all. But few nurse cows achieve this unless they are specially selected for heavy milking ability. A nurse cow is usually a cull from the dairy herd and may have lost a quarter ; or is a slow milker ; but she will usually make less " fuss " of her foster calves. A nurse cow is controlled in multiple suckling by being tied up under cover and the calves are then brought to her, otherwise she spends her time out-of-doors or in a yard.

Multiple suckling is a system that fits in well on arable farms where rearing begins in the autumn. One batch of calves is reared before Christmas, a second batch by March and a third during the early part of the grazing season. The biggest drawback to this system, where enough calves are not bred on the farm, is to find suitable calves when wanted. This is most difficult in the spring ; hence autumn suckling should be considered where calves have to be purchased. In any case the cost of, and the mortality rate in, such calves is usually less than in spring-born calves.

To introduce a calf to its foster mother requires patience, and the following technique is recommended. First ensure the cow is tied up and unable to hurt the calf by butting or kicking. Then insert a finger in the calf's mouth and guide the calf into the normal suckling position. Entice its head towards the teat by leading it with the finger being suckled until the cow's teat can be substituted for that finger. Do not push the calf's hindquarters or its head in the hope that it will find a teat to suckle. It may do so, but many nurse cows with pendulous udders require a more subtle technique.

Calves that have been bought in should not be allowed to gorge themselves when freshly introduced to their foster mother, even if it means milking out the cow by hand for the first couple of meals. The new calves should suckle first ; the older calves finish off the cow and wean themselves in the process.

Pail Feeding. The third, and in terms of numbers reared the most important, method of calf rearing is by pail feeding on limited quantities of whole milk supplemented by proprietary foods known as *milk equivalents* or *milk substitutes*.

This method requires more specialised accommodation in the way of calf pens and is more exacting in labour demands, but it does enable a considerable saving in food costs to be made. The level of nutrition is lower than in the previous methods and is more suited to rearing dairy type calves. It is important to see that feeding is regular and that feeding buckets are kept scrupulously clean and sterilised frequently, and that foods are mixed correctly.

Calves may have to be taught to drink from a bucket by inserting one finger into the calf's mouth after the finger has been dipped into the milk ; as it sucks, the calf's head should be lowered into the milk and the finger removed. A calf suckles more readily at knee height. Whilst the calf is being fed on whole milk, the dilution of milk with 25 per cent water will reduce the risk of digestive scour, as will the avoidance of giving too much milk at once.

This system of calf rearing is now practised in two forms :

(*a*) The feeding of milk substitute or milk equivalent feeds to an age of 3–4 months ; this method is losing favour on account of the labour involved, apart from the rearing of veal calves.

(*b*) Early weaning at about five weeks of age. In this system, milk substitute is fed at about eight per cent of the calf's birthweight to induce the early consumption of concentrates and hay. Weaning occurs in the fifth or sixth week. The concentrates offered must be very palatable and of high starch equivalent, and the hay of top quality. A typical concentrate mixture would be : flaked maize 20 per cent, crushed oats 20 per cent, rolled barley 30 per cent, linseed cake 20 per cent, dried skimmed milk 10 per cent, plus minerals with added synthetic vitamins A and D. " Early weaning " calf foods are available in cube form which increases palatability. A supply of clean water should always be available. Early weaning of calves stimulates development of the rumen and enables young calves to utilise dry food to replace milk with satisfactory live-weight gains, provided the nutrient intake is maintained by high quality palatable feeds. Dairy heifer calves should not in fact be reared on a high plane of nutrition.

A summary of the various systems of calf rearing is given in Table 8 on page 602.

Rearing Calves at Grass. To reduce calf rearing costs, the New Zealand practice of rearing calves at grass has been developed

TABLE 8

CALF REARING GUIDE

(Methods of Rearing)

Feeding.	Suckling.		Pail Feeding.	
	Single.	Multiple.	Late Weaning System.	Early Weaning System.
At 1 week	Suckling dam	Suckling nurse cow	Colostrum for 4 days then whole milk at 10 per cent of birth weight + 25 per cent added water (× 2 feeds). Gradual change to milk substitute— 450 g replacing 4·5 litres whole milk	Colostrum 8 per cent of birth weight then on milk equivalent made up by adding 450 g milk equivalent to 4·5 litres of water (× 2 feeds). Concentrates and hay offered *ad lib.*
4 to 5 weeks	Suckling dam	Suckling. Hay and concentrates offered as eaten	Milk substitute 560–675 g made up with 5·5–6·5 litres water. Hay and concentrates offered as eaten	Milk equivalent 450 g in 4·5 litres of water. Reduced end 5th week to nil. Early weaning concentrates and hay to appetite
3 months	Suckling dam. Concentrates offered in creep and/or grazing	Suckling. 1·4 kg hay; 1·4 kg concentrates	Milk substitute reduced from now onwards. 1·4 kg hay; 0·9 kg concentrates	Now consuming "Early Weaning" concentrates—1·4– 1·8 kg. Hay 1·4 kg
4 months	Suckling dam. Concentrates offered in creep and/or grazing	Suckling ceases. 1·4 kg hay; 1·8 kg concentrates	Weaned off milk foods. 1·8 kg hay; 1·4 kg concentrates	Now consuming 1·8 kg calf rearing concentrates; 2·7 kg hay
6 months	Suckling dam. Concentrates offered in creep and/or grazing	2·8 kg hay; 2·3 kg concentrates or grazing	2·8 kg hay; 1·4 kg concentrates or grazing	1·8 kg calf rearing concentrates; 3·6 kg hay or grazing
Liveweight gain per day from birth	0·9–1·0 kg	0·8–0·9 kg	0·6–0·7 kg	0·6–0·7 kg
Approximate milk consumption	1,125–1,350 litres	405–450 litres	45–65 litres + 50 kg milk substitute + 200 kg concentrates	Nil + 16 kg milk equivalent and 250 kg concentrates
System suited to rearing of	15–18 months beef and stud bulls.		Dairy heifers, 18 month beef and barley beef.	

in this country. Weaned calves can be successfully reared on grass during the summer, provided clean grazing at the 10–15 cm growth stage is available, and this can reduce hay and concentrate consumption markedly.

Features of the system are : 1. The calves are early weaned. 2. Rotational grazing is practised with groups of, say, 12 calves in 0·1 ha paddocks—0·12–0·16 ha per calf being required up to 6 months of age. 3. Some form of shelter is valuable to protect the calves not only from bad weather but also from hot sunshine.

When calves are reared on grass and exposed to a gradually increasing burden of worm infection (gastro-enteritis and husk) they tend to build up an immunity at an early age.

Management of Weaned Calves. At weaning time, there is always a slight check to growth rate which is minimised by good stockmanship. Calves, particularly suckled calves, should be encouraged (e.g. by creep feeding) to consume hay and concentrates whilst still suckling. A suitable concentrate mixture for creep feeding is barley or oats, and for pail fed calves over 4 months old is cereals 850 kg ; barley balancer pellet 150 kg for a 1 tonne mix ; crude protein 13 per cent ; plus mineral mixture. Suckled calves are best yarded when weaning occurs until they have forgotten their dependence on the suckler cows.

BEEF PRODUCTION

The current consumption of beef in the United Kingdom is over one million tonnes a year or 21·2 kg *per capita*, of which some 71 per cent is home produced ; prime beef, as distinct from cow beef and bull beef, accounts for some 78 per cent of home production, or 55 per cent of the total beef consumed. It is with this production of prime beef with which we are now concerned. Consumer demand in recent years has been for leaner and smaller joints, and this has conditioned the type of cattle used for beef production, the age and liveweight at which slaughtered, and the level of feeding or of nutrition during life.

Type of Cattle. Apart from the pure beef cattle, some 80 per cent of our prime beef comes from cattle bred from dairy cows by crossing with beef bulls. This beef-from-the-dairy-herd trend has made rapid strides in recent years. More than 700,000 dairy cows are now mated to beef bulls, principally Hereford and Aberdeen Angus, by A.I. every year. Both of these breeds, in common with the Galloway, " colour mark " their calves, the black coat colour or the white face of the Hereford being dominant characters, so that progeny from these breeds are easily identified. Beef bulls confer the characters of earlier maturity and deeper fleshing on calves bred from dairy cows ; but the pure Friesian steer in particular is now in great favour for beef production because it possesses a high rate of live-weight gain. The use of Charolais bulls is now popular, particularly for crossing with the lighter fleshed pure dairy breeds—Ayrshire, Jersey and Guernsey.

Level of Feeding. It is now generally accepted that in rearing beef animals a high level of nutrition is required in the first six months of life to produce early development of muscle (lean meat) and bone. If subsequent feeding is continued indoors at a very high level, thus telescoping the development of the animal into as short a time as possible, this gives an animal fit for slaughter at 400–425 kg at 11–12 months of age.   This " Barley Beef " system, developed by Dr. Preston at the Rowett Research Institute, has largely replaced the traditional " baby beef " which is confined mainly to early-maturing pure beef breeds where similar weights are achieved at 14–15 months of age.

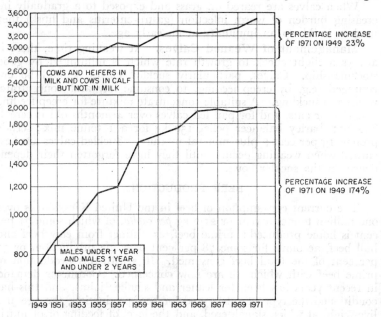

Graph illustrating the increasing importance of rearing for beef in   two recent decades. Figures are for June each year.   Source :   Ministry of Agriculture, England and Wales.

Feeding for baby beef normally includes a fair proportion of good quality roughages as well as concentrates.

Rearing after six months of age on a more moderate plane of nutrition permits full growth of muscle and bone to continue ; final fattening can then be delayed until the animals are either 11–12 months, 16–18 months or even 22–24 months old.   This final fattening period normally takes a further 5–6 months and may be carried out indoors over the winter period or on grass throughout the summer.

Finished beef of 400–450 kg at 16–18 months of age is becoming more popular and is ideally suited to autumn-born Friesian or Friesian-cross steer calves, where the middle period of their lives can be spent on good quality grazing to provide a moderate/high plane of nutrition and a reasonable daily liveweight gain.   The final fattening stage indoors can then be achieved through the use of good quality roughages plus some concentrate feed.

Older cattle are usually reared on more moderate grazing during the summer and on cheaper bulky-type foods during the winter : this system is common to large areas of upland country. These cattle are then sold as " stores " either in the spring or autumn to lowland farmers for a grass- or a yard-finishing period. The store cattle trade from Ireland into Great Britain has existed for many centuries.

Apart from the demands made by the trade and the consumer, which vary in different parts of the country, the level of feeding which the beef producer adopts is governed by the relative cheapness of concentrated foods compared with roughages and grazing. Thus in America, with its large grain surpluses, the trend has been towards a baby beef type of production ; whereas in Great Britain grazed and conserved grass together with cheaper roughages have a far greater economic value.

MANAGEMENT OF BEEF CATTLE

It is certainly advisable to breed naturally polled cattle for beef or to de-horn them as calves thus economising in accommodation when the animals are indoors and reducing the risk of injuries during handling and transport to market.

Cattle in Yards. Beef cattle are usually housed in yards (see p. 74) where adequate trough room to allow all to feed at once is essential except where self feeding on a 24-hour access is employed. From 60 to 70 cm of trough room per head is required. Dry lying is important, for " a dry back and a dry bed are worth an extra feed ". The cattle should be given every opportunity to rest, but housing need not be elaborate, for the margin of profit in beef production is too small for expensive or specialised accommodation. The stockman can contribute towards making the best use of the foods fed by ensuring clean feeding troughs, by even grading of the cattle in size to reduce bullying, and by quiet handling.

Cattle at Grass. The management of beef cattle at grass involves the gradual introduction of the cattle to full grazing in the first part of the grazing season, and this is achieved by an early turn-out and letting the grass " grow to the cattle ". Grazing is initially supplemented with a continuance of winter feeding, chiefly as hay or straw. If possible, shelter should be available in an open shed or a field with natural shelter (e.g. hedges) should be chosen. By such means the loss in liveweight, particularly with cattle that have been wintered indoors, is reduced to a minimum and the risk of bloat (see p. 679) largely eliminated.

Pastures are usually set-stocked with beef cattle although controlled grazing in smaller paddocks or by using electric fencing has become more common. Stocking rates vary from two beasts of 400–450 kg per hectare, or four yearlings of 250–300 kg, to double these rates under intensive systems.

The choice of pasture varies. For fattening purposes, pastures must produce a dense growth of grass over as long a grazing season as possible. This depends on rainfall and the natural level of the water table, so that most of the best fattening pastures are in low-lying areas such as the Norfolk marshes, the Welland valley and the river basins of the Trent, the Thames and the Severn. Less productive pastures are favoured for young cattle not being fattened that season, and these commonly carry sheep as companion stock to the cattle.

General Precautions. Each spring, usually before cattle are finally turned out to grass, they should be dressed against the Warble Fly with a derris preparation (see p. 449) if they were not dressed the previous autumn with a systemic phosphorus insecticide poured down the middle of the back. Otherwise except for daily observation to ensure that progress and general health are satisfactory, beef cattle should be disturbed as little as possible.

Labour requirements during the grazing season are extremely low. To assist in drafting out cattle, or for handling individual animals as in cases of lameness requiring veterinary attention, a stoutly built pen and cattle crush sited so as to be of easy access from as many grazing fields as possible is a very great asset. Shade is also appreciated by cattle in the hottest days of summer. A reliable water supply is essential. Access to lumps of rock salt or mineral mixtures is advised and the drinking trough is a convenient site for such purposes.

FEEDING SYSTEMS IN BEEF PRODUCTION

Against the general background of the management of beef cattle described above, examples are now given of the types of rations commonly fed in the production of beef. It must be emphasised how important economy of feeding is in its influence on profitability ; wherever possible, feeds grown on the farm either as hay, silage, cereals or arable crop by-products such as sugar beet tops, straw and pea haulm silage should be used as widely as possible, rather than the relatively more expensive purchased feeds. The margin between profit and loss in beef production is small so wasteful feeding of any kind must be avoided.

TABLE 9

FEEDING FOR DIFFERENT TYPES OF BEEF

	Barley Beef.		Baby Beef.		18 Month Beef.			Mature Beef.			
Age in Months	6	12	6	12	6	12	18	6	12	18	24
Plane of Nutrition	VH	VH	H	H	MH	H	H	H	L	L	H
Liveweight Gain/day in kg	1·23	1·32	0·9	1·0	0·77	0·9	0·8	0·9	0·45	0·45	0·77–0·9
Liveweight in kg	225	375–400	200	388	200	325	450	200	275	350	475–500
Typical Ration in kg/day											
Hay (good)	1	1	2·7	4·5	2·7			2·7			
,, (medium)						2·7			2·7	3·6	2·7–4·5
Straw						2·7	3·6		2·7	3·6	3·6
Silage						9·0	16·0		9·0	16·0	16·0–18·0
or											
Roots						16·0	25·0		16	25	25
Concentrates (cereals)	2·3	9·0–10·0	1·6	2·7	1·6	1·4	2·7	1·6			2·7
(high protein)	0·4	0·7	0·7	0·9	0·7	0·3	0·4	0·7			0·3
						or grazing			or grazing		

The table opposite shows the different demands made on feed supplies by the four systems of beef production previously described. Early maturing beef cattle reared on a continuously high plane of nutrition consume, relatively, a lower proportion of bulky type foods and preponderantly more concentrates. Higher costs of production per kg of liveweight are associated therefore with the more intensive systems of beef production, but such systems have the advantage of an increased rate of turnover owing to the early age of slaughter.

VEAL PRODUCTION

For veal production, beef × dairy calves are very suitable, as are pure bred calves of the dual purpose breeds, notably the British Friesian and South Devon. Calves capable of high live-weight gain and early fleshing are required, the aim being to finish calves of 150–168 kg liveweight at 13–14 weeks old. Such high rates of gain (1·15 kg/day from birth) are achieved by the feeding of specially compounded milk equivalent feeds based on a Dutch formula incorporating dried skimmed milk and animal fat, supplemented with vitamins and minerals. After the initial feeding on colostrum for four days the milk equivalent is fed at approximately the following daily rate when mixed with water at a dilution of 0·4 kg powder in 3 litres water.

In the early stages up to fourteen days of age, food consumption increases rapidly and a high level of stockmanship is required. At the

Age of Calf (days).	Milk Equivalent (litres).	Feeds per day.
7–14	2·8–5·0	2 or 3
15–24	5·7–8·0	2
25–34	8·5–11·4	2
35–59	12·0–14·8	2
60 onwards	15·4–17·0	2

No other food is given

slightest sign of scouring or loss of appetite the quantity of feed given is reduced until recovery occurs. Such setbacks are minimised by controlled ventilation ; some form of insulation may be necessary to maintain an even temperature of 14–15° C in the calf house up to four weeks of age. The calves are housed in individual pens—Dutch practice is to tether the calves in pens 1·44 m × 0·52 m with a floor of slate slats, 7·7 cm wide, spaced 2·6 cm apart, standing well clear of the concrete floor of the house. Straw may be used on the slatted floor to give extra warmth to the calves, but is not essential.

A high standard of general hygiene must be observed. Feeding buckets, for example, must be kept scrupulously clean and the feed properly mixed. To assist in this, an electrically heated domestic hot water boiler (45 litres capacity) in the mixing-room adjoining the calf house is a great advantage. Plastic buckets are recommended, being fitted easily into brackets to serve each pen, the calf feeding through a yoke. A well-designed calf house with pens on each side facing a central feeding passage greatly eases the labour of feeding which should not occupy more than 5 minutes per calf per day. A veal production unit could embrace up to 170 calves as a one-man enterprise.

Profit in veal production hinges largely on rearing the right sort of calf which should not be too expensive within current prices, and by good housing and management ensure that progress is rapid and continuous. In marketing, veal is not at present related to any guaranteed price and unless a return of at least 120p per kg (October 1976) deadweight is obtained (which makes veal one of the most expensive meats), profit margins are likely to be low—the feed cost alone, involving the consumption of around 175 kg of milk equivalent, is in the order of £57 per calf.

MARKETING OF BEEF AND VEAL

Sales of beef and veal can be on the deadweight or live-weight basis ; in the former, a cash sale is made through certified abattoirs, in the latter, through an auction market or by private treaty. In comparing returns from these alternative methods of marketing, the farmer needs to be able to estimate the killing-out percentage of his stock which is expressed as the percentage of carcase weight to live-weight. Offal includes the skin, head, feet, tail, intestines, heart and liver, but the greatest " loss " in weight from live to carcass weight is from the contents of the alimentary tract ; hence when weighed at markets, a deduction is often made to give the true or fasted liveweight.

Killing-out Percentages. Well-finished mature cattle, two years and upwards, give killing percentages of from 57 per cent to as high as 64 per cent : younger cattle kill out within the range 56–60 per cent, less well-finished cattle 54–6 per cent.

Veal calves have a killing percentage within the range of 58–62 per cent. A high killing-out percentage is of considerable importance to the butcher buying on liveweight, or to the farmer selling on a dead-weight basis, and varies from animal to animal dependent on the proportion of offal, degree of finish or fleshing, and conformation.

MILK PRODUCTION

Since the institution of organised marketing by the establishment of the Marketing Boards in 1933, milk production in Great Britain has forged ahead—the present rate of expansion in England and Wales alone is 2 per cent a year, and in 1974 production was a total of 11,265 million litres for milk sales off farms in England and Wales ; almost 4,546 million litres more than twenty-five years ago. Expansion has been based on increased consumption of liquid milk and an extension of milk manufacture into cheese, butter, condensed milk and cream. Liquid milk consumption is approximately 0·42 l per day per head of the population, and manufacturing milk represents some 40 per cent of production in 1974.

Within the industry itself certain significant trends are in evidence. The average size of herd is increasing from 17 cows in 1955 to 40 cows in 1974, and the number of herds of less than 20 cows has been halved in the same period. Larger herds save labour per cow, and developments in the housing and milking of cows have been prompted by the same aim to increase output per man. The loose housing of cows in yards and increasingly in kennels or cubicles, and parlour milking rather than housing and milking in cowsheds, is becoming more popular, the trend being assisted by Ministry capital grants.

Yield per cow has risen from an estimated 3,069 litres per cow per year in 1955 to 4,000 litres in 1974, as a result of better management of cows and grassland (from which the cheapest source of feed for dairy cows is obtained) and also by using cows of higher milking capacity either by change of breed or by better breeding methods.

THE REARING OF DAIRY HERD REPLACEMENTS

The average productive life of dairy cows today is between three and four lactations which requires the rearing of some 800,000 heifers

TABLE 10

DEVELOPMENT OF A BRITISH FRIESIAN DAIRY HEIFER

Age (months).	Live Weight (kg).	Typical Winter Rations (kg).	Summer Feeding (ha).	Remarks.
6	127	2·7 hay or 6·8 silage 0·9 concentrates	Grazing 0·08 per head	Vaccinated with S.19 at 3–4 months
12	229	5·4 hay or 13·6 silage	Grazing 0·16 per head	Vaccinated for Husk before turning out to grass
18	381	2·7 hay + 2·7 silage + 13·6 roots or 20·4 silage	Grazing 0·24 per head	Served
24	457	3·6 hay + 3·6 straw + 13·6 roots or 22·7 silage	Grazing 0·32 per head	Start concentrate feeding in winter 6 weeks before calving at 0·45 kg per day
27	533	4·5 straw + 4·5 hay + 15·9 roots or 27·2 silage	Grazing 0·40 per head	Concentrates fed, increasing by 0·45 kg per day per week up to calving. Maximum 3·6 kg per day (winter and early spring)

as dairy herd replacements every year. The methods of rearing such heifers has a significant effect on the cost of milk production, and the aim should be to rear as cheaply as possible without prejudice to future milking capacity.

Heifers intended for the dairy herd are reared on a significantly lower plane of nutrition than beef cattle, particularly after the first three months of age (see p. 606). A liveweight gain of 0·34 kg per day for Jerseys, 0·45 kg per day for Ayrshires and Guernseys, and up to 0·57 kg per day for Friesians and Dairy Shorthorns has been shown to give adequate skeletal development without the added cost of high condition or fleshing so desirable with cattle being reared for beef. Such a relatively low plane of nutrition permits the wider use of roughage and succulent feeds with minimal use of concentrates. Provided this restriction of nutrient intake is not too severe, saving in the overhead cost of rearing herd replacements can be achieved also by early calving, but early calving should not be achieved at the expense of body size. In-calf heifers should receive better food as calving approaches—with increasing length of pregnancy, so does the plane of nutrition rise. This is illustrated in Table 10 on page 609 which traces the development of a dairy heifer (British Friesian) from weaning to first calving.

THE MANAGEMENT OF MILKING COWS

Present methods of feeding, housing and milking of dairy cows are conditioned by the necessity for dairy farmers to reduce production costs. In the last ten years labour costs and the price of purchased concentrates have more than doubled, whereas milk prices have risen by about 70 per cent.

To reduce labour costs, herds are being increased in size ; bucket machine milking in cowsheds and parlours has now almost completely replaced hand milking ; housing in yards or in cubicles has gained in favour over cowshed housing because it facilitates the handling of dung and effluent by mechanical means (see p. 74). There has been an increase in self-feeding of silage, whereby cows help themselves from a vertical "grazing face" at the silo and also in the mechanical feeding of silage, thereby reducing labour in feeding. Feeding costs are being reduced by increasing the proportion of bulky foods such as hay, straw, silage, and forage crops at the expense of the more costly concentrated foods such as high protein cakes and some cereal grains.

Details of housing and the lay-out for milking parlours are to be found elsewhere (see Chapter III) and the principles of good grassland management and the production of arable crops for feeding to dairy cows in Chapters X, XI and XII.

FEEDING DAIRY COWS IN WINTER

The scientific background to the food requirements for milk production is given in Chapter XVIII. The translation of feeding standards to fit the circumstances of individual farms is based first of all on an estimate of the basic food requirements of cows for the winter in terms of the foods available on the farm. This will enable the dairy farmer to plan his cropping policy in relation to the size of his dairy herd. These requirements are set out in Table 11.

TABLE 11

WINTER FEED REQUIREMENTS OF DAIRY COWS
(October 1st–March 31st)

Breed of Cow.	Winter Requirements (as hay).		Daily Requirements.	
	Maintenance (M).	M + 5 kg Milk.	M.	M + 5 kg Milk.
	kg	kg	kg	kg
South Devon	1,829	2,480	10·88	14·72
British Friesian ⎫ Dairy Shorthorn ⎬ Red Poll ⎭	1,676	2,327	9·07	12·46
Guernsey ⎫ Ayrshire ⎭	1,372	2,023	7·26	11·13
Kerry ⎫ Jersey ⎭	1,067	1,718	5·44	9·31

It is then necessary to find out the " Hay Equivalent " of the bulk feeds ; this can be expressed as follows :

1 tonne medium hay equals
- 0·76 tonne very good hay or dried grass
- 2·29 tonnes grass silage (good)
- 3·05 tonnes grass silage (medium)
- 3·05 tonnes arable silage
- 4·06 tonnes kale, swedes or beet tops
- 5·08 tonnes mangels
- 3·05 tonnes fodder beet or potatoes
- 3·05 tonnes pressed beet pulp
- 0·51 tonne oat or barley straw + 3·05 tonnes kale or swedes

Thus if a dairy farmer plans to feed, say, his British Friesian dairy herd on kale and self-fed silage thereafter, he will need to grow for each dairy cow the equivalent of 2·33 tonnes of hay to feed for maintenance and the first 5 kg milk, or 1·68 tonnes for maintenance only, over 6 months of winter feeding. This would be met by growing :

3·15 tonnes kale = 0·77 tonne hay equivalent ⎫ for
4·73 tonnes silage = 1·55 tonnes hay equivalent ⎭ M + 5 kg milk
or
2·23 tonnes kale = 0·56 tonne hay equivalent ⎫ for
3·42 tonnes silage = 1·12 tonnes hay equivalent ⎭ M only

By a study of the above table, a dairy farmer is able to relate his stocking rate to food supplies. Young stock (or followers) can be reckoned to have half the feed requirements of mature cows, so that a dairy herd of 40 cows and 30 followers is equivalent to $40 + \dfrac{30}{2} = 55$ cows.

Choice of Crops. In his choice of crops to grow for a dairy herd, the farmer must be guided by the conditions on his own farm as determined by soil type, rainfall, and the equipment available for conservation. Where grass is the master crop, as on heavy land in high rainfall areas, winter feeding is likely to be largely based on silage, or haymaking where rainfall is lower. On arable farms a wider choice of crops is possible, including cereals and sugar beet.

Some guide to the average yield of fresh material and digestible organic matter per hectare of different forage crops is given in the following table :

TABLE 12

EXAMPLE YIELDS OF SOME FORAGE CROPS GROWN IN AVERAGE CONDITIONS

Crop.	Fresh Material (yield/ha) (tonnes)	Digestible Organic Matter (yield/ha) (tonnes)
Hay average (2 cuts)	9·3	4·5
Grass silage (unwilted 2 cuts) . . .	37·5	4·3
Kale	50·0	4·8
Mangels	100·0	7·5
Maize silage	42·5	5·8
Barley grain	4·0	2·9
Barley straw	2·5	0·8
Turnips	50·0	3·2
Swedes	62·5	6·1

When crops are grazed in the field considerable savings of harvesting cost can be made.   Grass is an outstanding example of this, nevertheless the more expensive conservation must be undertaken as a significant part of grassland husbandry for winter feeding. Productivity of grass is influenced largely by fertilisation, especially nitrogen, and by stocking rate.   A seasonal grazing utilisation of just over 10·1 tonnes dry matter per ha or 80 per cent of the estimated growth could be regarded as a satisfactory result.

Using Foods to Best Advantage.   The second need in feeding dairy cows is to ensure that the foods available are used to the greatest advantage.   In this context, good quality home-grown fodder (hay, silage and so on) has a protein-sparing function in that it reduces the need for protein in the concentrates fed in the production ration (see p. 578).   Economy in feeding concentrates is thus achieved as protein cakes are more expensive than starch- or energy-providing foods such as cereals.

The supplementation of silage or kale with cereal foods is now a recognised feeding practice.   Highly succulent foods such as kale or wet silage (below 22 per cent dry matter) tend to depress the dry matter intake, and supplementary feeding with cereal foods assists in raising the energy intake of diets largely composed of succulent feeds : this may avoid the risk of milk low in solids-not-fat.

Lastly, research on rumination has revealed the important role played by the micro-flora of the rumen in the fermentation of food during digestion.   This pre-digestion of food in the rumen is markedly affected by the physical condition of the foods eaten by the dairy cow. The cow needs fibre in her feeds in the " long " state.   There is no advantage in grinding hay or dried grass which are better fed long, and cereal grains should be rolled, rather than finely ground to a meal form.

WINTER RATIONS FOR DAIRY COWS

Some examples of suitable winter feeding rations for dairy cows are given below. Few farmers today aim at more than M + 10 kg from bulk feeds as this enables economy in feeding to be combined with maximum stocking ; to feed for M + 10 kg, for example, means that only 3 cows can be kept where 4 could be kept if fed to a maintenance level only on bulk feeds, with concentrates being fed entirely for production purposes. There is also the difficulty of getting cows to eat sufficient bulk feeds to provide for the production of more than 10 kg of milk owing to limitations of appetite.

This table emphasises that with the lower grade fodders such as

TABLE 13

EXAMPLES OF WINTER RATIONS FOR COWS
500 TO 570 kg LIVEWEIGHT
(British Friesian)

Ration per day. (kg)	Maintenance (M). (kg)	M + 5 kg (kg)	M + 10 kg (kg)	Remarks.
Mainly Grass Farms				
A. Hay (average) . .	7·6			Balanced concentrate mixture after first 5 kg milk
„ (good) . . .		7·6		
Cereals e.g.				
Oats				
Barley				
Beet pulp (dried) }		1·9		
B. Hay (average)	6·3	6·3	6·3	Balanced concentrate mixture after 10 kg milk
„ (good) . .				
Kale	6·7	25	25	
Cereals			1·9	
C. Silage (average) . .	23			Balanced concentrate mixture after 10 kg milk
„ (good) . . .		25	32	
Cereals		1·4	1·9	
Mainly Arable Farms				
A. Hay	3	3	3	Balanced concentrate mixture after 10 kg milk
Straw	3	3	3	
Sugar beet tops . .				
or Kale . . .	13	19	25	
Cereals		0·9	2·6	
B. Silage (average) . .	13	13	19	Balanced concentrate mixture after 10 kg milk
Straw	3	3	3	
Beet tops or Kale .	5·8	15	15	
Cereals		0·9	1·9	
C. Silage	17·5	17·5		Balanced concentrate mixture after 5 kg milk
Straw	3	3		
Cereals		1·9		

TABLE 14

Supplements for grass silage

Quality Digestibility.	Dig. Crude Protein. (g/kg DM)	Barley Fed. (kg)	Adequate for
Very high . . .	116	3	M + 15 kg milk
High	107	2	M + 10 kg milk
Moderate . . .	102	—	M + 5 kg milk
Low	98	—	M only

straw, or average silage and hay, the " saving " in protein concentrates is limited to the first 5 kg of milk. For yields in excess of 10 kg, usually a " balanced " mixture of concentrates is fed and this is discussed later.

Self-Feeding of Silage. This practice saves labour, the cows consuming silage from a vertical feeding face, usually at one end of the silo, with unrestricted access over the whole period of 24 hours. The cows eat to appetite which has been found to be of the order of 3·6–4·5 kg of silage per 50 kg of liveweight. Low levels of silage consumption result both from silage being too long, and here laceration or chopping during making is an advantage, and where the silage is too wet or unpalatable because of faulty fermentation. To safeguard butterfat production some long fodder, hay or straw is beneficial, particularly if the silage has been made from grass cut early in the 208–228 mm growth stage. Rarely will more than 2·7 kg of long fodder be eaten. The level of milk production achieved from self-fed silage is mainly dependent on the quality of the silage. Wilting the cut grass before collection to improve its dry matter content has become more common.

Feeding of Concentrates for Higher Levels of Production. Reference has already been made in the previous examples of winter feeding practices to the feeding of balanced mixtures of concentrates for yields in excess of 5 kg or 10 kg per cow. Such mixtures are compounded of cereal-type feeds mixed with high or medium type protein feeds and fed so as to provide, at defined rates of feeding per 5 kg, sufficient nutrients to meet production needs. (See p. 578 for a discussion on the application of feeding standards to dairy cow feeding.)

It is worth noting that the response in milk yield to increasing levels of concentrate feeding is subject to the Law of Diminishing Returns in that feeding of concentrates will increase milk yields at all levels, but at a diminishing rate as yield rises. This emphasises the need for controlled feeding of concentrates *according to the milk yield of the cow and her stage of lactation.* For example an increase of concentrate feeding to two cows both giving 15 kg of milk per day will give a greater response with one cow in early lactation, say one month calved, than with the other in late lactation, say seven months after calving. No cow will, however, produce milk beyond her genetical capacity ; adequate feeding enables that capacity to be fully revealed, and high yielding cows are distinguished by their ability to respond to rising levels of feeding without diminishing returns operating as soon as would be the case with low yielding (i.e. poorly bred) cows.

Present feeding standards are adequate for all cows in declining lactation, but in early lactation feeding at a level at least 10 per cent in excess of the standards (or 2·5 kg above what is actually being given) is justified, particularly to cows of high yielding potential. First-calf heifers also require more generous feeding to allow for development and better milk quality. Concentrates have a most important role today in raising milk yields per cow, provided feeding is done with care and discretion to match concentrate intake with the response in milk yield during lactation, and to build up the condition of the cow before calving—known as " steaming up ", or fitting the cow for calving.

Management of the Individual Cow. Cows should be given a dry period of at least 6–8 weeks between the end of the current lactation and the next calving. This rest period makes possible the repair and build up of udder tissue, full development of the unborn calf and the laying down of body reserves to be drawn upon during the next lactation, provided the level of nutrition is sufficiently high. To ensure this, cows during the winter period are fed concentrates, in addition to good quality fodder, from about six weeks before calving in increasing quantity reaching a level of 3·6–5·4 kg of concentrates per day at calving time. Accurate service records and calving dates must be known. During the grazing season steaming up can be at a reduced level, and with May-June calvings cut out altogether when grass is at its maximum feeding value (see p. 609 for steaming up of heifers).

As calving approaches the cow's appetite declines: concentrate or cake feeding is often then reduced and a bran mash or soaked beet pulp given to ensure that the cow is sufficiently laxative, a condition which assists parturition. Wherever possible, cows and heifers should be calved in a loose box, well littered with straw ; and if there is a smooth floor, it should be sprinkled with sand. Assistance at calving should be available if needed in the case of difficult or abnormal births. After calving, the cow is normally allowed to lick her calf dry and the calf remains with her for at least 24 hours or up to 96 hours, before the cow joins the milking herd. On farms where calf mortality is high due to bacterial infections it is recommended that the calves remain with the cow for the whole of the first four days of lactation, during which the first milk or colostrum is secreted which is so valuable to the calf for its nutritive and immunising properties.

Milk Fever, which may affect cows at, or soon after, calving, is a condition marked by failure to rise and a typical twisted posture in the neck. The condition is associated with a low calcium content in the blood stream, and is treated by injection of 200–400 ml of calcium borogluconate solution either intravenously or subcutaneously (see p. 678). Complications may occur and the advice of a veterinary surgeon should always be obtained ; meanwhile the affected cow should be propped up on her breast bone with straw bales ; if she is allowed to lie " flat out ", she may die from pressure of her rumen contents on the diaphragm causing suffocation.

Feeding after calving should be related to the cow's recovery of

Y

appetite.   Good quality hay or silage are safe feeds, and the quantity of cake fed is stepped up slowly over the first fortnight, when if recovery from calving is complete, feeding to yield can begin in earnest. Failure to consume food with relish after calving may be due to retention of the afterbirth, causing uterine inflammation (metritis) which is a subject for veterinary attention.   Once feeding and rumination are normal and the udder inflammation, common at calving time, has subsided, the cow enters into her full routine of milking and feeding for her next lactation.

FEEDING FOR MILK

During the first three months of lactation, a cow will normally give half of her total lactation yield, and it is during this period that concentrates are most justifiably employed to encourage and sustain a high level of milk production.   Rates of concentrate feeding per 5 kg during this period can, with justification, be raised from, say, 1·7 kg per 5 kg to 1·9 kg per 5 kg of milk (i.e. feeding in advance of yield) as long as milk yields continue to rise, thus matching output with greater nutrient input.   By the sixth week of lactation, yields tend to reach a maximum or peak limited by the frequency of milking.

Twice daily milking is now usual, but an increase of up to 20 per cent above the peak yield can be obtained by thrice milking, which on

TABLE 15

PRODUCTION STANDARDS PER kg OF MILK

Type of Milk.	Butter Fat. (g/kg)	Solids not fat. (g/kg)	ME allowance ML (MJ/kg)	Dig. Crude Protein. (g/kg)
Channel Island . . .	48	91	5·90	67
Shorthorn	36	87	4·98	52
Ayrshire	37	88	5·08	53
Friesian	35	86	4·88	51
Average	36	86	4·94	52
Solids corrected . . .	40 *	89 *	5·31 *	56

* Bulletin 33 HMSO

most farms today is uneconomic at present milk prices in relation to labour costs.

Once the peak yield is reached, the level of concentrate feeding is reduced to the conventional level, and concentrates should then be fed strictly to yield as production declines.   Decline in lactation is normally of the order of 10 per cent per month with cows and 7½ per cent with heifers in their first lactation.   If the rate of decline exceeds these levels in a herd, the most likely cause is poor quality fodder, e.g. poor silage or hay leads to shorter lactations and lean cows, unless concentrate feeding is increased or extra supplementary feed given.

Sugar beet pulp, stock feed potatoes or wet brewers' grains are suitable foods which can be purchased to boost the feeding value of low quality fodders.   For example, 3·2 kg of barley straw can be supplemented with 3·2 kg of wet brewers' grains to give the equivalent of

3·2 kg of good hay : or 3·2 kg of poor hay could be supplemented with 0·9 kg of dried beet pulp or 3·2 kg of potatoes. Such supplementation of fodders where quality is low is very important in avoiding low solids-not-fat percentage in milk, particularly in the late winter months (see also p. 622). The commonly accepted standards for milk production in terms of Metabolisable Energy and of digestible crude protein show variation according to the chemical composition of the milk (see Table 15). The Energy allowances for ruminants, as now defined in Metabolisable Energy terms, is the new and more precise method which can be used to determine the energy content of the diet and the needs of the animal. A full and comprehensive description of this is to be found in Technical Bulletin No 33, Energy Allowances and Feeding Systems for Ruminants issued by HMSO.

Production Mixtures and Rate of Feeding per 5 kg Milk
Mixture A, using mainly home grown cereals :

	kg	ME(MJ)	Dig. Crude Protein (g)
Rolled barley	1·0	11·8	70
Rolled oats	0·5	5·0	36
Decorticated ground nut cake	0·5	5·8	202
Total	2·0	22·6	308

5 kg of Friesian milk would require 24·4 ME(MJ) and 255 g D.C.P. and 2·16 kg of the mixture would be adequate for this quantity. However, 5 kg of Channel Island milk would require approximately 2·6 kg to supply its energy need.

Mixture B, using mainly purchased foods.

	kg	ME(MJ)	Dig. Crude Protein (g)
Rolled barley	1	11·8	70
Dried sugar beet pulp .	1	11·4	53
Flaked maize	1	13·5	95
Palm Kernel cake . . .	1	11·5	176
Soya bean cake . . .	1	12·0	409
Total	5	60·2	803

This ration fed at 2·5 kg for 5 kg milk would be adequate for Channel Island milk and fed at 2 kg for 5 kg of Friesian milk production.

By selecting feeds high in energy and/or protein it is possible to reduce the quantity of the mixture to be fed per 5 kg : and compound dairy cakes are sold with recommended rates of feeding per 5 kg according to milk quality.

Mineral Mixtures. The need for supplying minerals to dairy cows has already been discussed (see p. 539) ; one of the most convenient means of doing this is to add minerals to concentrate mixtures or compound cakes. Mineral mixtures in covered troughs can also be made available in the field. Deficiency in minerals may cause unthriftiness or infertility as well as lowered production. Trace element difficulties involving lack of cobalt (causing pine) or copper, or iodine, can be corrected by additions to the basic mineral mixture fed, which can be made up of limestone 9·0 kg, steamed bone flour 9·0 kg and common salt 9·0 kg added to each tonne of concentrates. Proprietary mineral mixtures fortified with trace elements are now widely used, and special mixtures containing up to 50 per cent of magnesium oxide (calcined magnesite) are also recommended in circumstances where hypomagnesaemia is likely to occur.

Feeding 56 g per day of calcined magnesite for 4–6 weeks before and after turning out to spring grass is a wise precaution (see p. 679).

Feeding for High Yields. In feeding for maximum yields, as for instance under test conditions in milking trials or in pedigree herds where cows are fed to capacity to determine their full milking potential in selection for breeding purposes, the limited appetite of the cow for dry matter consumption is overcome by concentrating the total diet. This practice of " bulk control " limits the intake of fodder or succulent feeds and maximises concentrate consumption, so that a highly nutritive low fibre diet is consumed. The maintenance ration is reduced in bulk by substituting such foods as dried beet pulp for hay and succulents, and by cutting out entirely low grade roughage foods such as straw. In order to maintain effective rumination the hay fed should not be less than 0·45 kg per 50 kg liveweight.

A typical diet then for a Friesian cow, giving say 47 kg of milk a day, might be : Hay 5·4 kg ; Beet pulp 1·8 kg ; Concentrates 13·2 kg (fed at 1·4 kg per 5 kg milk). Such a cow would be fed and milked three times a day and the order of feeding at each meal would be concentrates followed by hay. With increasing cost of concentrates in relation to milk prices and the additional labour cost and problems of organising three times milking, such methods are losing ground to systems of feeding and management involving lower cost feeding and lower labour input, even at the expense of reducing yields per cow. Such methods have already been described.

However, on small farms with limited land, heavy feeding of concentrates to cows of high milking capacity is still an important means of maximising production and increasing the volume of the farm business to an economic level. Such systems of high concentrate feeding also have greatest application to winter milk production, as in summer with lower milk prices, the substitution of grass by concentrate feeding is uneconomic.

The Summer Feeding of Dairy Cows. To exploit the full potential of grass as the cheapest food for dairy cows during the grazing season it is necessary to have cows calving mainly in the months of February and March so that the curve of production of milk follows the production curve for grassland in conjunction with controlled grazing. Such control involves paddock or strip grazing with rapid eating off of the grass by heavy stocking, followed by adequate rest periods of 20–25 days between grazings, these being determined by the rate of recovery of the sward. Increasing milk output per ha is correlated with increasing the stocking rate ; but over-stocking may reduce grass consumption to the point where milk yields suffer, so that stocking rate is also governed by the milk yields of the cows.

Grass during the early part of the grazing season is a high protein/low energy feed. Eating to appetite on May grass, a cow consumes sufficient digestible crude protein for maintenance and production of 33 kg to 38 kg milk, but the energy value is adequate only for maintenance and 19–23 kg milk. Supplementation of grass is economic only for those cows yielding over 19 kg milk. Restriction of grazing, and feeding of concentrates for yields below 19 kg milk is possible, but it would not pay to do this because grazed grass costs but little more than half of the cost of cereals on an ME basis. After the spring flush of grass, the case for supplementing grazing is stronger, but by using directly reseeded leys sown the same spring, or by grazing catch crops such as rape or Italian ryegrass and by good overall grassland management the summer grazing gap can be bridged with minimal use of concentrate supplements. Irrigation is another means of maintaining pasture growth.

If grazing is not maintained in periods of drought a low solids-not-fat percentage in the milk may well occur. On the other hand when the cows go out to grass on the early bite, low butterfat percentage is often the problem unless some hay or other roughage such as good oat or barley straw is provided after the morning milking.

As a general guide to summer feeding the following scheme is presented :

Time of year.	Growth stage of grass.	Supplementary feeding.
Early April	102–152 mm young and leafy (grazing controlled 2–4 hrs. daily)	Hay or other roughage 1·8–3·6 kg (for 2–3 wks.). Cereals at 1·9 kg per 5 kg milk over 14 kg or 19 kg milk
Mid-May .	200–250 mm long, still leafy (grazing to appetite)	No roughage needed. Cereals at 1·9 kg per 5 kg milk over 19 kg or 23 kg milk
Mid-July .	Grasses tending to flower (grazing to appetite	Cereals at 1·9 kg per 5 kg milk over 9 kg or 14 kg milk
Mid-August	Grazing of aftermaths, new seeds or forage crops, or green feed in dought conditions.	Balanced concentrates for yields over 9 kg or 14 kg milk
Mid-Sept.	ditto.	Balanced concentrates for yields over 9 kg

In interpreting this table, allowance must be made for local circumstances influencing grass growth, some of which are under the farmer's control, while others, such as rainfall or temperature (earliness of the season), are not. On the farm, close observation of the cows at pasture will show whether they are getting sufficient to eat, and their bodily condition and milk yields will indicate the quality of the food they do eat. A change in feeding is called for whenever yields start to decline too rapidly or the cows are in danger of getting too lean ; a change to new pasture, preferably young seeds or clean aftermaths, or to grazing a forage crop such as rape may obviate the need for more expensive hand-feeding.

A considerable amount of research work has been carried out into the question of grass utilisation by dairy cows both in terms of grazed material and conservation. Grazed material can be fed either " in situ " or by means of " zero grazing " whereby the grass is cut and carted to the cows. The advocates of zero grazing maintain a higher milk output per ha can be achieved compared with *in situ* grazing but the opponents of the system consider there is some evidence to show that infertility and other health problems tend to increase if cows are kept entirely " on concrete " throughout their lives. Slurry disposal also becomes a greater problem when zero grazing is adopted throughout the summer.

It is now appreciated that digestibility is a vital factor in determining milk output from grass, in addition to protein and dry matter content. As soon as grasses start to form stem, the digestibility starts to fall. Within the space of 12–14 days during the peak growing season, stem formation will lead to flowering heads if grazing has been uncontrolled : digestibility then falls markedly, leading to a big drop in milk output. Therefore, the whole aim must be to maintain grazed grass in a leafy state. Similarly, grass for conservation should be cut before flowering heads start to form. Fertiliser dressings, particularly nitrogenous fertiliser, have a very important part to play in achieving a high output from grass.

VARIATIONS IN THE COMPOSITION OF MILK

Milk may be considered to be a rather variable mixture of fat, non-fatty solids and water. Average commercial milk approximates to the composition : 87·7 per cent water, 3·6 per cent butterfat, 8·7 per cent solids-not-fat (including 4·6 per cent lactose, 3·3 per cent proteins and 0·8 per cent mineral matter and other substances), but these figures vary widely with individual herds and much more so with individual cows. The present state of legislation requires that genuine milk shall contain not less than 3 per cent butterfat and not less than 8·5 per cent solids-not-fat, but with entry into the EEC new regulations will come into being.

The composition of milk is influenced by many factors, genetic, physiological, seasonal and nutritional.

Influence of Breed on Milk Composition. Breed has a profound overall influence, since Channel Island herds normally produce milk of much higher levels of fat and non-fatty solids than indicated by the above figures. Many Jersey cows, for example, consistently produce milk containing 5 per cent or more butterfat and more than 9 per cent solids-not-fat (S.N.F.). Generally, breeds that tend to give the higher yields, such as Friesians, produce milk of lower compositional quality and vice-versa but, within the same breed structure, individual cows vary widely in the quality of the milk they produce.

To appreciate the variable quality of milk it is necessary to know how the quality of the milk of an average cow changes throughout her normal lactation of about ten months. The newly-calved animal produces for the first week or two milk of higher quality than the average for the lactation. As her output reaches its peak at about a month or six weeks after she has calved, butterfat levels tend to fall to a minimum and then slowly to rise towards the end of lactation, and very appreciably to rise as she is drying off. S.N.F. levels also begin high, dropping to a minimum at about eight to ten weeks after calving and remaining rather below the average for the greater part of the lactation. They usually rise steadily towards the end of lactation if the cow is in calf again, but may actually fall further if she fails to get in calf. Of the individual components of the S.N.F., the mineral matter tends to remain very steady all through, the lactose content very slowly and consistently falls over the lactation, whereas the protein levels drop very rapidly at first and then slowly begin to rise over the greater part of the lactation.

This typical picture of the behaviour of an individual cow reflects what happens under continuously maintained stall feeding conditions. The majority of cows calve over the autumn and winter periods and may pass from one-third to over one-half of their lactation period under stall conditions ; but mostly they are turned out on grass between March and May and spend the latter part of their lactation entirely on pasture. With the flush of spring grazing, yields are usually stimulated by about 15 per cent, and there is often an appreciable drop in the butterfat content of the milk and a rise in S.N.F. content.

The herd picture must necessarily be the resultant of the contribution made by the individual cows. Generally there is emphasis on autumn and winter calving, and the more nearly the cows are calving together the more closely the herd's milk compositional picture approaches that of the typical individual. Where the breed is one with the normally expected poorest compositional quality, danger periods for the milk to fail to reach the presumptive minimum legal standards are in the late autumn when the peak of post-calving production is being attained ; the late winter and mid summer periods (for S.N.F.) ; with also a temporary drop in butterfat level as spring grazing comes to its flush.

Influence of Age on Milk Composition. The effect of age on average butterfat levels is not very marked but, generally, cows tend to drop by about 0·03-0·05 per cent per lactation.

On the contrary, age has a profound effect on levels of S.N.F. content, the drop being of the order of 0·1 per cent per lactation for the first four or five lactations, slightly less thereafter. This means that an ageing herd is more likely to show poorer levels of S.N.F. in the milk than a younger herd. A heifer with a poor record for S.N.F. in her first lactation is almost certain to get worse in her subsequent lactations.

Solids-not-fat. The broad effects of underfeeding, i.e. supplying insufficient nutrients for maintenance and production, are loss of conditions in the cows, accompanied by a more rapid-than, normal falling away of yield. Underfeeding also has the effect of lowering the levels of S.N.F., and in this respect, although it is usually the protein content of milk that suffers most, deficiency of energy or starch equivalent intake has a more marked effect than shortage of digestible protein.

Where underfeeding has been sufficiently prolonged to cause the level of S.N.F. to drop below 8·5 per cent it is very difficult to effect any appreciable improvement by more liberal feeding with concentrates during the course of the remainder of the lactation. Underfeeding may take many forms other than an actual shortage of bulk of food, and possibly the most general form of underfeeding is substandard quality of the bulky foods, in particular hay and silage where maximum reliance is placed on these fodders.

The effect which turning the herd out to spring pasture has on the S.N.F. content of the milk is often dramatic, rises on average of as much as 0·4 per cent being attained within a month before the normal post-flush fall begins to set in again. The more liberally that a herd is fed before grazing commences the smaller is the rise in S.N.F. content of milk with grazing.

The effects of disease, especially of the udder, are likely to be adverse as far as S.N.F. are concerned. Damage by mastitis organisms to the internal tissues of the udder leads to a rise in the chloride content of the milk, and such a rise is offset by a tenfold drop in the lactose content, since milk is secreted at a constant osmotic pressure. Thus a rise of 0·1 per cent in the chloride content (doubling the normal value) is offset by a drop of 1 per cent in the lactose, the net effect being a fall of 0·9 per cent in the S.N.F. content of the milk. The after-effects of mastitis, successfully treated by the usual therapy, are likely to have a permanently lowering influence on the level of S.N.F. in the milk.

Butterfat. Until recently it was not thought possible to alter materially the butterfat content of milk by feeding methods. Some foods, such as palm kernel cake and coconut cake, were alleged to be able to raise the fat content, but experimental evidence was conflicting. Liberal doses of cod liver oil (170–220 g) per day could seriously depress fat levels, and the depression continued for some weeks after the cessation of the feeding of the oil.

Nevertheless, it was quite a common experience for farmers to face the possibility of having their milk rejected for low fat content shortly after turning out to spring grazing, and the effect of supplementing the grazing with concentrates often made matters worse. The discovery that long roughage, fed as hay and straw, in adequate amounts could prevent the trouble, or cure it after it had occurred, has had tremendous practical and scientific repercussions. If the same quantity of roughage, ground to a meal, were fed no improvement in the milk fat level resulted.

Spring grass causes a temporary stimulation of yield with often a corresponding fall in fat content of the milk, and the common practice of feeding concentrates (and often succulent foods in the form of mangels) to cows at grass often makes matters worse. It is sometimes difficult to persuade cows to eat roughage when newly turned out to grass, but a good practical measure is to put the roughage just under the electric fence controlling the area being strip-grazed, when it is found that the cows will eat sufficient to sustain adequate levels of butterfat. Too many farmers appear to be afraid to trust young spring grass to produce the amount of milk it is capable of, and the practice of continuing liberal feeding of concentrates is not only wasteful and unnecessary, but may actually jeopardise the sale of the milk by inducing low fat levels.

Flaked maize, although helpful in improving S.N.F. content in the winter months, can have a depressant effect on fat levels unless it is given in conjunction with adequate roughage. Since this food is sometimes included in coarse dairy rations, this provides a further argument against giving such rations to cattle when high quality grazing is available in the spring.

Underfeeding may have a slightly depressing effect on fat levels, although the principal effect of underfeeding is on yields and S.N.F. levels. Low butterfat levels may also occur in the early autumn when freshly calved cows have access to lush autumn grazing on which they are also fed liberal allowances of kale or beet tops, and are also being fed heavily on concentrates, with no roughage allowances. General experience suggests that about 2·7–3·2 kg of long roughage daily are required to sustain butterfat content of the milk at satisfactory levels.

MILKING METHODS AND CLEAN MILK PRODUCTION

Milking premises must be licensed by the Ministry of Agriculture through the County Agricultural Executive Committee, and milking methods must conform to the Milk & Dairies Regulations, 1959. Hand milking, except in very small herds, has now been superseded by either *bucket milking machines* where cows are milked and housed in cowsheds, or by *releaser milking machines* where cows are milked through a milking parlour. The latter system is gaining in favour (see p. 67 for a description of layouts for milking parlours).

The Milking Routine for machine milking is basically similar with all types of milking machine and incorporates operations designed to ensure clean milk production as well.

Preparing the Cow. Preparations of the cow for milking involves securing her in her stall—yoked or tied by the neck in a cowshed or confined to her milking stall in the parlour. Quiet handling of the cow is essential because if cows are nervous or frightened the *let down* of milk is inhibited. To encourage this " let down " of milk the udder and teats are washed with warm water (49° C) to which an antibacterial agent such as quaternary-ammonium compounds or hypochlorites may be added as a precaution against the spread of mastitis, and such washing if efficient reduces the risk of dirt contaminating the milk. Before udder washing the fore-milk is withdrawn from each teat on to a strip cup to reveal if any clots are present in the milk—a symptom of mastitis infection.

Applying the Teat Cups. Within one minute of washing the udder and teats the teat cups are applied, care being taken to see each teat cup is fitted snugly on to each teat. Release of milk (" milk ejection ") by the cow then proceeds over a milking time conditioned by the adequacy of milk let down by the cow, by the size of her teat orifices, by her milk yield, and by the degree of vacuum in the milk pipe line.

Degree of Vacuum. Most milking machines operate in the range of 380–457 mm of mercury vacuum, about half atmospheric pressure. Provided preparation of the cow has been satisfactory the majority of cows milk out in 4–6 minutes, dependent on yield. Some slow milking cows with abnormally small teat orifices take longer—up to even 10–12 minutes, and such cows are better culled. Increasing the vacuum will increase milking rate but, at over 457 mm of vacuum, udder injury may result. This is particularly the case if over-milking occurs when, as the milk pipe empties of milk, the teat cups tend to creep up the teats.

Machine Stripping. As soon as milk flow slackens and air bubbles appear in the sight glass or milk-flow indicator, pressure can be applied to the teat cup cluster with one hand, and the udder massaged downwards above each teat in turn with the other hand to " machine strip " the cow ; such last drawn milk is particularly rich in butterfat. Machine stripping is now becoming less common in larger parlour-milked herds. The teat cups are then removed by cutting off the vacuum to the teat cup cluster, which is then hung up ready for transfer to the next cow.

Time Taken in Milking. A summary of the time taken to perform this sequence of operations in a normal milking routine is :

	Seconds
Washing, udder massage and taking of fore-milk.	30
Attaching teat cups	10
Cow being milked	4–6 mins.
Machine stripping	20
Removal of teat cups	10
Changing cows	20
Total	90

The milking routine per cow allowing for contingencies is thus approximately 1½ minutes, so that up to 40 cows per hour can be milked by one man provided he has sufficient milking units to avoid idle time. By reducing the time devoted to operations in the milking routine, notably washing of the udder or machine stripping, this through-put can be exceeded.

In the maintenance of milking machines, vacuum pressure and pulsation rate (usually 50–60 per minute) should be periodically checked : teat cup liners should be changed approximately every six months.

Cooling. After milking, the milk has with bucket plants to be carried to the dairy whereas in releaser plants, conveyance is via a milk pipe-line direct under vacuum, so that releaser plants result in a 25 per cent saving in labour compared with bucket plants. In the dairy the milk is cooled either by a surface cooler, by in-churn immersion coolers operating on mains water or refrigeration or in a refrigerated bulk tank, to enhance milk keeping quality.

Milk sold for human consumption is subject to routine keeping quality tests, usually at the buyers' premises : since October, 1964, the methylene blue test has been superseded by the " Hygiene (2 hour) Resazurin " test. Where a consignment of milk fails to pass the quick resazurin or platform test on delivery it can be returned to the supplier as unmarketable. Poor keeping quality milk is most likely in hot summer weather or where faulty cleaning and inadequate sterilisation of the milking equipment has occurred.

Cleaning and Sterilising Equipment. Under the Milk & Dairies Regulations, 1959, steam sterilisation is no longer obligatory, and the use of hypochlorites approved by the Ministry of Agriculture is permissible. With parlour milking two developments to save labour in washing up equipment have occurred : first, with in-churn milking the milking clusters, teat cups and milk pipes are stored in caustic soda between milkings (*immersion cleaning*) ; second, a releaser plant is fitted up for circulation cleaning *in situ*, the detergent-sterilising solution being pumped through the plant under vacuum. In all cases the milking equipment must first have all milk residues removed by rinsing or by circulating cold water immediately milking has finished.

A further stage in labour saving has been achieved by the installation of storage and refrigeration tanks for the bulk collection of milk. This service is now operated in regional schemes throughout the United Kingdom, and such tanks, though expensive, fit in admirably with releaser type milking plants. This is the pattern of milk handling for the future.

MILK RECORDING

Official milk records are compiled under National Milk Records organised by the Milk Marketing Board since 1946. Commercial milk producers can also use the Farm Management Records operated by the Board to assist with their herd management, but such records are not accepted by Pedigree Breed Societies for official purposes. Under National Milk Records, weighing of milk is carried out with monthly weighings, either through Controlled Milk Recording (C.M.R.) or through Monthly Recording by Statement (M.R.S.). With both C.M.R. and M.R.S. the Board automatically issues certificates for completed lactations. A fully trained recorder visits the farm between three and twelve times a year, depending on which system is used, either to check accuracy or under M.R.S., to carry out weighing, sampling and recording. Butterfat testing is an integral part of National Milk Records, and a Protein Testing Service is available.

Membership costs vary markedly and depend upon the size of the herd and which recording system is adopted. The charges are reasonable and are deducted from the milk cheque.

Milk recording is of national importance as it provides data for the Bureau of Milk Records run by the M.M.B. to be used by Breed Societies and the A.I. service in bull selection and breeding projects, as well as in statistical investigations.

The standard lactation under N.M.R. is 305 days from the fourth day after calving, and herd averages are published on this basis. In

September 1973 there were 12,911 members of N.M.R. in England and
Wales, with a herd size averaging 52 cows (the average herd size for
all producers was then 30 cows.)   In Scotland, milk recording is run
by the Scottish Milk Marketing Board which offers both a butterfat
and total solids service to herds on the basis of individual cow records.

M.M.B. milk recording data is essential to the pedigree breeder.
With all milk producers it assists greatly in feeding cows more accurately particularly with concentrates, in culling of low-producing cows,
and in encouraging accurate breeding records in relation both to
management of the individual cow and regularity in breeding.   There
is a margin of some 468 kg per cow in milk recorded herds over cows
in non-recorded herds, which is a reflection of the higher management
standards in such herds.

REARING, LICENSING AND MANAGEMENT OF BULLS

The selection of bull calves for rearing as future herd sires is a
vital decision in breeding policy.   Such calves should be strong and
vigorous at birth, and they should preferably be by a progeny tested
sire out of a dam of proven merit either in milk production or in beef
qualities.   Bull calves are reared on a high plane of nutrition and by
the time they are ten months old must be licensed by the Ministry of
Agriculture.

Licensing.   The regulations were changed in September 1972
All potential breeding bulls are now examined and marked with a
" V " in the right ear at ten months old by a panel veterinary surgeon.
If the bull is sound, free of defects and healthy it will be eligible for
a licence, if not it must be slaughtered or castrated.   There is a right
of appeal against rejection of any bull.   Pedigree status in bulls
generally, and standards of suitability for dairy bulls, are no longer
grounds for refusing a licence and cross-bred bulls are also eligible for
a licence.   The Ministry of Agriculture still retains the responsibility
of issuing licences and dealing with applications for bull beef permits
and special experimental permits.   (See App. III.)

Once licensed, a bull can be used sparingly until mature when
twice a week service can be expected without loss of fertility providing
feeding is adequate.   Bulls should not be fed so as to become over fat
and sluggish in work.

Housing.   To maintain sound feet and an even temper, bulls
should be housed, preferably in a loose box provided with an outside
run, into which a service pen is incorporated with access from outside
to admit cows for service.

Bulls should be handled regularly with a firm hand ; workers
should not be exposed to avoidable risks by housing bulls in unsuitable buildings.   A properly constructed bull pen is a necessity if bulls
are to be kept long enough to assess their breeding worth (minimum
of 5 years) based on milk yields or liveweight-gain performance in
their progeny.   During the grazing season bulls can be tethered at
pasture, though beef bulls (and some dairy bulls) are often run with
the cows, with the disadvantage that service dates may be missed and
the risk to personnel is greater, but the conception rate is often higher

With suckler herds of beef cows, heat periods are often short and easily missed if reliance is placed on A.I. and the same is true of maiden heifers being served for the first time.

Feeding. In feeding bulls a fair assessment of their feed requirement is a full maintenance ration based on liveweight plus a " production ration " equivalent to that fed for 5 kg of milk (see p. 617). With bulls still growing the production ration is fed at the 10 kg milk level. The following illustrate suitable feeding schemes :

Age of Bull.	Live Weight. (kg)	Feeding.
18 months . . . (Immature)	500	Hay : 8 kg in 3 feeds Concs. : 3·5 kg
36 months . . . (Mature)	660	Hay : 11 kg in 3 feeds Concs. : 1·8 kg

Hay can be replaced by other feeds as desired and the concentrates can be of similar nature to a milk production mixture.

ARTIFICIAL INSEMINATION

An artificial insemination service is provided throughout Great Britain by the three Milk Marketing Boards and today three out of every five dairy cows are bred artificially and one of the three cows is inseminated by a beef bull (see beef from the dairy herd, p. 603). Bulls used in the A.I. stud are carefully selected both on pedigree and on their ancestors' performance. Young bulls are test mated and their relative value as sires obtained by the method of Improved Contemporary Comparison between their progeny and the daughters of other bulls in use in the same herds, which can be achieved on the wide scale of operations in the A.I. service, but would be impracticable to an individual breeder. The Commercial Service Fee is £1·65 for up to four inseminations per cow, but for the Nomination Service with choice of individual progeny tested bulls the fee is higher. In 1973–4 the total of all first inseminations of all breeds in England and Wales was 1,979,000 with a non-return rate of over 77 per cent.

In addition to the A.I. Centres run by the M.M. Boards there are six other privately run centres in England. Notification is usually made between 8 a.m. and 10.0 a.m. by telephone for inseminations the same day, later notification means insemination the following day. Insemination up to twelve hours after the end of the heat period is quite effective.

The A.I. Centres also offer private breeders a service whereby semen collected from the herd sire can be stored in deep freeze and used under licence in the owners' or other herds—such a bank of frozen semen is an insurance against death or injury to a bull. Semen has been stored for many years and still proved fertile. The deep freezing of semen also enables the Boards greatly to extend the influence of outstanding bulls in their own studs.

A.I. has already proved of tremendous benefit in livestock improvement but its future potential may well be even greater.

Chapter XX

SHEEP

In relation to its land area Great Britain carries more sheep than any country in the world except New Zealand ; more surprising still is the fact that this high sheep population is maintained in a thickly populated country which is highly industrialised.

In 1938 our sheep population exceeded 25 millions, only 1 million less than in 1898. A decline in numbers during the first world war was followed by an increase during the decade preceding the second world war.

Following a decline during the second world war there was a steady build up in sheep population reaching a peak of almost 30 million by 1966. There was a steady decline in numbers to around 26 million in 1971 in response to unfavourable economic returns but renewed confidence in the early 1970s has led to a recovery in the national flock to around 28 million in 1975.

Since the middle of the 18th century sheep in Britain have been bred primarily for their meat, but in the vast sheep-raising countries of the southern hemisphere wool production has been the main object though, since the development of refrigeration, there has been a marked tendency for the breeding of dual purpose (i.e. wool- and meat-producing) types. One of the reasons why the British sheep industry is so important is that every country which develops improved carcase-yielding types depends upon breeds of British origin for crossbreeding.

Britain's sheep are kept under very varied environmental conditions, on mountains, moorlands, uplands and lowlands. They are even to be found on farms mainly under arable cultivation. But during the past half-century there has taken place a marked shift of sheep from the south and east of Britain towards the north and west ; the greater part of our sheep population is now to be found on land of higher elevation. From this land sheep farming is not easily displaced by other kinds of husbandry.

BREEDS AND CROSSBREDS

No country can rival Britain in the variety of its sheep stock. There exist today more than thirty recognised pure breeds and certain well-defined local types which have not a flock book status. In addition, there are large numbers of crossbreds, which fall into two categories—those produced by systematic crossbreeding, and haphazardly bred " mongrels ".

The recognised breeds can be classified as follows :

Mountain and Moorland Breeds

Scotch Blackface	Herdwick
Swaledale	Whitefaced Dartmoor
Rough Fell	Welsh Mountain
Lonk	Exmoor Horn
Derbyshire Gritstone	Shetland
Limestone and Penistone	

Hill Breeds

Kerry Hill (Wales)
Clun Forest
Radnor Forest

North Country Cheviot
Cheviot

Short-woolled Breeds

(a) Whitefaced and Horned :
  Dorset Horn
  Wiltshire Horned or Western
(b) Whitefaced and Hornless :
  Ryeland
  Devon Closewool

(c) The Down Breeds :
  Southdown
  Shropshire
  Suffolk
  Dorset Down
  Hampshire Down
  Oxford Down

Long-woolled Breeds

Leicester
Border Leicester
Bluefaced or Hexham Leicester
Lincoln Longwool
Wensleydale

Kent or Romney Marsh
Devon Longwool
South Devon
Improved Dartmoor

In the following notes on individual breeds frequent reference will be made to size ; in the smaller breeds, mature ewes in good condition reach a live-weight of 32–36 kg ; in the medium-sized breeds the weight range in ewes is from 45 to 58 kg and in the large breeds from 58 to 76 kg.

MOUNTAIN AND MOORLAND BREEDS

The Scotch Blackface. Numerically this is one of the most important breeds in the whole of Britain ; it is without rival on the high, heather-clad moors of Scotland and northern England. Flocks are also to be found on Dartmoor and on the Cornish moors. Small in size, the breed is of exceptional hardiness, producing a long, coarse-textured fleece making it well equipped to endure snow and heavy rain. The wool is of value in the carpet-making industry. As in all breeds in this class the mutton is of excellent quality. Both ewes and rams are horned ; the face and leg colour may be black, or black mottled with white (Plate XLIIA).

The Swaledale. This breed of the northern Pennines is gaining in popularity at the expense of Scotch Blackface. It is slightly larger in size, carrying a shorter fleece in front. The face colour is a dark grey with a mealy nose.

The Rough Fell. This breed is found to the south and west of the Swaledale territory and it has a more localised distribution. The face colour is dark with a brownish tinge. Though having a shorter fleece it is claimed to be more hardy than the Swaledale.

The Lonk. Also a breed of rather localised distribution, this sheep is found mainly in the Pennine area of West Yorkshire and East Lancashire. Compared with the above-mentioned Blackfaced types it is larger and carries a finer, denser fleece. There is a considerable amount of white in the face markings.

The Derbyshire Gritstone. Again a breed somewhat local in its distribution—north-west Derbyshire. It is interesting as a transition type from the true Blackface, for it is much larger, hornless, and carries a much finer and softer fleece.

The Limestone and Penistone. Formerly of some localised importance, these are types indigenous to the Southern Pennines, but are not now numerically important.

The Herdwick. Throughout the Lake District this breed is predominant. Its endurance in severe storms and its ability to subsist on very scanty grazing prove that it is one of the hardiest types of sheep to be found in Britain. It must not be classed as a Blackface breed, although it has a coarse type of fleece, for some rams are hornless and, although the face colour in lambs is black, with advancing age the faces turn grey and finally white.

The Welsh Mountain. This is a breed of great numerical importance, for it has no rival on the extensive mountain land of Wales. Unlike the Cheviot, which is polled in both sexes, in this breed the rams are strongly horned. It is very small in body size and has a characteristic narrow-shouldered, long-necked appearance (Plate XLIIIA). The face and leg colour is a grey white, sometimes with tan markings. The wool is relatively soft and fine, though apt to contain hairy fibres (kemp). In its tolerance of high rainfall conditions and ability to exist on the sparse herbage of mountain land this breed is unrivalled.

The Whitefaced Dartmoor. A flock book has recently been started to revive this old type of Widecombe Dartmoor—one of the hardiest long-wooled breeds developed for moorland grazing.

The Exmoor Horn. This is a breed of considerable local importance on the high land of west Somerset and north-east Devonshire. Ewes and rams bear well-developed horns. The face and leg colour is a dull white and the dense fleece on a compact frame give to this breed a characteristically " chubby " appearance (Plate XLIIIB).

The Shetland. This interesting survival of the old Celtic fine-woolled type has been virtually driven out of Scotland by the Blackface. The very fine dense wool, removed by hand and not by clipping, is used in the manufacture of the famous Shetland shawls.

THE HILL BREEDS

The North Country Cheviot. This type has been developed from the Border Cheviot, and is very well established in Northern Scotland. It is larger than the Border Cheviot, with larger and less erect ears, and with a longer head than the south country type.

The Cheviot. This rivals the Scotch Blackface in being of very great numerical importance. Throughout Scotland it is found always on the " green " hill land, leaving the " black " or heather land for the Blackfaces. It has invaded Wales and even Exmoor—in fact the Cheviot can be found in many English counties on upland grazings of poorer quality.

A

B

PLATE XLII
A. Scotch Blackface shearling ram
B. Cheviot ewe

PLATE XLIII
A. Welsh Mountain ram
B. Exmoor Horn ram

PLATE XLIV
A. Kerry Hill ram
B. Clun Forest ram

PLATE XLV

A. Dorset Horn ram
B. Ryeland shearling ram

PLATE XLVI
A. Southdown ram lamb
B. Suffolk shearling ram

PLATE XLVII

A. Hampshire Down shearling ram
B. Oxford Down shearling ram

PLATE XLVIII
A. English Leicester shearling ram
B. Border Leicester shearling ram

PLATE XLIX
A. Lincoln Longwool ram
B. Kent or Romney Marsh ram

It is an impressive breed in appearance with its proud carriage of the head, its bright white face and leg colour, pricked ears, and distinctive ruff of wool behind the ears (Plate XLIIB). The wool quality is much higher than that of the Blackfaced types. The importance of this breed is still further enhanced by the fact that when crossed with the Border Leicester it produces in the " Half-bred " one of the most valuable types of crossbred ewe kept in Britain.

The **Kerry Hill (Wales).** Originating in the district whose name it bears, this breed is widely met with throughout the Welsh Border area and the Midlands of England. It is a medium-sized breed, polled in both sexes, with a dull white face and leg colour relieved by black markings round the muzzle and above the hoofs (Plate XLIVA). The wool is dense and always markedly coarser on the thighs and breech region. This breed is fairly prolific and is popular on lowland grass farms where a medium-sized breed is desirable. In some pure-bred flocks there is now a tendency to breed too closely towards Down breed standards of conformation ; this will ultimately lead to a falling-off in fertility and milking quality.

The **Clun Forest.** This breed has become very popular and is now widely distributed throughout the Midlands and also in southern England. It resembles the Kerry in size, in polled character, and in the carrying of coarser wool in the breech region. It differs from the Kerry in having a brownish black face and leg colour (Plate XLIVB). It has gained a good reputation as a type for mixed farms where it can be used to consume crops on arable land during the winter.

The **Radnor Forest.** This was a local type but is now a recognised breed. It is best described as an intermediate between the Welsh Mountain and the Kerry Hill.

THE SHORT-WOOLLED BREEDS

The **Dorset Horn.** Differing from all other British breeds this sheep has flesh-coloured lips and nostrils and produces a fine-textured wool which is very white in colour. It possesses another distinct character—it will breed earlier in the year than any other breed, for ewes will take the ram in May to lamb in October. This has made it a noted breed for the production of very early fat lamb. It is found mainly in south and west Dorset, Somerset and the Isle of Wight, usually on mixed and arable farms, for it is well adapted to feeding off arable land crops. It is very docile in grazing habit, resembling in this respect the Down breeds. Both ewes and rams bear well-developed, pale-coloured horns (Plate XLVA). While lacking the symmetry of carcase of the Down breeds it excels most of them in fertility rate and milking quality.

The **Wiltshire Horned (or Western).** Descended from a type once prevalent in Wiltshire, this breed is now found mainly in Northamptonshire and Buckinghamshire. Horned in both sexes, and with virtually no fleece, it is used mainly to provide rams for crossbreeding to produce early fat lambs.

The Ryeland. Originating in south Herefordshire this breed has now a scattered distribution. It is fairly small in size, very docile, and produces wool of good quality and uniformity. The face and leg colour is a dull white (Plate XLVB).

The Devon Closewool. Unlike the Ryeland, which is a breed of great antiquity, this is a relatively new breed. It owes its origin to crosses of the Devon Longwool with the Exmoor, in an attempt to produce a type suitable for habitat conditions intermediate between the higher uplands of Exmoor and the rich lowlands of Devon. It is polled in both sexes and bears a dense fleece of fairly long stapled wool.

THE DOWN BREEDS

The Southdown. No other breed in the whole world can challenge the Southdown in carcase quality ; in this breed the fineness of bone and texture of its dense flesh are supreme. In addition, it carries a short dense fleece of high-quality wool. It was the first of the Down breeds to be systematically improved. John Ellman in Sussex, towards the end of the 18th century, was the pioneer and his improved Southdowns were soon famed throughout England ; they were used to help in the establishment of the other Down breeds. In features the Southdown is quite distinctive ; it has a strikingly compact body on short legs, a mouse-grey face and leg colour, and its ears are small and rather rounded (Plate XLVIA). But it is unavoidable that in breeding for such a standard of excellence in flesh and wool production there is a sacrifice of the milking qualities and of prolificacy.

The Shropshire. This breed is now more widely known and more widely esteemed in the USA than in England. Though larger than the Southdown it does not approach in size to the other breeds in this class. The face and leg colour is black and it was in the past woolled down the legs and over the poll and cheeks to a greater extent than any other English breed.

The Suffolk. Though originally developed in East Anglia for arable farming conditions, the Suffolk can now claim to be the most widely distributed of all the Down breeds. It is extensively used in crossbreeding, its popularity being in part due to the absence of wool on the face. In sharp contrast to the other breeds in this class there is no wool on the head or on the legs and this clean-headed character has made it popular for mating with ewes of other breeds smaller in size (Plate XLVIB). The face and leg colour of the Suffolk is a glossy jet black ; in size it is larger than the Shropshire but inferior to the Oxford.

Both the Shropshire and the Suffolk are naturally more prolific than the rest of the breeds in this group.

The Dorset Down. This breed came into existence many years after the establishment of the Hampshire Down as a recognised breed. It is kept mainly in the counties of Dorset and Somerset. In many ways it is comparable with the Hampshire Down, though slightly smaller in size and having a lighter shade in face and leg colour.

The Hampshire Down. During the latter half of the last century arable-land sheep farming was extensively practised on the chalk uplands of Berkshire, Hampshire, Wiltshire and Dorset. For this system the Hampshire Down was pre-eminently suitable. It was accustomed to graze by day on downland pastures and it was penned at night in hurdle folds on arable crops. No other breed has such tolerance of living under close-folding conditions.

The arable sheep farming system has declined very greatly during the past fifty years and so this breed is numerically less important now. None the less, it is still an important breed since the rams are extensively used in crossbreeding for fat lamb production. It is a large-sized breed, capable of very rapid growth and early maturity if generously fed. It is woolled over the poll, on the cheeks and down the legs ; the face and leg colour is a very dark brown, almost black (Plate XLVIIA).

The Oxford Down. This has the largest-sized frame of all the Down breeds, owing its origin to crosses of the Hampshire with the large, long-woolled Cotswold. It was extensively kept on the arable farms of the limestone uplands in Gloucestershire and Oxfordshire. This again is an example of a breed developed for a system of farming now not extensively practised, but the Oxford is fairly widely distributed today because the rams are very popular in crossbreeding to produce store lambs for winter feeding. From its Cotswold ancestors it inherits a high bold head carriage ; it carries less wool on the legs and head than the Hampshire Down, though it has a comparable very dark brown face and leg colour (Plate XLVIIB). It must be ranked as one of the hardiest of the Down breeds in its tolerance of wetter soils and higher rainfall conditions.

THE LONG-WOOLLED BREEDS

As a class the long-woolled breeds are now much less numerous owing to the alteration in consumers' tastes which took place rapidly during the early part of the present century. The Longwools produce heavy fleeces of lustrous wool, coarser in fibre than the fleeces of sheep in the short-woolled group. The flesh is coarser in texture and there is some tendency to overfatness. With one or two exceptions, the Longwools have dull white face and leg colours and all are polled in both sexes.

The English Leicester. Systematically improved by Bakewell's careful breeding and selection, this must be regarded as the foundation breed of all other Longwools. Leicester rams were used to grade up other local types from which the rest of the breeds in this class were developed. Though well known abroad it is seldom met with except in north-east England (Plate XLVIIIA).

The Border Leicester. Like the Cheviot this breed is striking in its high head carriage and absence of wool on the legs and head (Plate XLVIIIB). It is a breed of first-class importance because the rams are extensively used to cross with Cheviot and Blackface ewes to produce the Half-bred and the Mule or Greyface, both of which are crossbreds of great economic value.

The Bluefaced or Hexham Leicester. In recent years rams of this breed have gained in popularity for crossing with Blackface and Swaledale ewes to produce the Greyface, Mule and Hexham crossbred types.

The Lincoln Longwool. The largest and heaviest fleeced breed in the whole of Britain, though today of greater economic importance overseas. It has contributed towards the formation of the Corriedale in New Zealand (Plate XLIXa).

The Wensleydale. In contrast to the other breeds in this class the Wensleydale has a bluish-grey face and leg colour with a fleece of silky and finely purled wool. It has not a wide distribution away from the district after which it is named.

The Kent or Romney Marsh. This breed is numerically one of the most important of the long-woolled breeds, and its habitat in the Romney Marsh area is the most densely sheep-stocked area in the whole of Britain. The wool is fairly fine in texture and of very good quality. It is highly esteemed in overseas countries which have specialised in wool production (Plate XLIXb).

The Devon Longwool. A breed of local importance in south-west England.

The South Devon. Rather similar in type to the Devon Longwool, but larger in size and slightly coarser in fleece. It is found mainly in Cornwall and south Devon.

The Improved Dartmoor. Another Longwool of very localised importance ; an interesting type, developed from the original moorland Dartmoor sheep by the use of Devon Longwool rams, and now kept under semi-lowland conditions.

CROSSBREEDING AND CROSSBRED TYPES

Systematic crossbreeding is widely practised by sheep breeders in this country. There also exists, to a regrettable extent, some crossbreeding which is unsystematic, resulting in numerous mongrel types which vary in wool and carcass quality. But when two breeds are crossed to yield a first-cross ewe, which is then mated to a ram of a third breed, this is definitely systematic since it results in the production of standardised commercial types.

Each year the older ewes in all the mountain and moorland flocks are drafted for sale to semi-lowland and lowland farms. Here they will live under more favourable habitat conditions—a higher plane of nutrition and less climatic severity. This environmental change brings about an increase in fertility rate and in potential milk yield. By mating these ewes with rams from breeds larger in body size and earlier in rate of maturity, the crossbred lambs produced have a capacity for rapid growth, for they can use to good effect the heavy milking capacity of their dams. Fundamentally the greatest reason for crossbreeding in sheep is to increase the progeny value of each ewe in the breeding flock.

For mating with mountain and moorland types of ewe both long-woolled and short-woolled types of ram are extensively used. In many cases from this type of crossing a valuable ewe for further breeding is produced. The following are good examples :

The Halfbred, from Border Leicester on Cheviot.
The Mule or Greyface, from Border Leicester on Blackface.
The Masham, from Wensleydale on Swaledale.
The Welsh Halfbred, from the Border Leicester on the Welsh Mountain ewe.

The wether lambs from these crosses are fattened for mutton, but the ewe lambs are kept for breeding to Down rams such as the Oxford and the Suffolk, yielding second-cross lambs much valued for winter feeding on roots, if not sold earlier off grass.

Another system of crossing, commonly practised to satisfy the considerable consumer preference for smaller and leaner joints, is the use of Down breed rams on the larger Longwoolled ewes. The following are examples of this :

Southdown on Kent or Romney Marsh.
Hampshire Down on South Devon and Devon Longwool.
Suffolk on Lincoln Longwool.

In early fat lamb production crossbreeding has always been a popular practice. Rams of many shortwool breeds—the Ryeland, the Western, the Suffolk, the Hampshire, the Dorset Down, the Southdown—have been widely used, also the Kerry Hill and the Clun. Some areas where the sheep population is high have always a surplus of ewe stock to export to other parts of the country ; most of these ewes are used to breed and fatten lambs for early disposal.

Since the 1960s interest has been shown in increasing the prolificacy of British sheep by cross breeding with certain European breeds. The introduction of the Finnish Landrace has shown how prolificacy can be improved, but whilst the crosses have shown great improvement in this respect, poor conformation, slow growth rates and high lamb mortality has prevented widespread adoption.

In crossbreeding in sheep there are two important considerations which govern the choice of the breed of ram to use : (1) To avoid excessive trouble in lambing, breeds of ram carrying much head wool should not be used on small-type ewes ; (2) The time of lamb disposal should also be considered ; lambs sired by smaller-type rams are usually more suitable for early slaughter, i.e. before weaning age. Crossbred lambs for sale as stores for winter feeding must be well grown in size of frame. To produce these it is obvious that rams from the larger-sized breeds will give the best results.

GENERAL FLOCK MANAGEMENT

The management of sheep is best dealt with by outlining the management of a breeding flock through the cycle of one complete year. In all breeding flocks the same sequence of operations must be carried out, but according to locality and breed the time of year of the lambing season varies considerably ; Dorset Horn flocks are lambed in the autumn, some Down flocks in January, and the latest breeding is found in the mountain breeds which are not lambed until April and early May.

Preparation for Mating. The shepherd's year begins when the breeding flock is made ready for mating. This means, in the first place, the sorting or culling of the flock ewes to ascertain those fit for retention for a further year's breeding and the addition of young ewes to make up the desired flock number.

Culling. Culling is an operation which must be carried out with great thoroughness, for age, fitness for further breeding, and type, must all be taken into consideration. In many flocks all ewes that have bred three crops of lambs are taken out for sale. Although they may be bred from for two or more seasons, they will if they are sound go to other farms which maintain " flying flocks ", i.e. flocks maintained by buying in for crossbreeding, no female progeny being retained for replacement.

In determining fitness for further breeding it is essential to examine the udder carefully to ensure that it is sound in both glands and teats. No ewes should be kept which show by poor condition and lack-lustre wool that they are unlikely to survive another pregnancy. In pure breeding flocks there should also be some selection for type and breed character, though the maintenance of these is governed largely by the quality of the young ewes drafted into the flock each year.

As a general rule ewes are mated for the first time to lamb their first crop at two years old ; a three-crop ewe is therefore four years old, showing a " full-mouthed " dentition, i.e. the eight permanent incisor teeth are up. In some late-maturing mountain breeds the first mating is postponed a further year. When culling is carried out all ewes should be " mouthed " to ascertain whether the incisor teeth are sound, because in the case of ewes which will have to consume roots during winter defective incisor teeth will be a severe handicap.

In the case of long-woolled types of ewe it is good practice to trim off some of the wool in the tail region before the rams are put with the flock.

Flushing. It is generally accepted that the potential breeding flock should be flushed before tupping. This implies a change to better grazing, or the provision of a special folding crop, and sometimes even the use of trough food. The purpose of flushing is two-fold : it may increase the percentage of twins conceived and reduce the number of barren ewes but, more important still, by the stimulation of ovarian activity it will bring the ewes into season more uniformly so that the subsequent lambing season will be of shorter duration. This latter is a very important consideration, because a long-drawn-out lambing season gives too wide a spread in age of the lamb crop, though this can also be controlled by limiting to six or eight weeks the period during which the rams are left with the ewes.

Under certain conditions of management and environment a good proportion of twins is desirable, though for mountain and moorland ewes one lamb reared per ewe is the maximum—often the lamb crop does not exceed 60 per cent of the total breeding flock.

But in measuring the fertility of a flock of ewes the figure taken is generally the number of lambs reared. If losses are heavy, even a high birth rate may result in an unsatisfactory rearage or final crop. Bad management and under-nutrition may lead to heavy losses in young lambs born too weak for survival, or receiving insufficient milk from their dams.

Number of Rams. The number of rams used is usually one for every 50–60 ewes in the flock, but there are two important considerations to be borne in mind—the age of the rams and the amount of range which the ewe flock will have. In the early maturing breeds most of the rams used are ram lambs of seven to nine months of age ; obviously older rams, shearlings and two shear, can be allowed to serve a larger number of ewes. But the amount of range over which the ewe flock can travel may render it necessary to turn out one ram to each 40 ewes else there may be a number of barren ewes, sterile because they were not found by the ram when they were in œstrum.

The œstrous cycle in the ewe is of 16–18 days in duration, the actual " heat period " lasting from 12 to 24 hours. It is a common practice to colour the rams on the breast when they are turned out so that all ewes served are marked ; by changing this colour after each 18-day period when the rams are with the ewes the progress of the flock mating is easy to observe ; if a large number of ewes is being marked more than once, a change of rams is essential. Another advantage of this marking is that it enables the ewes which will be the first to lamb to be picked out towards the advent of the lambing season.

Period of Gestation. The gestation period in ewes is normally 21 weeks, though a small proportion of the flock may produce lambs after 143 days of pregnancy. In large flocks the percentage of barren ewes may be 5 per cent, but this figure is often exceeded in flocks kept under environmental conditions where food is scanty and weather conditions unfavourable.

Precautions before Lambing. After the flock ewes are settled to the ram it is undesirable that they should be allowed to live too well, in fact during the first half of the gestation period no marked gain in live-weight should take place. But during the last eight weeks before lambing they should be managed well and adequately fed ; many losses occur before, during, and just after lambing because pregnant ewes have had unfavourable treatment. Undernourished ewes produce weakly lambs, which soon die from exposure, or such ewes will be short of milk and unable to rear their progeny. During the last weeks of pregnancy the feeding of excessive quantities of hay, chaff, and roots or kale will have a harmful effect, for as fœtal development in the uterus nears completion any overloading of the digestive tract with bulky foods should be avoided.

There are other ways in which lambing results are directly influenced by management. Ewes kept in folds on arable land may during wet weather have to lie in mud, and wet lying conditions are most unsuitable for ewes advanced in pregnancy. They should have access to drier ground at night even if they are feeding on a fold crop by day. Adequate access to supplementary forage and concentrates is essential particularly for the " shy " feeders which may otherwise be bullied out and be at risk from pregnancy toxaemia. For medium-sized ewes 150 mm of rack space should be adequate for access to ad-lib roughage but about 450 mm of trough space is desirable for concentrates. It is also most important that no sudden changes in the diet should be made during the latter part of the period of gestation. A progressive improvement in the nutritive quality of the diet, with a reduction in its fibrous bulk, will do much to ensure a trouble-free lambing with no lack of milk supply in the ewes.

Lambing. The majority of flocks in this country are lambed in the months of March and April, though in the ram-breeding Down flocks, and in flying flocks kept for early fat lamb production, January and February are the busiest months. Adverse weather conditions play a big part in determining the actual crop of lambs born alive and subsequently reared. Plans should be laid in advance for this critical period in the flock's history, the primary considerations being food supplies and shelter.

Shelter. In well-sheltered lowland areas the provision of artificial shelters can be reduced to a minimum, but in exposed situations and on arable farms lambing in January and February it is usual to erect large lambing pens, using timber, wire, straw and hurdles. Small pens are provided for ewes newly lambed, and large pens for ewes heavy in lamb and those with lambs at foot. This elaboration allows the shepherd to give very close supervision during lambing and to ensure that all lambs are properly " mothered ". Under conditions of extensive range such close supervision is not possible and much daily walking is necessary for the shepherd if he is to keep the newly born lambs under close supervision. Motherless lambs are a serious problem for the shepherd to face but, with patience and ingenuity, he can foster them on to ewes that are orphaned, or he can take a lamb from a ewe with a twin or a triplet to give to a ewe that has lost her lamb. A really efficient shepherd in charge of an intensively managed flock will see to it that every ewe with milk and a sound udder has a lamb to suckle.

Apart from a heavy mortality due to bad weather, serious losses in newly born lambs can occur from an outbreak of that serious infectious disease, lamb dysentery. This is largely controllable by good hygiene and there is now a reliable method of prevention by the use of serum inoculation. Mastitis is caused by several different organisms and can be a serious problem. If it occurs a veterinary surgeon is needed to decide what treatment to use.

It is rarely that a flock can be lambed entirely without some bad cases needing manual assistance, and it is in this direction that shepherding efficiency varies greatly. The skilled shepherd knows when a ewe needs aid and he is careful and patient in the way he handles a ewe.

Tailing. The tailing of lambs is an important routine operation except in flocks of mountain and moorland breeds which, living under high altitude conditions, need tails as some protection for the udder region of the ewes. It may also be omitted in the case of lambs destined for early slaughter. Tailing can be carried out at any age from one week to three months ; lambs under six weeks of age will not lose too much blood if the tail is severed with a sharp knife, but lambs that are older should be docked with a docking or tailing iron, which is used hot, so that the tail vein and stump can be seared to prevent excessive loss of blood. Tails may also be removed by placing a small highly tensile rubber ring on the tail. This operation must be carried out within seven days of birth.

Castration. The castration of all male lambs not required as rams is another routine task usually carried out when tailing is done There are three orthodox methods—drawing, castrating with clamp and the so-called bloodless method. The first of these is suitable for lambs under six weeks of age ; it involves cutting off the end of the scrotum and drawing each testicle away with a short portion of the spermatic cord attached. With older lambs this method may involve excessive bleeding and shock, so the use of a clamp to secure the ends of the spermatic cords after the testes have been cut off, and a hot iron to sear the cut surfaces, is the correct procedure. The " bloodless " method is now fairly widely practised ; in this a special type of pincer instrument is used to apply great pressure to the spermatic cords above the testes and no cutting of the scrotum is necessary. The pressure on the tissues of the cords destroys the connective tissues so that the testes gradually atrophy and disappear. This method is undoubtedly the most humane and the most hygienic, but unfortunately unless carried out by a skilled operator it is not always positive and many rig lambs may be left. If the bloodless method is practised a check examination of all wether lambs should be carried out after four weeks have elapsed so that further treatment can be given to those lambs which are rigs. Castration can also be effected by the rubber ring method : this involves placing the ring around the neck of the scrotum to cause atrophy of the entire scrotal sac. As with tailing this operation must be carried out within seven days of birth.

Weaning. Weaning is the next routine operation, and this usually takes place when the average age of the lambs is four months. It means nothing more than the separation of the ewes from their progeny and, if possible, segregation of the lambs out of earshot. The ewes should be relegated to poorer keep to hasten the drying off of milk secretion, and some observation of their udders is necessary at this time because some ewes may be overpent with milk. Ewes which after weaning develop udder trouble will have to be culled from the future breeding flock since their useful breeding life is over.

Shearing. One of the most important operations of summer sheep management is shearing or clipping. As a prelude to this in all well-managed flocks the ewes are crutched or tail trimmed ; this means the removal of all soiled wool from the tail and breech region. The time of shearing varies according to locality. In the south of England, and at low elevations, it is usually carried out in June, but on high land and in the north July and August are the busiest months for this work. It cannot be done until the yolk has risen in the fleece and this does not take place until warm weather promotes the rise in the wool. Machine shearing has almost totally replaced hand shearing and the use of New Zealand methods has substantially improved the rate of work. A skilled operator will shear 70–100 sheep in a day, though in the sheep-raising countries of the southern hemisphere, through greater specialisation of labour, a much faster rate of shearing obtains. It is most important to keep extraneous dirt and soiled locks from the shorn wool. The fleeces should be rolled and tied before packing in special wool sheets. In this country little attempt has been made to sort and class the wool before packing, whilst in the great wool-exporting countries this is done with great thoroughness.

The sale of all British wool is now controlled by a Wool Marketing Board. Prices have become more stabilised, and much progress has been made in the direction of eliminating the faults in handling and presenting wool for sale by the imposition of price reductions for tar stain, faulty fleeces, and the inclusion of tailings with sound wool in packing. The sale value of British wool can be much enhanced by the attention each individual wool producer gives to the handling and packing of wool at shearing time.

Whilst for most sheep their first shearing takes place in their second summer of life, some shearing or clipping of lambs is carried out in certain breeds and in the southern half of England.

Yield of Wool. The weight of wool yielded annually by adult sheep varies from 1·5 to 1·8 kg in the case of the smaller short-woolled breeds to 5·4–7·2 kg in the case of longwools. Well-fed rams will yield amounts of wool much in excess of these amounts.

After shearing or clipping has taken place the shepherd will find the supervision of the flock less arduous ; there is a reduced chance of maggot-fly strike in the shorn sheep, this being one of the major afflictions of sheep during the summer months, especially in flocks kept on land below 300 m in elevation (see p. 454).

Dipping. The next summer task is dipping, usually carried out about a fortnight after shearing. The re-appearance and rapid spread of sheep scab in the last three years has emphasised the importance of control, and all of the country is covered by regulations affecting movement and dipping. Several external parasites of sheep such as maggot-fly, the ked, the sheep tick and the sheep louse are all kept under control by dipping (see p. 455) ; a sound standard of flock hygiene cannot be maintained without the practice of dipping and without a sound standard of hygiene sheep flocks will not thrive.

In the 1960's alternative methods of wetting the fleece of shorn sheep by spraying or showering were developed. These methods require a suitably designed spray race or shower, a pump normally connected to the power take-off of a tractor, lengths of hose and a container such as a 180–litre drum for the spray fluid. The equipment is conveniently mobile and can be taken to a place near where the sheep are grazing if desired. Using a spray race or shower saves a certain amount of physical exertion and of sheep dip as compared with dipping. But it must be recognised that whereas careful spraying or showering will control maggot fly it is not acceptable at present under the regulations for the control of sheep scab and immersion dipping must be carried out at the appropriate times.

This is the last of the major operations in the routine management of a flock, though other tasks may be necessary to maintain the flock in good health. A shepherd should be always on his guard to prevent the spread of lameness through footrot; and periodic drenching to control parasitic worms may also have to be carried out at certain periods. In the management of sheep much depends upon the accurate daily observation of the flock by a shepherd or flockowner who is able to perceive any departure from normal health and well-being. The prompt diagnosis and treatment of certain ailments will prevent losses by death or lack of thriving.

On farms which carry medium- and large-sized permanent flocks, much time and labour can be saved in handling, drafting and dipping by the provision of fixed equipment of good design. Even the operation of dipping can be made possible for one man by the installation of a side slide-in device in lieu of the direct plunge method which involves handling individual sheep. In the large sheep-raising countries of the Southern Hemisphere, specialised sheds for shearing and handling wool are essential equipment. Flock sizes in this country do not always justify a heavy capital outlay in fixed equipment for handling sheep, but the rise in cost of labour now makes it imperative that some well-planned lay-out for handling and sorting sheep is available on farms which have a sheep-raising policy.

CAUSES OF LOSS IN SHEEP

Many diseases of sheep cause actual death, and while others may not be fatal they can and do cause loss of condition; more than this, they bring about the need for more labour in the tending of sheep. Blow-fly attack and footrot are examples of ailments peculiar to sheep which can greatly add to the labour of shepherding, and in the case of the former, failure to give effective treatment may ultimately result in death. The wide-scale use of the newer insecticidal compounds in dip preparations and as sprays makes the control of sheep maggot flies much easier than in the past.

Lameness through footrot is a very common affliction of sheep, especially in areas of low elevation and high rainfall. Research in Australia has shown that footrot is caused by a bacterium and vaccine has been developed to help its control; the disease is highly contagious but does not persist for any length of time in the soil. Certain animals which have a chronic degree of infection act as carriers and these should be eliminated from the flock. Footrot can be cured by paring the diseased tissues away from the hoof and passing sheep through a footbath containing some antiseptic preparation such as formalin or copper sulphate in solution.

The external parasites of sheep, in addition to the blow-flies, are lice, ticks, keds and the sheep scab mite. All these are easily and effectively controlled by sound dipping practice.

Sheep are subject to a wide variety of internal parasites. The liver fluke, stomach and intestinal worms, and husk and lung worms annually cause a heavy loss through unthriftiness and actual deaths. The liver fluke can be avoided by keeping sheep away from pastures in which there is stagnant water harbouring the small grey water snail, and dosage with an appropriate proprietary anthelminthic is effective in treating fluky sheep. The various types of roundworm causing parasitic gastritis and parasitic bronchitis become prevalent whenever sheep are grazed thickly and frequently without change of pasturage. Systematic management of grazing, the use of other classes of grazing stock, especially cattle, will do much to prevent serious worm infestation. Routine drenching with phenothiazine 'or the newer vermicides can do much to reduce losses from parasitic gastritis ; none the less, it should be remembered that the over-concentration of sheep on grazing will always result in re-infestation.

Sudden deaths in fast-thriving adult sheep and in young lambs can often be ascribed to bacterial activity in the bowel resulting in rapid toxin formation, causing death from entero-toxæmia in older sheep and from pulpy kidney in lambs. Such losses can now be prevented by vaccine treatment.

Many young lambs die each year from exposure and disease. Navel ill, tetanus and lamb dysentery are the more serious diseases ; there is now an effective serum treatment for lamb dysentery so that this disease is no longer a serious menace.

In breeding ewes losses may occur from pregnancy toxæmia (twin lamb disease), mastitis, milk fever and hypomagnesaemia. Sometimes septic metritis after lambing proves fatal. Correct nutrition and good hygiene in the lambing season will do much to counter these causes of loss.

Other sheep diseases not infrequently met with in some parts of Britain are braxy, scrapie, louping ill, and tick borne fever. But for these, serum or vaccine treatment can be used with considerable success. Further information about diseases of sheep can be found in Chapter XXIII.

It is necessary to note before leaving the subject of sheep diseases that mineral deficiencies in the herbage are all too common in some areas of high rainfall, especially on land of high elevation or on specific soil types. Undoubtedly these deficiencies predispose sheep to disease and ailments of many kinds. Research work has shown that the feeding of mineral mixtures containing copper salts has been effective in reducing losses from " swayback " in lambs in north Derbyshire. It is also well established that deficiencies of other elements such as copper, iron, manganese, cobalt and zinc are, in certain districts, the cause of ill-health in sheep—particularly in reducing their resistance to parasitic infestation. Much of our grazing land is very deficient in lime and phosphates ; in fact, these may be described as major deficiencies. Research work in the Border counties and in Scotland has now convincingly proved that the use of complete mineral mixtures has a marked influence on ovine health and vitality. No doubt our knowledge of this subject will be much advanced during the next decade ; a great deal of important work still remains to be done.

SYSTEMS OF SHEEP FARMING

The wide range of environmental conditions under which sheep can be kept in this country is reflected in the great variety of British breeds and crossbreds of commercial value. It is always possible to select a pure or crossbred type of sheep that will adapt itself to any farming system and to land of any level of fertility.

Farms on which sheep form either the whole or a part of the livestock enterprise can be classified as follows :

1. Mountain and moorland areas of high elevation.
2. Hill sheep farms.
3. Upland farms mainly pasture.
4. Upland farms mainly arable.
5. Lowland mixed farms.
6. Lowland farms mainly pasture.

1. Mountain and Moorland Farms. In Wales, northern England and Scotland there is much land over 450 m in elevation which is used to maintain breeds such as the Welsh Mountain, the Herdwick, the Swaledale and the Scotch Blackface. Individual sheep farms are often very extensive in area and boundary fences may be non-existent ; when a farm has a change of occupier the flock remains because it is important to maintain stock acclimatised to the area. Formerly some of the sheep stock on this high land consisted of two-, three- and four-year-old wethers which yielded an annual clip of wool to pay for their maintenance ; but now the sheep stock consists mainly of breeding ewes, with their followers during the summer months. After three breeding seasons the ewes are drafted and sold, passing to farms of lower elevation for a further year or two of breeding. The climatic conditions and sparse grazing on this high land make it essential that all lambs are brought away from the mountain land during autumn, the wether lambs being sold for feeding in the low-lands whilst ewe lambs which will be needed for flock replenishment have to be wintered away on better land, returning in the following spring.

The lamb crop on these farms is much influenced by weather conditions. Severe snowstorms in the winter may cause heavy losses in ewes and, although lambing does not take place until April and May, many lambs may be lost through exposure. In a good season the lamb crop per 100 ewes seldom exceeds 80 per cent ; in a bad year it may be as low as 40 per cent.

2. Hill Sheep Farms may be defined as those at an elevation of 270–365 m. Breeds such as the Cheviot, the Exmoor and the Kerry Hill may be regarded as typical for this class of land. Owing to a more favourable environment the yield of lambs on farms in this class is higher. A greater proportion of the grazing may be enclosed, affording the opportunity for better control of grazing, whilst supplementary food for winter-spring use can be more easily obtained and fed.

3. Upland Pasture Farms. On upland farms which are mainly pasture, sheep of the hill breed class have been extensively used. Farms in this category only differ from the hill farms in that all the area is enclosed and the land is at an elevation below 270 m which permits some arable cultivation ; this means that a small root break can be provided each year to produce roots for feeding in the early spring when pasture growth is scanty. During the decade prior to 1939 much land on the drier uplands of the oolitic and chalk geological formations was devoted to this type of sheep farming, pure and crossbred sheep of great variety being used, such as Cheviots, Exmoor, Kerry Hills, Ryelands and numerous crossbreds. The rate of stocking with sheep on such farms, if they maintained few or no cattle, was too high to ensure the maintenance of a satisfactory standard of health among sheep after two or three years. The drier uplands do not maintain a good type of pasture sward if grazed mainly by sheep, for herbage quality deteriorates owing to excessively hard grazing in the spring months and excessively close grazing during periods of summer drought inevitably leads to serious stomach worm infestation.

4. **Upland Arable Farms.** No system of sheep farming has shown a greater decline during the past half-century than the arable system, which was practised very extensively on the upland, mainly arable, farms on the chalk and limestone formations of the South and South Midlands and in East Anglia and on the Yorkshire Wolds. The system involves the intensive management of sheep with extensive arable farming. Throughout the whole year flocks lived on a succession of crops grown for them on the tillage area, each day a fresh fold or pen being set up with hurdles to allow the flock access only to an amount of crop which would be consumed daily. During winter, large amounts of hay and trough food were used to supplement the fold crop, and by this means a very high rate of sheep stocking per ha was maintained without risk of disease. On light and friable arable soils it was found that the treading and manuring by sheep resulted in a satisfactory level of yield from the subsequent cereal crops. When this arable system was at its zenith artificial fertilisers were not widely used but oilcakes were cheap and abundant—they were often wastefully used because of their manurial value to succeeding crops.

In their tolerance of close folding on arable crops our Down breeds and the Dorset Horn are quite unique. Although there has been such a marked decline in arable sheep farming for a variety of reasons—mainly economic—it is still carried out in several counties, particularly in pedigree flocks which specialise in ram breeding. The arable farm with its succession of tillage crops is able to cater for a ewe flock lambing in January to produce fat lambs for the Easter trade, and well-grown ram lambs for the summer and early autumn trade. A typical feeding programme, from January onwards, would be as follows : swedes ; swedes and kale ; winter barley or rye ; short-term ley ; aftermath ; stubbles and, in the autumn, back to forage crops.

The arable system with all-the-year-round folding is costly in labour, cultivations and equipment.

5. **Lowland Mixed Farms.** On lowland mixed farms sheep can be kept with profit if well managed, provided that they do not encroach on land suitable for dairying. Soils of the lighter type are desirable because during the winter months the flock can be penned on arable crops, sugar beet tops, turnips, swedes and kale. From the time when growth is abundant on leys and pastures, i.e. early May until the late autumn, no folding is necessary if all fields are fenced and watered and the flock can be grazed. This mixed-farm system of sheep management has much to commend it from the standpoint of disease control, since leys and pastures are rested from sheep grazing during several months of the year so that stomach and lung-worm infestation is to a large extent avoided.

6. **Lowland Grass Farms.** Except in localities unsuited for bullock feeding or dairying, sheep are not extensively kept on lowland grass farms. The reason for this is not hard to find—sheep are not so well suited to lowland grazing as are cattle. They are easily subject to footrot when kept on heavy damp soils, and in badly drained areas they may acquire liver fluke. In the whole of Britain there is only one large area of low elevation land heavily stocked with sheep—the Romney Marsh.

For lowland grazing on rich land the large long-woolled breeds or suitable crossbreeds are preferable to other types. It is one of the essential points in sound sheep management that the type of sheep kept should be correctly related to the productivity of the land. For mixed farms of average fertility the Half-bred (Border Leicester × Cheviot) or its crosses by a Down-breed ram will always maintain a high standard of lamb output. To put sheep of a large breed on poor land is a bad mistake, and it is equally wrong to put sheep of small size, accustomed to hard living, on to grazing of high productivity.

FLOCK SYSTEMS

Many farms maintain a permanent breeding flock of ewes in regular ages. This means that each year the oldest age group of ewes is drafted out and flock numbers are made up by introducing a sufficient number of young ewes to make up the flock number, these young ewes being either home bred or purchased. The draft ewes sold will, if sound, go to other farms for further breeding for one or more seasons. It is usual to cull out draft ewes from permanent breeding flocks after they have bred three or four crops of lambs. This practice is sound because the capital value of the flock is maintained so long as numbers are constant.

Another system is the maintenance of a flying flock, i.e. one which is bought in and then sold out, spending less than one year on the farm. Three kinds of flying flock can be kept : draft ewes for breeding ; store lambs for fattening ; or ewe lambs bought in to be grown on for subsequent sale as young breeding ewes. The flying breeding flock has often in the past been associated with dairy farms where the cattle are housed in winter, and the ewe flock is used to clean up the ungrazed growth on pastures in winter.

Although a buoyant market for lamb is assured the lowland and arable sections of the sheep industry are vulnerable to changes in EEC policy which could alter the relative profitability of the farm enterprises. However, on milk-producing farms and on arable farms producing cash crops such as sugar beet, a small flock of breeding ewes can add substantially to the gross farm revenue, without necessitating a heavy capital investment in specialised buildings.

SHEEP FEEDING

No other class of livestock is so difficult to feed according to exact feeding standards. It is possible, however, to state approximate quantities of food consumed by sheep of medium size. On grazing land five adult sheep can be considered equivalent to one bullock. With regard to hay, if no other food is given, an adult ewe of average size will consume 1·4–2·3 kg daily, but hay is generally fed as a supplement to other foods in the diet, allowing from 0·2–0·4 kg daily. When roots are fed the consumption will depend upon the amount of hay given and upon the trough food available; ewes of 63 kg live-weight will easily consume 6·3–9 kg of roots daily.

In the days when imported concentrates were plentiful and cheap they were extravagantly used in sheep feeding on some farms. Their liberal use can now only be justified in early, out-of-season, fat lamb production and to bring forward well-grown pedigree rams for the autumn sales. An allowance of over 0·5 kg of trough food per sheep per day must, under present-day conditions of price, be considered generous.

The problem of how to obtain an adequate supply of protein as a supplement to roots in winter feeding is now a serious one. It can be best solved by making hay of better quality, and by a greater extension of silage making to conserve for winter use the high protein content herbage from our pastures and leys, which is usually in plentiful supply during May and early June.

Information on feeding problems is given in Chapter XVIII.

THE NOMENCLATURE OF SHEEP

There is much variation in the terms by which sheep are described in different localities. The following table is a summary of the terms most commonly used :

MALES.		FEMALES.
Entire.	Castrated.	
Ram lambs	Wether lambs	Ewe ambs, Gimmer lambs, Chilvers
Hogg rams	Wether hoggs	Ewe hoggs
Ram hoggetts	Wether hoggetts	Ewe hoggetts
Teg rams	Wether tegs	Ewe tegs
Diamond tups		
Shearling rams	Shearling wethers	Shearling ewes
Shearling tups		Gimmers
		2-tooth ewes
Two-shear rams		One-crop ewes
Two-shear tups		4-tooth ewes
		Two-crop ewes
		6-tooth ewes
		Three-crop ewes
		Full-mouthed ewes

PLATE L
A. Large White boar
B. Large White gilt

PLATE LI

A. Camborough hybrid

B. Wall's cross-bred sow

PLATE LII

A. Welsh boar
B. British Saddleback sow

PLATE LIII

A. Landrace boar
B. Landrace sow

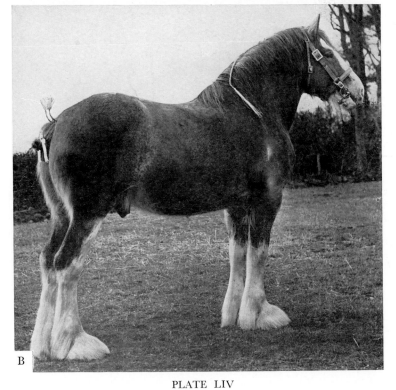

PLATE LIV

A. Shire mare
B. Clydesdale stallion

PLATE LV
A. Suffolk mare
B. Percheron stallion

PLATE LVI
A. Californian houses for 5,000 birds
B. Windowless broiler houses

PLATE LVII
A. Feeding caged layers
B. Breeding stock on deep litter

The teeth referred to are the permanent incisors of the lower jaw which replace the temporary incisors at approximately the following ages :

Age.	Eruption of Lower Permanent Incisors.	Definition of Ewe.
12–18 months	central (2)	2-tooth
18–24 months	outer middle (2)	4-tooth
30–36 months	outer (2)	6-tooth
36–48 months	corner (2)	Full mouth

The value of breeding ewes is much influenced by age since that denotes capacity for further breeding if the udder is sound. The dentition standard is a fairly reliable one when a seller is not prepared to give a warranty as to the number of seasons a ewe has been lambed (see also p. 502).

SHEEP HOUSING

Sheep are well adapted to an outdoor life but on many farms they can be managed more easily indoors during a period which according to the weather may begin as early as Christmas and end as late as 31 March. Heavy land, rain, snow, and a large flock on a small area of pasture are the conditions in which housing can be most useful. In these circumstances grazing provides little sustenance, supplementary fodder is a necessity and transport vehicles in addition to continual treading can do structural damage to the soil from which it is slow to recover.

Whether to provide winter housing is not entirely a husbandry question but also a question whether the improved output warrants the cost, bearing in mind that when sheep are indoors they are entirely dependent on the food which has to be carried to them. On the credit side pasture rested in winter produces earlier grass in the spring. On the debit side each ewe kept in will require about 35 kg more of hay or its equivalent than if she is on pasture, and will use about 35–50 kg of straw for bedding unless she is given a slatted floor, which is more expensive.

Large ewes require 1·2 m² each of floor space and double the area with lambs at foot. Very good ventilation is essential, but without floor draughts : this can be arranged by building around the pens to a height of 1·8 m with bales or corrugated iron, above which there can be either an opening or Yorkshire boarding to keep out driven snow. Large flocks are best split into lots of up to a hundred ewes and can be grouped according to probable date of lambing. Existing field hay racks can often be utilised to save the expense of purpose-made equipment.

Perhaps the chief advantage of winter housing is that the shepherd's work is made easier and he can therefore give more attention to a bigger flock.

z

Chapter XXI

PIGS

THE PIG INDUSTRY

IT is not generally realised how important pig production is in the United Kingdom. In 1976–7 the total value of pig meat " at the farm gate " was estimated at £596 million, 10 per cent of the total agricultural output. This compares with £1046 million for fat cattle and calves and £260 million for fat sheep and lambs. The volume of UK pigmeat production is, in thousand tonnes for April–March years :—

	Average of 1963–4 to 1965–6.	1971–2	1974–5	1976–7
Pork . .	592	631	666	588
Bacon . .	228	289	235	223
	820	920	901	811

In 1974–5 about 14·75 million pigs were slaughtered in the UK, and pigmeat represents about a third of the total meat consumed. The UK is virtually self-sufficient for pork and is about 47 per cent self-sufficient for bacon and ham, most of the remainder being imported from Denmark.

Increasingly, slaughterhouses and pig-processing factories have become manufacturing plants that divert the carcass to fulfil the currently most profitable market. Meat processing and marketing is becoming more highly specialised, with the number of slaughterhouses and butchers' shops declining fairly rapidly. Supermarkets and the prepacked meat trade are becoming increasingly important, and farmers are planning their production and marketing to meet the changing situation.

Over the years there has been some measure of Government control and support for the pig industry through the Fatstock Guarantee Scheme. This scheme ended in respect of pigs in July 1975 as the UK moved closer to the EEC system. A considerable proportion of pig carcasses are now classified by the MLC into the EEC grades, and the industry is moving into the position so that it may benefit from EEC Schemes.

Within the EEC, UK producers are part of a wider Community in which they have to compete equally with producers in other member states for the home market, but are able also to export freely to other member states. By agreement the Community veterinary rules are not yet applicable to the UK, which means that import of fresh pigmeat into the UK is still prohibited except from Denmark and the Irish Republic. This derogation is to be reviewed before the end of

1977. When the transitional period ends in December 1977, the full Community levies and restitutions will apply in UK trade with non-EEC countries.

The size of the UK pig industry has always been very sensitive to changes in both the cost of feed and the price of the product. Between 1973 and 1975 unfavourable feed/product price ratios led to a fall of some 25 per cent in the size of the industry. The structure of the industry has also changed markedly over the years and this can be illustrated by the breeding herd figures for England and Wales, collected annually by the Ministry of Agriculture.

Herd Size Range Sows and In-pig Gilts.	Percentage of Total Sows and In-pig Gilts.			Percentage of herds.
	1963	1968	1976	1976
1–19	53·1	34·1	13·3	65·8
20–49	25·8	28·2	17·8	17·1
50–99	12·2	20·1	23·5	10·1
Over 100	8·9	17·6	45·4	7·0
No. of sows and in-pig gilts (000's) .	702	714	745	
No. of holdings (000's) . . .	80	52	25	

During the thirteen-year period, which is a very short span in the lifetime of domestic agriculture, the proportion of total breeding pigs in herds of over 50 increased from 21 per cent to 69 per cent, and the number of pig breeders decreased from 80,000 to 25,000. The size of the " average " breeding herd ranges from 6 sows in Wales to 44 sows in the south-east of England.

Larger units can make better use of skilled labour, can justify more highly specialised buildings and can benefit from larger scale buying and selling as well as from implementing planned production schedules. Except, however, in the case of the most highly skilled labour there is probably little economy of scale above the 100 sow herd size. Large units, particularly on modest areas of farmland, can have serious effluent disposal problems. Notwithstanding this there is no reason why there should be a reduced return on capital, at least in the 100 to 500 breeding herd size range, so long as the unit is sensibly sited and manure handling and disposal taken into full account at the early planning stage.

Productivity has changed markedly in the pig industry, which has been quick to adopt new technology. An example can be taken from the results of the Cambridge University Pig Costings Scheme, which embraces about 140 farms in East Anglia.

Year.	Weaners per Sow per Year.	Kg of Feed (Sow + Litter) per 8-week-old Piglet.	Feed: Gain Ratio of finishing Pigs.
1962 . .	14	115	3·95
1966 . .	15·7	105	3·9
1969 . .	16·8	95	3·85
1976 . .	17·8	88	3·51

The greatest improvement has been in litters per sow per year, brought about by earlier weaning and better attention to management between weaning and service. This in turn, aided by better husbandry and housing, has reduced the total amount of feed used per weaner produced. Nationally the figures have not improved as far or as fast as in this example.

Breeds and Breeding. Pig production today is mainly based on the Large White and the Landrace, and on crosses between them. See Plates L–LIII. The Large White has slight advantage in litter size, efficiency of feed utilisation by growing pigs and in meat quality, whilst the Landrace scores on carcass grading. As with other species, however, there is great variation within the breeds and this can override the differences between them. For many characteristics the Welsh breed is comparable to the Large White and the Landrace and is used fairly widely. The British Saddleback is not widely used but still finds some favour either as a purebred or as a parent of crossbred sows destined to be kept under fairly rigorous conditions. Whilst renowned for its mothering qualities the breed lacks some merit in regard to feed : gain ratio and in carcass grading. The other British breeds are mainly kept on a very small scale and to fulfil intermittent export orders for breeding pigs. Of the more recently imported breeds the Hampshire has probably gained the most ground but fulfils a very minor role in the industry. The Pietrain failed generally on meat quality, and the small numbers of Durocs, Lacombes and Poland Chinas that have been imported have shown little potential when compared with Large Whites and Landrace. Maybe some of these breeds will ultimately find a place in hybrid breeding programmes.

The main improvement programme for pedigree pigs is the Pig Improvement Scheme organised by the Meat and Livestock Commission. Under this combined testing scheme two boars, one hog and one gilt all from the same litter (born in herds that are members of the Scheme) are consigned by the breeder to a MLC testing station (of which there are four) when about 9 weeks of age. The pigs are evaluated for efficiency in feed utilisation between about 30 kg and 90 kg liveweight. The hog and the gilt are slaughtered and the information regarding their efficiency of gain and carcass assessment supplements the information on feed : gain ratio and backfat thinness (measured ultrasonically) that is available for the two litter mate boars. Boars that achieve index scores that are better than just below average can be taken back by their breeders for sale or use as stock boars. The boars that are markedly below average are slaughtered. Herds that are full members of the Scheme are designated as " Nucleus herds ", and the other categories of herd in the Scheme are " Reserve Nucleus " and " Breeding Company ".

There is also a considerable amount of on-farm recording of weight for age and backfat thinness, and these observations are combined into an index and used particularly in the selection of gilts for retention in herds or for sale for breeding purposes.

Increasingly the industry has used crossbred sows, and at present about 70 per cent of sows used for commercial production are crossbred. The scientific justification for using crossbred sows is convincing :—

Average percentage superiority in performance of crossbred over purebred

	Crossbred litter.	Litter from crossbred sow.
Litter size at birth . .	2	5
Litter size at 3 weeks . .	4	7
Total litter weight at weaning	10	11

The quality of the parentage of crossbred sows remains very important, and since the early 1960's several Breeding Companies have developed that specialise in the production of maiden crossbred or " hybrid " gilts and purebred boars to mate with them. Most pig breeding companies are founded on groups of pedigree breeders and multipliers who co-operate with each other and with many of the breeders testing in the official MLC Scheme. Some breeding companies also operate their own boar performance testing units.

In an attempt to monitor the genetic quality of the boars and hybrid gilts offered for sale by breeding companies the MLC established a Commercial Product Evaluation station that takes in batches of boars and gilts from nine of the largest breeding companies, together with a group of " unimproved " breeding pigs for comparison. The trial runs over two litters per breeding company female and the evaluation is on the basis of litter size and feed utilisation by the sow and feed utilisation and carcass quality of her progeny. This must not be confused with a " which " type of test since the overall results are unlikely to be repeatable due to changes in breeding programmes in the companies over the time scale involved together with the difficulty in assembling a representative sample of animals for testing.

Selection of Breeding Animals. Some characters, for example litter size, are of low heritability (i.e. are not strongly inherited), others such as speed and economy of gain are of moderate heritability, whilst carcass length and quality are of higher heritability. Breeders bear these points in mind when making purchases. For factors of low heritability only slow, if any, improvement will be made by direct selection, although they can often be influenced by cross breeding or by better management. Selection pressure is likely to be applied most heavily for characters that have a high direct cash value and a reasonable level of heritability. Whilst it can be a wasted effort to select for characters of low heritability, the cash value of the possible improvement should be considered and the quick generation turnover of breeding pigs borne in mind. Thus, for example, improvement in litter size might be possible over several generations.

Purchases should be made from herds with sound health records, and preferably all purchases should be from the same source. The herd or company from which the pigs are bought should have a successful testing record and a high average performance in the required traits. The individuals selected should have good performance records for gain and leanness in relation to their contemporaries. This is particularly important for boars since they will make a genetic contribution equal to that of all the females, who perhaps outnumber them by 20 or 25 : 1. The boars should have been individually performance tested and, if from the MLC Scheme, should, at least for large herds where the cost can be easily recouped, have scored at least 145 points and so be in the top 10 per cent of the population being tested. It is important that pigs are tested and selected to meet the criteria of the market outlet.

Before purchase, breeding pigs should be inspected individually for absence of abnormalities, sound legs and feet and a good underline with preferably fourteen (minimum twelve) functional teats. If it is envisaged that the herd will continue to produce its own replacements, then care in the selection of foundation stock is particularly necessary. Whilst most of the breeding females used for commercial production are crossbred, in order to realise the benefits of heterosis, most of the boars used are still purebred. This is because real advantages in the use of crossbred boars have not been widely demonstrated and a tested purebred boar from nucleus stock can carry the maximum level of improvement.

The Commercial Breeding Herd. The objective of a breeding herd is generally to produce annually the maximum number of good quality weaners per sow at as low a cost per weaner, or per pound of weaner, as possible. Most costing schemes still interpret " weaner " as an 8-week-old piglet, irrespective of weaning age. The objective will be achieved by a combination of good stockmanship, correct feeding and efficient housing backed by a sound health and breeding policy.

The decision on weaning age is most important and will be governed to a degree by the housing available. Most herds now wean piglets at 5–6 weeks old but the successful trend is to wean earlier, and many are weaning between 18 and 24 days old. The figures in the table below illustrate that earlier weaning should produce slightly more weaners per sow per year. These figures are on a " model " basis, and good stockmanship, housing and luck will be needed for them to be exceeded under farm conditions.

Effect of weaning age on sow output

					10–12	18–24	32–38	56
Weaning age (days) .	.	.	.	.	10–12	18–24	32–38	56
Gestation (days)	.	.	.	.	115	115	115	115
Lactation (days)	.	.	.	.	11	21	35	56
Weaning to conception (days)	.	.	.	12	10	8	7
Farrowing interval (days) .	.	.	.	138	146	158	178
Farrowing Index (litters/year)	.	.	.	2·64	2·50	2·31	2·05
Pigs reared per litter.	.	.	.	.	9·0	9·3	9·5	9·5
Pigs reared per sow per year	.	.	.	23·8	23·3	22·0	19·5

Given suitable housing for the weaners then weaning at about 3 weeks of age has the additional advantage in that in general it is more efficient to feed the piglet direct than through the sow ; also housing and management problems can to a degree be simplified once weaning has occurred. The recent development of flat deck type pens has also enabled practical farmers to be aware of the efficient use that a young piglet can make of feed. The problems of weaning at below 2 weeks of age are not yet sufficiently resolved for this to be a practical consideration under most farm conditions.

Breeding gilts. Gilts generally first experience oestrus at 5–6 months of age and are normally served at about 110 kg liveweight when 7–8 months old. Increasingly they are being served when lighter and younger. Experimental work on this is not concluded, but some herds are mating gilts at about 100 kg, perhaps at second oestrus. Experience may show that this can be further reduced and selective breeding for early onset of oestrus may have a part to play. Feeding levels generally increase for gilts up to about 2·3 kg a day (depending on the energy value of the diet) and may be increased by a further 0·5 to 0·75 kg a day for the last 3 weeks before service in the hope of achieving a flushing effect. After service the feed level is generally reduced to about 2–2·5 kg a day depending on the feed analysis, housing conditions and the condition and growth of the gilts.

The Pregnant Sow and Gilt. A sow will increase her weight with succeeding pregnancies, quite apart from the products of conception and she will use feed more efficiently when pregnant than she would were she empty. Present indications are that an average liveweight gain of about 16 kg at each successive pregnancy up to the fourth or fifth is a satisfactory target for each sow. Her daily feed allowance should be adjusted with this in mind. Under good conditions of housing and individual feeding about 2 kg of feed will be adequate but careful watch should be kept on the body conditions and weight of the sows. Where individual feeders are not available it is necessary to allow about an extra 0·5 kg feed daily per sow to allow for bullied sows and slow feeders. Thus the provision of individual feeders can be very cost effective.

In large-scale pig units dry sows are increasingly housed in tie stalls or sow stalls. When well designed and maintained these can provide good environmental conditions for both sows and workers and can enable best use to be made of feed, since sows housed badly in cold surroundings will need additional feed to maintain body temperature and condition.

Farrowing. A sow is usually brought to her farrowing place a few days before she is due to farrow and if necessary is washed at the time of transfer. She should be in a good body condition and at this stage should be fed a light, laxative diet, which can be achieved by replacing half the diet with bran and feeding it wet. Overfeeding can increase udder and milk flow problems after farrowing. The level of feeding after farrowing will depend on the body condition of the sow, the proposed age at weaning for the litter, the size of the litter and the level of feeding during pregnancy.

Where dry sows have had a modest ration of perhaps 2·25 kg per day and farrow in average condition a useful basis is to feed sows with litters of eight at a rate of about 5·5 kg daily if they weigh about 135 kg or an additional 0·5 kg if they weigh 180 kg. For each piglet more or less increase or decrease by about 0·5 kg per piglet.

Where weaning is to be at 3 weeks of age suckling sows will not normally receive more than 4 kg of feed daily unless in poor condition at farrowing.

Overall, the level of feeding during both pregnancy and lactation is a matter of judgement and stockmanship. The best indicator of a good feeding régime is probably the number of good piglets reared per sow per year and a net gain in weight by the sow of about 16 kg in each reproductive cycle.

Piglets should be encouraged to eat solid feed as soon as possible in order to achieve good weight gains and to minimise post weaning checks. Proprietary creep feeds are usually used in preference to home mixing, for reasons of convenience and of acceptability to the piglet. Creep feed is offered fresh to piglets, in small quantities, from 10 days of age and the piglets should have good access to clean water. Freshness of the creep feed on offer is of paramount importance.

Sow's milk is deficient in iron and this may lead to anaemia in piglets. Mostly the iron supplement is now given by injection to piglets in their first few days of life or it can be provided in the form of paste for placement in the mouths of the piglets. In order to minimise tail biting later in life, the tips of tails are usually docked at the same time.

Housing the Sow and Litter. Something approaching 20 per cent of piglets die before weaning, mostly in the first 2 or 3 days of life. Considerable effort should be made to reduce such losses, since profit from the breeding herd is so closely related to piglets weaned per sow per year. Farrowing quarters should be warm, dry and free from draughts. A well-designed farrowing crate is recommended with a warm (often covered) creep area for the piglets. In larger herds farrowing crates can be in rooms managed on the batch principle, with the number of crates balanced to match the number of sows programmed to farrow at the planned intervals. Such a procedure gives good opportunity for cleansing, disinfection and drying out before the next batch. Given good housing and stockmanship piglet losses from birth to weaning should be contained at about the 10 per cent level.

One of the advantages of 3 week (17–24 days) weaning is that sows and litters can remain together in the farrowing crate until weaning, thus simplifying management. Even so, it is helpful for the pen floor to be partially slatted. In the case of units weaning at 5–6 weeks it is usual for the sow and her litter to be moved to cheaper accommodation, perhaps grouped in multiple occupation pens, when the piglets are 2–3 weeks old, and this becomes the weaner pool after weaning, the piglets remaining there up to about 35 kg liveweight.

The Newly Weaned Sow and the Boar. After weaning the sow must be quickly and happily got in pig again. At weaning sows should go to a specific service area or building, within smell and sound of boars and be comfortably housed and fed up to about 3·5 kg per day from weaning to service. Oestrus occurs some 5 days after weaning; sows are usually served about 24 hours after the onset of heat and again 12–18 hours later.

To ensure detection of any sows that return to service they should remain in the service area for a month after service. Gestation length is about 115 days.

Boar housing should be comfortable and well bedded, with facility for some exercise. A herd will usually carry at least 2 boars and large herds will have about one boar per 20–25 sows. As with sows, feed will be adjusted according to work and conditions, a rough guide being 3 kg of feed per day.

It is important that record cards are kept for sows showing litter sizes, service dates, service sires, etc. and a good simple system of identification maintained.

Artificial insemination is a most useful means of bringing safely the "blood" of some of the best boars available into a herd. It still needs further development, however, to become a widespread commercial proposition.

Growing and Finishing Pigs. The objective is to meet market requirements at minimum cost. The efficiency of the whole process is the overriding consideration, and the production factors need to be manipulated to meet changes in costs and in market requirements. In general, the relationship between the cost of feed per kg of gain and the value of the slaughter pig produced govern the profitability of the finishing enterprise. The main factors influenced by management are :—

Stockmanship and husbandry
Genetic quality and health status of the pigs
Feeding policy and practice
Pig environment and housing
Marketing policy

Feeding. The cost of feed represents about 75 per cent of the cost of pig production and it is therefore important that the feed is balanced correctly for nutrient content and fed accordingly. The background information on feeds and feeding is contained in Chapter XVIII, and the reader's attention is also drawn to the MAFF Advisory leaflet No. 104 (free).

Feed intake of finishing pigs above about 35 kg liveweight needs to be restricted, the level of restriction depending on the genetic make-up of the pig, the penalty against backfat imposed by the market outlet, the temperature in the piggery and the nutrient density of the diet. A rough guide would be a restriction of between 15 and 25 per cent of appetite. On many farms this is taken as increasing the feed given from about 1·4 kg daily at 35 kg to about 2·5 kg daily at 75 kg, and trial and error can indicate the best level for any production/marketing system.

Examples of currently recommended feed formulations are :—

	Suckling Sows.	Dry Sows.	Weaners above 16 kg.	Finishing Pigs above 36 kg.
Digestible Energy TDN	70	70	68–71	68–71
Crude Protein % . .	14	12	16·5–17·5	15–16
Lysine % . . .	0·6–0·7	0·6–0·7	0·75–0·85	0·6–0·7
Cystine + Methionine % . .	0·5	0·5	0·55	0·5

Plus minerals + vitamins at levels recommended by manufacturers.

The protein levels for the sows are below the levels adopted in the past and, in the case of the dry sows, it is assumed normal levels of feeding of not less than about 2 kg a day.

Swill and waste foods are fairly widely used on the fringes of large urban areas. Attention is drawn to the provisions of the Diseases of Animals (Waste Food) Order 1973.

The method of feeding finishing pigs will influence or depend on the design of the building. Where possible, a design suitable for the provision of troughs and wet feeding has much to commend it. This allows maximum flexibility to meet changing situations plus the possibility of best utilisation of feed by pigs weighing over 30 kg. Most experiments show that wet fed pigs give best feed : gain ratios, followed fairly closely by pellet fed pigs and with pigs fed dry meal on the floor the least efficient. Thus if floor feeding is adopted cubes should be used ; if it is meal (perhaps home mixed) it is best fed in the trough, wet.

Stockmanship and Husbandry. This represents the totality of the daily care of the pigs including the control and management of housing, feeding and attention to the detail of pig well-being and comfort. The difference between the profitability of pig units can often be attributed more to stockmanship than to any other single factor. Housing is a complex requirement that needs effective management. It involves consideration of living space, air temperature, ventilation and other factors and their impact on the pigs' behaviour pattern and feed utilisation as well as considerations of manure handling and worker satisfaction.

A good stockman must have time to stop and observe the animals and to check and reset carefully sited maximum and minimum thermometers. He needs to observe critically, draw sound conclusions and act quickly upon them, often by changing management practice to improve pig comfort or behaviour or to check or adjust ventilators or fan speed regulators according to the system adopted. It is also necessary to ensure that production is planned so that full use is made of the buildings available.

In connection with the maintenance of air temperature, the following is a guide to the pigs' requirements :

	°C
Sow in stall house	13–18
Farrowing house	13–18
Draught-free piglet creep	21–27
Weaners introduced to cages or flatdeck housing	21–24
Porkers	15–21
Baconers	13–18
Heavy hogs	10–15

The general air temperature within the piggery is often lower than that of the air closer to the pig, and care must be taken in the siting of thermometers. Also a deep bed of straw is worth at least 5° C for dry sows and finishing pigs. Excessive huddling and hairiness indicates that the pig finds the temperature cold. In practice the figures suggested above, particularly for finishing pigs, are likely to be exceeded on warm days. In some piggeries, particularly where floor feeding is practised this may lead to soiling of the floors if ventilation and air movement around the pigs is inadequate.

Cold pigs utilise feed inefficiently and even a moderately cold piggery can, over the finishing period, worsen the feed : gain ratio by 0·2, costing about £1 pig over the finishing period.

Increasingly the practice of castration is likely to be abandoned as prejudices are overcome, at least for pigs sold up to 75 kg liveweight. Boars are better feed converters than hogs and their carcasses carry much less fat.

In any pig unit regular consultation with the farm veterinarian is important in order to plan and maintain the health programme.

Chapter XXII

OTHER LIVESTOCK

HEAVY HORSES

The immense impetus given by the second world war to mechanical and scientific development in all aspects of human life has had, as can be seen in earlier chapters, its effects on the rural scene. The countryside is not what it was and many of the familiar sights and sounds have gone for ever. No longer can we see the teams of horses, steam rising from their backs in the wintry sunshine, stolidly pacing along the plough furrow and hear them being encouraged or admonished by the ploughman.

Sad it is, indeed, to have to relegate the farm horse from its previous position at the start of the chapters on livestock to a section of a chapter on " Other Livestock ". But facts have to be faced as technology moves on, and the hard fact is that the horse as a working animal has virtually disappeared from our farms. By 1965 the number of working horses in the United Kingdom had fallen to about 25,000 compared to a 1939 figure of about 750,000. In the last ten years working horses have ceased to be recorded in the annual Agricultural Returns. This decline is, of course, mainly due to the development of the tractor as a source of power but it has also been influenced by the reluctance of younger workers on specialised arable farms to undertake the week-long task of caring for horses. Whatever the reasons, the horse nowadays is used on only a few farms for the odd carting jobs.

There has, however, been a slight revival during the early 1970's in the use of heavy horses for purposes other than farming. For delivering heavy loads within a short radius of the source of the load and where frequent stops are necessary, as with the deliveries of brewers' products, a team of horses can be more economical and a great deal more eyecatching than an internal combustion engine.

A small export trade also helps to encourage heavy horse breeding on a limited scale.

There is no point in taking up space here on the detailed consideration of horse breeding, feeding and management. The reader who wishes to study these subjects can obtain all the information he needs and, possibly, introductions to some of the remaining breeders, from the Secretaries of the four Breed Societies concerned with heavy horses. They are :—

> Shire Horse Society, East of England Showground, Alwalton, Peterborough.
> Clydesdale Horse Society, 24 Beresford Terrace, Ayr KA7 2EL.
> Suffolk Horse Society, 6 Church Street, Woodbridge, Suffolk.
> Percheron Horse Society, Owen Webb House, 1 Gresham Road, Cambridge CB1 2ER.

LIGHT HORSES

While the heavy horse has gone into decline, there has been an upsurge in the number of riding horses, as people from both town and country seek a new pastime, as far removed as possible from the pace and pressures of 20th-century living, to fill their leisure hours.

At the beginning of 1976 it was estimated that about 2 million people rode regularly; approximately half of these are adults, and the growth rate of the sport is about 10 per cent per year.

The televising of show-jumping has played a considerable part in the rapid development of riding since it is essentially a spectator sport with clear-cut rules. Leading competitors have become household names, and dramatic jump-offs from the Horse of the Year Show at Wembley hold huge audiences captive and encourage many who have never ridden before to try their skill. By early 1976, the British Show Jumping Association had a membership exceeding 14,000.

Royal patronage of Horse Trials has undoubtedly contributed greatly to their continued increase in popularity, with thousands flocking annually to the two great three-day events, Badminton and Burghley.

This rapid development of riding for pleasure has created an increased demand for all breeds of horses, from heavyweight hunters, which are, as ever, scarce, to diminutive Shetland ponies. The part-thoroughbred hunter is extremely popular as it is most suitable for hacking and such forms of competition as the various Riding Club activities and Long Distance Rides.

Agricultural land is vital to the pastime of riding. Although the British Horse Society succeeded, in the High Court in 1975, in keeping open 3,200 km of bridleways, there are still not sufficient bridleways, and without the goodwill of farmers much more riding would take place on the often dangerous roads than is now the case. Also grass is essential for the breeding and rearing of horses, brood mares and their foals spend much of their first three formative years out in the fields. All horses benefit by being turned out for a rest—" Doctor Green " : many horses and, more especially, ponies are worked off grass, as can be seen by the equines of all shapes and sizes in the suburban fields and on the rural fringes of towns.

All these horses are dependent, in addition to their grazing, on the agricultural industry. The staple foods, of oats, bran and hay, as well as straw for bedding, form a valuable outlet for farm products, and those of top quality realise very high prices from racehorse trainers and breeding establishments.

Riding has, in fact, become a thriving industry which goes a long way to compensate for the disappearance of the heavy horse from the rural scene.

The British Horse Society, which has its headquarters at Stoneleigh, near Kenilworth, on the Royal Showground, in conjunction with the British Show Jumping Association, is run with the aim of promoting the interests of Horse and Pony Breeding, to further the art of riding and to encourage horsemastership and the welfare of horses and ponies.

A continually expanding organisation, it is responsible for all competitive and leisure riding with the exception of racing and show-jumping. Separate departments deal with the various facets, such as Combined Training, Combined Driving, The Pony Club, Riding Clubs, the Inspection of Riding Schools, Examinations and Training. A regional structure carries the work of the Society throughout the country. All details are available from The Secretary, National Equestrian Centre, Stoneleigh, Kenilworth, Warwickshire CV8 2LR.

POULTRY

Poultry, like horses, have left the farm but not for the same reason. The difference is that, as the horse has declined in numbers, poultry flocks have greatly increased in size and have developed into an independent industry, having little direct contact with the land. Practically all poultry are now housed in some form of intensive system and there is a sharp commercial distinction between fowls used for egg production and those destined for the table as meat. Here and there a few fowls can be seen scratching around farmyards and there is a limited trade in " free-range eggs ", but these are not a commercially significant part of the market for eggs. They do, however, provide a welcome reminder of the days when every farm had its flock of poultry, providing colour and characteristic noise in the countryside. There are a few small flocks from which freshly killed birds can be sold to discriminating members of the public, but even this form of sale is threatened by Common Market hygiene regulations.

Breeding and management of laying hens is a highly specialised business : the same applies to table poultry whether in the form of " broilers ", which have become an important source of animal protein for human consumption, ducklings or turkeys. There is also a small demand for fat geese : by contrast these can be kept more profitably on free range than under intensive conditions.

Sources of information about poultry keeping are :—

British Poultry Breeders and Hatcheries Association Ltd., 52–54 High Holborn, London WC1V 6SX.
British Poultry Meat Association, address above.
British Turkey Federation Ltd., The Bury, Church Street, Chesham, Bucks.

GOATKEEPING AND MILK PRODUCTION

Goatkeeping in this country has been encouraged by the British Goat Society which will celebrate its Centenary Year in 1979. During that time the standard of stock has increased with milk production going from a few litres soon after kidding to long lactation stock running through for two years and giving over 3,000 litres of milk in an extended lactation. A milk recording scheme is operated in conjunction with the Milk Marketing Board with many annual yields between 900 and 2,250 litres being recorded regularly.

In the world generally goats are reared for meat, but in this country the emphasis has been on milk production, and at present, in addition to supplies of fresh milk, large sales of cheese and yogurt are maintained from the farm gate, from health food stores and from supermarkets.

Information on goatkeeping and the various uses of milk can be obtained from the Secretary of the British Goat Society, at Rougham, Bury St. Edmunds, Suffolk IP30 9LJ.

MEAT RABBIT PRODUCTION

Introduced into Britain by the Normans and widely distributed in the countryside, the decimation of the wild rabbit population in the mid-1950's provided the incentive for commercial intensive rabbit production. United Kingdom consumption of rabbit meat slumped by 95 per cent after Myxomatosis but has since increased by 300 per cent, reaching 15,000 tonnes per annum by 1975.

Rabbit production in Britain is based on two main breeds, the New Zealand White and the Californian. Developed in America specifically for meat, the former produces meat rabbits at 2–2½ kg liveweight, the latter produces progeny weighing 1·8–2 kg, both breeds reaching meat weight in 9–11 weeks.

Rabbits are housed intensively in " all wire " cages and are fed pelleted rations containing 16–18 per cent protein and 12–14 per cent fibre. Rabbit fur production is uneconomic although the pelt is a valued by-product of meat production. Growing international interest in the meat production potential of the rabbit is stimulating much needed research into its requirements.

Further information is available from the Commercial Rabbit Association, Tyning House, Shurdington, Cheltenham, Glos., GL51 5XF.

GAME BIRDS

There are two ways of looking at shooting—as a leisure industry with an annual turnover of perhaps £100 million a year, if one includes everything from the sale of guns and cartridges to the income from letting grouse moors to overseas sportsmen. Or as a supplementary crop on farmland.

Good shooting country requires parcels of both arable and woodland in its make-up—the cover being interspersed throughout the area rather than in large blocks. Belts of trees, and woods shaped so that there is plenty of " edge ", will help provide the right conditions for wild pheasants. Vast hedgeless fields, monoculture, single species and even-aged plantations will restrict the abundance and the variety of game. And game in this context means most other forms of wildlife, because skilled shoot management equally benefits other birds, plants and butterflies.

One can assess the game potential of an area—rather like awarding two or three stars to hotels—by considering certain basic factors : the rainfall, the altitude, the nature of the soil (light soils being generally more favourable) and the type of farming. Mixed farming, with crop rotations, a proportion of *winter* cereals and preferably some undersowing, obviously produces more game than " prairie farming ", stock husbandry or market gardening, where there is always a high degree of disturbance.

Creating the right habitat for game by maintaining a chessboard pattern of crops, wiring off areas of rough cover which would otherwise be grazed down, planting up field corners and so on, is the first principle of wildlife management. Equally important is the careful use of farm chemicals. Partridges—useful barometers of wildlife stability—cannot live without a sufficiency of insects during their first three weeks of life. Insecticides tend to kill both the useful predatory insects as well as aphids and other pest species and herbicides will remove the plants upon which the insects live. It is said that a partridge today has to walk three times as far as a partridge ten years ago in order to find the same quantity of insects !

Predator control is also a vital part of game management. The surplus we produce by good conservation methods is for us to shoot and not to feed all the enemies of game—furred and feathered—that will be attracted to an area of easy pickings.

Covert feeding and winter feeding will always help to hold the game on the farm and ensure that fewer will stray over the boundaries. Food patches can also be extremely attractive—a few drills of maize left unharvested being one of the most useful crops of all.

Pheasant rearing is generally problem-free and well understood, but shoot owners and keepers tend to be less good at acclimatising their birds in the release pens, so that the maximum number is later shown to the guns. All pens should be fox-proof, carefully sited, and consist of open spaces for sunning, low ground cover for shelter and higher shrubs and trees to help them learn to roost.

There is no reason why good shooting and profitable farming should not go together, given a little compromise.

(Technical booklets on game conservation are available at low cost from the Game Conservancy, Fordingbridge, Hampshire.)

FUR FARMING

There are approximately 95 mink farms in the United Kingdom with a further 6 in Eire. They range from the largest, 18,000 breeding females producing some 70,000 skins per annum, down to 50 females and 150–200 skins.

The depressed market conditions over the decade 1965–75 forced mink farms throughout the world into bankruptcy or early retirement, resulting in a reduction of the world crop from 26 million skins per annum to 16 million. The United Kingdom suffered in loss of farms but not in number of skins. Although only one-sixth of the 650 farms in the UK during the 1960's remain they are a very skilled and professional group, producing skins of as good a quality as any in the world.

Mink are carnivores and as such no successful proprietary feeding stuffs have ever been available. Farmers, sometimes co-operating, but more often independently, find their own sources of raw materials which include fish offal, poultry offal, ox and sheep tripe and liver.

An ideal diet fed in the UK would be 30 per cent fish offal, 30 per cent poultry offal, 30 per cent meat offal and 10 per cent fortified cereal.

Labour varies from farm to farm, but a good stockman will look after 600–800 breeding females and their progeny. Costs, as in all farming, increase daily and today it costs between £7·00 and £8·00 to produce a pelt, the return on present (1976) market conditions being approximately £9·00–£10·50 dependent on quality.

Information about breeding and feeding of mink and fox can be obtained from the Secretary, Fur Breeders' Association, Beaver Hall, Garlick Hill, London EC4V 2AJ.

Chapter XXIII

ANIMAL HYGIENE

FARM animals, like all living creatures, are subject to ill health of various kinds. To a large extent the increasing intensity of livestock production has brought its own problems in this respect ; increased production per animal or per unit of land subjects the animal to stresses which may make it more susceptible to disease. With the current emphasis on efficient livestock production, the financial loss occasioned by ill health cannot be dismissed lightly and attention must be paid to increasing the health standards of farm animals.

The annual loss to the British farmer caused by disease is extremely difficult to estimate. Losses due to the death of animals are readily apparent but, more often, the animal survives but its production falls below full potential. Sub-clinical disease may cause losses of production in animals which appear perfectly healthy. These losses due to reduced production far outweigh those due to fatal disease and are virtually impossible to estimate in financial terms.

Many farmers tend to accept ill health in stock as a normal hazard of farming but it seems likely that much of the loss of production caused by diseases could be prevented by paying greater attention to their prevention and cure. Much research has been devoted to the study of ill health and an increasing knowledge of the causes and effects of ill health has contributed to the control of certain disease conditions. Also, many new drugs have been discovered in recent years which have greatly changed veterinary practice.

It is unreasonable to expect the farmer to keep abreast of these developments and to have the technical skill to administer drugs which, though they can effect remarkable cures, can be extremely dangerous if incorrectly used. However, the farmer can do much to assist in reducing the incidence of ill health by good management and good feeding of livestock with attention to hygienic measures. Further, when ill health does occur, the early recognition of an abnormal condition and recording of symptoms can be extremely useful to the veterinarian in making a rapid and accurate diagnosis and treating the condition.

Symptoms of Ill Health. Often the first sign of ill health is that the animal isolates itself from the herd or flock, but any departure from normal should act as a warning to the stockman. There may be changes in appearance, posture or movement, or abnormalities of respiration, pulse rate, temperature, appetite, rumination in cattle and sheep, defaecation, urination, the œstrous cycle or productivity.

When ill health is recognised the animal should if possible be isolated to facilitate observations and to prevent the spread of infectious disease. Records should be taken of respiration rate, pulse rate and body temperature (which is recorded by inserting a clinical thermometer into the rectum). If these records are to be of use the animal should not be frightened or excited during the period of examination. Normal values as given by Wooldridge are set out below :

	Respiration Rate (per minute)	Pulse Rate (per minute)	Body Temperature (Averages °C)
Horse	8–16	28–42	38
Foal (1st year) . . .	10–15	40–58	38
Cow	26–30	60–90	38
Calf (6 months) . . .	30	100	39
Sheep, goat	10–20	68–90	39
Pig	10–20	60–90	38·5
Dog	14–30	60–130	38
Cat	20–30	110–130	35·5
Fowl	15–48	120–160	41

It is evident that both respiration and pulse rates show a range of normal variation, but gross abnormalities may indicate disease conditions. For example, rapid panting may be caused by diseases of the lung such as pneumonia or may be associated with a feverish condition. The normal range of temperature variation is smaller and many diseases are associated with a definite rise of temperature.

Other symptoms may also indicate ill health. The eyes, which are normally bright, may be dull or there may be a drooping of the head (or wings and tail in birds). The coat of cattle is normally glossy, i.e. it has a good " bloom ", but in animals in poor condition may be dull or " staring " ; in advanced stages of wasting diseases the fleece of sheep may fall from the body. Animals may show abnormalities of movement ranging from an inability to move to violent convulsions. Abnormal colorations of the mucous membranes lining the lips and adjacent areas may be evident or there may be discharges from the nose. Diarrhœa, or the presence of blood in the fæces, may indicate specific diseases of the intestinal tract. Similarly, the urine which is normally a clear yellowish fluid, may be cloudy, brown or red. Sudden reductions in milk yield in dairy cattle may be early symptoms of ill health while an abnormal milk secretion, e.g. the presence of clots or of blood in the milk, is indicative of diseases of the mammary gland. Finally, there may be swellings of the body in different sites. These signs, occurring singly or in combination, can be symptomatic of specific diseases and therefore provide information which is of value to the veterinary surgeon in deciding on a course of treatment.

Occasionally more detailed examination of the animal is necessary, and this is normally carried out by the veterinary surgeon using special instruments and techniques. Often the exact cause of disease can only be established after laboratory examinations.

The Nature of Ill Health. Ill health may be the result of injuries or wounds, malnutrition, poisoning or attacks by specific infective agents. In some cases diseases are not readily transmissible from animal to animal and are described as being *non-infectious* ; in other cases disease may spread rapidly among a group of animals and is then described as being *infectious*. Disease is said to be *acute* when severe and of short duration, *chronic* when less severe but of long duration, *sporadic* when arising from occasional causes, and *epidemic* when widespread and arising from a general cause.

The causal agents of ill health are many and are listed below. (This list is not in order of importance).

(1) Injuries	(7) Helminths (worms)
(2) Poisons	(8) Acari (ticks and mites)
(3) Fungi	(9) Insects
(4) Bacteria	(10) Nutritional deficiencies
(5) Protozoa	(11) Metabolic disorders
(6) Viruses	(12) Mycoplasmas

Disease Resistance. It is a common observation that, when animals are exposed to an infectious disease, some contract it whereas others appear to be resistant. The role of the white cells or leucocytes in engulfing and destroying pathogenic micro-organisms has already been described (see p. 513), but the body has other means of defence against disease. Often, when an animal recovers from an infectious disease, there follows a period of immunity from further attacks. Clearly some reaction takes place in response to disease by which the animal develops an acquired immunity. This is due to the formation, from body proteins, of substances called *antibodies* which circulate through the body dissolved in the body fluids. The action of antibodies may be to kill or immobilise parasitic organisms, make them more susceptible to destruction by leucocytes or render their poisons innocuous. Antibody formation is particularly important in the development of resistance to attacks by bacteria and viruses, but antibodies are also produced in response to attacks by other parasites and, indeed, in response to the presence of almost any foreign substance in the body. The immunity is not always complete, for the supposedly immune animal may succumb if heavily challenged by disease or the immunity may fade with the passage of time.

Disease resistance has two important applications in the prevention of ill health. Firstly, it has been found that if a susceptible animal is dosed with blood serum from an animal which is immune to a certain disease, the antibodies contained in the immune serum confer resistance to that disease on the recipient. This is the principle of *immunisation* and immune sera are widely used to confer disease resistance on susceptible animals. The protection is only complete over a limited period of time after which the animal may again become susceptible. Secondly, it has been found that if the causal agent of a disease is inactivated (by treating it with heat or chemicals, or by irradiation) and injected into a susceptible animal, it stimulates the formation of antibodies without exposing the animal to the clinical effects of the disease. This is the principle of *vaccination* and the inactivated preparation of disease organisms is known as a vaccine.

In the case of immunisation, antibodies from the immune serum are responsible for the disease resistance, and immunity is acquired in a passive manner. However, when an animal is vaccinated immunity is acquired in an active manner, antibodies being produced by the animal in response to the presence of a foreign substance (the vaccine).

Legislation in Relation to Disease. Certain diseases have serious effects on the animal or may be a potential danger to humans eating livestock products from infected animals. Many such diseases have become subjected to legislative control measures, either to eradicate them completely from the country, or to control the incidence of sporadic diseases which might otherwise quickly overrun the country.

Under the Diseases of Animals Act, 1950 the outbreak of certain diseases must be notified to the police ; these notifiable diseases include Anthrax, Foot and Mouth Disease, Sheep Scab, Swine Fever Fowl Pest and more recently Swine Vesicular Disease.

THE CAUSES, PREVENTION AND TREATMENT OF ILL HEALTH

In the following description the various types of ill health in farm animals are classified under headings according to their causal agents. Under each heading the main types of ill health are listed and a selection of these is described in more detail both to show how they affect livestock production and to illustrate principles of prevention and control.

INJURIES

An injury or wound may directly affect the level of production of an animal or it may allow infectious organisms to gain entry to the body. Generally, simple injuries such as cuts and bruises can be readily treated with antiseptic dressings and recovery is rapid and complete. However, if the injury is severe, e.g. a fractured limb bone, it is probably best to slaughter the animal rather than to pursue a tedious course of treatment which may never give complete recovery ; the latter course may be followed if the animal concerned is valuable.

One troublesome condition in cattle is caused by the presence of foreign bodies, such as pieces of wire or nails, in the rumen or reticulum. Cattle graze extremely rapidly and, if metallic objects are left lying in fields, they may be eaten and can lead to sickness because of inflammation of the internal walls of the rumen or reticulum, or death if the object happens to pierce a vital organ or blood vessel. Much can be done to prevent trouble from this cause by ensuring that wire, nails, etc., are not left lying in the field. Affected animals may show a spontaneous recovery, but it may be necessary for the veterinary surgeon to operate on the animal and remove the offending object.

POISONS

Many pathogenic organisms cause ill health by the production within the body of toxic substances, but the animal may consume foods and other substances which are themselves poisonous and cause sickness. Some poisons produce specific symptoms in affected stock whereas others cause more general symptoms, usually including vomiting and diarrhœa.

On many farms, poisonous substances are in routine use and may be accidentally eaten by stock with disastrous results. Stock may also be poisoned if they have access to water supplies which are polluted by chemical poisons, or poisons (such as fluorine) present in factory smokes may be deposited on pastures as fine dusts. Poisoning from this latter cause may make it impossible to keep grazing stock in the neighbourhood of certain chemical works. A common form of poisoning in housed cattle (especially calves) arises from them gnawing at lead paint which is flaking from wood work, or licking new paint. Much can be done to prevent poisoning by the careful storage, handling and disposal of poisonous substances.

Several weed plants have been stated to be poisonous to stock eating them, but these claims are often difficult to substantiate. Cases of poisoning may occur only when pasture is in short supply and animals are forced to graze in the hedgerows and ditches. Plants which are definitely harmful to stock include ragwort, hemlock, horsetail and bracken ; some of these retain their poisonous properties when made into hay. Buttercups are probably only poisonous when eaten in large quantities. Ragwort is scheduled as an injurious weed and must be controlled either by cutting or by the application of weedkillers such as MCPA or 2–4 D. Bracken and horsetail are both difficult to kill in pastures, but horsetail shoots can be tackled with herbicides, making the pasture safer for grazing or haymaking.

Yew poisoning can occur with serious consequences, and it is important that yew trees should be fenced off from stock, or felled.

Some crop plants may, under particular circumstances, be poisonous to livestock. For example, where kale or rape are being extensively grazed by sheep or cattle poisoning may occur, the main symptom being the presence of blood in the urine. Similarly, freshly cut sugar beet tops may cause poisoning : they are quite safe for feeding if allowed to wilt for a few days.

In some areas of the world mineral substances such as molybdenum and selenium are present in relatively high concentrations in the soil. Under these conditions plants which are normally quite safe for feeding may accumulate excessive quantities of these substances and cause toxic effects in grazing animals. The only common type of poisoning of this sort in this country occurs on the so-called " teart " pastures of Somerset and Warwickshire where grazing animals fail to thrive because of the consumption of excessive quantities of molybdenum.

DISEASES CAUSED BY FUNGI

The only common condition resulting from fungal attack is Ringworm which, in farm animals, is most prevalent in calves and yearling cattle ; older cattle appear to be resistant.

The fungus grows in the superficial layers of the skin and results in the formation of bare circular patches, especially in the region of the head and neck. It may not have any obvious adverse effects on the animal but can cause severe irritation and leave ugly blemishes on the skin. Probably more important is the fact that the fungus can attack human beings in contact with infected stock in whom it produces highly irritant lesions.

Calves frequently show a spontaneous recovery from ringworm as they grow older, but the resistant spores produced by the fungus are harboured in buildings and infect subsequent occupants. The treatment of infected areas is now a thing of the past. Griseofulvin is now added to the diet and brings about a quick cure.

DISEASES CAUSED BY BACTERIA

Bacteria are small unicellular organisms belonging to the plant kingdom. Some types of bacteria may be present without causing disease and may be actually beneficial, e.g. bacteria in the rumen, but others may gain entry to the body where they multiply rapidly and cause disease. The diseases caused by bacteria include anthrax, bovine tuberculosis, brucellosis of cattle, infectious scours of calves, Johne's disease, mastitis, pulpy kidney and lamb dysentery of sheep, tetanus or lockjaw, braxy of sheep, foot rot of sheep and salmonellosis in poultry, pigs, calves and cattle.

Anthrax. This disease is caused by *Bacillus anthracis*, and all farm animals (except the hen) are susceptible. The disease is acute, and death is usually sudden, the body being grossly distended with gas and the limbs stiff. Anthrax is a notifiable disease, and whenever an animal dies suddenly showing the signs described above a police constable should be notified at once. It is extremely dangerous to open the body of an infected animal because the bacteria exposed to oxygen form resistant spores which can live in the soil for several years, thus forming a centre of infection for other animals. The only satisfactory method of disposing of the carcase is by complete burning. Apart from the effects it has on farm animals, the disease is important because it can be contracted by the human.

Bovine Tuberculosis. Tuberculosis has for many years been a most important disease causing a wasting condition of cattle with great losses of production. Further, pigs and, more important, humans consuming products from infected animals, may become infected. Tuberculosis is caused by *Mycobacterium tuberculosis*. The owner of an animal known to be shedding tuberculosis germs is legally bound to report this to the police.

Some control of the incidence of tuberculosis in humans has been achieved by the pasteurisation of milk, but the most important contribution to the control of the disease in both cattle and humans has been the eradication of tuberculosis from the cattle of this country. After several campaigns to eradicate bovine tuberculosis a complete eradication scheme was adopted in 1950 which planned to tackle the problem on an area basis. Infected cattle were detected by a tuberculin test which distinguishes between cattle infected with bovine tuberculosis and those having non-specific infections, e.g. the avian form of the disease and skin tuberculosis. Cattle infected with bovine tuberculosis were disposed of while those having non-specific infections could be retained in the herd. In the early stages of the scheme farmers were encouraged to co-operate by the payment of a bonus on tested animals or milk produced from them, but in the final stages of eradication from an area, testing became compulsory. The eradication scheme was completed in 1960 and has made an important contribution towards improving the health of both the cattle and human populations.

Brucellosis or Contagious Abortion. Brucellosis, caused by *Brucella abortus*, is a most important disease of cattle causing abortion in the fifth to seventh months of pregnancy; abortion may be restricted to odd animals in the herd but there may be " abortion storms " in which many cows abort over a short period. The disease in humans, once known as Undulant Fever, appears to be on the increase and lends urgency to the official cattle eradication scheme now in full swing. The disease can cause considerable financial loss to the dairy herd, not only because of the loss of calves, but also because of the delay in breeding and loss of milk production which it entails. Infected animals generally abort only once and then become resistant but continue to act as carriers of the disease ; infection appears to be spread from animal to animal from bacteria present in discharges from the genital tract, and from fœtus and after-birth being dragged across fields by dogs and foxes.

Infection can be detected by a blood test, but often the infection is so widespread in a herd that control of the disease by the disposal of infected animals is impracticable. Fortunately a vaccine, called Strain 19 or S.19, has been available since 1942 and the routine injection of this vaccine into all heifer calves in the herd between 3 and 6 months of age effectively suppresses the disease. In older cattle a vaccine known as 45/20 may be used, but veterinary consideration is essential in respect of each herd.

Infectious or White Scours of Calves. Young animals, whilst they are in the liquid feeding period, are frequently subject to digestive upsets which result in scouring. Scouring does not necessarily indicate the presence of pathogenic organisms—it may be merely a result of overfeeding or, in the case of bucket-fed calves, a poor feeding technique. However, pathogenic organisms may be the cause of digestive disturbances, and scouring due to infections with *Escherichia coli* can have serious consequences. Infectious scours are most prevalent in bucket-fed calves in which there develops a grey, evil-smelling diarrhœa and a rapid loss of condition frequently resulting in death. The disease can become serious where calf rearing takes place in unhygienic surroundings and where feeding equipment is not thoroughly cleaned after use. Much can be done to reduce the level of infection by thorough cleaning and disinfection of calf houses between batches of calves.

The first milk or colostrum produced by the cow contains antibodies which give the calf a measure of protection against infectious scours. It is therefore important to ensure that each calf receives colostrum at birth.

With good management and the use of antibiotics under veterinary direction (because of the danger of producing resistant strains) scours can be brought under control but it is well to ascertain if Salmonella bacteria are present. *Salmonellosis* is now known to cause scouring often with blood and other serious symptoms. So early diagnosis and isolation is important. The distribution of young calves can aggravate the situation spreading infection widely therefore the early use of the vaccine Mellovax is advised. The disease may be fatal in humans.

Foot Rot. Of the conditions causing lameness in sheep the commonest results from infections with *Fusiformis nodosus* which causes an evil-smelling putrefaction of the soft tissues of the feet. The lameness interferes with grazing and may therefore result in a loss of body condition.

The bacteria can be killed by treating infected feet with tinctures of chloromycetin, but this product is probably too expensive except for use in serious cases ; the most usual method of control is to walk infected sheep through a foot-bath containing a 5–10 per cent solution of formalin or copper sulphate. Whatever system of treatment is used it is important that it should be preceded by paring off infected tissue with a sharp knife so that the bactericidal agent can gain access to the sites of infection.

It is known that the bacterium cannot live in the soil for more than fourteen days but can be carried indefinitely on the feet of infected sheep. These facts make the eradication of the disease possible. The feet of all sheep in the flock should be examined, infected tissue pared off with a sharp knife and the sheep (including those which appear to be healthy) walked through a foot-bath containing a 10 per cent solution of formalin. After treatment the sheep should be moved on to pasture which has been kept clear of infected sheep for at least fourteen days. This treatment should be repeated at intervals of one to two weeks until no further cases of foot-rot are present. Once a flock is free of foot rot infection can only be contracted if infected sheep are brought on to the farm.

Mastitis. This is a disease which can be contracted by all lactating animals but is probably most troublesome in dairy cows. Among the organisms responsible are *Streptococcus agalactiae, dysgalactiae* and *uberis*, coliforms, staphylococci and pseudomonas. Cases are of varying severity showing just a few clots to a clear liquid with blood and a greatly swollen quarter. Infection spreads from cow to cow via the milkers' hands or the teat-cups of the milking machine. Leaving the latter on too long (over 4 minutes) can damage the teats and so predispose to mastitis. Dipping the teats after milking in an appropriate iodophor or chlorine solution greatly reduces the spread from animal to animal.

Treatment with an antibiotic is very effective except in chronic cases. When this type of therapy is used milk from affected quarters should not be marketed for 48 hours after treatment, or the presence of traces of antibiotic may have adverse effects on humans consuming such milk. Injection of the udder with a long-acting antibiotic at the time of drying off is very effective in reducing the incidence of new cases at calving.

The spread of infection in a herd can be checked by the use of a strip-cup, the cell count technique or the conductivity detector so that mastitis can be detected at an early stage.

A particularly serious form of mastitis occurs in heifers and dry cows, usually when the weather is hot and sultry. It is caused by *Corynebacterium pyogenes* and, because of its occurrence during the summer months, is called *Summer Mastitis*. The results of infection are serious ; there is an acute swelling of the infected quarter (which is usually permanently lost for milk production) and a feverish condition. Some cases may be fatal. Treatment introduced early enough may be effective. Flies play a part in the spread of infection and should be controlled.

DISEASES CAUSED BY PROTOZOA

Protozoa are microscopic, unicellular animals and cause two important diseases in farm animals ; firstly, coccidiosis which can occur in various classes of stock but is most important in poultry ; and, secondly, blackhead of turkeys.

Coccidiosis. This is caused by a protozoan parasite known as an *Eimeria* of which there are several species involved. These attack various parts of the chicken's intestine causing varying symptoms from the mild to the acute. Mortality is high in young chicks and unthriftiness and falling egg production characteristic in older birds.

A variety of effective remedies is now available and by giving half the curative dose (using as a coccidiostat) in the food over a period at selected stages of rearing, a useful degree of immunity is encouraged. The disease is also encountered in calves and pigs from time to time, but each sort of animal has its own coccidium which is not transferable to other species of animal.

Blackhead in turkeys can be controlled by drug therapy and by keeping them well apart from hens which harbour the parasite which affects the turkey's liver.

DISEASES CAUSED BY VIRUSES

There are several diseases in this group including three important notifiable diseases—foot and mouth disease, swine fever and fowl pest. Two other virus diseases, which occur in sheep, are orf (contagious pustular dermatitis), and enzootic abortion or kebbing.

Foot and Mouth Disease. Foot and mouth disease is highly infectious and can be contracted by cattle, sheep and pigs. Infected animals become feverish and develop blisters in the mouth which cause them to salivate abnormally freely, and on the feet, causing lameness. Death does not usually occur, but the disease is important because it leads to a tremendous loss of production and can quickly spread over large areas.

The disease occurs as sporadic epidemics in this country, being transported into Great Britain from countries where foot and mouth disease is endemic, either in imported meat or, perhaps, by migrating birds. The disease is notifiable and is controlled by the rigid enforcement of a slaughter policy ; all cattle, sheep and pigs on an infected premises, whether infected or not, are slaughtered and standstill orders are imposed on the movement of livestock within areas of possible infection. Under this policy the farmer is paid full value compensation for all animals slaughtered: both pedigree and potential are taken into account at valuation.

Vaccines as available abroad give some measure of protection against the disease and it has been argued that a permitted use of these vaccines should supplant the present slaughter policy. Unfortunately there are various strains of the virus causing foot and mouth disease, and it is difficult to produce a vaccine which is completely effective against all these strains. Also, the vaccines at present available are only effective over a relatively short period of time and periodic vaccination of all farm mammals would be necessary. It seems that the slaughter policy is less costly on a national scale than periodic vaccination. The development of more effective vaccines would, of course, completely alter the situation.

Any condition which resembles foot and mouth disease, foot rot in cattle and sheep and swine vesicular disease should be reported immediately to the police.

Swine Vesicular Disease. This virus disease was imported into this country in pig-meat from the Far East in 1970 and has caused widespread concern because the symptoms in the pig can only be differentiated from foot and mouth disease by the experienced expert assisted by laboratory techniques. Only pigs are affected. The disease is spread by careless swill feeders, by contaminated lorries and personnel which carry and handle affected animals. The virus is much more resistant to disinfectants and climatic conditions than that of F and M. Very rigorous restrictions and conditions have now been applied to swill feeding.

Swine Fever. This again is an infectious virus disease but occurs only in pigs. Swine fever may be accompanied by a wide range of symptoms, from an acute form in which sudden death occurs, to a chronic form where pigs are ill for long periods. Even in the acute form there are usually warning signs of the onset of disease. Affected animals become uneasy, fail to feed, have a catarrhal discharge from the eyes, and body temperature is elevated to 41 °C. In the acute form these signs are quickly followed by the development of a grey, evil-smelling diarrhœa and animals become abnormally thirsty ; there may be a reddening of the skin covering certain regions of the body. From this stage the pig declines rapidly in condition and dies within a few days. In the chronic form of swine fever the animals lose condition and have diarrhœa but may not die. Swine fever is still on the list of notifiable diseases, but as a result of a successful eradication campaign the country has been free since 1971.

Fowl Pest (Newcastle Disease). This virus disease is highly infectious and may occur in flocks where there has been no apparent contact with diseased stock. The disease occurs mainly in fowls but can also be contracted by turkeys.

Fowl pest may occur in two forms, an acute form and a mild form which in recent years has become more common than the typical acute form). In the acute form of the disease birds may be found dead without warning symptoms and over 90 per cent of an affected flock may die. Death may be preceded by symptoms including a loss of egg production, the development of a yellow diarrhœa, abnormal breathing with discharges from the nostrils and mouth, and sleepiness. The mild form of fowl pest resembles a nasal cold with the development of a watery diarrhœa. The infection spreads rapidly and there is a sharp fall in egg production though mortality (except in young birds) is low.

Following the introduction of the disease in pickled poultry products from central Europe immediately after the war the disease was made notifiable and compensation paid for compulsory slaughter. But, today vaccination of young birds by injection and aerosol is so effective that voluntary control is encouraged and the disease no longer comes under official control.

HELMINTH OR WORM PARASITES

The general term, helminths, covers a range of worms parasitic on animals. The most important are the *nematodes* or round worms which inhabit the stomach and small intestine giving rise to a wasting condition known as parasitic gastro-enteritis. Other parasitic worms include those which inhabit the lungs, e.g. nematodes causing husk in cattle ; *tapeworms* which inhabit the small intestine ; and *flatworms*, e.g. liver fluke.

Parasitic Gastro Enteritis. This condition has been most widely studied in relation to the sheep. In the sheep several species of nematode worms inhabit the stomach and small intestine, including *Haemonchus contortus*, *Ostertagia circumcincta* and various species of the Genera *Trichostrongylus* and *Nematodirus*. In infected sheep these worms give rise to a progressive loss of condition, diarrhœa and anæmia. In some cases they may cause the death of sheep. Parasitic gastro-enteritis is mainly a problem of sheep less than one year old, (for after this age sheep develop a resistance to worm infestation), and where sheep are heavily stocked on pasture.

A knowledge of the life cycle of parasitic nematodes and the method by which sheep become infected is extremely important, for it can indicate methods of controlling the level of worm infestation by grazing management. It is known that the lamb is born quite free from worms but becomes infected from worms present in its dam or other infected sheep. Adult worms lay eggs in the digestive tract which pass out on to the pasture with the fæces. After a time the eggs hatch to produce larvæ which can infect sheep if they are taken in through the mouth in herbage. In most worm species these infective larvæ take from three to fourteen days to develop from the egg, but worms of the genus *Nematodirus* lay eggs which do not usually develop immediately but overwinter on the pasture and give rise to infective larvæ in the following spring. If infective larvæ are not taken in by grazing sheep then they usually die within a few weeks : in exceptional circumstances they may live for several months.

These facts relating to the life cycle of nematodes are clearly important in relation to methods of worm control. For example, to control *Nematodirus* it is important not to put grazing sheep such as susceptible lambs on pastures grazed by infected sheep in the previous season. If the pasture is rested from sheep grazing, the infective larvæ when they develop will be unable to find a suitable host and will die. Other species of worms can be controlled by breaking the life cycle in a similar manner. This involves the rotational grazing of sheep around small paddocks with the lambs being encouraged to *creep-graze* either forwards or sideways of their infected dams. In this way the lambs are given access to the best pasture available. The sheep are moved to fresh paddocks every three to six days so that, when the infective larvæ develop on grazed areas, no sheep are present to act as hosts ; by the time sheep return to graze areas which they have grazed previously, the large majority of infective larvæ will have died. These systems are so effective in preventing the build-up of worm infestation in lambs that stocking rates of 7 to 9 ewes and their lambs per acre can be achieved without adverse effects on the level of fat lamb production.

A number of substances have been recently discovered which are very effective against the range of parasites referred to. These are given by mouth or injection and contribute greatly to the control of parasitism in the domestic animals if used at the correct time of year and age of animal.

Liver Fluke (Fasciola hepatica). This flatworm is a parasite of sheep and cattle and spends part of its life cycle in a small snail which inhabits damp or wet areas of fields. Immature stages of the parasite develop in the snail and then leave it and form small cysts on the pasture. When these cysts are taken in with the herbage by grazing animals, they develop to the adult stage. The adult worm becomes located in the liver where it lays eggs which pass out on to the pasture and re-infect the snail.

The damage to the liver caused by liver fluke may be relatively slight and have no apparent adverse effect on the host animal ; but if a large infestation is present the liver damage becomes extensive and may cause unthriftiness or death. The disease may occur in an acute form which can cause death, often with no preliminary symptoms of infestation ; or in a chronic form in which the animal becomes anæmic and unthrifty though it may recover spontaneously; occasionally there are epidemics of liver fluke disease with ruinous losses to the farmer. Apart from the direct effects of the disease on the animal there are indirect losses, because any liver in the abattoir showing signs of liver fluke infestation is condemned for human consumption. In cattle there may be a fall in milk production.

The life cycle of the liver fluke suggests two possible means of control ; either the adult stages can be killed in the liver or the snail can be eradicated, thus breaking the life cycle. There are now better and safer remedies than the popular carbon tetra-chloride which make it possible to control infection in the animal and reduce losses more effectively.

The snail breeds in damp parts of fields, in ditches or near any water and the best method of reducing the population of snails is to treat such areas with 30 kg of finely ground copper sulphate per ha. The treatment should be repeated two or three times during the summer months.

ACARI—TICKS AND MITES

Ticks and mites are troublesome external parasites affecting cattle sheep, pigs and poultry.

Ticks. Although these parasites cause irritation and loss of condition in infested animals, they are probably most important as disease carriers. Among the diseases carried are red-water disease of cattle and louping ill and tick-borne fever (two important virus diseases which are most prevalent in sheep).

Ticks are most prevalent in upland areas and their life cycle is prolonged, taking place on a range of hosts. Although ticks can be controlled on farm animals by using dips, sprays or dusts containing benzene hexachloride (BHC), reinfestation quickly takes place from birds and wild animals which harbour the parasite (see p. 459).

Mites. Mites are probably the most important external parasites of this group. Equine parasitic mange and sheep scab, caused by mites, are notifiable diseases, but fortunately both appear to have been eradicated from this country. Mange still occurs infrequently in cattle and can be troublesome in pigs. DDT and BHC have been of enormous value in reducing the incidence of mange conditions caused by mites (see p. 460).

INSECTS

In hot climates insects are probably most important as carriers of disease, but in this country, although insects undoubtedly act as disease carriers, they are more important for the direct effects which they have on animals.

Many insects, such as house-flies and stable-flies, have mainly an irritant effect though they may also cause slightly reduced productivity. Some insects cause actual harm to stock and the most important of these are lice, keds, blow-flies (causing blow-fly strike in sheep) and warble flies. For fuller descriptions of these and the harm they cause, reference should be made to Chapter XIV.

Lice. There are two distinct types of lice ; biting lice which live on scurf and hair, and sucking lice which live on blood. Both types cause intense irritation and may give rise to bare patches on animals and birds. In cattle, sheep and pigs, lice are effectively controlled by the use of dips, sprays or dusts containing BHC, but in poultry it is more usual to treat the poultry houses than the birds themselves.

Keds. Keds are external parasites of sheep. They are wingless flies which suck blood, and the whole of their life-cycle is spent on the sheep. The irritation caused by keds may lead to the loss of parts of the fleece. Modern sheep dips appear to give effective control of this parasite.

Blow-Fly Strike or Myasis. Certain blow-flies, particularly a green bottle, *Lucilia sericata*, lay their eggs in fleece, and when the eggs hatch the maggots live on the flesh of the sheep. Apart from the intense irritation caused, the sheep may be literally eaten to death within a few days. Sheep may be " struck " at any time from May until late-September, but the incidence is highest in warm, humid weather. During this period sheep should be observed carefully for any signs of irritation denoting the presence of maggots in the fleece. Sheep may be dipped as a routine measure before any are affected, but as soon as a sheep becomes parasitised all sheep should be dipped using one of the modern dipping preparations (which generally contain BHC). A single dipping gives protection against flies for several weeks but sheep may need to be dipped a second time later in the season. After a brief period of freedom from Scab it is now (1976) back and the entire country is under a compulsory Dipping Order for the time being.

Warble Flies. These flies are often called gad-flies though there are various other flies which cause cattle to " gad ". Adult warble flies appear in mid-summer and lay their eggs on the under-belly or on the legs of cattle. The eggs hatch to produce larvæ which burrow into the skin and wander through the body eventually becoming located beneath the skin of the back. In late winter the large grubs give rise to characteristic bumps on the backs of affected cattle and in spring or early summer the larvæ emerge through the skin and fall to the ground where they develop into adult flies.

The importance of these flies is first that the activities of the larvæ damage the skin of the back thus reducing the value of hides made from this skin : and second, the presence of the larvæ has a detrimental, if only slight, effect upon the growth and well-being of the infected animal.

The larvæ each make a small breathing hole in the skin, and the main method of control is to scrub the backs of affected cattle with a preparation containing derris which enters the breathing hole and kills the larva. This method of treatment does not eliminate the damage to the skin but reduces the population of adult flies in the following season. An order which required the dressing of all cattle was introduced some years ago but was revoked in 1964 as it was found quite impossible to police it. It is hoped that, with the development of more effective warble fly dressings, especially organo-phosphorus compounds which are effective when given as skin dressings, a more determined effort will be made to eradicate this pest and so eliminate the enormous spoilage of hides which occurs at present. Furthermore it has now been discovered that ruelene given as a drench will kill the larvae in the body at the time.

NUTRITIONAL DEFICIENCIES

The feeding of farm livestock can have marked effects on both health and level of production. The importance of feeding well-balanced rations in the right quantities has long been recognised.

Both under-feeding and over-feeding are likely to have adverse effects on the animal. There are many constituents of rations, such as vitamins and various mineral substances, e.g. copper and cobalt, which are only required in minute quantities but are essential for normal health and production. It is possible for rations which appear adequate to be deficient in one or more of these dietary essentials and animals fed on such rations develop typical deficiency symptoms. In this respect, particular care must be taken in formulating rations for animals, such as pigs and poultry, which may spend the whole of their life indoors with no access to natural sources of vitamins and minerals.

For information on the importance and results of deficiency of essential vitamins and minerals, the reader is referred to Chapter XVIII.

METABOLIC DISORDERS

Current systems of livestock production often extend animals to the limits of their productive capabilities while expensive feeds are carefully rationed to prevent wastage. Under such systems of management there may be temporary disruptions of the normal metabolic processes of the body with serious effects on production and health. These disturbances may be called metabolic disorders. The most important metabolic disorders are : milk fever, grass staggers, aceto-naemia, bloat and twin-lamb disease.

Milk Fever. Shortly after calving cows may become restless, pass into a comatose state and (if not treated) die. This condition, commonly called milk fever, is the result of an upset in calcium metabolism. In affected cows, blood calcium levels are sub-normal and so parturient hypocalcaemia is used as an alternative name for the disease. These low levels of blood calcium are not necessarily associated with low levels of calcium in the diet ; milk is rich in calcium and the sudden onset of lactation may use up the readily available body reserves of calcium whilst reserves of calcium in bone cannot be mobilised rapidly enough to maintain normal levels of calcium in the blood. The failure to mobilise calcium reserves from bone tissue is probably due to temporary malfunction of the parathyroid glands (see p. 522).

In former years milk fever was a common cause of death in dairy cattle but a completely effective remedy is now available. Affected cows are injected with a solution of calcium borogluconate and this effects a return to normal within a few hours.

Milk fever may also occur in lactating ewes and sows and the same method of treatment is used.

Grass Staggers, Hereford Disease or Hypomagnesæmia. This condition may occur in sheep or cattle and is characterised by a sharp fall in blood magnesium levels (hence the term hypomagnesæmia). Sudden death may be the first sign of the disorder, but usually the animal becomes restless with nervous twitching of the muscles and ears ; convulsive fits and death are later symptoms.

In many herds of cattle and flocks of sheep there is a steady fall of blood serum magnesium values during the late winter to critical levels (less than 2 mg per 100 ml) though animals show no outward sign of this. Grass staggers usually occurs when these animals are grazed on lush spring pasture though magnesium levels in the herbage are not necessarily sub-normal ; attacks are often precipitated by stress factors such as parturition, œstrus or spells of cold weather. Cases may also occur in autumn when pastures are sodden and the wet herbage causes animals to scour.

Formerly the occurrence of grass staggers was restricted to certain districts, being particularly prevalent in Herefordshire. In recent years cases have occurred in most areas, and the total number of cases, especially in dairy cows, has risen sharply. This increase seems to be associated with the intensification of grassland management, and it has been found that animals grazing pastures which have been dressed with potassic and/or nitrogenous fertilisers are particularly prone to attack ; the exact role of these fertilisers in relation to grass staggers is not yet fully understood.

In the early stages of an attack affected animals can be treated with injections of magnesium sulphate, and this effects a recovery in most cases. When animals are lactating it may be difficult to distinguish between the symptoms of grass staggers and milk fever, and for either condition it is usual to inject a preparation containing both calcium and magnesium. Preventative measures are of two kinds. Firstly, animals can be fed a supplement containing magnesium during the critical period just before being turned out and while they are grazing lush spring pasture. To do this, calcined magnesite should be fed at 50 g per day ; it may be necessary to mix the calcined magnesite with a little rolled barley or molassine meal to ensure that animals will eat it. Secondly, pastures may be top-dressed with dolomitic limestone (which contains magnesium) in an attempt to increase the magnesium content of the herbage. Both these measures markedly reduce the number of cases of grass staggers but do not necessarily eliminate the condition.

Acetonæmia (Ketosis). Acetonæmia occurs in the early weeks after calving and is generally restricted to high yielding cows which are being fed large quantities of winter rations ; it is rare in grazing cows. This is a disorder which is seldom fatal but which can cause a serious loss of milk production. The main symptom, apart from losses of milk yield and body condition, is a sickly smell of acetone in the breath. Acetonæmia is due to the accumulation of toxic ketone bodies in the blood ; these are produced as break-down products when fat is used extensively as a source of energy.

Cows normally show a spontaneous recovery after a short period, but this may be accelerated by oral dosing with 0·25 l per day of glycerine for four days ; it is claimed that oral dosing twice daily for three days with 25 g of potassium chlorate dissolved in 0·5 l of water gives an effective cure.

Bloat or Hoven. When ruminants are grazing on good pasture, particularly those in which the proportion of legumes is high, the rumen may become so distended with gas that the animal has a " blown-up " appearance. Some cases may be fatal due to pressure on the heart. It appears that the eructation or belching mechanism fails to function correctly and prevents the escape of gases produced as a result of bacterial fermentation in the rumen and the formation of a stable foam in which the gas is trapped. This is due to saponins in the young rapidly growing herbage.

Various agents may help in relieving the condition, including various vegetable oils, animal fats such as tallow, antibiotics and anti-foaming agents. These all appear to function by preventing the formation of a stable foam. In New Zealand some farmers spray the area to be grazed on a particular day with vegetable oil or liquid paraffins ; it is claimed that this prevents the occurrence of bloat, but the practice is expensive. In severely affected animals it may be necessary to release gas from the rumen with a stomach tube or by puncturing the rumen by inserting a knife through the abdominal wall.

Twin-Lamb Disease (Pregnancy Toxæmia). As its name suggests, this is a disease which may occur in ewes bearing twin (or triplet) lambs ; it occurs during the last two weeks before lambing and usually the ewe and her unborn lambs die. The course of the disorder is swift —the ewe refuses to eat, becomes weak, shows nervous tremors, may be blinded and finally becomes comatose and dies. However, if an affected ewe lambs, she may show a dramatic spontaneous recovery. Twin-lamb disease may occur in pregnant ewes which are in poor condition but more often it occurs in ewes in good or over-fat conditions which meet a sudden nutritional check, e.g. through a change of diet, snow covering the ground, etc.

Affected ewes may have mild ketosis but the only consistent finding is low blood sugar levels which are responsible for the symptoms of the disorder.

It is possible to prevent serious attacks by ensuring that ewes are in steadily improving condition during the later stages of pregnancy and that they suffer no nutritional checks ; this can be achieved by supplementary feeding with a concentrate ration for the last six to eight weeks of pregnancy. Where cases do occur the most usual method of treatment is to give the ewe an oral dose of 0·12 l of glycer-ine, but some ewes fail to respond to this treatment. Recently, treatment with hormonal substances such as cortisone and glucagon has given promising results in some cases.

HYGIENIC MEASURES IN LIVESTOCK MANAGEMENT

Many of the diseases described in the preceding sections are highly infectious. It may not be possible to prevent the occurrence of these diseases but, if they do occur, it is essential to prevent them from spreading rapidly throughout the herd or flock. Many specific measures designed to control the incidence and spread of disease have been described, but there are also many routine measures, some of which are not very obvious, which can materially reduce the incidence of disease.

The Design of Livestock Housing. All too frequently in the past, livestock accommodation has been designed with little thought for the comfort and health of the animals which live therein. Buildings, especially those for young animals and pigs, should be well ventilated without being draughty. In piggeries the direction of air flow should be out over the dung passage rather than in the reverse direction for this will ensure that the pigs breathe fresh rather than foul air and helps to prevent the spread of infection within the piggery (see p. 62).

Efficient drainage is an essential requirement of all livestock buildings.

Equipment. Milking machines, udder cloths, calf-feeding buckets drenching bottles, hypodermic syringes and other equipment in regular use can provide the means of spreading disease from one animal to another. It is essential that equipment of this sort should be cleaned and efficiently sterilised after use.

Drinking troughs and bowls should be regularly cleaned out.

Foodstuffs. Mice and rats can be carriers of certain diseases, and wherever possible foodstuffs should therefore be stored where they cannot be spoiled by these rodents. In old storage buildings and barns where this is not possible, a determined effort should be made to reduce the population of rats and mice.

Feeding troughs should be so arranged that stock are prevented from fouling their food.

Isolation of Animals. Any animals brought on to the farm should be isolated until it is certain that they are free from diseases which might infect existing stock. Animals which are sick should be isolated for the same reason and so that they may be given special attention to ensure a speedy recovery.

If possible, animals recovering from a period of illness should be given a period of convalescence until they are completely fit to rejoin the herd or flock. During this period the animal should be fed small quantities of a highly nutritious diet and, in the case of grazing animals, daily access to good pasture may be valuable. In most circumstances gentle exercise is beneficial to the recovering animal.

The location and design of loose-boxes to be used for isolation purposes is most important. These should be situated as distant from healthy stock as is practicable and should be well ventilated (though not draughty) and easily cleaned. Cleaning is facilitated if the walls have a smooth rendering of concrete on their surface to a height of some 1·5 to 2 m from the ground (see p. 78). So that animals can be easily examined and treated by the veterinary surgeon, it is essential to provide some means of restraining the animal.

Cleaning and Disinfection. It is necessary to clean thoroughly and disinfect isolation boxes immediately they become vacant ; the same is true of calving boxes, farrowing accommodation, permanent lambing pens and chick brooders and incubators. Diseases of calves such as infectious scours can be controlled by the regular cleaning and disinfection of calf houses.

In these operations cleaning must come first ; disinfectants are quickly rendered ineffective if they come into contact with large quantities of dirt. The bedding should be removed and, if an infectious disease has been present in the building, burnt. Then the walls and floors should be scrubbed and washed using plenty of clean water and a detergent ; special attention should be paid to cleaning fixed equipment in the building. When all dirt has been removed, disinfectant should be sprayed on the walls and floor. Many disinfectant materials are available including formalin, potassium permanganate, hypochlorites, carbolic acid, lysol, caustic soda and several proprietary disinfectants. Steam is used on many large poultry farms for the routine sterilisation of equipment.

If the disinfection of a premises is necessary after an outbreak of a notifiable disease, it is supervised by a veterinary officer of the Ministry of Agriculture, Fisheries and Food.

Stockmen. There is little point in employing hygienic measure on the farm if stockmen move at random between infected and healthy stock ; disease organisms can be carried on the person and clothing of stockmen. This method of spreading diseases is most difficult to overcome. The only general recommendation which can be given is that animals sick with infectious diseases should not be attended until work on healthy stock is completed. After attending to sick animals the operator should wash his hands and boots.

No vehicles carrying stock, feeding-stuffs or equipment which has visited other farms or markets, should be allowed beyond a fixed barrier at the entrance to or close to buildings where livestock are kept.

Chapter XXIV

FARM ORGANISATION AND MANAGEMENT

THE word " management " is now one of the more widely-used terms in agriculture, but unfortunately it has different meanings to differen-people. To some, especially practical farmers, it means the day-to-day organisation of work on the farm. Is the ground fit for hoeing—if so, who shall do it—and so on. These decisions are aspects of farm management, but they are the immediate ones, akin to the tactical decisions a soldier must take in battle. In this chapter the term is used in its broader and longer term sense, comparable to the strategic decision of the soldier when he decides where the battle will be, and what the objective is once the battle is won. The farmer's day-to-day decisions should depend upon his type of farm organisation (which in turn depends upon his long-term objective) and not the other way round. It is for this reason that this chapter concentrates upon the principles and techniques required to plan and run an organisation rather than upon the problems of day-to-day management, vital though the latter is.

The Objective. The objective of the farmer is, in general, to make the most money, but within the bounds set by good husbandry. There may well be other objectives, however. To build up capital at the expense of present profits, or to achieve a certain status—perhaps to have a farm of a certain size, even if he could manage a smaller one better and make more money thereon. The farmer may not keep the livestock or crops that would make the highest monetary return because he gets a greater satisfaction from some other line of production—say bullocks rather than dairy cows ; or he may accept a lower profit in return for an easier life.

All these are quite valid choices, but for this chapter it will be assumed that the *objective of the farmer is to produce the maximum profit (difference between returns and costs) with the resources available to him, and within the bounds of good husbandry.*

FACTORS OF PRODUCTION

The farmer starts out with his various resources, including the farm itself, the potential of the land to grow crops, the buildings for storage and livestock, and so on. There is the capital that he can command, either from his own pocket or by virtue of his credit-worthiness with banks and merchants. With this capital he can purchase different machines and livestock ; pay for all the productive resources he must buy (seeds, fertilisers and so on) ; and perhaps, most important of all, pay the wages of labour.

These form the three factors of production, *land, capital* and *labour.*

ORGANISATION AND CONTROL

The farmer has to *organise* the use of these factors to attain the objective set out above, and organisation is the first part of farm management.

But once he has organised his farm, he still has to see to it that the plan is carried out. This may seem so obvious as to be hardly worth saying, and so simple that there will be little chance of it not being done. But in many cases, failure to achieve a reasonable level of profit lies less with a poor plan than with poor execution of that plan. A typical example is the farmer who thinks that his dairy cows are being fed correctly just because he has produced a rationing scheme, without checking the actual amounts fed. Recent surveys have shown that only a minority of dairy farmers keep an accurate assessment of the concentrates fed. In other words, the best plan can be quite useless if no effective control is exercised during production.

Therefore the farmer sets out to farm his resources to the best of his ability, to produce as high a profit as possible, and he does this by organisation first, and control afterwards. In the following sections both these aspects are considered in some detail.

THE PRINCIPLES OF ORGANISATION

Land. The natural starting point for any farmer considering his plan will be the farm itself. Obviously, some farms are more suited to certain lines of production than others. Does a farmer choose a farm to suit a plan, or does he choose a plan to suit the farm ? While a man is looking for a farm, the first of these questions will be uppermost in his mind ; he will look for a farm that fits into the general pattern of what he wants to do, dairying, or extensive corn-growing, or mixed farming, perhaps. But once a farm has been selected, the detailed plan will be much more dependent on the potential of the land, its fertility and nature, the buildings and other facilities present. These aspects are mainly common sense, but a farmer should consider them in a logical order.

(*a*) **Soil and Climate.** In a temperate climate there is probably a wide range of crops or livestock that can be kept, but even so, some will be more suitable, in terms of physical yield at least, than others. Some soils are warm enough for early vegetables, others are not. Some soils are too heavy to have livestock treading upon them in winter, some are not.

(*b*) **Lay-out.** For some types of farming the lay-out of the holding is important. For arable cropping, provided there is reasonable access to fields, especially good roads in bad weather, the actual shape of the farm is not of prime importance. Indeed, many successful arable farms contain off-lying portions which are several miles away from the main steading. It is communications rather than proximity that matter most.

For a dairy farm, however, a compact lay-out is far more important. If some fields lie at a distance, or across busy roads, they will contribute less than they should to the general economy of the holding. On the other hand, communications in the form of hard roads are not nearly so important, for the livestock can move themselves across country with a fair degree of efficiency.

(*c*) **Buildings and other Facilities.** The type and condition of existing buildings may be very important in determining the plan, some systems, such as dairying, are much more likely to require specialist structures than others. In the increasingly competitive future, certain buildings are likely to play an even greater part than they do now, for they are one way in which a farmer may overcome the scarcity of one factor of production—skilled labour, by using another—capital. In recent years great strides have been made in building design to save labour on certain farm operations, milking cows and corn storage being two obvious examples.

Other facilities may also be important, including power and water supplies. Electricity may be a most useful source of power for certain types of production, and to produce it on the farm, particularly for high loads, can be very expensive.

Good water supplies are essential for dairying and also for irrigation of crops. Indeed, ability to meet this last requirement may well add thousands of pounds to the value of a farm.

(d) Markets. In this country, where communications are good and the country is densely populated, access to markets is not often given enough consideration. Occasions arise when a farm may be suited to the production of some commodity by virtue of its soil and lay-out but have such poor communications as to render this line of production uneconomic. An example would be an island farm and milk for sale on the mainland. Again, because of some obstacle, even of a political factor such as war or tariffs, it may be more profitable to grow something that the farm is not really suited for, rather than another crop which is a more natural one. A special case may also arise where the farmer has access to a limited market, perhaps owing to personal reasons, where a special price can be made. This occurs most frequently in " quality " production.

(e) Size of farm. The size of the holding will also determine what production pattern is followed. A plan suitable for a 320 ha arable farm would not necessarily be reasonable for a farm of 32 ha even in the same district. This aspect can be considered more logically in the section on combination of factors.

Labour. The second factor of production, labour, is in some ways the most difficult to deal with because, behind all the discussion of principles and results, the answers have to be translated into human terms. The " units of labour " in dairy work, for instance, refer essentially to a real cowman, with human qualities and failings. However, bearing this point in mind, it is still possible to make a logical survey of the problems and characteristics of this factor as they affect the general farm plan.

(a) Availability. It is first of all essential to decide if the labour necessary for the plan is available. Certain products, notably field-scale vegetable production, require large quantities of manual labour at specific periods of the year. Therefore, to produce a plan for a farm where soil and markets suit this type of production, but where the necessary labour force does not exist, is adding complications to the plan. This scarcity might be overcome in various ways : transport from a town or temporary immigrant labour are two ways, but they both add to costs and management problems.

Another example is the stockman. Good cottages are virtually essential to attract and retain the type of labour who must stay on the job all the year round. If there is no such accommodation, transport from town will not provide the answer. Accommodation must be built or the enterprise dropped.

(b) Methods of payment. The normal way of paying for most farm work is *time wage*, where the man is paid a set wage for a standard week, plus overtime. If the weather is too bad to work outside, the wage must still be paid and hence much " maintenance " work on some farms is really only disguised " idle-time ".

The other main method is by *piece-work*. Here the worker is paid per unit of output—so much a tonne picked or hectare worked, etc. The main advantage of piece-work is that it links effort with output and rewards it accordingly. However, it can cause work to be skimped and arguments can develop over appropriate rates. Also, there are many routine jobs which do not fit into this pattern for payment, especially in stock-rearing, where care is vital.

An attempt to get the best of both methods lies in paying a standard time wage, plus a bonus linked to output where possible. Unfortunately, such schemes can defeat themselves by becoming too complex or arbitrary. With the sort of numbers usually involved on a farm, it is probably better for the farmer to pay a good wage to get good men and rely on his own personality to see that proper work is done at the right time.

(c) Spasmodic requirement. With crops, especially those with a high value per ha such as roots and vegetables, the requirements for labour are spasmodic, as noted above. This means, perhaps, that many men may be required to harvest the crop, while relatively few are needed at other times. If casual labour is not available to supply the bulk of this requirement, then the farmer who intends to grow these crops must consider whether he can afford to keep sufficient men to cope with this peak demand. If he can find other crops with a high demand at other times in the year, he may be justified in keeping a big staff, as in the Fens in this country.

Alternatively, he may be able to mechanise the process at the peak period—capital again substituting for labour. Or he may have to reduce his labour requirement by reducing, or cutting out entirely, the area of the particular crop which causes the problem. On the whole it is the high-value crops which have these labour peaks, and this makes the problem increasingly complex, because giving up the crop means going without a lot of potential income.

Capital. Capital represents both the fixed equipment of the farm, other than the land itself, and the money required to purchase the ingredients of production—livestock, seeds, fertiliser, machinery or labour. The former is called *fixed* or *landlord's capital*. The latter is known as *working capital*. The aim of the farmer should be to keep a reasonable balance between the two.

On the whole, working capital is not only more immediately productive, but it can be released and turned to other uses more quickly. Capital in the form of buildings or land can usually be realised only by selling up the farm. If a farmer decides that another line of production would be more profitable than his present dairy herd, he can soon sell the cows and stop buying concentrates, and use the money released elsewhere ; but the money sunk in his cow-shed will be far more difficult to realise. It is for this reason that a depreciation allowance is charged in farm accounts on buildings and fixed equipment. This is not so much a wear-and-tear allowance as a way of gradually getting back the capital, out of profits, that has been put into the buildings, so as to use it again elsewhere. A realistic depreciation allowance is an essential piece of financial discipline in any farm plan requiring capital investment in fixed resources.

Sources of Fixed Capital. Money to purchase a farm may be borrowed by depositing the deeds with the lender as security. This is known as a mortgage, and the money can come from private in-dividuals, banks, the Agricultural Mortgage Corporation, the Land Improvement Company, etc. New buildings may also be financed from these sources when carried out by the owner. (A tenant obviously cannot borrow on mortgage, as he has no title deeds to deposit.)

Sources of Working Capital. Loans for working capital can be of two kinds, medium- and short-term capital, depending upon the needs of the farmer. For machinery or equipment, the loan may be a medium-term one, spread over several years. For short-term, for seed purchases for instance, the loan may nominally be for a few months, until harvest perhaps. However, the need will arise next year, when another loan may be required in the same way. The differences between the two, medium- and short-term, are really only of degree, and they will be treated as essentially similar in this section.

The main source of such credit is the banks, through overdraft or formal loans. In the former case the farmer is given permission to draw more money than he has in his account, the amount depend-ing upon the manager's estimate of his credit-worthiness and any security he may offer. He pays interest only on the amount actually over-drawn at any one time, rather than on the whole nominal value, as would be the case with a formal loan.

The other major source is merchant credit. Traders will supply seeds, etc., on credit, and charge interest for this service. There are drawbacks to this, in theory at least, as the farmer is then under obligation to the trader and the rate of interest may be high. But it is a simple system, and has worked surprisingly well over the years.

Hire purchase, especially for medium-term loans on equipment, is increasingly common, and is a form of merchant credit. Direct grants for part of the costs of specified improvement and guaranteed loans are available from government sources. Especially on small farms these can be a very important source of capital. The Agricul-tural Credit Corporation exists to underwrite loans on such farms for approved plans which will show an economic benefit.

Generally, sufficient credit is available to the prudent and estab-lished farmer. It is the newcomer lacking capital who finds it difficult to finance his ideas. But even here, there is a surprising number of ways in which a sound plan can be financed, if it is properly presented and supported by good estimates and figures.

COMBINATION OF ENTERPRISES

Having considered these various aspects of the three factors with which he has to work, the farmer must now organise his resources to give him the highest profit. To do this he must so juggle the combination of enterprises that the return to each factor tends to be equal. That is to say, he hopes to reach an organisation where any change in the balance of enterprises or resources would lead to less profit. This is known as the optimum solution, and is rather like the position of a traveller at the North Pole, where all directions are south. In the next section practical methods of implementing this concept are explained.

But before going on to consider them, there are two other aspects of organisation that require discussion. The first of these is size, both of farm and of enterprise.

Size of Farm. Taking the size of farm first, it was mentioned above that this will help to determine, to some degree, the type of enterprises that will be considered, as some are more suitable for large farms than small. Generally speaking, the larger the farm the larger the profit, because turnover, i.e. sales per annum, will be highest, and it is the volume of business that is most important in determining the profit.

A small farmer often overcomes this handicap of size by having intensive enterprises, that is enterprises with a high output per hectare or per man. Examples are market garden crops. Again, as he is short of land, he can operate " processing " enterprises, such as pigs or poultry, where the only farm resources needed are a little land as a site for buildings, and labour to mind the stock, while all feed is purchased and brought on to the farm from elsewhere. In this way he virtually adds a second storey to his farm.

But when all has been said, the advantage will normally lie with the bigger farm, as the capital and labour investment per hectare will usually be less, and it can more easily benefit from the economies of scale, that is the spreading of the cost of buildings or big equipment over more production. This assumes, of course, that the big and small farms are on similar soils and conditions. If one is comparing a small intensive farm in the Fens with a large hill farm, the turnover, and hence profit, is likely to be greater on the small farm. But this is an extreme case. The major drawback to a really large farm, apart from its cost, is the difficulty of supervision and management. It is on this point that many fail to reach their potential profit.

Size of Enterprise. When planning on any particular farm, the size of holding is normally given at the start. But there remains the question of size of individual enterprise on the holding. Should one have a number of small ones, providing a range of products which will spread the risk in face of failure or falling prices, or a few much larger ones ? The answer to this will depend upon a concept of economic theory which is quite vital to successful farm planning—the concept of *Fixed and Variable Costs*.

Fixed and Variable Costs. Under this concept, the costs of production can be divided into two classes. The first consists of those that must be incurred if there is to be any production at all, but they will not increase as the volume of production is altered. These are known as the *fixed costs* and are : regular labour ; machinery ; rent, or landlord's costs ; miscellaneous costs (telephone, etc., that cannot be allocated easily to enterprises). For instance, to harvest a crop of corn will require certain machinery—say a binder or combine. Once the machine is purchased, there is a wide range of production possible, without requiring a second machine ; yet no production is possible (in practice at least) without the machine.

Taking the combine as an example, with a capacity of 60 ha per annum, the capital cost to the farm of harvesting machinery will be the same whether 30 ha or the full 60 ha are cut each year. If the volume is stepped up to 80 ha, either a second similar machine will be required or the original one can be replaced by one with a bigger capacity.

Hence, though these costs are called fixed costs, they are only fixed for ranges in volume of production. Really large alterations will require a change in the fixed costs, but even then these changes occur spasmodically, and not smoothly, with gradual increases in production.

The other class consists of those costs which will change directly as the volume or pattern of production changes, and are known as *variable* costs. Examples are the cost of concentrates changing with milk production, fertiliser costs with ha of a particular crop, or even with yield per ha—and so on.

As an enterprise gets larger, fuller use is made of the fixed costs, which as a result become less per unit of product. This is the reason why large-scale production is usually cheaper. But there are drawbacks, or else all enterprises would be large-scale ones. For example, large-scale enterprises may have to use much more expensive equipment, the costs of which may work out less per unit of product, but still the capital cost proves a barrier to many farmers. Again, really large enterprises require a high grade of management, and this may also be difficult to provide, especially if more than one such enterprise is to be operated by a single manager.

Diversification versus Specialisation. In deciding the question of size of enterprises, these points have to be weighed up. In favour of *diversification* is the spreading of risk, possibly also a reduction in working capital, for while the variable costs will be the same for large or small enterprises no really big fixed equipment may be necessary.

On the other hand, *specialisation* implies not only that once equipment has been bought it will be economically used, and hence total costs should be lower per unit of product ; but also that management can concentrate on fewer lines and hence be more alive to new developments. With agricultural techniques changing as fast as they are today, this can be vital in the battle for the market.

Therefore it seems likely that reduction to three or four major enterprises, each of a reasonable size, may provide a practical compromise between the advantages of complete diversification and of specialisation, and might be termed *simplification*.

Profit. To get the biggest profit the farmer has to *organise* his resources of land, labour and capital so that the return to each is as high as possible. The plan he makes will be determined by the availability of resources and the productive enterprises open to him. Once the plan is made, he will have to ensure that it is carried out by exercising *control*. In the remaining sections various practical methods of attaining these objectives are described.

METHODS OF ORGANISATION

The first step in organisation is to make a list of the main resources that are to be used :

1. The area of the farm, including the amount that can be cultivated.
2. Any quota or husbandry restrictions, such as the biggest proportion of cereals that is acceptable, or the sugar beet quota that is available.
3. Capacity of buildings or facilities for livestock.
4. Regular labour and casual labour available, and for what tasks ; stockmen, tractor drivers, potato pickers, etc.
5. Total capital available to work the farm.

Next, a list of possible production enterprises is required giving the various kinds of livestock and crops that can be grown.

Planning. From this it is always possible to make up some sort of cropping and livestock plan on technical information alone, and then to cost out the various factors necessary, labour, feed, fertiliser, etc. A costing-out process of this sort is known as *budgeting* and such a budget is a *complete budget*. It is an essential preliminary before setting out to operate a farm organisation, because a plan, however sound technically, is of little use if it does not make the required profit. But here the first step is rather in the nature of an inspired guess.

There is another, more logical, method of obtaining guidance as to the type of plan to put together. This requires the planner to know *which enterprises give the highest return to his scarcest resource*. Suppose that land is the scarcest of the three factors, as it usually is in this country. Then one would plan to produce as much as possible of those enterprises that give the highest return to land—say potatoes or sugar beet, if they were on the list of " possibles ". But if too much of these crops were introduced into the plan, another resource —labour perhaps—would prove to be insufficient. Then the emphasis of the plan would be switched towards enterprises that might give lower returns to land, but which used much less labour, and so on until a complete organisation had been worked out.

There are mechanical and semi-mechanical methods for doing this, known as *programming*, which are beyond the scope of this section. But the basic idea underlying such methods can be used by anyone to arrive at a sound basis for his organisation. To do this it is necessary to write down the enterprises available in the order of their return to the resource believed to be the scarcest on the farm. If one mistakes the limiting factor, selecting land rather than labour, this will soon be made evident when the plan is drawn up, as one finds oneself with unused capacity.

These returns can be expressed in various forms : *Gross Income* (i.e. total sales) ; *Gross Output* (Gross Income less livestock purchases) ; *Net Output* (Gross Output less feed and seed purchases). Almost certainly the most useful is *Gross Profit* (sometimes called *Gross Margin*) which provides both a basis for planning and a shorthand for speedy budgeting. It is also the concept used in the computer programming techniques. In all the examples referred to below the figures relating to money are used solely to illustrate the points raised : they have no relation to current costs or returns.

Gross Profit or Gross Margin. Simply, this is the *total output less variable costs*. For example, for *wheat* this might be, per ha :

Variable Cost.	£	Gross Income.	£
Seed . . .	23	Sales 3·75 tonnes @ £78·0	
Fertiliser . .	30	(inc. subsidy) . .	292
Sprays, etc. . .	7		
	60		
Gross Margin .	232		
	£292		£292

To use this concept, the farmer must first split up his variable costs amongst the various enterprises, and deduct them from the relevant output to get the *gross margin* of each enterprise as in the example. The fixed costs (see p. 689) are not allocated but kept in a " pool " at first.

He must then consider various combinations of enterprises—so many hectares of cereals, potatoes, etc., which he thinks he can manage with the labour, machinery, etc., which comprise his fixed costs.

Once he has made such a plan, it is a simple matter to add up the gross margins and subtract the fixed costs to give the *net farm income* or profit of that plan. Another plan can then be considered, and its profit worked out in the same way until one is settled upon that will give the highest profit.

By using the gross margin as a measure of the return of each enterprise, an accurate indication is gained of the effect of expansion or contraction of any enterprise, and a great deal of laborious calculation of individual changes in variable costs is avoided.

The system is clearly suited to cash crops, but in addition the returns of grazing stock, such as dairy cows, can also be expressed in those terms. The gross profit is then related to the area of grass and other forage crops used, so that the gross profit per ha of dairy cows can be compared directly with that from cash crops.

PLANNING EXAMPLE

Data. 120 ha farm : 112 ha arable, 8 ha permanent grass, 12 ha sugar beet quota ; not more than 75 per cent in all cereals, and not more than 25 per cent in winter wheat.

Buildings etc. Suitable yard and parlour for up to 40 cows ; buildings for followers ; corn storage up to 180 tonnes ; combine of 88 ha capacity.

Labour. Accommodation for four men including the farmer; if necessary a fifth might be hired, living locally ; casual labour available for singling sugar beet but not for harvesting it.

Capital. £36,000 available from own resources and bank. Up to £9,000 more from merchant credit if necessary.

Possible enterprises.					Gross Margin per ha.
					£
Sugar beet	.	.	.	.	525
Wheat .	.	.	.	.	263
Dairy cows	.	.	.	.	249 (1 ha per cow and follower)
Barley .	.	.	.	.	203
Oats	.	.	.	.	180
Beans .	.	.	.	.	173
Clover .	.	.	.	.	112

If details of labour requirement per unit of each enterprise at various times in the year are available, it is possible to plan *concurrently* on land and labour use, as is done by programming. But space prohibits any demonstration of this here, and for simplicity's sake the plan will be set out, first, in terms of returns to land, and then the labour requirement checked : any necessary adjustments to fit into the labour available will be made later.

Returns to Land.

	Gross Margin £
The best crop is sugar beet—12 ha @ £525	6,300
Next best, wheat, up to 28 ha @ £263. . . .	7,364
40 ha used. . . .	13,664
Then dairy cows. Up to 40 + followers can be run in the buildings available. This means 40 × 1 ha = 40 ha @ £249 . .	9,960
80 ha used. . . .	23,624

The remainder, 40 ha, can go into barley, because the upper limit for this crop (in view of the wheat area) is 56 ha.

Barley 40 ha @ £203	8,120
Total	£31,744

Note that the last three crops listed as possible enterprises have not appeared at all. To introduce them would reduce the gross margin.

Labour Requirement. The labour requirement may be checked in one of several ways. The concept of *man-work units* may be used. A table of standard labour requirements is consulted and the figure for each enterprise multiplied by the number of ha or stock concerned. By adding up the results, the total man-work units can be found, and can be converted into the number of men needed by reference to the standard figure per man.

The system works well in organisations which have a steady labour demand throughout the year, as in livestock. With some crops such as cereals there is a peak demand, e.g. harvest, which must be met even if there is surplus labour throughout the rest of the year. In these cases it may be simpler to work out the number required at these peak months, depending upon the type of crop.

In our example there will be three peak periods ; sugar beet singling followed immediately by silage and hay making ; corn harvest ; sugar beet harvest and the cultivation and drilling of winter wheat.

For the first, casual labour is available for singling, and three men would be able to manage the silage and hay. For corn harvest, three men with a tanker combine could cope with 68 ha of cereals.

The third period, the autumn, presents the biggest problem. A sugar beet harvester requires three men to work efficiently, but can manage perhaps 20 ha or more in a season. At the same time there is the preparation of wheat land to be carried out. The answer is probably to get some early ploughing done after hay has been made, and then alternate periods of wheat land preparation and sugar beet harvesting until all the jobs are done. In this way, three men could manage the arable side. Without the harvester, however, a fourth man would be necessary or the area of beet must be reduced.

For livestock, provided the buildings and work routines were suitable, 40 cows + followers could be managed easily by one man.

The next step would be to ensure that this plan could be fitted into a reasonable cropping sequence. It might be necessary to amend the areas a little to fit into field sizes. In our case a 10-year rotation can be worked, though the proportion of wheat may have to be reduced a little to fit in.

Fixed Costs. Finally, the fixed costs would be budgeted. Labour costs can be estimated from the number of men required. A short cut for both machinery and miscellaneous costs would be to use standard rates for farms of similar types and size. Failing that, they can be estimated from scratch, a tedious task ; where an existing farm is being replanned, the current rates of expenditure can be used, altered only where obviously necessary—in our case for the purchase of a sugar beet harvester. The rent would be the contracted rate.

To finish off our example, the final budget would be :

Fixed Costs.	£		£
Labour—3 men . .	7,200	Gross Margin . . .	31,744
Machinery . . .	6,600		
Rent and miscellaneous .	4,500		
	18,300		
Farm Income . .	13,444		
or £112/ha			
£31,744			£31,744

Capital Requirement. Before finally accepting the plan, it is essential to check that sufficient capital is available. This will require a *capital budget*. There are two forms of such budgets, an annual one covering a whole year at a time ; or a consecutive series of budgets covering shorter periods, such as months or quarters, which give a *capital profile* over a given period.

Making an *annual* capital budget is really much the same as making a complete budget, except that certain items, such as machinery, are costed at full purchase price rather than on the basis of annual depreciation. Having produced a crop and livestock plan, the variable costs, rather than the gross profits, of each enterprise are entered in the budget, plus the fixed costs. This total is reduced by any sales that may be made during the year, as these receipts

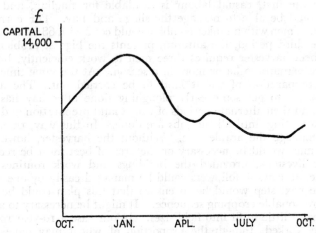

Capital profile—120 ha arable and dairy farm.

can be used to finance later expenditure. For instance, a milk cheque received in November can be used to pay for the concentrates used in December.

In such capital budgets, it is usual to assume that all bills are paid as soon as the items are received. Once the total capital requirement has been estimated, it is always open to the farmer to consider how much of this can be financed from merchant credit, i.e. by not paying bills at once.

When a new occupier is making a capital budget, additional items for living expenses during the first year, and tenant-right compensation, legal fees, etc., must also be taken into account.

The *capital budget* for the farm organisation considered in the previous section would be as follows, assuming a Michaelmas start.

Fixed Costs.	£	Contra Items.	£
Labour . . .	7,200	Milk (rec'd during 1st	
Machinery . .	18,000	year) . . .	14,400
Rent, etc. . .	4,500	Calves . .	900
Livestock purchase	15,000		
			15,300
		Net amount required .	43,710
Variable Costs.			
Wheat 28 ha @ £60 .	1,680		
Barley 40 ha @ £45 .	1,800		
Sugarbeet 12 ha @ £202·502	430		
Dairy Cows 40 ha @ £210	8,400		
	£59,010		£59,010

This is within the restriction set out in the example (£36,000 + £9,000 merchant credit).

If a growing sugar beet crop had been taken over, an ingoing valuation would have to be paid, but the sales would be available during the year to defray further expenditure, so that the total requirement in such a case would probably have been somewhat less.

A *capital profile* would be based upon the same figures, but all expenses and receipts would be allocated to the relevant periods. The result might be as shown in the diagram opposite.

CHANGING THE PLAN

However well the initial planning has been done, there will always come a time when a change has to be considered. This may come at the very beginning, when it is found that the proposed plan will use too much labour or capital. It may come after one or two seasons, when the farmer may want to try new enterprises or change some of the originals which have proved disappointing.

In all these cases a budget should be made for the proposed change, to see what the financial effect will be. Unless a major upheaval is contemplated, such as the complete elimination of a dairy herd and its replacement by cash crops, a *partial budget* will give sufficient guidance as to the benefit or otherwise of the proposal.

The Partial Budget. A partial budget only considers the costs that will alter with the change being considered, leaving the bulk of the farm's costs and returns untouched, hence its name. To carry out such a budget, four questions have to be answered :

Costs.	*Returns.*
1. What extra costs will be incurred ?	3. What extra sales will be made ?
2. What present income will be foregone ?	4. What present costs will be saved?

The balance between these two sides will show whether the proposal will result in a net increase in farm income or not.

If data about the farm are in gross profit terms, such a budget is a matter of minutes only, for the benefit will be the difference in gross profits, plus any change in the fixed costs. For instance, with a change of beans for barley, the gross profit or gross margin per ha for beans is £173, for barley £203. For every ha changed from

EXAMPLE OF A PARTIAL BUDGET

Extra costs.	£	£	Extra sales.	£	£
Beans			Beans		
Seed . . .	30				
Fertiliser . .	15				
Spray . .	22·50				
		67·50	2·60 tonnes @ £110		286·00
Income foregone			**Costs saved**		
Barley			Barley		
3·75 tonnes @ £80 .			Seed . . .	22·00	
			Fertiliser . .	22·00	
			Sprays . .	7·50	
		300			51·50
			Net loss . .		30·00
		£367·50			£367·50

barley to beans the loss is £30. If there is no saving in fixed costs, labour or machinery—and there does not seem to be any reason why there should be—the result would be a loss of income.

If data are not in gross profit form, then the full change in variable costs and sales will have to be set out, as shown at the bottom of page 695.

Clearly, having farm data in gross margin form means that these can be manipulated far more easily to estimate the effect of changes in plans, as well as providing guidance on the type of organisation to plan.

CONTROL—THE WHOLE FARM

Once the farm plan is under way, it is essential to keep a check on its progress to make certain that it is being carried out in the way intended. This is not just a question of a record of crop areas and livestock numbers ; it involves also costs and returns to make sure that commodities are not being produced at a loss.

There is also the diagnostic side, particularly important where a farmer is not satisfied with the existing profits of his farm, and wishes to know where the weaknesses of his organisation lie. In such a case a new organisation may be called for, but it may only be a question of tightening up on the technical husbandry of the existing plan.

Normal Farm Accounts. The easiest starting point in most cases is the farm account which, in this country, has in any case to be kept for taxation purposes. Over the last two decades, several methods of analysing the ordinary farm trading account to gain management information have been published. Here it is sufficient to describe the general principles underlying such analyses.

The basic idea in nearly all cases is to compare certain *efficiency factors*, calculated from the accounts, with standards drawn from the accounts of a sample of similar farms. This can be termed the *comparative method*. These efficiency factors are generally the expression of an *input-output* relationship, e.g. value of production per £100 labour costs, and so on. The basic framework can be described by the following diagram.

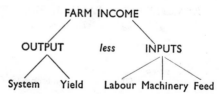

Farm Income. This is virtually the same as trading profit, though in some cases corrections are made for credit costs, owner-occupancy and farmer's manual work. If income is found to be low, then either output is low for the type of farm, or the inputs, especially the fixed costs, are too high.

Output. This can be measured by *Gross Output* (Sales, corrected for changes in valuation, less livestock purchases) ; *Net Output* (Gross Output less seed and feeding stuff purchases) ; *Gross Margin* (Gross Output less variable costs). There are arguments in favour of all these. Generally speaking, the more mixed the farming system is, with arable crops and several types of livestock, the more useful Net Output or Gross Margin are, because a proportion of the output is based on " imported acres " in the form of concentrates grown elsewhere than on the farm.

System. If the output is low, it may be due to the system followed, in that not enough of the resources are devoted to high potential enterprises. Dairy cows have a higher output per acre than beef, and a switch in this direction might be indicated. Various methods of calculating a System Index are published.

Yield. Alternatively, the pattern of farming may be reasonable, but the yield (output per unit of land or stock) may be low. Unless the farm system is a very simple one, or the accounts more detailed than is normal, additional records may be required to discover which enterprises are at fault.

Inputs. These costs are usually expressed in terms of costs per unit of output—Labour per £100 Gross Output, or Net Output, etc. A poor result may of course be due to output being low. But failing this there may be too much labour for the enterprises kept. This can be tested in several ways already discussed on p. 692, but the commonest is through *man work units*.

Machinery. Again expressed as costs per £100 output. To diagnose the cause of a poor result is much more difficult, for depreciation and repair costs are to some degree complementary, new machines having high depreciation costs but low repair costs, and older machines the reverse. To correct a poor result may be even more difficult, though if it is due either to extravagance in investment or bad technical management, then a correction can be made in time.

Feed. The cost of feed per unit of livestock output is often the weakest link in a mixed farm, especially where large quantities of homegrown cereals are used and not recorded. More data than are usually available from a trading account are required to find the weak enterprises, and these are considered in the final section of this chapter.

This type of analysis is the least that any informed farmer should carry out as a control method. Unfortunately, while it will give a reasonable indication of weaknesses, it is usually unable to show which particular enterprises are weak, or how best to remedy any fault in fixed costs. Additionally, when corrections come to be considered and partial budgets used to determine the financial value of such changes, quite different data have to be used. Therefore there is an interest in finding a different form of account-keeping which would make control easier and supply better data for budgeting.

Special Accounts. In addition to the ordinary farm trading account and balance sheet, there are special accounts of a more detailed nature which can be very useful. These all require rather more records to be kept than is necessary for the normal accounts, referring especially to the allocation of certain costs to various sections of the farm.

Full Costs. In manufacturing industry large sums are spent in keeping full cost accounts, so that the manager knows exactly the cost of each unit of product. Many attempts have been made to apply the same ideas to agriculture, but on the whole they have not been successful. There are three main reasons for this. In the first place, the fixed costs in farming represent perhaps two-thirds of the total costs, and therefore much of the apparent result will depend upon the method whereby these costs are allocated to individual enterprises. If based upon labour use, as in some cases, then one set of answers will be produced. If on capital, or even output, quite different ones. Secondly, a complex recording scheme is required to find out just where and when these fixed costs have been used, so that the sheer cost of running such accounts is high, and most agricultural businesses are not really large enough to carry such expenses. Finally, the management information derived from such accounts is not always useful. Because fixed costs represent so large a proportion of the total, reduction of output in one enterprise, or the elimination of an enterprise, may not affect costs very much if fixed costs are untouched —as is often the case.

Enterprise Costs. A variation of complete full costs is to keep the same sort of data for one or two major enterprises on a farm At first sight this seems a more feasible plan and certainly involves less work, for it is only necessary to record the use of fixed costs, labour and machinery, for these enterprises, rather than for the whole farm. But to make the scheme workable, certain standards have to be used, e.g. the cost of tractors per hour. This introduces an arbitrary element into what sets out to be an entirely objective record. More important, an arbitrary allowance for farm overheads may also affect the result. Then, there is no check that the recording has been done correctly, for the account does not tie up to any other accounts for the farm if as is usual, it is being carried out for only part of the farm organisation.

None the less, such enterprise accounts can be most useful, especially where livestock are concerned, for here they force the farmer to recognise the amount of feeding stuffs being used. However, information just as valuable could be achieved by simple feed recording, without going to the bother of allocating fixed costs at standard charges.

Intermediate Accounts. It should be possible to gain the advantages of enterprise costings without the disadvantages, by reorganising the farm accounts on the basis of fixed and variable costs. If the farm records were so organised as to show the allocation of variable costs—feed, seed, fertiliser etc.—to the various enterprises, and the value of sales were similarly allocated, then the gross profit or gross margin of each enterprise could be calculated. The fixed costs would be shown separately and would have to be defrayed before the profit were found. (For definition of gross margin, see p. 691.)

ANALYSIS

Enterprises. The pattern of analysis is much the same as that followed with the normal trading account. However, after looking at total output, attention is directed to the individual enterprises. It is a simple matter to compare the gross margin per acre of each enterprise with an acceptable standard, derived from the local advisers or from neighbours. If the result for any enterprise is low, the causes should be investigated, as described in the next section.

Fixed Costs. The fixed costs are much less amenable to alteration in themselves than the gross margins, and most corrections will take the form of altering the pattern of output rather than altering the fixed costs. Nevertheless, it is useful to have a quick check on the relationship between the fixed costs and output, as in the comparative method. Here the fixed costs can be expressed as a percentage of the total gross margin of the farm. The proportions will vary with the area and type of farm, but a very general rate has been found to be : Labour, $\frac{1}{3}$; Machinery, $\frac{1}{5}$; Others, 15 per cent of gross margin.

Under this system, any weak link amongst productive enterprises is located more readily. Disproportionate fixed costs are shown up, but here there are no particular advantages over the straight comparative method. More important is the fact that the data are in the form necessary not only to guide any reorganisation that may have to follow, especially to make better use of fixed costs, but also for budgeting such changes as are considered.

From what has been said it should now be clear that the first stage in controlling the farm organisation is to use the accounts to provide warning that all is not well. Various systems can be followed, the simplest being an analysis of the existing form of trading account as used for taxation returns. This has to be supplemented by additional information about individual enterprises. This may be done either as a separate operation, or by keeping the original accounts in the appropriate form.

CONTROL OF ENTERPRISES

The following sections are based upon the supposition that some form of analysis has been undertaken and that this has disclosed some weakness in the organisation. Depending upon the system of analysis adopted and the records kept, the farmer may also know in which enterprise or fixed cost this weakness lies. In all that follows, attention is mainly directed towards discovering the cause of such faults. In most cases the remedy will lie within the province of technical husbandry, and will not be discussed.

Crops. In almost all cases the reason for a low output from a crop enterprise will be yield rather than excessive inputs. The financial yield is made up of the physical yield × the price per unit. With some high value crops this fact may be vital ; a poorish physical yield may be more than compensated for by high prices. In these cases, marketing may be just as important as efficient technical production. This is so with many horticultural crops, where price received will depend upon many marketing factors, the state of demand, weather, other supplies, alternative products, presentation and packaging etc. But with most farm crops low output and low physical yield go together, so that it is essential to determine the cause of the latter, which will be a matter of husbandry. Therefore only such records should be kept as are necessary to help in such diagnosis. A field diary, in which a record of variable costs and yield is kept, together with a note of any special treatments, such as spraying, with all relevant technical information, is often of great service here.

Grazing Stock. With grazing stock, low output may be due to (i) Low physical or financial yield (mainly summer milk or small eggs are examples of the latter). (ii) Excessive concentrate feeding. (iii) Understocking of forage acreage. (iv) Too great a proportion of followers retained in dairy herds ; followers have a much lower output both per acre and per head than productive dairy cows.

Again enterprise records are necessary to determine which of these causes is operating. The field diary mentioned above can be expanded to take in the forage acreage, and a record kept of its use between livestock enterprises. In addition, a concentrate feed record is necessary, especially where home-grown cereals are used, and where there is more than one livestock enterprise.

Low yields can be checked from sales records, and excessive feeding from the feed record. Understocking can be checked from technical knowledge, or by the use of *livestock units*. These are a conversion factor, whereby all types of grazing stock can be reduced to a single common unit—equivalent to a medium-yielding dairy cow. By expressing all grazing stock on the farm in these terms and dividing the forage acreage by this total, the factor of *forage area per livestock unit* is calculated. The desirable figure will vary with the district, but a common one for lowland farms on fair soil is 0·6 ha per livestock unit.

Pigs and Poultry. Here there are no complications of forage acres and livestock density. Poor returns will be due to a low output from low yields or excessive concentrate feeding. But as these stock can be virtually divorced from the rest of the farm and they have a very high rate of turnover compared with other farm enterprises, more detailed recording is probably justified if the enterprise is of any size. The following records would be desirable on the majority of pig or poultry enterprises where the annual gross output exceeds £2,500.

Pigs. If both breeding and fattening are carried out on the same farm, they should be regarded as separate enterprises and the use of feeding stuffs recorded separately for each. The breeding herd should have litter records to show the average number of pigs farrowed and weaned per litter, the average number of litters per sow per year and the average weight per weaner. In addition, the total weight of food fed to the herd in a year can be divided by the number of weaners to give the weight and cost of food per weaner. For the fattening herd a record of all liveweight gained should be kept, by weighing all pigs on entry and despatch from the herd. The conversion factor, food per kg liveweight increase, can be calculated by dividing the food fed, and its cost, by the liveweight gain in the period. If pigs are being marketed to bacon, a record of gradings is also important.

Poultry. Simple egg yield over a year is really not sufficient. A calculation of the " percentage lay " should be done weekly, by taking the number of eggs laid as a percentage of the birds in the house. Once the rate falls below 60 per cent the margin left for profit must be getting very small. Food per bird per week should also be calculated and compared with accepted husbandry levels, as it is extremely easy to go astray in this type of intensive enterprise. One further aspect, at the end of the year, or when the flock is culled, deserves attention, and that is the difference between the value of the birds at point of lay and their sale value at the end. This depreciation can often be as much as 40–75p per bird, and may represent the gross margin on up to two-thirds of the bird's production. Therefore every effort should be made to close this gap, either by rearing more cheaply or attempting to get a better cull price.

The table bird industry is so very competitive that margins are very near the minimum, so there is little room for error. It should be possible to cost the food and sales of each batch with little difficulty. This should be done to keep a constant check on the profitability of the enterprise. With margins so low, to wait even two months before finding out that things are not going according to plan could be a disaster.

Summary. Keeping a check on crops presents little difficulty as physical yield, and in some cases prices, are all that is required, and the time element is not pressing. Once a year checks are sufficient.

For grazing stock, while continuous recording of feed is necessary, together with any sales, it is probably only with dairy cows that checks need to be made more often than once a year. Even with cows a quarterly check of feed use and milk output is probably sufficient.

With pigs and poultry, while there are really only two factors involved, concentrate feeding and production, yet because of the high rate of output constant checks are necessary, especially in the case of table birds, where costings should be on a batch rather than total flock basis.

CONTROL OF FIXED COSTS

In the discussion on organisation, the intractable nature of the fixed costs, especially machinery, was noted. As a result, no special records are generally necessary, for control purposes, for a farmer is hardly likely to have more men or tractors than he thought. Generally, the total cost of each of these factors, as recorded in the trading account, is the only financial record necessary.

However, there are two sets of records of a quasi-physical nature than can be kept to aid management, especially on really big farms : these are timesheets for labour, and log-books for tractors and large equipment.

Timesheets. These are often used to provide a basis for wage-payment. If the workers get in the habit of recording their tasks accurately, managers of large-scale farms often gain valuable insight into the amount of really productive work compared with the total hours worked, and can adjust their methods accordingly. If full-cost accounts are being kept, such records are essential, but the value of such accounts has already been discussed.

Log-books. These should be kept for all valuable machinery, not so much as a guide to the work that is done (though this is sometimes of great interest), but mainly to see that routine maintenance is being observed, and as a check on the rate of repairs. Again in full cost accounts they are essential.

Labour Costs. Though it has been stressed that the fixed costs are intractable, labour is relatively the easiest one to manipulate. It has been shown in the budgeting section that changes in organisation will not result in changes in labour costs unless the number of men is actually reduced. When a farmer is considering the possibility of reducing labour costs, he must first decide what are the enterprises or operations which cause him to keep his current labour force, by virtue of the peak labour demand. Once he has decided this, he has two courses open to him :

1. To eliminate or reduce the enterprise, thereby reducing demand. For this to be economically worthwhile, the value of wages saved must be greater than the loss in total gross margin.

2. To alter the techniques in the enterprises, so as to reduce labour. This normally means mechanisation, and here again the cost of mechanisation must obviously be less than the wages saved.

But on occasion a change in technique means doing the same task in a different way. Perhaps reorganising the lay-out of buildings or equipment, or rearranging men's tasks. It is here that *work-study* can play a vital part in examining current routines and pointing the way to better ones, that will reduce labour per unit of output. An outstanding example has been the use of work-study in milking routines.

A third type of alteration may be to vary the period in which certain tasks are performed, so as to reduce labour demand in bottle-neck periods. For instance, variations of drilling dates may serve to spread sugar-beet singling, and so allow fewer men to cover the same acreage.

SUMMARY

Farm Management means first of all planning the use of the farmer's resources—land, labour and capital—to give him the highest profit. A useful aid to this is the division of costs into two categories, fixed and variable. He should plan his farm so that as much of it as possible is devoted to those enterprises which give a high return (after defraying their respective variable costs) to whichever factor is scarcest. This means that on a small farm the plan should concentrate on those products that have a high return to land, whereas on a larger one, more products that give a lower return to land but require less labour can be considered.

Once the plan has been decided upon, it is essential to exercise control over it. The simplest way to do this is through the ordinary farm account, but usually this has to be supplemented with additional data, except in the simplest cases. As production continues, the plan should be constantly tested, to see whether improvements cannot be made.

But when all is said and done, the best plan will be unsuccessful if it is not supported by sound husbandry. The idea that a few abstruse calculations are sufficient to put a farmer on the road to success is a mistaken one. Farm Management is the economic expression of technical husbandry, and the two parts are vital to the sound working of a farm.

Chapter XXV

METRICATION

THE whole of British industry is in process of changing from the traditional Imperial system of units of measurement to that used in nine-tenths of the world. The new system, commonly called Metric, is the International System of Units, usually referred to, for clarity and brevity, as SI. The responsibility for supervising the change in the United Kingdom rests with the Metrication Board, but it should be emphasised that the Board is not the authority on the definition of SI units or on their application and use. Legal definitions of the most commonly used units and provisions for their use in trade are set out in the Weights and Measures Act, 1963.

The international authority for standardisation between countries is the International Organisation for Standardisation (I.S.O.) which is supported by sixty-two nations, including the United Kingdom.

The time-table for the change-over from Imperial units to SI varies from industry to industry. The changes affecting agriculture in the United Kingdom largely took place during 1976.

The seven basic SI units, common to all industrial programmes, are :

Quantity.	Unit.	Symbol.
length	metre	m
mass	kilogram	kg
time	second	s
electric current	ampere	A
temperature	kelvin	K
luminous intensity	candela	cd
amount of substance	mole	mol

Perhaps the most striking feature of this basic table to the farmer is the omission of the word " weight ". The conception of " mass " as the description of heaviness or lightness has replaced " weight ". It is important with all SI units that the symbols must always be used as set out in the tables. They should never be used in the plural, be followed by a full stop or written in capitals except when a capital letter is a symbol.

During the period of transition individual farmers will have to decide their own rate of change on the farm but, as the purchase and sale of different agricultural supplies and produce change to SI, farmers will need to be prepared to use the new units in all business transactions.

A comprehensive list of units likely to be used on farms follows this introduction to the subject.

Several leaflets are available from the Metrication Board, 22 Kingsway, London WC2B 6LE : probably the most useful for farmers is " Going Metric : Farming and Horticulture ". This contains some useful conversions which can be applied during the period of transition from Imperial to SI. There are also " Metric Memos " on the availability of metric weights for scales and on land measurement and maps.

SUMMARY OF UNITS USED ON THE FARM

TABLE A

COMMON METRIC WEIGHTS AND MEASURES FOR USE ON THE FARM

Unit of measurement.	Symbol.	Remarks.
LENGTH		
millimetre (one thousandth of a metre)	mm	about 25 to the inch (0·03937 in)
centimetre (one hundredth of a metre)	cm	less than half an inch (0·3937 in)
metre	m	just over a yard (1·094 yd)
kilometre (one thousand metres)	km	⅝ of a mile (0·6214 mile)
AREA		
quare centimetre (one ten thousandth of a square metre)	cm²	about 6 to the square inch (0·1550 in²)
square metre	m²	about 1¼ square yards (1·196 yd²)
hectare (ten thousand square metres)	ha	nearly 2½ acres (2·471 acres)
square kilometre (one million square metres)	km²	⅜ square mile (0·3861 mile²)
VOLUME AND CAPACITY		
millilitre (one thousandth of a litre)	ml	same as a cubic centimetre; about 30 to the fluid ounce (0·0610 in³ or 0·0352 fl oz)
litre	litre	about 1¾ pints; 4½ litres are about a gallon (1·760 pt or 0·2200 gal)
cubic metre	m³	about 220 gallons, 27 bushels or 1⅓ cubic yards (1·308 yd³)
WEIGHT		
gram (one thousandth of a kilogram)	g	about 450 to the lb (0·0353 oz)
kilogram	kg	nearly 2¼ lb (2·205 lb)
tonne	tonne	nearly a ton (0·9842 ton or 2205 lb)
TEMPERATURE		
degree Celsius (at present usually called *Centigrade*)	°C	100 steps of 1 °C between freezing and boiling

The symbol for litre is " l " and the symbol for tonne is " t " but the symbol " l " can be confused in type script with the figure " one ". Similarly the symbol " t " can be mistaken for an abbreviation for the imperial ton. It is customary, therefore, to write the units " litre " and " tonne " in full.

TABLE B

APPLICATIONS OF SOME METRIC WEIGHTS AND MEASURES

This table gives a selection of the units which will be encountered in various branches of farming.

Application and Units Used.	Symbol.	Remarks.
LINEAR MEASUREMENT OF LAND, BUILDINGS, FARMS, ETC.		
metre	m	See previous table for equiva-
centimetre	cm	lent imperial measures.
millimetre	mm	The building and construction industries have opted generally to use only the m and mm
FLOOR AND WALL LOADINGS		
kilograms per square metre	kg/m²	Equivalent to about ¼ lb/sq ft
SPACE REQUIREMENTS FOR ANIMALS (AREA AND VOLUME)		
hectares per animal	ha/animal	Nearly 2½ acres to the hectare
animals per hectare	animals/ha	
square metres per animal	m²/animal	Equivalent to about 10¾ sq ft
animals or birds per square metre	animals or birds/m²	
kilograms body weight per square metre	kg/m²	Equivalent to about ¼ lb/sq ft
kilograms body weight per cubic metre	kg/m³	16 kg/m³ is approximately 1 lb/cu ft
cubic metres per animal	m³/animal	About 35 cu ft
animals per cubic metre	animals/m³	
POWER		
kilowatts	kW	About 1⅓ HP
AIR AND FUEL CONSUMPTION		
litres per hour	litre/h	28 litres to the cubic foot or 4½ litres to the gallon
litres per minute	litre/min	
litres per second	litre/s	
cubic metres per hour	m³/h	35 cubic feet or 220 gallons to the cubic metre
cubic metres per minute	m³/min	
cubic metres per second	m³/s	
PRESSURE		
bar	bar	About 14½ lb/sq in
millibar	mbar	35 mbar is approximately 1 inch of mercury
SOWING; PLANTING; HARVESTING; FERTILISING; CROP YIELDS		
kilograms per sq metre	kg/m²	Nearly ¼ lb/sq ft or 2 lb/sq yd
kilograms per hectare	kg/ha	About 1 lb/acre
tonnes per hectare	tonne/ha	About 8 cwt/acre
litres per hectare	litre/ha	About 1/10 gallon/acre or 50 litre/ha is about 4½ gallons/acre
cubic metres per hectare	m³/ha	About 14 cubic feet or ½ cubic yard per acre

Application and Units Used.	Symbol.	Remarks.
CONCENTRATION		
kilograms per litre	kg/litre	About 10 lb/gallon
kilograms per kilogramme	kg/kg	lb/lb
grams per litre	g/litre	10 g/litre is about 1 lb/gallon
millilitres per litre	ml/litre	parts per 1,000
IRRIGATION		
millimetre depth	mm depth	1 mm = 1 litre/m² = 10 m³/ha
litres per square metre	litre/m²	About ⅕ gallon/sq yd

Other multiples, or divisions, of the units shown will also be met. For example, seed sowing of small areas may be shown in g/m² or mg/m² and these are related to the kg/m² :

1 kg/m² = 1000 g/m² ; 1 g/m² = 1000 mg/m².

Appendix I
SOWING AND YIELD DATA FOR CROPS

Crop.	Time of Sowing.	Time of Harvest.	Seed Rate. (kg/ha)	Yield of Grain. (tonnes/ha)	Yield of Straw or Feed (tonnes/h)
WHEAT:					
Winter varieties	Oct.–Jan.	Aug.–Sept.	150–220	3–5·5	3·8
Spring varieties	Feb.–April	Sept.	190–250	2·5–5·0	3·8
BARLEY:					
Winter varieties	Oct.–Dec.	July–Aug.	125–190	2·5–5·0	3·0
Spring varieties	Feb.–May	Aug.–Sept.	125–190	2·5–5·0	3·0
OATS:					
Winter varieties	Oct.–Nov.	July–Aug.	150–220	2·5–5·0	3·8
Spring varieties	Feb.–April	Aug.–Sept.	190–250	2·5–5·0	3·8
RYE:					
Winter varieties	Sept.–Oct.	July–Aug.	125–190	2·5–3·0	4·4
Spring varieties	Feb.–March	July–Aug.	190–250	2·5–3·0	4·4
BEANS:					
Winter varieties	Oct.–Nov.	Aug.–Sept.	190–250	2·5–3·8	3·0
Spring varieties	Feb.–April	Sept.–Oct.	220–290	1·9–3·0	3·0
PEAS	Feb.–May	July–Aug.	250–290	1·9–4·4	2·5
VETCHES	Oct.	July	150–250	1·9–2·0	3·0
MAIZE	May	Sept.–Nov.	30–40	3·8–5·0	50–60

				(seed)	
LUCERNE	April–May	June onwards	14–18	—	20–38
MANGEL	April–May	Nov.–Dec.	9–12	2·5	60
SUGAR BEET .	Mid-March–May	Nov.–Dec.	6–12	2·5	25–50
MARROW STEM KALE .	April–June	Nov.–Dec.	1–5	0·8	38–60
THOUSAND HEAD KALE	March–June	Nov.–Feb.	1–5	0·8	30–50
RAPE	Spring and Summer	Oct.–April	3–5	2–2·5	30–40
CABBAGE . . .	March	Nov.–Dec.	1–5	1·5	100–125
TURNIP	April	Aug.–Sept.	15–18	1·5	35
SWEDE	June	Nov.	1–5	1·0	38–60
MUSTARD . . .	July	Oct.	4–5	1–2·5	38–60
CARROT	April–May	Oct.–Dec.	2–5	—	25–30
LINSEED . . .	March–April	Aug.	60–90	1·5	5·6

Appendix II

SEEDS LEGISLATION

WITH the exception of provisions in respect of seed potatoes, the Seeds Act 1920 has been repealed by the Plant Variety and Seeds Act 1964.

(1) SEEDS ACT 1920

The purpose of the Act is to ensure that the farmer and horticulturalists are aware of the health status of the seed they are purchasing. The Seed Potatoes Regulations 1963–5 prescribe the criteria for the standard of seed potatoes and the particulars to be declared at the time of sale. These include :

(1) Name and address of the seller ; (2) Class ; (3) In the case of certified seed potatoes the reference letters and the number of the relative certificate ; (4) Variety ; (5) Size and Dressing ; (6) Where potatoes have been chemically treated, the names and types of treatment.

The certification of seed potatoes which is at the present time carried out under voluntary schemes within the UK are, following current discussions within the European Economic Community, expected to become compulsory and to include tuber standards. Farmers who are considering the growing of seed potatoes should consult the " Guide to the Seeds Acts and Regulations as regards seed potatoes " (Form SDS 6 (Seed Potatoes, Revised in July 1971)).

The Acts, Regulations and Orders which deal with seed potatoes are :—
The Seeds Act 1920
Seeds (Amendment) Act 1925
Agriculture (Miscellaneous Provisions) Act 1954 (Section 12 applies).
Agriculture (Miscellaneous Provisions) Act 1963 (Section 24 applies).
Seed Potatoes Regulations 1963 SIs (1963/1374).
Seed Potatoes (Amendment) Regulations 1964 (SI 1964/1174).
Seed Potatoes (Amendment) Regulations 1965 (SI 1965/1456).
Wart Disease of Potatoes (Gt Britain) Order 1973 (SI 1973 No. 1060).
Wart Disease of Potatoes (Gt Britain) (Amendment) Order 1974 (SI 1974/1159).
Potato Cyst Eelworm (Gt Britain) Order 1973 (SI 1973/1059).
Prevention of spread of Pests (Seed Potatoes) (Gt Britain) Order 1974 (SI 1974/1152).

(2) PLANT VARIETIES AND SEEDS ACT 1964

The purpose of the Act is to provide for the granting of proprietary rights to plant breeders of new varieties and the issue of compulsory licences in respect of those rights; to establish a Tribunal to hear appeals against the refusal to grant proprietary rights. For seeds, to establish an index of varieties and to provide protection for the farmer and horticulturalist against the sale of seeds of low vitality and contamination by other species and weed seeds.

(a) PLANT BREEDERS' RIGHTS

Part 1 of the Act introduced a system of plant breeders' rights in the United Kingdom under which the breeder of a new variety of plant may be given the right to license certain uses of his variety, i.e., the sale of seed, seed potatoes and other reproductive material and the production of seed, etc.,for sale. Royalties can be demanded by the breeder. Other uses such as production of a crop of wheat for milling or feeding or of a ware crop of potatoes do not require a licence. Many varieties are now subject to licensing and, generally, seed merchants, propagators and others will know which these are and the licence conditions.

Plant breeders' rights are at present (1976) being issued for :—

Wheat, oats, barley, lucerne, ryegrass, timothy, cocksfoot, tall and meadow fescue and red and white clover, French beans, peas, potatoes, runner beans and lettuces.

Carnations, perennial chrysanthemums, dahlias, perennial delphiniums, freesias, gladioli, herbaceous perennials, lilies, narcissi, rhododendrons, roses, trees, shrubs, woody climbers, cymbidiums, pelargoniums, and streptocarpus, conifers and taxads.

Apples, black currants, damsons, pears, plums, raspberries, rhubarb and strawberries.

(b) SEEDS—NATIONAL LISTS OF VARIETIES

Part II of the Plant Varieties and Seeds Act 1964 has been amended by the European Communities Act 1972 to provide for the introduction of the National List system for specified kinds of seed. The Regulations made under the Act prescribe that it is an offence, subject to certain exceptions, to offer or expose for sale or sell seed of a plant variety

(1) unless and until the name of that variety is included in the appropriate UK National List or EEC Common Catalogue of varieties (provided, of course, that the Common Catalogue entry has no note to the effect that marketing in the UK is prohibited) ; or

(2) under a name other than that which is given in a National List or Common Catalogue for that variety.

The following information is available from the Plant Variety Rights Office on request :—

The name of the breeder or organisation responsible for the maintenance of a variety.

Free leaflets : " Guide to Plant Breeders' Rights " and "Applying for a compulsory licence ".

Amendments to the National List are published in the Plant Variety Rights and Seeds Gazette.

Further information is available from the PVRO at the following address :—

Whitehouse Lane, Huntingdon Road, Cambridge CB3 0LF. Telephone number Cambridge 77151.

The Acts and Orders which are relevant to Plant Variety Rights or National Listing are :—

The Plant Varieties and Seeds Act 1964.
The Agriculture (Miscellaneous Provisions) Act 1968.

The Seeds (National List of Varieties) Regulations 1973 (SI 1973 No. 984).

The Seeds (National List of Varieties) (Fees) Regulations 1976 (SI 1976 No. 124).

(c) SEEDS

There have been many changes in respect of the marketing of seeds in the last few years and although the emphasis remains on consumer protection there is now a compulsory certification scheme for many farm and vegetable seeds. Certification includes the verification of trueness to variety and official testing of samples to ensure that analytical purity and germination standards, set for the European Economic Community as a whole, are maintained. Generation control has been introduced limiting the number of generations which may be bred from original stocks but this varies within species and genera. The certification scheme also ensures that seed is marketed by registered processors and seed merchants which has meant the elimination of farmer to farmer sales unless the seller is registered for that purpose. The Act and subsequent Regulations made thereunder prescribe the standards for both the crop in the field and the analytical purity in addition to sealing and labelling requirements for packages of seeds. A guide to the Seeds Regulations may be obtained from the Ministry of Agriculture, Fisheries and Food, PVRO and Seeds Division, Whitehouse Lane, Huntingdon Road, Cambridge CB3 0LF.

The Official Seed Testing Station for England and Wales is situated in Cambridge but additionally there are a number of licensed satellite testing stations throughout the country to assist with the volume of work entailed by compulsory certification. All samples tested at these stations are subject to the payment of a fee which is prescribed in Seeds (Fees) Regulations from time to time.

Field inspections are carried out both by officials of the Ministry and by officially licensed crop inspectors who are trained by the Certifying Authority.

Relevant legislation for the marketing of seeds is as follows :—

The Plant Varieties and Seeds Act 1964.
The Agriculture (Miscellaneous Provisions) Act 1968.
The Seeds (Registration and Licensing) Regulations 1974 (SI 1974/760).
The Seeds (Registration and Licensing) (Amendment) Regulations 1975 (SI 1975/720).
The Beet Seeds Regulations 1974 (SI 1974/898).
The Cereal Seeds Regulations 1974 (SI 1974/900).
The Fodder Plant Seeds Regulations 1974 (SI 1974/897).
The Oil and Fibre Plant Seeds Regulations 1974 (SI 1974/899).
The Vegetable Seeds Regulations 1975 (SI 1975/No. 1694).

Appendix III

BULL LICENSING REGULATIONS (ENGLAND AND WALES)

1. BULL LICENSING

THE licensing of bulls is carried out under the Improvement of Livestock (Licensing of Bulls) Act 1931 (as amended) and the Licensing of Bulls (England and Wales) Regulations 1972 which were made under the Act. These Regulations prohibit the keeping of a bull that has attained the age of 10 months unless a licence or permit is in force in respect of it. Only one type of licence, which is in the form of a veterinary certificate, is issued for bulls of any breed, including cross breds.

Licensing inspections are carried out by veterinary surgeons, who are members of a Panel established for bull licensing purposes by the RCVS/BVA acting jointly, who certify the bull's suitability for a licence on the basis of conformation and freedom from disease.

2. GENERAL PERMITS

A general permit is issued when a bull is kept entire to be fattened for slaughter and it is a condition of the permit that the bull must not be allowed to serve a cow or heifer.

Appendix IV

RECOMMENDED READING

CHAPTER I

Black, C. A., *Soil-Plant Relationships*, 2nd edn., John Wiley & Sons, Chichester, 1968.
Brade-Birks, S. G., *Good Soil*, Hodder & Stoughton for English Universities Press, Bickley, 1944.
Clarke, G. R., *The Study of Soil in the Field*, 5th edn., Clarendon Press, Oxford, 1971.
Davies, D. B., D. J. Eagle & J. B. Finney, *Soil Management*, 2nd edn., Farming Press, Ipswich, 1975.
Hall, D., *The Soil*, 5th edn., John Murray, London, 1945.
Hodgson, J. M., *Soil Survey Field Handbook*, Tech. Monograph No. 5 of " Soil Survey of England and Wales ", Harpenden, 1974.
Lyon, T. L., H. O. Buckman & N. C. Brady, *The Nature and Properties of Soils*, 5th edn., Macmillan, London, 1952.
MAFF, *Soil Physical Conditions and Crop Development*, Tech. Bulletin No. 29, HMSO, London, 1975.
Russel, E. W., *Soil Conditions and Crop Growth*, 10th edn., Longman, London, 1973.
Wallace, T., *Mineral Deficiencies in Plants*, HMSO, London 1961.

CHAPTER II

Agricultural Advisory Council, *Modern Farming and the Soil*, HMSO, London, 1970.
Berryman, C., et al., *Hydraulic Conductivity and Infiltration Rates*, FDEU Tech. Bulletin 74/4, 1974.
Carter, C. & B. D. Trafford, *Idealised Soil Profiles and Corresponding Drainage Designs*, FDEU Tech. Bulletin 73/12, 1973.
Croote, S., *The Use of Plastics in Field Drainage*, FDEU Tech. Bulletin 74/10, 1974.
" International Institute for Land Reclamation and Improvement ", *Drainage Principles and Applications*, Wageningen, Netherlands, 1973, vols. I and IV.
Luthin, J. N., *Drainage of Agricultural Lands*, American Society of Agronomy, Wisconsin, 1957.
Drainage Engineering, John Wiley & Sons, Chichester, 1966.
Massey, W., *Drainage Economics : change in land use as a result of drainage*, FDEU Tech. Bulletin 73/8, 1973.
Nicholson, H. H., *The Principles of Field Drainage*, Cambridge Univ. Press, Cambridge, 1953.
Rands, J. G., *Drainage Design and the Soils of Eastern Region*, FDEU Tech. Bulletin 73/3, 1973.
Schilfgaarde, J. van, *Drainage for Agriculture*, American Society of Agronomy, Wisconsin, 1974.
Trafford, B. D., *The Climatological Background to Field Drainage in General and FDEU Sites in Particular*, FDEU Tech. Bulletin 72/4, 1972.
The Effect of Waterlogging on the Emergence of Cereals, FDEU Tech. Bulletin 74/3, 1974.
The Evidence in Literature for Increased Yield due to Field Drainage, FDEU Tech. Bulletin 72/5, 1972.

"Field Drainage", *RASE Journal*, vol. 131, 1970.
Field Drainage Experiments in England and Wales, FDEU Tech. Bulletin 72/12, 1972.
A Guide to Drainage Design Technique Based on Scientific Methods, FDEU Tech. Bulletin 74/7, 1974.
Soil Water Régimes: what is known—the work which is in hand and suggestions for progress, FDEU Tech. Bulletin 74/13, 1974.

CHAPTER III

BOOKS : Sayce, R. B., *Farm Buildings*, Estates Gazette, London, 1967.
        Weller, J. B., *Farm Buildings*, vols. I and II, Crosby Lockwood Staples, London, 1965 and 1972.
LEAFLETS : Fixed Equipment of the Farm series, HMSO, London.
        MAFF, Advisory Drawings series, HMSO, London.
        Bulletins, HMSO, London.
PERIODICALS : *Farm Buildings Digest* (quarterly), FBC.
        Farm Building Progress (quarterly), SFBIU.
TECH. REPORTS : FBC Reports, Farm Buildings Information Centre, NAC, Kenilworth.
        SFBIU Reports, Scottish Farm Buildings Investigation Unit, Craibstone, Bucksburn, Aberdeen.

CHAPTER IV

Culpin, C., *Farm Machinery*, 9th edn., Crosby Lockwood Staples, London, 1976.

CHAPTER V

Hawkins, J. C., *Tractor Ploughing*, HMSO, London, 1967.

CHAPTER VI

MAFF, Mechanisation Leaflets, HMSO, London

CHAPTER VII

Cooke, G. W., *Fertilising for Maximum Yield*, Crosby Lockwood Staples, London, 1972.
Fertiliser Society of London, *Proceedings*, London.
MAFF, *Lime and Liming*, Tech. Bulletin No. 35, HMSO, London, 1973.
Fertiliser Recommendations, Tech. Bulletin No. 209, HMSO, London.

CHAPTER VIII

British Weed Control, *8th–12th Conference Proceedings*, British Crop Protection Council, Ombersley, 1966–74.
Fryer, J. D. & S. A. Evans (eds.), *Weed Control Handbook*, vol. I Principles, Blackwell Scientific Publications, Oxford, 1968.
Fryer, J. D. & R. J. Makepeace (eds.), *Weed Control Handbook*, vol. II Recommendations, Blackwell Scientific Publications, Oxford, 1973.
Gair, R., J. E. E. Jenkins & E. Lester, *Cereal Pests and Diseases*, Farming Press, Ipswich, 1972.
MAFF, *Agricultural Chemicals Approval Scheme: Approved products for farmers and growers*.
Irrigation, Tech. Bulletin No. 138, HMSO, London.
Modern Farming and the Soil, HMSO, London, 1970.

CHAPTER IX

Arable Farming, Farming Press, Ipswich.
Farmers' Leaflets : No. 8, *Cereals* ; No. 9, *Field Beans* ; Growers' Vegetable Leaflet No. 5, *Vining Peas,* National Institute of Agricultural Botany, Cambridge.
Millthorpe, F. L. (ed.), *Growth of Cereals and Grasses,* Butterworths, London, 1966.
Outdoor Vegetables : UK farming and the Common Market, NEDO, London, 1973.
Park, R. D. & M. Eddowes, *Crop Husbandry,* 2nd edn., Clarendon Press, Oxford, 1975.

CHAPTER X

British Sugar Beet Review, British Sugar Corporation, London.
Burton, W. G., *The Potato,* Veenman & Zon, Wageningen, Netherlands, 1966.
Farmers' Leaflets : No. 2, *Green Fodder Crops* ; No. 3, *Potatoes* ; No. 5, *Sugar Beet* ; No. 6, *Fodder Root Crops* ; No. 7, *Maize,* National Institute of Agricultural Botany, Cambridge.
MAFF, *Experimental Husbandry,* HMSO, London.
  Potatoes, Tech. Bulletin No. 94 ; *Sugar Beet Cultivation,* Tech. Bulletin No. 153 ; *Sugar Beet Diseases,* Tech. Bulletin No. 142 ; *Sugar Beet Pests,* Tech. Bulletin No. 162, HMSO, London.
Research Review, Agricultural Research Council, London.
Salaman, R., *The History and Social Influence of the Potato,* Cambridge Univ. Press, Cambridge, 1970.
Zaag, D. E. van der, *Potatoes and their Cultivation in the Netherlands,* Dutch Information Centre for Potatoes, The Hague, Netherlands.

CHAPTER XI

Clapham, A. R., T. G. Tutin & E. F. Warburg, *Flora of the British Isles,* Cambridge Univ. Press, 1962.
Davies, W., B.Sc., *The Grass Crop—Its Development, Use and Maintenance,* E. & F. N. Spon, Ltd., 1952.
Hubbard, C. E., *Grasses—A guide to their structure, identification, uses and distribution in the British Isles,* Penguin Books, 1954.
Robinson, D. H., Ph.D., B.Sc., N.D.A., *Leguminous Forage Plants,* Edward Arnold & Co., 1937.
Tansley, A. G., M.A., F.R.S., *The British Islands and their Vegetation,* Cambridge Univ. Press, 1939.
Grass and Clover Crops for Seed, Bulletin 204, MAFF, HMSO.
Grasses and Legumes in British Agriculture. Various authors and edited by C. R. W. Spedding and E. C. Diekmahns, Bulletin 49, Commonwealth Agricultural Bureaux, 1971.
Silage, Bulletin 37, MAFF, HMSO.
Current Literature
Grassland Practice Series—leaflets Nos. 1–10, MAFF, HMSO.
Farmers' Leaflets : No. 4, *Herbage Legumes* ; No. 16, *Grasses* ; No. 17, NIAB.
Grasses and Legumes for Conservation, NIAB.
Classified List of Herbage Varieties, England and Wales ; revised annually, NIAB.

CHAPTER XII

ADAS Profitable Farm Enterprises, Booklet No. 9, *Silage.*
Farmers' Leaflets : No. 4, *Recommended List of Legumes* ; No. 16, *Recommended*

List of Grasses; No. 17, Grasses and Legumes for Conservation, National Institute of Agricultural Botany, Cambridge.

MAFF, *Grass and Grassland*, Tech. Bulletin No. 154; *Quick Haymaking*, Tech. Bulletin No. 188; *Silage*, Tech. Bulletin No. 37, HMSO, London.

Raymond, F., G. Shepperton & R. Waltham, *Forage Conservation and Feeding*, Farming Press, Ipswich, 1972.

Watson, S. J. & M. J. Nash, *The Science and Practice of Conservation*, Oliver & Boyd, Edinburgh, 1960.

Chapter XIII

Ainsworth, G. C. & G. R. Bisby, *Ainsworth & Bisby's Dictionary of the Fungi*, 6th edn., C.M.I., Kew, 1971.

Ainsworth, G. C. & Kathleen Sampson, *The British Smut Fungi (Ustilaginales)*, C.M.I., Kew, 1950.

Alexopoulos, C. J., *Introductory Mycology*, 2nd edn., Wiley, New York, 1962.

Anon, *Plant Pathologist's Pocketbook*, C.M.I., Kew, 1968.

Anon, List of Approved Products and their Uses for Farmers and Growers. Published annually, MAFF.

Baker, J. J., " Report on Diseases of Cultivated Plants in England and Wales for the Years 1957–1968 ", Tech. Bull. 25, HMSO, London, 1972.

Baker, K. F. & W. C. Snyder, *Ecology of Soil-borne Plant Pathogens*, Berkeley : Univ. California Press, 1965.

Bawden, F. C., *Plant Viruses and Virus Diseases*, 4th edn., Ronald Press, New York, 1964.

Boyd, A. E. W., " Potato Storage Diseases ", *Rev. Pl. Path 51* (5) : 297–321, 1972.

Brooks, F. T., *Plant Diseases*, 2nd edn., Oxford Univ. Press, London, 1953.

Bruehl, G. W., " Barley Yellow Dwarf ", *Monograph. Amer. Phytopath. Soc. 1*, 1961.

Burton, W. G., *The Potato*, Chapman & Hall, London, 1948.

Butler, E. J. & S. G. Jones, *Plant Pathology*, Macmillan, London, 1949.

Cadman, C. H., " Modern Developments in Plant Virus Research with particular reference to Potatoes ", *Proc. 4th trienn. Conf. Eur. Ass. Potato Res. 1969 :* 51–64, 1970.

Chupp, C. & A. F. Sherf, *Vegetable Diseases and Their Control*, Ronald Press, New York, 1960.

Colhoun, J., " Club Root Disease of Crucifers caused by *Plasmodiophora brassicae* ", *Phytopath. Pap. 3*, Woron, C.M.I., Kew, 1958.

" Seed and Soil-borne Pathogens of Cereals ", *Proc. 5th British Insecticide and Fungicide Conf. :* 620–25, 1970.

Cox, A. E. & E. C. Large, " Potato Blight Epidemics throughout the World ", *Agric. Handbook* USDA, No. 174, 1960.

Dickson, J. G., *Diseases of Field Crops*, 2nd edn., McGraw-Hill, London, 1956.

Diercks, R., " Growth Regulators and Fungicides for Cereal Growing ", *Bayer. landw. Jb. 47* (3) : 283–300, 1970.

Dowson, W. J., *Plant Diseases due to Bacteria*, 2nd edn., Cambridge Univ. Press, London, 1957.

Gair, R., J. E. E. Jenkins & E. Lester, *Cereal Pests & Diseases*, Farming Press, Ipswich, 1972.

Garrett, S. D., *Pathogenic Root-infecting Fungi*, Cambridge Univ. Press, London, 1970.

Gerlagh, M., " Introduction of *Ophiobolus graminis* into new Polders and its Decline ", *Neth. J. of Pl. Path., 1974 Suppl. 2 :* 1–97, 1968.

Hawksworth, D. L., *Mycologist's Handbook : An Introduction to the Principles of Taxonomy and Nomenclature in the Fungi and Lichens*, C.M.I., Kew, 1974.

Horsfall, J. G., *Principles of Fungicidal Action*, Chronica Botanica Co., Waltham, Mass., 1956.

Hull, R., *Sugar Beet Diseases*, 2nd edn., MAFF, *Bull. 142*, 1960.
" Mycoplasma-like Organisms in Plants ", *Rev. Pl. Path. 50* (3) : 121–30, 1971.

Large, E. C. (reprinted 1962), *The Advance of the Fungi*, Jonathan Cape, London, 1940.

Manners, J. G., " The Rust diseases of Wheat and their Control ", *Trans. Brit. mycol. Soc. 52* (2) : 177–86, 1969.

Martin, H., *The Scientific Principles of Crop Protection*, 6th edn., Edward Arnold, London, 1973.
Insecticide and Fungicide Handbook for crop protection, edit. by Martin, H., 4th edn., Blackwell, Oxford, 1972.

McKay, R., *Crucifer Disease in Ireland*, Sign of Three Candles, Dublin, 1956.
Potato Diseases, Sign of Three Candles, Dublin, 1955.

Obst, A., " Glume Blotch of Wheat (*Leptosphaeria nodorum*)—Biology and Prevention ", *Bayer. landw. Jb. 46* (3) : 310–19, 1969.

Ogilvie, L., *Diseases of Vegetables*, 6th edn., MAFF, *Bull. 123*, London, 1969.

Rowell, J. B., " Chemical Control of the Cereal Rusts ", *A. Rev. Phytopath. 6* : 243–62, 1968.

Sampson, Kathleen & J. H. Western, *Diseases of British Grasses & Herbage Legumes*, 2nd edn., Cambridge Univ. Press, London, 1954.

Shipton, W. A., T. N. Khan & W. J. R. Boyd, " Net Blotch of Barley ", *Rev. Pl. Path. 52* (5) : 269–90, 1973.

Shipton, W. A., W. J. R. Boyd & S. M. Ali, " Scald of Barley ", *Rev. Pl. Path. 53* (11) : 839–61, 1974.

Slope, D. B., " The Benefits and Limitations of Break Crops and the Control of Soil-borne Diseases of Cereals ", *J. NIAB. 11* (Suppl.), 47–53, 1968.

Slykhuis, J. T., " Virus Diseases of Cereals ", *Rev. Appl. Mycol. 46* (8) : 401–29, 1967.

Smith, K. M., *A Textbook of Plant Virus Diseases*, 3rd edn., Longman, London, 1972.

Taylor, R. E., " Systemic Fungicides ", *Agriculture, Lond. 77* (7) : 310–12, 1970.

Walker, J., " Take all Diseases of *Gramineae* ", *Rev. Pl. Path. 54* (3) : 113–44, 1975.

Walker, J. C., *Plant Pathology*, 3rd edn., McGraw-Hill, London, 1969.
Diseases of Vegetable Crops, McGraw-Hill, London, 1952.

Western, J. H., *Diseases of Crop Plants*, Macmillan, London, 1971.

Whitehead, T., T. P. McKintosh & W. M. Findlay, *The Potato in Health and Disease*, 3rd edn., Oliver & Boyd, Edinburgh and London, 1953.

Wilson, M. & D. M. Henderson, *British Rust Fungi*, Cambridge Univ. Press, London, 1966.

CHAPTER XIV

Arthur, D. R., *British Ticks*, Butterworths, London, 1965.
Ticks and Disease, Pergamon Press, Oxford, 1962.

Edwards, G. A. & G. H. Heath, *The Principles of Agricultural Entomology*, Chapman & Hall, London, 1964.

Gair, R., J. E. E. Jenkins & E. Lester, *Cereal Pests and Diseases*, Farming Press, Ipswich, 1972.

Jones, F. G. W. & Margaret Hones, *Pests of Field Crops*, Edward Arnold, London, 1964.

Lapage, G., *Veterinary Parasitology*, 2nd edn., Oliver & Boyd, Edinburgh, 1968.

Martin, H., *Insecticide and Fungicide Handbook for Crop Protection*, 4th edn., Blackwell Scientific Publications, Oxford, 1972.

Monnig, H. O., *Veterinary Helminthology and Entomology*, 6th edn., Baillière Tindall, London, 1968.

CHAPTER XV

Akerberg, E., *et al.* (eds.), *Recent Plant Breeding Research : Sualöf 1946–61*, John Wiley & Sons, Chichester, 1963.

Allard, R. W., *Principles of Plant Breeding*, John Wiley & Sons, Chichester, 1966.

Bateson, W., *Mendel's Principles of Heredity*, Cambridge Univ. Press, Cambridge, 1909.

Bell, G. D. H. & F. G. H. Lupton, " The Breeding of Barley Varieties ", *Barley and Malt* (ed. A. H. Cook), Academic Press, New York, 1962.

Bell, G. D. H., " Cereal Breeding ", *Vistas in Botany*, vol. II Applied Botany (ed. W. B. Turrill), Pergamon Press, Oxford, 1963.

Biffen, R. H. & F. L. Engledow, *Wheat Breeding Investigations*, MAFF Research Monograph No. 4, HMSO, London, 1926.

Coffmann, Franklin A. (ed.), *Oats and Oat Improvement*, American Society of Agronomy, Wisconsin, 1961.

International Atomic Energy Agency, *Induced Mutations for Disease Resistance in Crop Plants*, Panel Proceedings series, Vienna, 1974.
Polyploidy and Induced Mutations in Plant Breeding, Panel Proceedings series, Vienna, 1974.

McLeish, John & Brian Snoad, *Looking at Chromosomes*, 2nd edn., Macmillan, London, 1972.

Moav, Rom (ed.), *Agricultural Genetics : selected topics*, John Wiley & Sons, Chichester, 1973.

Nelson, R. R. (ed.), *Breeding Plants for Disease Resistance*, Pennsylvania State U.P., University Park, 1973.

National Academy of Sciences, *Genetic Vulnerability of Major Crops*, Washington D.C., 1972.

Peterson, R. F., *Wheat Botany, Cultivation and Utilisation*, Leonard Hill Books, London, 1965.

Poehlman, John Milton, *Breeding Field Crops*, Holt, Rinehart & Winston, New York, 1959.

Quisenberry, K. S. & L. P. Reitz (eds.), *Wheat and Wheat Improvement*, American Society of Agronomy, Wisconsin, 1967.

Whitehouse, H. L. K., *The Matter of Mendelian Heredity*, Edward Arnold, London, 1965.

Williams, Watkin, *Genetical Principles and Plant Breeding*, Blackwell Scientific Publications, Oxford, 1964.

The published Annual Reports of the Plant Breeding Institute, Cambridge, the Scottish Plant Breeding Station and the Welsh Plant Breeding Station are relevant references for plant breeding work and supporting scientific research embracing the important crop plants of British agriculture. Commencing with the 1959–60 Report, the Plant Breeding Institute has included in its reports special feature articles dealing with individual crops and various aspects of supporting research. These three research centres can be consulted for further information.

The National Institute of Agricultural Botany, Cambridge, conducts official national and international crop testing. The Institute publishes *Recommended Lists* and *National Lists* for individual crops as well as a *Journal* and miscellaneous leaflets. It can be consulted also on specific problems relevant to its work.

On a broader front and relating to plant breeding in the European context, the *Proceedings* of the congresses held periodically by the European Association for Research in Plant Breeding (Eucarpia) contain a wide range of breeding topics, as does *Euphytica, the Netherlands Journal of Plant Breeding*.

Annual Reports of the Plant Breeding Institute, Cambridge

1959–60 *Developments in Wheat-Breeding Methods*, pp. 5–18. F. G. H. Lupton & R. N. H. Whitehouse.
1960–1 *Breeding Field Beans*, pp. 4–26. D. A. Bond & J. L. Fyfe.
1961–2 *Some Potato Breeding Problems*, pp. 5–21. H. W. Howard.
1963–4 *Breeding Sugar Beet*, 6–32. G. K. G. Campbell & G. E. Russell.
1964–5 *The Transfer of Alien Genetic Variation to Wheat*, pp. 6–36. Ralph Riley & Gordon Kimber.
1965–6 *Breeding Problems in Kale (B. oberacia) with Particular Reference to Marrow-stem Kale*, pp. 7–34. K. F. Thompson.
1966–7 *Applications of Physiological Analysis to Cereal Breeding*, pp. 5–26. F. G. H. Lupton & E. T. M. Kirby.
1968 *Barley Breeding at Cambridge*, pp. 6–29. R. N. H. Whitehouse.
1969 *Problems in Breeding for Resistance to Diseases and Pests*, pp. 6–36. H. W. Howard, R. Johnson, G. E. Russell & M. S. Wolfe.
1970 *Plant Breeding Institute 1960–70*, pp. 12–33. G. D. H. Bell.
1971 *Plant Breeding Research and Variety Production*, pp. 23–31. Ralph Riley.
1971 *Principles and Problems in Grass Breeding*, pp. 31–62. A. J. Thomson & A. J. Wright.
1972 *Aneuploidy in Wheat and its Uses in Genetic Analysis*, pp. 25–65. C. H. Law & A. J. Worland.
1973 *Maize in Britain—A Survey of Research and Breeding*, pp. 32–74. E. S. Bunting & R. E. Gunn.

Jubilee Report of the Welsh Plant Breeding Station, 1919–69, University College of Wales, Aberystwyth, 1970.
And subsequent annual reports.

CHAPTER XVI

Al-Murrani, " The limits to artificial selection ", *Animal Breeding Abstracts*, vol. 42, no. 12, 1974, p. 587.
Animal Breeding Research Organisation (Edinburgh), *Annual Reports*, 1963, *et seq.*
Barker, J. S. F. & A. Robertson, " Genetic and phenotypic parameters for the first three lactations in Friesian cows ", *Animal Production*, vol. 8, 1966, pp. 221–40.
Bowman, J. C. & J. S. Broadbent, " Genetic parameters of growth between birth and sixteen weeks in Down cross sheep ", *Animal Production*, vol. 8, 1966, pp. 129–35.
Bowman, J. C., J. E. Marshall & J. S. Broadbent, " Genetic parameters of carcass quality in Down cross sheep ", *Animal Production*, vol. 10, 1968, pp. 183–92.
Donald, H. P., " Perinatal deaths among calves in a crossbred dairy herd ", *Animal Production*, vol. 5, 1963, pp. 87–96.
Donald, H. P. & J. L. Read, " The performance of Finnish Landrace sheep in Britain ", *Animal Production*, vol. 9, 1967, pp. 471–6.
Donald, H. P., J. L. Read & W. S. Russell, " A comparative trial of crossbred ewes by Finnish Landrace and other sires ", *Animal Production*, vol. 10, 1968, pp. 413–22.
Donald, H. P. & W. S. Russell, " Some aspects of fertility in purebred and crossbred dairy cattle ", *Animal Production*, vol. 10, 1968, pp. 465–71.

Hinks, C. J. M., " The comparative milking performance of Canadian and US Holstein Cattle ", *British Cattle Breeders Club Digest*, no. 30, 1975, p. 84.
" Selection practices in dairy herds : 1. First lactation performance and survival to the second lactation ", *Animal Production*, vol. 8, 1966, pp. 467–80.
" Selection practices in dairy herds : 2. Selection patterns in later lactations ", *Animal Production*, vol. 8, 1966, pp. 481–8.
King, J. W. B., " New breeds for new purposes ", Paper presented at Summer Meeting of British Society of Animal Production, Aberdeen, 1968.
King, J. W. B. & L. Gajic, " The repeatability of Maternal Performance in inbred, outbred and linecross Large White sows ", *Animal Production*, vol. 11, 1969, pp. 47–52.
Kilkenny, J. B., " Estimates of heritability of weight at 400 days of age using field data ", *Animal Production*, vol. 12, 1970, p. 364.
Land, R. B., " Physiological studies and genetic selection for sheep fertility ", *Animal Breeding Abstracts*, vol. 42, no. 4, 1974, p. 155.
Lessells, W. J. & A. L. Francis, " The crossbred progeny test of beef bulls ", *Experimental Husbandry*, no. 16, 1968, pp. 1–10.
Lopez Fanjul, C., " Selection from crossbred populations ", *Animal Breeding Abstracts*, vol. 42, no. 9, 1974, p. 403.
Mason, I. L., " Comparative beef performance of the large cattle breeds of Western Europe ", *Animal Breeding Abstracts*, vol. 39, no. 1, 1971, p. 1.
M.M.B., *Breeding and Production Organisation Report*, nos. 18–24 incl.
Pearson, Lucia & R. E. McDowell, " Crossbreeding of dairy cattle in temperate zones ; a review of recent studies ", *Animal Breeding Abstracts*, vol. 36, 1968, pp. 1–15.
Robertson, A., " Optimum group size in progeny testing and family selection ", *Biometrics*, vol. 13, 1957, pp. 442–50.
British Cattle Breeders Club Digest, no. 22, 1966, p. 127.
" Biochemical polymorphisms in animal improvement ", *Proceedings*, Xth International Congress of Animal Blood Groups, Paris, 1967, pp. 35–42.
Robertson, A. & J. S. F. Barker, " The correlation between first lactation milk production and longevity in dairy cattle ", *Animal Production*, vol. 8, 1966, pp. 241–52.
Smith, C., " A comparison of testing schemes for pigs ", *Animal Production*, vol. 1, 1959, pp. 113–22.
" Results of pig progeny testing in Great Britain ", *Animal Production*, vol. 7, 1965, pp. 133–40.
Smith, C. & J. W. B. King, " Crossbreeding and litter production in British pigs ", *Animal Production*, vol. 6, 1964, pp. 265–72.
Smith, C., J. W. B. King & R. Gilbert, " Genetic parameters of British Large White bacon pigs ", *Animal Production*, vol. 4, 1962, pp. 128–43.
Smith, C. & G. J. S. Ross, " Genetic parameters of British Landrace pigs ", *Animal Production*, vol. 7, 1965, pp. 291–302.
Strang, G. S., " Litter productivity in Large White pigs : 1. The relative importance of some sources of variation ", *Animal Production*, vol. 12, 1970, pp. 225–34.
Strang, G. S. & J. W. B. King, " Litter productivity in Large White pigs : 2. Heritability and repeatability estimates ", *Animal Production*, vol. 12, 1970, pp. 235–44.
Wiener, G., " Genetic and mineral metabolism ", *Animal Breeding Research Organisation Report*, 1970, pp. 13–18.
Wiener, G. & M. R. Sampford, " The incidence of Swayback among lambs with particular reference to genetic factors ", *Journal of Agricultural Science*, vol. 73, 1969, p. 25.
Wijeratne, W. V. S., " The incidence and some causes of abortion in cattle

with particular reference to genetic factors ", *Animal Production*, vol. 12, 1970, p. 377.

Wijeratne, W. V. S., P. J. Crossman & C. M. Gould, " Evidence of a sire effect on piglet mortality ", *British Veterinary Journal*, vol. 126, 1970, pp. 94–9.

CHAPTER XVIII

MAFF, *Energy Allowances and Feeding Systems for Ruminants*, Tech. Bulletin No. 33, HMSO, London.

CHAPTER XIX

ADAS Profitable Farm Enterprises, Booklet No. 1, *Rearing Friesian Dairy Heifers* ; No. 2, *Beef Production Using Dairy Bred Calves* ; No. 4, *Suckler Beef*.

Barnard, C. S., R. J. Halley & A. H. Scott, *Milk Production*, Iliffe, London, 1970.

Barron, N., *Dairy Farmer's Veterinary Book*, 8th rev. edn., Farming Press, Ipswich, 1973.

Bowden, W. E., *Beef Breeding Production and Marketing*, Land Books, London, 1972.

MLC Handbook No. 2, *Cereal Beef*, Bletchley, 1974.

Russell, K., *Principles of Dairy Farming*, 7th rev. edn., Farming Press, Ipswich, 1974.

Tayler, J. C. & J. M. Wilkinson, *Beef Production from Grassland*, Butterworths, London, 1973.

Thomas, D. G. M. & W. I. J. Davies, *Animal Husbandry*, 2nd rev. edn., Cassell, London, 1971.

CHAPTER XX

Blood, D. C. & J. A. Henderson, *Veterinary Medicine*, Baillière Tindall, London, 1974.

Fraser, Allan & J. T. Stamp, *Sheep Husbandry and Diseases*, 5th rev. edn., Crosby Lockwood Staples, London, 1968.

MAFF, *Diseases of Sheep*, Tech. Bulletin No. 170, HMSO, London, 1966. *Sheep Breeding and Management*, Tech. Bulletin No. 166, HMSO, London, 1964.

Meat & Livestock Commission, *Feeding the Ewe*, Sheep Improvement Service : Tech. Report No. 2, 1973.

Spedding, C. R. W., *Sheep Production and Grazing Management*, 2nd rev. edn., Baillière Tindall, London, 1970.

CHAPTER XXI

Brent, G., D. Hovell, R. F. Ridgeon & W. J. Smith, *Early Weaning of Pigs*, Farming Press, Ipswich, 1975.

Cole, D. J. A., *Pig Production : Proceedings of the 18th School in Agricultural Science*, Butterworths, London, 1971.

Elsley, F. W. & C. T. Whittemore, *Practical Pig Nutrition*, Farming Press, Ipswich, 1975.

MAFF, *Blueprint for a 200-Sow Unit*, HMSO, London. *Farm Waste Disposal*, Short-Term Leaflet No. 67, HMSO, London. *Housing the Pig*, Tech. Bulletin No. 160, HMSO, London. *Pig Feeding*, Advisory Leaflet No. 104, HMSO, London. *Pig Husbandry and Management*, Tech. Bulletin No. 193, HMSO, London.

Sow Breeding Calendars and Records, Short-Term Leaflet No. 121, HMSO, London.
Weaning at 3 Weeks of Age, HMSO, London.
Thornton, K., *Practical Pig Production*, Farming Press, Ipswich, 1974.

CHAPTER XXIII

Barron, N., *Dairy Farmer's Veterinary Book*, 8th rev. edn., Farming Press, Ipswich, 1973.
Pig Farmer's Veterinary Book, 9th rev. edn., Farming Press, Ipswich, 1976.
Greig, J. R. (ed.), *Shepherd's Guide to the Prevention and Control of the Diseases of Sheep*, 3rd edn., HMSO, London, 1958.
Miller, W. C. & G. P. West (eds.), *Black's Veterinary Dictionary*, 10th rev. edn., A. & C. Black, London, 1972.

CHAPTER XXIV

Barnard, C. S. & J. S. Nix, *Farm Planning and Control*, Cambridge Univ. Press Cambridge, 1973.
Burr, H. & D. B. Wallace, *Planning on the Farm*, Report No. 60, Agricultural Economics Unit, Dept. of Land Economy, Cambridge University.
Sturrock, F. G., *Farm Accounting and Management*, 6th rev. edn., Pitman, London, 1972.
Thompson, M. C., *Farm Planning Data*, Report No. 60, Agricultural Economics Unit, Dept. of Land Economy, Cambridge University.

INDEX